T0206259

Biologics and Biosimilars

Biologics and Biosimilars

Drug Discovery and Clinical Applications

Edited by
Xiaodong Feng, Hong-Guang Xie, Ashim Malhotra, and Catherine F. Yang

CRC Press is an imprint of the
Taylor & Francis Group, an **informa** business

Cover image: © istock/Meletios Verras

First edition published 2022
by CRC Press
6000 Broken Sound Parkway NW, Suite 300, Boca Raton, FL 33487-2742

and by CRC Press
4 Park Square, Milton Park, Abingdon, Oxon, OX14 4RN

CRC Press is an imprint of Taylor & Francis Group, LLC

Library of Congress Cataloging-in-Publication Data

Names: Feng, Xiaodong, editor. | Xie, Hong-Guang, editor. | Malhotra,
Ashim, 1977- editor. | Yang, Catherine F., editor.
Title: Biologics and biosimilars : drug development and clinical affairs /
edited by Xiaodong Feng, Hong-Guang Xie, Ashim Malhotra, Catherine F. Yang.
Description: First edition. | Boca Raton : Taylor and Francis, 2022. |
Includes bibliographical references and index.
Identifiers: LCCN 2021056289 | ISBN 9781138594227 (hardback) |
ISBN 9781032262024 (paperback) | ISBN 9780429485626 (ebook)
Subjects: LCSH: Pharmaceutical biotechnology. | Drug development.
Classification: LCC RS380 .B536 2022 | DDC 615.1/9--dc23/eng/20211228
LC record available at https://lccn.loc.gov/2021056289

ISBN: 9781138594227 (hbk)
ISBN: 9781032262024 (pbk)
ISBN: 9780429485626 (ebk)

DOI: 10.1201/9780429485626

Typeset in Times LT Std
by KnowledgeWorks Global Ltd.

Contents

Preface

Distinct from the small-molecule chemical compounds that are synthesized chemically, biological products, composed of biologics and biosimilars, refer to a class of bioactive macromolecules that are manufactured in living systems. Since the first therapeutic protein product (Humulin) was approved by the US Food and Drug Administration in 1982, more than 350 biological products have been licensed in the United States (as of July 2021). In the past four decades, biological products have revolutionized the treatment of many chronic or life-threatening diseases and critically ill conditions, making more contribution to the global healthcare system due to the more active market access and more therapy options for life saving than we thought, particularly in patients who fail to respond to currently available therapy regimens or have no other medications to be used.

Recent advances in drug discovery have created such a new classification of biological drugs (relative to chemical drugs) and these new biodrugs have benefited patients and pharmaceutical markets the most. The book, entitled *Biologics and Biosimilars: Drug Discovery and Clinical Applications*, aims to systematically summarize the principles and practice of biological products and to further facilitate the use of these relevant biotechnology innovations and scientific breakthroughs in the process of new drug research and development as well as in clinical settings for better patient care. Accordingly, general aspects of the research and development of biological products are discussed first in Chapters 1–11, including but not limited to the well-defined concepts and essential principles, the entire process of manufacturing and its regulatory requirements, quality assurance and safety evaluation, pharmacological and clinical assessments, therapeutical delivery system, and the main strategies for the approval of biologics license application (BLA). In the second part of the book, clinical applications of biologics and biosimilars in patients are elaborated with the major focus on the following indications, such as cancer (see Chapters 12–17), diabetes (Chapter 18), asthma (Chapter 19), autoimmune renal disease (Chapter 20), hemophilia A (Chapter 21), inflammatory bowel disease (Chapter 23), neurological diseases (Chapter 24), and use of complement as a novel target of immunotherapy (Chapter 22). At the end of the book, other important relevant issues are also discussed, such as applications of modeling and simulation, machine learning, and artificial intelligence in the discovery and development of biological products (see Chapters 26 and 27), as well as use of some companion testing to predict the efficacy and safety of the biological drugs in clinical settings (Chapter 25).

This book is directed to those who are interested in the different aspects of drug design, utilization, and regulation, working in the biopharmaceutical industry, clinical care, scientific research, or the regulatory agency. The book is designed to provide readers not only with all aspects of biological drugs (from bench to bedside and back to bench) but also with insightful overview of key topics, such as development and emergency use authorization of vaccines in the time of the COVID-19 pandemic worldwide now (see Chapter 4).

All the chapters in the book have been written by the professional scientists working in the USA and China as well as the rest of the world. The editors would like to express their sincere appreciation to all the chapter authors, whose contributions to this book make it a comprehensive and valuable resource.

Xiaodong Feng, PharmD, PhD
Hong-Guang Xie, MD, PhD
Ashim Malhotra, PhD, FAPE
Catherine F. Yang, PhD

Editor Biographies

Dr Xiaodong Feng, PharmD, PhD, received his PhD in Cellular and Molecular Physiology from the Chinese Academy of Medical Sciences and his Doctor of Pharmacy degree from Albany College of Pharmacy and Health Sciences. He started his biomedical research career 25 years ago in the Wound Healing Center, School of Medicine at Stony Brook. After three years of fellowship, he continued research and teaching as an Assistant Professor in the Department of Dermatology, School of Medicine, State University of New York at Stony Brook for another four years. Currently he is teaching at California Northstate University College of Medicine as Professor of Pharmacology and Oncology. He has over 25 years of clinical and biomedical research experience in cancer, wound healing, and cardiovascular diseases. He also practices as a clinical pharmacist and oncology pharmacy specialist. His current research interests include drug discovery for anti-angiogenesis therapy and tumor metastasis, and the application of pharmacogenomics in patient care. Several US patents on strategies of anti-angiogenesis and cancer treatments have recently been issued for Dr Feng. Personalizing pharmacotherapy using pharmacogenomics has always been his professional passion. He also edited two textbooks on advanced pharmacology and the clinical application of pharmacogenetics in therapeutics. He has been serving as editorial board members and peer reviewers for several peer-reviewed biomedical and pharmaceutical journals.

Dr Feng is one of the founding faculty and administrators for California Northstate University College of Pharmacy and College of Medicine. He serves as a Professor of oncology and pharmacology at the College of Medicine and College of Pharmacy. He also assumes an administrative role as the Dean of College of Pharmacy and University Vice President for Admissions and Student Services. He previously served the College of Medicine as the Associate Dean of Student Affairs and Admission. By training he is an Oncology Clinical Pharmacist and practiced as a Clinical Pharmacist at Sutter Davis Hospital and an Oncology Pharmacy Specialist at Dignity Health Medical Foundation.

Dr Feng is a clinical pharmacist, biomedical scientist, pharmaceutical educator, academic administrator, community leader and an entrepreneur. He co-founded California Northstate University (CNU) College of Pharmacy which later evolved into one of the major medical universities in the country comprising six colleges: Pharmacy, Medical, Dental, Psychology, Graduate Studies and Health Sciences. Despite his commitment to CNU as a senior administrator, he has been actively involved in translational and educational research revolving around cardiopulmonary diseases, cancer and drug development. His integrated training, experience, and expertise in medical and pharmaceutical education, along with his clinical practice, community service and leadership, and productive biomedical research, position him to effectively deliver high quality patient care, medical education, and pharmaceutical development.

Hong-Guang Xie, MD, PhD, earned his bachelor's degree of medicine (MD equivalent), master's degree of medicine, and doctoral degree of medicine (PhD equivalent) in 1984, 1989, and 1995, respectively, from Central South University Xiangya (formerly Xiang-Yale) School of Medicine, Changsha, Hunan, China, where he was promoted as a lecturer and associate professor of pharmacology in 1989, and 1995, respectively, and subsequently selected as a mentor of graduate students for master's degree in pharmacology. He was appointed as the first associate director of the Pharmacogenetics Research Institute (currently Institute of Clinical Pharmacology), Central South University, China, as the co-founder of that institute in 1997. As a recipient of the Merck Sharp and Dohme (MSD) International Clinical Pharmacology Fellowship, Dr Xie joined Vanderbilt University School of Medicine (Nashville, Tennessee, USA) as a Research Fellow (postdoc) in 1997 and a Research Instructor (faculty) in pharmacology in 2002. After that, he joined the University of California at San Francisco (UCSF) Washington Center at Washington, DC, as staff (a Research Fellow) in 2008 and the US FDA in 2010 as an Oak Ridge Institute for Science and Education (ORISE) Fellow. In December 2011, Dr Xie joined Nanjing Medical University (Nanjing, Jiangsu, China) as a full Professor of pharmacology and a mentor of PhD students and was appointed as chief of General Clinical Research Center (GCRC), Nanjing First Hospital. In 2014, Dr Xie was awarded as a recipient of the Distinguished Medical Expert of Jiangsu Province, China. In 2015, he joined China Pharmaceutical University (Nanjing, China) jointly as a mentor of the PhD training program for clinical pharmacy.

Dr Xie is active in the research field of basic and clinical pharmacology, with at least 140 articles (original, review, and commentary) published, many of which published in some leading journals worldwide, such as *New England Journal of Medicine*, *Annual Review of Pharmacology & Toxicology*, *Advanced Drug Delivery Review*, *Pharmacology & Therapeutics*, *British Journal of Pharmacology*, *Clinical Pharmacology & Therapeutics*, *Biochemical Pharmacology*, *Journal of Thrombosis & Haemostasis*, *Drug Metabolism Review*, *Drug Metabolism and Disposition*, and more. In addition, Dr Xie has authored up to ten chapters in books published by Annual Reviews, Springer, Wiley, ASM Press, and CRC Press abroad, as well as the People's Medical Publishing House, Beijing, China, of which he served as a co-editor and associated editor for two books, respectively. Dr Xie has served on the editorial board of *BMC Medical Genomics* (associate editor), *Pharmacogenomics*, UK (2003–2011), *American Journal of Pharmacogenomics* (currently *Molecular Diagnosis and Therapy*), New Zealand, and *Journal of Geriatric Cardiology*, China. Dr Xie has also been invited as a manuscript reviewer by over 40 biomedical and pharmacological journals published predominantly in North America and Europe.

Ashim Malhotra, PhD, FAPE, is a pharmacist, an NYU Grossman School of Medicine-trained and extramurally funded molecular pharmacologist, and a pharmacy educator with over 15 years of academic experience.

He serves as Assistant Dean for Accreditation and Program Development, University Distinguished Teacher, and Associate Professor at the California Northstate University (CNU) College of Pharmacy (COP). He is Director of the CNU Institute of Teaching and Learning Excellence and also directs the University's interprofessional education and practice program. In these roles, he leads faculty development activities for CNU's six colleges of Medicine, Pharmacy, Health Sciences, Psychology, Graduate Studies, and Dental Medicine.

Dr Malhotra earned a Bachelor of Pharmacy degree in 2000, MS (2003), and PhD (2006) degrees specializing in molecular pharmacology from St John's University in New York City. He conducted postdoctoral work at the New York University (NYU) Grossman School of Medicine. Elected Lifetime Fellow of the American Academy of Pharmacology Educators of the American Society of Pharmacology and Experimental Therapeutics in 2020, his work has been recognized through institutional and national awards.

Dr Malhotra has a rich history of mentoring PharmD and undergraduate students and postdoctoral fellows in bench and pedagogy research. For seven continuous years (2014–2021), work conducted by his students was recognized through national awards such as the 2021 and 2019 AACP Walmart Scholar, 2015–2018 American Society of Pharmacology and Experimental Therapeutics (ASPET) travel awards, and 2017 first prize winner at the Dolores Shockley ASPET poster competition, among others.

Dr Malhotra has published over 44 peer-reviewed journal articles and book chapters and more than 80 conference abstracts, presented 32 invited national podium sessions, chaired an international conference on pancreatic cancer, edited two books, and garnered $374,000 in research funding with an emphasis on mitochondrial pharmacology and also the scholarship of teaching and learning. He keeps Pacific Standard Time and enjoys the snowy Sierras, theatre, poetry, and reading.

Catherine F. Yang, PhD, Vice President of Academic Affairs, Interim Dean, College of Medicine and Professor of Molecular Pharmacology/Medicinal Chemistry/Biomedical Sciences. Dr Yang received her Masters and PhD from Tufts University, followed by postdoctoral training in Molecular Pharmacology/Clinical Biochemistry from Harvard Medical School.

Dr Yang is a Professor of Molecular Pharmacology and Clinical Biochemistry at the Department of Basic Science of College of Medicine of California Northstate University (CNU). Before joining CNU in 2016, Dr Yang was a Professor of Biochemistry and Pharmacology at Rowan University and held a faculty appointment at its Cooper Medical School. She has also held research and faculty positions at Harvard Medical School, the American Health Foundation, Boston Biomedical Research Institute, Tokyo University of Medicine and Dentistry, University of Pennsylvania, and the Zhejiang University of Technology. Dr Yang has over 25 years of teaching experience in the areas of medicinal chemistry, molecular pharmacology, and clinical biochemistry.

Dr Yang's research has made strong contributions in elucidating mechanisms of tumor progression, the development of novel cancer drugs, and antibiotics. She has led research groups studying proteolytic regulatory mechanisms in the advanced stages of prostate cancer, lung cancer, and leukemia. Her in-depth research on type 2 diabetic metabolic regulation led to a dual function diabetes drug patent. Dr Yang's immunological research resulted in an allergy vaccine development that is currently under clinical trials at affiliated clinics.

Dr Yang has published more than 70 research papers and several biotechnology books and is an inventor of 10 patented inventions. She has also secured numerous grants from the NIH, NSF, Research Corporation, and New Jersey Health Foundation as well as funding from many pharmaceutical corporations and health foundations. She serves on various review boards of federal, private, and health foundation funding agencies.

Contributors

Sayeed Ahmad
School of Pharmaceutical Education and Research
Hamdard University
New Delhi, India

Fakhrul Ahsan
College of Pharmacy
California Northstate University
Elk Grove, California

R. Jayachandra Babu
Harrison School of Pharmacy
Auburn University
Alabama

Rinda Devi Bachu
College of Pharmacy and Pharmaceutical Sciences
The University of Toledo
Toledo, Ohio

Yahya B. Bello
College of Pharmacy
Chicago State University
Chicago, Illinois

Gewedy Berhe
College of Pharmacy
Chicago State University
Chicago, Illinois

Sai H.S. Boddu
Ajman University, UAE

Nivaz Brar
College of Pharmacy and College of Medicine
California Northstate University
Elk Grove, California

Saba Ghatrani Chalak
CVS Pharmacy
Los Gatos, California

Vaibhav Changedia
Dr Vedprakash Patil Pharmacy College
Aurangabad, India

Sylvia Nam-Phuong Dinh
College of Medicine
California Northstate University
Elk Grove, California

Qian-Ming Du
General Clinical Research Center
Nanjing First Hospital
Nanjing Medical University
Nanjing, China

Ahmed El-Shamy
College of Graduate Studies
California Northstate University
Elk Grove, California

Drew Fajardo
College of Medicine
California Northstate University
Elk Grove, California

Jiajun Fan
School of Pharmacy
Fudan University
Shanghai Hailu Biological Technology Co., Ltd.
Shanghai, China

Ahmed Abu Fayyad
College of Pharmacy
Chicago State University
Chicago, Illinois

Xiaodong Feng
College of Pharmacy and College of Medicine
California Northstate University
Elk Grove, California

Amy Ferrarotti
College of Pharmacy
California Northstate University
Elk Grove, California

Leo R. Fitzpatrick
College of Pharmacy
California Northstate University
Elk Grove, California

Caroline Goswami
College of Graduate Studies
California Northstate University
Elk Grove, California

Kanika Gulia
College of Graduate Studies – Pharmaceutical Sciences
California Northstate University
Elk Grove, California

Bang-Shun He
Nanjing First Hospital
Nanjing Medical University
Nanjing, China

DeVon Herr
Computer Science
Purdue University
West Lafayette, Indiana

Nazir Hossen
College of Pharmacy
California Northstate University
Elk Grove, California

Christiane How-Volkman
College of Graduate Studies
California Northstate University
Elk Grove, California

Tanvirul Hye
College of Pharmacy
California Northstate University
Elk Grove, California

Mohd Imran
Faculty of Pharmacy
Northern Border University
Rafha, Saudi Arabia

Satori Iwamoto
College of Graduate Studies
California Northstate University
Elk Grove, California

Simeon Kotchoni
College of Graduate Studies
California Northstate University
Elk Grove, California

Shikha Kumari
College of Pharmacy and Pharmaceutical Sciences
The University of Toledo
Toledo, Ohio

Uyen Minh Le
College of Pharmacy
California Northstate University
Elk Grove, California

Yubin Li
Perelman School of Medicine
University of Pennsylvania
Xinqiao Hospital, Third Military Medical University
Chongqing, China

Mengyao Liu
College of Graduate Studies
California Northstate University
Elk Grove, California

James Alexander Lugtu
College of Pharmacy
California Northstate University
Elk Grove, California

Ashim Malhotra
Institute of Teaching and Learning Excellence
Interprofessional Education and Practice Program
California Northstate University
Elk Grove, California

Priya K. Manhas
College of Graduate Studies
California Northstate University
Elk Grove, California

Ella Mokrushin
College of Pharmacy
California Northstate University
Elk Grove, California

Bahaar Kaur Muhar
College of Graduate Studies
California Northstate University
Elk Grove, California

Tung Hoang Ngo
Dr Tung Ngo, Inc.
Irvine, California

Anh Nguyen
College of Graduate Studies
California Northstate University
Elk Grove, California

Yashwant Pathak
Taneja College of Pharmacy
University of South Florida
Florida

Xinyu Pei
College of Medicine
California Northstate University
Elk Grove, California

Angela Penney
College of Graduate Studies
California Northstate University
Elk Grove, California

Mandeep Rajpal
College of Pharmacy
California Northstate University
Elk Grove, California

Dilip Shah
Cooper Medical School of Rowan University
Camden, New Jersey

Mohd Shahid
Department of Pharmaceutical Sciences
Chicago State University College of Pharmacy
Chicago, Illinois

Saif Anwar Shaikh
College of Pharmacy and College of Medicine
California Northstate University
Elk Grove, California

Anitha K. Shenoy
California Health and Science University College of Pharmacy
Clovis, California

Yihui Shi
College of Medicine
California Northstate University
Elk Grove, California

Jianbo Song
Fulgent Genetics
Temple City, California

Mohammad Tauseef
College of Pharmacy
Chicago State University
Chicago, Illinois

Amit K. Tiwari
College of Pharmacy and College of Medicine
Head, Cancer & Systems Therapeutics
College of Pharmacy and Pharmaceutical Sciences
The University of Toledo
Toledo, Ohio

Hongbin Wang
College of Graduate Studies
College of Pharmacy
California Northstate University
Elk Grove, California

Shaofei Wang
Department of Cellular and Genetic Medicine
School of Basic Medical Sciences
Fudan University
Shanghai, China

Wenjia Wang
College of Medicine
California Northstate University
Elk Grove, California

Tiffany R. Wong
College of Graduate Studies – Pharmaceutical Sciences
California Northstate University
Elk Grove, California

Tibebe Woldemariam
College of Pharmacy
California Northstate University
Elk Grove, California

Olivia Wu
College of Graduate Studies
California Northstate University
Elk Grove, California

Hong-Guang Xie
General Clinical Research Center
Nanjing First Hospital
Nanjing Medical University
Nanjing, China

Meng-Qiu Xiong
Nanjing First Hospital
Nanjing Medical University
Nanjing, China

Catherine F. Yang
College of Graduate Studies
College of Medicine
California Northstate University
Elk Grove, California

Erika Young
College of Pharmacy
Northeastern University
Boston, Massachusetts

Janie Yu
College of Pharmacy
California Northstate University
Elk Grove, California

Yingqi Zhang
Elixir Rx Solutions
Twinsburg, Ohio

James Zhou
College of Medicine
California Northstate University
Elk Grove, California

Biologics and Biosimilars 101 1

Principles and Practice

Hong-Guang Xie

Corresponding author: Prof. Hong-Guang Xie

Contents

1.1 INTRODUCTION

The documented history of use of the biological products can be traced back to an earlier vaccine story about Edward Jenner fighting against smallpox in 1796. However, the first therapeutic protein product (Humulin) was approved by the US Food and Drug Administration (FDA) in 1982,[1] which was recombinant human insulin developed by Genentech (later licensed to Eli Lilly) through using recombinant DNA technology to synthesize human insulin in bacteria. Since then, biotechnology innovation and scientific breakthroughs have resulted in the creation of a new class of drugs/medicines named biological products

DOI: 10.1201/9780429485626-1

that are composed of biologics, biosimilars, and interchangeable biosimilar products. As biotechnology constantly advances, new biological products are ceaseless occurrence in large numbers in the global biopharmaceutical marketplace due to the urgent need of the increased market and clinical access to life-saving medications.

In the past four decades of biological product development, there exist several milestone events worth mentioning. To increase more treatment options and reduce the healthcare cost of biologics through market competition, FDA established the 351(k) pathway, an abbreviated approval process for the biosimilars or interchangeable biosimilar products, as part of the response to the Biologics Price Competition and Innovation Act of 2009 (BPCI Act, signed in law in 2010).[2] After that, the first biosimilar filgrastim-sndz (recombinant human granulocyte colony-stimulating factor, or rhG-CSF) and the first interchangeable biosimilar product insulin glargine-yfgn (a long-acting insulin analog for diabetes) were approved by the FDA in 2015 and 2021, respectively (see Table 1.1). Since the approval of Humulin in 1982,[1] biologics and their derived biosimilars have gradually dominated new drug research and development (R&D) worldwide. As of 2021, more than 350 biological products have been approved by the FDA.[3,4] Clearly, these biological products represent the cutting-edge and frontier of biomedical and biopharmaceutical science research, providing more treatment choices and the most effective means to prevent, treat, and even cure a series of life-threatening diseases and critically ill conditions, some of which currently have no other available medications or therapy regimens (see below for their main indications).

As starting one of a series of the book chapters, this chapter briefly presents some essential knowledge about the biological products for the reader, such as their concise definitions, regulatory policy, justification of labeled dosing, main indications, the fate of the drugs in human body, and cost-effective strategies for their development and utilization.

1.1.1 How to Define the Term Biological Product?

As a new drug class, biologics, biosimilars, and interchangeable biosimilar products are all defined as the drugs or medicines that are manufactured from living organisms and distinguished from those that are from chemical synthesis. By definition, biological products contain an array of complex bioactive macromolecules (such as peptides or proteins) that are naturally occurring substances synthesized in living systems or produced by biotechnology and other cutting-edge technologies. They are frequently isolated from a variety of natural or biological sources, such as humans, animals, microorganisms, and even plants. In fact, as a group of the biological products that are highly similar to each other, biosimilars and their interchangeable products are almost the same as their corresponding biologics (currently defined as biological reference product or originator as a standard control due to the FDA approval first) in nature (at least bioactive principal components), with no clinically meaningful differences from their biologic reference product approved by the FDA first.

The term "biologic" is defined as a bioactive macromolecule that is manufactured in living systems for certain indications of patients. In general, biologics are more expensive than small-molecule chemical drugs because they demonstrate structural complexity and complicated manufacturing process.[5] This results in little market competition and almost exclusively market share, both of which are unfavorable conditions for increasing more treatment options and reducing healthcare cost. The high cost of the biologics also restricts their clinical access to a greater extent. On the other hand, because it is almost impossible to manufacture the "identical" biologics in living systems, there exists no "generic drug" for biologics. Therefore, the term "biosimilar" was proposed, which imitated or copied the concept of generic drug of chemically synthesized drugs. Correspondingly, the FDA established the 351(k) pathway as an abbreviated biologics license application (BLA) process for the biosimilars, which is similar to the abbreviated new drug application (ANDA) pathway for approval of the generic drugs. In other words, biologics are approved through the process of conventional BLAs, whereas biosimilars are through the abbreviated BLA process (i.e., the 351(k) pathway). As of July 2021, at least 30 biosimilar products have been licensed by the FDA (see Table 1.1).[6]

TABLE 1.1 A total of 30 approved biosimilars in relation to their biologics (reference products) as a group of the biological products licensed in the United States (as of July 2021)

BIOLOGIC (REFERENCE PRODUCT)	*NAMING OF BIOSIMILARS*	*TIME OF LICENSURE*
Filgrastim (Neupogen)	Filgrastim-sndz (Zarxio)	March 2015
	Filgrastim-aafi (Nivestym)	July 2018
Infliximab (Remicade)	Infliximab-dyyb (Inflectra)	April 2016
	Infliximab-abda (Renflexis)	May 2017
	Infliximab-qbtx (Ixifi)	December 2017
	Infliximab-axxq (Avsola)	December 2019
Etanercept (Enbrel)	Etanercept-szzs (Erelzi)	August 2016
	Etanercept-ykro (Eticovo)	April 2019
Adalimumab (Humira)	Adalimumab-atto (Amjevita)	September 2016
	Adalimumab-adbm (Cyltezo)	August 2017
	Adalimumab-adaz (Hyrimoz)	October 2018
	Adalimumab-bwwd (Hadlima)	July 2019
	Adalimumab-afzb (Abrilada)	November 2019
	Adalimumab-fkjp (Hulio)	July 2020
Bevacizumab (Avastin)	Bevacizumab-awwb (Mvasi)	September 2017
	Bevacizumab-bvzr (Zirabev)	June 2019
Trastuzumab (Herceptin)	Trastuzumab-dkst (Ogivri)	December 2017
	Trastuzumab-pkrb (Herzuma)	December 2018
	Trastuzumab-dttb (Ontruzant)	January 2019
	Trastuzumab-qyyp (Trazimera)	March 2019
	Trastuzumab-anns (Kanjinti)	June 2019
Epoetin-alfa (Epogen)	Epoetin-alfa-epbx (Retacrit)	May 2018
PEGfilgrastim (Neulasta)	Pegfilgrastim-jmdb (Fulphila)	June 2018
	Pegfilgrastim-cbqv (Udenyca)	November 2018
	Pegfilgrastim-bmez (Ziextenzo)	November 2019
	Pegfilgrastim-apgf (Nyvepria)	June 2020
Rituximab (Rituxan)	Rituximab-abbs (Truxima)	November 2018
	Rituximab-pvvr (Ruxience)	July 2019
	Rituximab-arrx (Riabni)	December 2020
Insulin glargine (Lantus)	Insulin glargine-yfgn (Semglee)[a]	July 2021

Source: Data from the FDA.[6]

Notes: Presented in parentheses is a brand name of a certain biologic or biosimilar product for its proprietary nature or exclusivity when patented or registered. The naming of biosimilars is based on the rule of the nonproprietary name of a biologic, followed by four letters, which is convenient for tracking and direct comparison. Among them, (peg) filgrastim is a recombinant human leukocyte growth factor with or without polyethylene glycol (PEG); infliximab is a purified recombinant DNA derived chimeric (human/mouse) IgG mAb that binds and neutralizes TNFα; etanercept is a fusion protein of TNF receptor to the Fc portion of the IgG1 antibody; adalimumab is a fully human mAb targeting TNFα; bevacizumab is an mAb targeting VEGF; trastuzumab is an mAb targeting HER2 receptor; epoetin-alfa is a recombinant human erythropoietin; rituximab is a chimeric mAb targeting CD20 antigen on B-cells; insulin glargine is a recombinant human insulin analog.

[a] The interchangeable product (a biosimilar itself).

In some cases, a biosimilar can be approved as an interchangeable product if its clinical efficacy and safety profile can be validated to be almost the same as that of an existing FDA-approved reference product in clinical settings. Switching may occur from a marketed reference product to its interchangeable biosimilar product via auto-substitution at some (if not all) retail pharmacies if needed or permitted.

Therefore, the term "interchangeable biosimilar product" refers, in particular, to a certain biosimilar that can directly substitute its biologic reference product as an alternative option due to its highly clinical similarity to that reference product. For the biologic and its derived biosimilar products, their naming just varies by the time sequence when they are approved as a group of drugs.

According to the well-recognized regulatory policy, a biosimilar is a certain biological product that is highly similar to its corresponding reference product in the molecular structure, quality control data, safety indicators, bioactivity measures, immunogenicity, disposition kinetics, selection of labeled dosing (desired dosage and route of administration), clinical efficacy (licensed indications), and more. In other words, a biosimilar is a highly biosimilar version of an FDA already approved biologic,[7,8] and biologics and biosimilars are brand name drugs and their generic versions (or generics) for the same group of the biological products, respectively. Different from chemical drugs that are identical chemically and easy-to-characterize, biologics and biosimilars are not chemically identical from one batch to another due to difficulty in projection and reproducibility that results from complexity and uncertainty of biological processes found in living organisms, in particular biosimilars that are allowed minor differences or variations in molecular structure and immunogenicity as well as clinically inactive components when compared with their biologic reference product.[8,9] In addition, biological products are more susceptible to exogeneous contamination and environmental temperature than chemical drugs during the period of their manufacturing, storage, and utilization. Thus, the biological products themselves and the whole process of their manufacturing and distribution are all regulated and approved by the regulatory agencies, such as the FDA and European Medicines Agency (EMA).

1.1.2 How to Support and Justify the Dose of a New Biological Drug in Labeling?

The dose proposed in product labeling is the result of a series of dose exploration and selection conducted in its dose-ranging studies (such as multiple-dose escalation), and then dose evaluation, adjustment, and even optimization (if requested for some, if not all, cases) assessed in its registrational clinical trials. Further, the dose that is tested in pivotal trials will be used to inform product labeling for the regulatory agency to evaluate a regimen proposed by its applicant. In general, registrational trials in patients with one or more than one intended indication (such as a serious disease or critically ill condition) are treated with a new biologic at only one dosage that is considered as the most appropriate for these trials.[10] Justification of the dose studied in clinical trials or proposed in product labeling is one of the critical steps in the development pathway or production timeline. In other words, if the dosage decided for a new biologic in a registrational trial is markedly inadequate or even wrong, the product would be at higher risk for toxicity (at a higher dose), lack of clinical efficacy (at a lower dose), or even facing an overall unacceptable risk-benefit imbalance (at an inadequate dose), all of which would jeopardize its development pathway and even market access.[10]

Ogasawara et al. systematically summarized the justification and rationale of dose selection and the labeled dose of 59 out of 79 BLAs reviewed by the Center for Drug Evaluation and Research (CDER) at the FDA between 2003 and 2016. They observed that clinical efficacy of a new biologic was used as the rationale for dose selection more frequently than its clinical safety (73% vs. 42% of BLAs) in registrational clinical trials, that dose-response (D-R) analyses were often (72%) used to determine the dose of the biological products whose dose appeared to be selected based on clinical efficacy, but that exposure-response (E-R) analyses for efficacy were used more frequently than D-R analyses (53% vs. 21%) to support the doses that were proposed in their product labeling.[10] Since the public release of guidance for industry on E-R relationships in May 2003,[11] E-R analysis is becoming preferred as a means to help justify dosing guidance in labeling. After 2012, as pharmacometrics advances,[12,13] E-R analysis and pharmacokinetics (PK)/pharmacodynamics (PD) modeling and simulation of clinical efficacy and/or safety, in combination with artificial intelligence (AI) and deep machine learning,[14–18] are all becoming more popular to support the labeled dose than before.[10]

In general, D-R relationship is the first step in evaluating the effects of different doses on the efficacy and safety when dose-ranging studies are conducted to clarify the range (or borderline) of the effective and safe doses, such as maximum tolerated dose (also known as MTD). In the past two decades, D-R analysis was still more frequently used for dose selection for registrational trials of FDA-approved biologics than E-R analysis. Despite this situation, E-R analysis is more commonly used for (1) evaluating the influence of varying levels of exposure parameters (such as trough concentration, average or median concentration, area under the plasma drug concentration-time curve or AUC at a steady state, and maximum plasma drug concentration or C_{max}) on product safety and/or efficacy; (2) identifying whether exposure variables are significant covariates in the modeling analyses of efficacy or safety; (3) informing clinical study design for recruited patients who are categorized into subgroups stratified according to the characteristics of the exposure parameters (such as median value or quartile); (4) confirming the appropriateness of the proposed dose in labeling; and (5) identifying individuals or subgroups who may need dose adjustment.[10] Increasing evidence suggested that E-R analysis could offer more scientifically sound basis and more convincing rationale for better dose selection and adjustment in late-stage registrational clinical trials than D-R analysis alone.[10,19,20] Despite these advantages over D-R analysis, widespread use of E-R analysis, however, seems not to be prevalent as expected due to more time to be spent for drug concentration measurements as well as use of the more and more complicated modeling and simulation technology.

It is estimated that 1 out of 6 products (17%) that failed at the first-round approval were declined by the FDA due to the failure of dose selection and the proposed dose in the product labeling.[21] For the drug that is approved with inadequate evidence supporting the labeled dose, the sponsor would receive an FDA-requested post-marketing requirement or commitment (PMR or PMC) to conduct additional E-R analyses for dose optimization at the stage of undergoing late-stage clinical trials. For example, a PMR is assigned to determine the dose with the optimal risk-benefit ratio for intended patients,[22] whereas a PMC is requested to optimize the regimen for a subgroup of patients with lower exposure at the approved dose,[23,24] or those with higher body weight (a significant covariate as assessed in modeling of clinical efficacy or safety that dramatically affects exposure parameters or drug concentration measurements).[25] The main explanations for these cases are that a proposed dose by the applicant might not be considered optimal by the regulatory agency and would need to be requested for further dose optimization if there was a significant difference in clinical efficacy or safety endpoints observed among different subgroups.[10] It is believed that the number of PMR/PMC focused on dose optimization may increase with either increased attention to E-R analysis or increased awareness of E-R relationships, but that the number of BLAs submitted with no E-R analysis may decline.

1.1.3 Who May Use the Marketed Biological Products?

Development and application of the biological drugs (also called biopharmaceuticals, or biodrugs) are aimed to prevent or treat some (if not all) chronic or life-threatening diseases or critically ill conditions, including but not limited to various cancers, diabetes, chronic inflammatory diseases (such as rheumatoid arthritis, ulcerative colitis, Crohn's disease, and ankylosing spondylitis), autoimmune-associated diseases (such as severe asthma, thrombocytopenia, systemic lupus erythematosus, myasthenia gravis, and pemphigus vulgaris), infectious diseases (such as AIDS, COVID-19, and viral hepatitis), hypercholesterolemia, multiple sclerosis, amyotrophic lateral sclerosis (ALS), psoriasis, severe chronic anemia, neutropenia, hereditary diseases (such as hemophilia, dwarfism, and gigantism), cystic fibrosis, age-related macular degeneration, cardiovascular diseases, rare diseases (such as spinal muscular atrophy or SMA, and growth hormone deficiency or GHD) treated with orphan drugs, and more. All these diseases and their complications are the clinical indications of the biological drugs (approved or underway). For the marketed biological products, their indications are each detailed in their product labeling and can also be found in the official website or URL link of the regulatory agencies, such as Drugs@FDA,[26] and in the Purple Book (a database of the approved biological products) launched by the FDA.[4]

In fact, BLAs for therapeutic use are a superfamily that is composed of various families and family members, including but not limited to an array of monoclonal antibody (mAb) drugs, peptides, and proteins for therapeutic use (such as insulin, hormone products, and enzymes), non-vaccine and nonallergenic immunomodulators, cytokines (such as interferons and interleukins), growth factors (such as VEGF, G-CSF, and EPO), and complement components (such as C3 and C5), all of which are reviewed and approved by the CDER at the FDA.[10] In addition, other relevant biological products include, but not limited to, human blood and blood components, plasma-derived products (such as albumin, immunoglobulin (Ig) proteins, or peptides), RhoGAM shot, and blood clotting factors, antitoxins, antivenins, animal venoms, cellular products, vaccines, allergenic products, and gene-therapy products, all of which are oversighted and approved by the Center for Biologics Evaluation and Research (CBER) at the FDA.[10] In addition to use of therapeutic proposes (prevention or treatment), some biological products can also be developed and used for molecular diagnostics or theranostics as a companion testing, such as measurement of HER2 protein over-expression/-amplification by immunohistochemistry (IHC) used for HER2-targeted therapy in patients with HER2-positive breast cancer (approximately 10–20% of all breast cancer cases), because these HER2-positive patients are much more likely to respond to treatment with drugs that target the HER2 proteins (receptors) expressed on breast cells, such as trastuzumab (Herceptin). And another similar case is to use KRAS companion diagnostic to accompany panitumumab.

The mechanisms of action of the biological products may vary by drugs themselves as well as their indications. For peptide- and mAb-type biologics, they exert their efficacy as an agonist/stimulator or antagonist/inhibitor that can specifically bind their targets. Many of the mAb drugs are antagonists or inhibitors, requiring high-dose administration in clinical settings. In contrast, as an efficient means to supplement or restore the insufficient amount of a certain endogenous substance important for a normal physiological or biochemical process for life, most peptides, such as insulin, growth hormone (GH), and coagulation factors VIII and IX, are used at very low dose to "rescue" these diseases due to insufficiency or deficiency via agonism or replacement therapy, such as diabetes, dwarfism, and hemophilia A and B, respectively.

The ADC (antibody-drug conjugate) is a new class of targeted drugs composed of a tumor antigen-specific mAb, a potent cytotoxic drug (also called payload), and a stable chemical linker that link these two components. In general, the ADCs exert their effects through the following four steps: (1) the circulating ADC binds to the antigen expressed on the target cell surface; (2) the ADC-antigen complex is internalized by endocytosis; (3) the ADC is degraded in the acidic lysosome; and (4) the cytotoxic payload (drug) is released to elicit target cell apoptosis.[27] For example, trastuzumab deruxtecan (T-DXd), a HER2-targeting mAb conjugated with a topoisomerase I inhibitor deruxtecan (DXd), was used to effectively treat patients with non-small cell lung cancer (NSCLC) harboring HER2-mutant alleles.[28] As of 2021, there are at least 11 FDA-approved ADCs.

Accordingly, new biological drugs would be developed based on identical, similar, or diverse therapeutic targets that may cause a certain disease, such as hormones, receptors, enzymes, cytokines, growth factors, signaling molecules, and more. As a result, one BLA can be labeled for more than one indication, indicating broad spectrum. For example, adalimumab (Humira) is a recombinant human mAb against TNFα (tumor necrosis factor-alpha), with at least 12 indications (autoimmune diseases) approved worldwide, including but not limited to rheumatoid arthritis (RA), Crohn's disease (CD), ulcerative colitis (UC), ankylosing spondylitis (AS), plaque psoriasis (PPs), psoriatic arthritis (PsA), polyarticular juvenile idiopathic arthritis (PJIA), and more. In addition, other TNFα-targeted biological drugs include infliximab (Remicade, an mAb against TNFα) and Enbrel (etanercept, a fusion protein against TNFα). This indicated that TNFα (an important inflammatory cytokine) is one of the shared therapeutic targets of various autoimmune-associated inflammatory diseases, and that these TNFα-targeted protein drugs are all effective to treat these diseases associated with autoimmune disorders and inflammation through inhibition of TNFα activity on the basis of the so-called class effect. On the other hand, one indication can be assigned with more than one BLA. For example, rheumatoid arthritis can be treated either with TNFα-targeted biological drugs, such as adalimumab, infliximab, and Enbrel or with rituximab

(Rituxan), which is an mAb against antigen CD20 on B lymphocytes. This clearly demonstrates (1) that a certain disease may have more than one therapeutic target, such as TNFα and CD20 for rheumatoid arthritis; (2) that concomitant use of biological drugs acting on different targets is expected to exert synergistic or additional efficacy for a certain disease; and (3) that the development of new biodrugs may follow the pipeline of multiple different therapeutic targets for the same indication or the same biodrug for the more than one indication.

If the patent expires, one indication that has been licensed for the reference product might be approved for its biosimilar or interchangeable product via another licensure approach named "indication extrapolation", which is a new concept proposed by the FDA as part of guidelines for biosimilar approval even if that biosimilar is not directly validated in that indication. In most cases, a manufacturer who plans to propose a biosimilar product may choose to only apply for one or part of all approved indications of interest. When the applicants or sponsors seek licensure of their proposed biosimilar product for one or more than one indication, rather than all indications, that its reference product was licensed for, they may still need to provide their own clinical safety and efficacy data acquired from the validation study of proposed appropriate indication(s) demonstrating biosimilarity to that reference product for each proposed indication in different patient populations for that indication extrapolation. However, some indications with orphan drug designation with exclusivity cannot be extrapolated due to the presence of remaining exclusivity period at the time of application.

1.1.4 How Does Neonatal Fc Receptor (FcRn) Work?

The term FcRn refers to neonatal Fc (fragment crystallizable) receptor or Fc receptor, which is the major histocompatibility complex (MHC) class I-related receptors, comprising an α-FcRn chain and β2 microglobulin components.[29–31] It is well known that FcRn functions as a recycling or transcytosis receptor that bidirectionally transports endogenous IgG and albumin across polarized cellular barriers,[32] facilitating their translocation, protecting these two ligands from intracellular catabolism (degradation),[33] and extending their circulating half-life (t½) as well as maintaining their homeostasis in the blood.[34,35] In addition, the FcRn also acts as an immune receptor through interacting with and facilitating antigen presentation of peptides derived from IgG immune complexes.[34] Based on this concept, use of the specific FcRn inhibitors or antagonism of the FcRn receptor is proposed to treat some autoimmune diseases,[36–38] such as myasthenia gravis. Now the FcRn comes of age.[39]

Fast-growing evidence indicated that FcRn is a crucial component with a direct link to the disposition and clearance of Fc, mAbs (via the Fc region), albumin, albumin-fused peptides, and fatty acid-conjugated peptides (acylated), facilitating the absorption of mAbs and fused/conjugated peptides from the subcutaneous space (and also presumably from the gut) via FcRn-mediated transcytosis,[40,41] regulating the systemic clearance and tissue distribution of mAbs and peptides.[41] For example, FcRn knock-out (KO) mice exhibited ~ three-fold lower subcutaneous bioavailability for an IgG1-type mAb than wild-type (WT) mice (control);[40] an abnormally short serum half-life of IgG was observed in β2-microglobulin-deficient mice;[42] the serum persistence of an IgG fragment was increased by random mutagenesis;[43] and serum half-life of albumin was extended by engineering FcRn binding.[44] Moreover, the clearance of 7E3 (an antiplatelet mAb; IgG1) increased with increasing doses in WT mice (due to binding saturation at high dose), but not in FcRn KO mice,[45] and the albumin-binding site of the FcRn was identified by structure-based mutagenesis.[46] All these results show rapid clearance of endogenous IgG and albumin in FcRn KO mice compared with WT mice, revealing that the FcRn functions to salvage IgG and albumin taken into cells, protecting them from elimination in the circulation. Further evidence included that uptake (internalization) of IgG and albumin into cells is FcRn-mediated,[47] and that the binding of FcRn to IgG (mediated through the Fc region) and albumin is pH-dependent within the endosomes, with a reduction in pH (~5 to 6) facilitating binding due to high affinity, followed by recycling of the FcRn-IgG or FcRn-albumin complex and exocytosis (or release) of bound molecules at a higher extracellular pH environment (usually ~7.4) due to pronouncedly lower FcRn affinity (see Figure 1.1 adapted from Ward et al.[48] for details).[39,48] On the other hand, IgG unbound to FcRn in the endosomes undergoes intracellular catabolism (proteolytic degradation) in lysosomes by various soluble or membrane-bound proteases.[35,48,49] Clearly, dynamics profile of IgG and albumin (bound vs. unbound) in their processed pathways (recycling vs. degraded) is the major determinant of the clearance, half-life, and thus the exposure of IgG and albumin species in the circulation. For IgG

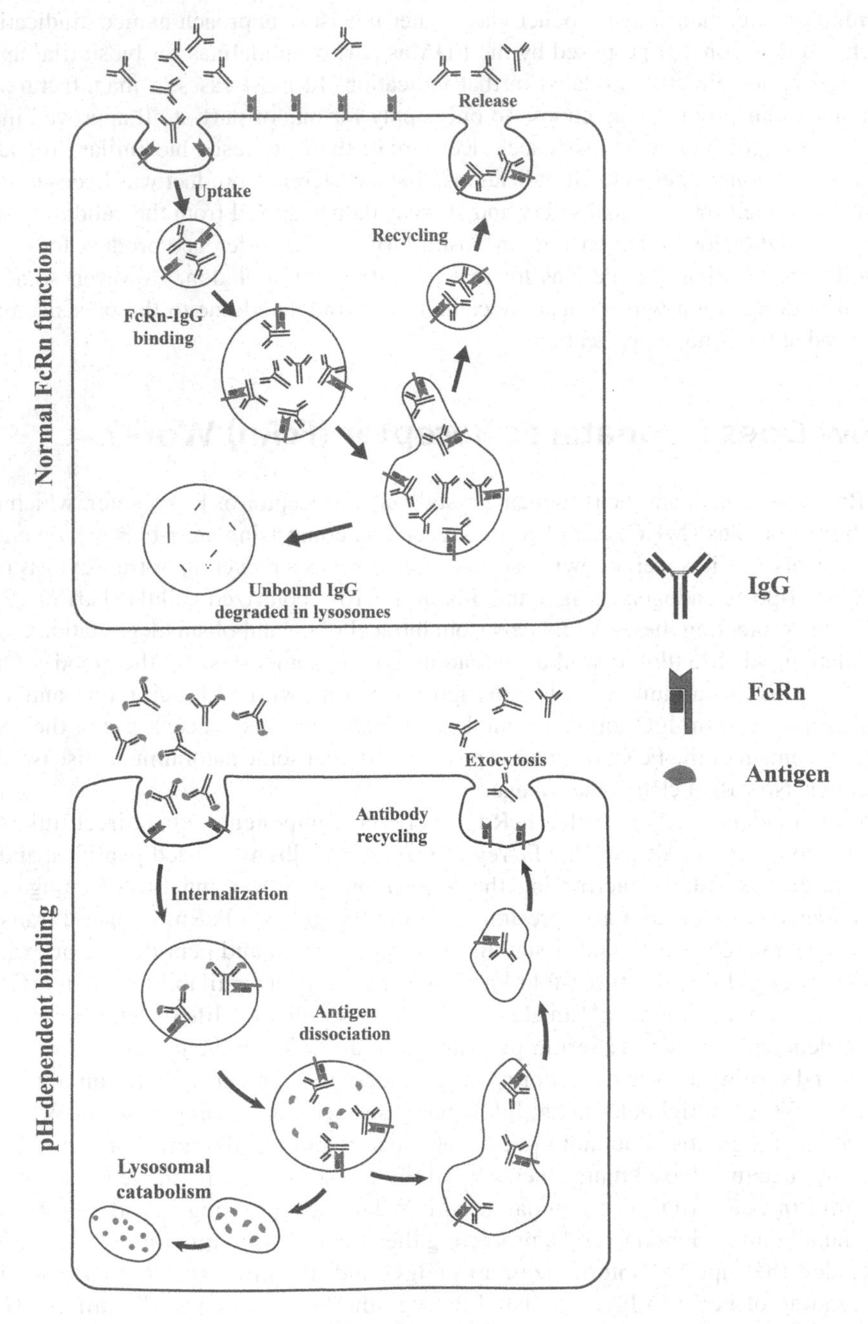

FIGURE 1.1 The FcRn-mediated recycling and release of the antibody IgG under physiological conditions (upper panel), and recycling (exocytosis) of the antibody when dissociated with its specific antigen within acidified endosomes (lower panel). (Adapted from Ward et al.[48])

antibodies, in addition to target-mediated elimination (see the content in the TMDD section), they also undergo FcRn-mediated elimination. However, the FcRn-mediated salvage pathways (recycling and exocytosis) are capacity limited and are expected to be saturated at very high doses (>100 mg/kg).[50] At the recommended therapeutic doses (usually <10 mg/kg), the nonlinear PK of an mAb is attributable to TMDD and to a less extent FcRn-mediated elimination due to less frequently seen saturation of FcRn and subsequent increase in the mAb clearance.[50,51] Please note that an Fc region or albumin also interacts with FcRn when it is fused/conjugated with a peptide or protein for increased molecular size to slow glomerular filtration and renal clearance (see below). In these cases, improved FcRn-mediated recycling (see Figure 1.1) is an important mechanism by which these fused moieties are also salvaged from intracellular catabolism for their slowed systemic clearance and extended circulation time or serum persistence. In addition, FcRn-mediated active transport could facilitate tissue distribution of mAbs and the fused peptides.

Please note that oxidation of methionine in the Fc region of mAbs and oxidation of methionine or tryptophan in the complementarity-determining region (CDR) of some mAbs were reported to reduce FcRn binding and the half-life of IgG and alumin,[52,53] or induce a loss of antigen binding and decreased potency,[54,55] with no reports currently available showing the effects of oxidation on the disposition and clearance of mAbs.[41]

The FcRn is a multitasking protein, transporting its ligand IgG within and across cells of diverse origins and regulating IgG homeostasis in the circulation. The main applications of FcRn-IgG interactions are focused on the following points. First, the FcRn helps to explain the much longer half-life of IgG compared with other Ig protein classes.[51] Second, the half-life extension of IgG by protein engineering may reduce the frequency for dosing. Third, the FcRn is a novel target for therapeutic antibodies and antibody engineering, exhibiting considerable promise for next-generation therapeutics and diagnostics.[56] For example, FcRn itself can be targeted by the FcRn inhibitors or antibodies to decrease circulating IgG concentrations, which could have implications in certain clinical settings. Currently, the development and use of the anti-FcRn antibodies exhibit clear prospects in the treatment of some diseases associated with elevated pathogenic IgG (auto)antibodies, such as pemphigus and pemphigoid disease,[57] primary immune thrombocytopenia,[58] organ transplantation (due to the occurrence of donor-specific IgG antibodies),[59] and neonatal diseases associated with transplacental transfer of maternal IgG antibodies.[60] Now, there are some promising BLAs, such as M281 (a fully human, aglycosylated IgG1 mAb), 4C9 (an anti-FcRn mAb), and 4470 (anti-mouse FcRn mAb), all of which could shorten the half-life of IgG and thus decrease pathogenic IgG levels in the blood through blockade of FcRn receptors and inhibition of the FcRn-mediated IgG recycling.[57–61]

1.1.5 What Is the Target-Mediated Drug Disposition (TMDD)?

TMDD is a term that was first raised by Dr. Levy in 1994,[62] describing a nonlinear PK behavior that is caused by the high-affinity binding of a compound to its pharmacological target. Clearly, the occurrence of TMDD is attributed to the complicated interactions of its pharmacological target with a drug itself (small or large molecule) at the sites of action (i.e., therapeutic targets) in the process of PD as a result of "PD affecting PK".[50] Increasing evidence demonstrated that, compared with a relatively low prevalence of TMDD in small-molecule drugs,[50,63] the prevalence of TMDD is high for large-molecule compounds (such as therapeutic protein drugs) since they usually have highly specific, high-affinity binding to their pharmacological targets (such as receptor, or target antigen), but concomitantly minimal binding to non-specific tissues and/or plasma proteins whose binding capacity is usually larger than we imaged.[50,64–70] Furthermore, a compound with low-capacity target binding seems to be more susceptible to nonlinear PK (which means a disproportional ratio of the dose to exposure) than that with high-capacity target binding because the low-capacity target is easily saturated at high doses. In other words, TMDD or nonlinear PK behavior of a compound (small or large) occurs usually at lower single dose due to highly specific binding to its low-capacity target with high affinity. Conversely, the target binding, even with high affinity, could be disguised easily by substantial nonspecific tissue binding.

Not all large-molecule compounds exhibit TMDD. In general, TMDD could occur when a large-molecule compound manifests nonlinear PK behavior across the dose range evaluated, whose nonlinearity is imparted by the specific binding of large molecules to its pharmacological targets with low capacity and high affinity. After a therapeutic protein (as a case of large-molecule compounds in this chapter) binds to its target to exert its efficacy, some (if not all) of the protein-target complex undergoes subsequent internalization, endocytosis, and degradation into amino acids for further recycling in a so-called mutual assured destruction (MAD) manner, suggesting that TMDD represents a major elimination pathway for most protein drugs. On the other hand, the remaining protein-target complex triggers cell proliferation and target synthesis on the surface of these target cells, leading to increased target binding capacity (due to expanded target "pool") and thus enhanced degradation of protein drugs (or elevated clearance as a result). Obviously, such an interplay between PD and PK is interchangeable and dynamic over time for some protein drugs (except for the antibody drugs; see below for details). For example, the serum concentrations of filgrastim, a recombinant human (rh) G-CSF, was found to be correlated inversely with the number of circulating neutrophils (part of G-CSF receptors are expressed on the peripheral neutrophils), suggesting that the saturable clearance pathway of filgrastim is mediated by neutrophils.[50] Furthermore, the clearance of rhG-CSF was also found to be closely associated with both the number of peripheral neutrophils and the percentage of G-CSF receptor-positive neutrophils in cancer patients,[71] demonstrating that the neutrophils-mediated clearance of rhG-CSF is mediated by G-CSF receptor on the surface of neutrophils in the blood and neutrophil precursors in the bone marrow (i.e., myeloid progenitor cells). Similar to filgrastim, plasma concentrations of pegfilgrastim (a PEGylated version of filgrastim; modified for half-life extension), was found to be decreased with multiple dosing in a time-dependent manner, whereas the absolute neutrophil count changed inversely, suggesting that its elevated clearance after repeated doses is also the result of expanded target "pool" (i.e., increased number of neutrophils and neutrophil precursors).[50] Moreover, this time-dependent nonlinearity of pegfilgrastim exposure was more profound at lower doses than at higher doses.[50] This is because neutrophil-mediated clearance of pegfilgrastim predominates at lower doses but gets attenuated at higher doses due to saturation of the target binding. Subsequent evidence indicated that the G-CSF receptor was identified as the target that specifically binds filgrastim or pegfilgrastim using G-SCF receptor KO mouse model, with linear PK and significantly decreased clearance of pegfilgrastim observed in KO mice, or nonlinear PK and increased clearance in WT mice when compared one with another.[72] This further confirmed that neutrophils-mediated clearance of pegfilgrastim is mediated by the G-CSF receptors expressed on the surface of neutrophils and their precursor cells, representing the major elimination pathway when the drug is given at lower doses. In other words, the target of these protein drugs destroys and clears themselves. A similar story occurred with the case of rhEPO (recombinant human erythropoietin), in which its clearance also decreases with increasing doses (or dose-proportional) but increases with multiple doses repeated (or time-dependent).[73–76]

Different from the above-mentioned two protein drugs (rhG-CSF and rhEPO products), the PK/PD interplay of various antibody (Ab) drugs, such as mAbs and even bispecific antibodies, is not strong as seen in the other protein cases because their Ab-target complex seems not to modulate dynamic change of the target expression and synthesis. However, these antibody drugs still exhibit TMDD and thus behave as nonlinear PK.[50,64,65,67,77] This is thought to be related to clearance of the antibody drugs via saturable target-mediated endocytosis and subsequent lysosomal degradation of antibody-receptor complex. The clearance of an mAb is relatively unaffected at its low exposure, and the target-mediated elimination represents the major clearance pathway. However, its clearance decreases with an increase in its exposure or doses because the target-mediated elimination pathway is saturated concentration-dependently, showing biphasic clearance or S-shaped mAb concentration-time curves when the TRX1 (anti-CD4 mAb) doses are increased gradually.[78] Of the antibody drugs, IgG antibody drugs also undergo FcRn-mediated elimination in addition to TMDD.[79] In terms of the fact that saturation of FcRn and subsequent increased clearance of IgG-type mAbs are less frequently seen within the labeled doses (usually <10 mg/kg) due to larger FcRn binding capacity,[51] nonlinear PK of these mAb drugs (IgG) is predominantly attributed to TMDD. Thus, how to determine whether nonlinear PK profile of a large-molecule compound is attributed to the TMDD is an important question to be answered.

1.1.6 How Does the Nontarget-Related Clearance Work?

In addition to TMDD and FcRn-mediated clearance (see above for details), many nontarget-related factors could affect the disposition and PK of the biological products, in particular mAbs and peptides,[41] leading to variations in the circulating exposure and consequently exposure-related response (i.e., efficacy). Further, thorough knowledge of nontarget-related clearance mechanisms would help enhance the druggability of therapeutic peptides and mAbs and improve the compliance and experience in patients when administered with these peptide- and mAb-based biologics.

Accumulating evidence indicated that there are a large number of causes that are involved in clinical attrition and even failure for biologics and biosimilars. Of them, poor relationship between exposure (PK) and response (efficacy, PD, or pharmacology) is considered as the major factor that precludes achieving market access and reaching patients.[41,80–82] Knowledge of various factors affecting the clearance of biologics and biosimilars from the body will further enhance the druggability of the biological products under development. Notably, optimizing their druggability would benefit patients the most.

Aim 1 is to extend elimination half-life of the biological products in the blood circulation by time-extension technology, reducing, in turn, their systemic clearance, the effective dosage required, and/or dosing frequency recommended as a result. In general, a peptide or small-size peptidic protein with a molecular weight (MW) of <2–50 kDa, rather than an mAb molecule (typically, MW ~150 KDa or above), is liable to renal filtration and consequently removal,[83,84] with some unmodified peptides exhibiting a shorter circulating half-life (usually in minutes) in the presence of normal renal function,[85] which could make product development and utilization very difficult. Clearly, increasing apparent MW of a peptide molecule will reduce its renal filtration and elimination. For example, conjugation of large polymers, such as polyethylene glycol (PEG), polysialic acid (PSA), hydroxyethyl starch (HES), to a peptide is used to increase its MW, slowing its glomerular filtration and renal clearance, extending elimination half-life, and thus slowing clearance.[86,87] An excellent case is head-to-head comparison of pegfilgrastim (PEGylated filgrastim) and its original version filgrastim, both of which are rhG-CSF, indicated for neutropenia (or low neutrophil count) and its complications in patients with severe chronic neutropenia or those receiving bone marrow transplantation or myelosuppressant chemotherapy. However, the only minor difference between the two molecules is that the former contains PEG moieties in its structure and thus exhibits half-life extension due to the shielding effect of the PEG moieties.[50,88,89] Other similar cases include, but not limited to, Mircera (methoxy PEGepoetin beta, approved in 2007), Plegridy (PEGinterferon beta-1α, approved in 2014),[10] and PEGvisomant (the GH receptor antagonist) indicated for acromegaly (a growth disorder caused by too much GH).[88] In addition, fusion of a peptide molecule to another endogenous larger species (such as circulating albumin, lipoproteins, or long-chain fatty acids) is also used to increase the apparent size of a peptide through noncovalent (direct or reversible) or covalent (or irreversible) interactions, slowing glomerular filtration and renal clearance of some peptides, in particular after increased binding of the two moieties.[90–92] Other large molecules, such as fatty acid long chain conjugated (or acylated, which facilitates interaction with albumin),[41] Fc domain (a protein fragment derived from an IgG1 antibody, or IgG-derived Fc),[93] and nanobody (a single domain antibody, such as engineered antibody targeting human serum albumin Fv domain),[94] have been developed for increased molecular size, leading to slowed renal filtration and elimination. Other examples include insulin,[95] rhG-CSF,[96] GH,[97] and interferon (IFN)-alpha or -beta,[98–100] etanercept (Enbrel, Fc-fused),[93] and more, all of which exhibit improved PK profiles when compared with its corresponding nonacylated or unfused counterparts, respectively. Among these cases, compared with albumin- or Fc-fused moieties, the interaction of an acylated peptide with FcRn is mediated through binding to albumin, and binding of a fatty acid-conjugated peptide to albumin may affect FcRn interactions within the acylated peptide bound albumin, leading to changes in FcRn affinity or binding due to varying saturation status of albumin with fatty acids, or the presence or absence of either a moiety bearing FcRn-binding peptide or an engineered albumin variant (harboring K573P) or YTE-bearing Fc variant.[41] Obviously, enhanced FcRn interactions with mAbs or fused/conjugated peptides is a robust strategy to improve the design of next-generation mAb and peptide therapeutics and their druggability. The updated knowledge will help better understand the role of FcRn in the disposition and clearance of various mAb and conjugated/fused peptide moieties in clinical settings.

Aim 2 is to modify the disposition and clearance of mAb and peptide molecules based on the minor changes of their structure and/or physicochemical properties via glycosylation, deamidation, oxidation, and isomerization of certain amino acid residues. In general, peptides and proteins are polar, poor membrane permeable, sensitive to temperature, and subject to enzyme degradation (or proteolysis). In addition to these physiochemical characteristics, for naturally glycosylated proteins, asialoglycoprotein receptor (expressed in the liver) and mannose receptor (predominantly expressed in immune cells) are each involved in clearing them (glycoproteins) from the blood circulation.[41] Endosomal acidic insulinase hydrolyzes internalized native human insulin in the liver at the aromatic locus Phe(B24)-Phe(B25)-Tyr(B26).[101] The mannose and galactose receptor in the liver is responsible for the clearance of tPA (tissue plasminogen activator).[102] For recombinant mAbs, the results are not always consistent. For example, a degalactosylated mAb showed slower clearance and longer half-life than its unmodified parent molecule,[103] but no differences in PK profiles were observed in mice for another mAb with high-mannose-enriched Fc,[104] implying that there may be mAb specificity due to diverse charge, hydrophobicity, solubility, stability, multidimensional (tertiary) structure, and dynamics in between. In contrast, glycosylation can significantly affect the disposition and clearance of peptide molecules because of increased molecular size, with the clearance slowed by the enhanced sialylation.[105] For asparagine and glutamine residues in the peptides (such as human GH), their deamidation is considered to alter their clearance due to increased overall net negative ionic charge on the surface of a moiety and altered local structural conformation presumably due to the charge-charge interactions or charge imbalance after creation of basic or acid moieties.[106] Similar to deamidation, isomerization of aspartate to isoaspartic acid in the CDR of mAb molecules seems to play an important role in changes of protein structure and function, such as decreased target-binding ability and mAb efficacy,[107–109] with no reports demonstrating the effects of isomerization on the disposition and clearance of mAbs and peptides.[41] Therefore, more evidence needs to be explored and accumulated to answer how to dissect and use the nontarget-related clearance mechanisms of the biological drugs for further improving their druggability under development and for better understanding their disposition and PK profiles when used for patient care.

1.1.7 How to Integrate All Acquired Information to Guide and Project R&D?

It is well-known that the entire process of the drug R&D, including chemical and biological drugs, is very complex, lengthy, and extremely expensive to achieve the requirements for approval based on systematic and rigorous evaluations by a certain regulatory agency, such as FDA or EMA. With accumulation and advances in knowledge and biotechnology, empirical guidance (such as trial-and-error approach) seems to be very limited and even impossible to do so, in particular for the development of biological drugs that are required to bind to their targets that are druggable or therapeutic. Thus, modeling and simulation are such indispensable and powerful tools that confirm predictions from the preclinical studies and guide optimal dosing regimens to be used for clinical trials.[110–112] Currently, there are many models available for the R&D, such as PK (what the body does to the drug),[113] PD (what the drug does to the body),[113] PK/PD, PK/TE/PD (PK/target engagement/PD), PBPK (physiologically based PK), popPK (population PK), QSP (quantitative systems pharmacology), and more. Of them, there are pronounced interplays between PK and PD, such as PK affecting PD and vice versa. Different from small-molecule chemical drugs, biological products demonstrate unique metabolism and elimination in the body, such as renal filtration and excretion (mostly for peptides of small molecular size), proteolysis (or enzyme degradation), TMDD (see above), recycling, and more, with most exhibiting nonlinear PK and half-life extension to a greater extent (usually from few days to few weeks) when saturated with target binding at the excess of the drug concentrations over the target amount.[114] In addition to other commonly used models, TMDD model is also a unique type of PK/PD model, which describes the nonlinear elimination and clearance most seen in mAbs to predict the PK, PD, and TE responses to diverse dosage regimens of interest and to elucidate the quantitative relationship between drug exposure (dose or plasma drug concentration) and response. To begin with the TMDD modeling, the first TMDD model was proposed in 2001,[115] and a tutorial on TMDD models (summarizing the various developed TMDD models) is currently available.[116] But not limited to this, the popPK analysis is an important component of 351(a) BLA submission. All information and parameters acquired from various modeling analyses will be integrated and analyzed for generating novel biological insights or achieving optimal expectation of clinical outcomes. All the incorporated efforts will help make informed decision when designing clinical trials, make database continuous to be updated and refined, and make simulations more accurate and precise.

There are tremendous multidimensional real-life data are generated and acquired due to biological complexity and uncertainty in the process of drug development, with big data and derived large-volume information composed of, but not limited to, quality-control data, *in silico* data, efficacy measurements (quantitatively), toxicological observations (qualitatively, such as safety signals), demographic data, blood biochemical testing data, routine testing data of liver and kidney function, imaging recognition dataset, omics data, biomarkers, confounding factors (variables, covariates, and noncompliance), study design, dosing regimen, algorithms designed for trials, missing data to be imputed, model scenario and building, EHR (electronic health records), and specific indication(s) to be prescribed. All generated and updated data are diverse and very complicated, with emerging at a speed of exponential growth. In these cases, AI and machine learning methods are needed to increase the accuracy of predictions on the efficacy and safety of drugs and diversify the pipelines and timelines of drug development as a highly efficient approach for speeding up discovery and preclinical stages by a factor of 15.[117] It is the case for the development of the biological products. In addition to modeling, simulating, machine learning, and AI, other new technologies also contribute to integrating and analyzing these big data, such as data mining, natural language processing, deep learning, neural network, bioengineering, imaging recognition and cluster analysis, and even digital cloud and blockchain technology.

1.1.8 How to Evaluate Cost-Based Value of the Biological Drugs?

The biologics and their derived biosimilars and interchangeable products constitute the biological products as a group of biodrugs. All marketed biologics are very expensive, whose major causes are structural complexity, complicated manufacturing process, and costly clinical trials.[5] Despite this, to further improve market access to and treatment options of the biological products, an abbreviated 351(k) licensure pathway and indication extrapolation have been used for the approval of biosimilars and interchangeable products at a significantly lower cost.

In general, the gold standard of rationalized medicine is accepted to be relatively safe, more effective, and less costly.[118] To acquire cost-normalized value of the biodrugs, a series of economic evaluations and comparisons [118] may be required to perform for each of the biological products and even for each indication (proposed or licensed), such as cost-effective analysis (CEA), cost-benefit analysis (CBA), cost-utility analysis (CUA), and cost-minimization analysis (CMA), with the major focus on the cost control of a marketed product, the total cost of each indication, the average cost per prescription or patient, cost per life-year gained, quality-adjusted life-year (QALY) gained, quality of life adjusted life expectancy, the number of the severe adverse medical events avoided, and the number of deaths averted. Clearly, these cost-related analyses would help further improve drug development and utilization, making manufacturing more cost-effective and thus benefiting patients the most.

1.2 SUMMARY AND FUTURE PERSPECTIVES

This year is the 100th anniversary of insulin since its discovery in 1921 by Frederick Banting and John Macleod.[119] The history of human fighting against diabetes mellitus is also the "evolutionary history" of the biological products, with the documented first recombinant human therapeutic protein product (Humulin in 1982),[1] insulin biosimilar (insulin glargine), and the first interchangeable biosimilar (insulin glargine-yfgn in 2021)[6] available in the market. In addition to insulin and its analogs, a class of the GLP-1 analogs have also been developed and marketed sequentially, such as exenatide (first-in-class), acylated liraglutide (conjugated with C16 or C18 fatty acid long chain that is bound to albumin; currently me-better or biobetter), dulaglutide (conjugated with IgG4-Fc), semaglutide (PEGylated and acylated; currently best-in-class), tirzepatide (a dual GLP-1/GIP agonist; currently under phase III clinical trials), and more to follow, with consequent improvements in marked half-life extension (from bid to once a day or week) and increased benefits (more potently glucose-lowing effect; and weight control to be added as another new indication). In the same class of biological products, me-better, biobetter, and best-in-class drugs would be evaluated and recognized after a series of the strict performance of noninferiority trials, equivalence trials, and superiority trials.

Most of currently available biological products are known to target more than one target or pathway involved in the process of diseases, resulting in more or less durable remission. Although they could help fight against life-threatening diseases or critically ill conditions, these biotherapies come at a higher price economically. However, the occurrence and development of biological products have increased one more option to treat diseases. It is anticipated that more and more new biological products would be developed and approved as the knowledge of biomedical sciences and biotechnology are renewed and advanced over time, but that a more complete and thorough understanding of currently available biological products could lead to the design of biological therapeutics (products and regimens) that would be much safer and more cost-effective.

In summary, biological products exert their clinical efficacy through specifically targeting their therapeutic target(s) of diseased cells and/or tissues in patients and thus, they have revolutionized the treatment of many chronic or life-threatening diseases and critically ill conditions and have become an increasingly more important contributor to our healthcare system due to the more active market access and more therapy options. This chapter helped the reader understand a series of important issues about the principles and practice of the biological products, with a major focus on the more cost-effective strategy for their research and development than ever. All these efforts made in this book are a fight worth fighting for taking human healthcare to the next higher level.

ACKNOWLEDGMENTS

This chapter was supported in part by grants No. 82073941, funded by the National Natural Science Foundation of China; No. BE2021603 by the Science and Technology Department of Jiangsu Province, China; and No. ZKX20035 by the Health and Human Service of Nanjing City, Jiangsu, China, all of which were awarded to Dr. Xie. In addition, Dr. Xie is a recipient of the distinguished medical expert of Jiangsu, China.

The author really appreciates PhD graduate student Li-Ping Jiang for preparing Figure 1.1 and all authors for sharing their articles cited in this chapter.

REFERENCES

1. Zaykov AN, Mayer JP, DiMarchi RD. Pursuit of a perfect insulin. *Nat Rev Drug Discov* 2016; 15(6):425–439.
2. FDA. Biosimilar and interchangeable products. https://www.fda.gov/drugs/biosimilars/biosimilar-and-interchangeable-products. Accessed: July 8, 2021.
3. FDA. Biological approvals by year. https://www.fda.gov/vaccines-blood-biologics/development-approval-process-cber/biological-approvals-year. Accessed: July 8, 2021.
4. FDA. FDA Purple Book. https://purplebooksearch.fda.gov/. Accessed: July 12, 2021.
5. Lexchin J. Affordable biologics for all. *JAMA Netw Open* 2020; 3(4):e204753.
6. FDA. Biosimilar product information. https://www.fda.gov/drugs/biosimilars/biosimilar-product-information. Accessed: July 12, 2021.
7. Belsey MJ, Harris LM, Das RR, Chertkow J. Biosimilars: Initial excitement gives way to reality. *Nat Rev Drug Discov* 2006; 5(7):535–536.
8. Nabhan C, Parsad S, Mato AR, Feinberg BA. Biosimilars in oncology in the United States: A review. *JAMA Oncol* 2018; 4(2):241–247.
9. Zangeneh F, Dolinar R. Biosimilar drugs are not generics. *Endocr Pract* 2016; 22(1):6–7.
10. Ogasawara K, Breder CD, Lin DH, Alexander GC. Exposure- and dose-response analyses in dose selection and labeling of FDA-approved biologics. *Clin Ther* 2018; 40(1):95–102.
11. FDA. Guidance for industry: Exposure-response relationships – Study design, data analysis, and regulatory applications. https://www.fda.gov/regulatory-information/search-fda-guidance-documents/exposure-response-relationships-study-design-data-analysis-and-regulatory-applications. 2003. Accessed: July 12, 2021.
12. Goldberger MJ, Singh N, Allerheiligen S, et al. ASCPT task force for advancing pharmacometrics and integration into drug development. *Clin Pharmacol Ther* 2010; 88(2):158–161.
13. Barrett JS, Fossler MJ, Cadieu KD, Gastonguay MR. Pharmacometrics: A multidisciplinary field to facilitate critical thinking in drug development and translational research settings. *J Clin Pharmacol* 2008; 48(5):632–649.
14. Koch G, Pfister M, Daunhawer I, Wilbaux M, Wellmann S, Vogt JE. Pharmacometrics and machine learning partner to advance clinical data analysis. *Clin Pharmacol Ther* 2020; 107(4):926–933.
15. Badillo S, Banfai B, Birzele F, et al. An introduction to machine learning. *Clin Pharmacol Ther* 2020; 107(4):871–885.
16. Bica I, Alaa AM, Lambert C, van der Schaar M. From real-world patient data to individualized treatment effects using machine learning: Current and future methods to address underlying challenges. *Clin Pharmacol Ther* 2021; 109(1):87–100.
17. Terranova N, Venkatakrishnan K, Benincosa LJ. Application of machine learning in translational medicine: Current status and future opportunities. *AAPS J* 2021; 23(4):74.
18. McComb M, Bies R, Ramanathan M. Machine learning in pharmacometrics: Opportunities and challenges. *Br J Clin Pharmacol* 2021; in press. https://doi.org/10.1111/bcp.14801
19. Pinheiro J, Duffull S. Exposure response – getting the dose right. *Pharm Stat* 2009; 8(3):173–175.
20. Overgaard RV, Ingwersen SH, Tornoe CW. Establishing good practices for exposure-response analysis of clinical endpoints in drug development. *CPT Pharmacometrics Syst Pharmacol* 2015; 4(10):565–575.
21. Sacks LV, Shamsuddin HH, Yasinskaya YI, Bouri K, Lanthier ML, Sherman RE. Scientific and regulatory reasons for delay and denial of FDA approval of initial applications for new drugs, 2000–2012. *JAMA* 2014; 311(4):378–384.
22. FDA. Yervoy (ipilimumab) injection, drug approval package. https://www.accessdata.fda.gov/drugsatfda_docs/label/2011/125377s0000lbl.pdf. Accessed: July 20, 2021.
23. FDA. Kadcyla (ado-trastuzumab emtansine) injection, drug approval package. https://www.accessdata.fda.gov/drugsatfda_docs/label/2020/125427s108lbl.pdf. Accessed: July 20, 2021.
24. FDA. Empliciti (elotuzumab), drug approval package. https://www.accessdata.fda.gov/drugsatfda_docs/label/2019/761035s010lbl.pdf. Accessed: July 20, 2021.
25. FDA. Cosentyx (secukinumab), drug approval package. https://www.accessdata.fda.gov/drugsatfda_docs/label/2021/125504s043lbl.pdf. Accessed: July 20, 2021.
26. FDA. Drugs@FDA: FDA approved drug products. https://www.accessdata.fda.gov/scripts/cder/daf/. Accessed: July 20, 2021.

27. Biopharma PEG. FDA approved antibody-drug conjugates up to 2021. https://www.biochempeg.com/article/74.html. Accessed: September 10, 2021.
28. Li BT, Smit EF, Goto Y, et al. Trastuzumab deruxtecan in HER2-mutant non-small-cell lung cancer. *N Engl J Med* 2022; 386(3):241–251. https://doi.org/10.1056/NEJMoa2112431
29. Brambell FW. The transmission of immunity from mother to young and the catabolism of immunoglobulins. *Lancet* 1966; 2(7473):1087–1093.
30. Burmeister WP, Huber AH, Bjorkman PJ. Crystal structure of the complex of rat neonatal Fc receptor with Fc. *Nature* 1994; 372(6504):379–383.
31. Oganesyan V, Damschroder MM, Cook KE, et al. Structural insights into neonatal Fc receptor-based recycling mechanisms. *J Biol Chem* 2014; 289(11):7812–7824.
32. Ghetie V, Ward ES. FcRn: The MHC class I-related receptor that is more than an IgG transporter. *Immunol Today* 1997; 18(12):592–598.
33. Andersen JT, Sandlie I. The versatile MHC class I-related FcRn protects IgG and albumin from degradation: Implications for development of new diagnostics and therapeutics. *Drug Metab Pharmacokinet* 2009; 24(4):318–332.
34. Pyzik M, Sand KMK, Hubbard JJ, Andersen JT, Sandlie I, Blumberg RS. The neonatal Fc receptor (FcRn): A misnomer? *Front Immunol* 2019; 10:1540.
35. Ward ES, Zhou J, Ghetie V, Ober RJ. Evidence to support the cellular mechanism involved in serum IgG homeostasis in humans. *Int Immunol* 2003; 15(2):187–195.
36. Wyckoff SL, Hudson KE. Targeting the neonatal Fc receptor (FcRn) to treat autoimmune diseases and maternal-fetal immune cytopenias. *Transfusion* 2021; 61(5):1350–1354.
37. Peter HH, Ochs HD, Cunningham-Rundles C, et al. Targeting FcRn for immunomodulation: Benefits, risks, and practical considerations. *J Allergy Clin Immunol* 2020; 146(3):479–491.
38. Gable KL, Guptill JT. Antagonism of the neonatal Fc receptor as an emerging treatment for myasthenia gravis. *Front Immunol* 2019; 10:3052.
39. Roopenian DC, Akilesh S. FcRn: The neonatal Fc receptor comes of age. *Nat Rev Immunol* 2007; 7(9):715–725.
40. Wang W, Wang EQ, Balthasar JP. Monoclonal antibody pharmacokinetics and pharmacodynamics. *Clin Pharmacol Ther* 2008; 84(5):548–558.
41. Datta-Mannan A. Mechanisms influencing the pharmacokinetics and disposition of monoclonal antibodies and peptides. *Drug Metab Dispos* 2019; 47(10):1100–1110.
42. Ghetie V, Hubbard JG, Kim JK, Tsen MF, Lee Y, Ward ES. Abnormally short serum half-lives of IgG in beta 2-microglobulin-deficient mice. *Eur J Immunol* 1996; 26(3):690–696.
43. Ghetie V, Popov S, Borvak J, et al. Increasing the serum persistence of an IgG fragment by random mutagenesis. *Nat Biotechnol* 1997; 15(7):637–640.
44. Andersen JT, Dalhus B, Viuff D, et al. Extending serum half-life of albumin by engineering neonatal Fc receptor (FcRn) binding. *J Biol Chem* 2014; 289(19):13492–13502.
45. Garg A, Balthasar JP. Physiologically-based pharmacokinetic (PBPK) model to predict IgG tissue kinetics in wild-type and FcRn-knockout mice. *J Pharmacokinet Pharmacodyn* 2007; 34(5):687–709.
46. Andersen JT, Dalhus B, Cameron J, et al. Structure-based mutagenesis reveals the albumin-binding site of the neonatal Fc receptor. *Nat Commun* 2012; 3:610.
47. Goebl NA, Babbey CM, Datta-Mannan A, Witcher DR, Wroblewski VJ, Dunn KW. Neonatal Fc receptor mediates internalization of Fc in transfected human endothelial cells. *Mol Biol Cell* 2008; 19(12):5490–5505.
48. Ward ES, Devanaboyina SC, Ober RJ. Targeting FcRn for the modulation of antibody dynamics. *Mol Immunol* 2015; 67(2 Pt A):131–141.
49. Ward ES, Martinez C, Vaccaro C, Zhou J, Tang Q, Ober RJ. From sorting endosomes to exocytosis: Association of Rab4 and Rab11 GTPases with the Fc receptor, FcRn, during recycling. *Mol Biol Cell* 2005; 16(4):2028–2038.
50. An G. Concept of pharmacologic target-mediated drug disposition in large-molecule and small-molecule compounds. *J Clin Pharmacol* 2020; 60(2):149–163.
51. Dirks NL, Meibohm B. Population pharmacokinetics of therapeutic monoclonal antibodies. *Clin Pharmacokinet* 2010; 49(10):633–659.
52. Wang W, Vlasak J, Li Y, et al. Impact of methionine oxidation in human IgG1 Fc on serum half-life of monoclonal antibodies. *Mol Immunol* 2011; 48(6–7):860–866.
53. Folzer E, Diepold K, Bomans K, et al. Selective oxidation of methionine and tryptophan residues in a therapeutic IgG1 molecule. *J Pharm Sci* 2015; 104(9):2824–2831.

54. Hensel M, Steurer R, Fichtl J, et al. Identification of potential sites for tryptophan oxidation in recombinant antibodies using tert-butylhydroperoxide and quantitative LC-MS. *PLoS One* 2011; 6(3):e17708.
55. Wei Z, Feng J, Lin HY, et al. Identification of a single tryptophan residue as critical for binding activity in a humanized monoclonal antibody against respiratory syncytial virus. *Anal Chem* 2007; 79(7):2797–2805.
56. Wang Y, Tian Z, Thirumalai D, Zhang X. Neonatal Fc receptor (FcRn): A novel target for therapeutic antibodies and antibody engineering. *J Drug Target* 2014; 22(4):269–278.
57. Kasprick A, Hofrichter M, Smith B, et al. Treatment with anti-neonatal Fc receptor (FcRn) antibody ameliorates experimental epidermolysis bullosa acquisita in mice. *Br J Pharmacol* 2020; 177(10):2381–2392.
58. Smith B, Christodoulou L, Clargo A, et al. Generation of two high affinity anti-mouse FcRn antibodies: Inhibition of IgG recycling in wild type mice and effect in a mouse model of immune thrombocytopenia. *Int Immunopharmacol* 2019; 66:362–365.
59. Manook M, Flores WJ, Schmitz R, et al. Measuring the impact of targeting FcRn-mediated IgG recycling on donor-specific alloantibodies in a sensitized NHP model. *Front Immunol* 2021; 12:660900.
60. Roy S, Nanovskaya T, Patrikeeva S, et al. M281, an anti-FcRn antibody, inhibits IgG transfer in a human ex vivo placental perfusion model. *Am J Obstet Gynecol* 2019; 220(5):498.
61. Coutinho E, Jacobson L, Shock A, Smith B, Vernon A, Vincent A. Inhibition of maternal-to-fetal transfer of IgG antibodies by FcRn blockade in a mouse model of arthrogryposis multiplex congenita. *Neurol Neuroimmunol Neuroinflamm* 2021; 8(4):e1011. https://doi.org/10.1212/NXI.0000000000001011
62. Levy G. Pharmacologic target-mediated drug disposition. *Clin Pharmacol Ther* 1994; 56(3):248–252.
63. An G. Small-molecule compounds exhibiting target-mediated drug disposition (TMDD): A minireview. *J Clin Pharmacol* 2017; 57(2):137–150.
64. Meijer RT, Koopmans RP, ten Berge IJ, Schellekens PT. Pharmacokinetics of murine anti-human CD3 antibodies in man are determined by the disappearance of target antigen. *J Pharmacol Exp Ther* 2002; 300(1):346–353.
65. Schropp J, Khot A, Shah DK, Koch G. Target-mediated drug disposition model for bispecific antibodies: Properties, approximation, and optimal dosing strategy. *CPT Pharmacometrics Syst Pharmacol* 2019; 8(3):177–187.
66. Abraham AK, Kagan L, Kumar S, Mager DE. Type I interferon receptor is a primary regulator of target-mediated drug disposition of interferon-beta in mice. *J Pharmacol Exp Ther* 2010; 334(1):327–332.
67. Kakkar T, Sung C, Gibiansky L, et al. Population PK and IgE pharmacodynamic analysis of a fully human monoclonal antibody against IL4 receptor. *Pharm Res* 2011; 28(10):2530–2542.
68. El-Komy MH, Widness JA, Veng-Pedersen P. Pharmacokinetic analysis of continuous erythropoietin receptor activator disposition in adult sheep using a target-mediated, physiologic recirculation model and a tracer interaction methodology. *Drug Metab Dispos* 2011; 39(4):603–609.
69. Bauer RJ, Gibbons JA, Bell DP, Luo ZP, Young JD. Nonlinear pharmacokinetics of recombinant human macrophage colony-stimulating factor (M-CSF) in rats. *J Pharmacol Exp Ther* 1994; 268(1):152–158.
70. Krzyzanski W, Smits A, van den Anker J, Allegaert K. Population model of serum creatinine as time-dependent covariate in neonates. *AAPS J* 2021; 23(4):86.
71. Terashi K, Oka M, Ohdo S, et al. Close association between clearance of recombinant human granulocyte colony-stimulating factor (G-CSF) and G-CSF receptor on neutrophils in cancer patients. *Antimicrob Agents Chemother* 1999; 43(1):21–24.
72. Kotto-Kome AC, Fox SE, Lu W, Yang BB, Christensen RD, Calhoun DA. Evidence that the granulocyte colony-stimulating factor (G-CSF) receptor plays a role in the pharmacokinetics of G-CSF and PegG-CSF using a G-CSF-R KO model. *Pharmacol Res* 2004; 50(1):55–58.
73. Chapel S, Veng-Pedersen P, Hohl RJ, Schmidt RL, McGuire EM, Widness JA. Changes in erythropoietin pharmacokinetics following busulfan-induced bone marrow ablation in sheep: Evidence for bone marrow as a major erythropoietin elimination pathway. *J Pharmacol Exp Ther* 2001; 298(2):820–824.
74. Chapel SH, Veng-Pedersen P, Schmidt RL, Widness JA. Receptor-based model accounts for phlebotomy-induced changes in erythropoietin pharmacokinetics. *Exp Hematol* 2001; 29(4):425–431.
75. D'Cunha R, Schmidt R, Widness JA, et al. Target-mediated disposition population pharmacokinetics model of erythropoietin in premature neonates following multiple intravenous and subcutaneous dosing regimens. *Eur J Pharm Sci* 2019; 138:105013.
76. D'Cunha R, Widness JA, Yan X, Schmidt RL, Veng-Pedersen P, An G. A mechanism-based population pharmacokinetics model of erythropoietin in premature infants and healthy adults following multiple intravenous doses. *J Clin Pharmacol* 2019; 59(6):835–846.
77. Ng CM, Stefanich E, Anand BS, Fielder PJ, Vaickus L. Pharmacokinetics/pharmacodynamics of nondepleting anti-CD4 monoclonal antibody (TRX1) in healthy human volunteers. *Pharm Res* 2006; 23(1):95–103.

78. Yan X, Mager DE, Krzyzanski W. Selection between Michaelis-Menten and target-mediated drug disposition pharmacokinetic models. *J Pharmacokinet Pharmacodyn* 2010; 37(1):25–47.
79. Peletier LA, Gabrielsson J. Dynamics of target-mediated drug disposition: Characteristic profiles and parameter identification. *J Pharmacokinet Pharmacodyn* 2012; 39(5):429–451.
80. Grilo AL, Mantalaris A. The increasingly human and profitable monoclonal antibody market. *Trends Biotechnol* 2019; 37(1):9–16.
81. Mohs RC, Greig NH. Drug discovery and development: Role of basic biological research. *Alzheimers Dement* 2017; 3(4):651–657.
82. Recio C, Maione F, Iqbal AJ, Mascolo N, de Feo V. The potential therapeutic application of peptides and peptidomimetics in cardiovascular disease. *Front Pharmacol* 2016; 7:526.
83. Haraldsson B, Nystrom J, Deen WM. Properties of the glomerular barrier and mechanisms of proteinuria. *Physiol Rev* 2008; 88(2):451–487.
84. Sarin H. Physiologic upper limits of pore size of different blood capillary types and another perspective on the dual pore theory of microvascular permeability. *J Angiogenes Res* 2010; 2:14.
85. Lau JL, Dunn MK. Therapeutic peptides: Historical perspectives, current development trends, and future directions. *Bioorg Med Chem* 2018; 26(10):2700–2707.
86. Patel A, Cholkar K, Mitra AK. Recent developments in protein and peptide parenteral delivery approaches. *Ther Deliv* 2014; 5(3):337–365.
87. Gregoriadis G, Fernandes A, Mital M, McCormack B. Polysialic acids: Potential in improving the stability and pharmacokinetics of proteins and other therapeutics. *Cell Mol Life Sci* 2000; 57(13–14):1964–1969.
88. Turecek PL, Bossard MJ, Schoetens F, Ivens IA. PEGylation of biopharmaceuticals: A review of chemistry and nonclinical safety information of approved drugs. *J Pharm Sci* 2016; 105(2):460–475.
89. van Witteloostuijn SB, Pedersen SL, Jensen KJ. Half-life extension of biopharmaceuticals using chemical methods: Alternatives to PEGylation. *ChemMedChem* 2016; 11(22):2474–2495.
90. Strohl WR. Fusion proteins for half-life extension of biologics as a strategy to make biobetters. *BioDrugs* 2015; 29(4):215–239.
91. Lim SI, Mizuta Y, Takasu A, Hahn YS, Kim YH, Kwon I. Site-specific fatty acid-conjugation to prolong protein half-life in vivo. *J Control Release* 2013; 170(2):219–225.
92. Chanson P, Timsit J, Harris AG. Clinical pharmacokinetics of octreotide: Therapeutic applications in patients with pituitary tumours. *Clin Pharmacokinet* 1993; 25(5):375–391.
93. Spencer-Green G. Etanercept (Enbrel): Update on therapeutic use. *Ann Rheum Dis* 2000; 59 (suppl 1):i46–i49.
94. Adams R, Griffin L, Compson JE, et al. Extending the half-life of a Fab fragment through generation of a humanized anti-human serum albumin Fv domain: An investigation into the correlation between affinity and serum half-life. *MAbs* 2016; 8(7):1336–1346.
95. Duttaroy A, Kanakaraj P, Osborn BL, et al. Development of a long-acting insulin analog using albumin fusion technology. *Diabetes* 2005; 54(1):251–258.
96. Halpern W, Riccobene TA, Agostini H, et al. Albugranin, a recombinant human granulocyte colony stimulating factor (G-CSF) genetically fused to recombinant human albumin induces prolonged myelopoietic effects in mice and monkeys. *Pharm Res* 2002; 19(11):1720–1729.
97. Osborn BL, Sekut L, Corcoran M, et al. Albutropin: A growth hormone-albumin fusion with improved pharmacokinetics and pharmacodynamics in rats and monkeys. *Eur J Pharmacol* 2002; 456(1–3):149–158.
98. Sung C, Nardelli B, LaFleur DW, et al. An IFN-beta-albumin fusion protein that displays improved pharmacokinetic and pharmacodynamic properties in nonhuman primates. *J Interferon Cytokine Res* 2003; 23(1):25–36.
99. Bain VG, Kaita KD, Yoshida EM, et al. A phase II study to evaluate the antiviral activity, safety, and pharmacokinetics of recombinant human albumin-interferon alfa fusion protein in genotype 1 chronic hepatitis C patients. *J Hepatol* 2006; 44(4):671–678.
100. Subramanian GM, Fiscella M, Lamouse-Smith A, Zeuzem S, McHutchison JG. Albinterferon alpha-2b: A genetic fusion protein for the treatment of chronic hepatitis C. *Nat Biotechnol* 2007; 25(12):1411–1419.
101. Authier F, Danielsen GM, Kouach M, Briand G, Chauvet G. Identification of insulin domains important for binding to and degradation by endosomal acidic insulinase. *Endocrinology* 2001; 142(1):276–289.
102. Smedsrod B, Einarsson M. Clearance of tissue plasminogen activator by mannose and galactose receptors in the liver. *Thromb Haemost* 1990; 63(1):60–66.
103. Newkirk MM, Novick J, Stevenson MM, Fournier MJ, Apostolakos P. Differential clearance of glycoforms of IgG in normal and autoimmune-prone mice. *Clin Exp Immunol* 1996; 106(2):259–264.
104. Millward TA, Heitzmann M, Bill K, Langle U, Schumacher P, Forrer K. Effect of constant and variable domain glycosylation on pharmacokinetics of therapeutic antibodies in mice. *Biologicals* 2008; 36(1):41–47.

105. Egrie JC, Dwyer E, Browne JK, Hitz A, Lykos MA. Darbepoetin alfa has a longer circulating half-life and greater in vivo potency than recombinant human erythropoietin. *Exp Hematol* 2003; 31(4):290–299.
106. Robinson NE. Protein deamidation. *Proc Natl Acad Sci USA* 2002; 99(8):5283–5288.
107. Dick LW Jr, Qiu D, Wong RB, Cheng KC. Isomerization in the CDR2 of a monoclonal antibody: Binding analysis and factors that influence the isomerization rate. *Biotechnol Bioeng* 2010; 105(3):515–523.
108. Wakankar AA, Borchardt RT, Eigenbrot C, et al. Aspartate isomerization in the complementarity-determining regions of two closely related monoclonal antibodies. *Biochemistry* 2007; 46(6):1534–1544.
109. Prueksaritanont T, Tang C. ADME of biologics – What have we learned from small molecules? *AAPS J* 2012; 14(3):410–419.
110. Bonifacio L, Dodds M, Prohaska D, et al. Target-mediated drug disposition pharmacokinetic/pharmacodynamic model-informed dose selection for the first-in-human study of AVB-S6-500. *Clin Transl Sci* 2020; 13(1):204–211.
111. Cai W, Leil TA, Gibiansky L, et al. Modeling and simulation of the pharmacokinetics and target engagement of an antagonist monoclonal antibody to interferon-gamma-induced protein 10, BMS-986184, in healthy participants to guide therapeutic dosing. *Clin Pharmacol Drug Dev* 2020; 9(6):689–698.
112. Chien JY, Friedrich S, Heathman MA, de Alwis DP, Sinha V. Pharmacokinetics/pharmacodynamics and the stages of drug development: Role of modeling and simulation. *AAPS J* 2005; 7(3):e544–e559.
113. Holford NH, Sheiner LB. Kinetics of pharmacologic response. *Pharmacol Ther* 1982; 16(2):143–166.
114. Krzyzanski W, Harrold JM, Wu LS, Perez-Ruixo JJ. A cell-level model of pharmacodynamics-mediated drug disposition. *J Pharmacokinet Pharmacodyn* 2016; 43(5):513–527.
115. Mager DE, Jusko WJ. General pharmacokinetic model for drugs exhibiting target-mediated drug disposition. *J Pharmacokinet Pharmacodyn* 2001; 28(6):507–532.
116. Dua P, Hawkins E, van der Graaf PH. A tutorial on target-mediated drug disposition (TMDD) models. *CPT Pharmacometrics Syst Pharmacol* 2015; 4(6):324–337.
117. Zhavoronkov A, Ivanenkov YA, Aliper A, et al. Deep learning enables rapid identification of potent DDR1 kinase inhibitors. *Nat Biotechnol* 2019; 37(9):1038–1040.
118. Xie HG. Chapter 11. Pharmacoeconogenomics: A good marriage of pharmacoeconomics and pharmacogenomics. In: Feng X, Xie HG, ed. *Applying Pharmacogenomics in Therapeutics*. 1st ed. Boca Raton, FL: CRC Press, Taylor & Francis Group; 2016. pp. 271–286.
119. Sims EK, Carr ALJ, Oram RA, DiMeglio LA, Evans-Molina C. 100 years of insulin: Celebrating the past, present and future of diabetes therapy. *Nat Med* 2021; 27(7):1154–1164.

Frontiers in Biopharmaceutical Development of Biologics and Biosimilars

2

Opportunities and Challenges

Yingqi Zhang, Nivaz Brar, Saif Anwar Shaikh, Drew Fajardo, and Xiaodong Feng

Corresponding author: Xiaodong Feng

Contents

DOI: 10.1201/9780429485626-2

2.1 INTRODUCTION

As the battle against chronic disease continues and cases rise globally, effective forms of treatment are in high demand. Biologics have revolutionized treatment for patients with chronic disease, autoimmune disorders, and cancers. Biosimilars are considered lower cost alternatives that are equally effective as reference biologics or originators, while they are not identical copies or generic versions of their originators. Biosimilars are widely used in Europe and have been proven to be safe and efficacious. Ever since the first biosimilar approved in the United States in 2015, the uptakes of biosimilars have been challenging. Most of the barriers stem from their likeliness to biologics, cost, and the complex U.S. healthcare system. To better understand biosimilars, it is essential to understand biologics first. This chapter will discuss the development process of biologics and biosimilars in detail, as well as the benefits and barriers of incorporating biosimilars to the U.S. market.

2.2 BIOLOGICAL PRODUCTS AND THEIR GROWTH IN MARKET SHARE

Before diving into biosimilar products, it is worth mentioning that biosimilars are also considered biological products or biologics. Biologics are generally large and complex molecules produced in living organisms, usually viruses or animal cells (also known as "cell line"). This advanced biotechnology allows biologics to target specific receptors, up- or down-regulating certain pathways in the human body. Promisingly, biologics have provided more treatment options for complex diseases, such as autoimmune diseases like rheumatoid arthritis and Crohn's disease.[1] Some examples of FDA-approved biologics include therapeutic proteins (e.g. insulin, filgrastim, and growth hormone), monoclonal antibodies (e.g. infliximab and adalimumab), and vaccines (e.g. COVID-19 vaccines).

Notably, the majority of the biological products can also be categorized as specialty medications, in comparison with nonspecialty medications. Nonspecialty medications, also known as traditional medications, are small, low-cost molecules that can be dispensed at retail pharmacies. A specialty medication is usually defined by one or more of the following characteristics: high cost, indication for complex or rare disease states, requiring special handling, requiring ongoing clinical assessment (e.g. Risk Evaluation and Mitigation Strategy or REMS program), or using a limited distribution network. For instance, medications indicated for multiple sclerosis, cystic fibrosis, HIV, hepatitis C, and various cancers are usually specialty medications. With constantly advancing technology, the market of biologics continues to grow. In 1990, there were only ten specialty drugs on the market.[2] As of 2021, over 350 biological products have been approved by the Food and Drug Administration (FDA).[3]

However, biologics or specialty medications are known to be expensive. According to the Express Script/Evernorth 2020 Drug Trend report,[4] although less than 2% of the population uses specialty medications, it accounts for more than half (51%) of all spending on prescription drugs. For example, specialty medications indicated for inflammatory conditions (e.g. rheumatoid arthritis, psoriasis) account for over 90% of medication spent for this drug class, with an average cost of at least $4,500 per prescription. A report published by IQVIA Institute[5] also mentioned that public and private insurers in the United States spent $211 billion on biologics in 2019 alone.

The high cost of biologics is due to several reasons. For one thing, it is extremely expensive to develop and launch a biological product. According to a study published in *JAMA Internal Medicine*, the estimated cost of pivotal trials for the approval of new drugs is $19 million. This is a modest amount compared to the development cost, which may be between $2 and $3 billion, and additional marketing and commercial expenses.[6] Moreover, there is no "generic drug" for biologics. Generic products exist for small molecule drugs for nonspecialty brand products. They are approved via the Abbreviated New Drug Application (ANDA) pathway by demonstrating bioequivalence. As biologics have inherent variations resulting from the choice of living organisms and manufacturing process, it is almost impossible to create the "identical" biologics. For these reasons, biological products tend to maintain little competition and huge market share.

In order to provide more treatment options, increase access to life-saving medications, and potentially lower health care costs through competition, on March 23, 2010, Congress passed The Biologics Price Competition and Innovation Act of 2009 (BPCI Act) as part of health reform (Affordable Care Act), which created an abbreviated licensure pathway for biosimilar products.[7]

2.3 THE BEGINNING OF BIOSIMILAR ERA AND THE U.S. REGULATIONS

Following the BPCI Act being signed in law in 2010, the U.S. FDA approved its first biosimilar filgrastim-sndz in 2015. However, biosimilars have been around for more than a decade. In 2006, the European Medicines Agency (EMA) approved the first biosimilar somatropin (a medication indicated for growth deficiency).[8] With the guidance of the EU and other countries' biosimilar regulation experience, the U.S. FDA established the 351(k) pathway, an abbreviated licensure pathway for biological products that are demonstrated to be biosimilar to or interchangeable with an FDA-approved biological product.[9]

Biologics that are already approved by the FDA with preclinical and clinical data, including full safety and effectiveness data, can be referred to as reference products or originators. These reference products are approved via the pathway known as Biologics License Applications (BLA).[10] As mentioned previously, biologics are produced from living organisms and as such may vary over time. This makes it nearly impossible to have "identical" compounds. Thus, the FDA 351(k) pathway maintains that while biosimilars may be inherently different from their biologic counterparts, they must be clinically the same. It is in this manner that the regulatory pathways for biosimilars differ so vastly from that of generic drugs. While reference biologics' approval process requires extensive clinical trials to demonstrate safety and efficacy, and traditional generic products' approval focuses on proof of bioequivalence to their reference products,[11] the 351(k) pathway requires manufacturers to provide evidence that their biosimilar products have high similarity with no clinically meaningful differences from an existing FDA-approved reference product.[12,13]

Specifically, the approval of biosimilar products focuses on two key components: being "highly similar" and having "no clinically meaningful differences" to the reference product.[9,14] A biosimilar product is required to have the same mechanism of action, route of administration, dosage form, and strength as the reference product. A manufacturer should conduct analytical studies demonstrating that the biological product is highly similar to the reference product, notwithstanding minor differences in clinically inactive components, including stabilizer and buffers. In addition, the manufacturer should also provide data from clinical studies, including assessments of immunogenicity and pharmacokinetics (PK) and/or pharmacodynamics (PD), demonstrating there are no clinically meaningful differences between the biosimilar and reference product in terms of safety, purity, and potency in an appropriate condition of use. Last but not least, data from animal studies, including the assessment of toxicity, should be provided.[15]

2.4 INTERCHANGEABLE PRODUCTS

If a biosimilar product meets additional requirements,[11,14,15] it can also be approved as an interchangeable product by FDA. While biosimilar products require specific prescriptions, an interchangeable product allows auto-substitution at the pharmacy level. In other words, if a patient brings in a prescription of a reference product, a pharmacist could substitute the interchangeable biosimilar products without consulting the prescriber, depending on the state's pharmacy laws. To be approved as an interchangeable product, a manufacturer would not only need to demonstrate the proposed product produces the same clinical results as the reference product in any given patient but also provide a risk assessment of switching between the products. On July 28, 2021, the FDA approved the first interchangeable biosimilar product Semglee® (insulin glargine-yfgn), a long-acting insulin analog, for the treatment of diabetes. It is expected that more interchangeable biosimilar insulins will be approved in the near future, signifying the FDA's goal to provide patients with additional safe, high-quality, and potentially cost-effective options for diabetic patients.[16]

2.5 NAMING OF BIOLOGICS

As part of the strategy to differentiate various biosimilars from their reference biologic (or originator), the FDA released the draft guidance of Nonproprietary Naming of Biological Products in 2017.[17] According to the guidance document, the FDA would designate a proper name that is a combination of the core name and a distinguishing suffix that is devoid of meaning and composed of four lowercase letters. This naming policy would be applied to any newly licensed originators, biosimilars, and interchangeable products. Adding unique suffixes to all biologics helps avoid creating a misimpression that products with such suffixes are inferior to those without. In 2019, an updated draft guidance published clarified that the FDA would not retrospectively modify the proper names of biological products that have already been licensed or approved without designated suffix.[18] The updated guidance also mentioned that the FDA would consider having a unique suffix that distinguishes an interchangeable product from other products sharing the same core name. However, if a biologic is first licensed as a biosimilar and later determined to be interchangeable, then it will keep its nonproprietary name after receiving a determination of interchangeability. Overall, the unique four-letter suffix allows biologics to be appropriately distinguished from one another at the pharmacy level and facilitate pharmacovigilance and adverse event report tracking.[19]

2.6 INDICATION EXTRAPOLATION

While reference products' manufacturers need to conduct extensive comparative clinical studies for each indication approval, a biosimilar product can be approved for an indication that is approved for the reference product, even if the biosimilar is not directly studied in that indication. This is achieved via a new concept called "extrapolation", introduced by the FDA as part of guidelines for biosimilar approval. According to the guidance document Scientific Considerations in Demonstrating Biosimilarity to a Reference Product, upon providing their own clinical safety and efficacy study data in one appropriate indication, the biosimilar manufacturer or applicant may seek licensure of the proposed product for one or more additional indications for which the reference product is licensed. The applicant would need to provide sufficient scientific justification for extrapolating clinical data to support a determination of biosimilarity for each indication, such as the mechanism(s) of action in each condition of use, the PK, biodistribution, and immunogenicity of the product in different patient populations. Of importance, indication extrapolation aligns with the purpose and goal of an abbreviated pathway, which is to improve access and options at a potentially lower cost.[13,14]

However, depending on biosimilar manufacturers' applications, scientific justification material provided, and FDA final decision, a proposed biosimilar product may still not be approved for all indications that the reference product is licensed. For one thing, a biosimilar manufacturer may choose to only seek one or part of currently approved indication(s). For another, certain indications, such as indication with orphan drug designation, may have different exclusivity periods and cannot be extrapolated at the time of application. For example, Avastin® (bevacizumab), a reference product used to treat various cancer diseases, has a total of seven approved indications. Among these indications, unresectable or metastatic hepatocellular carcinoma was granted with orphan drug designations with exclusivity until 2023. Upon FDA's final review, one of the Avastin®'s biosimilar products, Mvasi (bevacizumab-awwb)® was approved for five of the seven indications. The other biosimilar Zirabev® (bevacizumab-bvzr)® was approved for six of the seven indications. Neither Mavsi® nor Zirabev® was approved for hepatocellular carcinoma.[20–22]

2.7 PURPLE BOOK

To ensure biological products properly incorporate biosimilar and interchangeable information, the FDA launched the Purple Book, which is a database of licensed biological products.[23] The database allows users to search reference products, biosimilars, and interchangeable products. Each product listing also includes specific drug detail ranging from route of administration to license type. It is also important to note that insulin and other biologics that were approved under the traditional New Drug Application (NDA) have been transitioned to the BLA. Therefore, these products' information is also available in the Purple Book. Currently, there are over 20 biosimilars approved by FDA. Table 2.1 lists approved biosimilars with launch status.[24]

2.8 IMPACTS OF BIOSIMILARS

With the guidance of the abbreviated approval pathway, it is expected that the cost of biosimilar development and the timeline for approval would be reduced compared to reference biologics, which follow the standard approval pathway. The approval timeline for a reference biologic may take over 12 years, which typically spans from discovery phase to phase 3 clinical trials. The extensive research and clinical phases may cost a manufacturer over $1 billion for reference biologic development. On the contrary, biosimilar products already have reference biologics with known attributes. Therefore, the development of biosimilar products does not include the discovery phase or initial efficacy/safety phase but mainly focuses on comparative clinical studies instead. The approval timeline for biosimilar is estimated to be around eight years. As a result, the overall cost for biosimilar product development is about $100 to $200 million, significantly lower than the development cost for reference biologics.[35] In the long run, the shortened approval timeline and lower development cost may drive more manufacturers to enter the biosimilar market and subsequently increase patient access and market competition.

The introduction of biosimilar products also helps some biologic products re-align to the appropriate approval pathway. Prior to the establishment of the BLA, some biologic products, such as insulin and human growth hormone, were historically approved under the traditional NDA pathway. Since these proteins are actually large complex molecules developed in living organisms, it is almost impossible to develop their generic versions or exact copies, which limits market competition and patient access. However, by transitioning these products to the BLA pathway, they would be deemed as biologic products and opened up to potential biosimilar and interchangeable competition. With the BPCI Act signed in law for biosimilar products, it also clarified that the statutory category of "protein" should be categorized within the definition of biologic products. Consequentially, the FDA released a 10-year plan to transition these protein products from NDA to BLA pathway by March 22, 2020. The current Purple Book has already been updated to include these transitioned biologic products.[36–38]

TABLE 2.1 Currently FDA-approved biosimilars[25–34]

BIOSIMILAR NAME	*APPROVAL DATE*	*REFERENCE PRODUCT*	*LAUNCHED/ESTIMATED LAUNCH DATE*
Semglee[a] (insulin glargine-yfgn)	July 2021	Lantus (insulin glargine)	Launched
Riabni (rituximab-arrx)	December 2020	Rituxan (rituximab)	Launched
Hulio (adalimumab-fkjp)	July 2020	Humira (adalimumab)	Estimated launch date: July 31, 2023
Nyvepria (PEGfilgrastim-apgf)	June 2020	Neulasta (pegfilgrastim)	Launched
Avsola (infliximab-axxq)	December 2019	Remicade (infliximab)	Launched
Abrilada (adalimumab-afzb)	November 2019	Humira (adalimumab)	Estimated launch date: 2023
Ziextenzo (PEGfilgrastim-bmez)	November 2019	Neulasta (pegfilgrastim)	Launched
Hadlima (adalimumab-bwwd)	July 2019	Humira (adalimumab)	Estimated launch date: June 30, 2023
Ruxience (rituximab-pvvr)	July 2019	Rituxan (rituximab)	Launched
Zirabev (bevacizumab-bvzr)	June 2019	Avastin (bevacizumab)	Launched
Kanjinti (trastuzumab-anns)	June 2019	Herceptin (trastuzumab)	Launched
Eticovo (etanercept-ykro)	April 2019	Enbrel (etanercept)	Estimated launch date: 2028–2029
Trazimera (trastuzumab-qyyp)	March 2019	Herceptin (trastuzumab)	Launched
Ontruzant (trastuzumab-dttb)	January 2019	Herceptin (trastuzumab)	Launched
Herzuma (trastuzumab-pkrb)	December 2018	Herceptin (trastuzumab)	Launched
Truxima (rituximab-abbs)	November 2018	Rituxan (rituximab)	Launched
Udenyca (PEGfilgrastim-cbqv)	November 2018	Neulasta (pegfilgrastim)	Launched
Hyrimoz (adalimumab-adaz)	October 2018	Humira (adalimumab)	Estimated launch date: September 30, 2023
Nivestym (filgrastim-aafi)	July 2018	Neupogen (filgrastim)	Launched
Fulphila (PEGfilgrastim-jmdb)	June 2018	Neluasta (pegfilgrastim)	Launched
Retacrit (epoetin alfa-epbx)	May 2018	Epogen (epoetin-alfa)	Launched
Ixifi (infliximab-qbtx)	December 2017	Remicade (infliximab)	Estimated launch date: TBD
Ogivri (trastuzumab-dkst)	December 2017	Herceptin (trastuzumab)	Launched
Mvasi (bevacizumab-awwb)	September 2017	Avastin (bevacizumab)	Launched
Cyltezo (adalimumab-adbm)	August 2017	Humira (adalimumab)	Estimated launch date: July 01, 2023
Renflexis (infliximab-abda)	May 2017	Remicade (infliximab)	Launched
Amjevita (adalimumab-atto)	September 2016	Humira (adalimumab)	Estimated launch date: January 31, 2023
Erelzi (etanercept-szzs)	August 2016	Enbrel (etanercept)	Estimated launch date: 2028–2029
Inflectra (infliximab-dyyb)	April 2016	Remicade (infliximab)	Launched
Zarxio (filgrastim-sndz)	March 2015	Neupogen (filgrastim)	Launched

Note: PEG, polyethylene glycol; TBD, to be determined.

[a] The first FDA-approved interchangeable biosimilar.

In order to encourage the adoption of biosimilar products, the Centers for Medicare & Medicaid Services (CMS) has also made adjustments to their "buy-and-bill" reimbursement model in 2018. Prior to this, the "buy-and-bill" model was traditionally adopted by the medical benefit for provider-administered therapy. Under this model, the providers would purchase medicine at a negotiated price, usually at average sales price (ASP), and charge patients' insurers for the price of the drug plus a 6%–10% markup for administration. For example, a 6% markup with a $700 biologic would net $42, whereas a $500 biosimilar would net $30. Thus, providers would not have an incentive to purchase and administer biosimilars, as they are lower in cost and lower in markup. In 2018, the CMS changed the Medicare reimbursement policy under Part B so that biosimilars would be reimbursed at 100% of the individual biosimilar's ASP + 6% of the reference product's ASP. The above example would mean being paid $42 instead of $30 to administer the biosimilar. This way, providers are paid the same profit regardless of whether they choose the biologic or the biosimilar.[39–42]

Ever since the first biosimilar was approved in 2015, biosimilar products in the United States have demonstrated cost reduction and an increase in market uptake. A 2020 report has shown that biosimilars had an average of 30% reduction in ASP compared to reference biologic's ASP. The market share of the first approved biosimilar molecule, filgrastim, has grown from 25% to 80% after six years on the market. Overall, the biosimilars launched account for over 20% of competitive molecule volume. It is also projected that with the continuing increase in biosimilar adoption, savings related to biosimilar products may range from $69 to $140 billion over the next five years.[43]

2.9 UNCERTAINTIES OF THE BIOSIMILAR MARKET

The introduction of biosimilars into the U.S. market has brought in a new competitive environment as biosimilars are priced 15%–35% lower than the list prices of the reference products.[44] However, the volume and price dynamics of the biosimilar market remain volatile due to some significant uncertainties.

2.9.1 Patent Litigation

When a biologic is approved under the BLA pathway, it is granted 12-year market exclusivity. That means another manufacturer cannot launch its biosimilar product until 12 years of the reference product having been on the market. The manufacturers of the reference products also have the right to assert their patents before the market entry of a biosimilar, which is also known as "patent dance". To simplify the highly complex process of "patent dance", the reference product manufacturer must provide a list of potentially infringed patents to the biosimilar applicant(s) and state whether it is willing to license any of these patents. As a result, the launch of a biosimilar may be prolonged significantly.[45] For example, although there have been several biosimilars approved for the reference product Humira®, they cannot launch before 2023 due to their settlement over patent litigation. This allows reference products to retain market exclusivity for a longer period of time.[46]

2.9.2 Pricing Competition

Although biosimilars may be at a cheaper price compared to the reference products, manufacturers may negotiate with the payers or pharmacy benefit managers to achieve similar or lower net costs for reference products, subsequently maintaining their preference on formularies. For instance, a 2017 lawsuit by a biosimilar manufacturer alleged that the reference product manufacturer told the insurers to exclude biosimilars or they would withhold rebates on other products.[47]. It is not uncommon for a rebate to cover up to 50% of the biologic's list price. When a reference company withdraws the rebate, the price can double for the payer. This is referred to as a "rebate trap" and can drive the payer's cost to actually increase after the introduction of a cheaper biosimilar. A study published in 2018 showed that among 2547 Medicare Part D plans, only 10% of the plans covered biosimilar infliximab-dyyb, while 96% covered the reference product infliximab. The study concluded that the patients were paying higher out-of-pocket costs for the biosimilar than the reference product.[46,48] As a result of such a competitive dynamic, infliximab has the lowest biosimilar uptake at 6% share of molecule volume in 2020.[48] In addition to the rebate trap, reimbursement models are vastly different among Medicaid, Medicare, and commercial plans, making it difficult to streamline rebate or reimbursement strategies and encourage the biosimilar market.[39,49]

2.9.3 Difference in Physician Acceptance and Product Uptake

When biosimilar products first launched, some physicians were skeptical about prescribing biosimilars. A survey conducted in 2016 showed that about 55% of the U.S. physicians in specialties do not believe biosimilars are safe and appropriate for use in patients.[50,51] The low physician acceptance contributed to a rough start to the biosimilars market in its infancy phase. As more biosimilar products have launched to the market, their uptake has gradually increased with diverse patterns. For example, bevacizumab, a treatment of various cancers, is considered one of the fastest-growing biosimilar products on the market, and reached 42% of market volume in June 2020. However, its adoption varies significantly between healthcare providers. Many physicians prescribed bevacizumab biosimilar very sparingly or not at all, while some prescribed 100% biosimilar products.[43]

2.10 FUTURE PERSPECTIVES

Understanding the benefits of biosimilars and the challenges they are facing on the market, the government has taken several actions to address the barriers to development and increase utilization of biosimilars.

In February 2018, the President signed into law the Bipartisan Budget Act of 2018 (BBA of 2018), which included biosimilars in the Medicare prescription drug benefit's coverage-gap discount program starting 2019.[52].Once a beneficiary reaches the coverage gap (also known as the "donut hole"), the manufacturer and plan cover 75% of the cost, and the beneficiary would only need to pay 25% of the cost. Previously, biosimilars were not included in the coverage-gap discount program and the beneficiary would be responsible for 100% of coinsurance of the biosimilar.

In July 2018, the FDA published the Biosimilar Action Plan with the purpose of balancing innovation and competition among biologics and the development of biosimilars.[53] The action plan focuses on four key areas: (1) improving the efficiency of the biosimilar and interchangeable product development and approval process; (2) maximizing scientific and regulatory clarity for the biosimilar product development community; (3) developing effective communications to improve understanding of biosimilars among patients, clinicians, and payers; and (4) supporting market competition by reducing gaming of FDA requirement or other attempts to unfairly delay competition.

On February 3, 2020, the FDA and the Federal Trade Commission (FTC) signed a Joint Statement promoting competition in biologic markets, which included four key goals.[54] Firstly, the FDA and FTC will coordinate to promote greater competition in biologic markets. This would include development materials for consumers' and providers' educations, as well as public outreach. Secondly, the FDA and FTC will collaborate to deter behavior that impedes access to samples of reference products needed for the development of biologics, including biosimilars. Thirdly, the FDA and FTC will work together to address false or misleading communication about biologics, including biosimilars, with their respective authorities. That would include any false or misleading comparison between a reference product and a biosimilar in terms of safety or efficacy. The FDA is also publishing a draft guidance outlining considerations for FDA-regulated advertisements and promotional labeling that contains information about biologics. Finally, the FTC will also review patent settlement agreements involving biologics, including biosimilars, for anti-trust violation. The FTC may evaluate the agreements, including, among other things, anticompetitive reverse payments that slow or defeat the introduction of lower-priced medicines (e.g. biosimilars).

In addition to strengthening regulatory actions exerted by the FDA, payer and commercial insurers have started adding biosimilar products on the formularies, while excluding reference products. According to the Express Script 2020 National Preferred Formulary Exclusion list, reference products such as Avastin®, Heceptin®, and Rituxan® have been moved to formulary exclusion, while their biosimilar products are now considered formulary preferred alternatives.[55] With one of the largest Pharmacy Benefit Managers setting the example, the increase in biosimilar utilizations is imminent.

2.11 CONCLUSION

While bearing many uncertainties, the biosimilar market is expected to continue to grow and increase market share over time. It is also highly anticipated when biosimilars for Humira® and Enbrel® launch in 2023, as these biosimilars are under the highest spending drug category. In addition, regulation and price strategy implementation that incentivize biosimilars may help drive biosimilar adoption and utilization. Ultimately, patients are the beneficiaries as they will have more access to cost-effective biologic products.

REFERENCES

1. Center for Biologics Evaluation and Research. "What Are 'Biologics' Questions and Answers." FDA. Accessed February 28, 2019. https://www.fda.gov/about-fda/center-biologics-evaluation-and-research-cber/what-are-biologics-questions-and-answers.
2. AJMC. "The Growing Cost of Specialty Pharmacy – Is it Sustainable?". Accessed July 9, 2021. https://www.ajmc.com/view/the-growing-cost-of-specialty-pharmacyis-it-sustainable.
3. Center for Biologics Evaluation and Research. "Biological Approvals by Year." FDA. Accessed April 12, 2021. https://www.fda.gov/vaccines-blood-biologics/development-approval-process-cber/biological-approvals-year.
4. "2020 Drug Trend Report." Evernorth. Accessed July 9, 2021. https://www.evernorth.com/drug-trend-report.
5. IQVIA Institute for Human Data Science. "The Global Use of Medicine in 2019 and Outlook to 2023." Accessed July 9, 2021. https://www.iqvia.com/insights/the-iqvia-institute/reports/the-global-use-of-medicine-in-2019-and-outlook-to-2023.
6. Moore, Thomas J., Hanzhe Zhang, Gerard Anderson, and G. Caleb Alexander. "Estimated Costs of Pivotal Trials for Novel Therapeutic Agents Approved by the US Food and Drug Administration, 2015–2016." JAMA Internal Medicine 178, no. 11 (November 2018): 1451–57. https://doi.org/10.1001/jamainternmed.2018.3931.
7. Center for Drug Evaluation and Research. "FDA Webinar – Overview of the Regulatory Framework and FDA's Guidance for the Development and Approval of Biosimilar and Interchangeable Products in the US." FDA. Accessed February 9, 2019. https://www.fda.gov/drugs/biosimilars/fda-webinar-overview-regulatory-framework-and-fdas-guidance-development-and-approval-biosimilar-and.
8. European Medicines Agency. "Biosimilars in the EU – Information Guide for Healthcare Professionals". EMA. Accessed July 20, 2021. https://www.ema.europa.eu/en/documents/leaflet/biosimilars-eu-information-guide-healthcare-professionals_en.pdf.
9. Center for Drug Evaluation and Research. "Biosimilar and Interchangeable Products." FDA. Accessed February 9, 2019. https://www.fda.gov/drugs/biosimilars/biosimilar-and-interchangeable-products.
10. Center for Biologics Evaluation and Research. "Biologics License Applications (BLA) Process (CBER)." FDA. Accessed April 12, 2021. https://www.fda.gov/vaccines-blood-biologics/development-approval-process-cber/biologics-license-applications-bla-process-cber.

11. Center for Drug Evaluation and Research. "FDA Ensures Equivalence of Generic Drugs." FDA. Accessed November 3, 2018. https://www.fda.gov/drugs/resources-you-drugs/fda-ensures-equivalence-generic-drugs.
12. Office of the Commissioner. "Biosimilar and Interchangeable Biologics: More Treatment Choices." FDA. Accessed September 9, 2020. https://www.fda.gov/consumers/consumer-updates/biosimilar-and-interchangeable-biologics-more-treatment-choices.
13. Center for Drug Evaluation and Research. "Biosimilar Development, Review, and Approval." FDA. Accessed December 3, 2019. https://www.fda.gov/drugs/biosimilars/biosimilar-development-review-and-approval.
14. Center for Drug Evaluation and Research. "Scientific Considerations in Demonstrating Biosimilarity to a Reference Product." U.S. Food and Drug Administration. Accessed April 24, 2020. https://www.fda.gov/regulatory-information/search-fda-guidance-documents/scientific-considerations-demonstrating-biosimilarity-reference-product.
15. Center for Drug Evaluation and Research. "Considerations in Demonstrating Interchangeability with a Reference Product Guidance for Industry." U.S. Food and Drug Administration. Accessed May 6, 2020. https://www.fda.gov/regulatory-information/search-fda-guidance-documents/considerations-demonstrating-interchangeability-reference-product-guidance-industry.
16. Office of the Commissioner. "FDA Approves First Interchangeable Biosimilar Insulin Product for Treatment of Diabetes." FDA. Accessed July 30, 2021. https://www.fda.gov/news-events/press-announcements/fda-approves-first-interchangeable-biosimilar-insulin-product-treatment-diabetes.
17. Federal Register. "Nonproprietary Naming of Biological Products; Guidance for Industry; Availability." Accessed January 13, 2017. https://www.federalregister.gov/documents/2017/01/13/2017-00694/nonproprietary-naming-of-biological-products-guidance-for-industry-availability.
18. Federal Register. "Nonproprietary Naming of Biological Products: Update; Draft Guidance for Industry; Availability." Accessed March 8, 2019. https://www.federalregister.gov/documents/2019/03/08/2019-04242/nonproprietary-naming-of-biological-products-update-draft-guidance-for-industry-availability.
19. Office of the Commissioner. "Statement from FDA Commissioner Scott Gottlieb, M.D., on FDA's Steps on Naming of Biological Medicines to Balance Competition and Safety for Patients Receiving These Products." FDA. Accessed March 24, 2020. https://www.fda.gov/news-events/press-announcements/statement-fda-commissioner-scott-gottlieb-md-fdas-steps-naming-biological-medicines-balance.
20. Avastin (bevacizumab) [prescribing information]. South San Francisco, CA: Genentech; May 2020.
21. Mvasi (bevacizumab-awwb) [prescribing information]. Thousand Oaks, CA: Amgen Inc; June 2019.
22. Zirabev (bevacizumab-bvzr) [prescribing information]. New York, NY: Pfizer Labs; January 2020.
23. "FDA Purple Book." Accessed July 16, 2021. https://purplebooksearch.fda.gov/.
24. Center for Drug Evaluation and Research. "Biosimilar Product Information." FDA. Accessed December 17, 2020. https://www.fda.gov/drugs/biosimilars/biosimilar-product-information.
25. Insulin Glargine. Lexi-Drugs. Lexicomp. Wolters Kluwer Health, Inc. Riverwoods, IL. Accessed July 30, 2021. http://online.lexi.com.
26. Rituximab. Lexi-Drugs. Lexicomp. Wolters Kluwer Health, Inc. Riverwoods, IL. Accessed July 16, 2021. http://online.lexi.com.
27. Adalimumab. Lexi-Drugs. Lexicomp. Wolters Kluwer Health, Inc. Riverwoods, IL. Accessed July 16, 2021. http://online.lexi.com.
28. Pegfilgrastim. Lexi-Drugs. Lexicomp. Wolters Kluwer Health, Inc. Riverwoods, IL. Accessed July 16, 2021. http://online.lexi.com.
29. Infliximab. Lexi-Drugs. Lexicomp. Wolters Kluwer Health, Inc. Riverwoods, IL. Accessed July 16, 2021. http://online.lexi.com.
30. Bevacizumab. Lexi-Drugs. Lexicomp. Wolters Kluwer Health, Inc. Riverwoods, IL. Accessed July 16, 2021. http://online.lexi.com.
31. Trastuzumab. Lexi-Drugs. Lexicomp. Wolters Kluwer Health, Inc. Riverwoods, IL. Accessed July 16, 2021. http://online.lexi.com.
32. Entanercept. Lexi-Drugs. Lexicomp. Wolters Kluwer Health, Inc. Riverwoods, IL. Accessed July 16, 2021. http://online.lexi.com.
33. Filgrastim. Lexi-Drugs. Lexicomp. Wolters Kluwer Health, Inc. Riverwoods, IL. Accessed July 16, 2021. http://online.lexi.com.
34. Epoetin-alfa. Lexi-Drugs. Lexicomp. Wolters Kluwer Health, Inc. Riverwoods, IL. Accessed July 16, 2021. http://online.lexi.com.
35. Agbogbo, Frank K., Dawn M. Ecker, Allison Farrand, Kevin Han, Antoine Khoury, Aaron Martin, Jesse McCool, et al. "Current Perspectives on Biosimilars." Journal of Industrial Microbiology & Biotechnology 46, no. 9 (October 1, 2019): 1297–1311. https://doi.org/10.1007/s10295-019-02216-z.

36. Center for Drug Evaluation and Research. "'Deemed to Be a License' Provision of the BPCI Act." FDA. Accessed April 2, 2020. https://www.fda.gov/drugs/guidance-compliance-regulatory-information/deemed-be-license-provision-bpci-act.
37. Center for Drug Evaluation and Research. "List of Approved NDAs for Biological Products That Were Deemed to be BLAs on March 23, 2020." FDA. Accessed April 2, 2020. https://www.fda.gov/media/119229/download.
38. Office of the Commissioner. "FDA Works to Ensure Smooth Regulatory Transition of Insulin and Other Biological Products." FDA. Accessed March 24, 2020. https://www.fda.gov/news-events/press-announcements/fda-works-ensure-smooth-regulatory-transition-insulin-and-other-biological-products.
39. IQVIA Institute for Human Data Science. "Biosimilars in the U.S.: Reimbursement and Impacts to Uptake." Accessed July 27, 2021. https://www.iqvia.com/locations/united-states/library/white-papers/biosimilars-in-the-us-reimbursement-and-impacts-to-uptake.
40. "Part B Biosimilar Biological Product Payment and Required Modifiers – CMS." Accessed July 27, 2021. https://www.cms.gov/Medicare/Medicare-Fee-for-Service-Part-B-Drugs/McrPartBDrugAvgSalesPrice/Part-B-Biosimilar-Biological-Product-Payment.
41. The Center For Biosimilars. "CMS Biosimilar Reimbursement Shift: What You Need to Know." Accessed July 27, 2021. https://www.centerforbiosimilars.com/view/cms-biosimilar-reimbursement-shift-what-you-need-to-know.
42. Centers for Medicare & Medicaid Services. "Medicare Program; Revisions to Payment Policies under the Physician Fee Schedule and Other Revisions to Part B for CY 2018; Medicare Shared Savings Program Requirements; and Medicare Diabetes Prevention Program". Federal Register 82, no. 219 (November 15, 2017): 53349. http://www.gpo.gov/fdsys/pkg/FR-2013-11-05/pdf/2013-26670.pdf.
43. IQVIA Institute for Human Data Science. "Biosimilars in the United States 2020–2024." Accessed July 27, 2021. https://www.iqvia.com/insights/the-iqvia-institute/reports/biosimilars-in-the-united-states-2020-2024.
44. The Food and Drug Administration (FDA) and the Federal Trade Commission (FTC). "Joint Statement of the Food & Drug Administration and the Federal Trade Commission Regarding a Collaboration to Advance Competition in the Biologic Marketplace". FTC. Accessed February 3, 2020. https://www.ftc.gov/system/files/documents/public_statements/1565273/v190003fdaftcbiologicsstatement.pdf.
45. Aporte, Claire. "Manufacturers of Biosimilar Drugs Sit Out the 'Patent Dance.'" Health Affairs. Accessed March 10, 2017. https://healthaffairs.org.
46. Zhai, Mike Z., Ameet Sarpatwari, and Aaron S. Kesselheim. "Why Are Biosimilars Not Living up to Their Promise in the US?" AMA Journal of Ethics 21, no. 8 (August 1, 2019): 668–78. https://doi.org/10.1001/amajethics.2019.668.
47. Humer, Caroline. "Pfizer Files Suit against J&J over Remicade Contracts." Reuters. Accessed September 20, 2017, sec. Business News. https://www.reuters.com/article/us-pfizer-trial-johnson-johnson-idUSKCN1BV1S8.
48. Yazdany, Jinoos, R. Adams Dudley, Grace A. Lin, Randi Chen, and Chien-Wen Tseng. "Out-of-Pocket Costs for Infliximab and Its Biosimilar for Rheumatoid Arthritis Under Medicare Part D." JAMA 320, no. 9 (September 4, 2018): 931–33. https://doi.org/10.1001/jama.2018.7316.
49. The Center for Biosimilars. "CMS Payment Policy Plays Role in Biosimilar Uptake." Accessed July 30, 2021. https://www.centerforbiosimilars.com/view/cms-payment-policy-plays-role-in-biosimilar-uptake.
50. STAT. "Overcoming Challenges in the US Biosimilars Marketplace." Accessed July 30, 2021. https://www.statnews.com/sponsor/2018/11/13/challenges-in-the-us-biosimilars-marketplace/.
51. Cohen, Hillel, Donna Beydoun, David Chien, Tracy Lessor, Dorothy McCabe, Michael Muenzberg, Robert Popovian, and Jonathan Uy. "Awareness, Knowledge, and Perceptions of Biosimilars Among Specialty Physicians." Advances in Therapy 33, no. 12 (January 2017): 2160–72. https://doi.org/10.1007/s12325-016-0431-5.
52. Larson, John B. "H.R.1892 – 115th Congress (2017–2018): Bipartisan Budget Act of 2018." Legislation, February 9, 2018. https://www.congress.gov/bill/115th-congress/house-bill/1892/.
53. US Food and Drug Administration. "Biosimilars Action Plan: Balancing Innovation and Competition." FDA, July 2018. Accessed February 15, 2019. https://www.fda.gov/media/114574/download.
54. Federal Trade Commission. "FTC, FDA Sign Joint Statement Promoting Competition in Markets for Biologics." Accessed February 3, 2020. https://www.ftc.gov/news-events/press-releases/2020/02/ftc-fda-sign-joint-statement-promoting-competition-markets.
55. Express Script. "2021 National Preferred Formulary Exclusions". Accessed July 31, 2021. https://www.express-scripts.com/art/open_enrollment/DrugListExclusionsAndAlternatives.pdf.

Biologics and Biosimilars
Research and Development

3

Bahaar Kaur Muhar, Angela Penney, and Simeon Kotchoni

Corresponding Author: Simeon Kotchoni

Contents

3.1 INTRODUCTION

Current treatments against human illnesses have been dominated by small molecule drugs (molecular mass < 1000 g/mol).[1] However, large-scale biologics can be made cost-effectively. Biologics are complex molecule drugs that range from small components such as DNA and sugars to large entities such as whole cells generally used to treat or prevent illnesses.[2] Biologics are manufactured using living cells, and therefore, at the end of their production chains, biologics are unique entities, and two biologics cannot be identical because they are made from two different cells lines and/or

DOI: 10.1201/9780429485626-3

using different processes. With the exception of long-lived entities such as monoclonal antibodies, biologics are usually polar, heat-sensitive, membrane-impermeable, and readily degraded.[3] To mention just a few, common biologic examples include Lantus (insulin glargine) and Avastin (bevacizumab).

Biosimilars are "highly similar" versions of an innovator drug or biologic. Biosimilars are not to be confused with generics (generic drugs are identical to reference or brand drugs with only changes in the excipients).[4] Biosimilars utilize different cell lines and manufacturing processes, and/or raw material as the biologic developer.[4] However, the commonality between generic approved drugs and biosimilars includes the fact that both are marketed as cheaper versions, and are available when exclusive patent drugs expire, and are designed to have the same clinical effect as their pricier counterparts.[3] Biosimilars, therefore, offer an advantage of providing more affordable healthcare care by reducing the cost of the pharmaceutic for the consumer and the healthcare system as a whole (Table 3.1).[5]

This chapter will focus in detail on the industry aspects of biologics and biosimilars with regard to the whole process of development and research, as well as advantages, process time, and limitations.

3.2 DEVELOPMENTAL PROCESS

Every novel drug as well as vaccine candidate must be analyzed for safety, immunogenicity, and protective efficacy in humans before it may be used clinically. This can take on average 12–15 years and an investment of near one billion dollars.[20] Prior to regulatory approval, upon discovery, a biologic candidate usually undergoes different phases trials: pre-clinical, phase 1, phase 2, and phase 3 (Figure 3.1).[21] After the successful completion of phase 3, and following licensure of the product, phase 4 studies, also referred to as post-marketing surveillance studies (PMS) are used to continue to monitor the new drug and/or vaccine for safety and effectiveness in the population (Figure 3.1).[21]

Due to their high degree of complexity (Figure 3.2)[22], the development of biologics requires a comprehensive set of quantitative, accurate, and precise bioanalytical tools.[23] Biologic discovery and development can occur through various ways. Cell-based functional bioassays enable quantitative interrogation and validation of a biologic's target specificity.[23] Also, reagents that facilitate streamlined antibody purification and site-specific enzymes for mass spectroscopy-based analysis are crucial tools in biologics development workflows.[23]

In pre-clinical trials, biologics undergo *in vitro* and *in vivo* testing with the latter occurring first in animals first and then in humans. In phase 1 clinical trials, the biologic is introduced in the human subject for the first time and the prime focus is safety, followed by biological signs of efficacy. Phase 1 trials incorporate fewer than 100 healthy participants may be randomized and placebo-controlled and take several months to several years. Phase 2 trials study exclusively the dosing and safety of the biologic. Phase 2 trials include a larger participant size, 100–300, are randomized and placebo-controlled, and take a few years, averaging two years. Phase 3 studies focus on the efficacy and safety of the biologic. Phase 3 trials incorporate thousands of patients, are randomized and placebo-controlled, and last from one to four years. After successful completion of phase 3, the U.S. Food and Drug Administration (FDA) reviews the data and will move to the approval of the biologic's Biologics License Application (BLA). After a BLA has been approved, the biologic may then be marketed. Upon marketing, phase 4 or post-marketing surveillance continues during which the safety and efficacy of the biologic are monitored as well as any rare and long-term effects via post-marketing adverse events. Additionally, if a biologic has been given approval based on a surrogate endpoint, phase 4 studies will monitor the data for clinical efficacy confirmation.[24] In general, phase 4 studies are a minimum of one year.

TABLE 3.1 Average retail price comparison between approved biologics and biosimilars in USD. Biosimilars are on average 10%–37% cheaper than biologic products[5]

	BIOLOGIC (MANUFACTURER)	*AVERAGE RETAIL PRICE (USD)*	*DOSAGE FORM*	*BIOSIMILAR (MANUFACTURER)*	*DOSAGE FORM*	*AVERAGE RETAIL PRICE (USD)*	*REFERENCES*
Adalimumab	Humira (Abbvie)	$9,197.07	1 vial of 10 mL of 100 units	Hulio (Fujifilm)	–[a]	– [a]	[6]
Infliximab	Remicade (Janssen Biotech Inc.)	$5, 840.58	4 vials of 100 mg each	Avsola (Amgen)	3 vials of 100 mg each	$1,783.21	[7,8]
Rituximab	Rituxan (Genentech and Biogen)	$990	1 vial of 10 mg/mL	Riabni (Amgen)	100 mg	$716.80	[9,10]
Etanercept	Enbrel (Amgen & Wyeth)	$8,343.81	1 carton of 4 mini cartridges of 50 mg	Erelzi (Sandoz)	$6843.81[b]	1 carton of 4 mini cartridges of 50 mg	[11]
Insulin glargine injection	Lantus (Sanofi)	$123.99	1 pen of 3 mL	Basalgar[c] (Eli Lilly and Company)	$326.36	5 pack pens of 3 mL each	[12,13]
Bevacizumab	Avastin (Roche)	$842	4 mL	Zirabev (Pfizer)	$650	4 mL	[14,15]
Trastuzumab	Herceptin (Genentech)	$1,636.49	150 mg powder for injection	Kanjinti (Amgen)	$1,388	150 mg powder for injection	[16,17]
Pegfilgrastim	Neulasta (Amgen)	$6,231	1 injection of 6 mg/0.6 mL	Ziextenzo (Sandoz)	$4,108	1 injection of 6 mg/0.6 mL	[18,19]

[a] Hulio was approved by the U.S. Food and Drug Administration (FDA) in February 2021, so cost data is still awaiting.
[b] Erelzi pricing information is unknown at this time but is expected to be ~$1,500 less than the original biologic, Enbrel.
[c] Currently, there is no biosimilar for Lantus but Basalgar is categorized as its biosimilar as it uses the same active ingredient as Lantus.

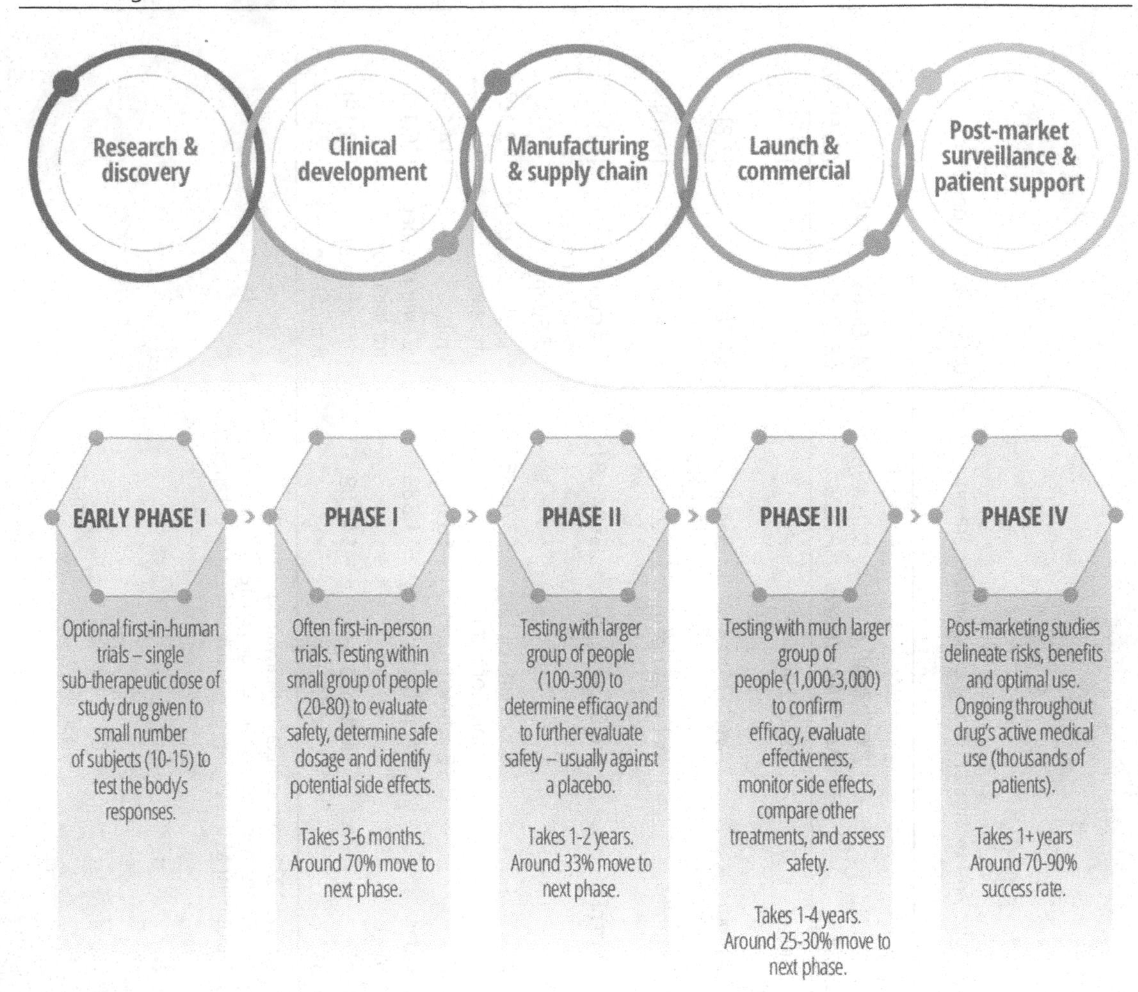

FIGURE 3.1 Biologic development timeline.[21]

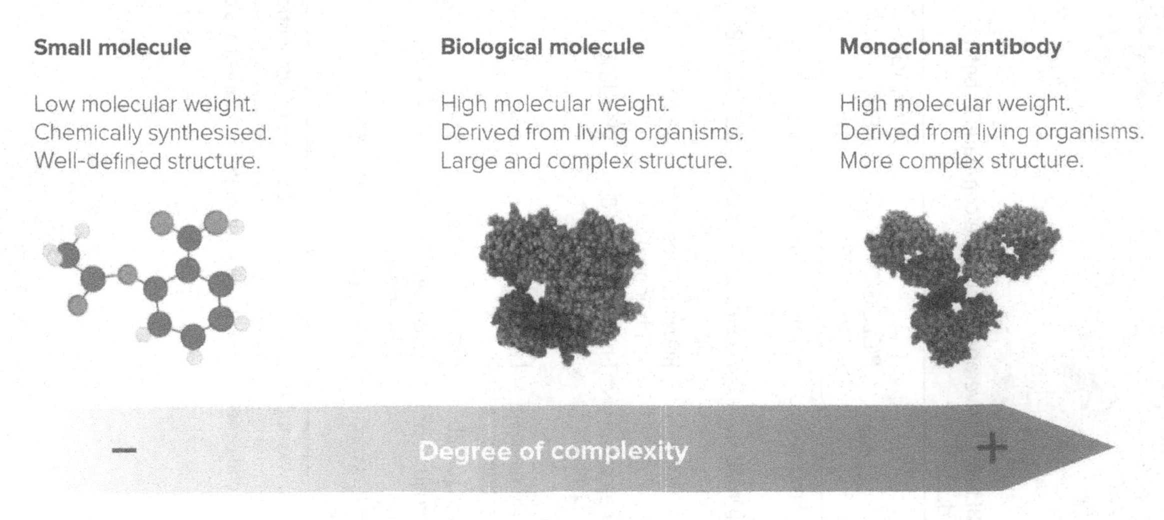

FIGURE 3.2 Complexity of drugs based on size and composition.[22]

TABLE 3.2 Key differences in the development of biologics and biosimilars[22]

	BIOLOGIC	*BIOSIMILAR*
Development cost (USD)	800 million	100–300 million
Time to market	8–10 years	7–8 years
Clinical studies	Phase 1–3 studies focusing on efficacy and safety	Phase 3 pharmacokinetic comparison studies
Patients/participant size	800–1000	100–500

Furthermore, biosimilar development is shorter than biologic development (Table 3.2),[22] (Figure 3.3),[25] but unlike generic drug development, biosimilars undergo more testing and trials, under an abbreviated approval pathway.[26] This is because biosimilars are very complex and process-specific and, therefore, it is not possible to produce biosimilars with identical characteristics of the standard product. Instead, the goal of a biosimilar development program is to demonstrate the biosimilarity between the proposed biosimilar product and the reference product, not to independently establish the safety and effectiveness of the proposed product.[27] Additionally, even with technological advances used to manufacture, purify and test biosimilars, the protocols or techniques may be different and sometimes inadequate in proving bioequivalence. Therefore, biosimilar development, as per U.S. FDA guidelines, consists of pre-clinical and clinical compatibility studies proving the biosimilar safety, purity, and efficacy.[27] This typically includes assessing immunogenicity, pharmacokinetics (PK), and, in some cases, pharmacodynamics (PD) and may also include a comparative clinical study.[27]

The development of biosimilars requires comparative *in vitro* non-clinical studies to confirm that the biosimilar matches the standard biologic.[26] Non-clinical animal studies may be avoided unless there are clear differences between the biosimilar and original biologic in terms of process related to impurities or important differences in the formulation that might cause potential efficacy or toxicity concerns.[26] The clinical trials for biosimilars are not standardized to those of biologics. Inclusion and exclusion criteria, statistical analyses, immunogenicity, and PK assays are just a few examples of aspects of clinical trial design that are not standardized between biologics and biosimilars.[26] If they were standardized, it would be easier to make indirect comparisons between biologics and different biosimilars. Overall, the most

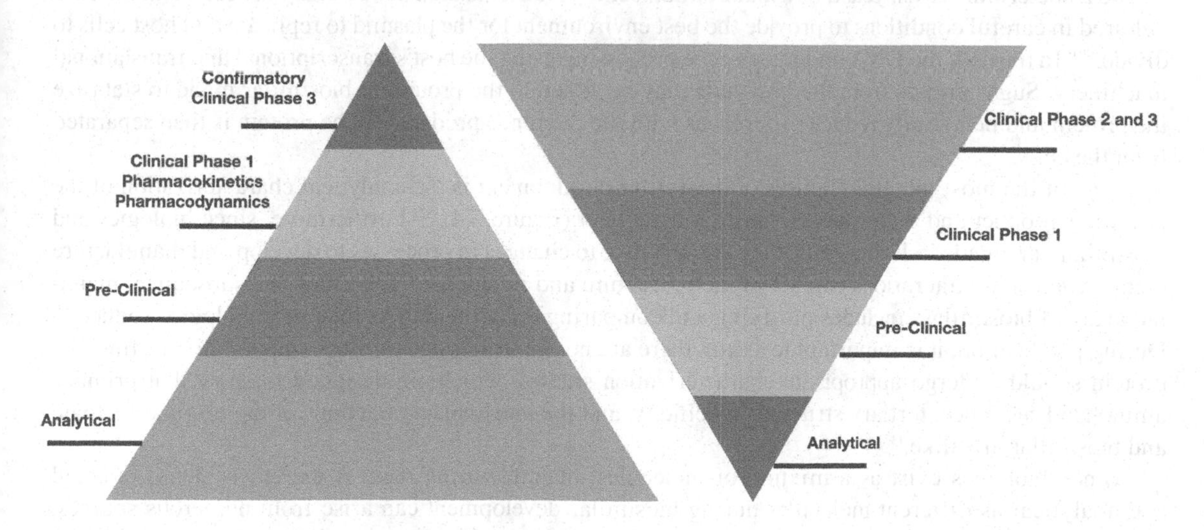

FIGURE 3.3 Contrast between the standard regulatory pathway to establish safety and efficacy of new biologic product versus biosimilar regulatory pathway to establish adequate comparability and clinical efficacy.[25]

prominent development concepts for the biosimilar approval process include design control with validation and verification studies, quality, statistical considerations for demonstration of analytical similarly, clinical aspects, and the FDA guidance of biosimilar labeling.[28]

3.3 RESEARCH

A biosimilar is researched based on the original biologic's patent, price, and mechanism of action. A biosimilar, or protein product, is a large molecule based on a sequence of amino acids folded in secondary and tertiary structures that undergo post-translational modifications to ultimately fold into a complete complex structure.[29] The post-translational modifications differ between the original biologics and biosimilars and this is because these modifications are a host cell attribute, and biologics and biosimilars do not use the same host cell line. Furthermore, this intricate process is challenging to reproduce even in the production of the reference biologic.[29] Upon altering the post-translational modifications by using a different host cell, culture conditions, or downstream processing like purification techniques or packaging, a bioequivalent formulation can be made. The selection of a biosimilar is complicated and more specific – especially in comparison to the selection of generic drugs.[29]

3.4 TECHNOLOGY AND MANUFACTURING

Biologics are made in living cells often by using recombinant DNA technology and require highly skilled technologically advanced capabilities.[30] The first step in manufacturing is to determine the biosimilar's genetic code, the DNA sequence of the desired protein, from the original biologic product.[30] Then, the specific DNA sequence is cloned into a vector using restriction enzymes as part of the recombinant DNA protocol.[31] The recombinant vector is then introduced into and taken up by host cells. Note, a host cell can be a bacterium, virus, yeast, or a mammalian cell.[30] Nutrients are added to the host cells and then are cultured in careful conditions to provide the best environment for the plasmid to replicate and host cells to divide.[30] In the cell, the DNA undergoes gene processing using the host's transcriptional and translational machinery. Sugar groups from the host cells may be added to the protein or biosimilar made to stabilize the protein and potentially reduce differences with the reference product.[30] The protein is then separated from the cells.

One of the most integral features of biosimilar development is the analytical characterization of the innovator product and the proposed biosimilar product (Figure 3.4).[32] Furthermore, since biologics and biosimilars are made in living cells, they are sensitive to changes in processes to develop and manufacture them – even small alterations can affect their structure and function.[33] Therefore, the subsequent manufacturing of biosimilars includes purifying and comparing the protein to its reference biologic product.[33] During purification, it is important to assure there are no impurities and during comparability testing, the protein should undergo appropriate characterization studies, which are designed to show that primary amino acid sequence, tertiary structure specificity, and the mechanism of actions of the original biologic and biosimilar are alike.[33]

Since biologics exist as a mixture of molecules, manufacturing them is extremely detail-oriented and analytical as different molecules during biosimilar development can arise from numerous sources, including N-terminal variants, post-translation modifications, glycoforms, and degradation products.[34] The high amount of heterogeneity in biosimilars makes it extremely necessary to demonstrate control

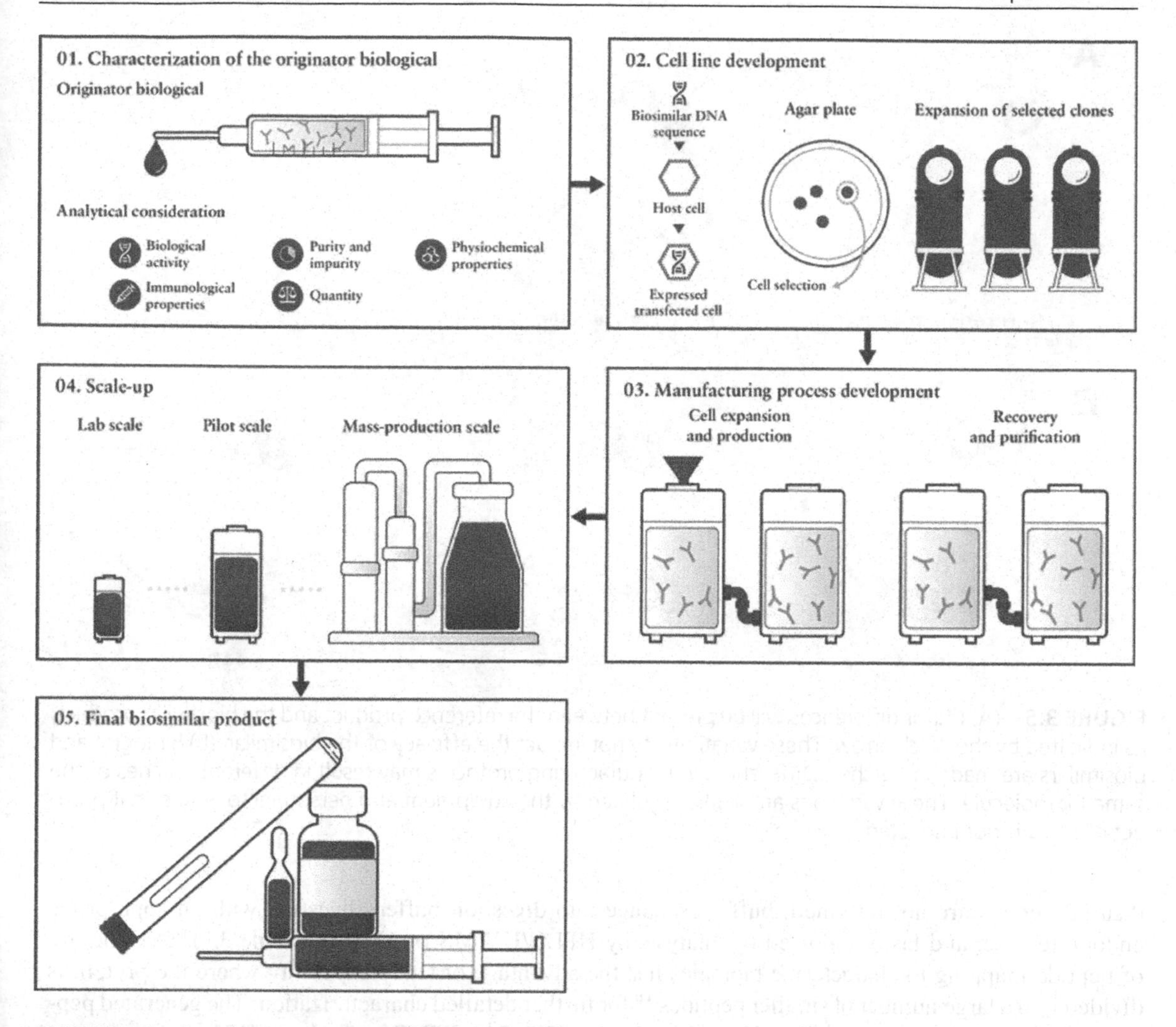

FIGURE 3.4 Development process of biosimilars.[32]

over the biosimilar manufacturing process (Figure 3.5).[35,36] This is done by using robust biochemical, biophysical, and function techniques requiring the manufacture to have analytical and characterization expertise and analytical techniques.[30]

One tool, intact mass analysis, assesses a protein's total molecular weight by mass spectrometry without prior digestion or fragmentation of the molecule of interest allowing for the observed mass to then be compared to the expected mass for a given amino acid sequence.[37] Intact mass analysis is a critical tool used for verifying that the purified biosimilar was successfully expressed and purified.[38] However, this technique only provides confirmation of protein molecular weight and not the location of any modifications that may be present.[38] In order to determine site-specific information (modification), peptide mapping with high-resolution mass spectrometry is used.[38] Ultra-high resolution mass spectrometry allows detection of mass analytes to the nearest 0.001 atomic mass units, and because this technique is very selective in measuring the exact mass of a compound, minor changes in structure may be easily distinguished.[38,39]

Peptide mapping is essential as it targets impurities or differences in the biosimilar that may reduce its efficacy.[38] Generally, the steps of peptide mapping include denaturing of the target protein, reduction of disulfide bonds, alkylation of the free thiol groups within the side chains of cysteine residues to ensure

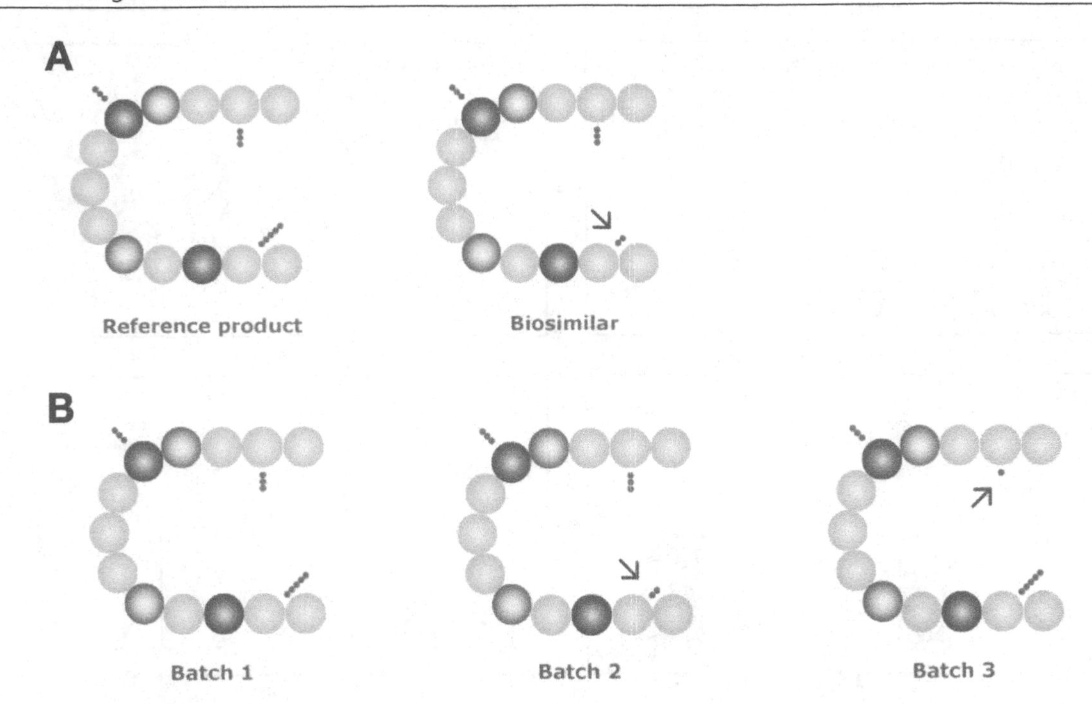

FIGURE 3.5 (A) Minor differences will be present between the reference product and the biosimilar product, as indicated by the black arrow. These variations do not impact the efficacy of the biosimilar. (B) Biologics and biosimilars are made in batches. Differences in manufacturing protocols may result in different batches of the same biomolecule. These variations are highly regulated by the equipment and personnel to assure that product efficacy is not impacted.[36]

disulfide bonds are not reformed, buffer exchange into digestion buffer, digestion with an appropriate endoproteinase, and lastly followed by analysis by HPLC/UV/MS or MS/MS (Table 3.3).[38,47] The use of peptide mapping to characterize biologics has the advantage of study capability where the protein is divided into a large number of smaller peptides[38] for further detailed characterization. The generated peptides can then be chromatographically separated using HPLC or UPLC technology.[38] The intact masses of the peptides can provide information regarding the type of modification made.[38] MS/MS sequencing of the peptides can provide the exact amino acid residue containing the modification (Figure 3.6).[38,40–46]

TABLE 3.3 Five steps for protein digestion[47]

PROCEDURE	*INTENDED EFFECT*	*GENERAL EXPERIMENT*
Sample preparation	Preparing sample for digestion	Depletion, enrichment, dialysis, desalting
Selection of cleavage agent	Specific cleavage requirement	–
Reduction and alkylation	Reduction reduces disulfide bonds Alkylation caps SH groups	Reduction: 1,4-dithiothreitol (DTT), 45 min, 60°C Alkylation: Iodoacetamide, 1 hr, in the dark
Digestion process	Cleavage of proteins	Digestion: pH 8, 37°C, overnight Quenching: Trifluoroacetic acid (TFA) addition
Enrichment/cleanup	Preparing sample for liquid chromatography (LC) or liquid chromatography/mass spectrometry (LC/MS) analysis	Concentrating, dialysis, affinity columns

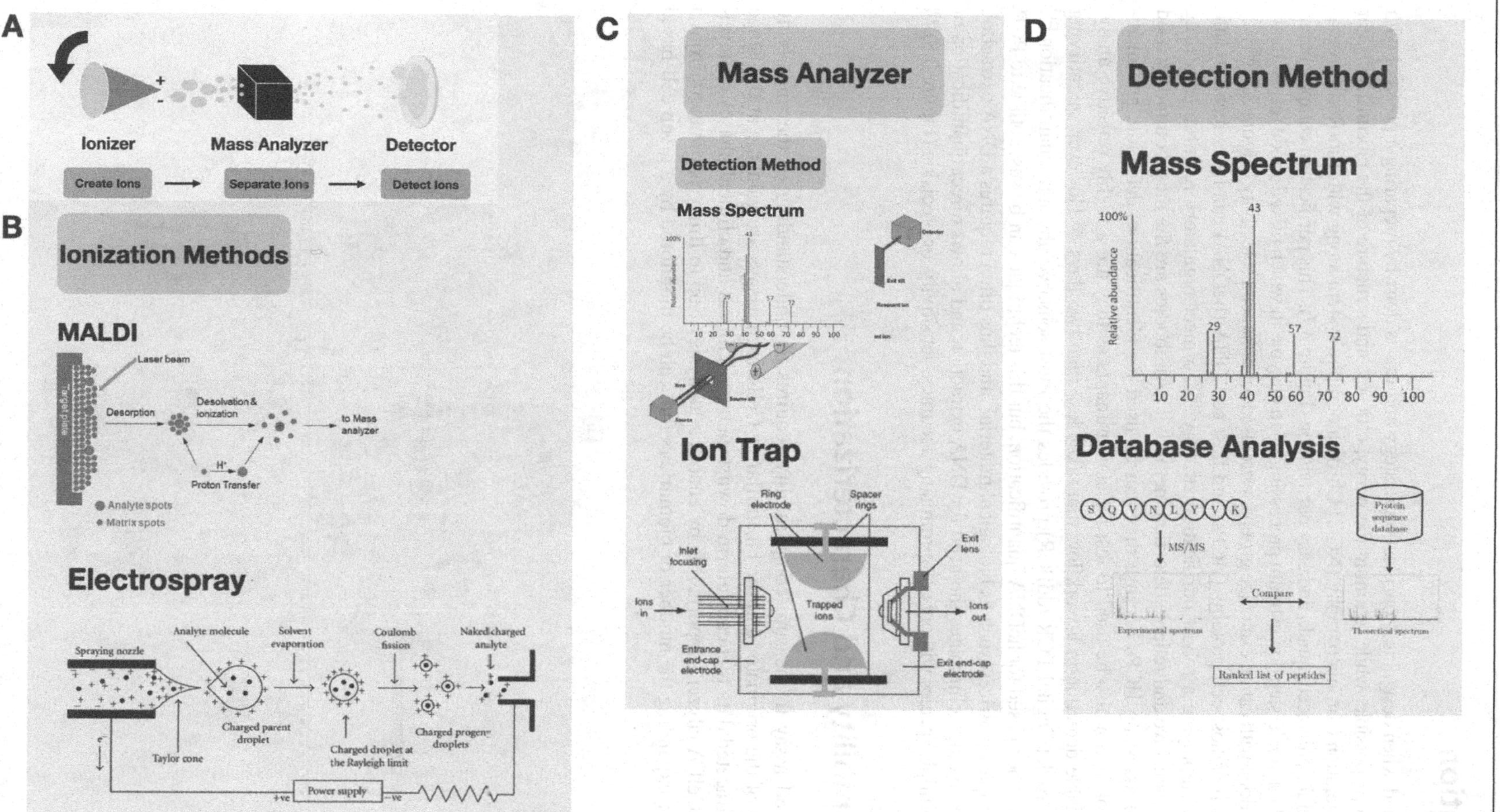

FIGURE 3.6 Mass spectrometry methodology. Mass spectrometry is an analytical tool useful for measuring the mass-to-charge ratio (m/z) of one or more molecules present in a sample. These measurements can often be used to calculate the exact molecular weight of the sample components as well. (A) The general steps consist of generating ions from organic or inorganic molecules, separating the ions based on the m/z to finally detect and analyze qualitatively and quantitatively. (B) Different ionization methods. (C) Different methods used to analyze mass and m/z. (D) Different methods or technology used to detect and compare.[40–46]

3.4.1 Purification

Costs must be controlled when designing a purification strategy and this is done by preparing a protocol with the least number of steps possible, while retaining the purity, efficacy, and structure of the biosimilar.[48] For example, the Next Generation Chromatography (NGC) Chromatography system along with Bio-Rad chromatography resins, is used for enhanced, quick, and easy purification (Figure 3.7). This purification equipment has been successfully used for single-step downstream processing of an enzyme in which mixed-mode resins, made of hydrophobic interactions with cation exchange groups, were used for enhanced selectivity (Figure 3.7).[49]

Contaminants such as host cell proteins (HCPs) and host cell DNA (hcDNA) can be reduced to safe (acceptable) levels by employing a robust purification strategy.[49] To meet regulatory requirements, drug manufacturers employ sensitive and reliable detection methods.[49] Antibodies are the most commonly used tool for detecting and assessing HCPs because they can be used for identification, detection, and quantification.[49] Western blots can also be used to test for contaminants as they are a highly sensitive analysis that allows for reliable detection in a workflow that takes less than two days.[49] However, quantifying residual hcDNA by Droplet Digital PCR (ddPCR) provides the most sensitive absolute quantification.[49] Traditionally, qPCR has been used for hcDNA quantification, but this technique can be susceptible to PCR inhibitors found in complex matrices, such as biological material, and thus often requires a DNA extraction step.[49] The ddPCR technology eliminates the need for DNA extraction and allows direct quantification of residual DNA from multiple species without compromising accuracy, sensitivity, or precision (Figure 3.8).[49]

3.4.2 Comparability and Characterization

Designing a cell-based assay to measure comparability requires an understanding of the underlying mechanism of action of the originator and the biosimilar.[49] Cell-based assays are valuable at this stage as they provide multifaceted data that can inform downstream pre-clinical and clinical phases of development.[49] The ZE5 Cell Analyzer supports this by enabling analysis of up to thirty parameters from a single sample, greatly expanding the number of biomarkers that can be monitored by a given cell-based

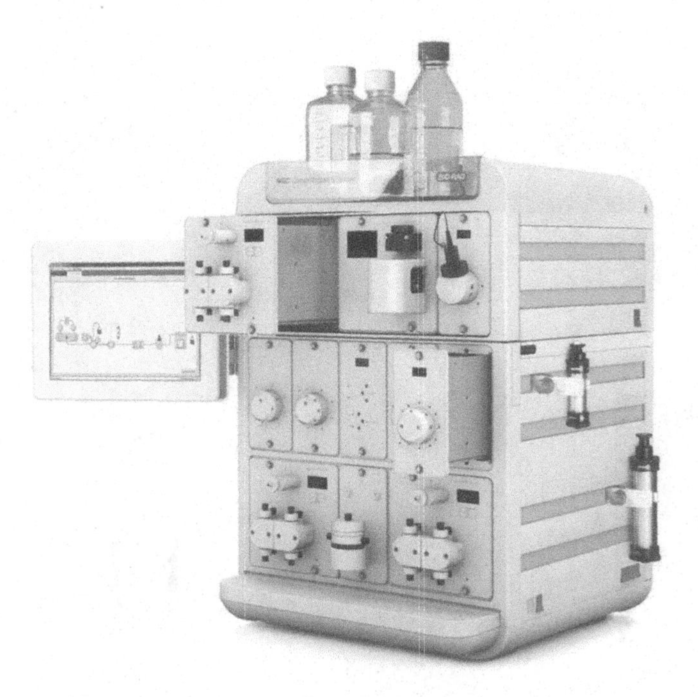

FIGURE 3.7 NGC™ chromatography system.[49] (Courtesy of Bio-Rad.)

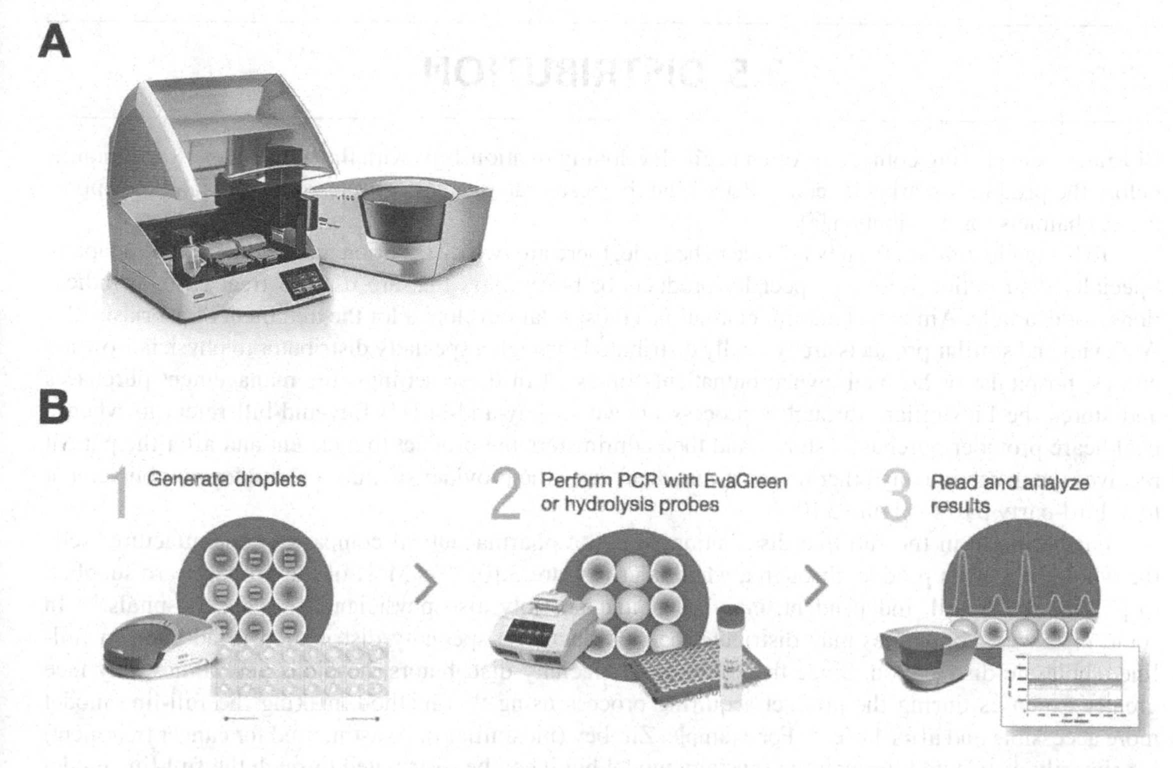

FIGURE 3.8 ddPCR technology by Bio-Rad.[49] (A) QX200 droplet digital PCR (ddPCR) equipment platform. (B) Steps of ddPCR workflow.

assay (Figure 3.9).[49] This is important when gathering data to provide as part of the totality of evidence required by regulatory agencies.[49] The ZE5 Cell Analyzer is also well-suited for measuring rare or transient cell populations due to its fast analysis speed (Figure 3.9).[49] As part of the biosimilar development process, when the biosimilar is compared to the original biologic, there must be no clinically meaningful differences.[49] Additionally, PK assays may be used for biosimilar comparability.[49] The assay is selected *in vitro* from the synthetic Human Combinatorial Antibody Libraries. Guided selection strategies enable the generation of fully human, highly specific inhibitory and non-inhibitory anti-idiotypic antibodies, and specialized drug-target complex binders.[49] These different types of antibodies enable the development of PK assays to detect free or total drug, or drug bound to its target.[49]

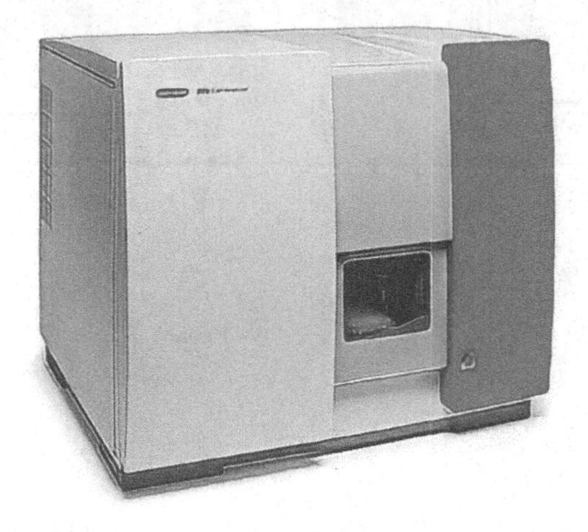

FIGURE 3.9 ZE5™ Cell Analyzer.[49]

3.5 DISTRIBUTION

Biosimilar developing companies often begin developing relationships with the wholesaler 18–24 months before the product's market launch to determine the best strategies for commercialization and the appropriate channels for distribution.[50]

To bring biosimilars from benchside to bedside, there are two distribution models that can be adapted. Specialty distribution refers to specialty products or biosimilars that are used to treat chronic indications; for example, Amjevita (adalimumab-atto), a biosimilar developed for the treatment of psoriasis.[50,51] Amjevita and similar products are typically distributed through a specialty distributor to physician-owned clinics, hospitals, or hospital-owned outpatient clinics.[50] In these settings, the management purchases and stores the biosimilars through a process known as buy-and-bill.[50] Buy-and-bill refers to when a healthcare provider purchases, stores, and then administers the product to a patient and after the patient receives the drug, and any other necessary medical care, the provider submits a claim for reimbursement to a third-party payer (Figure 3.10).[52,53]

Furthermore, in the full-line distribution model, a pharmaceutical company or manufacturer sells the whole line of its product through a wholesaler (Figure 3.10).[50,53] Most of the products are supplied to pharmacies (retail, independent, mail-order) and possibly also physician clinic and hospitals.[50] In some instances, companies may distribute their products via specialty distribution in addition to full-line wholesale distribution. Since there are fewer specialty distributors, hospitals and clinics may face greater expenses during the product acquiring process using this method, making the full-line model more accessible and affordable.[50] For example, Zirabev (biosimilar of Avastin; used for cancer treatment) is primarily distributed through the specialty model but it can be distributed through the full-line model for economic reasons.

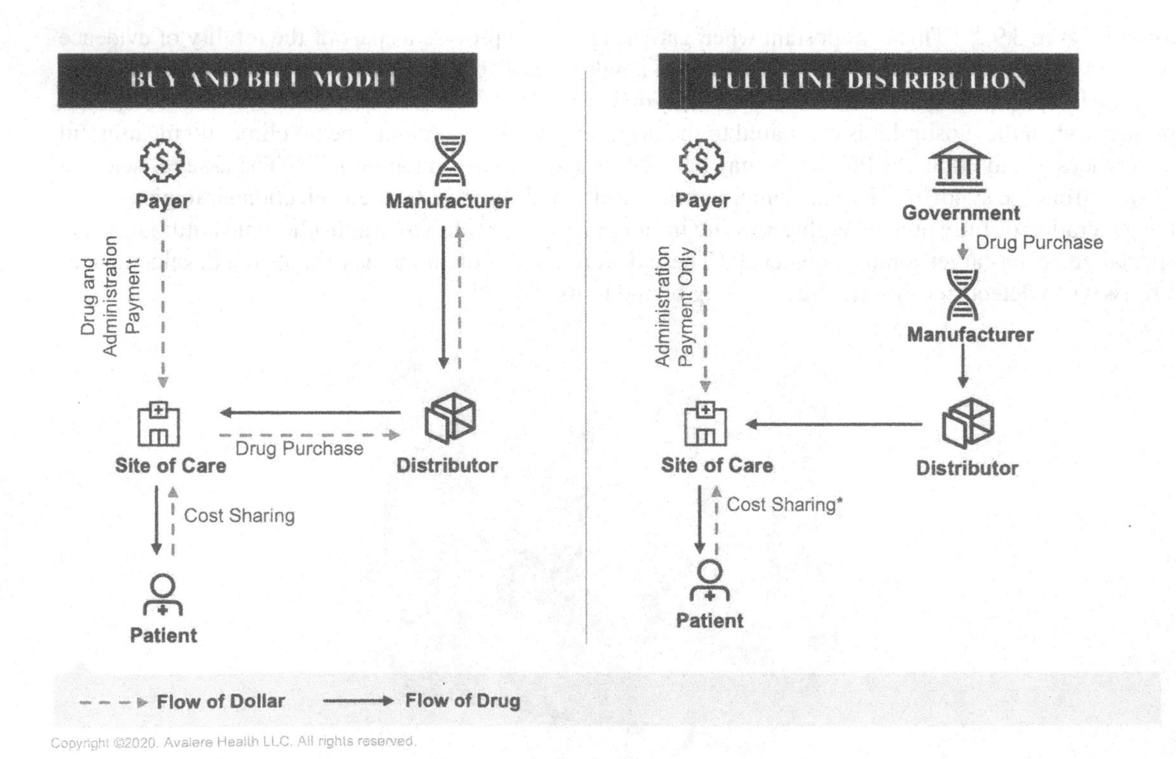

FIGURE 3.10 Flowchart describing the buy and bill model versus the full-line distribution model[53]

The best model may vary based on the biosimilar's distribution, cost, and/or product itself (Figure 3.10). Developing and manufacturing companies design their distribution and commercialization strategies prioritizing access and affordability for the patients.[50] Another factor considered includes how the company-wholesaler relationship can best deliver treatments to each individual patient, which often requires an understanding of the competitive market in which the biosimilar will compete and how the original biologic's product is being distributed.[50]

3.6 ADVANTAGES

The prime benefit that biosimilars have is that they are cheaper than biologics. They are not more efficacious or safe, if they were, then they would not constitute as a "biosimilar". On average, biosimilars are 10%–37% cheaper than biologic products.[5] Though, this does not translate to a major difference as seen in small molecule drugs and their corresponding generics, it is still cheaper, especially since biologics generally are expensive ($100s–$1000s). The reason that biosimilars do not have a dramatic decrease in price compared to generics is because manufacturing biosimilars is more extensive and difficult.[54] Additionally, biosimilars create additional treatment options for expensive brand-name products, thereby enhancing healthcare and delivery.[55]

3.7 CHALLENGES

During development, a prominent challenge biosimilars face is that the endpoints used in clinical trials may not be identical to those used in the biologic studies, thus making the comparability analysis more difficult.[56] Furthermore, clinicians, pharmacists, and even consumers must understand that the biosimilar development process is not to establish benefit but to prove bioequivalence.[57]

Although manufacturer, distributor, and payer strategies for managing biosimilars are evolving, providers and patients are slowly beginning to understand what biosimilars are and how they fit into healthcare options.[58] Therefore, lack of awareness and acceptance of biosimilars places a severe limitation on its use and marketing.[59] One way to conquer this challenge is by biosimilar manufacturers aligning with the supply chain channel strategy of the reference drug.[58] Another obstacle biosimilars face is that because they are not directly interchangeable with reference drugs, substitution becomes more complex, especially financially.[58] Providers or customers may feel more comfortable prescribing or being prescribed, respectively, a biologic that is more commonly known. Furthermore, the reduced cost of biosimilars may outweigh this concern. Also, if multiple biosimilars are launched for the same reference drug, there may be various fluctuations in market price.[58] In order to create a sustainable market for biosimilars where investment is attractive, stability and awareness are needed by manufacturers and providers, respectively.[58]

3.8 EXAMPLE: ZIRABEV

High pharmaceutical costs impact medication adherence.[60] For example, if a biologic is priced at $100 for ten doses and is not covered by insurance, the price adds up very quickly for a patient and, therefore, they may quit taking the pharmaceutical and thereby negatively impacting their health. However, with the reduced costs associated with biosimilars, the likelihood of patients discontinuing existing therapy drastically decreases. Avastin, a biologic used with chemotherapy as an anti-angiogenic therapy, costs $841.51 for 4 milliliters without insurance and was approved in 2004 by the U.S. FDA.[14] Zirabev, approved in the United States by FDA in 2019, is a monoclonal antibody biosimilar developed for Avastin that costs $650 for four milliliters without insurance.[15] The clinical trials for Zirabev studied clinically meaningful differences in safety, PK, immunogenicity across treatment groups and found none.[61] 45.3% of patients in the biosimilar arm and 44.6% in the reference arm achieved an objective response.[61] 96.6% of patients in the biosimilar arm and 96.9% of patients in the reference arm experienced at least one adverse event.[61] Furthermore, Avastin and Zirabev had similar serious adverse reactions (for example, gastrointestinal perforation and renal injury). Also, the incidence of immunogenicity was low since 1.5% of patients in the biosimilar arm and 1.4% of patients in the reference arm tested positive for antidrug antibodies.[61] The treatments were comparable.[61]

3.9 CONCLUSION

In summary, the greatest advantage of biosimilars over their bio-originators is their reduced cost in development, the improvements and increased efficiency of the production process, as well as the abbreviated clinical trial process.[27] During biosimilar development, the priority is to analyze comparability of the reference biologic to biosimilar of interest. Also, it is important to note, from a manufacturer perspective that developing a biosimilar is convenient and profitable as it is based on a biologic that is known to produce a pharmacological response safely and effectively and, therefore, the risk of failure is already known and are expected to be lower compared to its reference biologic.[62]

REFERENCES

1. Gurevich, E.V. and V.V. Gurevich, *Beyond traditional pharmacology: new tools and approaches*. Br J Pharmacol, 2015. **172**(13): p. 3229–41.
2. Lee, B., *What Are Biologics? 5 Examples of Biological Drugs You May Already Be Taking*. GoodRx, 2018.
3. Charles Oo, S.S.K., *Leveraging the attributes of biologics and small molecules, and releasing the bottlenecks: a new wave of revolution in drug development*. Taylor & Francis Online, 2016. **9**: p. 747–49.
4. Mora, F.d., *Biosimilar: what it is not*. Br J Pharmacol, 2015. **80**(5): p. 949–56.
5. Chase, L., *A Guide to Biosimilar Prices: How Much They Cost and How You Can Save*. GoodRx, 2020.
6. GoodRx, *Humira Prices, Coupons & Savings Tips*. GoodRx Inc., 2021.
7. GoodRx, *Remicade Prices, Coupons & Savings Tips*. GoodRx Inc., 2021.
8. GoodRx, *Avsola Prices, Coupons & Savings Tips*. GoodRx Inc., 2021.
9. Delgado, D.C., *FDA Approves Riabni, 3rd Rituxan Biosimilar to Treat GPA and MPA*. ANCA Vasculitis News, 2021.
10. Drugs.com, *Rituxan Prices, Coupons and Patient Assistance Programs*. Price Guide, 2021.
11. GoodRx, *Enbrel Prices, Coupons & Savings Tips*. GoodRx Inc., 2021.
12. GoodRx, *Lantus Prices, Coupons & Savings Tips*. GoodRx Inc., 2021.
13. SingleCare, *Lantus Solostar Coupons & Prices*. SingleCarer, 2021.
14. Drugs.com, *Avastin Prices, Coupons and Patient Assistance Programs*. Drugs.com, 2021.
15. Drugs.com, *Zirabev Prices, Coupons and Patient Assistance Programs*. Drugs.com, 2021.
16. Drugs.com, *Herceptin Prices, Coupons and Patient Assistance Programs*. Drugs.com, 2021.
17. Drugs.com, K*anjinti Prices, Coupons and Patient Assistance Programs*. Drugs.com, 2021.
18. Neulasta, *Prescribing Information*. Neulesta, 2021.
19. Drugs.com, *Ziextenzo Prices, Coupons and Patient Assistance Programs*. Drugs.com, 2021.

20. Sullivan, T., *A Tough Road: Cost to Develop One New Drug Is $2.6 Billion; Approval Rate for Drugs Entering Clinical Development Is Less Than 12%*. Policy & Medicine, 2019.
21. AXXELIS, *COVID-19 Impact: Rethinking How to Plan for and Run Clinical Trials*. Latest News, 2020.
22. mAbxience, *Generics, Biologics, Biosimilars: Who's Who?* mAbxience, 2021.
23. Promega. *Biologics Drug Discovery*. Promega Corporation. 2021.
24. Kraus, V.B., L.S. Simon, J.N. Katz, T. Neogi, D. Hunter, A. Guermazi, and M.A. Karsdal, *Proposed study designs for approval based on a surrogate endpoint and a post-marketing confirmatory study under FDA's accelerated approval regulations for disease modifying osteoarthritis drugs*. Osteoarthr Cartil, 2018. **27**(4): p. 571–79.
25. Cohen, H., *Biosimilar development – approval of biosimilar medicines through totality of the evidence*. Drug Dev Deliv, 2019. **19**(5): p. 40–44.
26. Isaacs, J., et al., *The biosimilar approval process: how different is it?* Considerations Med, 2017. **1**(1): p. 3–6.
27. FDA, *Biosimilar Development, Review, and Approval*. Biosimilars, 2017.
28. Aziz, K., *Biosimilar Development – Biosimilar Biological Products: Development & Applications*. Drug Dev Deliv, 2020. **20**(2): p. 24–29.
29. Niels Boone, H.v.d.K., Mike Scott, Jill Mairs, Irene Kramer, Arnold Vulto, Rob Janknegt, *How to select a biosimilar* 20, 2013(275–286).
30. Pfizer, Technology and Manufacturing Biosimilars. 2020.
31. Lodish H, B.A., Zipursky SL, et al., *Section 7.1 DNA Cloning with Plasmid Vectors*. Molecular Cell Biology. 4th edition., 2000.
32. Louis Bridges, S. Jr., D.W. White, A.B. Worthing, E.M. Gravallese, J.R. O'Dell, et al. The Science Behind Biosimilars. *Arthritis Rheumatol*, 2018. **70**(3): p. 334–344.
33. Berkowitz, R.P., *2012 updated consensus guidelines for the management of abnormal cervical cancer screening tests and cancer precursors*. Obstet Gynecol, 2013. **122**(2 Pt 1): p. 393.
34. Carol, F.K, X.Z.M. Wang, H.D. Conlon, S. Anderson, A.M. Ryan, A. Bose. *Biosimilars: key regulatory considerations and similarity assessment tools*. Biotechnol Bioeng, 2019. **114**(12): p. 2696–705.
35. Jaquez, A.G.V.a.O.A., *The process defines the product: what really matters in biosimilar design and production?* Rheumatology (Oxford England), 2017. **56**: p. iv14–iv19.
36. *Biosimilars in the EU: Information Guide for Healthcare Professionals*. European Medicines Agency (EMA), 2017.
37. Cecco, M.D., *Characterization of Protein Structure: The Value of Intact Mass Analysis*. Satorius Blog, 2021.
38. Boomershine, W., *High-Resolution Mass Spectrometry – A Map to Biologics*. Drug Dev Deliv, 2018. **18**(6): p. 44–48.
39. Cook-Botelho, J.C., L. Bachman., and D. French, Chapter 10 – Steroid hormones, *Mass Spectrometry for the Clinical Laboratory. Elsevier*, 2017: p. 205–230.
40. Kashyap, A., *Proteomics and its applications in phytopathology*. Slideshare, 2015.
41. Creative Proteomics, *MALDI-TOF Mass Spectrometry*. Creative Proteomics. 2021.
42. Mazumdar, S.B.a.S., *Electrospray ionization mass spectrometry: a technique to access the information beyond the molecular weight of the analyte*. Int J Anal Chem, 2012.
43. Santoiemma, G., *Recent methodologies for studying the soil organic matter*. Appl Soil Ecol, 2017. **123**: p. 546–50.
44. Yinon, J., *Detection of Explosives by Mass Spectrometry*. Counterterrorist Detection Techniques of Explosives, 2007.
45. Saxton, C., *How Does Mass Spectroscopy Work?* BiteSizeBio, 2021.
46. Lahesmaa-Korpinen, A.-M., *Computational approaches in high-throughput proteomics data analysis*. Helsinki University Biomedical Dissertations, 2012. **No. 169**.
47. Agilent Technologies Inc., *Peptide Mapping: Application Compendium*. Agilent Biocolumns, 2018.
48. Eva Rahman Kabir, S.S.M. and M.K. Sharif Siam, *The breakthrough of biosimilars: a twist in the narrative of biological therapy*. Biomolecules 2019. **9**(9): p. 410.
49. Bio-Rad, *A Biosimilar Workflow: From Purification to Characterization*. Bio-Rad Laboratories, Inc., 2021: p. 1–11.
50. Welch, A.R., *The Wholesaler's Guide to Biosimilar Distribution*. Biosimilar Development, 2018.
51. FDA approves Amjevita, *a biosimilar to Humira*. FDA News Release, 2016.
52. Fein, A.J., *Follow the Vial: The Buy-and-Bill System for Distribution and Reimbursement of Provider-Administered Outpatient Drugs*. Drug Channels, 2016.
53. Sloan Tande, K.O., J.C. Neal, L. Grady, A. Moorman, *CMS Takes Steps to Expand Access to COVID-19 Antibody Treatment*. Avalere, 2020.

54. Cronstein, B.N., *The benefits and drawbacks of biosimilars.* Clin Adv Hematol Oncol, 2015. 13(10): p. 639–41.
55. Diner, E.J., *Great Minds Think Similarly: The Pros and Cons of Biosimilars.* CGLifeAgency, 2021.
56. Mark, A.S., G. Curigliano, I. Jacobs, B. Gumbiner, J. MacDonald, and D. Thomas, *Clinical considerations for the development of biosimilars in oncology. MAbs*, 2015. **7**(2): p. 286–93.
57. Richard, M., J. Lui, M. Ramchandani, D. Landa, T. Born, and P. Kaur, *Developing the totality of evidence for biosimilars: regulatory considerations and building confidence for the healthcare community.* BioDrugs, 2017. **31**: p. 175–87.
58. Syrop, J., *Report: Biosimilars Will Play Increasingly Important Role in US Biologic Market.* AJMC, The Center for Biosimilars, 2017.
59. Cohen, H., D. Beydoun, D. Chien, T. Lessor, D. McCabe, M. Muenzberg, R. Popovian, and J. Uy, *Awareness, knowledge, and perceptions of biosimilars among specialty physicians.* Adv Ther, 2016. **33**: p. 2160–72.
60. Zullig, L.L., B.B. Granger, H. Vilme, M.M. Oakes, and H.B. Bosworth, *Potential impact of pharmaceutical industry rebates on medication adherence.* Am J Manag Care, 2019. **25**(5).
61. Reinmuth, N., M. Bryl, I. Bondarenko, K. Syrigos,V. Vladimirov, M. Zereu, A.H. Bair, F. Hilton, K. Liau, and K. Kasahara, *PF-06439535 (a bevacizumab biosimilar) compared with reference bevacizumab (Avastin®), both plus paclitaxel and carboplatin, as first-line treatment for advanced non-squamous non-small-cell lung cancer: a randomized, double-blind study.* BioDrugs, 2019. **33**(5): p. 555–70.
62. Erwin, A.B. and P.F. Joseph, *The economics of biosimilars.* Am Health Drug Benefits, 2013. **6**(8): p. 469–78.

Vaccines as Biologics in the Era of Pandemic

4

Priya K. Manhas, Christiane How-Volkman, Olivia Wu, Satori Iwamoto, Caroline Goswami, and Ahmed El-Shamy

Corresponding author: Ahmed El-Shamy

Contents

DOI: 10.1201/9780429485626-4

4.1 WHAT IS A VACCINE?

Humanity has been no stranger to the disease. As societies faced the increasing plight of infectious diseases, it became paramount to attempt to alleviate the populations afflicted and find a source of primary prevention. In order to better the health of populations, primary prevention strategies take the prophylactic rather than therapeutic approach– one such route being vaccination [1, 2].

The term vaccination originates from the Latin phrase "vacca", meaning cow [3]. Naming the term vaccination after a cow pays homage to Edward Jenner's discovery of cowpox exposure conferring immunologic protection to smallpox infection. Smallpox loomed over societies for nearly 3,000 years and it was not only until the end of the 18th century that Edward Jenner, the Father of Immunology, would identify a phenomenon, sparking the formation of a hypothesis that would later curtail smallpox [3, 4]. In 1796, Jenner discovered that milkmaids who had been infected with cowpox were shielded from the perils of smallpox, and Jenner was motivated to further investigate these findings [4, 5]. Drawing on his experience observing the milkmaids, Jenner predicted cowpox could hold the key to salvation. To test Jenner's predictions, Jenner injected components of a cowpox blister into an adolescent named James Phipps [4, 5]. Subsequently, Jenner exposed Phipps to variola virus (smallpox) multiple times and noted that Phipps never developed the disease that annihilated so many [4, 5]. Left with no symptoms, James Phipps gained full immunological protection to fend against smallpox infection and Jenner had performed the first account of contemporary vaccination [5]. Today, we can credit Jenner's unethical experiment to the development of modern-day vaccination.

Vaccination is an active form of immunization in which an antigen is introduced to the body to provoke an immune response, and optimistically manifests protection and long-term memory against the foreign antigen. Vaccines undoubtedly have been a remarkable tool since the 18th century warding off and closely eradicating infectious diseases, such as smallpox. Vaccination saves approximately 2 to 3 million lives annually, according to the Centers for Disease Control and Prevention (CDC) [6].

It is important to note that vaccines fall under the category of biologics rather than a typical drug. Biologics are classified as stemming from a living biological system and are often composed of larger, complex molecules such as saccharides, proteins, nucleic acids, cells, tissues, or an assortment of these entities [7]. Examples of biologics include vaccines, blood, and recombinant therapeutic proteins [7]. There is an underlying theme of mystery regarding all the components contained in biologics as not all attributes may be characterizable, making the need for biologic manufacturing to be more surveillant and controlled [8].

In this chapter, we will discuss the basis of vaccines, provide examples demonstrating the various vaccine types, examine factors affecting vaccine accessibility, and will explore possible routes of vaccine evolution in relation to the concept of biosimilarity. Although there is currently active discussion regarding how to properly define biosimilarity in regard to vaccines, for the purpose of this chapter, we will define biosimilarity as a medicinal biological product with an active substance that closely resembles an original or a reference product, in terms of molecular composition and bioequivalence in pharmaceutical and clinical settings [9]. The dawn of the COVID-19 pandemic demonstrated the strong need for innovative vaccine technology. We will be hypothesizing how biosimilar vaccines may bring about a solution to the many barriers to absolute health equity we see today.

4.2 HOW DO VACCINES WORK?

The purpose of vaccines is to provide a protective effect against pathogens by stimulating an immune response and subsequently generating immunological memory. Pathogens vary from bacteria, viruses, parasites, and fungi that cause disease within the body [10]. Pathogens tend to have unique subparts that are often referred to as antigens, because they illicit an immune response through activation of humoral and cellular immunity. The humoral immunity, which is responsible for antibody production, is a key contributor to the body's defensive system. Furthermore, cell-mediated immunity tackles infected cells with the use of T-cells and initiates a response to destroy the microorganisms through cell signaling and lysis [11].

In general, time is necessary for the immune system to respond to a new antigen and produce specific humoral and cell-mediated immunity. Vaccines manipulate immune responses by taking advantage of pathogens' antigens. With the use of an inactivated or weakened disease-causing microorganisms, the body can initiate an immune response that would mimic a natural response to infection. Vaccines incorporate the antigens of pathogens that would not cause disease but will trigger in an immune response and antibody production [12]. When antibodies are present, the antigens will be neutralized and destroyed by other immune cells. This response permits for an immunological memory against the pathogen, and enables the immune system to recognize and respond in both cell-mediated and humoral-like fashions if ever exposed to the disease-causing pathogen at a later time [13].

Vaccines have been evolving in the past 300 years. Traditional and older vaccines were often constructed from whole pathogens that were killed or live-attenuated products or subunits of viral particles [12]. However, these conventional vaccine approaches did not meet the demands of evolving pathogens, like influenza, Ebola, and Zika viruses. Therefore, newer vaccines have recently been developed and produced using recombinant viral vectors to deliver the antigens more locally for replication and amplification within the recipient's cells [14]. Antigens are not the only component to the formulation of vaccines; adjuvants are substances that are also added in the formulation to help promote the immunogenic profile of the antigens formulated in vaccines. Adjuvant incorporation is critical in vaccine development for vaccines that lack potent immunogenicity [15]. As a relevant example during the COVID-19 global pandemic, Novavax COVID-19 vaccine's incorporation of the saponin-based adjuvant, Matrix-M, helped direct immune responses to humoral immunity, preferentially the helper T cell type 1 (Th1) subset [16].

4.3 BENEFITS OF VACCINATION

Vaccination has greatly reduced the burden of infectious diseases [17]. Given their less reactogenic property, independent experts and the World Health Organization (WHO) have deemed vaccines far safer than therapeutic medicines [18]. With the push for global coverage against childhood infectious diseases, vaccinations have greatly contributed to the betterment of global health. In just the United States between 1993 and 2014, the Vaccines for Children (VFC) program prevented approximately 322 million illnesses, 21 million hospitalizations, and 732,000 premature deaths among children born, at cost savings to society of $1.38 trillion [19].

There is a long-standing list of benefits due to vaccinations. First, the disease control benefits through eradication and elimination, as seen with rinderpest and smallpox [20]. At the same time, polio is on the brink of eradication. Vaccines are able to control mortality, morbidity, and complications of diseases by mitigating their severity or preventing them as a whole [17]. In the United States, there has been a 99% decrease in incidence for the nine diseases for which vaccines have been recommended for decades, accompanied by a similar decline in mortality and disease sequelae [21]. Some infective agents can cause cancer, but vaccines against these agents can prevent the incidence of the disease and their progression to such cancers. For example, chronic hepatitis B virus infection that leads to liver cancer, or human papillomavirus (HPV) leading to cervical cancer [21].

Vaccines help protect against disease-causing pathogens. Herd immunity is a type of protection proposing that a sufficient proportion of the population needs to be immune (through overcoming natural infection or through vaccination) to an infectious agent for it to stop generating large outbreaks, especially in the presence of those unimmune or unvaccinated groups [22]. Though this effect is dependent on the efficacy and coverage of the vaccine, vaccination dramatically reduces the spread of the infectious agent by reducing the number of transmitting variables. The greatly decreased incidence of measles, mumps, polio, and varicella are greatly related to the establishment of herd immunity. Outbreaks for these diseases tend to occur when there is lower vaccine coverage in particular communities [23]. Hence, the roll-out of vaccines is quintessential to the containment of disease-causing pathogens, as seen through the rapid efforts for the COVID-19 vaccine delivery.

There are also societal benefits provided by vaccination and their programs. The quality of health care increases by providing a stable infrastructure to give opportunities for a better healthcare systems for both individuals and their communities, especially during critical periods of their lives [24]. Vaccinations also extend life expectancies by protecting against pathogens that could harm weaker individuals like the elderly, infants, or those with pre-existing health conditions. By protecting individuals on a health basis, the ripple effect allows for a larger societal impact for factors not always thought brought to the table like safer travel and mobility, protection against bioterrorism, and promoting economic growth [17].

4.4 CHALLENGES OF VACCINATIONS

Although vaccine refusal is complex and there are many reasons behind such reservations, there are legitimate failures to vaccinations. There are two major reasons for failure. First, the delivery system of the vaccine may fail to meet the therapeutic threshold needed by the individual(s). For example, some vaccines require multiple doses over a period of time because of the failure to produce a sufficient immune response in the initial dose(s). However, the risk is minimized when offered a subsequent dose [12]. Second, the immune system may fail to respond, whether it is due to inadequacies of the vaccine or factors related to the individual. Inadequate vaccines factors may include incorrect immunization route or schedule, or failures in delivery due to storage and distribution conditions. Moreover, host-related factors for nonresponsiveness are often difficult to define but may be associated with individuals' immune statuses, health status, age, genetic factors, or predispositions [25]. There are numerous factors that may contribute to failures in the vaccine, including improper use, efficacy at the time of use, or even lack of knowledge pertaining to the vaccine(s). Yet, despite the shortcomings of vaccines, the cumulative advantages of vaccines greatly outweigh their disadvantages.

4.5 VACCINE TYPES

Now that the need for vaccines has been established, how can scientists go about creating them and overcoming challenges to face the onslaught of ever-evolving, pathogenic microorganisms (microbes) like virulent bacteria, exotoxins, protozoa, and viruses? The answer, which is still being molded and perfected today, is the creation of different vaccine types. Ideally, vaccines would mimic wild-type infections via cell-mediated immunity through the action of cytotoxic T cells, and humoral immunity via antibody

TABLE 4.1 Unique advantages and disadvantages of vaccine type in the fight against COVID-19

VACCINE TYPE	*MAIN COMPONENT*	*MAIN ADVANTAGES*	*MAIN DISADVANTAGES*	*VACCINES AGAINST COVID-19*
Inactivated	Killed whole pathogen	Mimic wild-type infections	Poor immunogenicity	• Bharat Biotech • Chumakov Center • Kazakhstan RIBSP • Minhai Biotechnology Co • Sinopharm Sinovac
Live attenuated	Weakened whole pathogen	Effective against microbes with antigenic variability	Risk of virulence secondary to mutation	• None
Subunit	Antigenic microbial component	Low risk of virulence	Poor immunogenicity and need for adjuvant	• Anhui Zhifei Longcom • FBRI
DNA	Specific pathogenic genes	Activates cell-mediated and humeral immunity	Requirement of large bolus dose and risk of VITT	• Oxford/AstraZeneca
mRNA	Specific pathogenic mRNA	Safe and activates cell-mediated and humoral immunity	Unstable with the need for strict storage conditions	• Pfizer/BioNTech • Moderna • Takeda
Recombinant	Genetically engineered genes on a plasmid	Allow for modulation of immune response	Cost of manufacturing	• Cansino • Gamaleya • Johnson & Johnson • Serum Institute of India

generation, all without harming the host. Different vaccine types offer unique advantages and disadvantages when tackling the array of disease-causing microbes, especially SARS-CoV-2, the infectious virus behind COVID-19 (Table 4.1). Amid a pandemic, vaccine types are being explored and improved to accelerate population herd immunity.

4.5.1 Inactivated Vaccines

The goal of a vaccine is to generate protective immunity without the distressful symptoms and risks of infection [26]. Injection of a live, virulent pathogen would generate the desired immunity, but would also (and very likely) result in infection. One way to circumvent the dangers of infection while safeguarding protective immunity is the use of an inactivated or killed vaccine [27].

Inactivated vaccines are created by taking the disease-causing pathogen and "killing" it with heat, radiation, or fixation with formalin [28]. The process of inactivation prevents replication of the disease-causing pathogen, while stimulating the humoral immune response effectively creating memory [29]. Though the pathogen contained in the vaccine can no longer infect the host, it contains components (i.e. a capsid, cell wall, etc.) that are recognized as foreign by the immune system. The generated immunological memory results in greater survival when the vaccinated individual encounters the pathogen in the community [28]. This memory allows the immune system to "skip" the lag time needed for the adaptive immune system to become activated. The faster and more robust response results in fewer to no symptoms of disease upon infection [30].

In practice, inactivated vaccines have been successful. These vaccines have prevented bacterial diseases, such as typhoid caused by *Salmonella typhimurium* (*S. typhi*) [28]. But, they have also been developed for a variety of detrimental and deadly viral diseases, such as rabies, which can cause neurological damage; hepatitis A, which can cause liver failure; and salk poliomyelitis, which can cause paralysis [27, 31, 32]. Inactivated vaccines are still in development for some protozoan diseases, such as toxoplasmosis caused by *Toxoplasma gondii* (*T. gondii*) [33]. They have also been used with mild success against SARS-CoV-2, which has been seen in the inactivated vaccines made by Sinovac Biotech, Bharat Biotech, Chumakov Center, Kazakhstan RIBSP, Minhai Biotechnology Co, and Sinopharm approved for emergency use in many countries [34, 35].

Inactivated vaccines are primed to combat pathogens with low antigenic variability [36] and are deemed safe, since the pathogenic microbe is inactivated and can no longer cause infection. However, in circumstances of mutations, such as antigenic shift or antigenic drift, inactivated vaccines may be inadequate. For example, with influenza viruses, inactivated vaccines tend to fail to induce specific neutralizing antibodies against mutated viral antigens, hemagglutinin, and neuraminidase [37]. Thus, protection and immunological memory of the vaccinated individual may be insufficient and oftentimes requires numerous booster injections and adjuvants [27].

4.5.2 Live Attenuated Vaccines

To ensure vaccination compliance among the community, vaccines need to be safe, effective, and administered in the fewest inoculations [27]. In contrast to inactivated vaccines, live attenuated vaccines often require fewer booster injections and have longer-lasting immunological memory [38]. As the name suggests, live attenuated vaccines are created by using the live pathogen, or a similar pathogen, and weakening it by aging or altering growth conditions [27]. The pathogen is still viable but should no longer be virulent and dangerous.

According to the host's immune system, this weakened pathogen is identical to its virulent, wild-type counterpart. Like inactivated vaccines, live attenuated vaccines prime the host to generate immunological memory so when the virulent pathogen is encountered, the adaptive immune system will not delay and produce a robust immune response to neutralize the threat before the host experiences any symptoms. However, unlike inactivated vaccines, live attenuated vaccines stimulate both the cell-mediated and humoral branches of the immune system resulting in a stronger immune response [39], which accounts for the superior effectivity and fewer booster injections of the vaccine.

Live attenuated vaccines currently dominate vaccination strategies; since in theory it should stimulate the same immune response as the virulent wildtype. Thus, they are often utilized as protection against a variety of viruses due to their highly mutagenic nature. Notable live attenuated vaccines include those against influenza A, deadly smallpox, highly infectious measles, mumps, rotavirus, and yellow fever [40, 41]. In addition, the Bacille Calmette-Guérin (BCG) vaccine protects against tuberculosis (TB) [42], a bacterial infectious disease that is second only to COVID-19 as the top infectious disease cause of mortality globally [43].

For all its merits, live attenuated vaccines also carry a large risk: an unprecedented return to a virulent form. Since the vaccine contains a live pathogen, there is a risk of the pathogen mutating to a virulent form that can cause harm to the host. Though the risk is small, live attenuated vaccines are often contraindicated for immunocompromised individuals [37], which can reduce the accessibility of the vaccine and resultant herd immunity.

4.5.3 Subunit and Conjugate Vaccines

Both inactivated and live attenuated strategies utilize the entire infectious pathogen within the vaccine. These methods mimic the infectivity of the wild-type pathogen; however, they often have trade-offs with safety concerns [27]. In contrast, subunit strategies use only the antigenic components of the pathogen within the vaccine to generate an immune response [44]. Since only components of the pathogen are used, there is little risk of virulence as the pathogen is not actually present within the vaccine, ultimately making subunit vaccines more accessible for immunocompromised individuals. Likewise, subunit vaccines can be mass-produced and ensure consistency between lots [44].

However, these vaccines depart from the mimicry of wild-type infection, as seen with whole pathogen vaccines and require extensive knowledge of resultant immune responses. This can lead to decreased immunity compared to live attenuated vaccines due to their inability to stimulate pattern recognition receptors (PRRs) of the innate immune system and effective cytotoxic T cell responses, resulting in a reduction of cell-mediated immunity [45]. Thus, they often require adjuvants for increased immunogenicity and booster injections [44]. Coupling microbial components with antigenic protein carriers result in conjugate vaccines, which also carry risks regarding adjuvant safety, including allergy concerns [27]. Common types of subunit and conjugate vaccines include those based on antigenic exotoxins, proteins, and polysaccharides.

4.5.3.1 Toxoid vaccines

Pathogenic bacteria can cause harm to a host by infection as well as by the release of exotoxin. Exotoxin, toxins released by bacteria, cause damage by disrupting metabolism and impairing cellular membranes [46]. Toxoid vaccines attempt to prevent bacterial infection by combining a toxoid, an inactivated toxin, with adjuvant to induce an immune response [27].

Toxoid vaccines achieve immunity without the need for a natural infection. A common toxoid vaccine is diphtheria-pertussis-tetanus (DPT), which combines diphtheria and tetanus toxins with acellular pertussis as an adjuvant to stimulate an immune response [46]. Without toxoid vaccines, diphtheria can spread through respiratory droplets and cause a serious respiratory infection. Likewise, tetanus is a deadly disease that can cause neurological and breathing complications [47]. A toxoid vaccine is also being explored against exotoxin released by *Clostridium difficile* (*C. difficile*), which causes nosocomial diarrhea and can result in life-threatening colitis [48].

Even though toxoid vaccines are deemed safe and necessary for high-risk individuals [48], they still suffer the pitfalls of subunit vaccines. Both the vaccines for DPT and *C. difficile* require booster injections due to poor immunogenicity [27, 48].

4.5.3.2 Protein subunit vaccines

Protein subunit vaccines consist of an antigenic protein from the pathogen of interest coupled to an adjuvant to induce an immune response [27] and subsequently offer protection to the host from the wild-type pathogen. Choice of the microbial protein requires identification of the microbial components that are responsible for protective immunity. Likewise, when constructing the adjuvant, special care is taken to identify proper stabilizers, an effective delivery system, and optimal immunogenicity. Following stringent purification processes, the immune response can be "customized" by optimization of the vaccine delivery systems. A thorough understanding of the tailored immune response to the wild-type pathogen is required, otherwise, resultant immunity could be inadequate. For example, alum is a commonly used adjuvant due to its safety; however, it is a relatively weak inducer of the immune response [44].

Though protein subunit vaccines may produce reduced immune responses, their safety compared to live attenuated vaccines have led to their use and subsequent success in preventing prominent viral diseases. The hepatitis B vaccine utilizes the hepatitis B surface antigen (HBsAg) and the HPV vaccine utilizes the HPV-6 L1 capsid protein to generate protective immunity in the host [36]. In more recent news, protein subunit vaccines synthesized by Anhui Zhifei Longcom and FBRI utilize the SARS-CoV-2 spike protein and have been approved for use in at least two countries to help prevent COVID-19 [34].

4.5.3.3 Polysaccharide subunit vaccines

When constructing a subunit vaccine, or any vaccine for that matter, it is important to consider the stage of the infection in which to target with an immune response. To minimize the host cell death, many vaccines attempt to induce an immune response against pathogenic entry mechanisms [49–51]. This method helps minimize symptoms when the host encounters the wild-type pathogen, since the protective immunity prevents the pathogen from infecting host cells. Many microbes express surface polysaccharides found on the cell wall, which make for selective targets as they are typically not found on mammalian cells [49, 51]. Thus, antigenic polysaccharides can be formulated into a subunit vaccine to offer protective immunity against virulent bacteria and viruses [49–51].

Like most subunit vaccines, polysaccharide subunit vaccines require an adjuvant for immunogenicity [51]. Construction of the adjuvant is a balancing act between ensuring optimal immunogenicity without any harm to the host. Successful polysaccharide vaccines have prevented deadly bacterial diseases such as meningococcal disease caused by *Neisseria meningitis* (*N. meningitis*) and pneumococcal pneumonia caused by *Streptococcus pneumoniae* (*S. pneumoniae*) [49, 50].

4.5.4 Nucleic Acid Vaccines

The central dogma of molecular biology states that genetic information travels from DNA to messenger RNA (mRNA) and finally to protein. This idea holds for most microbes with the exception of RNA viruses that have their genetic information stored in an RNA genome [52]. Nucleic acid vaccines capitalize on the central dogma and differ from their subunit counterparts by utilizing specific parts of the microbial genome (rather than already transcribed and translated proteins) within the vaccine.

This method allows host cells to internalize the viral nucleic acids, express the viral peptides, and display them on their surface via their major histocompatibility complex I (MHC I). Specific immune and nonimmune cells, named antigen-presenting cells (APCs), also display viral peptides on another surface MHC molecule called MHC II. With the expression of viral peptides on both MHC molecules, both the cell-mediated and humoral branches of the adaptive immune system can be activated, unlike subunit vaccines which typically only stimulate humoral immunity. With the activation of both branches, a stronger immune response is created, resulting in longer-lasting immunological memory [26].

Like subunit strategies that use a specific component of the pathogen, nucleic acid vaccines contain only specific genes or transcribed products of RNA, rather than the entire viral genome. Thus, they should be safe for immunocompromised individuals as they carry little risk for virulence compared to live attenuated methods [53].

In theory, nucleic acid vaccines combine effectiveness with safety. However, nucleic acids, especially RNA, are inherently unstable and require regulated storage and transportation, which greatly limits the accessibility of the vaccines to rural areas which may lack storage facilities. In addition, extensive knowledge of the tailored immune response to the wild-type virus is needed or else vaccine efficacy could be reduced [45]. There are two main types of nucleic acid vaccines: DNA vaccines and mRNA vaccines.

4.5.4.1 DNA vaccines

DNA vaccines utilize specific genes of the pathogen within the vaccine. The antigen-encoding gene is inserted into a bacterial plasmid and then administered to the host. Addition of different motifs to the plasmid can also modulate the immune response. These motifs can act as pathogen-associated molecular patterns (PAMPs), which activate the innate immune system via PRRs and eventually induce adaptive immune responses such as the release of Th1 cytokines [26].

In practice, DNA vaccines are effective when vaccinating small animals [53], but have struggled when used to vaccinate humans. Though DNA vaccines are easier to store than their mRNA counterparts due to their stability and succeed in generating both cell-mediated and humoral immune responses, they tend to require large bolus doses since they do not elicit an adequate antibody response [53]. Additionally, they carry risks of vaccine-induced thrombotic thrombocytopenia (VITT), as seen during the trials of the Oxford-AstraZeneca COVID-19 vaccine [54].

4.5.4.2 mRNA vaccines

Within the central dogma of molecular biology, DNA gets transcribed into mRNA, which then gets translated into protein [52]. The transition from DNA to mature mRNA requires different regulatory processes and splicing mechanisms. Unlike DNA vaccine strategies, mRNA vaccines are already in a processed form, thus preventing any uncontrolled splicing by the host cell. In practice, mRNA vaccines utilize host translational machinery to produce the target antigen, which will subsequently initiate an adaptive immune response. After inoculation, the mRNA vaccine is internalized by resident nonimmune cells at the injection site. The mRNA is then expressed, which stimulates select PRRs of the innate immune system, such as RIG-1 and MDA5. This stimulation starts a domino effect which leads to the upregulation of signaling molecules, like cytokines and chemokines, that recruit immune cells. This process leads to the activation of cytotoxic T cells through MHC-I and helper T cells and B cells through MHC-II. Thus, mRNA vaccines can stimulate both cell-mediated and humoral immunity [45].

Currently, there are two types of mRNA vaccines: (1) conventional, non–self-amplifying mRNA and (2) self-amplifying mRNA [45]. Conventional mRNA vaccines only encode the antigen of interest, while self-amplifying mRNA vaccines have auto-replicative activity based on the RNA viral vector [55]. Even though self-amplifying mRNA vaccines have the potential to induce the host to produce high levels of mRNA based on the antigenic gene, the mRNA does not contain any structural components, thus it is incapable of producing infectious virions making the vaccine safe and accessible to healthy hosts [45]. mRNA vaccines have proven successful in the plight against COVID-19 as the vaccines created by Pfizer/BioNTech, Moderna, and Takeda have been approved for emergency use in a number of countries [34]. These vaccines induce cell-mediated and humoral immunity while having intrinsic adjuvant properties reducing the need for external excipients [45]. However, mRNA is much less stable compared to DNA, and is only viable in specific storage conditions, which limits its accessibility. The Pfizer/BioNTech vaccine requires storage in a –70°C ultra-cold freezer and once thawed must be used within five days. Likewise, the Moderna vaccine requires storage within a –20°C freezer but can be used within 30 days after thawing [56]. In addition, knowledge of the tailored immune response to the wild-type pathogen is required as mRNA vaccines can stimulate an innate antiviral response, which reduces vaccine effectiveness via type I interferon (IFN) responses [45].

4.5.5 Optimization with Recombinant Vaccines

Each vaccine type has unique advantages and disadvantages. Whole pathogen vaccines, like inactivated and live attenuated, mimic the natural infection of the pathogen but lack immunogenicity or carry the underlying risk of harmful mutation, respectively. Subunit vaccines have a low risk of harm and can be administered to immunocompromised individuals but often rely on adjuvants due to their low immunogenicity. Nucleic acid vaccines balance effectivity and safety but sacrifice accessibility due to strict storage requirements.

How can scientists pick and choose only the beneficial attributes of each vaccine type? Through genetic engineering, scientists can optimize tailored immune responses via recombinant vaccines [26, 36, 44, 53, 57, 58]. Recombinant vaccines rely on gene manipulation *in vitro* on recombinant vectors or viral vectors and use bacterial or yeast expression systems for any required post-translational modifications [36]. Thus, pathogenic genes can be modified in the lab to modulate the host immune system.

Recombinant strategies break past many limitations suffered from predecessor vaccines. A single recombinant vaccine can contain multiple, modified antigens on a single plasmid, effectively inducing immunity against different pathogens in a single inoculation [57]. Recombinant strategies are also combined with other vaccine types to improve safety and immunogenicity. Viral vectors are used as delivery systems for nucleic acid vaccines, most notably those against COVID-19. The COVID-19 vaccines created by CanSino, Gamaleya, Johnson & Johnson, Oxford/AstraZeneca, and the Serum Institute of India utilize nonreplicating viral vectors to induce immunity [34]. In addition, recombinant strategies are also being explored with the SARS-CoV-2 spike protein embedded on the yellow fever viral vector [59]. Even viruses that are controlled with inactivated vaccines, like rabies virus, are being explored with recombinant methods, as seen with the rabies surface glycoprotein embedded within the vaccinia viral vector to improve efficacy [58].

Though recombinant strategies improve many pitfalls held by other strategies, a major barrier that remains is the cost of manufacturing resulting in accessibility issues [36]. With many infectious diseases at large, including the continued pandemic of COVID-19, access to effective vaccines is a global priority. Recombinant strategies may offer the key to creating biosimilar vaccines, which could aid in the accessibility of COVID-19 vaccines in rural areas.

4.6 FACTORS IMPACTING THE AVAILABILITY OF VACCINES

Since the dawn of vaccine development, we have seen the production of multiple generations of vaccines for various diseases, but these discoveries were not realized without considerable obstacles and controversies. Some issues and concerning the pre-development, development/production, and post-development/marketing processes are outlined below.

4.6.1 Pre-Development Process

4.6.1.1 Discrepancies on the definition of biosimilars concerning vaccines limit the number available on the market

Vaccines themselves are considered a biologic and can enter the market under a Biologics License Application (BLA). Correspondingly, biosimilars are biologic products that have been proven to be biosimilar or interchangeable with an already FDA-approved biologic. However, vaccine biosimilars are seemingly rare and may be due in part to the strict and ill-defined guidelines that are more applicable to biologics and biotherapeutics [9]. (More details will be discussed under Section 4.7.1.)

4.6.1.2 Barriers to vaccine innovation

4.6.1.2.1 Identifying priorities for vaccine production

For many people, it seems that the success of vaccines no longer warrants the need for new and more effective preparations [60]. However, many vaccines that currently exist on the market are not optimal and would benefit from safer versions that can be produced. Additionally, the lack of vaccines for many diseases in such as HIV/AIDS and malaria underscore the need for targeted vaccine production, especially in tropical areas where parasitic diseases are of high concern.

4.6.1.2.2 Determining the viability of vaccine production

Basic funding for research on infectious diseases and vaccine development, especially from federal sources, is difficult to obtain and requires a solid determination of the infectious agent, host response to that agent, and the pathobiology within human hosts [60]. Additionally, information regarding the quantity of serotypes, antigen determination, and epidemiology of the disease are required to develop an effective vaccine, as well as one that is capable of remaining stable in relevant storage environments. A lack of understanding in any of these areas may delay or hinder vaccine development when resources and funding are already limited.

4.6.1.2.3 Economic disincentives for pharmaceutical companies

The issues related to the development of a vaccine may be enough to discourage pharmaceutical manufacturers from producing a vaccine, even if a need has been established and a means of production is feasible due to the large cost-benefit ratio. For example, while a process for development may have been outlined, the complexity, length, and quality control methods may vastly outweigh the benefits associated with the production. Similarly, the cost of research and development of such a vaccine may fall significantly short of projected sales and lead to a net loss for the company.

4.6.1.2.4 Market and commercial influences

One of the main requirements of a vaccine is that it provides immunity for an extended period of time, if not for an entire lifetime – this requirement runs contrary to the prospect of multiple product sales within a company [60]. Export out to foreign countries is difficult because manufacturers in the United States have stricter guidelines in comparison to other competing countries.

4.6.1.3 FDA incentives

The FDA established the Office of Orphan Products Development (OOPD) Grants Program to encourage the development of products to treat rare diseases or conditions where no current therapy exists, or the new product will be superior to existing therapy. Any proposed orphan drug can be a vaccine as long as it meets one of two following criteria: the disease affects less than 200,000 Americans or the disease affects more than 200,000 Americans but has no potential recovery costs after sales [61]. Obtaining this incentive allows for market exclusivity for seven years and a 50% deduction of tax credit for clinical trial expenses. While many medicinal products have gained orphan designation, there exist obstacles to the process itself. Funding, correct information on the epidemiology of the disease, disease severity, and the effect of the disease on public health may be lacking when required for final authorization. Additionally, the lack of a streamlined process for final authorization and exemptions from some or all parts of the necessary registration fee can decrease overall cost, reduce the number of staff required, and make the product more attractive to the sponsor. Because of the increase in patent protection and the establishment of market exclusivity, sponsors and manufacturers experience a decrease in investment risk and may ultimately garner the favor of companies looking to benefit from an ethical business image.

4.6.2 Development/Production Process

4.6.2.1 High cost of production

The production of a vaccine is challenging due to the complexity of the production process, the number of different biological starting materials that can be used, and dependency on a specific biological process [62]. The entire process of vaccine production must be safe, effective, and consistent as well as replicable, especially since regulatory authorities license both the biologic and the process in which it is produced. Because vaccines use different starting materials, microorganisms from which a product is extracted or purified, environmental conditions in which the microorganism was kept, and experiences of the manufacturing technician in performing the purification, many avenues exist for variability in the end product. If a manufacturer cannot meet these demands and criteria, the product may end up recalled and/or suspended and penalties may arise if a quota cannot be met. Additionally, changes in the production process may require new and costly clinical trials to prove the new product is bioequivalent to the original product [62]. Facilities for manufacture can generate exponentially rising costs, especially if changes occur to a previously streamlined process. Eventually, this can impact immunization programs and national public health outcomes if the vaccine is utilized in a nationwide program.

4.6.2.2 High cost of testing

Due to the irregularity in tests available, there is a need for tests that are capable of detecting slight but biologically and therapeutically significant differences in vaccines without arduous, expensive, and time-consuming clinical trials [9]. Additionally, *in vitro* testing may not detect changes in upstream production process alterations and further clinical trials may be required to validate the process and obtain final approval for a license [62]. The difficulties in cost of overall production help, in part, to explain why few vaccine manufacturers exist when the demand for vaccines is so high [62]. This has led many vaccine patents to include the manufacturing process itself rather than the specific antigen that is created, unlike those of small molecule pharmaceuticals [62].

4.6.2.3 Supply of vaccines

Due to the inherent variability in vaccine product make-up, problems can arise that impact the supply of a vaccine such as changes in potency, stability, variations in population responses to inactivation of the vaccine, excessive undesirable biological activity, and contamination [60]. If a manufacturer is contracted to produce a specific number of vaccines and encounters any one of these issues, they will be unable to meet the agreed-upon quota and can subsequently affect immunization programs.

4.6.2.4 Reliance on foreign manufacture and lack of manufacturing capabilities in countries lacking the proper infrastructure to support a production system

The lack of manufacturing companies in the United States due to the aforementioned issues may lead to reliance on foreign manufacturers to fulfill the need for new vaccines. This could potentially cause a variety of issues ranging from licensing issues to discrepancies in communication, issues with vaccine stability over a long period of time, political issues due to demands in the foreign country to deliver the product to their citizens before an export, and an increase in cost due to the potential need for more laboratory and regulatory staff resulting from more stringent U.S. standards. Many of the diseases that currently do not have a patented vaccine for their prevention tend to exist in third-world countries where these products are needed the most [61]. These locations often experience poverty, a lack of education, cold-chain issues, and a shortage in vaccine transport facilities [63]. The economic impact and strain caused by diseases like HIV/AIDS further decrease the possibility for these affected countries to allocate funds for immunization programs.

4.6.3 Post-Production/Marketing Process

4.6.3.1 Vaccine utilization

Factors that affect provider administration
Health care providers can influence the utilization of vaccines as they are generally the ones who are administering the product. Such use depends on those giving the vaccine, the vaccine itself, and the setting in which the vaccine is introduced [60]. For example, providers may be concerned about the efficacy of the vaccine, the way in which it is administered, the complexity of vaccination schedules, cost, and the risk that can be caused onto patients. Additionally, the environment and culture in which the vaccine is being introduced into may hold opposing beliefs, attitudes, and norms toward vaccinations and should be taken into consideration.

Factors that affect client compliance and use
While a vaccine may enter the market and be approved for administration, the general public's decision to receive the vaccine will determine the ultimate outcome of an immunization program. Those who hold

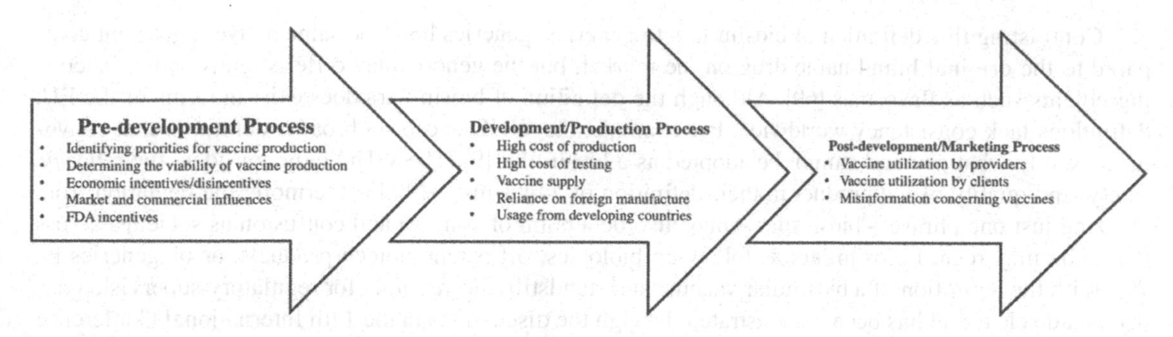

Pre-development Process
Identifying priorities for vaccine production
Determining the viability of vaccine production
Economic incentives/disincentives
Market and commercial influences
FDA incentives
Development/Production Process
High cost of production
High cost of testing
Vaccine supply
Reliance on foreign manufacture
Usage from developing countries
Post-development/Marketing Process
Vaccine utilization by providers
Vaccine utilization by clients
Misinformation concerning vaccines

FIGURE 4.1 Overview of the vaccine production process and factors that can affect the cost, efficiency, and time required to develop a vaccine.

the belief that vaccines do not cause adverse side effects, have general positive attitudes about vaccination, have positive vaccine recommendations, and perceive fewer difficulties of vaccination are more likely to receive a vaccine and encourage others to also receive a vaccine [64]. These patients also hold knowledge about a particular vaccine, social influences and trust in the healthcare profession, and information about the vaccination process [64]. Other factors such as previous vaccination experiences, beliefs and attitudes about health, knowledge and awareness of vaccines, trust in the healthcare system, and risk vs. benefit outcomes can influence individual decisions to vaccinate [65]. On a more macro scale, social involvement regarding vaccines can be wide and varied, including media involvement, influential public leaders, historical influences, religious influences, cultural influences, socio-economic influences, political figures, geographic barriers, and perceptions of the pharmaceutical industry. One or more of these sources can influence and/or pressure the general public's decision to vaccinate.

Vaccine hesitancy fueled by misinformation about vaccines and their efficacy can lead to tragic outcomes that could have been avoided [66]. A recent example concerns the COVID-19 vaccine. Social media remains a quick and efficient way to share unreliable information and can influence public decision to get vaccinated. Companies like Twitter and Facebook have been mounting efforts to address false news regarding the vaccine. Many deaths in vulnerable populations may have been avoided if eligible patients were vaccinated.

Numerous factors starting from addressing an unmet vaccine need for a disease to the production process all the way down to ensuring the healthcare sector and the public are adequately informed about a marketed vaccine can affect widespread immunization efforts to eradicate a disease (Figure 4.1). Each step of the process requires an efficient and streamlined process to minimize extraneous costs and maximize administration of the vaccine. Ultimately, an increase in efforts in each of these steps can potentially allow for the efficient development of vaccines and vaccine biosimilars.

4.7 INTRODUCTION OF BIO-SIMILARITY CONCEPT IN VACCINOLOGY

With the emergence of biotechnology products and the novel concept of biosimilarity, there have been highlighted needs for regulation and a well-defined definition for biosimilars regarding vaccines. In 2004, the European Medicines Agency (EMA) pioneered and laid the foundation for the inception of "similar biological medicinal products", most commonly known as biosimilars, and the European Union (EU) soon approved their first biosimilar in 2006 [67, 68]. The EU defines a biosimilar as a biologic medicine that would fit the profile of being comparable to another reference product, contingent upon the fact of analogous biological activity, efficacy, immunogenicity, structure, and safety profiles between the two [68].

Contrasting this definition of biosimilars to generics, generics have the same active ingredient compared to the original brand-name drug on the market, but the generic may differ slightly in the inactive ingredients such as flavorings [69]. Although the definition of biosimilars does exist in terms of the EU, definitions lack consistency worldwide. For example, the WHO accepts a broader definition that convolutes exactly what can and cannot be adopted as a biosimilar [9]. The WHO only considers the efficacy, safety, and quality of the product in their definition of a biosimilar [9]. Furthermore, the deviation from utilizing just one phrase – biosimilar – may also be a point of concern and confusion as societies across the globe may regard biosimilars as follow-up biologics, off-patent biotech products, or biogenerics [9, 70]. With the definition of a biosimilar vaccine at a standstill, the yearning for regulatory supervision has been made clear and has been demonstrated through the discussions in the 12th International Conference of Drug Regulatory Authorities (ICDRA) [71]. In the ICDRA, the WHO has been urged to provide regulation management for biological products as there is an expected increase of biosimilar infiltration to the market [67, 71]. As the biosimilarity approach gains popularity and applicable use, the world may need to better define a biosimilar vaccine.

Generally, biosimilars, not directly related to vaccines, have been looked to as a potential for lowering the cost of treatment options. Biosimilars were predicted to save Americans 4.5 billion dollars if they had access to biosimilars [72]. To the best of our knowledge, there are not any available biosimilar vaccines that fit the EU definition. So, why are we not using the concept of biosimilarity in vaccines more if there are public health cost benefits? Are there current hindrances to vaccine availability that block this action? Could biosimilarity be a possible hope to a future of improved health outcomes or a mirage of delusion?

4.7.1 Issues That Prevent the Emergence of Biosimilar Vaccines [9]

First, there is a lack of a universal definition for biosimilarity that is specific to vaccines (Table 4.2).

Second, there need to be guidelines or minimal criteria that (1) take into account the specific properties of vaccines, which differentiate them from therapeutic products, and also (2) justify the acceptance of the product as biosimilar. For example, immunogenicity is generally an undesirable property in therapeutics, whereas it is a central requirement for vaccines. In addition, vaccines are administered to healthy individuals, unlike therapeutics, and there is a greater need for stringency in ensuring safety and efficacy because of no trade-off between relief of symptoms and adverse effects as in the case of therapeutics.

TABLE 4.2 Variation in regulatory approval of biosimilar [73]

REGULATORY AGENCY	*DEFINITION*
EMA	A biological medicinal product that contains a version of the active substance of an already authorized product in the European Economic Area (EEA).
FDA	A biological product that is highly similar to a US-licensed reference product notwithstanding minor differences in clinically inactive components, and for which there are no clinically meaning differences between the biological product and the reference product in terms of safety, purity, and potency of the product.
WHO	A biotherapeutic product that is similar in terms of quality safety and efficacy to an already licensed reference product.
Japan PMDA	A biotechnological product developed by a different company to be comparable to an approved biotechnology-derived product (hereinafter "reference product") of an innovator. It can generally be developed on the basis of data that demonstrates the comparability between the biosimilar product and the reference product with respect to quality, safety, and efficacy, or other relevant data.

Third, a phase II level noninferiority comparison in relation to safety and immunogenicity parameters should be demonstrated because there need to be minimal clinical data requirements for the acceptance of biosimilar vaccines. Even if a certain type of vaccine formulation and production method have been approved, a hypothetical biosimilar vaccine that uses the same method of production, and claims to be noninferior, must be supported by evidence and have side-by-side comparison data made available. This is because vaccines are diverse in composition, and so even if they may have similar efficacy, for example, they could differ in adverse event profile or vice versa.

Fourth, the inability to define and characterize the active substance at the molecular level hinders in demonstrating that new preparations are similar to the reference product. The composition of active substances cannot be defined for many types of vaccines. Methods must be developed to identify the molecular composition of the active substance. This will support the development of test requirements to compare safety, quality, and efficacy between superficially similar products (i.e. same strain and production yet different biological properties).

4.7.2 Overview of Current Vaccines for Biosimilarity Potential

As a concept, some vaccine subtypes, if applied for biosimilarity, may meet the definition for biosimilars, while others will not (Table 4.3).

TABLE 4.3 Current vaccines that may or may not fit the definition for biosimilars

	VACCINE SUBTYPES	*VACCINE TYPE*	*EXAMPLES*	*MEETS BIOSIMILAR DEFINITION?*	*REASONS*
Bacterial	Whole-cell bacterial	Inactivated	Pertussis, oral cholera, and typhoid	No	Composition of active component cannot be defined. May produce inconsistent performance for products prepared from same strains by the same process.
	Bacterial	Live attenuated	Bacillus Calmene-Guerin (BCG) and oral typhoid	No	Different substrains vary in biological properties.
	Polysaccharide	Purified component	Typhoid Vi, meningitis	Yes	Properties can be predicted through physical and chemical analysis. Active component is characterizable.
Viral	Whole virus	Inactivated	Poliomyelitis, hepatitis A, influenza, and rabies	No	Composition of active component cannot be defined. Batch-to-batch variation in biological properties may occur.
	Viral	Live attenuated	Measles, mumps, rubella, varicella, oral polio, and yellow fever	No	Different strains may give different results.
	Influenza vaccine	Viral subunit	Influenza	Maybe, but already treated as one	Produced and distributed quickly on an annual basis for new strains and this gives little time for any clinical evaluations.
	Recombinant viral	Viral subunit	Human papillomavirus and hepatitis B	Yes	Slight variation in active components still maintains biological properties.

4.7.3 Existing Vaccines That Fit the Definition for Biosimilars [9]

Polysaccharide vaccines can meet the definition of biosimilar because the activity of polysaccharide antigen is related to the integrity of chemical composition and molecular size. This means that the vaccine properties can be predicted on the grounds of physical-chemical analysis. Despite some occurrence of microheterogeneity, or a variation in the chemical structure of a substance, having small differences within the products or between batches has not been correlated with vaccine failure, and thus acceptable as biosimilar [74–76]. Studies of two different typhoid Vi vaccines, for example, from two separate populations had similar efficacy: 72% in Nepal and 69% in South Africa for one vaccine; 69% in China for the other vaccine [75]. The molecular composition is well characterized and produced by similar processes. Even with a slight difference in impurity profile, if these products meet pharmacopeia requirements, they would most likely produce similar clinical data and act as biosimilars.

Influenza vaccines do not necessarily fit the definition for biosimilars, but are treated as one, mainly because of the need for an updated version of the vaccine on an annual basis due to the nature of constant evolution of the antigenic properties of the virus. The hemagglutinin and neuraminidase surface glycoproteins mutate frequently such that they are unrecognizable to the host immune [77]. The WHO annually identifies which mutant strains to be placed into vaccine production for that year, and its production and distribution must be done rapidly. There is very little time for clinical evaluation and thus the vaccines essentially have to be treated as biosimilars [78].

For recombinant viral vaccines, one example from this type of vaccine that fits biosimilarity definition is the HPV vaccine. For the HPV vaccine, there are two products that differ in the number of purified protein antigens, where one has HPV genotypes 16 and 18, while the other has 6, 11, 16, and 18. Both vaccines prevent cervical cancer caused by HBV, which corresponds to genotype 16 and 18, while the latter has the additional benefit of preventing genital warts, which corresponds to genotype 6 and 11 [79, 80]. The two extra antigen genotypes differentiate the vaccines but they are still similar in composition in relation to proteins efficacious in the prevention of cervical cancer. Thus these meet the criteria for biosimilars.

4.7.4 Existing Vaccines That Fail to Fit the Definition for Biosimilars [9]

Both types of the inactivated vaccines do not fit the definition for biosimilars because of variability in performance and lack of defined composition. The former include notable batch-to-batch variation, even though products are being prepared from using the same strain and identical process. Pertussis vaccine, for example, showed variable biological properties in clinical trials from Italy, Germany, Sweden, and Senegal, where some showed the efficacy of over 80%, while others were below 40% [81–83]. The protective components are uncertain and other parts that contribute to its efficacy are undefined, making these kinds of vaccines to be excluded from biosimilars.

The attenuated bacterial vaccines, specifically the BCG vaccines, cannot be assumed for biosimilarity among different products because the production process creates extensive genetic variation and the various substrains greatly differ from the original. The genetic variation among substrains produced differing biological properties. For example, "Strong" strains of Danish and Pasteur were thought to have higher levels of residual virulence and show higher frequency and severity of local reaction than the "weak" strains of Japan and Glaxo [84]. Even wide differences in the protective efficacy against pulmonary TB resulting from reactivating or reinfection have been reported [85]. Thus, the differing substrains being created during the manufacturing process create a differing biological property and this makes these vaccines fail to fit the definition as a biosimilar.

There is a special case for attenuated viral vaccines that do not meet the biosimilar criteria, and this is due to the fact that even when the same strains are used, the vaccine produced by different manufacturers may not behave in the same way. This has been the case, for example, with measles-mumps-rubella vaccines when mumps strains from Jeryl Lynn, Urabe and Leningrad-Zagreb were compared. Upon closer look at the effects of mumps strain, which implicated post-vaccination aseptic meningitis, it has been reported that Jeryl Lynn strain was less likely to produce aseptic meningitis than the other two, although it also may have produced less durable immunity [86–88]. In addition, it has also been reported that vaccines made from Jeryl Lynn strain from different companies did not always have identical properties [89]. When producing certain attenuated viral vaccines, caution must be used when licensing what may seem like identical vaccines, and this makes these vaccines to be excluded from the potential to be categorized as a biosimilar.

4.7.5 The First "Biosimilar" Vaccine [90]

In India, the first "biosimilar" or what they call "similar biologic" that was approved and marketed was their hepatitis B vaccine, *Biovac-B* by Wockhardt, in 2000. However, no specific guidelines for "biosimilars" existed in India until 2012. This has been the case even though the requirements for granting regulatory approval for "biosimilars" required more data than for generics for their drug application. In other words, the "biosimilars" approved in India might not have been authorized following as strict a regulatory process as is required for approval of biosimilars in the EU. The EMA regulatory requirements ensure the same high standards of quality, safety, and efficacy for biosimilars as for originator biologicals, and also include a rigorous comparability exercise with the reference product [90]. Thus this vaccine does not fit our definition of biosimilar and will be excluded.

4.7.6 COVID-19 Vaccines as Biosimilar

Making its first appearance in December 2019, COVID-19 has taken the world by storm [91]. Totaling COVID-19 cases up to June 2021, there have been as many as 172 million confirmed cases and 3.7 million confirmed deaths from COVID-19 [92]. With the increasing devastation and human tragedy that disturbed societies across the globe, researchers and pharmaceutical companies raced to produce a vaccine. The first vaccine to gain Emergency Use Authorization (EUA) in the United States was the Pfizer-BioNTech vaccine in December 2020, which was closely followed by Moderna [93].

At the time of writing this chapter, three COVID-19 vaccines are released under EUA, but none of them are fully approved by the FDA. Thus, without a reference product, there would be no possibility of developing a biosimilar anytime soon. The current pandemic, however, may potentially change that.

COVID-19 pandemic has brought world leaders together and allocated resources to those countries in need to save as many lives across the globe. The United States has proposed to WTO a waiver to the intellectual property rights of the COVID-19 vaccines to allow middle- and low-income countries to have access to their formulation data in order to produce their own vaccines in pharmaceutical companies that provide for those regions [94]. If passed, in theory, this movement would utilize the concept of biosimilarity because this would help solve the issue of availability and accessibility by lowering the cost of production and thus the price of the vaccines. Hypothetically, if Pfizer releases its detailed information about its mRNA vaccine technology, then low- and middle-income country's pharmaceutical companies would be able to develop a similar product much sooner with a variant mRNA being the difference in active component, and submit its application to a regulatory agency as a biosimilar. In addition, a development of a booster shot for the upcoming season may also qualify as biosimilar, based on its definition and through similar reasoning as the influenza vaccine.

As the world prepares to further eradicate the disease that human society just met two years ago, the prospective avenue of taking a biosimilar approach in vaccine development and production is promising. Biosimilars with their potential to lower production cost, and to increase accessibility, availability, and treatment options, may pave the way to better health not only for those who can afford and access healthcare but also those who may depend on it for survival.

REFERENCES

1. Organization WH: EPHO5: Disease prevention, including early detection of illness. Geneva: World Health Organization 2018.
2. Primary, secondary and tertiary prevention [https://www.iwh.on.ca/what-researchers-mean-by/primary-secondary-and-tertiary-prevention] April 2015.
3. The history of vaccines [https://www.historyofvaccines.org/content/jenner].
4. History of smallpox [https://www.cdc.gov/smallpox/history/history.html].
5. Riedel S: Edward Jenner and the history of smallpox and vaccination. In: *Baylor University Medical Center Proceedings: 2005*: Taylor & Francis; 2005: 21–25.
6. Organization WH: Ten health issues WHO will tackle this year. World Health Organization [https://www who int/news-room/feature-stories/ten-threats-to-globalhealth-in-2019 2019].
7. Food U, Administration D: What are "biologics" questions and answers. In: About the Center for Biologics Evaluation and Research [https://www.fda.gov/about-fda/center-biologics-evaluation-and-research-cber/what-are-biologics-questions-and-answers;] 2018.
8. How do drugs and biologics differ? [https://archive.bio.org/articles/how-do-drugs-and-biologics-differ#:~:text=Drugs%20versus%20Biologics&text=Most%20biologics%20are%20very%20large, ingredients%20in%20an%20ordered%20process].
9. Corbel MJ, Cortes Castillo MdLA: Vaccines and biosimilarity: a solution or a problem? *Expert Review of Vaccines* 2009, 8(10):1439–1449.
10. How do vaccines work? [https://www.who.int/news-room/feature-stories/detail/how-do-vaccines-work].
11. Marshall JS, Warrington R, Watson W, Kim HL: An introduction to immunology and immunopathology. *Allergy Asthma Clinical Immunology* 2018, 14(Suppl 2):49.
12. David Salisbury MR, Karen Noakes (ed.): *Immunisation against Infectious Disease, 3rd edn*. London: TSO; United Kingdom: Stationery Office; 2006.
13. Janeway C: *Immunobiology 5: The Immune System in Health and Disease*, 5th edn. New York: Garland Publishing; 2001.
14. Ewer KJ, Lambe T, Rollier CS, Spencer AJ, Hill AV, Dorrell L: Viral vectors as vaccine platforms: from immunogenicity to impact. *Current Opinion in Immunology* 2016, 41:47–54.
15. Citera G, Pra FD, Waimann CA, Ficco HM, Alvarellos T, Mas LA, Cerda OL, Paira S, Pellet AC, Secco Aet al.: Frequency of human leukocyte antigens class II-DR alleles (HLA-DRB1) in Argentinian patients with early arthritis. *Clinical Rheumatology* 2019, 38(3):675–681.
16. Keech C, Albert G, Cho I, Robertson A, Reed P, Neal S, Plested JS, Zhu M, Cloney-Clark S, Zhou H: Phase 1–2 trial of a SARS-CoV-2 recombinant spike protein nanoparticle vaccine. *New England Journal of Medicine* 2020, 383(24):2320–2332.
17. Andre FE, Booy R, Bock HL, Clemens J, Datta SK, John TJ, Lee BW, Lolekha S, Peltola H, Ruff TAet al.: Vaccination greatly reduces disease, disability, death and inequity worldwide. *Bulletin of the World Health Organization* 2008, 86(2):140–146.
18. Folb PI, Bernatowska E, Chen R, Clemens J, Dodoo AN, Ellenberg SS, Farrington CP, John TJ, Lambert PH, Macdonald NEet al.: A global perspective on vaccine safety and public health: the Global Advisory Committee on Vaccine Safety. *American Journal of Public Health* 2004, 94(11):1926–1931.
19. Whitney CG, Zhou F, Singleton J, Schuchat A: Benefits from immunization during the vaccines for children program era – United States, 1994–2013. *Morbidity and Mortality Weekly Report* 2014, 63(16):352–355.
20. Yamanouchi K: Scientific background to the global eradication of rinderpest. *Veterinary Immunology and Immunopathology* 2012, 148(1–2):12–15.
21. Centers for Disease Control and Prevention: Impact of vaccines universally recommended for children – United States, 1990–1998. *Morbidity and Mortality Weekly Report* 1999, 48(12):243–248.
22. John TJ, Samuel R: Herd immunity and herd effect: new insights and definitions. *European Journal of Epidemiology* 2000, 16(7):601–606.
23. Papaloukas O, Giannouli G, Papaevangelou V: Successes and challenges in varicella vaccine. *Therapeutic Advances in Vaccines* 2014, 2(2):39–55.
24. Shearley AE: The societal value of vaccination in developing countries. *Vaccine* 1999, 17(Suppl 3):S109–S112.
25. Wiedermann U, Garner-Spitzer E, Wagner A: Primary vaccine failure to routine vaccines: why and what to do? *Human Vaccines & Immunotherapeutics* 2016, 12(1):239–243.
26. Cui Z: DNA vaccine. *Advances in Genetics* 2005, 54:257–289.

27. Okonek BAM, Peters PM: Vaccines – How and why? *Access Excellence Classic Collection* 2009:1–3.
28. Herzberg M, Nash P, Hino S: Degree of immunity induced by killed vaccines to experimental salmonellosis in mice. *Infection and immunity* 1972, 5(1):83.
29. Krahenbuhl JL, Ruskin J, Remington JS: The use of killed vaccines in immunization against an intracellular parasite: *Toxoplasma gondii. The Journal of Immunology* 1972, 108(2):425–431.
30. Plotkin SA: Vaccines: past, present and future. *Nature Medicine* 2005, 11(4):S5–S11.
31. Yang D-K, Kim H-H, Lee K-W, Song J-Y: The present and future of rabies vaccine in animals. *Clinical and Experimental Vaccine Research* 2013, 2(1):19.
32. Innis BL, Snitbhan R, Kunasol P, Laorakpongse T, Poopatanakool W, Kozik CA, Suntayakorn S, Suknuntapong T, Safary A, Tang DB: Protection against hepatitis A by an inactivated vaccine. *JAMA* 1994, 271(17):1328–1334.
33. Waldeland H, Frenkel J: Live and killed vaccines against toxoplasmosis in mice. *The Journal of Parasitology* 1983:60–65.
34. COVID-19 vaccine tracker [https://covid19.trackvaccines.org/vaccines/].
35. Palacios R, Patiño EG, de Oliveira Piorelli R, Conde MTRP, Batista AP, Zeng G, Xin Q, Kallas EG, Flores J, Ockenhouse CF: Double-blind, randomized, placebo-controlled phase III clinical trial to evaluate the efficacy and safety of treating healthcare professionals with the adsorbed COVID-19 (inactivated) vaccine manufactured by Sinovac–PROFISCOV: a structured summary of a study protocol for a randomised controlled trial. *Trials* 2020, 21(1):1–3.
36. Nascimento I, Leite L: Recombinant vaccines and the development of new vaccine strategies. *Brazilian Journal of Medical and Biological Research* 2012, 45(12):1102–1111.
37. Lanthier PA, Huston GE, Moquin A, Eaton SM, Szaba FM, Kummer LW, Tighe MP, Kohlmeier JE, Blair PJ, Broderick M: Live attenuated influenza vaccine (LAIV) impacts innate and adaptive immune responses. *Vaccine* 2011, 29(44):7849–7856.
38. Maiztegui JI, McKee Jr KT, Oro JGB, Harrison LH, Gibbs PH, Feuillade MR, Enria DA, Briggiler AM, Levis SC, Ambrosio AM: Protective efficacy of a live attenuated vaccine against Argentine hemorrhagic fever. *Journal of Infectious Diseases* 1998, 177(2):277–283.
39. Barrett AD: Yellow fever live attenuated vaccine: a very successful live attenuated vaccine but still we have problems controlling the disease. *Vaccine* 2017, 35(44):5951–5955.
40. Minor PD: Live attenuated vaccines: historical successes and current challenges. *Virology* 2015, 479:379–392.
41. Hoft DF, Lottenbach KR, Blazevic A, Turan A, Blevins TP, Pacatte TP, Yu Y, Mitchell MC, Hoft SG, Belshe RB: Comparisons of the humoral and cellular immune responses induced by live attenuated influenza vaccine and inactivated influenza vaccine in adults. *Clinical and Vaccine Immunology* 2017, 24(1):e00414–16.
42. Andersen P, Doherty TM: The success and failure of BCG – implications for a novel tuberculosis vaccine. *Nature Reviews Microbiology* 2005, 3(8):656–662.
43. Chakaya J, Khan M, Ntoumi F, Aklillu E, Fatima R, Mwaba P, Kapata N, Mfinanga S, Hasnain SE, Katoto PD: Global Tuberculosis Report 2020 – Reflections on the global TB burden, treatment and prevention efforts. *International Journal of Infectious Diseases* 2021, 113(Suppl 1):S7–S12. doi: 10.1016/j.ijid.2021.02.107.
44. Moyle PM, Toth I: Modern subunit vaccines: development, components, and research opportunities. *ChemMedChem* 2013, 8(3):360–376.
45. Iavarone C, O'hagan DT, Yu D, Delahaye NF, Ulmer JB: Mechanism of action of mRNA-based vaccines. *Expert Review of Vaccines* 2017, 16(9):871–881.
46. Middlebrook JL, Dorland RB: Bacterial toxins: cellular mechanisms of action. *Microbiological Reviews* 1984, 48(3):199.
47. McQuillan GM, Kruszon-Moran D, Deforest A, Chu SY, Wharton M: Serologic immunity to diphtheria and tetanus in the United States. *Annals of Internal Medicine* 2002, 136(9):660–666.
48. Kotloff KL, Wasserman SS, Losonsky GA, Thomas Jr W, Nichols R, Edelman R, Bridwell M, Monath TP: Safety and immunogenicity of increasing doses of a *Clostridium difficile* toxoid vaccine administered to healthy adults. *Infection and Immunity* 2001, 69(2):988.
49. Greenwood B, Bradley A, Blakebrough I, Whittle H, de C. Marshall T, Gilles H: The immune response to a meningococcal polysaccharide vaccine in an African village. *Transactions of the Royal Society of Tropical Medicine and Hygiene* 1980, 74(3):340–346.
50. Shapiro ED, Berg AT, Austrian R, Schroeder D, Parcells V, Margolis A, Adair RK, Clemens JD: The protective efficacy of polyvalent pneumococcal polysaccharide vaccine. *New England Journal of Medicine* 1991, 325(21):1453–1460.
51. Verez-Bencomo V, Fernandez-Santana V, Hardy E, Toledo ME, Rodríguez MC, Heynngnezz L, Rodriguez A, Baly A, Herrera L, Izquierdo M: A synthetic conjugate polysaccharide vaccine against *Haemophilus influenzae* type B. *Science* 2004, 305(5683):522–525.

52. Crick F: Central dogma of molecular biology. *Nature* 1970, 227(5258):561–563.
53. Smith TR, Patel A, Ramos S, Elwood D, Zhu X, Yan J, Gary EN, Walker SN, Schultheis K, Purwar M: Immunogenicity of a DNA vaccine candidate for COVID-19. *Nature Communications* 2020, 11(1):1–13.
54. McGonagle D, De Marco G, Bridgewood C: Mechanisms of immunothrombosis in vaccine-induced thrombotic thrombocytopenia (VITT) compared to natural SARS-CoV-2 infection. *Journal of Autoimmunity* 2021, 121:102662–102662.
55. Blakney AK, Ip S, Geall AJ: An update on self-amplifying mRNA vaccine development. *Vaccines* 2021, 9(2):97.
56. Holm MR, Poland GA: Critical aspects of packaging, storage, preparation, and administration of mRNA and adenovirus-vectored COVID-19 vaccines for optimal efficacy. *Vaccine* 2021, 39(3):457.
57. Stover C, De La Cruz V, Fuerst T, Burlein J, Benson L, Bennett L, Bansal G, Young J, Lee M, Hatfull G: New use of BCG for recombinant vaccines. *Nature* 1991, 351(6326):456–460.
58. Brochier B, Kieny M, Costy F, Coppens P, Bauduin B, Lecocq J, Languet B, Chappuis G, Desmettre P, Afiademanyo K: Large-scale eradication of rabies using recombinant vaccinia-rabies vaccine. *Nature* 1991, 354(6354):520–522.
59. Sanchez-Felipe L, Vercruysse T, Sharma S, Ma J, Lemmens V, Van Looveren D, Javarappa MPA, Boudewijns R, Malengier-Devlies B, Liesenborghs L: A single-dose live-attenuated YF17D-vectored SARS-CoV-2 vaccine candidate. *Nature* 2021, 590(7845):320–325.
60. National Research Council (US) Division of Health Promotion and Disease Prevention. Vaccine Supply and Innovation. Washington (DC): National Academies Press (US); 1985, 3, Vaccine Availability: Concerns, Barriers, and Impediments. Available from: https://www.ncbi.nlm.nih.gov/books/NBK216807.
61. Lang J, Wood SC: Development of orphan vaccines: an industry perspective. *Emerging Infectious Diseases* 1999, 5(6):749.
62. Plotkin S, Robinson JM, Cunningham G, Iqbal R, Larsen S: The complexity and cost of vaccine manufacturing – an overview. *Vaccine* 2017, 35(33):4064–4071.
63. Arya SC: Human immunization in developing countries: practical and theoretical problems and prospects. *Vaccine* 1994, 12(15):1423–1435.
64. Smith LE, Amlôt R, Weinman J, Yiend J, Rubin GJ: A systematic review of factors affecting vaccine uptake in young children. *Vaccine* 2017, 35(45):6059–6069.
65. What influences vaccine acceptance: a model of determinants of vaccine hesitancy [https://www.canvax.ca/what-influences-vaccine-acceptance-model-determinants-vaccine-hesitancy].
66. Misinformation fueling vaccine hesitancy, PAHO Director says [https://www.paho.org/en/news/21-4-2021-misinformation-fueling-vaccine-hesitancy-paho-director-says].
67. Knezevic I, Griffiths E: Biosimilars – global issues, national solutions. *Biologicals* 2011, 39(5):252–255.
68. European MedicinesAgency: Biosimilar medicines: Overview. 2018. [https://www.ema.europa.eu/en/human-regulatory/overview/biosimilar-medicines-overview]
69. U.S. Food & Drug Administration: Generic drugs: questions & answers. 2017.
70. United States Food & Drug Administration: Pfizer-BioNTech COVID-19 Vaccine EUA Letter of Authorization. Collegeville, Pennsylvania; 2021, May 10.
71. Organization WH: *Proceedings of the Twelfth International Conference of Drug Regulatory Authorities (ICDRA)*, Seoul: Republic of Korea, 3–6 April 2006.
72. Biosimilars [https://www.pfizer.com/science/research-development/biosimilars].
73. Kirchhoff CF, Wang XZM, Conlon HD, Anderson S, Ryan AM, Bose A: Biosimilars: key regulatory considerations and similarity assessment tools. *Biotechnology and Bioengineering* 2017, 114(12):2696–2705.
74. Heath PT, Feldman RG: Vaccination against group B streptococcus. *Expert Review of Vaccines* 2005, 4(2):207–218.
75. Levine MM: (2018) Typhoid fever vaccines. In: *Plotkin's Vaccines* (7th ed.). Philadelphia, Pennsylvania: Elsevier; 2018: 1114–1144.
76. MacDonald NE, Halperin SA, Law BJ, Forrest B, Danzig LE, Granoff DM: Induction of immunologic memory by conjugated vs plain meningococcal C polysaccharide vaccine in toddlers: a randomized controlled trial. *JAMA* 1998, 280(19):1685–1689.
77. Treanor J: Influenza vaccine – outmaneuvering antigenic shift and drift. *New England Journal of Medicine* 2004, 350(3):218–220.
78. Jefferson T, Smith S, Demicheli V, Harnden A, Rivetti A, Pietrantonj CD: WHO position paper on influenza vaccines: selected references. *The Lancet* 2005, 365(9461):773–780.
79. Jansen KU, Shaw AR: Human papillomavirus vaccines and prevention of cervical cancer. *Annual Review of Medicine* 2004, 55:319–331.

80. Pagliusi SR, Aguado MT: Efficacy and other milestones for human papillomavirus vaccine introduction. *Vaccine* 2004, 23(5):569–578.
81. Gustafsson L, Hallander HO, Olin P, Reizenstein E, Storsaeter J: A controlled trial of a two-component acellular, a five-component acellular, and a whole-cell pertussis vaccine. *New England Journal of Medicine* 1996, 334(6):349–356.
82. Heininger U, Cherry JD, Christenson PD, Eckhardt T, Göering U, Jakob P, Kasper W, Schweingel D, Laussucq S, Hackell JG: Comparative study of Lederle/Takeda acellular and Lederle whole-cell pertussis-component diphtheria-tetanus-pertussis vaccines in infants in Germany. *Vaccine* 1994, 12(1):81–86.
83. Simondon F, Preziosi M-P, Yam A, Kane CT, Chabirand L, Iteman I, Sanden G, Mboup S, Hoffenbach A, Knudsen K: A randomized double-blind trial comparing a two-component acellular to a whole-cell pertussis vaccine in Senegal. *Vaccine* 1997, 15(15):1606–1612.
84. Trunz BB, Fine P, Dye C: Effect of BCG vaccination on childhood tuberculous meningitis and miliary tuberculosis worldwide: a meta-analysis and assessment of cost-effectiveness. *The Lancet* 2006, 367(9517):1173–1180.
85. Bertolli J, Pangi C, Frerichs R, Halloran ME: A case-control study of the effectiveness of BCG vaccine for preventing leprosy in Yangon, Myanmar. *International Journal of Epidemiology* 1997, 26(4):888–896.
86. da Cunha SS, Rodrigues LC, Barreto McL, Dourado I: Outbreak of aseptic meningitis and mumps after mass vaccination with MMR vaccine using the Leningrad–Zagreb mumps strain. *Vaccine* 2002, 20(7–8):1106–1112.
87. Jefferson T, Price D, Demicheli V, Bianco E, Project ERpfIVSS: Unintended events following immunization with MMR: a systematic review. *Vaccine* 2003, 21(25–26):3954–3960.
88. Phadke M, Patki P, Kulkarni P, Jadhav S, Kapre S: Pharmacovigilance on MMR vaccine containing L-Zagreb mumps strain. *Vaccine* 2004, 22(31–32):4135–4136.
89. Bonnet M-C, Dutta A, Weinberger C, Plotkin SA: Mumps vaccine virus strains and aseptic meningitis. *Vaccine* 2006, 24(49–50):7037–7045.
90. Derbyshire M: "Similar biologics" approved and marketed in India. *Generics and Biosimilars Initiative Journal* 2013, 2(1):50–52.
91. Coronavirus disease (COVID-19) pandemic [https://www.euro.who.int/en/health-topics/health-emergencies/coronavirus-covid-19/novel-coronavirus-2019-ncov].
92. WHO: Coronavirus (COVID-19) data [https://covid19.who.int/] 2021.
93. US Food and Drug Administration: FDA takes additional action in fight against COVID-19 by issuing emergency use authorization for second COVID-19 vaccine. In: *Action Follows thorough Evaluation of Available Safety, Effectiveness, and Manufacturing Quality Information by FDA Career Scientists, Input from Independent Experts*; 2021.
94. Zarocostas J: What next for a COVID-19 intellectual property waiver? *The Lancet* 2021, 397(10288):1871–1872.

5 Therapeutic Delivery Systems for Biologics and Biosimilars

Tiffany R. Wong and Catherine F. Yang

Corresponding author: Catherine F. Yang

Contents

Since the discovery of recombinant DNA technology by Boyer and Cohen in 1973 and the first recombinant DNA-based protein drug approval by the Food and Drug Administration (FDA) of Humulin in 1982 (Rohrer, Lupo, and Bernkop-Schnürch 2018), biological therapy research and development in the last several decades has revolutionized medicine for treating many complex diseases including some of the most debilitating and chronic diseases such as cancer, autoimmune diseases, diabetes, HIV, neurogenic disorders, and many cardiovascular-metabolic illnesses (Chan and Chan 2017; Rohrer, Lupo, and Bernkop-Schnürch 2018).

Biological therapy is also referred to as biopharmaceuticals or biotherapeutics and includes drugs based on biologics and biosimilars. Biologics are large complex molecules manufactured from living systems. However, a biosimilar is "a biological product that is highly similar to and has no clinically meaningful differences" from an existing FDA-approved reference product, the originator biologic (Kabir,

DOI: 10.1201/9780429485626-5

Moreino, and Siam 2019). Biosimilars are not generics that are chemically synthesized and identical to the reference product. Instead, they are copies of reference products made from living cell systems, and since no living cell systems are identical, natural variations exist. Therefore, biosimilars clinically have high similarity to an approved biologic product, including safety, purity, potency, and efficacy but are not identical.

The process for the bench to bedside journey for a biological drug starts from its discovery, followed by a drug approval process comprised of development, pre-clinical, and clinical phases I, II, and III. This typically takes an average of 12 years from discovery to clinical trials. In contrast, a biosimilar approval starts from analytical studies to demonstrate similarity, animal toxicity studies, clinical pharmacology studies (to assess the immunogenicity and to generate pharmacokinetics and pharmacodynamic profiles), and finally clinical studies. This typically takes approximately eight years. Therefore, a biosimilar has a shortened approval pathway compared to a biologic (Agbogbo et al. 2019; Kabir, Moreino, and Siam 2019). Examples of some biological medications include insulin, hormones, growth factors, cytokines, blood products, gene and cellular therapy, monoclonal antibodies (mAbs), fusion proteins, soluble receptors, and vaccines.

Biotherapeutic delivery can be invasive or noninvasive, each with its mechanisms, advantages, and disadvantages (Anselmo et al. 2019). The predominant delivery for biological therapy is the invasive parenteral route, with the most common form being intravenous injection due to its excellent bioavailability and reproducibility as well as the nature of these drugs' large complex molecules. The less invasive routes, including subcutaneous and intramuscular injections, have also been used. Given the rapidly increasing usage and success of biotherapeutics in medicine, especially for severe and/or chronic diseases in a target-directed fashion with favorable safety and pharmacokinetics, noninvasive delivery strategies for biologics have been researched and developed to overcome challenges including inherent bioavailability barriers (e.g., epithelial and mucosal) for complex macromolecules, biodegradation, drug clearance, and targeting, as well as to increase patient compliance and comfort (Anselmo et al. 2019; Durán-Lobato, Niu, and Alonso 2020; Mantaj and Vllasaliu 2020). The typical noninvasive administration routes for biotherapeutics include oral, transdermal, inhalation, intranasal, and buccal administration (Durán-Lobato, Niu, and Alonso 2020; Mantaj and Vllasaliu 2020; Škalko-Basnet 2014).

5.1 NANOMEDICINE

Since biologics are produced from living systems, immunogenicity can potentially occur and cause allergic reactions or alter intended biological activities. In addition, biological therapy is challenging to deliver since the large complex molecules tend to be unstable and can be easily degraded chemically and enzymatically *in vivo*. To address these challenges, nanomedicine (also known as nanoparticles or nanocarriers) has been used to improve the delivery of biological therapy in various routes, including both invasive and noninvasive delivery (Škalko-Basnet 2014; Wahlich et al. 2019).

Nanomedicine is defined as the branch of therapeutics that employs particulate material between 1 and 100 nm (Zizzari et al. 2021). The advantages of nanocarriers to improve bioavailability and to prevent rapid drug clearance include versatile delivery routes, variable nanoscale sizes and shapes, adjustable charges, high service-to-volume ratios, loading modulation ability, controlled release capability, site-specific targeting modification, and carrier degradation (Wahlich et al. 2019). In general, the overall biologic pharmacokinetics can be controlled by adjusting the nanoparticle's size, hydrophilicity, and crystallinity (Škalko-Basnet 2014; Tran et al. 2017).

Nanocarrier drug delivery systems, which can vary in size, shape, and surface characteristics such as charge and surface coating, are commonly classified into several types including (Cullis and Hope 2017; Durán-Lobato, Niu, and Alonso 2020; Tran et al. 2017; Wahlich et al. 2019):

1. Liposomes are spherical nanoparticles that encapsulate active molecules in vesicles composed of lipid bilayers that form in the shape of a hollow sphere containing an aqueous compartment, allowing the biologics to be encapsulated within an aqueous compartment if hydrophilic or within a lipid bilayer if lipophilic. Liposomes typically consist of 50–500 nm spheres (Tran et al. 2017). Liposomes can protect biologics from physical degradation and facilitate their absorption into cells.
2. Lipid nanoparticles (LNPs) are liposome-like lipid nanoformulation structures by design but can take various types of forms and shapes in their applications. As such, LNPs are commonly used to encapsulate a wide variety of nucleic acids (RNA and DNA), making them the lead non-viral gene delivery systems (Cullis and Hope 2017). LNPs are primarily composed of cationic lipids with an anionic nucleic acid. Other lipid ingredients typically include neutral phospholipid molecules such as phosphatidylcholine class and sterols such as cholesterol. Coating LNPs with polyethylene glycol (PEG) – a hydrophilic and neutral polymer to help increase its stability and solubility – has been commonly used in nanomedicine including gene therapy. In gene therapy, segments of siRNA can be designed to silence any gene, and encapsulation of siRNA segments by LNPs provides protection for them until they reach the intended target destination and help facilitate their delivery. Since PEG has a neutral charge, it does not disrupt the function of any charged molecules such as DNA. As a result of PEGylation, the PEGylated LNPs as nanocarriers help increase biologic half-life circulating in the body and its bioavailability, creating the so-called stealth effect (Park 2019).
3. Polymer nanocarriers such as polymeric micelle, polymeric nanoparticle, nanoparticle-albumin bound technology (*Nab*), dendrimers (branched polymeric molecules), and hydrogel. They are typically composed of polymers (synthetic or naturally occurring) with dense matrices and well-known degradation patterns, thus allowing easier customizations of size, hydrophilicity, and biodegradability (Tran et al. 2017). A hydrogel is a three-dimensional network of cross-linked water-soluble polymers. Additional components can be added to the hydrogel to create a degradable nanocarrier. In addition, by controlling the amount of cross-linking in the matrix, the drug loading and release rate can be controlled (Park 2019; Tran et al. 2017).

Drug-conjugate nanocarriers such as protein-polymer conjugate, drug-polymer conjugate, and drug-antibody conjugate, or various combinations of the above as the hybrid nanocarriers (Durán-Lobato, Niu, and Alonso 2020; Mantaj and Vllasaliu 2020; Tran et al. 2017).

1. Chitosan-based nanoparticles are one of the examples. Chitosan is derived from chitin, a naturally occurring polysaccharide, cationic, mucoadhesive biocompatible polymer. Its positive surface charge and muco-adhesiveness allow it to adhere to mucous membranes and release drugs in a sustained released fashion, making it very versatile in many applications in not only intravenous but also in noninvasive delivery for any mucosal targets such as brain, head and neck, GIT, lungs, and cancer cell.
2. Another successful example of a hybrid nanocarrier is polylactide (PLA)/polylactide-co-glycoside (PLGA) PEGylated NPs. This nanocarrier formulation is a lipid polymer hybrid with a hydrophobic PLGA core and a hydrophilic PEGylated-lipid shell. By using different oil-in-water emulsion techniques, PLGA can form nanoemulsion, typically a transparent, monophasic, isotropic colloidal dispersion composed of oil, water, surfactants, and cosurfactant.
3. Inorganic nanoparticles such as metals (e.g., gold and silver), silica nanoparticles, and hafnium oxide nanoparticle. Inorganic nanoparticles are mostly used in diagnosis rather than drug delivery (Tran et al. 2017).

5.2 POTENTIAL ROUTES FOR DELIVERY OF BIOLOGICS/BIOSIMILARS

5.2.1 Oral Delivery

Though oral delivery is the most convenient and preferred method of medication administration by patients, it remains a challenge for biotherapeutics. As large complex molecules, biotherapeutics are highly susceptible to the relatively harsh physical and chemical environment of the gastrointestinal tract (GIT). Additionally, oral biotherapeutics are significantly affected by the first-pass metabolism, when the concentration of the drug is greatly reduced during the process of absorption in the liver before it reaches systemic circulation (Anselmo et al. 2019). Though the first-pass effect occurs mainly in the liver, it can also occur in the remaining GIT and other metabolically active tissues in the body (Herman and Santos 2021). There are multiple physiological and chemical barriers as well as micro-environmental challenges to limit biotherapeutic absorption via oral administration, including acidic pH-induced proteolysis of proteins in the stomach, bile salt effect, proteolytic enzymes in the intestinal lumen at the brush border membrane, efflux pump P-glycoprotein, microbiome-drug interaction, and intestinal epithelium various physical barriers including tight Junction (Anselmo et al. 2019).

Many strategies implemented to improve oral delivery of biologics include (1) protection from stomach acid degradation by using an enteric-coated delivery system, and coadministration of protein and/or peptide drugs with protease inhibitors, chemical modifiers, or cyclic peptides to improve stability in GIT (Anselmo et al. 2019); (2) prolongation of the contact time between the delivery system and the gastric mucosa by using natural mucoadhesive polymers such as chitosan, pectin, gelatin, or synthetic mucoadhesive polymers such as cellulose derivatives and vinyl alcohol (Mantaj and Vllasaliu 2020; Wahlich et al. 2019); (3) improvement in the mucosal permeability which will enhance the absorption of the drug by using surfactant or permeation enhancers which open the intracellular tight junction and facilitate paracellular transport of proteins and peptides, leading to enhanced absorption (Mantaj and Vllasaliu 2020; Škalko-Basnet 2014; Zizzari et al. 2021). Surfactants have both hydrophilic and hydrophobic components, and the most commonly used surfactants for the development of oral peptide formulations are based on medium-chain fatty acid, bile salts, and acylcarnitine. For tight junction-opening enhancers, chitosan-based nanoparticles are probably the most commonly studied products, which will be discussed in detail later; (4) deployment of nanoparticles that increase the transepithelial passage of biologics through carrier-mediated transport mechanisms (Zizzari et al. 2021); and finally, (5) the use of "smart" ingestible devices such as microneedle or ionic liquid (Durán-Lobato, Niu, and Alonso 2020; Mantaj and Vllasaliu 2020; Zizzari et al. 2021).

In general, the mucosal permeation enhancers enable the desired molecules to pass through the mucus layer, cell membranes, or tight junctions. Salcaprozate sodium (SNAC) and sodium caprate (C10) are two of the best-known intestinal permeation enhancing surfactants studied for oral delivery of biological drugs (Twarog et al. 2019). SNAC, a synthetic N-acetylated amino-acid derivative of salicylic acid, has been used for oral vitamin B12 administration and to increase oral insulin and octerotide-calcitonin bioavailability. The first oral glucagon-like peptide-1 analog, semaglutide, was first approved by FDA for type II diabetes mellitus (DM) treatment in 2019 using SNAC technology (Zizzari et al. 2021). C10 is a soluble anionic surfactant that has been studied in the oral delivery of small molecules such as zoledronic acid and macromolecules such as oral insulin, desmopressin, and some oligonucleotides (Twarog et al. 2019). Their exact mechanisms of action are uncertain. It is thought that C10 acts via openings of epithelial tight junctions and/or membrane perturbation to enable paracellular flux and transcellular perturbation, while SNAC is thought to increase passive transcellular permeation across small intestinal epithelia and to modulate solubility or possibly tight junction opening. An additional mechanism of SNAC, including a transient increase of pH, complex formation, and pepsin-inhibition effect in the stomach, which results in an increased concentration-dependent flux of semaglutide across the gastric mucosa by using a transcellular mechanism, was also observed (Twarog et al. 2019).

Medium-chain fatty acids such as C10 have also been utilized in one of the permeation enhancer systems – Gastrointestinal Permeation Enhancement Technology (GIPET) (Zizzari et al. 2021). It is a micro-emulsion system comprised of enteric-coated tablets or soft-gel capsules with a pH-sensitive coating, a medium-chain fatty acid such as C10, and the desired drug in selected ratios. A randomized, double-blind, phase II German trial has shown oral insulin 338 is safe to use with no difference in efficacy compared to commonly used subcutaneous insulin for insulin naïve type II DM (Halberg et al. 2019). However, further development of this particular oral insulin product was discontinued as the production of quantities required for wide public use was deemed not commercially viable.

For tight junction-opening enhancers, the best-known examples are chitosan-based nanoparticles as their mucoadhesive features enable them to mediate the opening of tight junctions (Chen et al. 2013). Tight junctions consist of a complex combination of transmembrane integral proteins, including claudins, occludin, and junctional adhesion molecules alone with several additional proteins to help anchor the transmembrane proteins to the actin cytoskeleton. Transmembrane proteins, especially claudin, are known to play major roles in forming the seal between adjacent cells. After interacting with chitosan-based nanoparticles, redistribution of claudin from cell membrane to cytosol occurs, which leads to its degradation in lysosomes, resulting in subsequent decreases in tight junction strength and ultimate increases in paracellular permeability. A study has also shown the effects of chitosan on opening epithelial tight junctions and paracellular transport, observed by *in vivo* fluorescence-microscopic data and verified by computed-tomography following the oral administration of isotope-labeled nanoparticles in Caco-2 cell monolayers and animal models (Sonaje et al. 2012).

One of the new microneedle enhancer devices is called a self-orienting millimeter-scale applicator (SOMA). This novel system delivers biologics orally using a tortoise-shaped device and low centered gravity to self-orient in the correct position and physically insert a biodegradable microneedle through the gastric mucosa to deliver the desired biologics (Durán-Lobato, Niu, and Alonso 2020; Mantaj and Vllasaliu 2020; Zizzari et al. 2021). This ingestible device consists of a core of stainless steel and low-density polycaprolactone, which automatically delivers milliposts loaded with biotherapeutic peptides into the gastric mucosa and is currently being studied for the delivery of insulin (Zizzari et al. 2021). Within the shell, the SOMA contains a 7 mm millipost consisting of a 1.7 mm compressed tip with 0.5 mg of insulin mixed under high pressure with poly(ethylene)oxide and a shaft made of biodegradable polymers (Cully 2019).

In addition to various classes of nanocarriers, robotic microneedle devices such as SOMA and robotic capsule, self-emulsifying drug delivery system (SEDDS) with mixtures of lipids, surfactant, and cosolvents that improve peptide bioavailability, permeation enhancer systems such as POD (Protein Oral Delivery), and structural modifiers such as cyclic peptides have also been used for oral delivery (Mantaj and Vllasaliu 2020; Zizzari et al. 2021). When SEDDS are administered orally, they form emulsions or microemulsions in the GIT to enhance drug delivery and absorption by improving permeation through the intestinal mucus layer and by providing protection against the biodegradation (Zizzari et al. 2021). The best-known drug using SEDDS is cyclosporin to prevent transplant rejection or Graft versus Host disease. For POD, it is a system incorporating various types of strategies to provide protection during GI passage and to enhance absorption within the same formulation. SEDDS has been utilized for oral insulin through strategies including encapsulation, a permeation enhancer, a chelating agent, and a protease inhibitor (Zizzari et al. 2021). For structural modifiers, peptide cyclization can also help improve stability from GI enzymatic degradation and enhanced membrane permeability. In addition to SOMA device, other robotic pills using microneedle technology typically involve an enteric-coated capsule utilizing a carbon dioxide inflated small balloon that pushes out sucrose-based needles, or a compressed spring enclosed in an enteric-coated capsule driven by osmotic controlled released to propel drug-loaded microneedles into the small intestine (Durán-Lobato et al. 2019; Zizzari et al. 2021).

5.2.2 Subcutaneous Delivery

For minimally invasive biologic delivery, the subcutaneous route is commonly used for many hormones, proteins, and antibodies. The advantages are self-administration by the patient and the avoidance of first-pass metabolism. The disadvantages are patient compliance and potential painful injection perceived by some patients based on their individual pain threshold. Anatomical challenges for this delivery modality include the presence of an extracellular matrix barrier and limited space for injected volumes to achieve therapeutic efficacy (Anselmo et al. 2019).

One approach to ensure the adequate volume is injected for biologics to achieve therapeutic efficiency and bioavailability is the use of enzymes called hyaluronidases which aid in increasing the biodegradation of the matrix, thereby increasing the absorption and injectable volume of the biologics. However, using naturally derived hyaluronidases can lead to increased immunogenicity as these products are all derived from a living system, which can induce allergic reactions and/or potentially alter the biological activity with clinical consequences. To account for the increased hypersensitivity and immunogenic reactions to the naturally derived hyaluronidases, recombinant human hyaluronidases PH20 were developed to address the issues (Locke, Maneval, and LaBarre 2019). ENHANZE® is one of the most recent subcutaneous drug delivery technology currently using recombinant human hyaluronidase PH20 enzyme for therapy of breast cancer, lymphoma, and primary immunodeficiency disorders (Locke, Maneval, and LaBarre 2019).

5.2.3 Transdermal Delivery

The advantages of transdermal delivery are painless administration, control of timed biologic release, avoidance of first-pass metabolism, and ease of therapy termination (Peña-Juárez, Guadarrama-Escobar, and Escobar-Chávez 2021). It can also serve as a local treatment for dermatological diseases. Through this route, passive diffusion of molecules across the skin layers may happen, however, substantial barriers exist for this route since skin primarily serves as a physical protective barrier for humans to the external world. In particular, stratum corneum, the topmost skin layer as a physical barrier, cell shedding, and skin metabolism are considered major drawbacks for transdermal biologic delivery (Peña-Juárez, Guadarrama-Escobar, and Escobar-Chávez 2021). Some advances have been made to address these issues. Transdermal patches have been used for several decades for opioids, hormones, small molecules of anesthetics, and many other biologics (Benson et al. 2019). Commonly used transdermal delivery enhancers include microneedles, thermal ablation, microdermabrasion, laser-assisted delivery, and colloidal nanocarriers such as nanoemulsion, liposome, or flexible vesicles, or various lipid nanoparticle carriers (Benson et al. 2019). Electroporation, iontophoresis, and sonophoresis, magnetophoresis, and nanocarrier-assisted microjets have also been explored (Benson et al. 2019). Smart wearable technology using microneedle patches for disease diagnosis and treatment such as monitoring and managing diabetes is currently undertaken, providing a noninvasive personalized feedback system for combining glucose monitoring and flexible responsive insulin delivery (Benson et al. 2019).

5.2.4 Inhalation Delivery

The inhalation route offers a noninvasive way for both local and systemic delivery, with additional advantages of rapid and large surface area for absorption, favorable pharmacokinetics, and avoidance of first-pass metabolism (Škalko-Basnet 2014). Commonly used inhaler delivery include metered-dose inhalers, nebulizers, and dry powder inhalers. For biologics, since most nebulizers and metered dose inhalers typically use aqueous suspension that is not ideal for biologics, dry powder inhalers are generally preferred devices for inhaled biotherapeutics (Liang et al. 2020).

The main challenges in this delivery route are the presence of physicochemical barriers such as muco-constricted bronchioles, mucociliary and macrophage clearance, pulmonary surfactant, airway epithelium, and enzymatic degradation, as well as the variability in dosing of the lung tissues, which depends on types of inhaler techniques and patient training (Liang et al. 2020). Drug formulation such as particle surface modifications (e.g., size, shape, charge, porosity), the use of mucoadhesive, mucopenetrative additives, or chemical or permeation enhancers (such as surfactants and polymers), and drug encapsulation to overcome the physicochemical barriers have been explored to minimize clearance and to help increase absorption or diffusion of biologics via inhalation delivery (Liang et al. 2020; Anselmo et al. 2019).

5.2.5 Intranasal and Buccal Delivery

Intranasal and buccal delivery offer advantages of noninvasive delivery such as easy self-administration, rapid absorption, and the avoidance of first-pass metabolism (Anselmo et al. 2019; Škalko-Basnet 2014). Disadvantages for both routes include potential mucosal irritation and low surface area for absorption with total dose limitation (Anselmo et al. 2019; Škalko-Basnet 2014). The main challenges are mucus and epithelial cells. Nasal sprays, dry nasal powder, nasal drops, buccal soluble film, and nano-formulated buccal sprays are currently used for vaccines and hormones (Anselmo et al. 2019). A buccal insulin delivery system via Rapidmist device using surfactants to form insulin-containing micelles as absorption enhancers has been studied in a phase III trial for type I DM (Zizzari et al. 2021). Additional ways to improve delivery such as permeation enhancers, mucolytic agents, enzymatic inhibitors, and mucoadhesive agents have also been explored and used (Rohrer, Lupo, and Bernkop-Schnürch 2018).

5.2.6 Blood–Brain Barrier (BBB): Trojan-Horse Technology (THT)

Though there are many debilitating neurogenic disorders, most available drugs cannot cross the blood–brain barrier (BBB). This fact also applies to biologic drugs. Intrathecal (IT) injection of biotherapeutics is useful when the target is on the surface of the brain or spinal cord, thus in close proximity to the cerebrospinal fluid (CSF) flow (Pardridge 2017). The IT route, however, is not ideal for the delivery of biologics into deeper tissues of brain parenchyma, which is mostly needed for treating neurodegenerative disorders or aggressive brain cancers. To overcome the challenge, the initially so-called molecular Trojan-Horse technology (THT) has emerged (Pardridge 2017). The technology allows for fusing the biologic pharmaceutical to a mAb which can recognize endogenous BBB receptors such as insulin receptors and/or transferrin receptors. The mAb, therefore, acts as a molecular Trojan Horse to deliver the biologic pharmaceutical across the BBB (Pardridge 2017). Recently, it has been reported that neurotransmitter-lipidoids was used as a THT carrier to deliver a diverse range of biotherapeutic products into the brain neural cells of mice – a major breakthrough and a promising beginning to treat potentially many chronically debilitating neurogenic disorders and central nervous system diseases (Ma et al. 2020). Though mAb-based targeted delivery has been used extracranially, this technology is relatively new for the central nervous system since most available drugs to date cannot cross the BBB adequately. We will await its wider promising use for many chronically debilitating neurogenic diseases.

5.2.7 Vaccine Delivery

For the year 2020, the pandemic of the novel coronavirus SARS-CoV-2 has universally hit the globe without exception. Given that SARS-CoV-2 is a virus that humans have never been exposed to in the past, there was no vaccine available when the human transmission was first reported in late 2019.

In general, there have been many different vaccine platforms, including live attenuated, inactivated, recombinant, viral vector, DNA, RNA, and protein-based vaccines (Pollard and Bijker 2021).

These platforms are currently being utilized to develop various SARS-CoV-2 vaccines. Once a vaccine is developed, it must go through a rigorous testing process from extensive pre-clinical studies on small animals to clinical phases I, II, and III, where safety, efficacy, and immunogenicity are being tested.

5.2.8 Types of Vaccine Products

1. Viral-based vaccines:
   a. Inactivated vaccines are manufactured by cultivating viruses and subsequently inactivating them using heat or chemicals to prevent the pathogen's ability to replicate and cause disease while still being recognizable by the host immune system (Vetter et al. 2018). This type of a vaccine is typically administered intramuscularly and is generally considered a safer vaccine type compared to live attenuated vaccines as there is no risk of the virus reverting to a disease-causing capable state (Vetter et al. 2018). Another benefit of the inactivated vaccine is that it would target multiple components of the virus, not just the spike protein. There are several types of this form of SARS-CoV-2 vaccines being developed around the world, and quite a few of them are in the late-stages of clinical trials (Edwards and Orenstein 2021; Gomez, Robinson, and Rogalewicz 2013; Plotkin et al. 2017).
   b. Live-attenuated vaccines are genetically or conditionally weakened wild-type viruses capable of generating an immune response within the body but not causing the diseases themselves (Vetter et al. 2018). Common methods to achieve attenuation are genetically altering the virus or growing the virus in an unfavorable environment until its virulence, that is, its ability to cause disease, is lost while still maintaining its immunogenicity or an ability to elicit an immune response to protect the host against future infection (Gomez, Robinson, and Rogalewicz 2013; Plotkin et al. 2017). Common examples of this type of vaccine in the United States include the varicella vaccine, the influenza nasal spray vaccine, MMR combined vaccine, and the rotavirus vaccine (Plotkin, Orenstein, and Offit 2012). One theoretical advantage of producing a live attenuated SARS-CoV-2 vaccine is that it may potentially stimulate both cellular and humoral immunity against various components of the whole attenuated virus (Pollard and Bijker 2021). When administered intranasally, it may offer another potential advantage of inducing mucosal immune responses at the viral entry site of the upper respiratory tract (Edwards and Orenstein 2021). One major safety concern of using live attenuated vaccines is the possibility of the virus reverting to the wild-type virus that can cause disease (Edwards and Orenstein 2021).
2. Protein-based vaccines:
   a. Protein-based subunit vaccines include recombinant protein, purified protein, peptide, polysaccharide, or virus-like particle vaccines (Pollard and Bijker 2021). Recombinant protein vaccines are produced through the insertion of a gene encoding the viral protein into another system or producer cell to serve as a carrier. The vaccine protein is formed once the carrier expresses the viral protein without replication of the live virus. The immune system will then recognize the protein and protect against the virus. A common example of this vaccine type is the Hepatitis B vaccine or human papillomavirus vaccine (Pollard and Bijker 2021). There are three types of recombinant SARS-CoV-2 vaccines in development: spike protein, recombinant receptor-binding domain, and virus-like particle vaccines (World Health Organization 2021).
3. Viral vector vaccines: This platform uses a vector as the skeletal backbone. A vector is a different and typically less harmful virus than the one the vaccine is created for – to express the viral protein to trigger the development of an immune response. There are three types of vectors:
   a. Non-replicating vector vaccine: These utilize a genetically engineered vector virus that does not replicate *in vivo* but expresses the intended viral protein target (Zhu et al. 2020). The majority of this type of vaccine in development for SARS-CoV-2 are made to express

the spike protein using intramuscular administration (World Health Organization 2021; Zhu et al. 2020). Johnson & Johnson's Janssen SARS-CoV-2 vaccine platform is a non-replicating adenovirus vector that expresses a stabilized spike protein from the DNA of COVID-19 (Sadoff et al. 2021; Stephenson et al. 2021).

b. Replicating vector vaccine: Compared to the non-replicating vector vaccines, the replicating ones typically result in a more robust immune response as they are made from attenuated or vaccine-strains of viruses and can replicate within the vaccinated host for an innate immune response (Edwards and Orenstein 2021). There are multiple vaccines in development for SARS-CoV-2, which target the spike protein and can be given intranasally (Case et al. 2020; Sun et al. 2020; World Health Organization 2021).

c. Inactivated virus vector vaccine: Inactivated virus vector vaccines are created to express the protein of interest but inactivated, and thus are safer, even in immunocompromised hosts. This type of vaccine to target spike proteins for SARS-CoV-2 is still in pre-clinical development as of late 2020 (World Health Organization 2021).

4. DNA and RNA vaccines:

a. DNA and RNA are considered genetic vaccines that use part of the virus' own genes to stimulate an immune response in the vaccinated recipient. In DNA vaccines, plasmid DNA is precipitated onto an inert particle (generally gold beads) and forced into the cells with a helium blast. Transfected cells then express the antigen encoded on the plasmid, resulting in an immune response (Abbasi 2020). DNA vaccines are specifically composed of the target gene's plasmid DNA, which gives a large manufacturing advantage as plasmid DNA is relatively stable and has the same straightforward encoding process for most genes (Liu 2019). However, their use is limited because DNA vaccines tend to have lower immunogenicity and require the use of specific delivery devices (Yu et al. 2020). Johnson & Johnson's Janssen SARS-CoV-2 vaccine utilizes a piece of DNA that encodes the spike protein; however, the DNA is added to an adenovirus vector, which is why it is also discussed in the viral vector vaccine section above (Sadoff et al. 2021; Stephenson et al. 2021).

b. RNA vaccines work by introducing a piece of mRNA to the recipient that corresponds to a viral protein, usually a smaller piece of a protein found on the virus' outer membrane. RNA vaccines have an advantage over DNA vaccines because the mRNA remains in the cytoplasm and does not enter into the nucleus, nor does it interact or integrate into the recipient's DNA. Once administered, the mRNA vaccine is directly translated into its target protein and can elicit an immune response (Walsh et al. 2020). There are two types of RNA viruses: positive- and negative-sense or strand. Positive-strand RNA is similar to and functions as mRNA, and it can be directly translated within the body. Negative-strand RNA, however, is a complimentary template to mRNA and has to be converted to positive-strand RNA before translation to the target protein (Payne 2017). RNA vaccines have several benefits compared to other types of vaccines, including shorter manufacturing times and no risk of causing disease in the vaccinated recipient. The first SARS-CoV-2 vaccines that were developed for mass production are Pfizer-BioNTech's and Moderna's mRNA vaccines; both are given intramuscularly (Walsh et al. 2020).

5.3 CHAPTER SUMMARY

Future medical use and applications of biological therapy will continue to increase and expand, ultimately leading to well-targeted, personalized medicine. Further research and development for its delivery, including noninvasive routes for patient ease, and compliance while maintaining excellent bioavailability and reproducibility is warranted, ultimately leading to highly targeted and personalized medicine.

REFERENCES

Abbasi, Jennifer. 2020. COVID-19 and mRNA vaccines: First large test for a new approach. *Journal of the American Medical Association* 324 (12):1125–1127.

Agbogbo, Frank K., Dawn M. Ecker, Allison Farrand, et al. 2019. Current perspectives on biosimilars. *Journal of Industrial Microbiology & Biotechnology* 46 (9):1297–1311.

Anselmo, Aaron C., Yatin Gokarn, and Samir Mitragotri. 2019. Non-invasive delivery strategies for biologics. *Nature Reviews Drug Discovery* 18 (1):19–40.

Benson, Heather A. E., Jeffrey E. Grice, Yousuf Mohammed, Sarika Namjoshi, and Michael S. Roberts. 2019. Topical and transdermal drug delivery: From simple potions to smart technologies. *Current Drug Delivery* 16 (5):444–460.

Case, James Brett, Paul W. Rothlauf, Rita E. Chen, et al. 2020. Replication-competent vesicular stomatitis virus vaccine vector protects against SARS-CoV-2-mediated pathogenesis. bioRxiv:2020.07.09.196386.

Chan, Juliana C. N., and Anthony T. C. Chan. 2017. Biologics and biosimilars: What, why and how? *ESMO Open* 2 (1):1–2.

Chen, M. C., F. L. Mi, Z. X. Liao, et al. 2013. Recent advances in chitosan-based nanoparticles for oral delivery of macromolecules. *Advanced Drug Delivery Reviews* 65 (6):865–879.

Cullis, Pieter R., and Michael J. Hope. 2017. Lipid nanoparticle systems for enabling gene therapies. *Molecular Therapy* 25 (7):1467–1475.

Cully, M. 2019. Tortoise-inspired device for oral delivery of biologics. *Nature Reviews Drug Discovery* 18:254.

Durán-Lobato, Matilde, Zhigao Niu, and María José Alonso. 2020. Oral delivery of biologics for precision medicine. *Advanced Materials* 32 (13):1901935.

Edwards, Kathryn M., and Walter A. Orenstein. 2021. COVID-19: Vaccines to prevent SARS-CoV-2 infection. UpToDate, April 2, 2021]. Available from https://www.uptodate.com/contents/covid-19-vaccines-to-prevent-sars-cov-2-infection.

Gomez, Phillip L., James M. Robinson, and Joseph A. Rogalewicz. 2013. 4 – Vaccine manufacturing. In *Vaccines (Sixth Edition)*, edited by S. A. Plotkin, W. A. Orenstein, and P. A. Offit. London: W.B. Saunders.

Halberg, Inge B., Karsten Lyby, Karsten Wassermann, Tim Heise, Eric Zijlstra, and Leona Plum-Mörschel. 2019. Efficacy and safety of oral basal insulin versus subcutaneous insulin glargine in type 2 diabetes: A randomised, double-blind, phase 2 trial. *The Lancet Diabetes & Endocrinology* 7 (3):179–188.

Herman, T. F., and C. Santos. 2021. First Pass Effect. In *StatPearls. Treasure Island (FL): StatPearls Publishing Copyright © 2021.* StatPearls Publishing LLC.

Kabir, Eva Rahman, Shannon Sherwin Moreino, and Mohammad Kawsar Sharif Siam. 2019. The breakthrough of biosimilars: A twist in the narrative of biological therapy. *Biomolecules* 9 (9):410.

Liang, Wanling, Harry W. Pan, Driton Vllasaliu, and Jenny K. W. Lam. 2020. Pulmonary delivery of biological drugs. *Pharmaceutics* 12 (11):1025.

Liu, Margaret A. 2019. A comparison of plasmid DNA and mRNA as vaccine technologies. *Vaccines* 7 (2):37.

Locke, Kenneth W., Daniel C. Maneval, and Michael J. LaBarre. 2019. ENHANZE® drug delivery technology: A novel approach to subcutaneous administration using recombinant human hyaluronidase PH20. *Drug Delivery* 26 (1):98–106.

Ma, Feihe, Liu Yang, Zhuorui Sun, et al. 2020. Neurotransmitter-derived lipidoids (NT-lipidoids) for enhanced brain delivery through intravenous injection. *Science Advances* 6 (30):eabb4429.

Mantaj, Julia, and Driton Vllasaliu. 2020. Recent advances in the oral delivery of biologics. *The Pharmaceutical Journal* 304 (7933). doi:10.1211/PJ.2020.20207374

Pardridge, William M. 2017. Delivery of biologics across the blood–brain barrier with molecular Trojan Horse technology. *BioDrugs* 31 (6):503–519.

Park, Jin Seon. 2019. The impact of nano-technology on biologic drug development. *Pharmaceutical Outsourcing* 5 (13):1575–1581.

Payne, Susan. 2017. Introduction to RNA viruses. *Viruses*: 97–105. https://www.sciencedirect.com/science/article/pii/B9780128031094000106?via%3Dihub

Peña-Juárez, Ma Concepción, Omar Rodrigo Guadarrama-Escobar, and José Juan Escobar-Chávez. 2021. Transdermal delivery systems for biomolecules. *Journal of Pharmaceutical Innovation*. https://link.springer.com/article/10.1007/s12247-020-09525-2#article-info

Plotkin, Stanley, James M. Robinson, Gerard Cunningham, Robyn Iqbal, and Shannon Larsen. 2017. The complexity and cost of vaccine manufacturing – An overview. *Vaccine* 35 (33):4064–4071.

Plotkin, Stanley A., Walter A. Orenstein, and Paul A. Offit. 2012. *Vaccines: Expert Consult – Online and Print* (6th ed.). Philadelphia: Saunders.

Pollard, Andrew J., and Else M. Bijker. 2021. Publisher Correction: A guide to vaccinology: From basic principles to new developments. *Nature Reviews Immunology* 21 (2):129–129.

Rohrer, Julia, Noemi Lupo, and Andreas Bernkop-Schnürch. 2018. Advanced formulations for intranasal delivery of biologics. *International Journal of Pharmaceutics* 553 (1):8–20.

Sadoff, Jerald, Mathieu Le Gars, Georgi Shukarev, et al. 2021. Interim results of a phase 1–2a trial of Ad26.COV2.S Covid-19 vaccine. *New England Journal of Medicine* 384 (19):1824–1835.

Škalko-Basnet, N. 2014. Biologics: The role of delivery systems in improved therapy. *Biologics: Targets and Therapy* 8:107–114.

Sonaje, Kiran, Er-Yuan Chuang, Kun-Ju Lin, et al. 2012. Opening of epithelial tight junctions and enhancement of paracellular permeation by chitosan: Microscopic, ultrastructural, and computed-tomographic observations. *Molecular Pharmaceutics* 9 (5):1271–1279.

Stephenson, Kathryn E., Mathieu Le Gars, Jerald Sadoff, et al. 2021. Immunogenicity of the Ad26.COV2.S Vaccine for COVID-19. *Journal of the American Medical Association* 325 (15):1535–1544.

Sun, Weina, Stephen McCroskery, Wen-Chun Liu, et al. 2020. A Newcastle Disease Virus (NDV) expressing a membrane-anchored spike as a cost-effective inactivated SARS-CoV-2 Vaccine. *Vaccines* 8 (4):771.

Tran, Stephanie, Peter-Joseph DeGiovanni, Brandon Piel, and Prakash Rai. 2017. Cancer nanomedicine: A review of recent success in drug delivery. *Clinical and Translational Medicine* 6 (1):e44.

Twarog, C., S. Fattah, J. Heade, S. Maher, E. Fattal, and D. J. Brayden. 2019. Intestinal permeation enhancers for oral delivery of macromolecules: A comparison between Salcaprozate Sodium (SNAC) and Sodium Caprate (C(10)). *Pharmaceutics* 11 (2):78.

Vetter, Volker, Gülhan Denizer, Leonard R. Friedland, Jyothsna Krishnan, and Marla Shapiro. 2018. Understanding modern-day vaccines: What you need to know. *Annals of Medicine* 50 (2):110–120.

Wahlich, John, Arpan Desai, Francesca Greco, et al. 2019. Nanomedicines for the delivery of biologics. *Pharmaceutics* 11 (5):210.

Walsh, Edward E., Robert W. Frenck, Ann R. Falsey, et al. 2020. Safety and immunogenicity of two RNA-based Covid-19 vaccine candidates. *New England Journal of Medicine* 383 (25):2439–2450.

World Health Organization. *COVID-19 vaccine tracker and landscape*. World Health Organization 2021 [cited February 7, 2021]. Available from https://www.who.int/publications/m/item/draft-landscape-of-covid-19-candidate-vaccines.

Yu, Jingyou, Lisa H. Tostanoski, Lauren Peter, et al. 2020. DNA vaccine protection against SARS-CoV-2 in rhesus macaques. *Science* 369 (6505):806–811.

Zhu, Feng-Cai, Yu-Hua Li, Xu-Hua Guan, et al. 2020. Safety, tolerability, and immunogenicity of a recombinant adenovirus type-5 vectored COVID-19 vaccine: A dose-escalation, open-label, non-randomised, first-in-human trial. *The Lancet* 395 (10240):1845–1854.

Zizzari, Alessandra T., Dimanthi Pliatsika, Flavio M. Gall, Thomas Fischer, and Rainer Riedl. 2021. New perspectives in oral peptide delivery. *Drug Discovery Today* 26 (4):1097–1105.

6 Legislative and Regulatory Pathways for Biosimilars

Erika Young, Kanika Gulia, and Catherine F. Yang

Corresponding author: Erika Young

Contents

6.1 INTRODUCTION

Biological products or biologics are large and complex molecules that are used to diagnose, prevent, treat, and cure diseases and medical conditions (1). They are produced through biotechnology in a living system, such as a microorganism, plant cell, or animal cell. Therefore, biologics are different from small molecule drugs in that they are difficult to characterize because of their complex nature and the variations that can result from the manufacturing process. The Public Health Service Act (PHSA) also establishes the definition of a biological product as "*a virus, therapeutic serum, toxin, antitoxin, vaccine, blood, blood component or derivative, allergenic product, or analogous product … applicable to the prevention, treatment or cure of a disease or condition of human beings*" (2). Most biological products regulated under the PHSA also meet the definition of a drug under Section 201(g) of the Federal

DOI: 10.1201/9780429485626-6

Food, Drug & Cosmetic Act (FFDCA) as *"articles intended for use in the diagnosis, cure, mitigation, treatment, or prevention of disease in man or animals; and articles (other than food) intended to affect the structure or any function of the body of man or other animals"* (3). Since biological products are a subset of drugs, both are subject to approval under the FFDCA. However, only biologics are licensed under PHSA (4).

The Food & Drug Administration (FDA)'s Center for Biologics Evaluation and Research (CBER) regulates biologics, such as vaccines, blood and blood components, allergenic patch tests and extracts, human immunodeficiency virus (HIV) and hepatitis tests, gene therapy products, cells and tissues for transplantation, and new treatments for cancers and other serious diseases (5). The CBER assesses the biologics regarding aspects of safety, purity, potency, and effectiveness. As part of its review, CBER ensures that the biologics are manufactured following Current Good Manufacturing Practice (CGMP) regulations to help ensure that the aspects of safety, purity, potency, and effectiveness are in place and that the manufacturers produce biological products with consistent clinical performance.

6.2 HISTORICAL CONTEXT FOR EVOLUTION OF LAWS PERTINENT TO BIOLOGICS

The regulation of biologics by the federal government began with the Biologics Control Act (BCA) of 1902 (also known as Virus-Toxin law) as a response to two incidents that stemmed from inadequate controls during the manufacture of biologics, leading to the death of children caused by diphtheria antitoxin serum, which was made from the blood of horses contaminated with tetanus, and contaminated smallpox vaccine (6, 7). The later incident involved the death of children from tetanus after receiving a contaminated smallpox vaccine. Before the BCA was legislated, there were no regulations overseeing the manufacturing establishments. Therefore, the law gave the government the first control over establishment inspections and mandated routine annual inspection of manufacturing facilities before biologic licensure to ensure that safety, potency, and purity of biologics are in place.

The BCA has since undergone revision and been recodified after the PHSA was passed in 1944 to give the FDA authority over manufacturers of biologic products (8). The PHS Act, among other jurisdictions, laid out the provisions for new biological product licensure process, and currently the regulations (21 CFR Parts 600–680) can be implemented to get approval for licensing.

An emphasis is placed on appropriate manufacturing processes due to the complexity of manufacturing and characterizing a biologic. Any adjustments in the manufacturing process can disrupt the integrity of the biological molecule and, therefore, may profoundly alter the safety and efficacy profile. Such changes are often undetectable by standard techniques and require highly sophisticated technologies. Furthermore, in many cases, instruments with limited capabilities cannot identify clinically active components of a biological product. Therefore, biological products must be evaluated based on their specific manufacturing process (9). The PHSA, among other jurisdictions, laid out the provisions for new biological product licensure process, and currently the regulations (21 CFR Parts 600–680) can be implemented to get approval for licensing. An emphasis is placed on appropriate manufacturing processes due to the complexity of manufacturing and characterizing a biologic.

The FDA Modernization Act of 1997 (FDAMA) has amended several sections of the PHSA; one of which is Section 506(a)(2) that added provisions for fast-track designation for products intended to treat serious or life-threatening conditions to facilitate their development and expedite their review. In addition, the FDAMA amended the PHSA to eliminate the Establishment License Application (ELA). Biologics now obtain approval through a biologics licensure application (BLA) via 351(a) under the PHSA, which is the traditional pathway for approval of biologics. The submitted application must contain all the information regarding the safety, purity, potency, and effectiveness of the product and is also known as a "stand alone" application as it does not depend upon any other biological product.

6.3 FDA AUTHORITY OVER BIOLOGICS

The FDA oversees the safety and effectiveness of biologics, drugs, and devices and has held regulatory responsibility for biologics since 1972 (10). The CBER and Center for Drug Evaluation and Research (CDER) are responsible for reviewing oversight of biological products.

Following initial laboratory and animal testing data that establish safety for human studies, biological products may first be studied in clinical trials under an investigational new drug application (IND) process as outlined under 21 CFR Part 312 (11). There are three types of INDs – an investigator IND, an emergency use IND. A treatment IND. An Investigator IND is submitted by a physician who initiates and leads the investigation, to either study an unapproved drug, or an approved one with a new indication. An emergency use IND is submitted with the intention of using an experimental drug that does not permit the time needed to submit an IND in accordance with 21 CFR Section 312.23 or Section 312.20. A treatment IND requests permission for use of an experimental drug that has already shown efficacy in clinical testing but has not yet completed the final clinical assessments (12). In general, the IND application must contain information regarding animal pharmacology and toxicology studies, manufacturing information, as well as clinical protocols and investigator information (13). If the outcomes from these studies evidence safety and efficacy for its intended use, the data are then submitted via BLA under PHSA Section 351(a).

The FDA approval to market a biologic is granted by issuance of a biologic license, which is determined by the quality of the product, manufacturing process, and manufacturing facilities in the context of appropriate requirements that ensure continued safety, purity, and potency of the product. Particular to biologics, a potency assay is required due to the complexity and heterogeneity of biologics. To obtain licensure, the sponsor must establish in the BLA that the biologic, as well as the facility in which it is manufactured, meets standards to assure safety, purity, and potency. Any subsequent change to the approved manufacturing process requires a demonstration that the product still maintains safety and effectiveness.

6.4 LEGISLATION AND REGULATORY PATHWAY FOR BIOSIMILARS

Biosimilar products are biological products that need to fulfill all biological products' regulatory requirements, while also providing evidence to prove similarity to and no clinical meaningful difference from the reference product. The Biologics Price Competition and Innovation Act (BPCIA) of 2009, enacted as Title VII of the Affordable Care Act (ACA), revised the PHSA to establish an abbreviated licensure pathway, 351(k), for follow-on biologics shown to be "highly similar" (biosimilar) to or "interchangeable" with an FDA-licensed biologic product (13). The abbreviated 351(k) pathway reduces costs of the development process and streamlines licensure by allowing publicly available data regarding the reference product to which the biosimilar is compared as reliable data to determine safety, purity, and potency. The PHSA defines the "reference product" for a 351(k) application as the "single biological product licensed under Section 351(a) against which a biological product is evaluated" (14).

Since the primary goal is to establish biosimilarity between product and reference product, rather than re-establishing safety and effectiveness, the data required to bolster biosimilarity is quite extensive (15). Information demonstrating biosimilarity must include, among other things, adequate data from analytical studies (structural and functional tests), animal studies (including assessment of toxicity), and clinical studies including pharmacokinetics and pharmacodynamics to demonstrate safety, purity, and potency in the indicated conditions for use in which the reference product is licensed for. Approval of a biosimilar product is based on a comprehensive review of evidence submitted by the applicant to provide an overall assessment that the proposed product is biosimilar to or interchangeable with the reference product (16).

6.5 BIOSIMILAR PRODUCT DEVELOPMENT

Biosimilar product development (BPD) has a stepwise approach to eliminate residual uncertainty regarding quality attribute similarity to the reference product. Overall, the amount of nonclinical and clinical data required depends on the residual uncertainty remaining after evaluating nonclinical and the chemistry (CMC) data (17). Since safety and effectiveness are proven in the reference product's development program, it is assumed that the biosimilar product is equally as safe and effective as the reference product, given that biosimilarity is demonstrated. Thus, the goal of BPD is to demonstrate biosimilarity to the reference product and show that there are no significant clinical differences from the reference product (8).

The BPCI Act changed the biological product approval pathway to ensure that all biological products, except chemically synthesized polypeptides, are subject to BLA under the PHSA (18). The BPCIA also specifies that biosimilar sponsors applying for a BLA under 351(k) need to meet the following conditions (19).

1. The biosimilar product's similarity to the reference product must be supported by analytical, animal, and clinical studies.
2. The reference and biosimilar need to have the same mechanism of action.
3. The conditions of use, for labeling purposes, must be the same as those used previously by the reference product.
4. The biosimilar and reference product administration route, dosage form, and strength need to be the same.
5. The manufacturing facility must meet cGMP standards.

BPD components include three main types of studies:

1. Analytical studies demonstrating similarity to reference product.
2. Animal studies which include toxicity studies.
3. Clinical studies and other studies for immunogenicity, pharmacodynamics, and pharmacokinetics in order to demonstrate that the biosimilar product's safety, purity, and potency are similar to that of the reference product.

The International Conference on Harmonization (ICH) aims to achieve global harmonization to ensure safe, effective, and high-quality medicines. The ICH provides guidelines for product development and its guidelines for biological products also apply to BPD. For biosimilars, one important ICH guideline is ICH Q5E: *Comparability of Biotechnological/Biological Products Subject to Changes in Their Manufacturing Process*. This states that since the biosimilar sponsor has no insight on the reference product license holder's manufacturing process, it is important to address the impact of manufacturing differences on the safety, purity, and potency of the biosimilar through the use of analytical tests, functional assays, animal and clinical studies (20).

6.6 PERIODS OF EXCLUSIVITY

The periods of exclusivity for biologics under the PHSA are generally longer than those for chemical drugs under the FFDCA, which detail a five-year new chemical entity exclusivity and three-year new clinical study exclusivity. The BPCIA imposes two periods of exclusivity applicable to brand-name biologics to regulate the ability for competitors to reference the data generated by name-brand manufacturers. Section 351(k)(7) of the PHSA, entitled "Exclusivity for Reference Product", instructs that the BLA for a biosimilar or interchangeable must wait until four years after the licensure of the reference product before submission, and that the FDA may not approve a BLA for a biosimilar or interchangeable product until 12 years after the date of licensure of the reference product (21).

Biologic products submitted between the 10-year transition period between BPCIA enactment on March 23, 2010, and March 23, 2020, are not eligible for the 12-year exclusivity period because they were licensed under Section 505 of the FEDCA without intention to obtain a 12-year exclusivity upon subsequent license approval under Section 351(a) of the PHSA.

An additional six-month period of exclusivity may be added to the 12- and 4-year periods if the sponsor conducts pediatric studies that meet the requirements for pediatric exclusivity period pursuant to Section 505A of the FFDCA (22). Additionally, a biosimilar product with a reference product with a protected orphan indication may not obtain licensure until after the expiration of the 7-year orphan drug exclusivity period or the 12-year exclusivity period, which is later (14).

6.7 PURPLE BOOK

Originally, the Purple Book consisted of two lists of FDA-licensed biological products regulated by the CDER and the CBER. This resource has since transitioned into an online database that can be accessed through the FDA's website. The database contains information about biological products, biosimilars, and interchangeable biological products licensed by the FDA under the PHSA, along with their BLA Type, licensure and patent information, and product labeling and naming details. Currently, the Purple Book database contains information on all FDA-licensed biological products regulated by CDER and FDA-licensed allergenic, cellular and gene therapy, hematologic, and vaccine products regulated by CBER.

6.8 LABELING OF BIOSIMILARS

As a part of the BLA, sponsors are required to submit a draft labeling, along with the proposed nonproprietary name (23). For biosimilar labeling, details can be found within the FDA's *Guidance for Industry: Labeling for Biosimilar Products.*

The product label should include all relevant data and information from the reference product label, including any clinical data that supports the findings of safety and effectiveness for the reference product (24). The label needs to use the proprietary name when referring to the biosimilar, explain the meaning of a biosimilar, include a statement that indicates that the product is a biosimilar, and name the reference product (8). Since the benefit-risk profile for the reference product is relevant to the biosimilar, the core name (INN) followed by "products" should be included in the benefit-risk profile (25). It is also important to be clear about the reference product indication for which biosimilar has been approved and include any differences in administration, preparation, storage, or safety that do not impact the biosimilarity demonstration (26). Any immunogenicity information should be included in the "Adverse Reactions" subsection titled "Immunogenicity". It should be noted that clinical studies of the proposed biosimilar are not included within the label because these studies compare the safety and efficacy of the biosimilar to the reference product, rather than demonstrating safety and efficacy (27).

6.9 NAMING OF BIOSIMILARS

For biosimilars, product naming is the same as biological products and more details can be found within the *Guidance for Industry: Nonproprietary Naming of Biological Products.* The nonproprietary naming system creates a pharmacovigilance system, which allows product tracking back to the manufacturer for adverse events and the detection of safety signals in the product life cycle. It also prevents accidental switching between reference and biosimilar products (including off-label use for unapproved indication), and it ensures patient safety and safe use of the products.

The nonproprietary name consists of a combination of the core name and a hyphenated distinguishable suffix, and this applies to products under BLA Type 351(a) or 351(k). The core reflects certain scientific product characteristics to help healthcare providers identify the drug substance (*Guidance for Industry: Nonproprietary Naming of Biological Products*) (28).

6.10 BIOSIMILAR USER FEE

The Biosimilar User Fee Amendments of 2017 (BsUFA), enacted as Title IV of Food and Drug Administration Safety and Innovation Act (FDASIA), authorized the FDA to assess and collect user fees to expedite the review of biosimilar product applications submitted under Section 351(k) of the PHSA (29). The biosimilar user fee program was renewed (BsUFA II) to authorize fee collection from October 2017 through September 2022. Under BsUFA II, BPD fees include the initial BPD fee, the annual BPD fee, and the reactivation fee (30).

6.11 EMERGENCY USE AUTHORIZATION

Emergency Use Authorization (EUA) provides a mechanism to facilitate the availability and use of medical countermeasures during public health emergencies. Under Section 564 of the Federal Food, Drug, and Cosmetic Act (FEDCA), when the Secretary of Department of Health and Human Services (HHS) declares an EUA is appropriate, the FDA can authorize the use of unapproved medical products or unapproved uses of approved medical products to diagnose, treat, or prevent serious or life-threatening diseases or conditions when certain criteria are met, including that there are no adequate, approved, and available alternatives (31).

The HHS declaration to support a EUA must be justified based on one of four determinants: (1) A domestic emergency or potential for domestic emergency determined by the Secretary of Homeland Security; (2) A material threat pursuant to Section 319F-2 of the PHSA sufficient to affect national security identified by the Secretary of Homeland Security; (3) A military emergency or potential for such determined by the Secretary of Defense; (4) A public health emergency or potential for such determined by the Secretary of HHS. In the context of the recent COVID-19 pandemic, several EUAs have permitted the use of biological products to treat COVID-19 and serious conditions caused by COVID-19, which include bamlanivimab, remdesivir, convalescent plasma, and others that may be found on the FDA website (32).

The EUA was initially introduced after the events of September 11, 2001 and subsequent anthrax mail attacks, when the Project Bioshield Act of 2004 was passed (33). This Act called for emergency countermeasures for a bioterror attack and to be able to act rapidly in an emergency; it allowed the FDA to authorize formally unapproved products for emergency use against a threat to public health and safety. FDA's EUA authority was used relatively sparingly at first, in which its most extensive use was in combating the H1N1 swine flu pandemic of 2009 by authorizing medical equipment and existing influenza drugs such as Tamiflu, peramivir, and Relenza (34, 35). It was also used to authorize occasional countermeasures in anticipation of MERS, Ebola, Zika, and other epidemics.

Recently, the FDA Advisory Committee has convened multiple times to assess EUA for vaccines for the prevention of coronavirus disease 2019 (COVID-19). On December 11, 2020, the FDA issued a EUA for the Pfizer-BioNTech COVID-19, allowing the vaccine to be distributed in the United States. Following this, EUAs were issued for the Moderna and Janssen COVID-19 vaccines and on December 18, 2020, and February 27, 2021, respectively.

To issue a EUA for a vaccine, there needs to be adequate manufacturing information to ensure quality and consistency and the FDA needs to determine whether the known and potential benefits outweigh the known and potential risks. A EUA request for a vaccine can be submitted on a final analysis of phase 3 clinical efficacy trial. A EUA submission needs to include over 3,000 vaccine recipients, representing a high proportion of participants enrolled in the phase 3 study, who have been followed for serious adverse events for at least one month after completion of the vaccination regimen. The FDA's evaluation of a EUA request for a vaccine also includes evaluation of the chemistry, manufacturing, and controls information and sufficient data should be submitted to ensure the quality and consistency of the product.

REFERENCES

1. Food and Drug Administration (2018). "Definition of the Term 'Biological Product' Proposed Rule Preliminary Regulatory Impact Analysis". Silver Spring, MD: U.S. Food and Drug Administration. https://www.fda.gov/about-fda/reports/economic-impactanalyses-fda-regulations
2. Section 351(i)(1) of the Public Health Services Act, 42 U.S.C. § 262. Accessed 22 June 2021. https://www.govinfo.gov/content/pkg/USCODE-2010-title42/html/USCODE-2010-title42-chap6A-subchapII-partF-subpart1-sec262.htm
3. Section 201(g) of the Federal Food, Drug, and Cosmetic Act, 21 U.S.C § 321. Accessed 22 June 2021. https://www.govinfo.gov/content/pkg/COMPS-973/pdf/COMPS-973.pdf
4. "Assignment of Agency Component for Review of Premarket Applications", Final Rule, Federal Register, vol. 56, no. 225, November 21, 1991, pp. 58754–58758.
5. Science and the Regulation of Biological Products. https://www.fda.gov/about-fda/histories-product-regulation/science-and-regulation-biological-products
6. David, M. Dudzinski, "Reflections on Historical, Scientific, and Legal Issues Relevant to Designing Approval Pathways for Generic Versions of Recombinant Protein-Based Therapeutics and Monoclonal Antibodies", Food and Drug Law Journal, 2005, vol. 60, pp. 143–260.
7. Anna Wilde, Mathews and Jeanne, Whalen, "FDA Clears Copycat Version Of Human Growth Hormone", The Wall Street Journal, June 1, 2006, and Anna Wilde Mathews, "FDA Is Ordered to Rule on Generic Biotech Drug", The Wall Street Journal, April 15, 2006.
8. Section 351(i)(2) of the Public Health Services Act, 42 U.S.C. § 262. Accessed 22 June 2021. https://www.govinfo.gov/content/pkg/USCODE-2010-title42/html/USCODE-2010-title42-chap6A-subchapII-partF-subpart1-sec262.htm
9. Section 505(j) of the Federal Food, Drug, and Cosmetic Act, 21 U.S.C § 321. Accessed 22 June 2021. https://www.govinfo.gov/content/pkg/COMPS-973/pdf/COMPS-973.pdf.
10. Shuren, J. "Drug and Biological Product Consolidation," Federal Register, June 26, 2003, vol. 68, no. 123, pp. 38067–38068. Accessed 22 June 2021. https://www.federalregister.gov/documents/2003/06/26/03-16242/drug-and-biological-product-consolidation
11. Investigational New Drug (IND) Application. https://www.fda.gov/drugs/types-applications/investigational-new-drug-ind-application
12. Procedural Guidance for Industry. Reference Product Exclusivity for Biological Products Filed Under Section 351(a) of the PHS Act. August 1, 2014.
13. Section 351(m) of the Public Health Services Act, 42 U.S.C. § 262. Accessed 22 June 2021. https://www.govinfo.gov/content/pkg/USCODE-2010-title42/html/USCODE-2010-title42-chap6A-subchapII-partF-subpart1-sec262.htm
14. Section 7002(h) of the Patient Protection and Affordable Care Act, 42 U.S.C. § 1301. Accessed 22 June 201. https://www.govtrack.us/congress/bills/111/hr3590/text.
15. Section 7002(f) of the Patient Protection and Affordable Care Act, 42 U.S.C. § 1301. Accessed 22 June 201. https://www.govtrack.us/congress/bills/111/hr3590/text.
16. FDA, Interpretation of the "Deemed to be a License" Provision of the Biologics Price Competition and Innovation Act of 2009 Guidance for Industry, December 2018, pp. 9–10, https://www.fda.gov/downloads/Drugs/ GuidanceComplianceRegulatoryInformation/Guidances/UCM490264.pdf
17. FDA, Guidance for Industry: Scientific Considerations in Demonstrating Biosimilarity to a Reference Product. Accessed 22 June 2021. https://www.fda.gov/media/82647/download

18. PHSA. Sec. 351(a). Section 351(a) of the Public Health Services Act, 42 U.S.C. § 262. Accessed 22 June 2021. https://www.govinfo.gov/content/pkg/USCODE-2010-title42/html/USCODE-2010-title42-chap6A-subchapII-partF-subpart1-sec262.htm.
19. Biologics Price Competition and Innovation Act of 2009, Section 351(k). Available from the FDA website https://www.fda.gov/media/78946/download. Accessed 30 June 2021.
20. Guidance for Industry: Q5E Comparability of Biotechnological/Biological Products Subject to Changes in Their Manufacturing Process. Accessed 22 June 2021. https://www.fda.gov/media/71489/download.
21. Section 7002(b)(3) of the Affordable Care Act, adding Section 351(i)(4) of the PHS Act. "Compilation of Patient Protection and Affordable Care Act" Section 7002(a) of the Affordable Care Act, Public Law 111-148, June 9, 2010, pp. 750, adding Section 351(k)(7) of the Public Health Services Act, 42 U.S.C. § 262. Accessed 22 June 2021. https://www.hhs.gov/sites/default/files/ppacacon.pdf.
22. Section 7002(g) of the Affordable Care Act, adding Section 351(m) of the PHS Act. "Compilation of Patient Protection and Affordable Care Act" Section 7002(g) of the Affordable Care Act, Public Law 111-148, June 9, 2010, pp. 762–763, adding Section 351(m) of the Public Health Services Act, 42 U.S.C. § 262. Accessed 22 June 2021. https://www.hhs.gov/sites/default/files/ppacacon.pdf.
23. PHSA. Sec. 351(a)(1) of the Public Health Services Act, 42 U.S.C. § 262. Accessed 22 June 2021. http://www.govinfo.gov/content/pkg/USCODE-2010-title42/html/USCODE-2010-title42-chap6A-subchapII-partF-subpart1-sec262.htm.
24. PHSA. Sec 351(k)(2)(iii). Biologics Price Competition and Innovation Act of 2009, Public Health Services Act Section 351(k)(2)(iii), 42 U.S.C. § 262(k). Accessed 22 June 2021. https://www.fda.gov/media/78946/download.
25. PHSA. Sec 351(k)(5)(C). Biologics Price Competition and Innovation Act of 2009, Public Health Services Act Section 351(k)(5)(C), 42 U.S.C. § 262(k). Accessed 22 June 2021. https://www.fda.gov/media/78946/download.
26. PHSA. Sec 351(k)(4). Biologics Price Competition and Innovation Act of 2009, Public Health Services Act Section 351(k)(4), 42 U.S.C. § 262(k). Accessed 22 June 2021. https://www.fda.gov/media/78946/download
27. FDA, Guidance for Industry: Labeling for Biosimilar Products, July 2018, pp. 8–9. Accessed 22 June 2021. https://www.fda.gov/media/96894/download.
28. FDA, Guidance for Industry: Nonproprietary Naming of Biological Products, January 2017, pp. 4–10. Accessed 22 June 2021. https://www.fda.gov/media/93218/download.
29. Thaul, Susan, Bagalman, Erin, Corby-Edwards, Amalia K., Glassgold, Judith M., Johnson, Judith A., Lister, Sarah A., and Sarata, Amanda K. "Title IV of the Food and Drug Administration Safety and Innovation Act (FDASIA, P.L. 112–144)", CRS Report R42680, February 4, 2013, p. 32.
30. Sarata, Amanda K., Dabrowska, Agata, Johnson, Judith A., and Thaul, Susan. "FDA Reauthorization Act of 2017 (FDARA, P.L. 115-152)", CRS Report R44961, September 21, 2017, pp. 9–10. Accessed 22 June 2021. https://sgp.fas.org/crs/misc/R44961.pdf
31. Azar II, Alex M. "Emergency Use Authorization Declaration", Federal Register, vol. 85, April 1, 2020, pp. 18250–18251. Accessed 22 June 2021. https://www.federalregister.gov/documents/2020/04/01/2020-06905/emergency-use-authorization-declaration
32. Azar II, Alex M. "Determination of Public Health Emergency", Federal Register, vol. 85, February 4, 2020, pp. 7316–7317. Accessed 22 June 2021. https://www.federalregister.gov/documents/2020/02/07/2020-02496/determination-of-public-health-emergency
33. Crawford, Lester M. and Shuren, Jeffrey. "Authorization of Emergency Use of Anthrax Vaccine Adsorbed for Prevention of Inhalation Anthrax by Individuals at Heightened Risk of Exposure Due to Attack with Anthrax; Availability", Federal Register, February 2, 2005. Accessed 22 June 2021. https://www.federalregister.gov/documents/2005/02/02/05-2028/authorization-of-emergency-use-of-anthrax-vaccine-adsorbed-for-prevention-of-inhalation-anthrax-by
34. Horowitz, David. "Authorization of Emergency Use of the Antiviral Product Peramivir Accompanied by Emergency Use Information; Availability", Federal Register, November 2, 2009. Accessed 22 June 2021. https://www.federalregister.gov/documents/2009/11/02/E9-26291/authorization-of-emergency-use-of-the-antiviral-product-peramivir-accompanied-by-emergency-use
35. Lutter, Randall W. "Authorizations of Emergency Use of Certain Antiviral Drugs – Zanamivir and Oseltamivir Phosphate; Availability", Federal Register, August 4, 2009. Accessed 22 June 2021. https://www.federalregister.gov/documents/2009/08/04/E9-18568/authorizations-of-emergency-use-of-certain-antiviral-drugs-zanamivir-and-oseltamivir-phosphate

7 The Development of Biologics from the Perspective of the Pharmaceutical Industry

Mohammad Tauseef, Yahya B. Bello, Ahmed Abu Fayyad, and Mohd Imran

Corresponding author: Mohammad Tauseef

Contents

7.1 INTRODUCTION

Biological preparations or biologics are therapeutic entities made from living organisms such as humans or animals or microorganisms, including bacteria, fungi, and viruses (1–3). These therapeutics substances are also called vaccines (3, 4). Biologics are complex biological substances, and chemically they are composed of proteins, nucleic acids, carbohydrates, or even a combination of them. Besides vaccines, biologics also include cells and tissues used in transplantation purposes (2, 3, 5).

Biosimilars or follow-on biologics are therapeutic entities, which are like biologics in terms of clinical efficacy; however, their structures could be different from the brand name biologics (3, 6). Brand name

DOI: 10.1201/9780429485626-7

biologics are also termed innovators or reference products. It is important to understand here that biosimilars should not be confused with generic drugs, which have a simple structure, and chemically they are the exact replica of the already-approved brand therapeutic drug (5). Since biosimilars are much more economical compared to their brand biologics, the global market for biosimilars is continuously growing, and it is expected to expand further by about 25% by 2026 compared to its value in 2020 (5).

Since biologics and biosimilars are of biological origin, they require special handling and processing to prevent their degradation and contamination from microbes (7, 8). More importantly, strict procedures and guidelines must be followed due to the reason that they are administered directly into the bloodstream of the patients via infusion or injections (7). Minute variability in their manufacturing, storing, or handling during the administration may lead to severe consequences, including disability and death in the patients. Therefore, biologics are referred to as specialty medications in the clinical setting (9).

The cost of biologics is very high. For example, the cost of the Soliris® (eculizumab) and Vimizim® (elosulfase alfa) in the United States is more than 250,000 dollars per patient per year (1, 10). The cost of biologics is higher in the United States compared to that in Canada or Europe. The reason for price reductions in Europe is the introduction of biosimilars. For instance, in 2018, AbbVie launched several biosimilars of its top-selling biological antibody-based drug, Humira, in Europe. That led to about 80% price reduction in the drug compared to its brand medication (1, 10).

7.2 REGULATION OF BIOLOGICS BY THE FDA

The National Institutes of Health (NIH) originally regulated biologics (1). However, in 1972, the regulatory responsibilities were transferred to the Food and Drug Administration (FDA) (1, 11). Under the Public Health Service Act (PHSA), biologics and biosimilars can be approved for US marketing (11, 12) following pathways outlined in §351(a) and §351(k), respectively. The biotech industry submits the Biologics License Application (BLA) to the FDA under regulatory mechanisms outlined in 21 Code of Federal Regulations Parts 600-800 (21 CFR 600-680). Within the FDA, the responsibilities to regulate biologics are shared by both the Center for Biologics Evaluation and Research (CBER) and the Center for Drug Evaluation and Research (CDER) (11, 13). More importantly, traditional biologics such as blood and blood products, vaccines, allergenic extracts, test kits, and certain devices are regulated by CBER. Besides, CBER is also responsible for regulating other specialty products, including cellular therapy products, gene therapy products, human tissue used in transplantation, and the tissue used in xenotransplantation – the transplantation of nonhuman cells, tissues, or organs into a human (14). Furthermore, in addition to regulating prescription brand-name and generic drugs and over-the-counter drugs, CDER also regulates most therapeutic biologics (14, 15). In 2003, the CBER transferred its responsibility to regulate the therapeutic biologics to CDER (14, 15). Here is a brief description of the therapeutic biologics which are currently regulated by the CDER (11, 15).

i. *Monoclonal antibodies*: Monoclonal antibodies are proteins that have a binding affinity to a specific substance in the body or on a specific cell (16–18). A monoclonal antibody may sometimes carry a drug or toxin to kill tumor cells, which are otherwise hard to target (16, 17). An example of the most common monoclonal antibody drug available in the US market is infliximab. Infliximab is a paradigm shift in the management of Crohn's disease, rheumatoid arthritis, ulcerative, psoriasis, and colitis (17, 19).
ii. *Growth factors*: Growth factors such as hormones are made by the body and they control cell division and cell survival (20). An example of a growth factor is the human growth hormone somatropin (21).
iii. *Cytokines*: Cytokines are proteins that activate or inhibit the immune system (22). They are proinflammatory and some are autoinflammatory in nature (21). They also fight cancer and infections. Interleukins, interferons, and colony-stimulating factors, such as filgrastim, are a few examples of cytokines (21).

iv. *Immunomodulators*: As the name suggests, immunomodulators are substances that stimulate or suppress the immune system (22, 23). By modulating the immune system, they prepare the body to fight cancer, infection, or other inflammatory diseases (22).
v. *Enzymes*: They are also proteins that catalyze the biochemical reactions in the living system (24). Enzymes control prominent cell functions and processes, such as cell growth, cell signaling, and cell division (25). In the treatment of certain types of cancers, inhibiting the activity of the enzymes blocks cancer growth and metastatic (24). Enzyme inhibitors are the emerging therapeutic to manage cancers (24, 25).

Chemical drugs are approved for marketing by the FDA via new drug application (NDA) or abbreviated new drug application (ANDA) (26, 27). Likewise, most biologics are also regulated and licensed for marketing by the FDA (18, 27, 28). However, it is done via a BLA under the PHSA (18, 28). To obtain the licensure, the manufacturer must demonstrate that the biologic product and the facility in which the biologic will be manufactured, packaged, and stored meet the standards and the product will remain safe, pure, and potent (18). As consistent with other FDA-approved products for the marketing, if the manufacturer does any changes in the manufacturing process, such as changes in the supplier of the raw material, or installation or replacement of the manufacturing equipment, it requires to demonstrate the FDA that the product quality and effectiveness are same before and after any changes in the manufacturing processes (28).

Before 1941, certain biological products were regulated as drugs and were approved by the FDA through NDA under the Federal Food, Drug, and Cosmetic Act (FFDCA) rather than as a biological product under the PHSA (29, 30). However, in 1941, Congress gave authority to the FDA to market insulin, a natural biological product, as it was extracted from animal sources (30). Even though there was a similarity between insulin and other biologics, the FDA regulated insulin as a drug. Biologics include hormones such as human growth hormone, hormones to treat infertility, glucagon, and hormones used clinically for the management of osteoporosis menopause (30). Started in late 70s and early 80s, there was development if biologics were for human therapeutic use, such as recombinant proteins and monoclonal antibodies (31). Some of these products, like insulin and growth hormones (created by recombinant technology), were regulated by the FDA under the FFDCA, while others were regulated by FDA's PHSA (31). Examples of PHSA-regulated biologics are blood factors, proteins involved in the immune response, and cytokines (31, 32). Till March 2020, most biologics were regulated by the PHSA, and some were approved as drugs under the FFDCA (1). After March 2020, when applications for biologics were approved under the FFDCA, they were deemed to be licensed under the PHSA (1, 30, 31).

In 2010, the Biologics Price Competition and Innovation Act (BPCIA) was passed as Title VII of the Patient Protection and Affordable Care Act (ACA, P.L. 111-148) (33). This act, for the first time, created a pathway under the PHSA for the approval of biosimilars (33, 34). As we mentioned earlier, a biosimilar is a biologic that is clinically highly similar but not identical compared to the FDA-approved biological product or a reference product (34). Generic drugs have been marketing in the United States since 1984 (35). This happened due to the enactment of the Drug Price Competition and Patent Term Restoration Act of 1984 (P.L. 98-417), which is famously also called the Hatch-Waxman Act (35). Due to this act, there was an establishment of the abbreviated approval pathway for generic chemical drugs (35). Thus, ultimately, this led to the lowering of the prices of the medications by providing the viable alternative to expansive brand medications. And it boosted the expansion and market of generic drugs in the United States (36). The significant price reduction (60%–90%) to generic drugs happened due to the manufacturers saving a huge amount of money by avoiding the initial drug research and development cost as well as the expensive clinical trials incurred on the development of the brand drug products (37). To obtain the approval on generic drugs, manufacturers need to submit to the FDA an ANDA, which demonstrates that the generic products are pharmaceutically equivalent to the brand name drug and at the same time are demonstrated to be bioequivalent to the brand drug (37). Furthermore, the generics should also have the same active pharmaceutical ingredient, dosage form, route of administration, as well as strength compared to the reference product (38–39).

It is also worth mentioning here that the Hatch-Waxman Act also established a second abbreviated pathway, called the 505(b)(2) pathway (36). This pathway is used to approve biosimilars that had received approval under the FFDCA rather than licensure under the PHSA (36). Now the question arises, how a 505(b)(2) NDA differs from ANDA? A 505(b)(2) NDA contains complete details on the safety and effectiveness of the products, and some of the information that it is based upon for approval comes from other studies, which were not investigated by or for the manufacturer of the medication (40). In contrast to ANDA, A 505(b)(2) NDA may rely on published articles and/or on FDA's finding of safety and effectiveness for the brand drug product (41).

The main purpose of the Hatch-Waxman Act is to provide a process for the approval of the generic medications as well as some biosimilars under FFDCA (36). It is important to note that this act does not provide any mechanism of approval on biosimilars under the PHSA (36).

In June 2006, the FDA announced the first approval of a biosimilar of Omnitrope via the mechanism of 505(b)(2, 42, 43). This FDA decision was followed by the ruling of the US District Court for the District of Columbia. In its ruling, the court said that the FDA should move forward on the application submitted by Sandoz for the approval of the Omnitrope. Omnitrope is the biosimilar of Pfizer's Genotropin (42–44).

7.3 NEW REGULATORY PATHWAY FOR THE APPROVAL OF THE BIOSIMILARS

In the year of 2010, the BPCIA enacted the Title VII of the ACA and thereby set up a regulatory authority for the FDA and formed an abbreviated licensure pathway in Section 351(k) of the PHSA (45, 46). This pathway solely maintained the approval of the biosimilars, which are the products highly like the FDA-licensed brand biologics (35, 45). Besides, the ACA also instructed the FDA to submit the Congress recommendations for a user fee program to back the review of biosimilar product applications, which are submitted under Section 351(k) of the PHSA (46) – the Biosimilar User Fee Act of 2012 (BsUFA), which is passed as the Title IV of Food and Drug Administration Safety and Innovation Act (FDASIA, P.L. 112-144) (47). This act authorizes the FDA to assess and collect the fee for reviewing biosimilar applications (35, 46, 47).

Under Section 351(k) of the PHSA, when a biopharmaceutical or a biotechnology company wants to market a biosimilar product in the United States, the prerequisites are that the company should submit a list of documents to the FDA (48–50). These requirements are toxicological and clinical studies and biosimilarity data based on structural and functional studies. Based on the submitted data, the FDA will decide if the product is "highly similar" compared to the reference brand biological in terms of there being no clinically meaningful differences between them (49). The agency may accept negligible differences in the clinically inactive ingredients that have no effect on the efficacy of the product (49). However, the biosimilar must demonstrate no variability in terms of safety, purity, and potency with reference to the reference product (48, 50). Furthermore, to be considered interchangeable, the biosimilar should produce the exact same clinical effect in the patient compared to the reference brand biological product (50). The interchangeable product must be substituted for the brand product by the pharmacist without any intervention from the prescribing healthcare provider (48, 50).

7.4 BIOSIMILARS: APPROVAL AND MARKETING

The market of biosimilars is expanding rapidly (30, 51). Although about 2% of Americans use biologics, in terms of total spending, these products represent about 40% of total prescription drugs (48). As of July 2020, 28 biosimilars have been approved by the FDA (30, 52). Even after being approved by the FDA, many of these biosimilars are still not available to treat patients in the clinic due to ongoing litigation and settlement agreements (48). After approval and launching of the biosimilars in the market, there will always be factors identified which potentially limit biosimilar competition (51, 52). These potential factors include conventions of biosimilar naming and labeling, requirements of interchangeability, and even access to samples for biosimilars testing. The FDA has shed light on how to address these issues in their 2018 Biosimilars Action Plan (30, 48).

7.5 PATENT LITIGATION AND SETTLEMENTS

It has been regularly observed that the launch of several biosimilars is delayed due to the ongoing patent litigation and settlements (53, 54). This has been happening between the brand biological products and the biosimilar companies. For instance, in the recent past, AbbVie has been the subject of Congressional inquiry for its use of the "patent thicket" to protect its biological product named Humira (adalimumab) from biosimilar competition (1, 55). Data showed that AbbVie filed 247 applications for patents regarding Humira. However, it got 132 patent approvals (1). Humira was licensed in the United States in the year of 2002 (55). Besides, the FDA approved three biosimilars of Humira – Sandoz's Hyrimoz, Boehringer Ingelheim's Cyltezo, and Amgen's Amjevita (1, 56). But none of them are available to the United States (1). AbbVie also delayed the biosimilars developed by various other companies of the Humira to be available in the United States until 2003 (1). Later, concerns were raised on the transparency of granting the patents on biologics and biosimilars as well as on the patents thickets that were designed to delay the entry of the biosimilars in the market by the brand biologics companies (57, 58). For those drug products which were granted approval under FFDCA, pursuant to the Hatch-Waxman Act, it was required that when a manufacturer submit an NDA to the FDA and, as a part of the NDA, the manufacturer must also provide any information that any patent that claims the drug (58). The FDA then later will publish that information in the Approved Drug Products with Therapeutic Equivalence Evaluations, also called an Orange Book (1, 59). The Orange Book is publicly available and is revised every 30 days (1, 59). The FDA also publishes voluntarily (although not required by the law) the information on biologics and biosimilar in the Purple Book (60). The Orange Book is available in paper form as well as in a searchable electronic database. However, the Purple Book is available in two formats (60): one for the biological drugs, which also includes biosimilar and interchangeable products, licensed by CDER, and the other list includes the drugs licensed by the CBER. (59, 60). Furthermore, for the brand-name biological products, the list also identified the drug by the date when they were licensed, and if the FDA assessed the medication for reference drug exclusivity, the date the exclusivity will expire (1).

In 2018, the FDA mentioned that it would further add the information in the Purple Book about the approval of biologicals and this would happen under the FDA's "Biosimilars Action Plan" (30, 61). Besides, the FDA urged the public what further steps the agency should take to provide more information on biologics in Purple Book so that the Purple Book will become more useful to healthcare providers, patients, pharmacists, and manufacturers (61, 62). The FDA received the comments to include detailed and comprehensive information on biologics in the Purple Book, update the book more often, and arrange the Purple Book as a single searchable database for its use by the stakeholders (62). Thus, in the 116th Congress, bipartisan legislation has been introduced to make the Purple Book a single searchable database just like the Orange Book. It was also approved to include the patent information of the biologics in the Purple Book. This provided more patent transparency and thus enhanced the biosimilar competition (1).

7.6 NAMING OF THE BIOSIMILAR

When the patent litigation of the brand biological product is resolved, the biosimilar is introduced into the market. One critical factor which has been identified is the naming of the biosimilar (63, 64). This naming is specifically regarding the nonproprietary name of the biosimilar compared to the brand biological. The nonproprietary name or a proper name of the biosimilar is used on the label of the biosimilar product (64). It is also used for regulation and in the scientific literature to determine the active ingredient of the pharmaceutical product (65). This is a name a drug company will use to market its products and generally, it is capitalized, which is followed by a superscript R in a circle (®). For instance, the product Neupogen® is the proprietary name for the biologic filgrastim, the nonproprietary name for the active substance (65, 66).

The FDA released the detailed guidance in the form of a draft on the nonproprietary naming of the biological products in August 2015 (1, 64, 66). Later, this draft was finalized on January 12, 2017 (1). In March 2019, the FDA further issued revised guidance (67). As per the naming convention, "the nonproprietary name assigned for each the reference product, which is also called an originator biological drug, related biological drug, and a biosimilar drug will be a proper name (1). According to the rule, the proper name will be formed by joining the core name with a distinctly unique suffix that has no meaning and is composed of four lowercase letters (68). In contrast to the suffix, the core name indicates the part that is shared among a reference or originator biological drug and any related biosimilar, biological, or interchangeable product as part of the proper name (68). This concept of naming can be better understood by an example of a core name, filgrastim. Here, there is a core name followed by a hyphen and then a suffix, for instance, filgrastim-abxy. The naming convention discussed here is applied to previously licensed and future biologics and biosimilars. In 2019, the FDA then released its revised guidelines on the naming of the products; they advised that it is no longer mandatory to change or modify the proper name of the biological products if they were previously approved under the PHSA without the FDA-mandated suffix in their name. It was stated that the FDA was also intended to modify the name of the transition biologics such as insulin (69). However, the FDA is now required to apply the naming convention to the biological drugs when they are licensed. This is basically the opposite of the agency's earlier mandate, where it was mentioned to give proper names even to those products which were approved on previous dates. Regarding the interchangeable drugs, the FDA's revised guidelines say that it is appropriate to provide a unique suffix that distinguishes the products from the other products that share the common name (1).

7.7 LABELING OF THE BIOSIMILARS

Next to the naming, which affects the launching of the biosimilars in the market, is the labeling of the biosimilars. Labeling is done based on the policy of the FDA, more specifically, the inclusion part of a statement of a biosimilarity (1, 70). Labeling the drugs is necessary as it provides information to the healthcare providers about the safety and effectiveness of the medication in the patients. This helps the providers to decide if these products can be prescribed to a patient/s or not. In 2006, the FDA issued its final guidelines on the format as well content of the labeling for the prescription medications as well biological drugs (1, 70). The FDA mandated that the labeling should begin with the warning information on the medications. Labeling should also provide detailed information on usage, indications, dosage, dosage form, contraindications, strength of the medication, precautions, warnings, clinical pharmacology, toxicity, use of the drug in a specific population, adverse drug reactions, drug abuse, overdosage, drug dependence, references, patient counseling, and storage conditions of the medication.

In March 2016, the FDA initially released a draft of guidelines on the labeling of biosimilars. Later, in July 2018, the agency finalized the guidelines (71). In these labeling guidelines, the FDA recommended that labeling contains a "Biosimilarity Statement". This statement explains the relationship of a biosimilar drug to its reference drug. For instance, "Nivestym® (filgrastim-aafi)" is a biosimilar product to NEUPOGEN (filgrastim) (1, 70, 72). This will be followed by the following statement:

> *Biosimilar means that the biological product is approved based on data demonstrating that it is highly like an FDA-approved biological product, known as a reference product, and that there are no clinically meaningful differences between the biosimilar product and the reference product. Biosimilarity of [BIOSIMILAR PRODUCT'S PROPRIETARY NAME] has been demonstrated for the condition(s) of use (e.g., indication(s), dosing regimen(s)), strength(s), dosage form(s), and route(s) of administration described in its Full Prescribing Information* (1, 71, 72).

7.8 SUBSTITUTION AND INTERCHANGEABILITY

A biosimilar product cannot instantly be chosen for the reference or brand biological product at the pharmacy, thus adding another layer that may affect the uptake of the biosimilar medications (73). In other words, to choose the biosimilar over the brand biologic at the pharmacy, the biosimilar should be interchangeable with the brand product. An interchangeable drug product must be producing exactly the same clinical outcome compared to its brand or reference product in patients (73). Moreover, for the biological product will be administered in the patients more than once, the risk of switching between the biosimilar and its brand product must be more than if the patient will be receiving the brand biological drug. Therefore, it must be clear that the intertangle drug should be substituted by the pharmacist with the involvement of the physician who prescribed the biological drug product (1, 33, 73).

Generic medications are considered interchangeable for their corresponding brand drug products. And as per the states' laws, a pharmacist can substitute generic medications for their brand medication. By following the Hatch-Waxman Act, the pharmacist can easily switch the generic medication for the brand medication to save the cost without the intervention of the prescribing healthcare professional. However, this is not the case with biosimilars because they are not structurally the same as their corresponding reference or brand products. In this case, the interchangeability is a completely different and different process (73).

In the United States, although the FDA regulates the drug products, the pharmacies and the practice of the pharmacy are controlled by the State Governments. As per the National Conference of State Legislatures (NCSL), since October 2018, almost 49 states have considered substituting biosimilars for their brand biologicals through their legislation (1). NCSL further stated that 45 states as well as Puerto Rico now have passed the legislation (74).

REFERENCES

1. Dabrowska A. Biologics and Biosimilars: Background and Key Issues. CRS Report. 2019;1:1–30.
2. Kesselheim AS, Avorn J, Sarpatwari A. The High Cost of Prescription Drugs in the United States: Origins and Prospects for Reform. JAMA. 2016;316(8):858–71.
3. Gotham D, Barber MJ, Hill A. Production Costs and Potential Prices for Biosimilars of Human Insulin and Insulin Analogues. BMJ Glob Health. 2018;3(5):e000850.
4. Morrow T, Felcone LH. Defining the Difference: What Makes Biologics Unique. Biotechnol Healthc. 2004;1(4):24–9.
5. Kallberg C, Hudson J, Salvesen Blix H, Ardal C, Klein E, Lindbaek M, et al. The Effect of Generic Market Entry on Antibiotic Prescriptions in the United States. Nat Commun. 2021;12(1):2937.
6. Raimond VC, Feldman WB, Rome BN, Kesselheim AS. Why France Spends Less Than the United States on Drugs: A Comparative Study of Drug Pricing and Pricing Regulation. Milbank Q. 2021;99(1):240–72.
7. Oliva A, Llabres M. New Quality-Range-Setting Method Based on Between- and Within-Batch Variability for Biosimilarity Assessment. Pharmaceuticals (Basel). 2021;14(6):527.
8. Peeters M, Planchard D, Pegram M, Goncalves J, Bocquet F, Jang H, et al. Biosimilars in an Era of Rising Oncology Treatment Options. Future Oncol. 2021;17(29):3881–3892.
9. Barszczewska O, Piechota A. The Impact of Introducing Successive Biosimilars on Changes in Prices of Adalimumab, Infliximab, and Trastuzumab-Polish Experiences. Int J Environ Res Public Health. 2021;29;18(13):6952.
10. Coghlan J, He H, Schwendeman AS. Overview of Humira(R) Biosimilars: Current European Landscape and Future Implications. J Pharm Sci. 2021;110(4):1572–82.
11. Kuehn CM. A Proposed Framework for Patient-Focused Policy at the U.S. Food and Drug Administration. Biomedicines. 2019;7(3):64.

12. Federal Register, Biological Products Regulated Under Section 351 of the Public Health Services Act; Implementation of Biologics License; Elimination of Establishment License and Product License–FDA. Proposed Rule. Fed Regist. 1998;63(147):40858–71.
13. FDA website; Vaccines, Blood & Biologics, Accessed June 2021.
14. Alexander GC, Ogasawara K, Wiegand D, Lin D, Breder CD. Clinical Development of Biologics Approved by the US Food and Drug Administration, 2003–2016. Ther Innov Regul Sci. 2019;53(6):752–8.
15. Carlson B. FDA: Change Is Good. Biotechnol Healthc. 2004;1(1):27–30.
16. Ovacik M, Lin K. Tutorial on Monoclonal Antibody Pharmacokinetics and Its Considerations in Early Development. Clin Transl Sci. 2018;11(6):540–52.
17. Ecker DM, Jones SD, Levine HL. The Therapeutic Monoclonal Antibody Market. MAbs. 2015;7(1):9–14.
18. Kaplon H, Reichert JM. Antibodies to Watch in 2021. MAbs. 2021;13(1):1860476.
19. Khanna R, Sattin BD, Afif W, Benchimol EI, Bernard EJ, Bitton A, et al. Review Article: A Clinician's Guide for Therapeutic Drug Monitoring of Infliximab in Inflammatory Bowel Disease. Aliment Pharmacol Ther. 2013;38(5):447–59.
20. Kraemer WJ, Ratamess NA, Hymer WC, Nindl BC, Fragala MS. Growth Hormone(s), Testosterone, Insulin-Like Growth Factors, and Cortisol: Roles and Integration for Cellular Development and Growth with Exercise. Front Endocrinol (Lausanne). 2020;25(11):33.
21. Devesa J, Almenglo C, Devesa P. Multiple Effects of Growth Hormone in the Body: Is It Really the Hormone for Growth? Clin Med Insights Endocrinol Diabetes. 2016;9:47–71.
22. Savic S, Caseley EA, McDermott MF. Moving Towards a Systems-Based Classification of Innate Immune-Mediated Diseases. Nat Rev Rheumatol. 2020;16(4):222–37.
23. Turner MD, Nedjai B, Hurst T, Pennington DJ. Cytokines and Chemokines: At the Crossroads of Cell Signalling and Inflammatory Disease. Biochim Biophys Acta. 2014;1843(11):2563–82.
24. Duronio RJ, Xiong Y. Signaling Pathways That Control Cell Proliferation. Cold Spring Harb Perspect Biol. 2013;5(3):a008904.
25. Acharya S, Yao J, Li P, Zhang C, Lowery FJ, Zhang Q, et al. Sphingosine Kinase 1 Signaling Promotes Metastasis of Triple-Negative Breast Cancer. Cancer Res. 2019;79(16):4211–26.
26. Chazin H, Woo J, Han J, Grosser S, Luan J. FDA's Generic Drug Program: Decreasing Time to Approval and Number of Review Cycles. Ther Innov Regul Sci. 2020;54(4):758–63.
27. Ciociola AA, Cohen LB, Kulkarni P, Gastroenterology FD-RMCotACo. How Drugs Are Developed and Approved by the FDA: Current Process and Future Directions. Am J Gastroenterol. 2014;109(5):620–3.
28. Krause S, Lacana E, Welch J, Shapiro M, Downey C, Chung J, et al. PDA Biosimilars Workshop Report (September 27–28, 2018)-Getting It Right the First Time for Biosimilar Marketing Applications. PDA J Pharm Sci Technol. 2019;73(4):401–16.
29. George K, Woollett G. Insulins as Drugs or Biologics in the USA: What Difference Does It Make and Why Does It Matter? BioDrugs. 2019;33(5):447–51.
30. Gherghescu I, Delgado-Charro MB. The Biosimilar Landscape: An Overview of Regulatory Approvals by the EMA and FDA. Pharmaceutics. 2020;13(1):48.
31. Kesik-Brodacka M. Progress in Biopharmaceutical Development. Biotechnol Appl Biochem. 2018;65(3):306–22.
32. Lu RM, Hwang YC, Liu IJ, Lee CC, Tsai HZ, Li HJ, et al. Development of Therapeutic Antibodies for the Treatment of Diseases. J Biomed Sci. 2020;27(1):1.
33. Koyfman H. Biosimilarity and Interchangeability in the Biologics Price Competition and Innovation Act of 2009 and FDA's 2012 Draft Guidance for Industry. Biotechnol Law Rep. 2013;32(4):238–51.
34. Ventola CL. Biosimilars: Part 1: Proposed Regulatory Criteria for FDA Approval. P T. 2013;38(5):270–87.
35. Darrow JJ, Avorn J, Kesselheim AS. FDA Approval and Regulation of Pharmaceuticals, 1983–2018. JAMA. 2020;323(2):164–76.
36. Gupta R, Kesselheim AS, Downing N, Greene J, Ross JS. Generic Drug Approvals Since the 1984 Hatch-Waxman Act. JAMA Intern Med. 2016;176(9):1391–3.
37. Gronde TV, Uyl-de Groot CA, Pieters T. Addressing the Challenge of High-Priced Prescription Drugs in the Era of Precision Medicine: A Systematic Review of Drug Life Cycles, Therapeutic Drug Markets and Regulatory Frameworks. PLoS One. 2017;12(8):e0182613.
38. Al-Jazairi AS, Bhareth S, Eqtefan IS, Al-Suwayeh SA. Brand and Generic Medications: Are They Interchangeable? Ann Saudi Med. 2008;28(1):33–41.
39. Dunne S, Shannon B, Dunne C, Cullen W. A Review of the Differences and Similarities between Generic Drugs and Their Originator Counterparts, Including Economic Benefits Associated with Usage of Generic Medicines, Using Ireland as a Case Study. BMC Pharmacol Toxicol. 2013;14:1.

40. Freije I, Lamouche S, Tanguay M. Review of Drugs Approved via the 505(b)(2) Pathway: Uncovering Drug Development Trends and Regulatory Requirements. Ther Innov Regul Sci. 2020;54(1):128–38.
41. Salminen WF, Wiles ME, Stevens RE. Streamlining Nonclinical Drug Development Using the FDA 505(b)(2) New Drug Application Regulatory Pathway. Drug Discov Today. 2019;24(1):46–56.
42. Bhardwaj KK, Bangarurajan K, Naved T, Rajput SK. Perspective, Perceptions, and Promulgation of Biosimilars: A Questionnaire-Based Study to Assess and Understand the Current Challenges of Biosimilars to the Potential and Intended Users. J Pharm Bioallied Sci. 2020;12(2):124–30.
43. Barlas S. Opening the Door to Follow-on Proteins? Biotechnol Healthc. 2006;3(5):47–54.
44. Saenger P. Current Status of Biosimilar Growth Hormone. Int J Pediatr Endocrinol. 2009;2009:370329.
45. Heled Y. The Case for Disclosure of Biologics Manufacturing Information. J Law Med Ethics. 2019;47(4_suppl):54–78.
46. Paradise J. Foreword: Follow-on Biologics: Implementation Challenges and Opportunities. Seton Hall Law Rev. 2011;41(2):501–10.
47. Sharfstein JM. Reform at the FDA-In Need of Reform. JAMA. 2020;323(2):123–4.
48. Federal Register, Biological Products Regulated Under Section 351 of the Public Health Service Act; Implementation of Biologics License; Elimination of Establishment License and Product License; Public Workshop–FDA. Proposed Rule; Notice of Workshop. 1998;63(154):42773–4.
49. Lemery SJ, Ricci MS, Keegan P, McKee AE, Pazdur R. FDA's Approach to Regulating Biosimilars. Clin Cancer Res. 2017;23(8):1882–5.
50. Ho RJ. Midyear Commentary on Trends in Drug Delivery and Clinical Translational Medicine: Growth in Biosimilar (Complex Injectable Drug Formulation) Products within Evolving Collaborative Regulatory Interagency (FDA, FTC, and DOJ) Practices and Enforcement. J Pharm Sci. 2017;106(2):471–6.
51. Bennett CL, Schoen MW, Hoque S, Witherspoon BJ, Aboulafia DM, Hwang CS, et al. Improving Oncology Biosimilar Launches in the EU, the USA, and Japan: An Updated Policy Review from the Southern Network on Adverse Reactions. Lancet Oncol. 2020;21(12):e575–e88.
52. Moore TJ, Mouslim MC, Blunt JL, Alexander GC, Shermock KM. Assessment of Availability, Clinical Testing, and US Food and Drug Administration Review of Biosimilar Biologic Products. JAMA Intern Med. 2021;181(1):52–60.
53. Zhai MZ, Sarpatwari A, Kesselheim AS. Why Are Biosimilars Not Living up to Their Promise in the US? AMA J Ethics. 2019;21(8):E668–78.
54. Pagani E. Why Are Biosimilars Much More Complex Than Generics? Einstein (Sao Paulo). 2019;17(1):eED4836.
55. Lee CC, Najafzadeh M, Kesselheim AS, Sarpatwari A. Cost to Medicare of Delayed Adalimumab Biosimilar Availability. Clin Pharmacol Ther. 2021;110(4):1050–6.
56. Blauvelt A, Leonardi CL, Gaylis N, Jauch-Lembach J, Balfour A, Lemke L, et al. Treatment with SDZ-ADL, an Adalimumab Biosimilar, in Patients with Rheumatoid Arthritis, Psoriasis, or Psoriatic Arthritis: Results of Patient-Reported Outcome Measures from Two Phase III Studies (ADMYRA and ADACCESS). BioDrugs. 2021;35(2):229–38.
57. Druedahl LC, Almarsdottir AB, Kalvemark Sporrong S, De Bruin ML, Hoogland H, Minssen T, et al. A Qualitative Study of Biosimilar Manufacturer and Regulator Perceptions on Intellectual Property and Abbreviated Approval Pathways. Nat Biotechnol. 2020;38(11):1253–6.
58. Dougherty MK, Zineh I, Christl L. Perspectives on the Current State of the Biosimilar Regulatory Pathway in the United States. Clin Pharmacol Ther. 2018;103(1):36–8.
59. Rumore MM, Randy Vogenberg F. Biosimilars: Still Not Quite Ready for Prime Time. P T. 2016;41(6):366–75.
60. Osemene NI. The Purple Book: A Compendium of Biological and Biosimilar Products. US Pharm. 2015;40(6):22–9.
61. News in Brief, FDA Promotes Efficient Biosimilar Approval. Cancer Discov. 2018;8(10):1200.
62. Sabatelli AD, Alpha-Cobb CJ. Will Purple Become the New Orange? The New FDA Purple Book for Biologics: What Does the Future Hold? Pharm Pat Anal. 2015;4(2):63–8.
63. Golden L. Peters EKH. Naming of Biological Products. US Pharm.;45(6):33–6.
64. Socal MP, Garrett JB, Tayler WB, Bai G, Anderson GF. Naming Convention, Interchangeability, and Patient Interest in Biosimilars. Diabetes Spectr. 2020;33(3):273–9.
65. Harvey RD. Science of Biosimilars. J Oncol Pract. 2017;13(9_suppl):17s–23s.
66. Niederwieser D, Schmitz S. Biosimilar Agents in Oncology/Haematology: From Approval to Practice. Eur J Haematol. 2011;86(4):277–88.
67. Kabir ER, Moreino SS, Sharif Siam MK. The Breakthrough of Biosimilars: A Twist in the Narrative of Biological Therapy. Biomolecules. 2019;9(9).

68. Barbier L, Ebbers HC, Declerck P, Simoens S, Vulto AG, Huys I. The Efficacy, Safety, and Immunogenicity of Switching Between Reference Biopharmaceuticals and Biosimilars: A Systematic Review. Clin Pharmacol Ther. 2020;108(4):734–55.
69. FDA Website, Nonproprietary Naming of Biological Products: Guidance for Industry, 2017. Accessed June 2021.
70. FDA Website, Labeling for Biosimilar Products: Guidance for Industry, 2018. Accessed June 2021.
71. Celia Lu ECJ. Biosimilars: Not Simply Generics. US Pharm. 2019;44(6):36–9.
72. American Pharmacists Association (APhA). Re: Labeling for Biosimilar Products; Draft Guidance for Industry [Docket No. FDA2016-D-0643-0001]. 2016.
73. McKinley L, Kelton JM, Popovian R. Sowing Confusion in the Field: The Interchangeable Use of Biosimilar Terminology. Curr Med Res Opin. 2019;35(4):619–21.
74. Cauchi R. State Laws and Legislation Related to Biologic Medications and Substitution of Biosimilars. NCSL. 2018.

The Manufacturing and Quality Assurance Process for Biologics and Biosimilars

8

Vaibhav Changedia and Yashwant Pathak

Corresponding author: Vaibhav Changedia

Contents

DOI: 10.1201/9780429485626-8

8.1 INTRODUCTION

The biopharmaceutical market is a rapidly growing class of therapeutics, showing significant potential in oncology, diabetes, and other disease areas. Unlike conventional chemically synthesized pharmaceuticals, biopharmaceuticals – also known as biologics – are derived from living organisms, typically using biotechnology. Examples of biologics include hormones, blood products, cytokines, monoclonal antibodies (mAbs), and vaccines, as well as gene transfer, cell therapy, and tissue-engineered products. There are more than 300 mAbs, more than 250 vaccines, and more than 100 other biologics – including cell and gene therapies – are currently in clinical development. The global biologics market is expected to reach around $291 billion in 2020, and by 2022, 50% of the pharmaceutical market share is expected to be in biologics. But the future of biologics won't be focused solely on the discovery of new therapeutics (1). There is also a significant market for biosimilars, biologic drugs that demonstrate high similarity to an already approved biologic standard drug, and can in turn serve as an alternative to it. Biosimilars are different from generics, which are synthetic chemical copies of their standard drugs and are identical in active ingredients, strength, dosage form, and route of administration. There are significant benefits to biosimilars. Biosimilars are a more cost-effective way to manufacture biologics, increasing the affordability of life-saving biologics to patients. To be approved, biosimilars need to prove a high degree of similarity in biophysical properties, safety, and effectiveness to the marketed standard product. The agent must demonstrate no significant difference in immune response and pharmacokinetic (PK)/pharmacodynamics (PD) outcomes in a clinical trial. However, lower immune response and better safety are acceptable (2, 3).

8.1.1 Biological Product

The term "biological product" means a virus, therapeutic serum, toxin, antitoxin, vaccine, blood, blood component or derivative, allergenic product, protein except for any chemically synthesized polypeptide applicable to the prevention, treatment, or cure of a disease or condition of human beings.

8.1.2 Biosimilar or Biosimilarity

Biosimilarity means there are no clinically significant differences between the biological product and the standard product in terms of the safety, purity, and potency of the product.

TABLE 8.1 Drugs vs. biological products – generally

SMALL MOLECULE DRUGS	*BIOLOGICAL PRODUCTS*
Fewer critical process steps	Many critical process steps
Usually made by organic or chemical synthesis	Made with/from live cells/organisms Inherent and contamination risk
Generally low molecular weight	Generally high molecular weight
Homogeneous drug substance	Heterogeneous mixtures may include variants
Known structure	Structure may or may not be completely defined or known
Well-characterized	Less easily characterized

8.1.3 Standard Product

Standard product means the single biological product, licensed under Section 351(a) of the Public Health Service (PHS) Act, against which a biological product is evaluated in an application submitted under Section 351(k) of the PHS Act.

Biosimilar medicines are a unique type of potent medicine that have revolutionized the way physicians are now able to treat cancers, diabetes, autoimmune disorders, and other conditions. The outcomes of biosimilar medicines for patients have been fantastic. Today, there are over 200 biosimilar products, including vaccines in the global market and most of these products are therapeutic proteins. These are proteins that are applied in the treatment of autoimmune disorders. Currently, greater than 1000 more biosimilar medicines and vaccines are under research and development (4).

Biosimilar medicines are more complex in nature as they are derived from living organisms; they cannot be synthesized by a chemical process as compared to traditional drugs. Biosimilar medicines or biosimilars, which may also be known as follow-on biologics, are developed to be highly similar versions of approved biologics. Biosimilars medicines are more cost-effective and may promote a challenging environment for further development of biologic medicines. Biosimilar medicines have demonstrated similarity in terms of safety, efficacy, and potency as compared to the marketed biologic product. The biosimilar medicines, as compared with marketed products, usually have structural similarities, but minor differences in the clinically inactive components exist in them. Biosimilars medicines are not generic drugs because generic drugs are exactly copies of brand name drugs, have the similar active component, and are precisely the same as their brand name drugs in quality, route of administration, dosage form, safety, strength, performance, etc. as shown in Table 8.1.

Biosimilar medicines have demonstrated improved health outcomes for patients. In the coming days, the patents protection of many synthetic drug products will soon expire, and biosimilar medicines are only an alternative treatment choice. In this chapter, we will discuss what makes biosimilar medicines different from other synthetic drugs products and the mechanism of their action in our bodies and fighting against diseases. We will also discuss the manufacturing procedure, challenges, and obstacle that arises in the regulation of these types of biosimilar medicines. Our aim is to provide recent updated information on biosimilar medicines that can help patients and the healthcare system to make correct decisions concerning the applications of biosimilars medicines (5, 6).

8.2 MANUFACTURING PROCESS CONSIDERATIONS

Different manufacturing processes lead to changes in a protein product in a way that could affect the safety or efficacy of the product. For example, differences in biological systems used to manufacture a protein product may cause different post-translational modifications, which in turn may affect the safety

TABLE 8.2 Difference between biological products, biosimilars, and generic

PROCESS	*BIOLOGICAL PRODUCT*	*GENERIC*	*BIOSIMILAR*
Manufacturing	Produced by biological processes in host cell lines. Sensitive to changes in the production process: Expensive and specialized production facilities. Reproducibility is difficult to establish.	Produced by using chemical synthesis. Less sensitive to changes in the production process. Reproducibility is easy to establish.	Produced by biological process in host cell lines. Sensitive to changes in the production process: expensive and specialized production facilities. Reproducibility is difficult to establish.
Clinical Development	Extensive clinical trials, including phase I–III. Pharmacovigilance and periodic security updates required.	Often only phase 1 study. Short term for approval.	Extensive clinical trials, including phase I–III. Pharmacovigilance and periodic security updates required.
Regulation	Need to demonstrate "comparability" regulatory path. Currently, no automatic substitution is foreseen.	You must show bioequivalence. Abbreviated registration procedures. Automatic replacement is allowed.	Need to demonstrate "comparability" regulatory path. Currently, no automatic substitution is foreseen.

and/or effectiveness of the product. Thus, when the manufacturing process for a marketed protein product is changed, the application holder must assess the effects of the change and demonstrate through appropriate analytical testing, functional assays, and/or in some cases, animal and/or clinical studies that the change does not have an adverse effect on the identity, strength, quality, purity, or potency of the product as they relate to the safety or effectiveness of the product. The International Conference on Harmonisation (ICH) guidance for industry Comparability of Biotechnological/Biological Products Subject to Changes in Their Manufacturing Process describes scientific principles in the comparability determination for manufacturing changes (7, 8).

Demonstrating that a proposed test product is biosimilar to a standard product typically will be more complex than assessing the comparability of a product before and after manufacturing changes made by the same manufacturer. This is because a manufacturer that modifies its own manufacturing process has extensive knowledge and information about the product and the existing process, including established controls and acceptance parameters. By contrast, the manufacturer of a proposed test product is likely to have a different manufacturing process (e.g., different cell line, raw materials, equipment, processes, process controls, and acceptance criteria) from that of the standard product and no direct knowledge of the manufacturing process for the standard product as shown in Table 8.2.

8.3 DEVELOPMENT CONCEPTS

8.3.1 Stepwise Approach to Generate Data in Support of a Demonstration of Biosimilarity

The purpose of a biosimilar development program is to support a demonstration of biosimilarity between a proposed test product and a standard product, including determination of the effects of any observed differences between the products, but not to independently establish the safety and effectiveness of the proposed test product. As per the Food and Drug Administration (FDA) guidelines, sponsors should use a stepwise approach to develop the data and information needed to support a biosimilarity. At each step, the sponsor should evaluate the extent to which there is any uncertainty about the biosimilarity of the proposed test product and identify the next steps to address that uncertainty. To demonstrate biosimilarity, studies should be designed to maximize their contribution. The stepwise approach should be initiated with extensive structural and functional evaluation of the proposed test product and the standard product. The more comprehensive and comparative structural and functional evaluation of the extent to which these studies are able to identify qualitative as well as quantitative differences in relevant product attributes between the proposed test product and the standard product, the more useful such characterization will be in determining what additional studies may be needed. It may be useful to further determine the similarity or differences between the two products using a meaningful fingerprint-like analyses algorithm that covers a large number of additional product attributes and their combinations with high sensitivity using orthogonal methods. Such a strategy may further reduce the possibility of undetected structural differences between the products and lead to a more selective and targeted approach to animal and/or clinical testing. The sponsor should then conduct comparative human PK/PD studies and compare the clinical immune response of the two products in an appropriate study population. After performing structural evaluation, functional evaluation, animal testing, human PK and PD studies, and the clinical immune response determination, the sponsor should then consider what additional clinical data may be needed to adequately address that uncertainty. The FDA encourages sponsors to consult extensively with the agency after completion of comparative structural and functional analyses (before finalizing the clinical program) and throughout development as needed. The FDA recognizes that some of the aforementioned investigations could be performed in parallel; however, the agency recommends that sponsors use a stepwise approach to better address residual uncertainty about biosimilarity that might remain at each step and incorporate The FDA's advice provided after the FDA review of data and information collected at certain milestones (9–11).

8.3.2 Use Totality-of-the-Evidence Approach

In evaluating a sponsor's demonstration of biosimilarity, the FDA will consider the totality of the data and information submitted in the application, including structural and functional evaluation, nonclinical evaluation, human PK and PD data, immune response data, and comparative clinical studies data. The FDA intends to use a risk-based approach to evaluate all available data and information submitted in support of the biosimilarity of the proposed test product. Thus, a sponsor may be able to demonstrate biosimilarity even though there are formulation or structural differences, provided that the sponsor provides sufficient data and information demonstrating that the differences are not clinically significant and the proposed test product meets the regulatory requirement for biosimilarity. Information provided by the sponsor shows that the proposed test product is highly similar to the standard product, notwithstanding minor differences in clinically inactive components and that there are no clinically meaningful differences between the products in terms of safety, purity, and potency (12).

8.4 STEPWISE EVIDENCE DEVELOPMENT

8.4.1 Extensive Structural and Functional Characterization

The PHS Act requires that the application should include information demonstrating biosimilarity based on data derived from, among other things, analytical studies that demonstrate that the biological product is highly similar to the standard product unless the FDA determines that an element is unnecessary in a 351(k) application. The FDA expects that first, a sponsor will extensively characterize the proposed test product and the standard product with state-of-the-art technology because extensive characterization of both products serves as the foundation for a demonstration of biosimilarity. Additionally, sponsors should consider all relevant characteristics of the protein product (e.g., the primary, secondary, tertiary, quaternary structure and post-translational modifications, etc.) to demonstrate that the proposed test product is highly similar to the standard product. The more comprehensive and comparative structural and functional evaluation is, the stronger the scientific justification for a selective and targeted approach to animal and/or clinical testing (13).

Sponsors should use suitable analytical techniques with sufficient sensitivity and specificity for structural characterization of the proteins. Generally, the proposed test product and the standard product should be compared in terms of the following parameters:

- Primary structures, such as amino acid sequence.
- Higher-order structures, including secondary, tertiary, and quaternary structures.
- Enzymatic post-translational modifications, such as glycosylation and phosphorylation.
- Chemical modifications, such as PEGylation.

Sponsors should conduct extensive structural characterization of both the proposed test product and the standard product in multiple representative lots to understand the lot-to-lot variability of both products in the manufacturing processes. Lots used for the analyses should support the biosimilarity of both the clinical material used in the clinical studies intended to support a demonstration of biosimilarity, and the to-be-marketed proposed test product, to the standard product. In addition, the FDA recommends that sponsors analyze the finished dosage form of multiple lots of the proposed test product and the standard product, assessing excipients and any formulation effect on purity, product and process-related impurities, and stability. Differences in formulation between the proposed test product and the standard product are among the factors that may affect the extent and nature of subsequent animal or clinical testing. The nature and extent of the changes may determine the extent of the analytical similarity and comparability studies and any necessary additional studies. If the standard product or the proposed test product cannot be adequately evaluated with state-of-the-art technology, the application for the proposed test product may not be appropriate for submission under Section 351(k) of the PHS Act; and the sponsor should contact the FDA for appropriate guidance on the appropriate submission pathway (14, 15).

8.4.2 Functional Assays

The pharmacologic activity of protein products should be evaluated by *in vitro* and/or *in vivo* functional assays. *In vitro* assays may include, but are not limited to, biological assays, binding assays, and enzyme kinetics. *In vivo* assays may include the use of animal models of disease (e.g., models that exhibit a disease state or symptom) to evaluate functional effects on PD markers or efficacy measures. A functional evaluation comparing a proposed test product to the standard product using these types of assays is also an important part of the foundation that supports a demonstration of biosimilarity and may be used to scientifically justify a selective and targeted approach to animal and/or clinical testing.

Sponsors can use functional assays to provide additional evidence that the biologic activity and potency of the proposed test product are highly similar to those of the standard product and/or to support a conclusion that there are no clinically significant differences between the proposed test product and the standard product. Such assays may also be used to provide additional evidence that the mechanism of action of the two products is the same to the extent to which the mechanism of action of the standard product is known. Functional assays can be used to provide additional data to support results from structural analyses, investigate the consequences of observed structural differences, and explore structure-activity relationships. These assays are expected to be comparative so they can provide evidence of similarity or reveal differences in the performance of the proposed test product compared to the standard product, especially differences resulting from variations in structure that cannot be detected using current analytical methods. The FDA also recommends that sponsors discuss limitations of the assays they used when interpreting results in their submissions to the FDA. Such discussions would be useful for the evaluation of analytical data and may guide whether additional analytical testing would be necessary to support a demonstration of biosimilarity (16, 17).

8.5 ANIMAL STUDIES

8.5.1 Animal Data

The PHS Act also requires that a 351(k) application include information demonstrating biosimilarity based on data derived from animal studies (including the determination of toxicity), unless the FDA determines that such studies are not necessary in a 351(k) application. Results from animal studies may be used to support the safety evaluation of the proposed test product and more generally to support the demonstration of biosimilarity between the proposed test product and the standard product (18).

8.5.1.1 Animal toxicity studies

As a scientific matter, animal toxicity data are considered useful when, based on the results of the extensive structural and functional evaluation, uncertainties remain about the safety of the proposed test product that needs to be addressed before initiation of clinical studies in humans. The scope and extent of any animal toxicity studies will depend on the availability of studies data about the standard product, studies data about the proposed test product, and the extent of similarities or differences between the two. Safety data derived from animal toxicity studies are generally not expected if clinical data using the proposed test product are available (with the same proposed test route of administration and formulation) that provide sufficient evidence for its safe use.

Animal toxicity studies are generally not useful if there are no animal species that can provide pharmacologically relevant data for the product. However, there may be some instances when animal data from a pharmacologically nonresponsive species (including rodents) may be useful to support clinical studies with a proposed test product that has not been previously tested in human subjects. Data that was obtained using studies on human cells can provide important information between the proposed test product and the standard product regarding potential clinical effects. In general, nonclinical safety pharmacology, reproductive and developmental toxicity, and carcinogenicity studies are not required when the proposed test product and the standard product have been proved to be highly identical through extensive structural and functional evaluation and animal toxicity studies (19, 20).

8.5.1.2 Inclusion of animal pharmacokinetic and pharmacodynamic measures

Under certain circumstances, a single-dose study in animals comparing the proposed test product and the standard product using PK and PD measures may contribute to the totality of evidence that supports a demonstration of biosimilarity. Specifically, sponsors can use results from animal studies to support the degree of similarity based on the PK and PD profiles of the proposed test product and the standard product.

8.5.1.3 Interpreting animal immune response results

Animal immune response studies are conducted to help in the interpretation of the animal study results and generally do not predict potential immune responses to protein products in humans. However, when differences in manufacturing (e.g., impurities or excipients) between the proposed test product and the standard product may result in differences in immune response, measurement of anti-therapeutic protein antibody responses in animals may provide useful information (21).

8.5.1.4 Study population

The choice of study population should allow the determination of clinically meaningful differences between the proposed test product and the standard product. Often the study population will have characteristics consistent with those of the population studied for the licensure of the standard product for the same indication. However, there are cases where a study population could be different from that in the clinical studies that supported the licensure of the standard product. For example, if a genetic predictor of response was developed following licensure of the standard product, it may be possible to use patients with the response marker as the study population (22, 23).

8.5.1.5 Sample size and duration of study

The sample size for and duration of the comparative clinical study should be adequate to allow for the detection of clinically meaningful differences between the two products.

8.5.1.6 Study design and analyses

A comparative clinical study for a biosimilar development program should be designed to investigate whether there are clinically meaningful differences between the proposed test product and the standard product. The design should take into consideration the nature and extent of residual uncertainty that remains about biosimilarity based on data generated from comparative structural and functional characterization, animal testing, human PK and PD studies, and clinical immunogenicity determination. Generally, the FDA expects a clinical study or studies designed to establish statistical evidence that the proposed test product is neither inferior to the standard product by more than a specified margin nor superior to the standard product by more than a (possibly different) specified margin. Typically, an equivalence design with symmetric inferiority and superiority margins would be used. Symmetric margins would be reasonable when, for example, there are dose-related toxicities (24).

In some cases, it would be appropriate to use an asymmetric interval with a larger upper bound to rule out superiority than a lower bound to rule out inferiority. An asymmetric interval could be reasonable, for example, if the dose used in the clinical study is near the plateau of the dose–response curve and there is little likelihood of dose-related effects (e.g., toxicity). In most cases, the use of an asymmetric interval would generally allow for a smaller sample size than would be needed with symmetric margins. However, if there is a demonstration of clear superiority, then further consideration should be given as to whether the proposed test product can be considered biosimilar to the standard product. In some cases, depending on the study population and endpoint(s), ruling out only inferiority may be adequate to establish that there are no clinically meaningful differences between the proposed test product and the standard product (25). For example, if it is well established that doses of a standard product pharmacodynamically saturate the target at the clinical dose level and it would be unethical to use lower than clinically approved doses, a noninferiority (NI) design may be sufficient. A sponsor should provide adequate scientific justification for the choice of study design, study population, study endpoint(s), estimated effect size for the standard product, and margin(s) (how much difference to rule out). Sponsors should discuss their study proposal(s) and overall clinical development plan with the FDA before initiating the comparative clinical studies) see Table 8.3.

TABLE 8.3 Study design considerations in biosimilar development

	STANDARD PRODUCT	*BIOSIMILAR*
Comparator	Placebo or active comparator	Active comparator study – standard product
Endpoint	Outcome by which the effectiveness of treatment in a clinical trial is evaluated	Traditional efficacy endpoints not sensitive to detect differences between similar active products Endpoints should reflect the activity of the product
Timepoint for assessing endpoint	Adequate time for product to take and maintain the clinical effect	Timepoint(s) when most likely to detect differences between products, e.g., ascending portion of the dose–response curve (activity) rather than the efficacy, look for similarity between "activity" responses
Patient population	Disease population for which licensure is sought	Same or different from the standard product Should be sensitive to detect differences; for example, populations in early or late-stage disease which may not be confounded by concurrent or previous therapy
Dose	Objective is to obtain clinical efficacy as efficiently and safely as possible	May be a therapeutic dose, or a lower dose (if ethical) Dose should produce an effect over a time period that is conducive to detecting differences between products, e.g., therapeutic dose may reach a plateau before one can assay for differences between products
Sample size	Powered to demonstrate efficacy by detecting treatment difference	Based on the selected endpoint and margins (generally equivalence) under the chosen study conditions
Duration	Adequate to assess the efficacy and reasonable safety follow-up	Driven by study design (e.g., endpoint and timepoint); generally same or shorter duration because not independently establishing safety and efficacy of the product

8.6 CLINICAL STUDIES – GENERAL CONSIDERATIONS

The sponsor of a proposed test product must include in its submission to the FDA information demonstrating that "there are no clinically meaningful differences between the biological product and the standard product in terms of the safety, purity, and potency of the product". The nature and scope of the clinical study or studies will depend on the nature and extent of residual uncertainty about biosimilarity after conducting structural and functional characterization and, where relevant, animal studies (26).

8.6.1 Human Pharmacology Data

Functional assays and/or animal studies alone cannot be adequately predicted human PK and PD profiles of a protein product. Therefore, human PK and PD studies comparing a proposed test product to the standard product generally are fundamental components in supporting a demonstration of biosimilarity. Both PK and PD studies (where there is a relevant PD measure(s)) generally will be expected to establish biosimilarity, unless a sponsor can scientifically justify that such a study is not needed. Even if relevant PD measures are not available, sensitive PD endpoints may be assessed if such determination may help reduce residual uncertainty about biosimilarity (27).

Sponsors should provide a scientific justification for the selection of the human PK and PD study population (e.g., patients versus healthy subjects) and parameters, taking into consideration the relevance and sensitivity of such population and parameters, the population and parameters studied for the licensure for the standard product, as well as the current knowledge of the intra-subject and inter-subject variability of human PK and PD for the standard product. For example, comparative human PK and PD studies should use a population, dose(s), and route of administration that are adequately sensitive to allow for the detection of differences in PK and PD profiles. The FDA recommends that, to the extent possible, the sponsor select PD measures that (1) are relevant to clinical outcomes (e.g., on mechanistic path of mechanism of action or disease process related to effectiveness or safety); (2) are measurable for a sufficient period of time after dosing to ascertain the full PD response and with appropriate precision; and (3) have the sensitivity to detect clinically meaningful differences between the proposed test product and the standard product. The use of multiple PD measures that assess different domains of activities may also be of value.

When there are established dose–response or systemic exposure–response relationships (response may be PD measures or clinical endpoints), it is important to select, whenever possible, a dose(s) for study on the steep part of the dose–response curve for the proposed test product. Studying doses that are on the plateau of the dose–response curve is unlikely to detect clinically meaningful differences between the two products. Sponsors should predefine and justify the criteria for PK and PD parameters for studies included in the application to demonstrate biosimilarity.

A human PK study that demonstrates similar exposure (e.g., serum concentration over time) for the proposed test product and the standard product may provide support for a demonstration of biosimilarity. A human PK study may be particularly useful when the exposure correlates with clinical safety and effectiveness. A human PD study that demonstrates a similar effect on a relevant PD measure(s) related to effectiveness or specific safety concerns (except for immune response, which is evaluated separately) represents even stronger support for a biosimilarity determination (28).

In certain cases, establishing a similar clinical PK, PD, and immune response profile may provide sufficient clinical data to support a conclusion that there are no clinically meaningful differences between the two products. PK and PD parameters are generally more sensitive than clinical efficacy endpoints in assessing the similarity of two products. For example, an effect on thyroid-stimulating hormone (TSH) levels would provide a more sensitive comparison of two thyroxine products than an effect on clinical symptoms of thyroidism.

In cases where there is a meaningful correlation between PK and PD results and clinical effectiveness, convincing PK and PD results may make a comparative efficacy study unnecessary. Even if there is still residual uncertainty about biosimilarity based on PK and PD results, establishing a similar human PK and PK profile may provide a scientific basis for a selective and targeted approach to subsequent clinical testing (29).

8.6.2 Clinical Immune Response Determination

The goal of the clinical immune response determination is to evaluate potential differences between the proposed test product and the standard product in the incidence and severity of human immune responses. Immune responses may affect both the safety and effectiveness of the product by, for example, altering PK, inducing anaphylaxis, or promoting the development of neutralizing antibodies that neutralize the product as well as its endogenous protein counterpart. Thus, establishing that there are no clinically meaningful differences in immune response between a proposed test product and the standard product is a key element in the demonstration of biosimilarity. Structural, functional, and animal data are generally not adequate to predict immune response in humans. Therefore, at least one clinical study that includes a comparison of the immune response of the proposed test product to that of the standard product will be expected. The FDA encourages that, where feasible, sponsors collect immune response data in any clinical study, including human PK or PD studies.

The extent and timing of the clinical immune response determination will vary depending on a range of factors, including the extent of analytical similarity between the proposed test product and the standard product, and the incidence and clinical consequences of immune responses for the standard product. For example, if the clinical consequence is severe (e.g., when the standard product is a therapeutic counterpart of an endogenous protein with a critical, nonredundant biological function or is known to provoke anaphylaxis), a more extensive immune response determination will likely be needed to support a demonstration of biosimilarity. If the immune response to the standard product is rare, a premarketing evaluation to assess apparent differences in immune responses between the two products may be adequate to support biosimilarity. In addition, in some cases, certain safety risks may need to be evaluated through postmarketing surveillance or studies.

The overall immune response determination should consider the nature of the immune response (e.g., anaphylaxis, neutralizing antibody), the clinical relevance and severity of consequences (e.g., loss of efficacy of life-saving therapeutic and other adverse effects), the incidence of immune responses, and the population being studied. The FDA recommends the use of a comparative parallel design (i.e., a head-to-head study) in treatment-naïve patients as the most sensitive design for a premarketing study to assess potential differences in the risk of the immune response. However, depending on the clinical experience of the standard and proposed test products (taking into consideration the conditions of use and patient population), a sponsor may need to evaluate a subset of patients to provide a substantive descriptive determination of whether a single cross-over from the standard product to the proposed test biosimilar would result in a major risk in terms of hypersensitivity, immune response, or other reactions. The design of any study to assess immune response and acceptable differences in the incidence and other parameters of immune response should be discussed with the FDA before initiating the study. Differences in immune responses between a proposed test product and the standard product in the absence of observed clinical sequelae may be of concern and may warrant further evaluation (e.g., extended period of follow-up evaluation).

The study population used to compare immune response should be justified by the sponsor and agreed to by the agency. If a sponsor is seeking to extrapolate immune response findings for one condition of use to other conditions of use, the sponsor should consider using a study population and treatment regimen that are adequately sensitive for predicting a difference in immune responses between the proposed test product and the standard product across the conditions of use. Usually, this will be the population and regimen for the standard product for which the development of immune responses with adverse outcomes is most likely to occur.

The selection of clinical immune response endpoints or PD measures associated with immune responses to therapeutic protein products (e.g., antibody formation and cytokine levels) should take into consideration the immune response issues that have emerged during the use of the standard product. Sponsors should prospectively define the clinical immune response criteria, using established criteria where available, for each type of potential immune response and should obtain agreement from The FDA on these criteria before initiating the study. The duration of follow-up evaluation should be determined based on (1) the time course for the generation of immune responses (such as the development of neutralizing antibodies, cell-mediated immune responses) and expected clinical sequelae (informed by experience with the standard product), (2) the time course of the disappearance of the immune responses and clinical sequelae following cessation of therapy, and (3) the length of administration of the product (30).

8.6.3 Comparative Clinical Studies

As a scientific matter, a comparative clinical study will be necessary to support a demonstration of biosimilarity if there is residual uncertainty about whether there are clinically meaningful differences between the proposed test product and the standard product based on structural and functional characterization, animal testing, human PK and PD data, and clinical immune response determination. A sponsor should provide a scientific justification if it believes that a comparative clinical study is not necessary (30).

The following are examples of factors that may influence the type and extent of the comparative clinical study data needed:

- The nature and complexity of the standard product, the extensiveness of structural and functional characterization, and the findings and limitations of comparative structural, functional, and nonclinical testing, including the extent of observed differences.

- The extent to which differences in structure, function, and nonclinical pharmacology and toxicology predict differences in clinical outcomes, in conjunction with the degree of understanding of the mechanism of action of the standard product and disease pathology.
- The extent to which human PK or PD is known to predict clinical outcomes (e.g., PD measures known to be relevant to effectiveness or safety).
- The extent of clinical experience with the standard product and its therapeutic class, including the safety and risk-benefit profile (e.g., whether there is a low potential for off-target adverse events), and appropriate endpoints and biomarkers for safety and effectiveness (e.g., availability of established, sensitive clinical endpoints).
- The extent of any other clinical experience with the proposed test product (e.g., if the proposed test product has been marketed outside the United States).

The differences between conditions of use with respect to the factors described above do not necessarily preclude extrapolation. A scientific justification should address these differences in the context of the totality of the evidence supporting a demonstration of biosimilarity.

In choosing which condition of use to study that would permit subsequent extrapolation of clinical data to other conditions of use, the FDA recommends that a sponsor consider choosing a condition of use that would be adequately sensitive to detect clinically meaningful differences between the two products.

8.7 QUALITY, SAFETY, AND EFFICACY

The quality, safety, and efficacy of a biosimilar product must be approved by the relevant regulatory body before marketing approval can be gained, which requires an appropriate comparability exercise. The European Medicine Agency (EMEA) requires a comparison of the biosimilar product with the innovator product to determine the absence of any detectable differences. The quality comparison between the biosimilar and the innovator product is crucial because the quality of a protein product affects its safety and efficacy. It is known that biopharmaceutical manufacturing is a multistep process, involving cloning of the appropriate genetic sequence into a carefully selected expression vector, selection of a suitable cell expression system, and scale-up and purification, up to the formulation of the end product (Figure 8.1). Toward the particular manufacturing process used, biopharmaceuticals exhibited great sensitivity, and variation in product quality was commonly observed, even when the exact same process of manufacturing was used. The challenge then remains to assess and quantify these differences, and determine whether the new product is as safe and efficacious as the innovator product. Further, variability of the source material has also been known to affect product quality. Thus the product is affected both by the host cell and the processing steps that follow. In addition, protein molecules can be degraded during processing steps and impurities created in these steps can contribute to decreased potency and/or increased immune response. With a large number of quality attributes, acquiring a complete knowledge of the impact of each of the attributes on clinical safety and efficacy is not feasible (31).

Quality by design and process analytical technology initiatives from the FDA has improved understanding of the impact of manufacturing processes and their starting materials on product quality. Biochemical characterization of the protein product requires sophisticated analytical tools to detect the possibilities of changes to the product. Further, the characterization of the product requires a variety of methods for different attributes or, alternatively, with orthogonal methods for the characterization of a given attribute, thus developing a comprehensive finger-printing of a protein product. However, key challenges remain that continue to require attention, primarily because of the high complexity of the products, processes, and raw materials that are part of the manufacturing of biotechnology products. Virtually all therapeutic proteins induce some level of antibody response. The immune reaction can vary from low titer, low affinity, and transient IgM antibodies to a high-titer, high-affinity IgG response, with

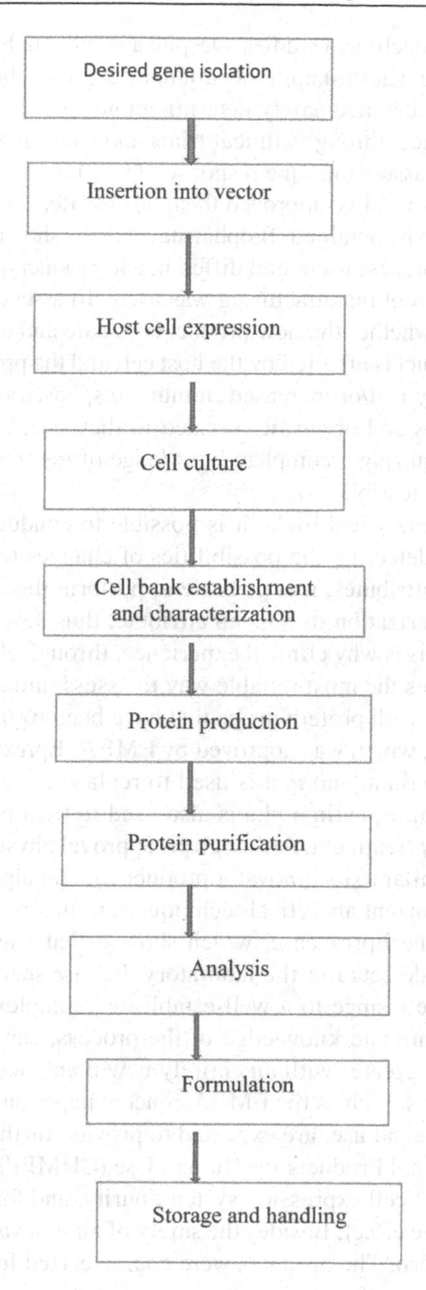

FIGURE 8.1 Typical steps in manufacturing a biologic product.

consequences ranging from none to severe or life threatening. Many factors determine the appearance of immune response, including patient characteristics and disease state, and the therapy itself influences the generation of an immune response. Product-related factors such as the molecule design, the expression system, post-translational medications, impurities, contaminants, formulation and excipients, container, closure, as well as degradation products are all implicated. It is fundamental to conduct preclinical and clinical studies to understand the safety, efficacy, and quality of both the innovator product and biosimilar medicines. Preclinical studies are not yet capable of assessing the clinically relevant immune response potential of these product-related factors. Understandably, most of the focus has been on assessing immune response of protein products by nonclinical studies. Despite advances in biochemical characterization and other nonclinical methods for the determination of immune response, the unpredictability of the human immune system still necessitates detailed safety determinations, which will rely heavily on clinical trials. This is why clinical experience, through clinical trials and extensive pharmaco-vigilance programs, remains the most reliable way to assess immune response (32, 33).

The relevant regulatory body must be approved the quality, safety, and efficacy of a biosimilar product before marketing approval can be obtained. Biopharmaceuticals showing great sensitivity with regards to the particular manufacturing process used, and difference in product quality was commonly observed, even when the exact same process of manufacturing was used. To assess and quantify these differences, challenges remain to determine whether the new product is as safe and efficacious as compared with the innovator product. Thus the product is affected by the host cell and the processing steps that follow. There are chances of decreased potency and/or increased immune response because protein molecules can be degraded during processing steps and impurities created in these steps can contribute to. With a large number of quality attributes, acquiring a complete knowledge of the impact of each of the attributes on clinical safety and efficacy is not feasible.

By applying sophisticated analytical tools, it is possible to conduct biochemical characterization of the protein product requires detecting the possibilities of changes to the product. By implementing different methods for different attributes, it is possible to perform the characterizations of the product required or alter for the characterization of a given attribute, thus developing a comprehensive fingerprinting of a protein product. This is why clinical experience, through clinical trials and extensive pharmacovigilance programs, remains the most reliable way to assess immune response. The best example of unexpected things happening with protein drugs that have been well characterized is Hospira's biosimilar epoetin zeta (Retacrit®), which was approved by EMEA. Eprex® (Amgen/Johnson & Johnson) is synthetic erythropoietin (epoetin alpha) and is used to replace the erythropoietin that is lacking in people who cannot make enough. Epoetin alpha is also used to treat people with cancer who develop anemia because of chemotherapy treatment. Although preapproval physicochemical, *in vivo*, and animal testing showed that it was biosimilar to its innovator product epoetin alpha (Eprex), it had lower potency in clinical trials. Accordingly, current analytical techniques are unable to assess immune response and potency. This is evident from the Eprex case, which showed that one protein can be different from another in ways that cannot be detected in the laboratory, but are seen only by the body's exquisitely sensitive immune system. If one change to a well-established complex manufacturing process, made by the manufacturer who has intimate knowledge of the process, can cause a problem with immune response, surely the risk is even greater with an entirely new manufacturer and process – as with biosimilar. Recently launched efforts, such as the EMEA concept paper on immunogenicity determination of mAbs intended for *in vivo* clinical use, are expected to provide further clarity on this topic. Further, when the Committee for Medicinal Products for Human Use (CHMP)/EMEA evaluations to date were examined, any difference in host cell expression system, purity, and formulation appears acceptable if the clinical data show no negative effect. Besides the safety of an innovator product, evidence suggested that efficacy can also be a concern. The products were characterized for similarity in the types of glycoforms present, the relative degree of unfolding, *in vitro* potency, presence of covalent aggregates, and presence of cleavage aggregates. The biochemical discrepancies between the different copy products were most likely caused by the differences in the cell lines and the manufacturing process (34).

8.8 CHALLENGES FACING THE BIOSIMILAR MARKET

Despite its potential in cost savings, the biosimilar market still faces hurdles. There are gaps in knowledge and understanding of biosimilars among the public, and that can lead to challenges with physicians' and patients' perception of the safety and efficacy of the product. Worldwide there is also a high variability in regulations regarding the approval of biosimilars. In Europe, the regulatory pathway and litigation procedures are much clearer and more defined. In the United States, however, there are very broad IP and patent laws, which can make patent infringements a major obstacle in effective marketing. This can impact patient access and increase the cost. Health plans and pharmacy benefit managers (PBMs) can also contribute to increased costs, as some require high rebates that block biosimilar access and uptake. Finally, multiple companies also often compete to make the same biosimilar of a biological drug that has recently come off patent, causing stiff competition across the market (30).

8.9 OVERCOMING THESE OBSTACLES

To stand out from the competition, companies that make biosimilar products can consider lowering prices or offering high rebates to PBMs. However, this approach is best suited in developing countries with high price sensitivity. In more established markets for biologics, such as the United States, Europe, and Japan, pricing alone is not sufficient, as they are typically no significant price variety among similar therapeutics. It is also possible to attempt to differentiate the molecule itself by providing better tolerability. However, regulatory agencies will not typically allow bio superiority to be claimed on a label on a biosimilar product.

One potentially effective way to differentiate a biosimilar is through a novel delivery device or container closure system, which improves convenience, ease of use, or patient acceptability of the therapeutic from its standard product. Some design differences in the delivery device or container closure system used with the proposed test biosimilar product also may be acceptable for regulatory approval of a biosimilar. For a proposed test biosimilar product in a different delivery device or container closure system, the presentation must be shown to be compatible for use with the final formulation of the biological product through appropriate studies including, for example, extractable/leachable studies and stability studies. For certain design differences in the delivery device or container closure system, performance testing and a human factors study may be needed (35).

8.10 BARRIERS TOWARD ACCESS TO BIOSIMILARS

It is possible to expand access to high-quality treatment options for doctors and patients, as well as reduce costs for the healthcare system using biosimilars. All those involved (health professionals, payers, legislators, politicians, patient advocates, and patients) need a clear, impartial, and evidence-based source of information to better understand biosimilars and make wise decisions about politics and use. Patients who suffer from many complex and chronic diseases have promised new treatment options in biologics and specialized medicines. These drugs are extremely expensive, which creates problems for patients' access. Now, thanks to biological drugs known as biosimilars, these advanced treatments are available at a lower cost for millions of patients in the United States living with cancer, rheumatoid arthritis, Crohn's disease, anemia, psoriasis, and other conditions. Biosimilar manufacturers seeking to launch in the nascent US market, face challenges related to scientific development, regulatory approval, and manufacturing and production. The following are some of the important factors that act as barriers to access to biosimilars (34, 35).

8.10.1 Pricing

The price of biosimilars is lower than the original organic product like generic medicines. However, cost of a generic drug could cost 70% less than a brand-name drug; biosimilar manufacturers were not able to keep much lower prices than the original organic because the costs of manufacturing biosimilars are much higher and manufacturers have to invest in marketing and sales.

8.10.2 Lack of Information, Experience Education, among Prescribers and Patients

Prescribers' knowledge and attitude regarding biosimilars will be a critical decision of uptake as more products enter the market. In the United States, prescribers simply write recipes for the products they are more used to. In addition to this, there are relatively low levels of biosimilar knowledge among many providers. In a recent medical survey, less than half (45%) believed that "biosimilars would be safe and appropriate for use in both untreated and existing patients", while more than one-third (36%) believed "that a biosimilar would be less safe than the standard biologic". Even more than doctors, patients lack awareness and education on biosimilars. According to a recent survey, 70% of the US respondents in the general population had never heard of biosimilars, compared to 57% who had never heard of biologics. Also, among patients diagnosed with Crohn's disease, ulcerative colitis, rheumatoid arthritis etc.; 54% had never heard of biosimilars, compared to 33% of people diagnosed with these diseases who had never heard of organic products. The WHO, like many other organizations and agencies, has provided some materials such as questions and answers, frequently asked questions, etc., to explain some basic facts about biosimilars. But it seems that there is still a need to do more in terms of key players in education.

8.10.3 Patent Abuses

Patent abuse is an important element of innovation in drug development and new effective treatments to meet unmet medical needs. However, the patent system is increasingly being used to unjustly prolong the monopoly of a brand-name drug and delay the access of patients to cheaper biosimilar medicines. Even though the Biologics Price Competition and Innovation Act of 2009 (BPCIA) grants brand-name organic drug manufacturers a 12-year market exclusivity period to ensure return on investment in new drugs (more than anywhere else in the world), these patent abuses are proliferating and delaying competition.

8.10.4 Litigation

There may be delay in the development and marketing of biosimilars due to the legal dispute, which generally focuses on patent infringement among biopharmaceutical companies. One of the major legal aspects is the "dance of patents", which colloquially describes a provision in the BPCIA that requires the manufacturer of biological standard products to share information about the patents with the biosimilar producers. The reluctance to fully comply with this provision greatly slows down the development of biosimilars. Some producers of biopharmaceutical products, or their subsidiaries, produce biological standard products, as well as biosimilars for competitive products, which complicates matters relating to organic products and biosimilars. Requests related to biosimilars are usually addressed to two biopharmaceutical manufacturers in the courtroom. In a recent lawsuit against Johnson & Johnson and Janssen, two pharmaceutical companies in the pharmaceutical chain (Walgreens and Kroger) claimed that biopharmaceutical companies practiced anticompetitive practices to prevent insurers from covering biosimilars infliximab.

8.10.5 Substitution – Interchangeability

The BPCIA amended the PHS Act, creates an approval path for the biosimilars referred to in Section 351(k). Similar to generic applications filed under Abbreviated New Drug Applications (ANDA), per the

Hatch-Waxman Act of 1984, biosimilars approved under Section 351(k) are based on the safety and efficacy data of the standard product. BPCIA created two levels of biosimilar medicines: (1) a biosimilar that would not be substitutable at the pharmacy level without the prescriber's participation and (2) an interchangeable biosimilar that could be replaced without the prescriber's intervention (similar to generic small molecule drugs). An interchangeable biosimilar is a biosimilar that can be expected to produce the same clinical result in a given patient and for which there are no further risks to safety or efficacy with repeated changes compared to the constant use of the standard product. Interchangeability would allow pharmacists to replace biosimilars without the intervention of a prescriber. This is the next major hurdle in the path of biosimilars, as it will create a real challenge for the biosimilars product markets.

8.10.6 Stringent Regulatory Regulations and Requirements for Biosimilars

Congress passed the BPCIA in 2010 as Title VII of the Patient Protection and Affordable Care Act (ACA). Under the BPCIA, there must be "no clinically meaningful" differences in terms of the safety, purity and potency. A biosimilar product application must include data demonstrating biosimilarity to the standard product. This usually includes data from analytical studies demonstrating that the biological product is highly similar to the standard product and animal studies including determination of toxicity.

8.10.6.1 Biosimilarity: A novel approach to develop therapies and vaccines for COVID-19

The coronavirus disease 2019 (COVID-19) pandemic has created a rush to develop diagnostic, treatment, and prevention modalities that is unique in the past 100 years. As of the end of August, there were more than 3000 studies underway in 115 countries and in every state in the United States, where there were 685 studies in progress. The US federal funding had gone to 79 studies in 23 countries. There were 108 vaccines trials underway. To manage this process effectively, the FDA has developed the Coronavirus Treatment Acceleration Program (CTAP), and this has reported more than 570 potential COVID-19 treatment development plans. So far, the FDA has reviewed more than 270 trials and authorized emergency use of dexamethasone and remdesivir; no other COVID-19 treatment approvals are imminent.

Vaccines development is a challenging task because they typically require years of research and testing before they reach the market. The work to create COVID-19 vaccines began in January 2020 with the decoding of the SARS-CoV-2 genome. As of this writing, China has approved Sinovac vaccine for limited use (August 31), and Russia's Gamaleya has moved to phase 3 testing of a vaccine called Gam-Covid-VacLyo, a combination of adenoviruses engineered with a coronavirus gene. The vaccine, Sputnik V, has conditional approval in Russia pending results of the phase 3 testing. There are no data available to evaluate the safety and efficacy of either of these vaccines. In June 2020, the FDA issued detailed guidance on the development of COVID-19 vaccines, requiring that a vaccine protect at least 50% of the recipients.

The guidelines regarding to vaccine mandates testing a new vaccine first on cells and then in animals, such as mice or monkeys, to see if it produces an immune response resulting in the production of virus-specific antibodies. In phase 1 trials, the vaccine is given to a small number of healthy volunteers to test safety and dosage as well as to confirm that it stimulates the immune system. In phase 2, the trial is expanded to hundreds of volunteers split into groups, such as children and the elderly, to see if the vaccine acts differently in them. These trials further test the vaccine's safety and ability to stimulate the immune system. In phase 3 trials, thousands of volunteers receive the vaccine, and investigators wait to see how many become infected compared with volunteers who receive a placebo. These trials can determine if the vaccine protects against the coronavirus.

On August 30, the FDA announced that it would allow review and approval of vaccines based on preliminary phase 3 trials data. The rushes to develop by investigators and the desire to approve by the regulatory agencies have brought us to the brink of decision-making that can have serious consequences. At the same time, a new drug approved for a disease or condition demonstrates the risk based on the number of patients exposed. However, with a vaccine that is given to billions of healthy people and, in the case of COVID-19, a majority of the world's population, a large number of subjects can be adversely affected even if the percentage of events is small; such risk does not apply to therapeutic products developed to treat patients because of the limited use only in patients. COVID-19 is a challenge that requires novel approaches to overcome such pandemic. One such strategy is using the principles of establishing biosimilarity to assess the safety and efficacy of both treatments and vaccines. Biosimilars are macromolecules that include antibodies, and their safety and efficacy are tested first by comparing the biosimilar candidate with the standard product to ensure that there is no clinically significant difference between them. The analytical science has advanced well over the past few decades to enable us to compare the primary, secondary, and tertiary structures of proteins of all sizes with a high degree of accuracy.

The FDA suggested that it require a test comparing the structure of antibodies produced in patients with COVID-19 (serving as an equivalent of a standard product) with the antibodies produced using a vaccine (as the proposed test biosimilar product). This test can be conducted within a few days of administering the first dose of the vaccine. This testing can also be a preamble to full-blown testing to further reduce the risk to study subjects. The testing will also include binding or functionality tests and any other novel testing methods that the developers can propose to assure that the antibodies produced by a vaccine are highly similar to the antibodies produced in patients. Another test for structural similarity can be a PK study in animal species such as monkeys, where three to four animals can be administered the antibodies to determine if there are any differences in their disposition kinetics profile.

8.10.6.2 Abbreviations

EMEA – European Medicine Agency
PHS – Public Health Service
FDA – Food and Drug Administration
BPCIA – Biologics Price Competition and Innovation Act
PK – Pharmacokinetics
PD – Pharmacodynamics
PBM – Pharmacy benefit Manager
ANDA – Abbreviated New Drug Applications

8.11 CONCLUSION

Biosimilar products are the biosystems-based products that are being approved based on the similarity to the biological products released by The FDA known as a standard product. Productions of the similar copies of the standard products are not possible therefore the procedures used to manufacture the generic medicines are not applicable to these biosimilar products due to their complex nature. Because biosimilar products are very complex molecules, factors such as the robustness of the manufacturing process, structural similarity to the parent molecule, mechanism of action, quality of PD assays utilized, demonstrated comparability in PK and immunogenicity, quantity and quality of clinical data, and the innovator's experience with the parent product needs to be considered critically before marketing. Biosimilars products and reference products are generated in the living cells and require trained expertise's, as well as technology in manufacturing biosimilar products, are usually highly effective compared to small molecule drugs. The FDA's regulatory authority for the approval of biosimilars products is under PHS Act however that are also suggested to regulate under the Federal Food, Drug, and Cosmetics Act (FD&C).

REFERENCES

1. Kefalas CH, Ciociola AA. The food and drug administration's generic-drug approval process: Similarities to differences from brand-name drugs. Am J Gastroenterol. 2011; 106:1018–1021.
2. European Medicines Agency. Guideline on Similar Biological Medicinal Products [draft]. http://www.ema.europa.eu/docs/en_GB/document_library/Scientific_guideline/2013/05/WC50 (last accessed 31 March 2021).
3. Mellstedt H.et. al. The challenge of biosimilars. Annals of Oncology. 2008; 19: 411–419.
4. Camacho LH, Frost CP, Abella E. Biosimilars 101: Considerations for U.S. oncologists in clinical practice. Cancer Med. 2014; 3:889–899.
5. Gascon P. Presently available biosimilars in hematology-oncology: G-CSF. Target Oncol. 2012; 7:29–34.
6. Reichert JM. Trends in US approvals: New biopharmaceuticals and vaccines. Trends Biotechnol. 2006; 24:293–298.
7. Crommelin DA. Shifting paradigms: Biopharmaceuticals versus low molecular weight drugs. Int J Pharm. 2003; 266: 3–16.
8. Bennett CL, Chen B, Hermanson T. Regulatory and clinical considerations for biosimilar oncology drugs. Lancet Oncol. 2014; 15(13):594–605.
9. Kozlowski S, Woodcock J, Midthun K, Sherman RB. Developing the nation's biosimilars program. N Engl J Med. 2011; 365:385–388.
10. http://www.Food and Drug Administration.gov/downloads/Drugs/Guidance Compliance Regulatory Information/UCM21614 6.pdf (last accessed 31 May 2021).
11. Mulcahy AW. Perspective: Expert insights on a timely policy issue. The Cost Savings Potential of Biosimilar Drugs in the United States. Rand Corporation. 2014. https://www.rand.org/pubs/perspectives/PE127.html
12. Di Masi JA, Hansen RW, Grabowski HG. The price of innovation: New estimates of drug development costs. J. Health Econ. 2003; 22:151–185.
13. Henry D, Taylor C. Pharmacoeconomics of cancer therapies: Considerations with the introduction of biosimilars. Semin Oncol. 2014; 41(3):13–20.
14. Rovira J, Espin, J. Garcia L, Labry A. 2011. The impact of biosimilars' entry in the EU market. Andalusian School of Public Health, Granada, Spain. European Generic and Biosimilar Medicines Association. Biosimilars Handbook. 2011.
15. Patro SY, Freund E, Chang BS. Protein formulation and fill-finish operations. Biotechnol Annu Rev. 2002; 8:55–84.
16. Rathore AS. Follow-on protein products: Scientific issues, developments and challenges. Trends Biotechnol. 2009; 27:698–705.
17. Rathore AS, Winkle H. Quality by design for biopharmaceuticals. Nat Biotechnol. 2009; 27:26–34.
18. http://www.Food and Drug Administration.gov/downloads/Drugs/guidance compliance regulatory information/guidances/ucm070305.pdf. September 2004.
19. Chirino AJ, Mire-Sluis A. Characterizing biological products and assessing comparability following manufacturing changes. Nat Biotechnol. 2006; 22:1383–1391.
20. Kozlowski S, Swann P. Current and future issues in the manufacturing and development of monoclonal antibodies. Adv Drug Deliv Rev. 2006; 58:707–722.
21. Sharma B. Immune response of therapeutic proteins. Part 3: Impact of manufacturing changes. Biotechnol Adv. 2007; 25:325–331.
22. Swann PG, Tolnay M, Muthukkumar S. Considerations for the development of therapeutic monoclonal antibodies. Curr Opin Immunol. 2008; 20:493–499.
23. De Groot AS, Scott DW. Immune response of protein therapeutics. Trends Immunol. 2007; 28:482–490.
24. De Groot AS, Moise L. Prediction of immune response: State of the art. Curr Opin Drug Discov Dev. 2007; 10:332–335.
25. Zuniga L, Calvo B. Biosimilars: Pharmacovigilance and risk management. Pharmacoepidemiol Drug Saf. 2010; 19:661–669.
26. Thijs J Giezen, Sabine MJM, Straus A. Pharmacovigilance of biosimilars from a regulatory point of view: Is there a need for a specific approach? Int J Risk Saf Med. 2009; 21:53–58.
27. Pavlovic M, Girardin E, Kapetanovic L, Ho K. Similar biological medicinal products containing recombinant human growth hormone: European regulation. Horm Res. 2008; 69:14–21.

28. Schellekens H. Assessing the bioequivalence of biosimilars: The Retacrit case. Drug Discov Today. 2010; 14:495–499.
29. Park SS, Park J, Ko J. Biochemical determination of erythropoietin products from Asia versus US Epoetin alfa manufactured by Amgen. J Pharmaceut Sci. 2009; 98:1688–1699.
30. Büttel IC. Taking immune response determination of therapeutic proteins to the next level. J Int Asso Biol Standard. 2011; 39:100–109.
31. http://www.ema.europa.eu/docs/en_GB/document_library/Scientific_guideline/2009/09/WC50 0003953.pdf (last accessed 31 May 2021).
32. http://www.ema.europa.eu/ema/index (last accessed 31 May 2021).
33. http://www.Food and Drug Administration.gov/downloads/Drugs/Guidance Compliance Regulatory Information/Guidances/ (last accessed 31 May 2021).
34. Camacho LH, Frost CP, Abella E, Morrow PK, Whittaker S. Biosimilars 101: Considerations for U.S. oncologists in clinical practice. Cancer Med. 2014; 3(4):889–99.
35. Schellekens H, Moors E. Clinical comparability and European biosimilar regulations. Nature Biotechnol. 2010; 28:28–31.

9 Biopharmaceutics, Pharmacokinetics, and Pharmacodynamics of Biological Products

Nazir Hossen, Tanvirul Hye, and Fakhrul Ahsan

Corresponding author: Fakhrul Ahsan

Contents

DOI: 10.1201/9780429485626-9

9.1 INTRODUCTION

Biopharmaceutics (BP), pharmacokinetics (PK), and pharmacodynamics (PD) are three subdisciplines of pharmaceutical sciences that revolve around the physicochemical properties of the drugs, the fate of the drugs in the body, and their clinical efficacy and safety profiles. BP explains the influence of the physicochemical properties of drugs or drug formulations on absorption, distribution, metabolism, and elimination (ADME) of drugs, while PK describes the rate and extent of ADME of drugs. More specifically, PK explains the fate of a drug in the body, and PD, on the other hand, delineates the relationship between drug concentrations in the blood or site of action and therapeutic or toxicological effects elicited by the drugs. Unique to the sciences of BP, PK, and PD is that the scientific principles surrounding the fate of drugs are intricately intertwined with the ADME of drugs and thus play an important role in the process of drug development and discovery. As such, the principles of BP, PK, and PD are the basis for the rationale design of dosing regimen, therapeutic monitoring, and clinical application of both small- and large-molecule drugs, such as biologics and biosimilars.

Various BP, PK, and PD properties of chemical drugs that comprise 90% of commercially available therapeutic agents (1) are well known or have been extensively studied. In fact, the science of BP, PK, and PD of therapeutic substances has been developed based on or evolved from the data generated using chemical drugs are the basis of the science behind ADME of biologics. While existing principles of BP, PK, and PD can be used for studying and predicting ADMEs of chemical drugs, the same principles often do not apply or cannot be utilized for studying the ADMEs of biological products that are usually macromolecular.

Unlike chemical drugs, biologics are obtained from biological sources, such as human cells, microorganisms, whole blood, or plasma. Most common biologicals are genetic materials, such as DNA and RNA, or proteins such as hormones or antibodies. The molecular weights (MW) of these biological substances are much greater than those of traditional chemical drugs. In fact, a single molecule of a biological therapeutic agent may have 200 to 50,000 atoms (3). As such, large-scale production, characterization of physicochemical properties, optimization of formulations of biologicals are far more complex than those of chemical drugs. Unlike chemical drugs, biologics undergo degradation in the gastrointestinal (GI) tract because of the acidic environment of the GI tract and the presence of various types of proteolytic enzymes. Large molecular size of biologics is a major barrier for biologicals to transport cross the GI epithelium and subsequently enter the systemic circulation. A small fraction of orally administered biologics that crosses the GI epithelium undergoes degradation in the liver because of first-pass metabolisms. For these reasons, biologics are currently available as intravenous (IV), subcutaneous (SC), or intramuscular (IM) injection (4, 5). Since biologics share many structural features with those of native proteins and genetic materials, they are more specific in eliciting pharmacological effects and work at a reduced dosing rate when compared with chemical drugs.

Due to their large MW and three-dimensional (3D) structures, ADMEs of biologics have little or no relevance with those of chemical drugs. In fact, most biologics, upon IV administration, travel across the vasculature, enter the interstitial space and traverse cell membranes (5, 6). In the case of SC or IM administration, biologics chiefly undergo absorption via the lymphatic system, although the extent of absorption via the lymphatic system varies depending on the size of the biological substance. Proteins with an MW larger than 16 kDa, for example, undergo absorption via the lymphatic system (6, 7). Factors that influence ADME of biologicals include target-mediated clearance (CL), neonatal Fc receptor (FcRn, also known as Brambell receptor) recycling for Fc-containing proteins, immunogenicity, isoform heterogeneity, and metabolic stability (8). As biologics have limited tissue distribution, understanding the mechanisms of tissue penetration and the relationship between tissue concentration and efficacy is important for studying the PK/PD of biologics. Since the CL of biologics such as target- and antidrug-antibody-mediated CL, immunogenicity, FcRn/IgG interactions of biologics vary depending on the species, data obtained using common laboratory animals such as rodents and dogs may not accurately reflect the ADME of biologics in humans. Quantitation of biologics in plasma or tissue samples is also a major challenge for studying ADME of biologics. Indeed, because of the similarity between the structures of endogenous proteins and therapeutic biologics, excessive hydrophilicity of therapeutic macromolecules, extraction and purification of these large MW drugs from the blood and tissue samples and subsequent quantitation are very challenging. Without purification and enriching via immunocapturing, existing technology for quantitation cannot detect or measure biologicals in the presence of biological matrices (9). Thus, to identify structures and quantify biological therapeutics, samples of biologics often require complex processing such as enzymatic digestion. In fact, sample recovery of therapeutic biologics from their matrices is very challenging because of possible loss during the digestion process and interference with endogenous proteins (10). However, modern instrumental techniques with many innovative features allow easy handling of samples and high-efficiency detection and quantitation. Overall, owing to the dissimilar biopharmaceutical properties of biologics versus chemical drugs, PK/PD of biologicals are distinctly different from those chemical drugs. Here in this chapter, we sought to summarize the BP, PK, and PD of therapeutic peptides, proteins, and monoclonal antibodies (mAbs).

9.2 BIOPHARMACEUTICAL PROPERTIES OF BIOLOGICS AND BIOSIMILARS

9.2.1 Biopharmaceutical Properties of Biologics

Many biopharmaceutical properties of biologics are distinctly different from those of chemical drugs. Although peptides are made of small amino acid chains with minimal tertiary structures, proteins often have complex structures, including folded 3D structures that must remain unaltered for eliciting their pharmacological effects. Therapeutically active proteins are generally polar, sensitive to temperature, have poor membrane permeability, and are subject to enzymatic degradation. Because protein therapeutics tend to be unstable or have unfavorable physicochemical properties, their formulation and manufacturing are challenging. Physical properties that should be considered for the formulation of protein therapeutics include unfolding, misfolding, and aggregation. Similarly, chemical properties that adversely affect the stability of biologics include oxidation, deamidation, racemization, hydrolysis, disulfide exchange, and carbamylation (11). The physicochemical stability and structural integrity of protein therapeutics can be improved by conjugating them with hydrophilic polymers such as polyethylene glycol (PEG), which increases the hydrodynamic diameter and thus reduces glomerular filtration or elimination rates. PEGylation of biologics also increases the circulation time in the blood by reducing the uptake by the reticuloendothelial system (RES). In fact, many PEG-modified protein therapeutics are now commercially available (12, 13). Other hydrophilic polymers that have been used to extend the biological half-life of biologics are sialic acid, hyaluronic acid, and hydroxyl ethyl starch (14–16). Although PEGylation decreases protein immunogenicity, increases solubility, helps maintain a steady drug level in the blood for an extended time, the pattern of changes of drug level in the blood follows first-order kinetics (13, 17, 18). However, PEGylation of biologicals is reported to interfere drug–receptor interactions (13, 17, 18), which can potentially be addressed by optimizing the molecular size and design of PEG-biologic conjugates. Recent advances in polymer chemistry and site-specific conjugation may address some challenges associated with polymeric-macromolecule–based long-circulating biologicals. The MW of PEG is reported to influence the site-specificity of PEGylated biologicals. Longer oligomers, for example, are reported to reduce the folding rate and reduce the rate of protein unfolding (14). Further, poly(2-oxazoline), which are synthesized using a method called living polymerization, allow controlling MW, modulate hydrophobicity and thus increase cellular uptake of poly(2-oxazoline) conjugated macromolecules (14).

Moreover, FcRn recycling pathways have also been used to extend the half-life of macromolecules (14). Using this technology, one can develop fusion proteins of macromolecules and the Fc region of IgG. Amgen Inc. (Thousand Oaks, CA) used this protein engineering technology for developing long-acting etanercept (Enabrel®). Several fusion proteins are now under various phases of clinical development (14). Similarly, albumin-based fusion protein has been used to develop biologics with extended half-life and thus to reduce the dosing frequency. An albumin fusion of GLP1, for example, is now commercially available for the treatment of adults with Type 2 diabetes (15). An albumin fusion of Factor IX is undergoing Phase III trial for the treatment of bleeding episodes in pediatric and adult patients with severe hemophilia B (16). However, fusion proteins prepared with nonnative macromolecules have safety concerns. Likewise, fusion of native human protein may cross react with endogenous homologues and could be reasons for long-term safety issue. Both PEGylated-biologics and fusion proteins are known to lose their long-circulating properties upon repeated administration (14).

9.2.2 Biopharmaceutical Properties of Biosimilars

In the United States, the Waxman-Hatch law, enacted in 1984, has been the chief driver for the development of generic drugs and industries surrounding generic drug development. Under this law, any commercial entity can make a copy, commonly known as generic drug products, of an innovator's product without submitting a full application required for new therapeutic agents. For generic versions of chemical drugs, the Food and Drug Administration (FDA), with a few exceptions, requires a comparative study that can demonstrate the plasma concentration-time profiles of the imitator's product are not statistically different from those of the innovator's product. However, the term, generic drug products, is not used to describe or denote biologicals that are a copy or therapeutically equivalent to innovator's biological products. The regulatory authorities call imitator's products that are of biological origin as biosimilars and thus distinguish generics of chemical drugs from those of biologics. This distinction was made because, unlike generics of chemical drugs, the production and formulation of an exact replica of innovators' biologics are unfeasible. As such, biosimilars are very similar to innovators' biological products, but with little differences in the structures that are unlikely to affect the therapeutic efficacy of the biologics. Because biologics are produced in live organisms or recombinant DNA technology that involve complex manufacturing process, a small change in the structures of biologics may trigger immune reactions. Thus, comparative plasma profiles of imitator's biologics versus innovator's reference products are not used to demonstrate equivalence of therapeutic agents that are derived from organisms or by recombinant technology. Instead, equivalency between imitator's biologics versus innovator's reference products is required to be demonstrated by structural analysis, functional assays, animal data, clinical studies, clinical immunogenicity assessment, and sometimes postmarketing safety monitoring. These studies are performed to ensure the identity, strength, quality, purity, and potency of biosimilars (19, 20). While the methodology for producing biosimilars may not be the same as that of the reference product, the manufacturing process of biosimilars must not adversely affect the identity, strength, quality, purity, and potency of the product. Similarly, analytical methods and the characterization procedure should have the precision and sensitivity for demonstrating equivalency of imitator's products with those of reference product (19). Other requirements for demonstrating the equivalency include studying lot-to-lot variability, using the functional assays in support of biologic activity and potency of the biosimilars, and population PK (popPK) parameters. Similarly, postmarketing safety monitoring is conducted to rule out any possible adverse effects of biosimilars (20). Overall, unlike comparative bioavailability study required for chemical drugs, a series of experimental evidence are required for demonstrating the similarity between imitator's and innovator's products.

The biopharmaceutical properties of biological products such as proteins, antibodies, and their biosimilar products that influence drug formulations and absorption are MW, structure, solubility, permeability, and 3D structures. In fact, these properties can not only influence PK/PD of biological products and their biosimilars but also affect their safety and efficacy.

9.3 PHARMACOKINETICS OF BIOLOGICAL PRODUCTS

9.3.1 Absorption

Major biopharmaceutical properties that influence the ADME of chemical drugs are solubility, partition coefficients, membrane permeability, and presence or absence of influx and efflux transporters at the site of absorption. Further, depending on the route of administration, the extent of availability of chemical drugs in the blood may vary between imperceptibly low to 100%. However, unlike chemical drugs, biologics are available only in injectable forms and are chiefly administered via one of the three major parenteral routes of administration: IV, SC, or IM routes. Like chemical drugs, IV administered biologics are 100% bioavailable. When biologics with MW of larger than 16 kDa are administered via the SC or IM routes, they must traverse one or more biologic membranes for absorption to occur and consequently the bioavailability of SC or IM biologics often vary from 20% to 95% (4, 5, 21). Because some biologics are absorbed via the lymphatic system (Figure 9.1) after SC or IM injections, the rate of absorption is very

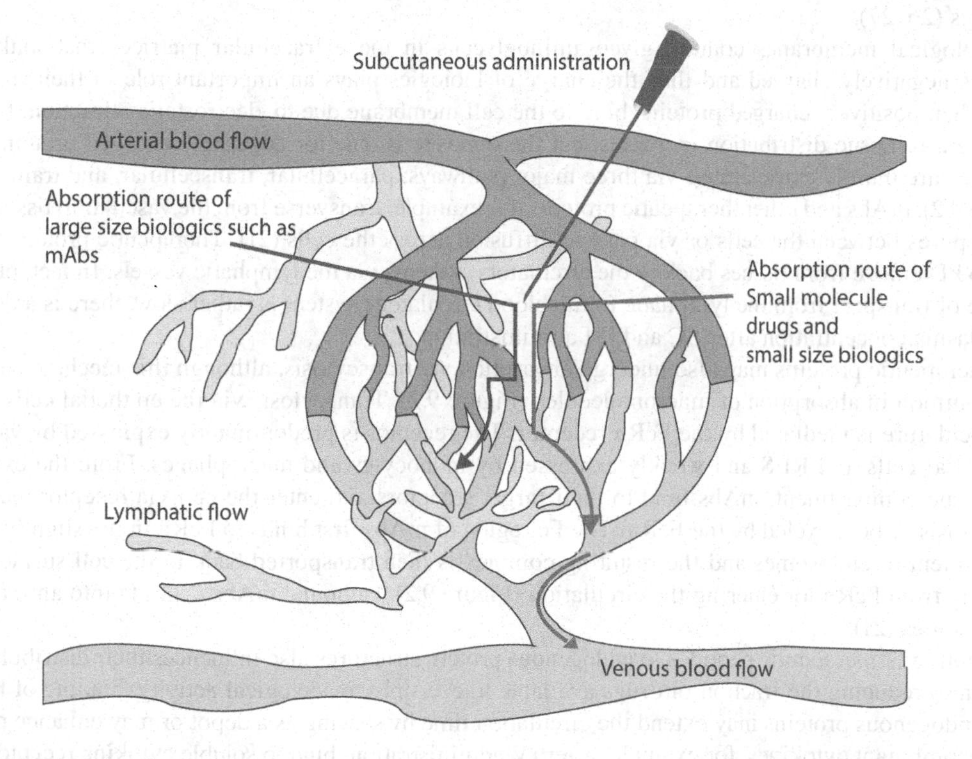

FIGURE 9.1 Molecular weight of biologics dictates the route by which they may move into the blood circulation or lymphatic system upon subcutaneous administration. Larger therapeutic proteins such as monoclonal antibodies (mAbs) with a molecular weight greater than 20 kDa generally traverse from the subcutaneous site to the lymphatic vessels and then to the systemic circulation. Smaller proteins such as insulin enter the bloodstream via the capillaries (redrawn based on ref. 23).

slow, which may significantly delay the time for maximum plasma concentration. The absorption of biologics is slow because of absorption-limited elimination and degradation of the drugs by the proteolytic enzymes in the lymphatics. This phenomenon is common in the case of antibody-based biologics (22, 23). However, biologics with MW below 16 kDa have no such issues with absorption. Recombinant human insulin and insulin aspart, for example, reach the peak plasma concentrations within hours (24). Insulin glargine, a derivative human insulin whose solubility at physiological pH is lower than that of native insulin, develop micro-precipitates in the SC tissue and release insulin in a pattern similar to basal insulin secretion for an extended period of time (25).

9.3.2 Distribution

In the case of chemical drugs, absorption and distribution of drugs occur concurrently after administration via all routes of administration except IV route. Regardless of the routes of administration, the drug moves from the circulatory system to tissues and then to the sites of action. Chemical drugs can also bind to plasma proteins, although only unbound drugs produce therapeutic effects. The distribution of chemical drugs depends on physical properties such as lipophilicity, extent of plasma protein binding, properties of biological membranes, and blood perfusion of the organs. Similarly, physicochemical properties of the biologics such as charge, lipophilicity, MW, structure, and properties of the capillaries influence the distribution of biologics. Since biologics are large MW drugs, the rate and extent of their distribution vary according to the rates of extravasation and partitioning into the interstitial space. In fact, because of the large size of biologics, their apparent volume of distribution (V_D) is typically between plasma volume (0.04 L/kg) and the volume of the extracellular space (0.23 L/kg) and often smaller than the V_D of chemical drugs (26, 27).

Biological membranes contain glycosaminoglycans in the extracellular matrices that make cell surfaces negatively charged and thus the charge of biologics plays an important role in their distribution. When positively charged proteins bind to the cell membrane due to electrostatic attraction, the rate and extent of tissue distribution increases, but the opposite is true for negatively charged proteins (28). Biologics are mainly extravasated via three major pathways: paracellular, transcellular, and transcytosis (Figure 9.2). mAbs and other therapeutic proteins, for example, transverse from the vascular to tissue sides via the pores between the cells or via passive diffusion across the cells (21). Therapeutic proteins larger than 16 kDa move from tissues back to the circulatory systems via the lymphatic vessels. In fact, because the rate of transport from the lymphatic to the blood circulatory system is rather slow, there is a delay to peak plasma concentration after SC and IM administration.

Therapeutic proteins may also undergo absorption via transcytosis, although this mechanism is not very common in absorption of macromolecules (Figure 9.2). Transcytosis via the epithelial cells lining the vasculature is mediated by the FcRn receptor. The receptor is predominantly expressed by vascular endothelial cells and RES and weakly expressed by monocytes and macrophages. From the extracellular tissue compartment, mAbs bind to their target receptors and enter the cell via receptor-mediated endocytosis or be recycled by the FcRn. The Fc region of mAbs first binds to FcRn in the slightly acidic environment of endosomes and the resulting complex is then transported back to the cell surface, dissociating from FcRn for entering the circulation (Figure 9.2). Unbound mAbs degrade into amino acids by lysosomes (21).

Binding of therapeutic peptides to endogenous protein structures also influences their distribution. In addition to reducing the fraction of drugs available to exert pharmacological activity, binding of biologics to endogenous proteins may extend the circulation time by serving as a depot or may enhance protein CL. Recombinant cytokines, for example, after IV administration, bind to soluble cytokine receptors and anti-cytokine antibodies and thereby either prolong the cytokine circulation time or expedite the rate of elimination. Like chemical drugs, peptides and proteins may also nonspecifically bind to plasma proteins: the extent of binding of metkephamid, a metenkephalin analog, to albumin is 44%–49% that of octreotide, a somatostatin analog, to lipoproteins is 65% (21).

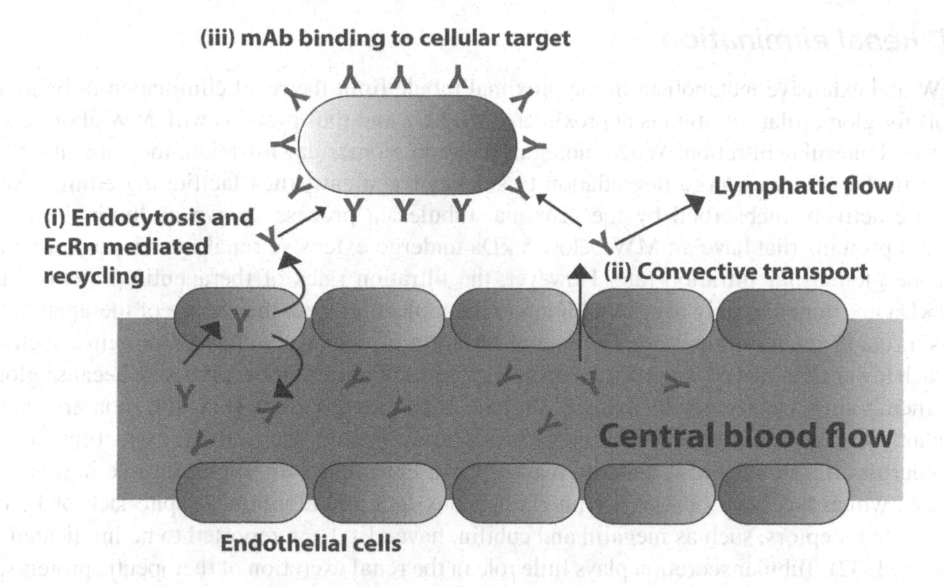

FIGURE 9.2 Distribution of biologics from the circulatory systems into tissues or cells: (i) Monoclonal antibodies (mAbs) are first taken up by vascular endothelial cells through endocytosis and then enter the interstitium or recycled back to the bloodstream via FcRn mediated exocytosis. (ii) mAbs move across intercellular pores or paracellular routes via convective transport resulting from the pressure gradient. mAbs thus transported may undergo lymphatic elimination. (iii) Binding of mAbs to the cellular target (redrawn based on ref. 5).

9.3.3 Clearance

For elimination, therapeutic proteins generally undergo degradation into small peptide fragments or amino acids by the lysosomal proteolytic enzymes. The resulting peptide fragments or amino acids are excreted via the kidneys or recycled for de novo protein/peptide synthesis via the endogenous amino acid pool (29). Unlike chemical drugs, biologics are not usually metabolized by the liver enzymes, and they break into smaller fragments by nonspecific peptidases, which is the chief reason for poor oral bioavailability of biologics. Of the two major pathways of metabolism – specific and nonspecific – therapeutic proteins are chiefly eliminated via nonspecific proteolytic catabolic pathways within the plasma and tissues because of the absence of fragment of antigen-binding (Fab) and Fc region. Furthermore, the lack of FcRn in proteins leads to substantially shorter half-lives compared to that of mAbs. The elimination of mAbs occurs due to the proteolysis by the RES. Binding of mAbs to target epitopes on cell surfaces leads to phagocytosis by the RES and then they are eliminated via intracellular catabolism. This type of endocytosis and elimination is a form of target-mediated drug disposition (TMDD), where the interaction of the drug and its pharmacological target serves as a major contributor in the kinetics of antibody distribution and elimination. In fact, the rate of uptake into cells and the elimination of mAbs by TMDD pathways are significantly influenced by the dose size, the level of expression of the target antigen or endogenous chemical, and the rate of receptor-mediated endocytosis and lysosomal catabolism. The rate of lymph flow and protein catabolism by the tissue also influences the elimination rate from the tissues. Because these are relatively slow processes and recycling of mAbs via FcRn pathway, mAbs have relatively long half-lives and small V_D. This slow process, in turn, decreases the frequency of dosing and increases the time to reach the steady-state concentration.

9.3.3.1 Renal elimination

Large MW and extensive metabolism in the proximal tubule limit the renal elimination of biologics. The MW cutoff for glomerular filtration is approximately 60 kDa and thus proteins with MW above 60 kDa do not undergo glomerular filtration. When biologics undergo glomerular filtration, they are either excreted unchanged in the urine, undergo degradation to smaller fragments, thus facilitating elimination in the urine, or are actively reabsorbed by the proximal tubules, a process known as luminal endocytosis. Peptides and proteins that have an MW below 5 kDa undergo extensive renal filtration at a rate that can approach the glomerular filtration rate. However, the filtration rates of therapeutic proteins with MW above 30 kDa are imperceptibly low. In addition to the molecular size, the charge of therapeutic proteins also plays a role in renal elimination. The rate of filtration of negatively charged proteins, such as albumin, is much lower than that of cationic and neutral proteins of similar molecular size because glomerular capillary membranes are negatively charged. Proteins that undergo glomerular filtration are catabolized predominantly by exopeptidases present on the brush border membrane of the proximal tubule. Proteins in the glomerular filtrate may also undergo reabsorption. FcRn plays an important role in this reabsorption process, which has been observed with endogenous IgG and albumin (despite lack of Fc domain) (30). Endocytic receptors, such as megalin and cubilin, have also been reported to be implicated in renal elimination (31, 32). Tubular secretion plays little role in the renal excretion of therapeutic proteins. Unlike small-molecule drugs, there are not many studies demonstrating the effects of renal impairment on the PK of therapeutic proteins. The total CL of protein drugs that are mainly eliminated by the renal route, such as anakinra, an interleukin-1 receptor antagonist with an MW of 17 kDa, may decline in patients with renal impairment (33). Further, the total CL of the anti-topotecan antibody 8C2, a 14 kDa murine antibody, was reported to be double in a murine model of diabetic nephropathy compared to that in control animals with urinary albumin excretion rate (34).

9.3.3.2 Hepatic elimination

In addition to metabolizing peptide and protein drugs, the liver plays an important role in the removal of proteins from the systemic circulation by eliminating drugs via receptor-mediated endocytosis and nonselective pinocytosis. In the case of hepatic elimination, endopeptidases attack the proteins by acting on the mid-section of the protein, and the resulting oligopeptides are then further degraded by exopeptidases. Amino acids and dipeptides and major metabolites of proteins finally join the endogenous amino acid pool. The rate of hepatic metabolism varies depending on specific amino acid sequences of the proteins. Substrates for hepatic metabolism include insulin, glucagon, and tissue plasminogen activator (t-PA, 65 kDa). An acidic endopeptidase, called endosomal acidic insulinase, mediate internalized insulin proteolysis at several sites (35). Specifically, an aspartic acid protease, called cathepsin D, is responsible for the proteolysis of insulin. Similarly, membrane-bound cathepsins B and D are responsible for the proteolysis of glucagon. For hepatic metabolism, proteins and peptides are first taken up hepatocytes. Small hydrophobic peptides enter the hepatocyte membrane via passive diffusion and larger proteins, such as t-PA, enter hepatic cells by FcRn-mediated transport processes. Radio-iodinated t-PA studies suggest that mannose and asialoglycoprotein receptors in the liver may have a role in facilitating t-PA uptake and CL (36). Low-density lipoprotein receptor-related protein, a hepatic membrane receptor may also contribute to the CL of t-PA.

9.4 PHARMACODYNAMICS OF BIOLOGICAL PRODUCTS

Biologics may exhibit either linear or nonlinear PK characteristics. Biologics with high molecular weight and large dose follow first order kinetics. Depending on the molecular size, biologics are absorbed via tissue penetration or moves from lymphatic vessels into the systemic circulation. With the increase in MW, absorption via the lymphatic system increases. Because absorption pathways involve both lymphatic and circulatory systems, biologics, such as belimumab, blisibimod, atacicept, and AMG 811, when given by the SC routes, show highly variable absorption profiles with a T_{max} of approximately one day to two weeks and an absolute bioavailability of 30%–82% (37). The absorption kinetics from the depot to the central compartment follows first-order kinetics for AMG 811 and the estimated absorption rate constant was 0.178 per day, which is close to those obtained for mAbs used in other disease types (37).

The rate elimination of biologics could be linear, nonlinear, or both. The nonantibody therapeutic proteins usually exhibit first-order kinetics because they lack specificity. The mechanism of nonspecific cellular elimination involves intracellular proteolysis after endocytosis by RES and FcRn recycling. This phenomenon tends to be linear as the usual doses of biologics lead to therapeutic concentrations that do not saturate FcRn on the cell surface. FcRn, present on the cell surface of vascular endothelial cells and RES cells, is responsible for the recycling of biologics with a functional Fc region. FcRn binds to the Fc region of mAb or its derivative, protects it from degradation in cellular lysosomes, and returns it to the cell surface where the biologic is dissociated from FcRn and re-enters the circulation. CDP7657, for example, does not have an Fc region and is a monovalent PEGylated protein that shields it from RES uptake or immunologic recognition (37).

MAbs that are highly specific for their target generally exhibit nonlinear eliminations and first bind to target antigen via the Fab region to form the drug-target complex for elimination to occur. The kinetics of the macromolecule-target complex is described by the TMDD model, which is a function of target amount, target turnover rate, the reversible binding between the drug and the target, and the elimination rate of the complex (2, 38). The factor that chiefly affects the overall CL in target-mediated CL is the target amount. The extent of expression of target antigens may dictate the nonlinear relationship between dose and systemic exposure. For example, in individuals with a high burden of disease, TMDD-mediated elimination is high and this form of CL changes as the burden of disease declines in case of successful treatment. The nonlinear behavior is observed at lower doses in the body, when membrane-bound or circulating targets remain unoccupied (2). As elaborated below (Figure 9.3), this nonlinear phenomenon in which the PK of biologics are affected due to their binding to drug receptors involves a high-affinity binding of biologics to low-capacity receptors and is more frequently observed in mAbs compared to nonantibody therapeutic proteins and chemical drugs due to the presence of Fab in the former that specifically and strongly binds

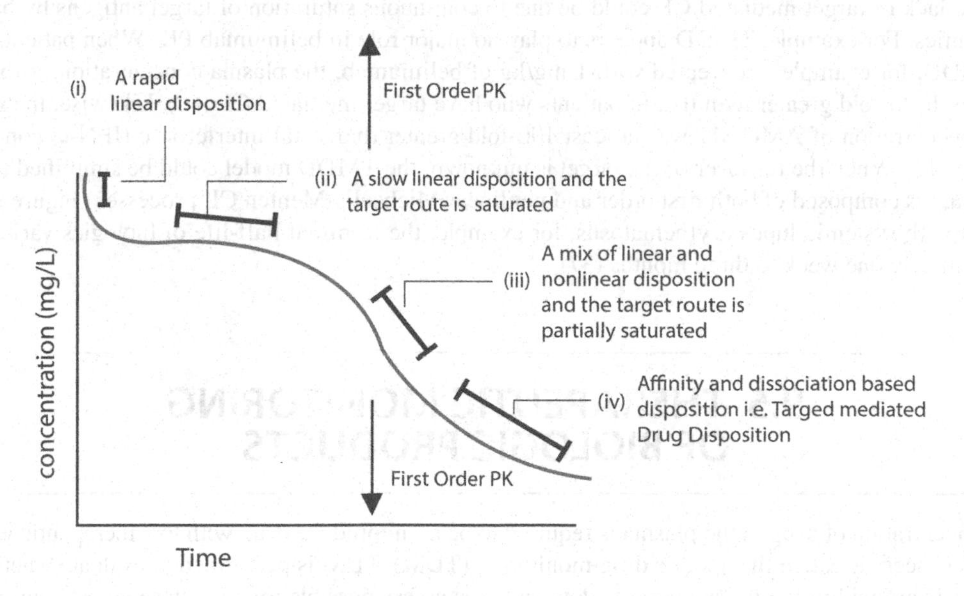

FIGURE 9.3 A log concentration of ligand versus time curve for target mediated drug disposition model showing the four phases disposition: Biologics first show a sharp linear decline in disposition (i); then a less than proportional disposition as the target route becomes saturated (iii); which is followed by a mixed first order and nonlinear disposition, signaling partial saturation of the target route; (iv) and they finally undergo an affinity- and dissociation-based disposition (redrawn based on ref. 21).

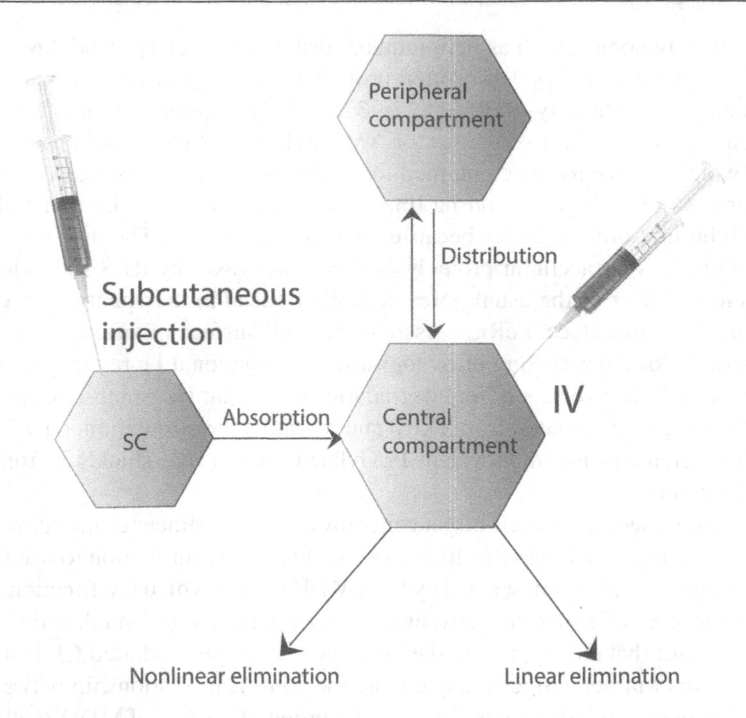

FIGURE 9.4 Monoclonal antibodies (mAbs) follow nonlinear two-compartmental pharmacokinetic profiles. They are eliminated from the central compartment. The total clearance is the sum of linear clearance plus nonlinear Michalis-Menten kinetics.

to its target. At low systemic concentrations of mAbs (39), TMDD accounts for a significant fraction of total mAb CL, whereas at high mAb concentrations, target-mediated elimination becomes saturated and total mAb CL decreases, approaching linear elimination (2).

The lack of target-mediated CL could be due to continuous saturation of target antigens by biologic therapeutics. For example, TMDD appears to play no major role in belimumab PK. When patients without TMDD, for example, are treated with 1 mg/kg of belimumab, the plasma concentration of the drug becomes 1000-fold greater than that in patients who have target-mediated CL (40). Likewise, the steady-state concentration of AMG 811 was at least 100-fold greater than total interferon-c (IFN-c) concentrations (41, 42). When the turnover of the target is unknown, the TMDD model could be simplified to a PK model that is composed of both first order and nonlinear Michaelis-Menten CL processes (Figure 9.4). In patients with systemic lupus erythematosus, for example, the terminal half-life of biologics varies from approximately one week to three months (37).

9.5 THERAPEUTIC MONITORING OF BIOLOGIC PRODUCTS

The concentration of drug in the plasma is required to be monitored for drug with low therapeutic window and this process is called therapeutic drug monitoring (TDM). TDM is performed to evaluate whether the drug levels are within the therapeutic window and thus assess possible toxicity, dose adjustment, or help develop a patient-specific therapy. Chemical drugs that are often monitored include carbamazepine, valproate, digoxin, and anticonvulsants (43). TDM of biologics or biosimilars has recently allowed for improved treatment plans and better clinical outcomes. The infliximab biosimilar CT-P13 has recently been approved for the treatment of autoimmune diseases due to the rising cost of healthcare and patent expirations for anti-TNF-alpha biologics. The common concerns of using biosimilar include immunogenicity, safety, and efficacy. Thus, sensitive assays are required to help define the activity profile. TDM assays, such as infliximab Drug Level ELISA®, can be used to quantitate biosimilars and can potentially be used for TDM of biosimilars for autoimmune diseases, leading to better clinical outcomes. With the approval of CT-P13, it is important to develop methods for TDM of biosimilars for better clinical outcomes for individuals with rheumatoid arthritis (RA) and inflammatory bowel disease (IBD). For patients with IBD and RA, adequate serum concentrations of biologics have been linked with sustained clinical responses; growing evidence suggests that TDM is beneficial in psoriasis as well. In the case of biosimilars, however, reports concerning the use of TDM are very scarce. The published TDM data focuses on the use of infliximab biosimilars in IBD and RA, with studies suggesting that biosimilar switching is feasible with few adverse events.

9.6 COMPUTATIONAL MODELS FOR PK/PD OF BIOLOGICS AND BIOSIMILARS

Over the past two decades, computational models or modeling and simulation have been deployed in various stages of discovery, development, and optimization of small molecules, biologics, and biosimilars. Modeling and simulation integrate and analyze all available information within a mathematical framework and help address issues concerning PK, PD, and toxicity of the most promising candidates and thus expedite the development of biologics. Computational models offer insights into the mechanism of action, biomarkers, and selection of first-in-human dose of biologics for translating animal data to humans (44). In the early stages of biologics development, target identification is explored through a systems pharmacology model to validate the target's role in disease pathology. These models help identify lead biologics that can inhibit or stimulate the target with high selectivity and potency. Once a lead compound with ideal affinity is selected for clinical development, different PK and PK/PD models can be applied to predict human PK, clinical dose, and dosage regimen, here we will discuss the use of computational models in studying PK of mAbs in humans using animal data (45).

Predicting human PK of mAbs from animal species is quite complex. In fact, cynomolgus monkeys have been used as the standard animal model for conducting nonclinical PK studies. This is because of remarkable similarities between monkeys and humans in several parameters: similar affinity constants for antigen, similar binding to FcRn, and identical elimination pathways (46).

As discussed above, biologics, such as mAbs, often follow the complex distribution and elimination pathways. At reduced doses, mAbs that exhibit higher selectivity for antigen binding follow the nonlinear distribution and elimination, i.e., TMDD. At a higher dose, the target is often saturated and mAbs follow linear elimination (Figure 9.4). Moreover, in the absence of dominant target expression (low endogenous level of antigen), mAbs display linear elimination with a longer half-life due to FcRn-mediated recycling (46). Thus, PK analysis of mAbs is divided into two broad categories based on their elimination kinetics: (1) mAbs that follow linear elimination kinetics and (2) mAbs that follow nonlinear elimination. Various commercial software can conduct both types of analysis, including NONMEM, R, MATLAB, Phoenix WinNonlin, SimBiology, gPKPDSim, Simulation Plus (47). Each type of software has its pros and cons, requiring expert training for successful execution and analysis.

9.6.1 mAbs with Linear PK

Usually, noncompartmental analysis (NCA) is commonly used for nonclinical PK studies of mAbs. NCA analysis assumes that there is no compartment in the body, drug elimination occurs exclusively from the sampling compartment, and that the mAbs follow linear PK (Figure 9.4). NCA assumes that TMDD is insignificant in overall elimination (46). NCA analysis receives input data from the monkey plasma concentration-time plot and then calculates the output in the form of the area under the concentration-time curve (AUC), C_{max}, T_{max}, half-life ($t_{1/2}$), CL, and V_D. For precise estimation of the parameters mentioned above, a more extensive data set per sample is needed for NCA analysis (48). However, compared to compartmental analysis, NCA is an easy, time-efficient, and cost-effective method since it is based on simple algebraic equations to estimate PK parameters, such as trapezoidal rule to estimate AUC. The next step of mAb development is to translate monkey PK data to humans to assess how a mAb might behave in humans before any clinical studies are conducted. Two empirical models are available for scaling monkey data to humans: (i) allometric scaling and (ii) species-invariant time methods (39).

9.6.1.1 Allometric scaling

Allometric scaling uses mathematical models to describe physiological, anatomical, and biochemical changes in animals as their size changes (44). Allometric scaling reasonably predicts two PK parameters such as CL and V_D from monkey to human using equation $\log Y_{human} = \log Y_{monkey} + b \log (W_{human}/W_{monkey})$, where Y = CL or V_D, W = body weight, and b = scaling factor. Allometric scaling assumes a linear correlation between W and Y after log-transformation. To estimate a CL value, b usually is 0.8–0.9 for estimation of V_D and b = 1. The allometry predicted CL and V_D could be verified by comparing the CL and V_D from clinical data (49).

9.6.1.2 Species-invariant time methods

This model, a modified version of allometric scaling for predicting human concentration-time plots from monkey data, assumes that the metabolic rate is the same in all mammals. The variation of physiological processes across the mammal stems from the biological clock: animals with shorter life spanning, such as cynomolgus monkey is 25–30 years old, and that for humans is 79 years. That means that PK profile in the animals is different from that in humans. Moreover, body weight is linked to life span, and therefore, an exponential weight factor is considered for optimal prediction of human concertation-time curves (49). $Time_{human} = Time_{monkey} \times (W_{human}/W_{monkey})^{1-b}$; $Concentration_{human} = Concentration_{monkey} \times (Dose_{human}/Dose_{monkey})$.

9.6.2 mAbs with Nonlinear PK

A more sophisticated compartmental analysis is performed for studying PK of mAbs exhibiting nonlinear kinetics. Compartmental models of mAbs can successfully simulate dose-dependent kinetics and provide insights into a mechanistic understanding of PK. This approach helps find an optimal first-in-human dose of mAbs, ensuring the safety and efficacy of the clinical studies. Several commonly used computational models are: (i) empirical models: two compartmental models with linear and nonlinear elimination, and a sub-model of popPK model, (ii) mechanistic models or TMDD model, (iii) physiologically based PK (PBPK) model, (iv) condensed PBPK model, and (v) quantitative systems pharmacology (QSP) modeling that integrates system biology approaches with traditional PK/PD approaches to understand the whole system.

9.6.3 Two Compartmental Models with Linear and Nonlinear Elimination

This model assumes linear and nonlinear elimination from the central compartment. The Michalis-Menten equation explains nonlinear elimination and estimates several PK parameters, such as maximal elimination velocity (V_{max}) and the drug concentration at half V_{max} (Km). The first-order kinetic process helps find the linear CL value. The total CL of mAbs depends on population estimates of linear and nonlinear CL (Figure 9.3). Then the application of allometric scaling predicts the human PK from this monkey model PK parameters (46).

9.6.4 Mechanistic Models

These models are more complex than the empirical model but provide a comprehensive understanding of mAb PK/PD. For example, a PBPK model answers the disposition of mAbs at physiological levels, such as nonspecific endothelial uptake, pH-dependent FcRn binding to the endosome, lysosomal degradation, the release of mAbs to plasma or interstitial fluids (50). In the condensed PBPK model of mAbs, the tissue compartment is connected to two central compartments: lymph nodes and plasma. The tissue compartment is further subdivided into vascular, endothelial, and interstitial space (51). The PBPK models integrate all these physiological parameters to derive the PK profile of mAbs. Scaling to human parameters (e.g., lymph flow rate) can be obtained from the literature or direct allometric scaling.

TMDD model is best suited for the mABs that exhibit extensive binding to a pharmacological target and display a dose-dependent PK profile (52). When the dose of mAbs is greater than the concentration of endogenous antigen/receptor target, it shows a multiphasic concentration-time plot (53, 54): (i) the mAb concentration declines readily due to rapid target binding and peripheral distribution; (ii) the target is saturated with mAb and mAbs follow a linear elimination (nontarget routes); (iii) the target is unsaturated, and mAbs is eliminated both from target- and nontarget-based routes (nonlinear elimination predominates); and (iv) due to extremely low target binding, target-based elimination of mAbs predominates, and the PK becomes linear (Figure 9.3). Scaling of TMDD model-derived PK parameters to humans is more complex, and simple allometry may be used from monkey data or considered similar to data obtained in monkey models.

9.7 SUMMARY AND FUTURE PERSPECTIVES

BP, PK, and PD of biologics and biosimilars do not always follow the principles of BP, PK, and PD of chemical drugs. In fact, there are both similarities and differences between the ADME of biologics and those of chemical drugs. Major factors that influence the ADME of biologics include MW, route of administration, transport mechanism, and metabolic pathways. Further, ADME of biologics is more specific to biological molecules, such as mAbs, than to large MW proteins or small peptides. To elucidate the BP, PK, and PD of biologics, various computational models have shown that PK of biologics follows a very complex pathway, which is distinctly different from that of chemical drugs. We believe that micro-engineered models, such as organ-on-chip, can potentially be used for studying the ADME of biologics and biosimilars and thus reduce or eliminate the use of animal models.

ACKNOWLEDGMENT

The authors thank Dr. Xiadong Feng, Dean of the California Northstate University College Pharmacy, for his inputs throughout the writing process of the manuscript.

REFERENCES

1. Gurevich EV and Gurevich VV. Therapeutic potential of small molecules and engineered proteins. Handb Exp Pharmacol 2014; 219:1–12. https://www.ncbi.nlm.nih.gov/pubmed/24292822
2. An G. Concept of pharmacologic target-mediated drug disposition in large-molecule and small-molecule compounds. J Clin Pharmacol 2020; 60(2):149–163. https://www.ncbi.nlm.nih.gov/pubmed/31793004
3. Small Molecules, Large Biologics and the Biosimilar Debate by AZBio. https://www.azbio.org/small-molecules-large-biologics-and-the-biosimilar-debate. Accessed date: February 18, 2021.
4. Mould DR. The pharmacokinetics of biologics: A primer. Dig Dis 2015; 33(Suppl 1):61–69. https://www.ncbi.nlm.nih.gov/pubmed/26367860
5. Lobo ED, Hansen RJ and Balthasar JP. Antibody pharmacokinetics and pharmacodynamics. J Pharm Sci 2004; 93(11):2645–2668. https://www.ncbi.nlm.nih.gov/pubmed/15389672
6. Porter CJ and Charman SA. Lymphatic transport of proteins after subcutaneous administration. J Pharm Sci 2000; 89(3):297–310. https://www.ncbi.nlm.nih.gov/pubmed/10707011
7. Supersaxo A, Hein W, Gallati H and Steffen H. Recombinant human interferon alpha-2a: Delivery to lymphoid tissue by selected modes of application. Pharm Res 1988; 5(8):472–476. https://www.ncbi.nlm.nih.gov/pubmed/3244653
8. Ezan E, Becher F and Fenaille F. Assessment of the metabolism of therapeutic proteins and antibodies. Expert Opin Drug Metab Toxicol 2014; 10(8):1079–1091. https://www.ncbi.nlm.nih.gov/pubmed/24897152
9. Xu X and Vugmeyster Y. Challenges and opportunities in absorption, distribution, metabolism, and excretion studies of therapeutic biologics. AAPS J 2012; 14(4):81–91. https://www.ncbi.nlm.nih.gov/pubmed/22864668
10. Roopenian DC and Akilesh S. FcRn: the neonatal Fc receptor comes of age. Nat Rev Immunol 2007; 7(9):715–725. https://www.ncbi.nlm.nih.gov/pubmed/17703228
11. Parkins DA and Lashmar UT. The formulation of biopharmaceutical products. Pharm Sci Technol Today 2000; 3(4):129–137. https://www.ncbi.nlm.nih.gov/pubmed/10754542
12. Kontermann RE (2012). Half-life modulating strategies - An introduction. In Kontermann R (Ed.) Therapeutic Proteins: Strategies to Modulate Their Plasma Half-Lives (pp.1–21). https://onlinelibrary.wiley.com/doi/book/10.1002/9783527644827
13. Pasut G and Veronese FM. State of the art in PEGylation: The great versatility achieved after forty years of research. J Control Release 2012; 161(2):461–472. https://www.ncbi.nlm.nih.gov/pubmed/220.94104
14. Mitragotri S, Burke PA and Langer R. Overcoming the challenges in administering biopharmaceuticals: Formulation and delivery strategies. Nat Rev Drug Discov 2014; 13(9):655–672. https://www.ncbi.nlm.nih.gov/pubmed/25103255
15. Bush MA, Matthews JE, De Boever EH, Dobbins RL, Hodge RJ, Walker SE, et al. Safety, tolerability, pharmacodynamics and pharmacokinetics of albiglutide, a long-acting glucagon-like peptide-1 mimetic, in healthy subjects. Diabetes Obes Metab 2009; 11(5):498–505. https://www.ncbi.nlm.nih.gov/pubmed/19187286
16. Santagostino E, Negrier C, Klamroth R, Tiede A, Pabinger-Fasching I, Voigt C, et al. Safety and pharmacokinetics of a novel recombinant fusion protein linking coagulation factor IX with albumin (rIX-FP) in hemophilia B patients. Blood 2012; 120(12):2405–2411 https://www.ncbi.nlm.nih.gov/pubmed/22859609
17. Constantinou A, Epenetos AA, Hreczuk-Hirst D, Jain S, Wright M, Chester KA, et al. Site-specific polysialylation of an antitumor single-chain Fv fragment. Bioconjug Chem 2009; 20(5):924–931. https://www.ncbi.nlm.nih.gov/pubmed/19402707
18. Mero A, Pasqualin M, Campisi M, Renier D and Pasut G. Conjugation of hyaluronan to proteins. Carbohydr Polym 2013; 92(2):2163–2170. https://www.ncbi.nlm.nih.gov/pubmed/23399272
19. FDA. Quality Considerations in Demonstrating Biosimilarity to a Reference Protein Product. 2012. https://www.fda.gov/media/135612/download
20. FDA. Scientific Considerations in Demonstrating Biosimilarity to a Reference Product. 2012. https://www.fda.gov/media/135612/download
21. Mclachlan AJ and Adiwidjaja J (2021). Pharmacokinetics of biologics. In Ramzan I (Ed.) Biologics, Biosimilars, and Biobetters: An Introduction for Pharmacists, Physicians and Other Health Practitioners (pp. 125–146). In Tech. DOI: 10.1002/9781119564690
22. Lin JH. Pharmacokinetics of biotech drugs: Peptides, proteins and monoclonal antibodies. Curr Drug Metab 2009; 10(7):661–691. https://www.ncbi.nlm.nih.gov/pubmed/19702530
23. Jones GB, Collins DS, Harrison MH, Thyagarajapuram NR, Justin M and Wright JM. Subcutaneous drug delivery: An evolving enterprise. Sci Transl Med. 2017; 9:eaaf9166. https://pubmed.ncbi.nlm.nih.gov/28855399/

24. Osterberg O, Erichsen L, Ingwersen SH, Plum A, Poulsen HE and Vicini P. Pharmacokinetic and pharmacodynamic properties of insulin aspart and human insulin. J Pharmacokinet Pharmacodyn 2003; 30(3):221–235. https://www.ncbi.nlm.nih.gov/pubmed/14571692
25. Morello CM. Pharmacokinetics and pharmacodynamics of insulin analogs in special populations with type 2 diabetes mellitus. Int J Gen Med 2011; 4:827–835. https://www.ncbi.nlm.nih.gov/pubmed/22267935
26. Datta-Mannan A. Mechanisms influencing the pharmacokinetics and disposition of monoclonal antibodies and peptides. Drug Metab Dispos 2019; 47(10):1100–1110. https://www.ncbi.nlm.nih.gov/pubmed/31043438
27. Glassman PM, Abuqayyas L and Balthasar JP. Assessments of antibody biodistribution. J Clin Pharmacol 2015; 55(Suppl 3):S29–S38. https://www.ncbi.nlm.nih.gov/pubmed/25707961
28. Tibbitts J, Canter D, Graff R, Smith A and Khawli LA. Key factors influencing ADME properties of therapeutic proteins: A need for ADME characterization in drug discovery and development. MAbs 2016; 8(2):229–245. https://www.ncbi.nlm.nih.gov/pubmed/26636901
29. Tang L, Persky AM, Hochhaus G and Meibohm B. Pharmacokinetic aspects of biotechnology products. J Pharm Sci 2004; 93:2184–2204. https://pubmed.ncbi.nlm.nih.gov/15295780/
30. Sand KM, Bern M, Nilsen J, Noordzij HT, Sandlie I and Andersen JT. Unraveling the interaction between FcRn and albumin: Opportunities for design of albumin-based therapeutics. Front Immunol 2014; 5:682. https://www.ncbi.nlm.nih.gov/pubmed/25674083
31. Christensen EI, Birn H, Storm T, Weyer K and Nielsen R. Endocytic receptors in the renal proximal tubule. Physiology (Bethesda) 2012; 27(4):223–236. https://www.ncbi.nlm.nih.gov/pubmed/22875453
32. Chadha GS and Morris ME. Monoclonal antibody pharmacokinetics in type 2 diabetes mellitus and diabetic nephropathy. Curr Pharmacol Rep. 2016; 2:45–56. https://link.springer.com/article/10.1007/s40495-016-0048-z
33. Yang BB, Baughman S and Sullivan JT. Pharmacokinetics of anakinra in subjects with different levels of renal function. Clin Pharmacol Ther 2003; 74(1):85–94. https://www.ncbi.nlm.nih.gov/pubmed/12844139
34. Engler FA, Zheng B and Balthasar JP. Investigation of the influence of nephropathy on monoclonal antibody disposition: A pharmacokinetic study in a mouse model of diabetic nephropathy. Pharm Res 2014; 31(5):1185–1193. https://www.ncbi.nlm.nih.gov/pubmed/24203494
35. Authier F, Danielsen GM, Kouach M, Briand G and Chauvet G. Identification of insulin domains important for binding to and degradation by endosomal acidic insulinase. Endocrinology 2001; 142(1):276–289. https://www.ncbi.nlm.nih.gov/pubmed/11145591
36. Smedsrod B and Einarsson M. Clearance of tissue plasminogen activator by mannose and galactose receptors in the liver. Thromb Haemost 1990; 63(1):60–66. https://www.ncbi.nlm.nih.gov/pubmed/2160132
37. Yu T, Enioutina EY, Brunner HI, Vinks AA and Sherwin CM. Clinical pharmacokinetics and pharmacodynamics of biologic therapeutics for treatment of systemic lupus erythematosus. Clin Pharmacokinet 2017; 56(2):107–125. https://www.ncbi.nlm.nih.gov/pubmed/27384528
38. Mager DE and Jusko WJ. General pharmacokinetic model for drugs exhibiting target-mediated drug disposition. J Pharmacokinet Pharmacodyn 2001; 28(6):507–532. https://www.ncbi.nlm.nih.gov/pubmed/11999290
39. Kamath AV. Translational pharmacokinetics and pharmacodynamics of monoclonal antibodies. Drug Discov Today Technol 2016; 21–22:75–83. https://www.ncbi.nlm.nih.gov/pubmed/27978991
40. Struemper H, Chen C and Cai W. Population pharmacokinetics of belimumab following intravenous administration in patients with systemic lupus erythematosus. J Clin Pharmacol 2013; 53(7):711–720. https://www.ncbi.nlm.nih.gov/pubmed/23681782
41. Chen P, Vu T, Narayanan A, Sohn W, Wang J, Boedigheimer M, et al. Pharmacokinetic and pharmacodynamic relationship of AMG 811, an anti-IFN-gamma IgG1 monoclonal antibody, in patients with systemic lupus erythematosus. Pharm Res 2015; 32(2):640–653. https://www.ncbi.nlm.nih.gov/pubmed/25213774
42. Dirks NL and Meibohm B. Population pharmacokinetics of therapeutic monoclonal antibodies. Clin Pharmacokinet 2010; 49(10):633–659. https://www.ncbi.nlm.nih.gov/pubmed/20818831
43. Ghiculescu RA. Therapeutic drug monitoring: Which drugs, why, when and how to do it. Aust Prescr 2008; 31:42–44. https://www.nps.org.au/australian-prescriber/articles/therapeutic-drug-monitoring-which-drugs-why-when-and-how-to-do-it
44. Deng R, Iyer S, Theil FP, Mortensen DL, Fielder PJ and Prabhu S. Projecting human pharmacokinetics of therapeutic antibodies from nonclinical data: What have we learned? MAbs 2011; 3(1):61–6. https://www.ncbi.nlm.nih.gov/pubmed/20962582
45. Dixit S (2015). Computational Methods in the Optimization of Biologic Modalities. In Kumar S and Singh SK(Eds.) Developability of Biotherapeutics Computational Approaches (pp. 35–60). Boca Raton, FL: CRC Press.
46. Dong JQ, Salinger DH, Endres CJ, Gibbs JP, Hsu CP, Stouch BJ, et al. Quantitative prediction of human pharmacokinetics for monoclonal antibodies: Retrospective analysis of monkey as a single species for first-in-human prediction. Clin Pharmacokinet 2011; 50(2):131–142. https://www.ncbi.nlm.nih.gov/pubmed/21241072

47. Hosseini I, Gajjala A, Yadav DB, Sukumaran S, Ramanujan S, Paxson R, et al. gPKPDSim: A SimBiology((R))-based GUI application for PKPD modeling in drug development. J Pharmacokinet Pharmacodyn 2018; 45(2):259–275. https://www.ncbi.nlm.nih.gov/pubmed/29302838
48. Kuester K and Kloft C (2006). Pharmacokinetics of Monoclonal Antibodies. In Meibohm B (Ed.) Pharmacokinetics and Pharmacodynamics of Biotech Drugs (pp. 45–91). Weinheim, German: Wiley-VCH GmbH and Co. KGaA.
49. Wang J, Iyer S, Fielder PJ, Davis JD and Deng R. Projecting human pharmacokinetics of monoclonal antibodies from nonclinical data: Comparative evaluation of prediction approaches in early drug development. Biopharm Drug Dispos 2016; 37(2):51–65. https://www.ncbi.nlm.nih.gov/pubmed/25869767
50. Jones HM, Zhang Z, Jasper P, Luo H, Avery LB, King LE, et al. A physiologically-based pharmacokinetic model for the prediction of monoclonal antibody pharmacokinetics from in vitro data. CPT Pharmacometrics Syst Pharmacol 2019; 8(10):738–747. https://www.ncbi.nlm.nih.gov/pubmed/31464379
51. Li L, Gardner I, Dostalek M and Jamei M. Simulation of monoclonal antibody pharmacokinetics in humans using a minimal physiologically based model. AAPS J 2014; 16(5):1097–1109. https://www.ncbi.nlm.nih.gov/pubmed/25004823
52. Luu KT, Bergqvist S, Chen E, Hu-Lowe D and Kraynov E. A model-based approach to predicting the human pharmacokinetics of a monoclonal antibody exhibiting target-mediated drug disposition. J Pharmacol Exp Ther 2012; 341(3):702–708. https://www.ncbi.nlm.nih.gov/pubmed/22414855
53. Peletier LA and Gabrielsson J. Dynamics of target-mediated drug disposition: Characteristic profiles and parameter identification. J Pharmacokinet Pharmacodyn 2012; 39(5):429–451. https://www.ncbi.nlm.nih.gov/pubmed/22851162
54. Gabrielsson J and Peletier LA. Pharmacokinetic steady-states highlight interesting target-mediated disposition properties. AAPS J 2017; 19(3):772–786. https://www.ncbi.nlm.nih.gov/pubmed/28144911

Understanding Variability, Stability, and Immunogenicity of Biosimilars

10

Tibebe Woldemariam and Janie Yu

Corresponding author: Tibebe Woldemariam

Contents

10.1 INTRODUCTION

Biosimilars have been developed and marketed as lower-cost alternatives to reference products. Biosimilarity refers to a product's similarity to its reference product only and does not indicate similarity to other biosimilar products. Biosimilars have a non-proprietary name plus a suffix designated

DOI: 10.1201/9780429485626-10

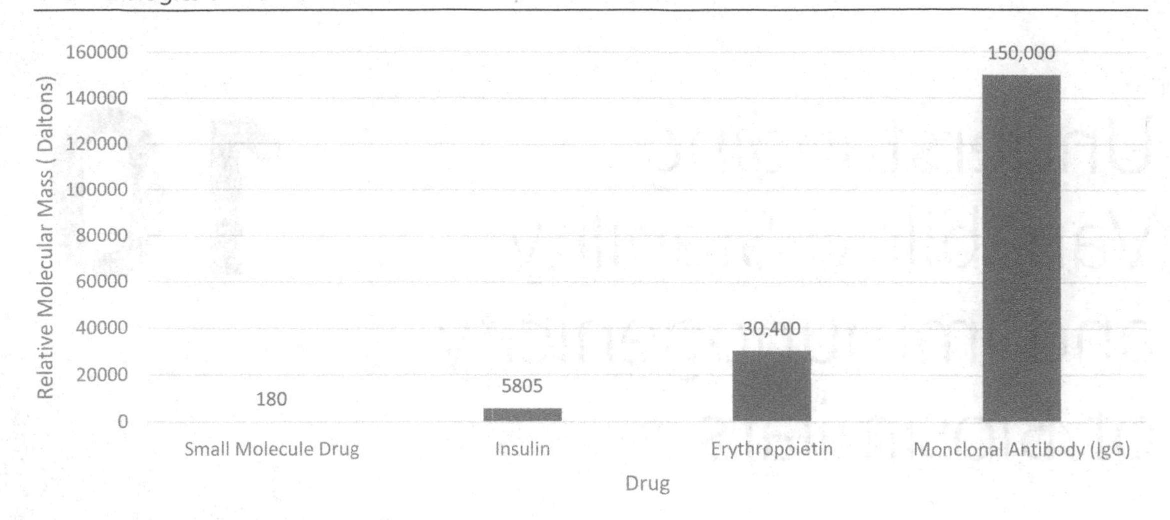

FIGURE 10.1 Difference of size between small molecule drugs and larger molecule biosimilars.[5]

by the US Food and Drug Administration (FDA) consisting of four lower-case letters that have no intended meaning.

The first biologic drug approved by the FDA was a monoclonal antibody, ORTHOCLONE OKT3® (muromonab), in 1982.[1] Biologics are recombinant therapeutic protein drugs produced to combat difficult to treat diseases using biotechnology.[2] Biologics are termed the "originators" or "reference product" of recombinant protein drugs that paved the way for improving treatment guidelines and drug development.[2] Biologics could be an antibody or non-antibody drugs. Non-antibody drugs include insulin, and antibody drugs include monoclonal antibodies such as Rituximab.[3]

Biosimilars are essentially biological products that are highly similar to the reference product with no clinically meaningful differences between active ingredients and minor differences in inactive ingredients when comparing the purity, potency, and safety profile.[4]. They are a cheaper alternative to biologics, whose patents have ended or are coming to an end.[2]

One of the first biosimilars to be approved by the FDA in 2015 was ZARXIO® (filgrastim-sndz), a biosimilar for the reference product Neupogen.[4] Unlike small-molecule drugs, biosimilars are more difficult to reproduce in terms of ensuring bioactivity, stability, and immunogenicity due to their large complex structure and different batch variabilities (Figure 10.1).[5] The goals of biosimilars are to provide cost saving alternative drugs with a similar biologic efficacy and safety profile when treating diseases.

10.2 VARIABILITY OF BIOSIMILARS

The goal of biosimilar development is to generate a biologic drug product that is highly similar to the reference biologic product with no clinically meaningful differences in terms of safety and efficacy. [6,7] There is a strong relationship between the manufacturing process and the characteristics of the final biosimilar.[8] Even small changes in manufacturing can result in altered protein stability and impact post-translational modifications such as glycosylation (i.e. the addition of glycans [carbohydrate groups] to the structure of a mAb).[8,9] Glycans can impact biologic activity and, in so doing, affect overall efficacy, safety, and immunogenicity.[8]

Given the delicate and sensitive nature of the manufacturing process and that each manufacturer is required to develop a new process for each biosimilar, it is necessary that appropriate safeguards be established to protect patients.

Biosimilars can never be identical to originators. They are "living" molecules, and there is variability among lots, even in the reference products. The FDA's approval criteria require a biosimilar to be highly similar to its original biologic and show "no clinically meaningful differences." They are precisely engineered to match an existing reference biologic medicine in all relevant attributes. For example, REMICADE® (infliximab) has had several manufacturing changes and other changes over its lifespan, yet these changes have not had a negative impact on its efficacy or safety.

There is an intra- and inter-variability in the biosimilar manufacturing processes in demonstrating biosimilarities to a reference product. Biosimilars are not usually indistinguishable from the originator compound because of molecular complexity and production using recombinant DNA techniques in living organ systems.

Marginal clinical studies (or perhaps even no clinical studies) might be sufficient for the approval of the biosimilar if biochemical, biophysical, and biological data (structural and functional analysis) can demonstrate that the innovator drug (also known as the reference product) and its biosimilar are identical (or similar enough) and that there is no effect of any difference in the mode of formulation, container closure as well as handling and administration so that equivalence in clinical performance (pharmacokinetics/pharmacodynamics (PK/PD) and immunogenicity) can be secured.[10]

The evaluation of biosimilarity between a proposed biosimilar product and its reference product involves the vigorous characterization of the proposed biosimilar product, including comparative physicochemical and functional studies.

The key requirements for biosimilarity are matching structure and function; similar PK/PD, clinical efficacy and safety; dose (strength), and method of administration. Biosimilars must have certain relationships with their reference medicines in terms of structure. The amino acid sequence must be identical, while secondary, tertiary, and quaternary folding patterns must be indistinguishable across multiple redundant methods of analysis. Owing to biosynthesis in living cells, post-translational modifications such as glycosylation and sialylation demonstrate a certain degree of variability for biologic medicines, and each batch of a given biologic can be differentiated from other batches using robust and refined analytical methods.[11]

Manufacturing changes may result in significant shifts in characteristics, which in some cases may be associated with changes in biological function. For example, the antibody-dependent cellular cytotoxicity activity of reference RITUXAN® (rituximab) was going through an obvious shift following a change in the manufacturing processes.[12] Nevertheless, this variability is tightly controlled within acceptable limits to ensure that such shifts have no relevant clinical impact.

Stable manufacturing of biosimilars is safeguarded through control of the manufacturing process, including control of raw material, process design, in-process testing, control of process parameters, and final dosage form.

Biosimilars are not identical to the reference product because variability is natural in all biologics. The variability starts as early-stage chain amino acids are folded and transformed into final stage 3-D proteins. Variability also results from frequent changes or minor inconsistencies in biologic manufacturing processes. Variations occur between individual production lots, when production volume increases, when new technologies are employed, or when manufacturing changes are made. When biosimilar sponsors are applying for approval, they must measure and report the variations in different lots of the original biologic. The series of variations in the characteristics of the original biologic then becomes the limit for acceptable variations of the biosimilar.

10.3 BIOPHARMACEUTICAL STABILITY STUDIES OF BIOSIMILARS

Biopharmaceutical stability studies for biologic therapeutics are vital to evaluating sensitivity to factors that could cause aggregation and degradation, which impacts biologic activity, product safety, and quality. To reduce the risk of degradation and maintain the biological activity of the product, suitable conditions for storage and shelf life must be established. By understanding degradation routes, it is possible to establish which critical quality attributes (CQAs) are more susceptible to change throughout the lifespan of the biologic.[13]

Variations in biopharmaceutical stability due to degradation, chemical or physical instability can change protein folding and the three-dimensional protein structure, thus affecting efficacy. The use of multiple justified and validated analytical methodologies may be required to determine stability. Biological activity assays and the quantitative detection of degradation products are also required to meet the requirements of the current Good Manufacturing Practice (cGMP) compliant stability program. Environmental factors (such as temperature, exposure to oxygen or changes in pH), adsorption onto surfaces, and interactions with excipients or inactive substances that serve as the vehicle or medium drug in question, can all affect stability. In order to demonstrate the stability, forced degradation studies should be designed and conducted to determine stability-indicating methods suitable for ongoing stability studies.[13]

It is essential to conduct biopharmaceutical forced degradation, accelerated stability studies, short and long-term stability studies for biosimilars, as it is done with antibodies and other biologics. Forced degradation is a degradation of new drug substance and drug product at conditions more severe than accelerated conditions. Forced degradation studies show the chemical manners of the molecule, which supports the development of formulation and packaging. It is required to perform stability studies of new drug moiety before filing in the registration dossier. The stability studies include long-term studies (12 months) and accelerated stability studies (6 months). However, intermediate studies (6 months) can be performed at conditions milder than that used in accelerated studies.[14]

Stability-indicating analytical method developments provide confirmation that any changes in physicochemical properties, structure, aggregation, biological activity, visual appearance, impurities, excipient degradation, and container/closure interactions will be detected. The knowledge from forced degradation studies, in conjunction with a detailed understanding of the product and process, helps to establish CQAs.

External factors such as interaction with formulation components and storage conditions can impact biophysical behavior and higher-order structure, therefore, it is important to characterize the impact of these factors on the structure and conformational stability. Some of the excipients which may also be susceptible to degradation could also react with the main biologic product.

Extractables and leachables studies confirm that the product container/packaging system demonstrates sufficient stability over the relevant lifecycle of the product in question.

Regulatory approval for a biosimilar product is provided on the basis of its comparability to an originator product. A thorough physicochemical and functional comparability implementation is a key element in demonstrating biosimilarity. Studies have been reported for therapeutic proteins that have been mixed using magnetic stir bars, and it was demonstrated that this type of mixing may induce aggregates.[13]

Many proteins are glycosylated, and some have other post-translational modifications such as phosphorylation, which also affects their potential degradation pathways as well as the kinetics of their degradation. Proteins are typically sensitive to slight changes in solution chemistry. They remain compositionally and conformationally stable only within a relatively narrow range of pH and osmolarity, and many require additionally supportive formulation components to remain in solution, particularly over time.[14]

Advances in analytical chemistry have identified many degradation pathways that can occur in recombinant protein therapeutics over time. These pathways generate either chemical or physical instability. *Chemical instability* refers to the formation or destruction of covalent bonds within a polypeptide or protein structure. Chemical modifications of proteins include oxidation, deamidation, reduction, and hydrolysis.[15] Unfolding, dissociation, denaturation, aggregation, and precipitation are known as *conformational* or *physical* instabilities.[15] In some cases, protein degradation pathways are synergistic: A chemical event may trigger a physical event, such as when oxidation is followed by aggregation.

There are several protein degradation events: oxidation, photodegradation, disulfide scrambling, deamidation, aggregation, precipitation, dissociation, and fragmentation.

Primary and higher-order protein structures of biosimilars can be analyzed using a variety of methods that include high-performance liquid chromatography electrospray ionization mass spectrometry

(HPLC-ESI-MS), peptide mapping with UV and MS detection, circular dichroism (CD), Fourier transform infrared (FTIR) spectroscopy, hydrogen deuterium exchange (HDX) MS, 1D 1H nuclear magnetic resonance (NMR) spectroscopy, X-ray crystallography, and differential scanning calorimetry (DSC). Charge and amino acid modifications are assessed using cation exchange chromatography (CEX) and peptide mapping using reversed-phase (RP) HPLC.

To ensure product safety and efficacy, protein therapeutics must meet defined quality characteristics after manufacture as well as at the end of their designated shelf lives. Many physical and chemical factors can affect the quality and stability of biopharmaceutical products, particularly after long-term storage in a container–closure system, which is likely to be subject to variations in temperature, light, and agitation with shipping and handling.

Compared with traditional chemical pharmaceuticals, proteins are considerably larger molecular entities with inherent physiochemical complexities. Proteins are typically sensitive to slight changes in solution chemistry. They remain compositionally and conformationally stable only within a relatively narrow range of pH and osmolarity, and many require additionally supportive formulation components to remain in solution, particularly over time. Even lyophilized protein products experience degradation.

Proteins and peptides are susceptible to oxidative damage through the reaction of certain amino acids with oxygen radicals present in their environment.[16] Oxidation can alter a protein's physiochemical characteristics (e.g. folding and subunit association) and lead to aggregation or fragmentation. It can also induce potential negative effects on potency and immunogenicity depending on the position of oxidized amino acids in a protein relative to its functional domain.

Photooxidation can change the primary, secondary, and tertiary structures of proteins and lead to differences in long-term stability, bioactivity, or immunogenicity.[17] Exposure to light can trigger a chain of biochemical events that continue to affect a protein even after the light source is turned off. These effects depend on the amount of energy imparted to a protein and the presence of oxygen.

With many recombinant proteins, changes in peptide and protein structure are observed through the non-enzymatic deamidation of glutamine and asparagine residues. This can have varying effects on their physiochemical and functional stability.[18] It has been observed that deamidation of hGH alters proteolytic cleavage of the human growth hormone.[18]

10.3.1 Aggregation and Precipitation

Aggregated proteins are a significant concern for biopharmaceutical products because they may be associated with decreased bioactivity and increased immunogenicity. Macromolecular protein complexes can trigger a patient's immune system and mount an antigenic response.[19] Large macromolecular aggregates also can affect fluid dynamics in organ systems such as the eyes.[20]

Aggregation is a common problem encountered during the manufacture and storage of proteins.[20] The potential for aggregated forms is often enhanced by exposure of a protein to liquid–air, liquid–solid, and even liquid–liquid interfaces.[21] Mechanical stresses of agitation (shaking, stirring, pipetting, or pumping through tubes) can cause protein aggregation. Freezing and thawing can promote it as well. Solution conditions such as temperature, protein concentration, pH, and ionic strength can affect the rate and amount of aggregates observed. The formulation in sucrose can increase aggregation over time because of protein glycation when sucrose is hydrolyzed.[22] The presence of certain ligands, including certain ions, may enhance aggregation. Interactions with metal surfaces can lead to epitaxic denaturation, which triggers aggregate formation. Foreign particles from the environment, manufacturing process, or container–closure system (e.g. silicone oil) can also induce aggregation.[23]

Multimeric proteins with two or more subunits can become dissociated into monomers, and monomers (or single peptide chain proteins) can degrade into peptide fragments. Non-enzymatic fragmentation usually proceeds by hydrolysis of peptide bonds between amino acids, releasing polypeptides of lower molecular weight than the intact parent protein. Peptide bonds of Asp–Gly and Asp–Pro are most susceptible to hydrolytic protein cleavage.[24]

10.3.2 Protein Aggregation Analysis

A major concern in manufacturing protein biopharmaceuticals is their propensity to form aggregates. These undesirable associated states of the monomeric form can be reversible or irreversible and can range in size from a dimer to particles that may contain trillions (or more) of monomer units that can be visible to the naked eye. In general, aggregation can be a problem for any protein biopharmaceutical. Beyond the obvious detrimental impact of reducing the actual dosing concentration of the drug (as most aggregates have little or substantially reduced drug activity in comparison with the monomeric form of the drug), by far the greatest concern surrounding the presence of aggregates is their unpredictable ability to give rise to adverse toxicological and immunological responses, which in extreme cases can result in severe responses that can be life-threatening.[25,26]

As a result, the area of aggregation has attracted considerable amounts of research attention. Weak evidence has mounted over the years pointing to factors such as the amount, size, and native-like repeating array structure of these aggregates as potential key attributes associated with the adverse effects.[9] Hence, there is considerable scrutiny and interest in how the biopharmaceutical industry monitors and assesses protein biopharmaceutical aggregation in terms of its detection, quantification, and characterization threatening.[27,28]

Protein aggregation characterization is key to successful biopharmaceutical development and manufacturing. Protein aggregates can form at any stage of the development or manufacturing process including bioprocessing, purification, formulation, and packaging, and also during storage. Protein aggregates can be considered an impurity, or a molecular variant, resulting from changes that take place over time and/or by the action of physical factors such as light, temperature, pH, water, shear-forces, or by reaction with an excipient in the formulation, concentration of buffers or interaction with the container/closure system (e.g. elastomer seals or glass delamination). Aggregates include reversible non-covalent and irreversible covalent bonded species, dimers, oligomers, and higher multiples of the desired protein product and can be present as small soluble particles ranging in size from a few nm to large sub-visible/visible particles up to microns.

Protein aggregation is a complicated phenomenon, which is sensitive to solvent conditions, sample history, protein sequence, and so on. The understanding aggregation will depend on the identification of patterns within this vast parameter space. These patterns can be described using five categories: size, reversibility/dissociability, conformation, chemical modification, and morphology, to consistently describe protein aggregates. Characterizations of aggregates or improperly associated species that affect product safety or efficacy are a huge task and require highly skilled personnel.[29]

Aggregation of a protein therapeutic can have serious implications for patient safety, biologic product stability, potency, biological activity, quality, and efficacy. As aggregation has been reported to lead to adverse immune reactions in patients, it is important to mitigate health risks during drug development through a comprehensive understanding of the biomolecule's propensity to aggregate and characterization of the aggregation state. Analytical ultracentrifugation (AUC) allows for the assessment of homogeneity of proteins/peptide solutions and to qualitatively assess the molecular weight and presence of aggregates over a broad range of molecular weights ranging from a few kDa to MDaltons.

For submicron aggregates, size-exclusion chromatography (SEC) is routinely used to detect and quantify irreversible aggregates from oligomers through to ultra-high order aggregates. This technique is used to establish the molecular weight of observed aggregates and is typically coupled with multi-angle static light scattering (SEC–MALS).

It is worth mentioning one example of a study that looked at the long-term stability of CT-P6, a trastuzumab biosimilar referencing HERCEPTIN® (a monoclonal antibody used to treat breast cancer and stomach cancer), to see if it had the same stability as the reference product.[30] This study examined the extended stability of CT-P6 by several complementary methods, both in reconstituted vials at a concentration of 21 mg/mL and after dilution to bracket concentrations of 0.8 and 2.4 mg/mL in polyolefin bags, and after storage at 4°C and 22°C for up to three months. After 90 days of storage at both 4°C and

22°C, no signs of physical instability, such as the formation of aggregates or oligomers, were observed, regardless of the antibody concentration. After 90 days at 4°C, there was no change in the distribution of the seven ionic variants of the compound. The tertiary structure of the compound was unaltered after storage for 28 days at 4°C. However, when stored at 22°C for 28 days, the tertiary structure was slightly altered, and there were signs of hydrolysis but no observable aggregate formation or significant signs of thermodynamic destabilization. The same conclusions can be made for reconstituted vials at 21 mg/mL.

10.4 IMMUNOGENICITY OF BIOSIMILARS

10.4.1 What Is Immunogenicity?

Immunogenicity is simply the body's ability to stimulate an immune response, also known as the body's adaptive immunity.[31,32] In response to diseases and infections, the immune system is activated to combat foreign antigens. Initially, an antigen or foreign substance enters the host and is processed by an antigen-presenting cell (APC) such as a dendritic cell or B-cell.[3,33] The APC will then present the antigen peptide that is bound to the MHC class II molecule to a T helper cell.[8] B-cells are activated with the help of T helper cells by a series of two interactions.[8] The first interaction requires the MHC class II molecule to interact with the TCR complex on the T helper cell, which is followed by an interaction between costimulatory molecules such as CD-80/86 on the APC and CD-28 on T helper cells.[33]

The interaction between the APC and T helper cell will activate T cells to secrete inflammatory cytokines such as IL-4, IL-5, and IL-6, and B-cells to produce antibodies, such as IgG, IgM, IgA, IgE, and IgD.[3,33] Initially, IgM is produced during the primary immune response when exposed to a new antigen.[3] After an initial production of IgM, IgG is produced when IgM declines, which is followed by isotype class switching of antibodies into IgD, IgE, or IgA for a more diverse response against a wide variety of antigens.[34]

During the secondary immune response, antibodies that are re-exposed to the same antigen will rapidly produce more IgG antibodies to eradicate the antigen that is causing the disease or infection.[33] These antibodies are generated to combat the antigens through antibody-mediated immunity mechanisms, such as neutralization, complement activation, opsonization, and antibody-mediated cellular toxicity.[35] Specifically, neutralization by binding of antibodies to antigen will inactivate the antigen so that it cannot cause an infection or disease.[35]

10.4.2 Biologic Induced Immunogenicity

Immunogenicity, however, can act against biosimilars and hinder the efficacy of biosimilars by forming anti-drug antibodies (ADAs), rendering the disease difficult to treat.[3] ADAs can be categorized into two different types of antibodies, neutralizing ADAs and non-neutralizing ADAs.[12] Neutralizing antibodies (NAbs) can reduce the efficacy of the biosimilar by directly binding to sites of the biosimilar and impairing its function, which is seen in biosimilar products such as streptokinase.[3] NAbs not only neutralize certain therapeutic sites of the biosimilars, but they can also affect other endogenous components of the biosimilars such that it leads to a serious adverse event, such as anaphylaxis.[3]

Non-neutralizing antibodies (NNAbs) are normally produced in the presence of non-monoclonal antibody therapeutic proteins that have the ability to bind to biosimilar therapeutic sites and not affect its therapeutic effects; however, it may promote the clearance of certain biosimilars and decrease its efficacy.[3] For example, biosimilars such as insulin may promote the production of NNAbs and not have an effect on the clinical efficacy of the drug, therefore, clinicians will still continue therapy despite the presence of such ADAs.

The immunogenicity of biosimilars is affected by different biological, patient, and disease factors.[2] Some immunogenic effects of biosimilars that may lead to negative clinical effects in patients include hypersensitivity reactions, anaphylaxis, serum sickness, and decreased efficacy of the drug, as well as reduced half-life IgD.[3,33] Therefore, immunogenicity is an important factor when assessing patient condition and safety; and extensive research on biological, patient, and disease factors are necessary to develop ways to better produce biosimilars.

10.4.3 Replication of Biosimilar-Induced Immunogenicity in Different Disease States

Different diseases will activate the immune system and cause a cascade of interleukin and cytokine release. In a normal immune response, an epitope of a foreign antigen binds to an antibody, which then activates effector mechanisms, such as neutralization or destruction of the antigen and upregulation or downregulation of other immune components, such as interleukins and cytokines.[32] Biosimilars can elicit such an immune response due to the foreign component of the biosimilar. For example, a biosimilar of rituximab contains murine protein components as it is not completely humanized, therefore, these murine peptides can form aggregates and stimulate a release of cytokines, causing an immune reaction in different disease states (Figure 10.2).[36] Moreover, the depletion of CD-20 B cells due to rituximab may lead to the formation of ADAs, which produce an immune response to counteract the biosimilar.[36]

10.4.4 Biological Factors Affecting Immunogenicity in Biosimilars

Different biological factors that can affect the immunogenicity of biosimilars include the chemical structure of the biosimilar, which can affect its physical degradation and decomposition through oxidation.[2] Post-translational modifications such as deamination, oxidation, and glycosylation of amino acid side chains can have a direct or indirect effect on immunogenicity.[3]

Biosimilars may have different variations in their amino acid sequence, which can determine the glycosylation of antibodies.[2] Glycosylation is the addition of carbohydrate molecules, also known as glycans, to protein surfaces.[3] Glycosylation of biosimilars may attribute negatively to immunogenicity as high levels of glycosylation make biosimilars more prone to degradation, which may lead to aggregate formation.[2]

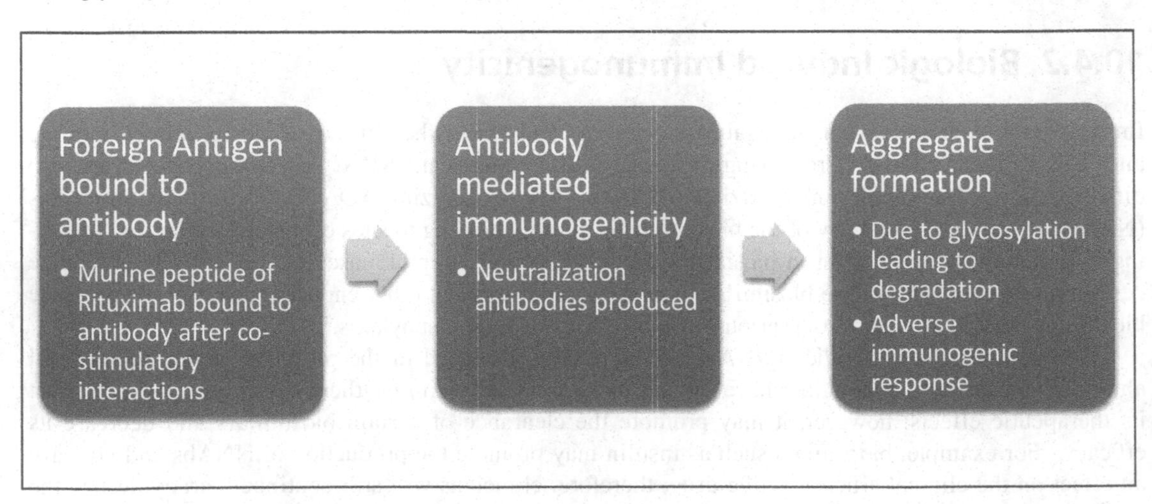

FIGURE 10.2 Schematic view of an immunogenic response induced by monoclonal antibodies.[2,36]

The aggregate formation may then lead to an immunologic effect against the drug.[2] One example of this is interferon-alpha, which has a higher tendency to form aggregates under certain conditions, for example, lower pH conditions.[2]

Moreover, the glycan itself can have an effect on immunogenicity, or it can cause the therapeutic protein to become immunogenic.[3] The glycan can also affect the physicochemical properties such as solubility, stability, mass, size, electrical charge, and folding, as well as the biological properties such as activity, cell receptor function, and half-life.[12] This, in turn, can affect the purification process of manufacturing a biosimilar, which may affect its physiological effects and have a direct or indirect impact on the immunogenicity of biosimilars.[37]

10.4.5 Route of Administration Affecting Immunogenicity in Biosimilars

Other patient factors that could affect immunogenicity include the route of administration of biosimilars. Since the body is composed of barriers such as the skin and mucosa membranes, it is susceptible to immunogenic responses, especially after injections of therapeutic proteins.[3] According to Singh,[38] subcutaneous injection of biosimilars have the highest probability of immunogenic response, followed by intramuscular, intranasal, and intravenous routes of administration.

Immunogenic responses are elicited after subcutaneous injection due to the localization and prolonged exposure to therapeutic protein in the subcutaneous tissue, which is located adjacent to lymph nodes, where there are B and T cells.[3] Therapeutic proteins exposed to lymph nodes can increase the activity of APCs and activate B and T cells for an antibody-related immunogenic response, leading to adverse events, such as serious injection reactions.[3] Serious injection reactions can occur from the use of biosimilars such as rituximab.[39,40] Therefore, it is important to premedicate patients on rituximab with anti-histamines and acetaminophen to prevent life-threatening injection reactions.[40]

10.4.6 Patient and Disease Factors Affecting Immunogenicity in Biosimilars

Patient factors that affect immunogenicity include patient age, response to biosimilar, drug burden, and disease burden.[5] Patients with an immunocompromised condition, for example, early or late-stage cancer, may have increased difficulty in producing antibodies to combat antigens, thus re-exposure to therapeutic antibodies that the patient cannot independently produce may lead to the formation of NAbs.[3] Older patients may also have a decreased ability to tolerate medications used in chemotherapy due to normal degenerative state, patient comorbidities, or organ failure.[33] As a result, the degenerative immune systems of older patients or immunocompromised patients may respond negatively to biosimilar therapy.[41]

Patient chemotherapy regimens or concomitant medications may affect immunogenicity as well.[5,41] For example, in a study done by Jahn et al., patients treated with infliximab alone for psoriatic arthritis were five times more likely to develop ADAs compared to patients on infliximab and methotrexate, which may be due to methotrexate's immunosuppressive abilities.[42] The production of ADAs may indicate that the drug is not efficacious and is eliciting an immune response.[42] Hence, these patient and disease characteristics may impact a patient's ability to respond appropriately to the biosimilar.

10.4.7 Biosimilar Development for Different Indications

Biosimilars are normally approved through an abbreviated clinical trial program or a few comparative Phase 3 trials and may not be tested for all indications of the reference product, therefore, extrapolation of indications is used to determine its use.[5] According to the European Medicines Agency 2012 guidelines, extrapolation of data from biologic indication of one medical condition can be used to infer indications for other medical conditions, given there is great evidence within prior clinical trials.[5,43] However, the extrapolation of indications is a form of inference and may pose a great concern for immunogenicity because there is no clear clinical marker to measure safety and efficacy.[5]

Other concerns of extrapolation include using certain population studies, which can oversimplify the mechanism of action of complex biosimilars because different population factors can affect immunogenicity.[5] For example, patients with different stages of cancer may differ in disease burden, chemotherapy, concurrent medications, and immune response, which may be challenging to extrapolate.[5] Moreover, although the same therapeutic protein is used for different indications, the dosing and duration of therapy may differ, which makes extrapolating of previous clinical data for new indications challenging.[5,44] Extrapolation during different stages of cancer or disease burden makes it difficult to generalize the indication to the patient population affected because of specific patient characteristics that are altered during disease progression.[5,44] Therefore, there is a need to develop clinical markers to determine the safety and efficacy of biosimilars.

10.5 SUMMARY AND FUTURE CONSIDERATIONS

Since biologic patents are coming to an end for many new drugs on the market, biosimilars have become the new generic alternatives. New recombinant protein drugs such as biosimilars have future implications on providing therapeutic benefit to patients with difficult to treat diseases. Although large molecules, such as biosimilars, are difficult to replicate compared to small molecule drugs, biosimilars can greatly benefit patients by improving treatment modalities as well as providing cost saving alternatives to biologics.

Currently, there is no formal assessment of immunogenicity for newly developed biosimilars. The current standard of practice is to ensure that the properties of biosimilars are similar or equivalent to the biologic or originator drug, which may be done with or without another clinical trial. Therefore, it is important to understand the different factors of immunogenicity that affect the efficacy and safety of biosimilars in order to better develop a production protocol/regulation of biosimilars to improve patient outcomes.

REFERENCES

1. Kinch M. An overview of FDA-approved biologics medicine. Drug Discovery Today (2015) 20(4):393–398. https://doi.org/10.1016/j.drudis.2014.09.003.
2. Covic A, Kuhlmann MK. Biosimilars: Recent development. International Urology and Nephrology (2007) 39:261–266. doi: 10.1007/s11255-006-9167-5.
3. Kuriakose A, Chirmule N, Nair P. Immunogenicity of biotherapeutics: Causes and association with post-translational modifications. Journal of Immunology Research (2016) 2016:1–18. http://dx.doi.org/10.1155/2016/1298473.
4. Dabrowska A. Biologics and biosimilars background and key issues. Congressional Research Service (2019) R44620(12):1–27.
5. Mellstedt H. Clinical considerations for biosimilar antibodies. European Journal of Cancer Supplements (2013) 11(3):1–11. https://doi.org/10.1016/S1359-6349(13)70001-6.
6. U.S. Food and Drug Administration. *Biologics Price Competition and Innovation*. 2009; H.R. 3590–3697. Cited: November 4, 2016. http://www.fda.gov/downloads/Drugs/GuidanceComplianceRegulatoryInformation/UCM216146.pdf

7. U.S. Food and Drug Administration. *Scientific Considerations in Demonstrating Biosimilarity to a Reference Product: Guidance for Industry.* 2015. Cited: November 4, 2016. http://www.fda.gov/downloads/Drugs/.
8. Mellstedt H, Niederwieser D, Ludwig H. The challenge of biosimilars. Annals of Oncology (2008) 19:411–419.
9. U.S. Food and Drug Administration. Quality considerations in demonstrating biosimilarity of a therapeutic protein product to a reference product: Guidance for industry. 2015. Cited: November 4, 2016. http://www.fda.gov/downloads/drugs/guidancecomplianceregulatoryinformation/guidances/ucm291134.pdf.
10. US Food and Drug Administration, Center for Drug Evaluation and Research (CDER), Center for Biologics Evaluation and Research (CBER). Scientific Considerations in Demonstrating Biosimilarity to a Reference Product. Guidance for Industry. April 2015. https://www.fda.gov/downloads/drugs/guidancecomplianceregulatoryinformation/guidances/ucm291128.pdf. Accessed June 25, 2018.
11. Schiestl M, Stangler T, Torella C, Cepeljnik T, Toll H, Grau R. Acceptable changes in quality attributes of glycosylated biopharmaceuticals. Nature Biotechnology (2011) 29(4):310–312.
12. Lee KH, Lee J, Bae JS, Kim YJ, Kang HA, Kim SH, Lee SJ, Lim KJ, Lee JW, Jung SK, Chang SJ. Analytical similarity assessment of rituximab biosimilar CT-P10 to reference medicinal product. MAbs. 2018 April 10(3):380–396. Published online 2018 March 6. doi: 10.1080/19420862.2018.1433976.
13. Kiese S, Papppenberger A, Friess W, Mahler H-C. Shaken, not stirred: Mechanical stress testing of an IgG1 antibody. Journal of Pharmaceutical Sciences (2008) 97:4347–4366.
14. Wang W. Instability, stabilization, and formulation of liquid protein pharmaceuticals. International Journal of Pharmaceutics (1999) 185:129–188.
15. Manning MC, Patel K, Borchardt RT. Stability of protein pharmaceuticals. Pharmaceutical Research (1989) 6:903–918.
16. Li S, Schoneich C, Borchardt RT. Chemical instability of protein pharmaceuticals: Mechanisms of oxidation and strategies for stabilization. Biotechnology and Bioengineering (1995) 48:490–500.
17. Kerwin BA, Remmele RL. Protect from light: Photodegradation and protein biologics. Journal of Pharmaceutical Sciences (2007) 96:1468–1479.
18. Lewis UJ. Altered proteolytic cleavage of human growth hormone as a result of deamidation. Journal of Biological Chemistry (1981) 256:11645–11650.
19. Rosenberg AS. Effects of protein aggregates: An immunologic perspective. The AAPS Journal (2006) 8:E501–E507.
20. Kahook MY. 2010. High-molecular-weight aggregates in repackaged bevacizumab. Retina (2010) 30:887–892.
21. Philo JS, Arakawa T. Mechanisms of protein aggregation. Current Pharmaceutical Biotechnology (2009) 10:348–351.
22. Banks DD. The effect of sucrose hydrolysis on the stability of protein therapeutics during accelerated formulation studies. Journal of Pharmaceutical Sciences (2009) 98:4501–4510.
23. Jones LS, Kaufmann A, Middaugh CR. Silicone oil induced aggregation of proteins. Journal of Pharmaceutical Sciences (2005) 94:918–927.
24. Smith BJ. Chapter 6: Chemical Cleavage of Proteins. *New Protein Techniques Methods in Molecular Biology*, Springer, Heidelberg; 1988. 71–88.
25. Bucciantini M 1, Giannoni E, Chiti F, Baroni F, Formigli L, Zurdo J, Taddei N, Ramponi G, Dobson CM, Stefani M. Inherent toxicity of aggregates implies a common mechanism for protein misfolding diseases. Nature (2002) 416:507–511. [PubMed: 11932737]
26. Filipe V, Hawe A, Schellekens H, Jiskoot W. *Aggregation of Therapeutic Proteins.* Wang W, Roberts CJ, editors. John Wiley and Sons; 2010. 400–433.
27. Carpenter JF, et al. Overlooking subvisible particles in therapeutic protein products: Gaps that may compromise product quality. Journal of Pharmaceutical Sciences (2009) 98:1201–1205. [PubMed: 18704929]
28. Philo JS. A critical review of methods for size characterization of non-particulate protein aggregates. Current Pharmaceutical Biotechnology (2009) 10:359–372. [PubMed: 19519411]
29. Narhi LO, Schmit J, Bechtold-Peters K, Sharma D. Classification of protein aggregates. Journal of Pharmaceutical Sciences (February 2012) 101(2):1–6.
30. Paul M, Astier A, Vieillard V. Extended stability of a biosimilar of trastuzumab (CT-P6) after reconstitution in vials, dilution in polyolefin bags and storage at various temperatures. GaBI Journal (2018) 7(3):101–110. doi: 10.5639/gabij.2018.0703.022.
31. Locatelli F, Roger S. Comparative testing and pharmacovigilance of biosimilars. Nephrology Dialysis Transplantation (2006) 21(Suppl 5):v13–v16. doi: 10.1093/ndt/gfl47.
32. Pradeu T, Carosella E. On the definition of a criterion of immunogenicity. Proceedings of the National Academy of Sciences of the United States of America (2006) 103(47):17858–17861. https://doi.org/10.1073/pnas.060868310.

33. Barbosa M. Immunogenicity of biotherapeutics in the context of developing biosimilars and biobetters. Drug Discovery Today (2011) 16(7/8):345–353. doi: 10.1016/j.drudis.2011.01.011.
34. Durandy A. Activation-induced cytidine deaminase: A dual role in class-switch recombination and somatic hypermutation. European Journal of Immunology (2003) 33(8):2069–2073. doi: 10.1002/eji.200324133.
35. Casadevall A, Pirofski, LA. A new synthesis for antibody-mediated immunity. Nature Immunology (2011) 13(1):21–28. doi: 10.1038/ni.2184.
36. Sauna ZE, Lagasse D, Pedras-Vasconcelos J, Golding B, Rosenberg AS. Evaluating and mitigating the immunogenicity of therapeutic proteins. Trends in Biotechnology (2018) 36(10):1068–1084. https://doi.org/10.1016/j.tibtech.2018.05.00.
37. Van Beers MM, Bardor M. Minimizing immunogenicity of biopharmaceuticals by controlling critical quality attributes of proteins. Biotechnology Journal (2012) 7(12):1473–1484. doi: 10.1002/biot.201200065.
38. Singh AK. Impact of product-related factors on immunogenicity of biotherapeutics. Journal of Pharmaceutical Sciences (2011) 100(2):354–387. doi: 10.1002/jps.22276.
39. Genentech. Rituxan (Rituximab) [package insert]. U.S. Food and Drug Administration website. A11 above Kinch M. An overview of FDA-approved biologics medicine. Drug Discovery Today (2015) 20(4):393–398. https://doi.org/10.1016/j.drudis.2014.09.003. Available from: https://www.accessdata.fda.gov/drugsatfda_docs/label/2010/103705s5311lbl.pdf. Revised February 2010. Accessed February 27, 2020.
40. U.S. Food and Drug Administration. Center for Drug Evaluation and Research. Biosimilars – Biosimilar and Interchangeable Products. Available from: https://www.fda.gov/Drugs/DevelopmentApprovalProcess/HowDrugsareDevelopedandApproved/ApprovalApplications/TherapeuticBiologicApplications/Biosimilars/ucm580419.htm#biosimilar. Cited: October 23, 2017 Assessed February 27, 2020.
41. Lichtman SM, Reske T, Jarobs IR. Biosimilars and cancer treatment of older patients. Journal of Geriatric Oncology (2016) 7(1):S1–S8. https://doi.org/10.1016/j.jgo.2016.01.002.
42. Jahn EM, Schneider CK. How to systematically evaluate immunogenicity of therapeutic proteins – regulatory considerations. New Biotechnology (2009) 25(5):280–286. https://doi.org/10.1016/j.nbt.2009.03.012.
43. Rugo H, Linton K, Cervi P, Rosenberg J, Jacobs I. A clinician's guide to biosimilars in oncology. Cancer Treatment Review (2016) (46):73–79. http://dx.doi.org/10.1016/j.ctrv.2016.04.003.
44. Weise M, Bielsky MC, De Smet K, et al. Biosimilars: What clinicians should know. Blood (2012) 120:5111–5117. doi: 10.1182/blood-2012-04-425744.

Expanding the Product Shelf

11

Taking Biologics and Biosimilars from the Bench to the Bedside

Saba Ghatrani Chalak, Tung Hoang Ngo, and Uyen Minh Le

Corresponding author: Uyen Minh Le

Contents

DOI: 10.1201/9780429485626-11

11.1 INTRODUCTION

Biologic therapy has completely revolutionized the treatment of several chronic diseases such as cancer, autoimmune disorders, diabetes, chronic anemia, etc. [1–4]. Biological drugs can have significantly fewer side effects making them much more tolerable for patients, as these therapies are manufactured from living systems and are not chemical drugs. They are specifically targeting diseased tissues rather than normal tissues, and their clinical safety and efficacy have been proven to be lifesaving for patients worldwide who have these devastating chronic diseases. The main issue with cancer chemotherapy, for example, is the intolerable serious and life-threatening side effects it has on cancer patients. Biological therapy is not without side effects, but they are generally more tolerable than chemotherapy, and by being targeted therapy, toxicity can be minimized to just tumor cells rather than both tumor and normal cells as in the case of cancer chemotherapy [5]. The discovery and application of biologic therapy of chronic diseases started almost two decades ago and required sophisticated recombinant molecular biological laboratory set up. Biologic medications are produced using recombinant DNA techniques of tissue culture of living cell lines for the expression, translation, and biochemical engineering of large complex macromolecules. Vaccines, therapeutic proteins, blood and blood components, and tissues are biological products. Biologics are not chemically synthesized products. They are derived from living materials from human, animal, microorganisms, or plant source. The three-dimensional structure of biological drug products targets specific receptors or cell surface particles and exerts an inhibitory effect on those diseased tissues, leaving surrounding normal cells intact. The safety and efficacy of biological therapy have significantly changed to practice medicine and its use has exploded in the past decade, improving patient care, quality of life, and life expectancy. Well-known examples of biologic drugs currently available include Enbrel, NovoLog, Lantus, Neulasta, Humira, Rituxan, Remicade, Avastin, Herceptin, to name a few. Cells of this biological product in 2016 were more than $100 billion [6]. The latest commercial value of biologics according to EvaluatePharma, World Preview 2018 is estimated to be $194 billion. The high costs of biological medicines show that these are very expensive drugs. Brand-name biological medicines are very expensive and could cost a patient hundreds of thousands of dollars over a lifetime of treatment, limiting patients' access to these life-saving therapies. There is hope that the costs of biological treatment could be lowered, and their access significantly expanded through the production of smaller biological molecules that has the translated protein sequence similar to that of the brand name larger, more complex biological macromolecule.

Numerous advanced biologic and biosimilar products have been generated, contributing to the treatment of various diseases that have no other treatment available from conventional chemical products. FDA manages diligent and conscientious evaluation on the biologic, biosimilar, and interchangeable drug products to ensure the approval of a high standard of efficacy and safety products.

The Biologics Control Act, also known as the Virus-Toxin Law, was first passed in 1902 after the diphtheria antitoxin tragedy with 13-children death occurred in 1901 in S. Louis, Missouri [7]. The acts gave path to further regulations in control over the production process of biologic products. In 1903, for the first time, manufacturers of vaccines were required to obtain annual licenses and inspections for the production and sale of products which are vaccines, antitoxins, and serum. Since then, additional legislations, including the Federal Food and Drugs Act, then Food, Drug, and Cosmetics Acts, etc., have been passed that commit the consistency, safety, and protection to the recipients of the biologic products, which is also the protection of American public health.

According to the Food and Drug Administration (FDA) [8], "Biological products include a wide range of products such as vaccines, blood and blood components, allergenics, somatic cells, gene therapy, tissues, and recombinant therapeutic proteins. Biologics can be composed of sugars, proteins, or nucleic acids or complex combinations of these substances, or may be living entities such as cells and tissues". Unlike chemical products, biologic products are composed of complex mixtures whose structures and characterization are not well identified. In addition, the products are sensitive to heat and likely contaminated by microorganisms. Therefore, additionally care and aseptic requirements are crucial in the manufacturing of the products.

With the growth of biotechnology industry, resemble products to the original biologic products have been developed, attempting to achieve similar efficiency with lower cost. In 2009, an abbreviated licensure has been created by the Congress to allow biosimilar products in the treatment option. The permit helps to provide more medication choices and potentially reduce healthcare cost in general. According to the FDA [9], "a biosimilar is a biological product that is highly similar to and has no clinically meaningful differences from an existing FDA-approved reference product". The reference product is the approved biological products by the FDA. This product passed FDA's comprehensive requirements for efficacy and safety. The similarity is defined in terms of product' structure, function, characteristics, purity, identity, potency, and bioactivity; however, minor difference in excipients is allowable. A biosimilar product can be used interchangeably with the reference product only when it further meets the similarity requirement of both efficacy and safety in the same way as for the reference product. Any substitution of reference to a biosimilar product will require a prescription, but substitution with the interchangeable product may not require the prescription, dependent on the State law of pharmacy.

The FDA has established guidance and regulatory on biologic and biosimilar products for the therapeutic applications of monoclonal antibodies, cytokines, growth factors, immunomodulators, thrombolytics, enzymes, proteins, and other non-vaccine therapeutic immunotherapies, etc. This chapter focuses on the regulatory aspect of biologic products and further expands to biosimilar products.

11.2 BIOLOGICS AND FDA APPROVAL PROCESS FOR BIOLOGIC PRODUCTS

11.2.1 Process for Drug Approval from FDA on Biologic Products

Biologic, biosimilar, and interchangeable products approved by the FDA are contained in the FDA Purple Book database. Two centers regulating and oversighting biologic products are the Center for Biologics Evaluation and Research (CBER) and the Center for Drug Evaluation and Research (CDER). CBER oversights vaccines, allergenic products, human blood products, immunoglobulin, proteins, peptides, animal venoms, etc., whereases CDER is responsible for naturally occurring substance, nonhuman animal/solid human tissues, monoclonal antibodies, protein therapeutics, growth factors, cytokines, antibiotics, hormone products, etc. Being a subset of drugs, biological products are also regulated under the provisions of the FDC Act.

The process for a biologic product to obtain FDA approval is mainly similar to the one for any New Drug Application (NDA) (Figure 11.1) [10]. The first three initial steps that the product must undergo are pre-clinical testing, Investigation New Drug (IND) application, and clinical testing. In pre-clinical testing, there is a series of initial laboratory and animal testing that will determine if potential testing on humans is reasonably safe. The testing also includes information on manufacturing, clinical protocols, and information about the investigator. Clinical testing comprises three phases to ensure the potency, purity, and safety of the biological products on human beings [11]. In the test, phase I, on 20–100 healthy volunteers, will look for the optimal dose with minimal side effects. Phase II, on several hundreds of people, further evaluates the safety on specific disease conditions. Phase III will continue assess the efficacy and safety of the biological products on a large group of patients (300–3000). After getting FDA approval from the clinical testing results, the product sponsor can implement the next step for the products.

However, unlike other chemical new drugs, the new biologic product approval process will not finalize with NDA but with Biologic License Application (BLA). NDA and BLA are both steps for approval before the drug can come to the market; however, they are different in the content and submission requirements. BLA requires that biological products be not only safe and potent but also highly purified.

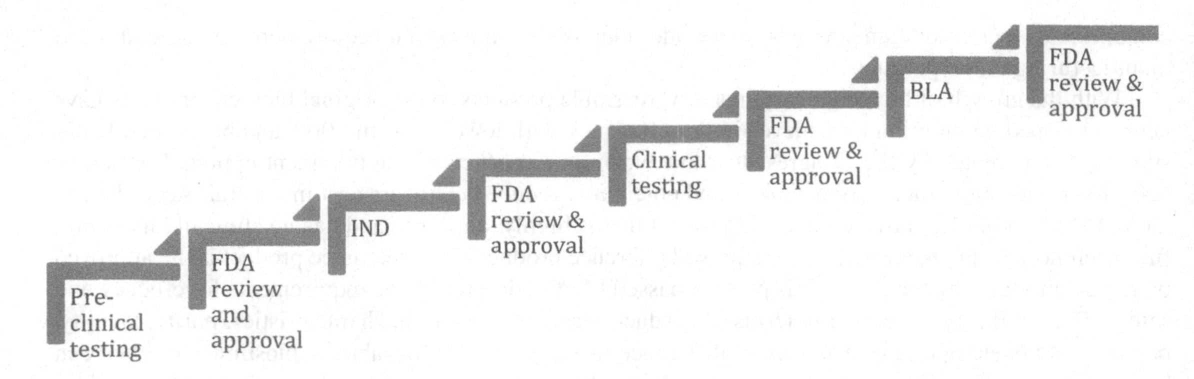

FIGURE 11.1 Process for drug review and approval from FDA on biologic products.

Since producing biological products is complicated, an inspection of the facility is normally required prior to the BLA approval. The BLA requirement can be summarized in Table 11.1 [12].

In general, it can take FDA 6–10 months to review a BLA after a 60-day filing submission. Averagely, the time for a biological product to start from the clinical testing to submission of BLA can be 81 months [13]. The number of BLA approvals is different from year to year (Figure 11.2) [14]. For the period from 1983 to 2018, there were legislation changes that impacted how drugs were approved by the FDA [15]. As a result of these changes, there was an increase in the number of BLAs that were approved. From 1990 to 1999, 34 BLAs were approved, 154 BLAs from 2000 to 2009, and 167 BLAs from 2010 to 2020.

After receiving the BLA, FDA will review and determine if the application is complete. That will include an initial review on the standard operating procedures and data validation. If any missing data or information, FDA will have a filing meeting to identify any issues, which can result in a Refuse to File (RTF). In case all data and information are completely reviewed and approved, FDA will grant a license to the applied biological product.

Biologic products can come in different routes of administration, such as oral, nasal, intra-nasal, inhaler, topical, or parenteral. Many types of biologic products have been approved for the diagnosis, prevention, treatment, and/or cure of diseases. In this chapter, major therapeutic biologic products, consisting of therapeutic proteins, vaccines, and cellular and gene therapy products are further discussed.

11.2.2 Therapeutic Proteins

Therapeutic proteins have emerged approximately 40 years ago. They include proteins that are developed for pharmaceutical use, which can be a replacement for abnormal or deficient protein in certain diseases or enhancement of the body's supply of necessary protein to decrease the impact of disease [16]. The first

TABLE 11.1 Requirements for Biologic License Application (BLA)

REQUIRED ITEMS	*DESCRIPTION*
Application form	People involved in manufacturing or responsible for biologic regulatory compliance
Product/manufacturing information	Quality, manufacturing, controls of the product, validation methods
Pre-clinical studies	*in vitro* and *in vivo* data; pharmacology and toxicology
Clinical studies	Safety and efficacy data of the product, pharmacokinetics and bioavailability, other clinical data
Addition	Post-marketing plan, proposed proprietary/nonproprietary name, priority review request if applicable

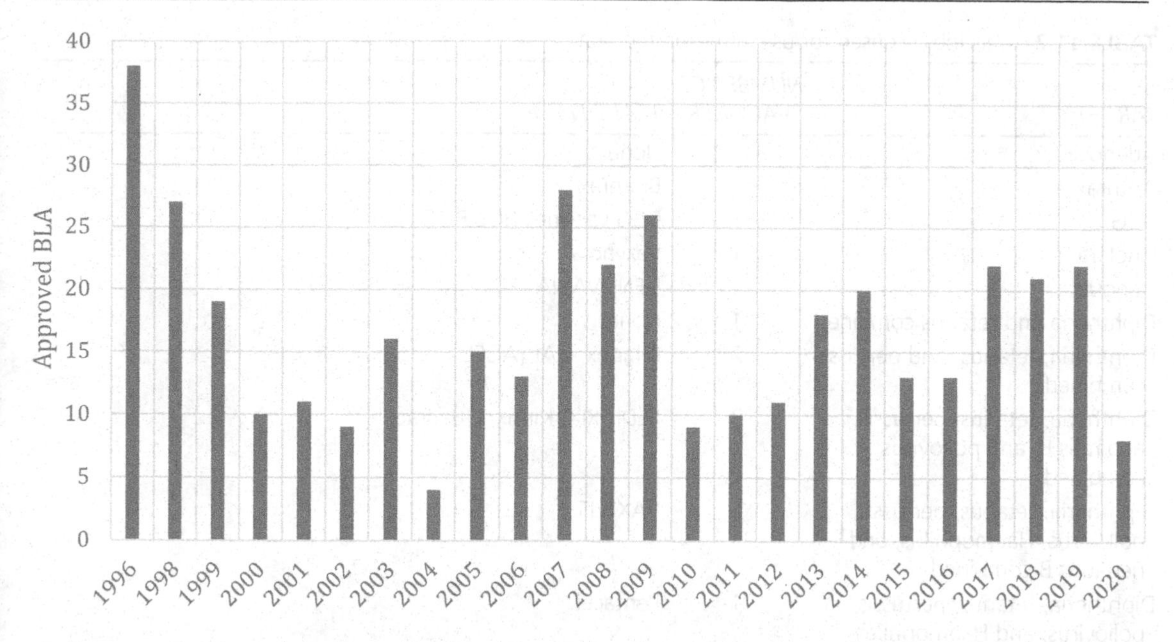

FIGURE 11.2 Biological License Application (BLA) approvals in 1996–2020.

therapeutic protein product approved by the FDA in 1982 was recombinant insulin (Humulin). Humulin was developed by Genentech but licensed to Eli Lilly. Since then, there have been hundreds of protein-based medicinal products have been approved and released to the market. Therapeutic proteins have been applied in various diseases, such as cancers (e.g., monoclonal antibodies, interferons), cardiovascular diseases (e.g., blood factors, enzymes), diabetes (e.g., insulin), anemia (e.g., erythropoietin), etc. If based on the structure, therapeutic proteins can be grouped into antibody-based products, Fc fusion proteins, enzymes, blood factors, hormones, growth factors, interferons, interleukins, and thrombolytics.

To be reviewed and approved by the FDA, therapeutic proteins must pass full requirement for biologic products (Table 11.1). Among assessments, immunogenicity is specifically important in assessing the safety of the product. In 2020, additional draft guidance of drug-drug interaction assessment was introduced for candidates of the Investigational New Drug Application (INDA) and Biologic License Application (BLA) to provide "a systemic and risk-based approach" on new therapeutic proteins. The guidance is added to the 2020 FDA guidance on "In vitro Drug Interaction Studies- Cytochrome P450 Enzyme and Transporter-Mediated Drug Interactions" and "Clinical Drug Interaction Studies – Cytochrome P450 Enzyme- and Transporter-Mediated Drug Interactions".

11.2.3 Vaccines

Vaccines' story must have begun far longer before the first one by Edward Jenner in 1796 to fight against smallpox. After that, the rabies vaccine (1885) by Louis Pasteur and other vaccines against cholera, anthrax, tetanus, diphtheria, plague, typhoid, tuberculosis, etc., were well developed through the 1930s. The research and vaccine industry has strongly grown since the 1940s for the target of measles, rubella, mumps, etc. Recently, with the outbreak of coronavirus diseases (COVID-19), vaccines research and development and the vaccine market have been more powerful than ever. Vaccine research has adopted innovative techniques, which drives the trend more to recombinant DNA technology and new delivery systems and new directions [17].

As other biologic products, vaccines must go through very careful review and examination from the FDA to "ensure the safety, efficacy, purity and potency of these products" [18]. Currently, FDA has approved approximately 80 vaccines for use in the United States (Table 11.2).

TABLE 11.2 Vaccines licensed for use in the United States

TARGET	*NUMBER OF VACCINES*	*TRADE NAME*
Adenovirus	1	None
Anthrax	1	Biothrax
BCG	2	BCG vaccine, TICE BCG
Cholera	1	Vaxchora
Dengue	1	DENGVAXIA
Diphtheria and tetanus combined	1	None
Diphtheria, tetanus, and pertussis combined	2	Infanrix, DAPTACEL
Diphtheria, tetanus, pertussis, hepatitis B, and poliovirus combined	3	PEDIARIX, Kinrix, Quadracel
Diphtheria, tetanus, pertussis, poliovirus, Haemophilus, and hepatitis B combined	1	VAXELIS
Diphtheria, tetanus, pertussis, poliovirus, and Haemophilus combined	1	Pentacel
Ebola		ERVEBO
Haemophilus B	3	PedvaxHIB, ActHIB, HIBERIX
Hepatitis A V	2	HAVRIX, VAQTA
Hepatitis A and hepatitis B combined	1	TWINRIX
Hepatitis B	3	Recombivax HB, Engerix-B, Heplisav-B
Human papillomavirus	3	GARDASIL, GARDASIL 9, Cervarix
Influenza A (H1N1)	5	No trade name
Influenza (H5N1)	2	No trade name, AUDENZ
Influenza	18	Fluad Quadrivalent, Fluad, Afluria Quadrivalent, Afluria Quadrivalent Southern Hemisphere, Flucelvax Quadrivalent, Afluria, Afluria Southern Hemisphere, FLULAVAL, FLUMIST, FLUARIX, Fluvirin, AGRIFLU, Fluzone, Fluzone High-Dose and Fluzone Intradermal, FLUCELVAX, Flublok, Flublok Quadrivalent, FluMist Quadrivalent, FLUARIX QUADRIVALENT, Fluzone Quadrivalent, FLULAVAL QUADRIVALENT
Japanese encephalitis virus	1	Ixiaro
Measles, mumps, and rubella combined	1	M-M-R II
Measles, mumps, rubella, and varicella	1	ProQuad
Meningococcal	6	Menactra, Bexsero, TRUMENBA, Menomune-A/C/Y/W-135, MenQuadfi
Plague	1	No trade name
Pneumococcal	1	Pneumovax 23, PREVNAR 13
Poliovirus	2	POLIOVAX, IPOL
Rabies	2	IMOVAX, RabAvert, No trade name
Rotavirus	2	ROTARIX, RotaTeq

(*Continued*)

TABLE 11.2 Vaccines licensed for use in the United States (*Continued*)

TARGET	*NUMBER OF VACCINES*	*TRADE NAME*
Smallpox and monkeypox combined	1	JYNNEOS
Smallpox	1	ACAM2000
Tetanus and diphtheria combined	2	TDVAX, TENIVAC
Tetanus, diphtheria, and pertussis combined	2	Adacel, Boostrix
Typhoid	2	Vivotif, TYPHIM VI
Varicella	1	VARIVAX
Yellow fever	1	YF-VAX
Zoster	2	Zostavax, Shingrix

The process of development for a vaccine is long and complex. Averagely, it may take 10–15 years following all stages from the drug discovery to market release [19]. Brief standard steps involved in vaccine development can be summarized in Figure 11.3 and Table 11.3.

Like other drugs, vaccines are developed and undergone steps of testing and regulating to ensure their purity, potency, and safety. However, the testing for vaccines is even more thorough than non-vaccine products due to a greater number of human subjects involved in vaccine clinical trials. Additionally, the monitoring for vaccines after licensure approval is closely supervised and examined by the Centers for Disease Control and the FDA.

In emergency, e.g., pandemic, disaster, etc., the regular standard process of approval for a vaccine can be in fast-track, which provides immediate response to the needs of population but still maintains a rigorous examination on the vaccine safety, effectiveness, and quality. In 2019, when the coronavirus (COVID-19) pandemic announced expanding from Asia to the rest of the world, different vaccine

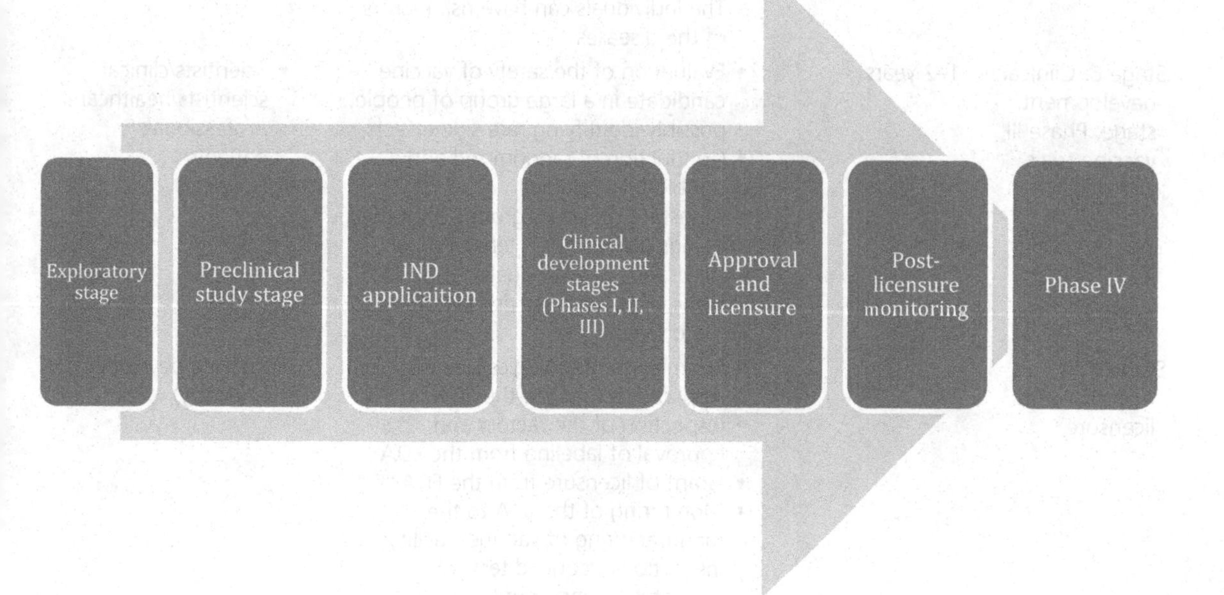

FIGURE 11.3 Regular process of development, review, and approval for a new vaccine.

TABLE 11.3 Summarized steps involved in regular vaccine's development, review, and approval

STAGES	*APPROXIMATE TIME*	*DESCRIPTION*	*PERSONNEL/PLACE*
Stage 1: Exploratory stage	2–4 years	• Fundamental laboratory research • Discovery of agents (particles, viruses, bacteria, toxins, pathogen derivatives, etc.) that might have effect on diseases	• Scientists from academia, government, industry, etc.
Stage 2: Pre-clinical study stage	2 years	• Determination of safety and immunogenicity • Improvement of vaccine candidates • Conduction of cell culture, tissue-culture, and animal studies	• Scientists in academia, government, private industry
Stage 3: Investigational New Drug (IND) Application	0.5–1 year FDA to respond within 30 days	• Submission of IND to the FDA • Describing laboratory reports, processes, proposed studies, manufacturing plans, proposed clinical trials, etc.	• Sponsors (private companies
Stage 4: Clinical development stage: Phase I vaccine trials	1–2 years	• Evaluation of the safety, side effect, and immune response type that the vaccine can produce • Conduction of open-label studies • Assessing on small groups of 20–100 healthy volunteers	• Scientists/clinical scientists/healthcare professionals/healthy volunteers • Hospitals
Stage 5: Clinical development stage: Phase II vaccine trials	1–2 years	• Evaluation of the safety and immunogenicity, of vaccine candidate on the proposed doses, immunization schedules, and delivery method • Conduction of well-controlled and randomized studies • Assessing on larger groups of volunteers (several 100 individuals). The individuals can have risk factors of the diseases	• Scientists/clinical scientists/healthcare professionals/ patients • Hospitals
Stage 6: Clinical development stage: Phase III vaccine trials	1–2 years	• Evaluation of the safety of vaccine candidate in a large group of people, possibly identifying rare side effects • Conduction of randomized and double-blind studies • Assessing on larger groups of human beings (thousands – tens of thousands of individuals). The individuals can have risk factors of the diseases	• Scientists/clinical scientists/healthcare professionals/ patients • Hospitals
Stage 7: Approval and licensure		• Submission of Biologics License Application (BLA) to the FDA • Inspection of the factory and approval of labeling from the FDA • Grant of licensure from the FDA • Monitoring of the FDA to the manufacturing of vaccine, facility inspections, required tests on vaccine's potency, purity, and safety	• Vaccine developer (industry)

(*Continued*)

TABLE 11.3 Summarized steps involved in regular vaccine's development, review, and approval (*Continued*)

STAGES	*APPROXIMATE TIME*	*DESCRIPTION*	*PERSONNEL/PLACE*
Stage 8: Post-licensure monitoring of vaccines		• Monitoring of vaccine' safety and adverse effects after the products are approved • Undergo phase IV trials, Vaccine Adverse Event Reporting System, and the Vaccine Safety Datalink	• Vaccine Adverse Event Reporting System and the Vaccine Datalink • Centers for Disease Control and the FDA
Stage 9: Phase IV trials		• Continued evaluation of the safety, efficacy, and other potential uses of the vaccine • Optional studies which are post-market release of the vaccine	• Drug company

candidates for the prevention of the disease have been developed, improved, and submitted for licensing. Under the Emergency Use Authorization (EUA), which "facilitates the availability and use of medical countermeasures, including vaccines, during public health emergencies, such as the current COVID-19 pandemic" (ref), FDA may accept the use of vaccine candidates, which have not yet been unapproved, in the emergency for the prevention and/or treatment of COVID-19. For a candidate to get to the EUA, it must demonstrate "potential benefits outweigh the known and potential risks of the vaccine" [20].

Under the EUA, the vaccine candidates still must be investigated throughout the Clinical Development Stages (Stage 4–6: Phases I, II, and III) to ensure the safety and effectiveness before being approved. On December 11, 2020, Pfizer-BioNTech's COVID-19 vaccine has marked its impression in the vaccine history as the first product approved by the FDA, under EUA, to prevent against the COVID-19 disease caused by severe acute respiratory syndrome coronavirus 2 in individuals of 16 years old and older. Following Pfizer-BioNTech's COVID-19 vaccine, Moderna's COVID-19 vaccine and Janssen COVID-19 vaccine have been approved for emergency use by the FDA on December 18, 2020 and February 27, 2021, respectively [21]. Since the approval was based on EUA, the use permit of the vaccine is effective only during the assigned emergency time. Vaccine manufacturers would have to gather comprehensive data to file the Biologics License Application (BLA) to the FDA for full regulatory approval.

11.2.4 Cellular and Gene Therapy Products

The Center for Biologics Evaluation and Research (CBER) regulates cellular and gene therapy products. Moreover, the products are oversighted and evaluated based on the Public Health Service Act and the Federal Food Drug and Cosmetic Act. Cellular and gene therapy products are composed on the following items:

Cellular immunotherapies
Hematopoietic stem cells
Adult and embryonic stem cells
Cancer vaccines

Although the research in this area has strongly developed for decades, only 19 products have been approved to market by the FDA within the last ten years (Table 11.4) [22]. The first cellular and gene

TABLE 11.4 List of approved cellular and gene therapy products (2010–2021)

PRODUCT	*AREA*	*ADMINISTRATION AND DOSAGE FORMS*	*TRADE NAME*	*MANUFACTURER*	*INITIAL US APPROVAL*
BREYANZI (lisocabtagene maraleucel)	Treatment of lymphoma	Suspension for intravenous use	BREYANZI	Juno Therapeutics, Inc., Bristol-Myers Squibb	2021
TECARTUS (brexucabtagene autoleucel)	Treatment of lymphoma	suspension for intravenous infusion	TECARTUS	Kite Pharma	2020
ZOLGENSMA (onasemnogene abeparvovec-xioi)	Treatment of spinal muscular atrophy	Suspension for intravenous infusion	ZOLGENSMA®	AveXis	2019
HPC, cord blood – MD Anderson Cord Blood Bank	Transplantation	Suspension for intravenous use	None	MD Anderson Cord Blood Bank	2018
KYMRIAH (tisagenlecleucel)	Treatment of leukemia	Suspension for intravenous use	KYMRIAH	Novartis	2017
Luxturna	Treatment of biallelic RPE65 mutation-associated retinal dystrophy	Intraocular suspension for subretinal injection	LUXTURNA	Spark Therapeutics	2017
YESCARTA (axicabtagene ciloleucel)	Treatment of lymphoma	Suspension for intravenous infusion	YESCARTA	Kite Pharma	2017
CLEVECORD (HPC Cord Blood)	Transplantation	Suspension for intravenous use	CLEVECORD	Cleveland Cord Blood Center	2016
HPC, cord blood – Bloodworks	Transplantation	Suspension for intravenous use	None	Bloodworks	2016
MACI (autologous cultured chondrocytes on a porcine collagen membrane)	Treatment of full-thickness cartilage defects of the knee	Cellular sheet for autologous implantation	MACI	Vericel Corporation	2016
IMLYGIC (talimogene laherparepvec)	Treatment of lesions in melanoma patients	Suspension for intralesional injection	IMLYGIC	Amgen	2015
ALLOCORD (HPC Cord Blood)	Transplantation	Suspension for intravenous use	ALLOCORD	SSM Cardinal Glennon Children's Medical Center	2013
HPC, cord blood – LifeSouth	Transplantation	Suspension for intravenous use	None	LifeSouth Community Blood Centers	2013
HPC, cord blood	Transplantation	Suspension for intravenous use	None	ClinImmune Labs, University of Colorado Cord Blood Bank	2012

(Continued)

TABLE 11.4 List of approved cellular and gene therapy products (2010–2021) (*Continued*)

PRODUCT	*AREA*	*ADMINISTRATION AND DOSAGE FORMS*	*TRADE NAME*	*MANUFACTURER*	*INITIAL US APPROVAL*
DUCORD (HPC Cord Blood)	Transplantation	Suspension for intravenous use	DUCORD	Duke University School of Medicine	2012
GINTUIT (Allogeneic Cultured Keratinocytes and Fibroblasts in Bovine Collagen)	Wound healing	Cellular sheet for topical oral application	GINTUIT	Organogenesis Incorporated	2012
HEMACORD (HPC, cord blood)	Transplantation	Suspension for intravenous use	HEMACORD	New York Blood Center	2011
LAVIV (Azficel-T)	Treatment of nasolabial fold wrinkles	Suspension for intradermal injection	Azficel-T	Fibrocell Technologies	2011
PROVENGE (sipuleucel-T)	Treatment of metastatic and resistant prostate cancer	Suspension for intravenous infusion	PROVENGE	Dendreon Corporation	2010

therapy product was PROVENGE (sipuleucel-T), by Dendreon Corporation, approved in 2010 for the treatment of prostate cancer. Since then, additional 18 products have been licensed for the treatment of lymphoma, spinal muscular atrophy, leukemia, cartilage defect, melanoma, wound, nasolabial fold wrinkles, retinal mutation, or transplantation.

Guidance for the manufacturing of cellular and gene therapy products was first released in 1998, which focused on human somatic cell therapy and gene therapy. Nine years later, in 2017, another guidance for the industry was presented to the public, instructing the "Eligibility Determination for Donors of Human Cells, Tissues, and Cellular and Tissue-Based Products".

From 2018 to 2021, FDA has released 25 guidance on different aspects and instructions for specific products and management [23]. The following products have been instructed in detail for the development, application, review, and approval:

Human somatic cell therapy
Allogeneic pancreatic islet cells
Cell therapy for cardiac disease
Tests for cellular and gene therapy products
Therapeutic cancer vaccines
Knee cartilage repairment or replacing
Allogeneic placental/umbilical cord blood
Gene therapies, vectored vaccines, and recombinant viral for microbial products
Virus or bacteria-based gene therapy and oncolytic products
Regenerative medicine therapies
Therapies for retinal disorders, rare disease, or hemophilia
Cellular and gene therapy products for COVID-19 public health emergency

The process of development, review, and approval for cellular and gene therapy products follow the general guideline for biological products.

11.3 BIOSIMILARS AND FDA APPROVAL PROCESS FOR BIOSIMILAR PRODUCTS

11.3.1 Biosimilar Products

Biosimilar drugs may offer as a version of brand name drug but at a more affordable cost. To prevent duplicate costly clinical trials, biosimilar products are approved through the shorter pathway of approval than reference product, but it is important to note that biosimilar product are not generic, and there is a meaningful difference between biosimilar and generic medication. Manufacturers are responsible to provide data and information to evaluate the risk of alternation between the products and make sure that interchangeable products produce the same clinical result as the reference product if the interchangeable drug is administered to patients more than once.

Biosimilars have demonstrated a remarkable cost-saving for the health care system [24]. In 2020, many biological products expired their patents in the United States. That facilitates biosimilar to play a key role in contributing billions of dollars to the global market [25]. However, due to strict requirements in the development, manufacturing, and licensing for a biosimilar product, the cost saving is expected to be approximately 15%–30%, which is much lower than that for a generic product (i.e., 80% for generics) [26]. Despite the low expected cost-saving, biosimilars still help to significantly reduce the health care cost for patients in general. For example, a 20% reduction in the price of six expired-patent biologic product would save billions of dollars, which would expand more treatment access to patients [26]. In March of 2015, FDA approved the first biosimilar product, Filgrastim-sndz (Zarxio), whose reference is Neupogen (filgrastim), a leukocyte growth factor indicated for the infection in patients with nonmyeloid malignancies after receiving myelosuppressive anticancer drugs [27]. This product was approved previously (2009) by the European Union. The approval of this biosimilar drug encourages other research companies to follow. Up to April 2021, 29 biosimilar products have been permitted for use by the FDA [28]. There was only one approved biosimilar product in 2015, two in 2016, but then 5–9 approved biosimilar product per year within the period of 2017–2019. However, in the year 2020, the number of approved biosimilar products reduced to only 3. The most recent approved biosimilar product that took place in December 2020 was Riabni (rituximab-arrx), whose reference is Rituxan (rituximab), a cytolytic antibody indicated for the treatment of lymphoma. In general, the approval rates for biosimilar drugs are significantly less than brand biological products. Biosimilar products are mostly in oncology drugs and injection administration.

11.3.2 Process for the Development, Review, and Approval of Biosimilar Products

To reach the market, biosimilar product candidates must go through a rigorous approval process by the FDA. This is to ensure public safety and any adverse effect of the drugs are known. A single biological product is referred to a reference product that is already approved by the FDA. The reference product is then benchmarked against single biological products which has been approved by the FDA. The application of an approved reference product, and on the standalone application, it must contact all the data and information necessary to illustrate the safety and effectiveness of the drug. We typically see the results of the clinical trials included in the paperwork to support the safety and effectiveness of the product. It is important to note the benefit and risk profile of a biosimilar drug is consistent with a reference product. The basis for this conclusion is through analytical studies, which included animal studies, clinical pharmacology, and the like. An important item to note is animal studies examined the level of toxicity while the clinical studies demonstrated purity, safety, and potency of the biosimilar product. The results of these studies are consistent and prove that biological products and biosimilar products are similar.

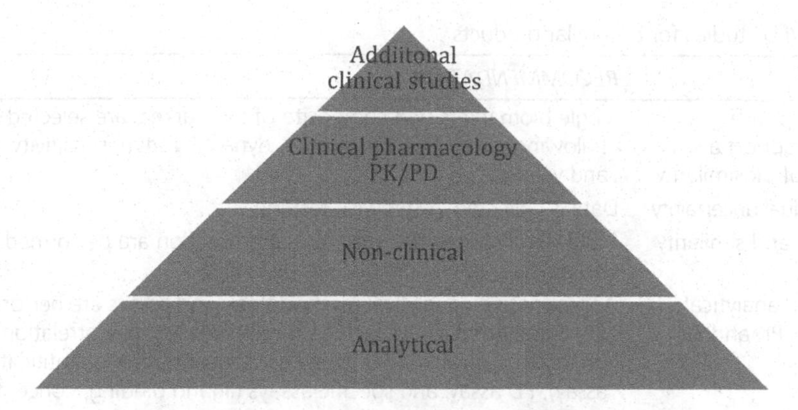

FIGURE 11.4 Required studies in drug development, review, and approval for a biosimilar product.

Unlike the reference product, biosimilar candidates do not undergo investigation for a full profile of non-clinical and clinical data. Hence, they will save the cost through by-passing the lengthy clinical trials. Instead, the candidates will show in part the determination of effectiveness and safety for the approved reference product and supplemental therapeutic options. Require studies for drug development, review, and approval is described in Figure 11.4 [29]. The detail of the data is clarified in terms of analytical studies, non-clinical studies, Clinical Pharmacology in Pharmacokinetics and Pharmacodynamics (PK/PD), and possible additional clinical studies.

11.3.2.1 Analytical studies

The studies have a high impact and consume most time in the process. They must establish and demonstrate that the biological candidates are highly like the reference product although including insignificant significance in excipients. The key is to understand the quality between attributes and the respective relationship along with the clinical safety and efficacy of the drug. It provides critical information to predict the expected clinical similarity.

11.3.2.2 Non-clinical studies

Those important experiments are the use of studying drugs on animals, checked for the assessment of toxicity, and the impact to the animal. This is useful when there are uncertainties regarding safety and efficacy. The data can provide information on how safe the product is prior to the initial clinical studies. The animal toxicity for the biosimilar product will be evaluated based on available information of the referenced product. Also, a comparison between the biosimilar and referenced product is conducted when it comes to toxicity and other information to better draw our conclusions. However, if no available animal species can provide relevant pharmacology data, then the results from animal toxicity studies are generally not warranted. In this case, *in vitro* testing, using human cells or tissue, is encouraged.

11.3.2.3 Clinical pharmacology in pharmacokinetics and pharmacodynamics (PK/PD)

Besides safety and immunogenicity are examined, critical stepwise development of a proposed biosimilar product is assessed based on PK/PD's response assessment, evaluation of residual uncertainty, and expectation of analytical quality and similarity. Description of PK/PD studies are summarized in Table 11.5 [30].

In the development of biosimilar products, PK/PD plays a crucial role in assessing the biosimilarity between a proposed product and the reference product in a stepwise manner. An approved biosimilar product needs to have supportive evidence of no clinically meaningful difference between itself and the reference product.

TABLE 11.5 PK/PD studies for biosimilar products

CRITERIA	*RECOMMENDATION*
Exposure and Response Assessment to support a demonstration of biosimilarity	Single biomarkers or a composite of biomarkers are selected based on the following properties: onset time, dynamic range, sensitivity, relevance, and validity.
Evaluation of residual uncertainty	Data is collected in stepwise fashion.
Analytical quality and similarity	Studies in comparative structure and function are performed. Methods' capabilities and limitations are described.
Integrity of the bioanalytical methods used in PK and PD studies	Appropriate bioanalytical methodology and assays are performed. Recommended assays include general PK assays (correlation of pharmacological and PD activities, concentration quantification, analytical assay), PD assay, and specific assays (ligand binding, concentration, and activity).
Safety and immunogenicity	Assessment of immunogenicity from loss of PD effect/efficacy or immune-mediated toxicity is conducted. Comparison of the data with available published information is necessary.
Study design	Single dose, randomized, and crossover study design is applied for PK similarity assessments. Whereas multiple-dose design is recommended for PD similarity assessments. Parallel group design is appropriate for products that have long half-lives and produce immunogenic responses.
Reference product	US-licensed reference product or non-US-licensed product can be used as a reference product, dependent on specific situation.
Study population	Healthy subjects are included in clinical PK/PD studies if the product is safe for their administration. If any immunogenicity of known toxicity occurs, patients are included instead of healthy subjects. Demographic group and number of subjects must be considered to ensure the best similarity and sufficient data.
Dose selection	For healthy volunteers, selected dose is mostly based on approved dose of the approved reference product. For patients, alternative dosing regimen can be applied to enhance the sensitivity of difference.
Route of administration	Same route of administration or the most sensitive for clinical detection as compared to the reference product should be applied.
PK measures	Values of peak concentration (C_{max}) and area under the curve (AUC) need to be obtained.
PD measures	Similarity on mechanism of drug action is recommended. PD biomarkers contain timing of the PD response after administration, half-life, duration of effect, area under the effect curve (AUEC), drug concentration measurements, etc.
Statistical analysis	Before any statistical analysis, a log-transformation of the exposure measures should be performed. Statistical approaches including 90% confidence interval (CI) and acceptable limit need to be qualified. Regularly, CI is recommended within 80%–125%.
Modeling and simulation	The PK/PD studies can be applied using modeling and simulation in the selection of optimal dosing regimen.

11.3.2.4 Additional clinical studies

After all required studies, such as structural/functional characterization, animal testing, are completed, if there is any uncertainty about the biosimilar results, the manufacture will be requested to conduct further comparative clinical testing to ensure the biosimilarity in terms of safety, purity, and potency of the proposed biosimilar product as compared to the reference drug. Moreover, the studies will be benchmarked and compared to the referenced product. Typically, this includes assessment on immunogenicity, pharmacokinetics, and, at times, pharmacodynamics. The nature and scope of additional clinical studies can depend on the extent of residual uncertainty about the biosimilarity of the two products. Based on the relevant animal studies, extensive structural and functional characteristics of the drug can be determined. Those studies can be shorter than full clinical studies as compared to the reference products.

11.4 IMPORT FOR BIOLOGIC AND BIOSIMILAR PRODUCTS

To our knowledge, so far FDA has not approved the import of any biosimilar product to United States. For a biologic product, the importer or his/her representative has the responsibility to file an entry notice and an entry bond with customs to make sure that the FDA is aware of all regulated products being imported into the United States. FDA relay on coordination with Customs to enforce inspection and enforcement procedures for imports. The Customs will notify FDA about the product's entry, and FDA will decide its admissibility. If FDA does not grant the entry admission, the product cannot move to the United States commerce.

U.S. Customs and Border Protection (CBP) administered the process of importing into the United States, which is governed by customs laws and regulations. For an article to be subject to FDA review, it must fall under FDA's jurisdiction. Section 801 of the Federal Food, Drug, and Cosmetic Act (21 USC 381) sets out basic standards and procedures for FDA review of imports under its jurisdiction. Section 801 (a) provides both examination of imports and authorizes FDA refusal. If the item being imported violates FDA requirements based on section 801 (a) FDA can refuse admission of imports. For importing procedures of human tissues, FDA regulations at 21 CFR 1271.420 set out the basic import standards and procedures.

11.5 CONCLUSION

Biological sciences have greatly developed over the past decades. Numerous challenging diseases have been cured or effectively treated by different biological products. To reduce the cost, biosimilar products have been developed and distributed, which improves the treatment access to a larger population. FDA's CBER and CDER play crucial roles in examining and regulating biological and biosimilar products to ensure their purity, potency, and safety. The process of drug development, review, and approval for those products is lengthy and costly, but the regulatory is essential to ensure the high quality and safety of products before being released to patients. With the continued growth of biological sciences and research, more advanced biological products will be developed, which challenges FDA not only on the regulatory but also on possible ethical and legal.

REFERENCES

1. Sun, W., et al., *Advances in the techniques and methodologies of cancer gene therapy.* Discov Med, 2019. **27**(146): p. 45–55.
2. Rosman, Z., Y. Shoenfeld, and G. Zandman-Goddard, *Biologic therapy for autoimmune diseases: An update.* BMC Med, 2013. **11**: p. 88.
3. Chellappan, D.K., et al., *Gene therapy and type 1 diabetes mellitus.* Biomed Pharmacother, 2018. **108**: p. 1188–1200.
4. Zhang, Q.S., *Stem cell therapy for Fanconi anemia.* Adv Exp Med Biol, 2018. **1083**: p. 19–28.

5. Nagasaka, M. and S.M. Gadgeel, *Role of chemotherapy and targeted therapy in early-stage non-small cell lung cancer.* Expert Rev Anticancer Ther, 2018. **18**(1): p. 63–70.
6. Calo-Fernandez, B. and J.L. Martinez-Hurtado, *Biosimilars: Company strategies to capture value from the biologics market.* Pharmaceuticals (Basel), 2012. **5**(12): p. 1393–408.
7. Bren, L., *The Road to the Biotech Revolution0 Highlights of 100 Years of Biologics Regulation.* FDA Consum., 2006. 40(1): p. 50–7
8. FDA. *Definition of the Term "Biological Product" Final Regulatory Impact Analysis.* [cited 2021 04/06]; Available from: https://www.fda.gov/about-fda/economic-impact-analyses-fda-regulations/definition-term-biological-product-final-regulatory-impact-analysis.
9. FDA. Biosimilars. [cited 2021 04/06]; Available from: https://www.fda.gov/drugs/therapeutic-biologics-applications-bla/biosimilars.
10. FDA. *Development & Approval Process (CBER).* [cited 2021 04/06]; Available from: https://www.fda.gov/vaccines-blood-biologics/development-approval-process-cber.
11. FDA. Step 3*: Clinical Research.* [cited 2021 04/06]; Available from: https://www.fda.gov/patients/drug-development-process/step-3-clinical-research.
12. FDA. *Biologics License Applications (BLA) Process (CBER).* Available from: https://www.fda.gov/vaccines-blood-biologics/development-approval-process-cber/biologics-license-applications-bla-process-cber.
13. DiMasi, J.A., H.G. Grabowski, and R.W. Hansen, *Innovation in the pharmaceutical industry: New estimates of R&D costs.* J Health Econ, 2016. **47**: p. 20–33.
14. FDA. *Biological Approvals by Year.* [cited 2021 04/06]; Available from: https://www.fda.gov/vaccines-blood-biologics/development-approval-process-cber/biological-approvals-year.
15. Darrow, J.J., J. Avorn, and A.S. Kesselheim, *FDA approval and regulation of pharmaceuticals, 1983–2018.* JAMA, 2020. **323**(2): p. 164–176.
16. Dimitrov, D.S., *Therapeutic proteins.* Methods Mol Biol, 2012. **899**: p. 1–26.
17. Liu, M.A., *A comparison of plasmid DNA and mRNA as vaccine technologies.* Vaccines (Basel), 2019. **7**(2): p. 37. doi:10.3390/vaccines7020037
18. Cope, J.U., et al., *FDA safety reviews on drugs, biologics, and vaccines: 2007–2013.* Pediatrics, 2015. **136**(6): p. 1125–1131.
19. The College of Physicians of Philadelphia. *Vaccine development, testing, and regulation.* 2018; Available from: https://www.historyofvaccines.org/content/articles/vaccine-development-testing-and-regulation.
20. FDA. *Emergency Use Authorization.* [cited 2021 04/08]; Available from: https://www.fda.gov/emergency-preparedness-and-response/mcm-legal-regulatory-and-policy-framework/emergency-use-authorization.
21. FDA. *COVID-19 Vaccines.* [cited 2021 04/08]; Available from: https://www.fda.gov/emergency-preparedness-and-response/coronavirus-disease-2019-covid-19/covid-19-vaccines.
22. FDA. *Approved Cellular and Gene Therapy Products.* [cited 2021 04/08]; Available from: https://www.fda.gov/vaccines-blood-biologics/cellular-gene-therapy-products/approved-cellular-and-gene-therapy-products.
23. FDA. *FDA In Brief.* [cited 2021 04/08]; Available from: https://www.fda.gov/news-events/fda-newsroom/fda-brief.
24. Henry, D. and C. Taylor, *Pharmacoeconomics of cancer therapies: Considerations with the introduction of biosimilars.* Semin Oncol, 2014. **41**(Suppl 3): p. S13–S20.
25. IMS Institute for Healthcare Informatics. *Global Medicines Use in 2020.* [cited 2021 04/08]; Available from: https://www.iqvia.com/-/media/iqvia/pdfs/institute-reports/global-medicines-use-in-2020.
26. Konstantinidou, S., A. Papaspiliou, and E. Kokkotou, *Current and future roles of biosimilars in oncology practice.* Oncol Lett, 2020. **19**(1): p. 45–51.
27. Awad, M., P. Singh, and O. Hilas, *Zarxio (Filgrastim-sndz): The first biosimilar approved by the FDA.* P T, 2017. **42**(1): p. 19–23.
28. FDA. *Biosimilar Product Information.* [cited 2021 04/08]; Available from: https://www.fda.gov/drugs/biosimilars/biosimilar-product-information.
29. FDA. *Biosimilar Development, Review, and Approval.* [cited 2021 04/08]; Available from: https://www.fda.gov/drugs/biosimilars/biosimilar-development-review-and-approval#:~:text=All%20FDA%2Dapproved%20biological%20products,efficacy%2C%20safety%2C%20and%20quality.
30. FDA. *Clinical Pharmacology Data to Support a Demonstration of Biosimilarity to a Reference Product.* 2016 [cited 2021 04/08]; Available from: https://www.fda.gov/regulatory-information/search-fda-guidance-documents/clinical-pharmacology-data-support-demonstration-biosimilarity-reference-product.

Hydrogel-Based Delivery of Biologics in Cancer and Cardiovascular Diseases *Proof-of-Concept*

12

Amy Ferrarotti, James Alexander Lugtu, and Ashim Malhotra

Corresponding author: Ashim Malhotra

Contents

DOI: 10.1201/9780429485626-12

12.1 INTRODUCTION

12.1.1 What Are Hydrogels?

Hydrogels are three-dimensional structures composed of hydrophilic polymers such that the hydrogel can imbibe water and retain it within its structure without dissolving. They offer a means of embedding in tissue structures since they have flexibility in their ability to absorb or release water and release drug cargo that may be dissolved within. Of particular help is their closeness to the solid state, which resembles tissue architecture. Hydrogels can be manufactured using a wide variety of polymers such as natural or synthetic polymers, which may be chemically or physically cross-linked. Hydrogels may be designed to evince overall anionic, cationic, ampholytic, or non-ionic properties and may be formulated in different sizes such as macrogels, microgels, and nanogels [1].

Of particular relevance is the ability of hydrogels to transition from a more fluid state, such as that of a liquid (called "sol") to a more solid phase ("gel"). Such sol-gel transformation offers incredible advantages for the manufacture of drug-cargo-loaded hydrogels, which may be implanted during surgery for the continued release of drugs. Thus, hydrogels may be gainfully employed for the delivery of therapeutic molecules including biological drugs products. If intended for such biotherapeutic purposes, polymer selection becomes an important strategy to ensure minimal toxicity and irritability to surrounding tissues.

12.1.1.1 The use of hydrogels for the delivery of biologics

Hydrogels are becoming a useful tool for controlled release drug delivery and are an ongoing area of research interest. Hydrogels can be used for several routes of drug administration including oral, rectal, vaginal, transdermal, ocular, and implants. They can control the release of medications by responding to different stimuli such as temperature, pH, glucose, electric signals, light, and other stimuli. The ability of hydrogels to respond to these chemical and/or physical stimuli renders them "smart" or "intelligent", making them ideal delivery vehicles for drugs and the new-age biologic products that are rapidly gaining attention in the treatment of a variety of human diseases.

With the advent of modern pharmaceutical science and discovery, biologics have quickly gained a considerable share of the drug market. Biologics are loosely defined as products obtained from cells, tissue, or whole organs of animals and humans and include products such as whole blood, cytokines, interleukins, antibodies, proteins biosynthesized and secreted by the tissues, and recently, recombinant proteins. In addition to their rapid research and development, biological drugs such as monoclonal antibodies offer the distinct advantage of targeted therapy and a relatively ameliorated toxicity and adverse events profile. Drug targeting can be further improved by the use of specialized formulations or vehicles of delivery. Hydrogels offer many advantages for the delivery of biologics in a variety of disease states. However, the specific advantages and disadvantages of hydrogels for the delivery of biologic cargo depend upon the type of hydrogel, which is discussed below.

12.1.2 Types of Hydrogels and the Polymers Used for Their Manufacture

12.1.2.1 Thermosensitive (temperature-sensitive) hydrogels

Temperature-sensitive hydrogels respond to temperature changes in the environment and can be particularly relevant for the delivery of biologic cargo. These types of hydrogels are composed of polymers with multiple hydrophobic groups such as methyl, ethyl, and propyl groups that allow the hydrogel to swell or shrink in response to changes in the local environment. The swelling and shrinking behavior of the hydrogel is what allows for the release of the drug. Temperature-sensitive hydrogels may be further subclassified as (1) negatively thermosensitive, (2) positively thermosensitive, and (3) thermally reversible hydrogels [1].

12.1.2.1.1 Negatively thermosensitive, positively thermosensitive, and thermally reversible hydrogels

Temperature changes regulate the hydrophobicity index of this type of hydrogel. Typically, negatively thermosensitive hydrogels have a lower critical solution temperature (LCST). This means that the hydrogel becomes less hydrophobic and expands when the temperature decreases past this point, and this "expansion" allows delivery of the drug cargo. Conversely, as the temperature increases, the hydrophobic interactions between the hydrophobic sections on the polymer chain increase, and the hydrogen bonding between the hydrophilic segments becomes weaker, causing the hydrogel to shrink. Generally, the LCST can be adjusted by adjusting the hydrophobic to the hydrophilic ratio of the polymer [1].

On the other hand, positively thermosensitive hydrogels have an upper critical solution temperature (UCST), where the hydrogel becomes less hydrophobic and expands when the temperature increases beyond this point. Again, it is the expansion of the hydrogel that ultimately leads to drug delivery.

Interestingly, thermally reversible hydrogels can be either positively or negatively thermosensitive. Mechanistically, for example, thermally reversible hydrogels do not undergo swelling-shrinking transitions. Instead, they undergo solution-gel transitions, changing from a solution to a gel at the LCST or UCST, depending on the properties of the hydrogel. This transition is reversed when the temperature decreases or increases, and such reversible transitions are due to the composite polymer chains not being covalently cross-linked. A good way to visualize this is to think about gelatin. At high temperatures, it is a liquid; when it cools down, it becomes has a gel-like consistency. This process can be repeated over and over, and the gelatin would continue to transition between these two states, which makes gelatin reversible. This is similar to what happens with thermally reversible hydrogels. This property helps control the delivery of drugs.

12.1.2.1.2 Disadvantages of temperature-sensitive hydrogels
There are some safety concerns with the clinical applications of temperature-sensitive hydrogels, especially when it comes to poly(N-isopropylacrylamide) (PINAAPm), which is the most commonly used polymers in temperature-sensitive hydrogels. There is the possibility that these could be toxic, carcinogenic, teratogenic, or cause other problems due to the incompatibility of the monomers and cross-linkers used in the production of the hydrogels. In addition, these particular hydrogels are not yet biodegradable. Further studies are warranted to better understand the efficacy and safety of these hydrogels, in addition to making them biodegradable.

12.1.2.1.3 Clinical application of temperature-sensitive hydrogels for delivering biologic cargo
Pan et al. [2] devised a poly(lactic-co-glycolic acid) iPLGA-PEG-PLGA (MW: PEG=1500Da, MPLGA: MPEG=3:1)-based hydrogel for the delivery of collagenase and the trastuzumab monoclonal antibody for the treatment of breast cancer using a peritumoral delivery system. Breast cancer is a serious disease that occurs commonly and with high frequency across the world. A specific type of breast cancer that overexpresses the human epidermal growth factor receptor 2 (HER2) is particularly recalcitrant to traditional anti-cancer therapeutic approaches. Thus, the development of a HER2-specific monoclonal antibody, trastuzumab, heralded the age of promise for the treatment of this type of breast cancer. Unfortunately, in vivo studies demonstrated that only about 0.01% of trastuzumab was bioavailable at breast tumor sites following intravenous injection [3]. This was because of the well-known concept of the paucity of drug penetrance into tumors which are often surrounded and protected by dense cellular mass in the extracellular matrix surrounding the tumors. Thus, Pan et al.'s approach of implanting a thermosensitive hydrogel system in a murine model of $HER2^+$ breast cancer that would locally release collagenase at the tumor site to enhance drug penetrance and delivery and subsequently increase the amount of trastuzumab delivered to the site was justified.

The authors demonstrated that the implanted hydrogel delivered its biologic cargo in a biphasic release pattern over 9 days in the $HER2^+$ mouse, with NIR imaging demonstrating a 20-day retention of the hydrogel at the tumor sites. The hydrogel treatment not only increased apoptotic cell death in the tumors compared to controls, but it was also able to achieve this at a much lower volume of only a 1/4th dose of intravenous trastuzumab.

Thus, long-term and sustained localized delivery of biologic cargo is possible with thermosensitive hydrogels for the potential innovative treatment of tumors and cancers to augment the delivery of targeted therapy while alleviating the challenge of drug penetrability.

12.1.2.2 pH-sensitive hydrogels

Hydrogels that are pH sensitive are commonly used for drugs given by the oral route. This use of hydrogels is based on the chemistry of environment pH alterations driving changes in the ionization/charge distributions of the polymer chains of hydrogels. Such a change in pH-based ionization, in turn, causes the hydrogels to shrink or swell. It is primarily the swelling of the hydrogels is that allows for the delivery of the drug. However, the shrinking of the hydrogels can also facilitate drug delivery through a squeezing mechanism. Controlling the swelling and shrinking of the hydrogels is what will ultimately lead to optimal drug delivery [4].

Properties of pH-sensitive hydrogels
Various properties of the polymer and the swelling medium control the swelling and shrinking of the hydrogel. Polymeric characteristics that aid in the design of pH-sensitive hydrogels for the delivery of biologic cargo include polymeric charge, concentration, and the overall pK_a of the polymer's ionizable groups, which influences the degree of ionization. Additional factors that regulate outcomes include the cross-linking density and overall balance between the hydrophilicity and hydrophobicity of the hydrogel. Properties of the swelling medium that are important to this type of hydrogel design include pH, ionic strength, and the counterion and its valency. In addition, the characteristics of the buffer solutions also impact the swelling and shrinking of hydrogels. Together, these properties play a vital role in optimizing drug delivery with hydrogels.

Cationic and anionic pH-sensitive hydrogels: clinical applications for the delivery of biologics
Polycationic hydrogels, such as chitosan and poly(ethylene imine), swell at an acidic pH due to the protonation of amine or imine groups. These positively charged groups repel each other, which in turn causes the hydrogel to swell and the drug to be released. An example of the delivery of biotherapeutic agents dependent upon the swelling of these hydrogels in an acidic environment makes them ideal for localized drug delivery in the stomach for infections such *H. pylori* [5].

Conversely, anionic hydrogels, such as poly(acrylic acid) and carboxymethyl chitosan, do not swell in a low pH but rather in a higher pH, such as in the intestine. In this case, it is the negatively charged polymer chains that repel each other that cause the hydrogels to swell. This property makes it effective for drug delivery in the intestines, and this delivery mode has been studied for drugs such as clonidine and prednisolone.

In an example of the clinical application of the use of pH-sensitive hydrogels for the delivery of biologic cargo such as monoclonal antibodies, Bellingeri et al. reported that the delivery of egg yolk immunoglobulin for the treatment of several gastrointestinal pathogens was improved by the use of pH-sensitive hydrogel systems. The authors demonstrated that gastric acid and gastric pepsin-driven degradation of orally delivered egg yolk antibody could be reduced by the use of encapsulation of this specific antibody in acrylamide and acrylic acid-based pH-sensitive hydrogels that possessed a superior drug-loading percentage (99.2±12.9 mg IgY/mg hydrogel) at pH 7.4. This hydrogel system retained the monoclonal antibody drug within at the highly acidic stomach pH while releasing the biologic drug cargo slowly and continuously at the more alkaline intestinal pH [6].

Disadvantages of pH-sensitive hydrogels
Despite the promising advantages of pH-sensitive hydrogels, there are also some limitations. The main disadvantage of the pH-sensitive hydrogels is their lack of biodegradability, especially observed with the synthetic pH-sensitive polymers. While this is generally not an obstacle for the delivery of biological cargo by the oral route, it is a hindrance for implants and other routes, as these polymers have to be removed from the body once the drug has been delivered. Currently, several different polypeptides are being studied and used for the synthesis of biodegradable hydrogels including dextran, poly(hydroxyl-l-glutamate), poly(l-ornithine), poly(aspartic acid), poly(l-lysine), and poly(l-glutamic acid). However, further research and development are needed to make pH-sensitive hydrogels biodegradable.

12.1.2.3 Electrosensitive hydrogels

A special type of hydrogel possesses the ability to respond to changes in electrical resistance in its environment. These so-called electrosensitive hydrogels are composed of polyelectrolytes that grow or diminish in size or experience both growth and shrinkage simultaneously on different portions of the hydrogel upon the passage of an electric current. The change in shape in this hydrogel system is dependent on the point of contact of the electric field and the presence of electrolytes in the aqueous phase. Commonly, electrohydrogels do not contain any salts [7].

Partially hydrolyzed polyacrylamide hydrogels that are in contact with both the anode and cathode undergo volume collapse by an infinitesimal change in electric potential across the gel. Application of an electric potential to hydrogel results in the movement of hydrated H^+ ions migrating toward the cathode. This causes a loss of water at the anode side and, concurrently, electrostatic attraction of negatively charged acrylic acid groups to the anode surface, creating uniaxial stress along the gel axis, primarily near the anode side. The result of these events is a decrease in the size of the hydrogel at the anode side.

For example, a sodium acrylic acid-acrylamide copolymer hydrogel placed in an aqueous solution of an acetone-water mixture under an electric field without any physical contact with the electrodes undergoes shape change dependent on the electrolyte concentration. In the presence of an electric field, low to zero concentrations of electrolytes result in a decrease in size of the hydrogel due to the movement of Na^+ to the cathode, while high concentrations of electrolytes drive an increase in the size of the hydrogels, most noticeably near the anode causing bending of the hydrogels, due to an increase in the concentration of Na^+ in the hydrogel. In the presence of a cationic surfactant such as n-dodecylpyridinium chloride, swelling occurs at the cathode side of the hydrogel due to the migration of positively charged surfactant molecules to the cathode, forming a complex with negatively charged polymer chains on the anode side. In the presence of an electric field, the absence of salts ameliorates the size of electrohydrogels due to electroosmosis and electrophoresis from the hydrogel to the cathode. This is commonly used for modulated drug delivery by the "on-off" switching of the accompanying electric field [7].

Electrosensitive hydrogels are most commonly used in controlled drug delivery. For example, electrosensitive hydrogels composed of poly (2-acrylamido-2-methylpropane sulfonic acid-c-*n*-butylmethacrylate) are capable of releasing edrophonium chloride and hydrocortisone in a pulsatile manner under an electric field. Specifically, Zhao et al. reported the development of a novel microporous hydrogel that was responsive to and could release biologic drug cargo in response to a magnetic field. Control of "on-off" drug release was achieved through varied electric stimulation intensity in distilled-deionized water. Edrophonium chloride is a positively charged drug that controls its release through an ion exchange between the positively charged solute and H^+ ion from the electrolysis of water. The authors tested the ability of their subcutaneously implanted macroporous ferrogel constructs to deliver stained murine mesenchymal stem cells *in vivo* under magnetic field and reported a "burst" of stem cell release upon variation of the magnetic field [8].

In another example, the chemomechanical shrinking and swelling of PMA hydrogels under an electric field was used for the pulsatile delivery of pilocarpine and raffinose. In the presence of an electrical current, microparticles of PAA hydrogel rapidly exhibited a decrease in size. In the absence of an electrical current, they returned to their original size. Because of this, it is believed that an electrical current induces a change in the size of the microparticles, which will cause the "on-off" release profiles [9].

In addition to drug delivery, electric signal-sensitive hydrogels play a role in other biomedical processes. Electrosensitive hydrogels are similar in structure to pH-sensitive hydrogels. Because of this, they are also able to convert chemical energy, such as action potentials, into mechanical energy, such as muscle contractions, allowing them to act as actuators or artificial muscles in many applications. This has been shown using weakly cross-linked poly(2-acrylamido-2-methylpropanesulfonic acid) hydrogels. When positively charged surfactant molecules interacted with the surface of the polyanionic hydrogel facing the cathode, the result was a reduction in the overall negative charge. This negative charge caused a local decrease in the size of the hydrogel, "bending" the hydrogel. Applying an oscillating electrode polarity resulted in the same worm-like contractile motion in the hydrogel. For example, electrosensitive conducting hydrogels composed of conducting polymers are used for the delivery of various types of cells in *in vivo* experimental animal models. Examples of electrical conducting polymers that may be used for the design of such hydrogels include polypyrrole (PPY), polythiophene (PT), poly(3,4-ethylene dioxythiophene) (PEDOT), and polyaniline (PAni) [10].

Disadvantages of electrosensitive hydrogels electrosensitive hydrogels in drug delivery are beneficial because their drug release rate is controlled by the modification of the electrical current flowing through the hydrogel. Some common issues with all hydrogels are the slow response of the hydrogels to the electrical current and the need for a controlled voltage source for the controlled drug delivery to work. As most electrosensitive hydrogels function without electrolytes, there may be some difficulties in the future creating drug delivery systems based on physiological conditions.

12.1.2.4 Glucose sensitive hydrogels: clinical applications of the types of hydrogel systems described above

Diabetes is a condition where the pancreas does not produce enough insulin or the human body becomes incapable of employing the insulin appropriately. Diabetic patients often require insulin therapy to keep their blood sugar at an appropriate level. Hydrogels combined with a glucose sensor are currently being studied as a way to make insulin delivery more effective and convenient for the patient. The three main systems that are currently being studied for insulin delivery are pH-sensitive membrane systems, ConA-immobilized systems, and sol-gel phase reversible systems.

First example: pH-sensitive membrane systems for diabetes
As previously mentioned, pH-sensitive hydrogels swell or shrink in response to changes in the environment. When it comes to insulin delivery, glucose oxidase is the most commonly used enzyme to sense changes in glucose. Glucose oxidase converts glucose into gluconic acid, making the environment more acidic. This allows the hydrogels to swell or shrink, which allows for the delivery of insulin.

One example of such a stimulus-responsive hydrogel system arises from the use of poly(N, N′-diethylaminoethyl methacrylate) (PDEAEM) for its construction. With PDEAEM, the lower pH causes ionization which leads to ionization, which ultimately causes the hydrogel to swell and release insulin. On the contrary, hydrogels made with polyanions, such as poly(methacrylic acid–co-butyl methacrylate), collapse when glucose is converted to gluconic acid due to the protonation of the carboxyl groups of the polymer. The shrinking of the hydrogel opens the pores of the porous filter, which allows for the delivery of insulin. Marek and Pappes studied insulin-release dynamics from PDEAEM-based hydrogel delivery systems. They reported that "the optimal conditions for loading insulin into PDEAEM microparticles were found to be at a loading pH of 5.6, particle to insulin mass ratio of 7:1, a concentration of 1.0 mg/mL insulin, and a collapsing pH of approximately 9.5. Microparticles exhibited a responsive (pH) or intelligent (glucose) release of insulin from a stimulus. Microparticles that had a nominal crosslinking ratio of 10% released a third of the insulin payload after a single stimulus, compared to nearly 70% for microparticles with a 3% crosslinking ratio". This leads them to conclude that "PDEAEM microparticles of 150 μm diameter showed promise as components of a system of automated, intelligent delivery method for insulin to type I diabetics" [11].

Second example: ConA-immobilized systems
Concanavalin A (ConA) is a tetravalent binding protein that belongs to the family of carbohydrate-binding proteins and supports the ability to bind glucose. ConA is derived from the jack bean plant called *Canavalia ensiformis*. Pai et al. devised a hydrogel system in which ConA was bound and immobilized on microspheres. In this system, p-succinylamidophenyl-alpha-D-glycopyranoside-insulin (SAPG-insulin) was cross-linked to a concanavalin A-based hydrogel system. Glucose molecules interrupted interactions between insulin molecules and their carriers or supports, usually through the addition of functional groups onto the insulin molecules. The authors reported that "as a new device, ConA MSs loaded with SAPG-insulin was enclosed in a heat-sealed, surface-modified, porous poly(vinylidene difluoride) membrane pouch. This system showed a pulsatile release pattern for insulin with a short lag time in response to glucose challenges of 50-500 mg/dL. The pattern of release of SAPG-insulin from the devices was studied with varying design parameters, such as surface area, membrane pore size, and loading content of SAPG-insulin" [12].

However, it is important to note that for these glucose-sensitive hydrogels to be successful, they need to be able to respond to changes in glucose quickly and deliver the appropriate amount of insulin rapidly. Currently, these hydrogels respond slowly to changes in glucose. In addition, they revert to their original state slowly and their actions are not as reproducible as they need to be. Finally, not all of the components of these hydrogels are biocompatible. These hydrogels are promising for patients who are dependent on insulin. However, further research and improvements are needed in order for these hydrogels to be successfully implemented into clinical practice.

12.2 CARDIOVASCULAR APPLICATIONS FOR HYDROGEL-BASED DELIVERY OF BIOLOGICS

Cardiovascular diseases continue to be a primary cause of death worldwide. Many cardiovascular disease patients undergo various pharmacological and surgical interventions, which can significantly improve and lengthen their lifespan. However, these interventions are costly and invasive. In the search for more cost-effective and less invasive options, hydrogels may be used either as scaffolds or continuous drug delivery systems for the treatment of ischemic diseases and arrhythmias.

12.2.1 Ischemic Heart Disease

Ischemic heart disease occurs when the arteries cannot deliver an adequate oxygenated blood supply to the heart, resulting in tissue neurosis and cell death. This is usually caused by the buildup of plaque which leads to the narrowing and hardening of the arteries. In addition, plaque can attract platelets and blood clots, which can lead to further blockage of the arteries. This condition can be particularly dangerous as it is often asymptomatic until there are severe complications, such as myocardial infarction. Standard treatment for ischemic heart disease usually includes lifestyle modifications, pharmacotherapy, and procedures such as cardiac bypass or other coronary interventions. Hydrogels are an ongoing area of research for their therapeutic potential in treating ischemic heart disease.

12.2.2 The Potential Use of Hydrogels for the Treatment of Ischemic Heart Disease

For hydrogels to be effective for cardiac tissue repair in ischemic disease, they must be biocompatible and biodegradable. In addition, they must remain functional long enough to promote sufficient new tissue generation and tissue repair. For these purposes, hydrogels can be administered as an injection or a patch. However, injectable hydrogels are generally preferred because it is a more convenient route of drug delivery. Hydrogels may be injected either as a liquid that transitions to a gel state (the sol-gel systems described above) in response to certain stimuli such as temperature and pH or in a partially cross-linked gel form. These hydrogels can be made from natural materials or can be made from synthetic materials.

12.2.3 The Use of Natural Polymer-Based Hydrogels in the Treatment of Cardiovascular Diseases

Natural hydrogels are usually biocompatible, bioactive, and many of them are found naturally in the human body. This makes it easier for them to mimic natural tissues. However, natural hydrogels can induce an immune response and usually have low mechanical strength. In addition, their structure is usually fragile and complex, making modification difficult. Several natural hydrogels are used for cardiac repair, including collagen, gelatin, hyaluronic acid, fibrin, alginate, chitosan, keratin, Matrigel, and decellularized ECM.

12.2.3.1 Collagen-based hydrogels for cardiovascular diseases

Collagen is a natural protein found in the extracellular matrix in various parts of the body, including bone, cartilage, blood vessels, skin, and other tissues. Collagen is widely used because of its biocompatibility, biodegradability, and its strong cellular activities. However, collagen has low physical strength and a high synthesis cost and can potentially elicit an immune response. Despite this, collagen is effective in aiding in the repair of damaged, ischemic tissues. For example, Dai et al. investigated the use of collagen hydrogels in a rat model of myocardial infarction to reduce the thickening of the infarcted heart wall. The authors reported an improvement in stroke volume, ejection fraction, and wall thickness in the collagen hydrogel group compared to the controls [13], suggesting the continued utility of hydrogels for the delivery of biological cargo such as collagen, which shows promise as an effective therapy for cardiac tissue repair in ischemic heart disease.

12.2.3.2 Gelatin-based hydrogels for the delivery of biologics for treating myocardial infarction

Gelatin is a natural polymer and derivative of collagen. In particular, gelatin is useful for the delivery of various growth factors, including basic fibroblast growth factor (bFGF) at the site of cardiac tissue death. It has been shown that gelatin loaded with bFGF is effective at inducing angiogenesis and improving left ventricle systolic and diastolic function at the site of the damaged cardiac tissues. For example, Tabata and Ikada prepared biodegradable gelatin hydrogels employing the technique of chemical cross-linking by using glutaraldehyde to cross-link gelatin with an isoelectric point of 5.0 and 9.0. The authors reported that "when implanted subcutaneously into the mouse back, bFGF-incorporating acidic and basic gelatin hydrogels with higher water contents were degraded with time faster than those with lower water contents. Significant neovascularization was induced around the implanted site of the bFGF-incorporating acidic gelatin hydrogel. The induction period was prolonged with the decrease in hydrogel water content. On the other hand, such a prolonged vascularization effect was not achieved by the bFGF-incorporating basic gelatin hydrogel and the hydrogel initially exhibited a less enhanced effect, irrespective of the water content. These findings indicated that the controlled release of biologically active bFGF is caused by biodegradation of the acidic gelatin hydrogel, resulting in induction of vascularization effect" [14].

12.2.3.3 Hyaluronic acid-based injectable hydrogels for myocardial infarction

Hyaluronic acid is a derivative of glycosaminoglycan that plays a role in tissue structure and function. However, hyaluronic acid does have weak mechanical properties, similar to other natural hydrogels. This can be overcome by modifying the molecular structure and composition with various functionalization. Though it has been used in wound dressings and tissue engineering, it has also shown promise for cardiac repair. In one study, Ifkovits et al. compared two methacrylated hyaluronic acid macromers to evaluate the associated effects of intramyocardial hydrogel injection on the remodeling response based on hydrogel mechanics using sheep models post-myocardial infarction [15]. Treatment with both hydrogels significantly increased the wall thickness in the apex and basilar infarct regions compared with the control infarct. However, only the higher-modulus (MeHA High) treatment group had a statistically smaller infarct area compared with the control infarct group. These results seem to indicate that hyaluronic acid hydrogels show promise for use in cardiac tissue repair.

Expanding the premise above, Abdalla et al. showed that hyaluronic acid-based hydrogels may be used for enhancing rat cardiac function in models of myocardial infarction. Specifically, the authors showed by hyaluronic acid-based hydrogels showed an 18.2% ($P < 0.01$) increase in ejection fraction in gel-injected groups when compared with the control group, almost returning the ejection fraction to baseline levels. The authors proposed the use of such hydrogels not only by themselves but as potential scaffolds for the delivery of biological agents for the sustained treatment of myocardial infarction [16].

12.2.3.4 Fibrin hydrogels for the delivery of transplanted cells into myocardial infarcts

Fibrin has been approved by the Food and Drug Administration (FDA) for wound healing and the formation of blood clots, and it plays a huge role in these applications. In addition, fibrin is biocompatible, biodegradable, and can facilitate cell adhesion. It can also be naturally synthesized from the host's blood without eliciting an immune response. This makes it highly suitable for use as a tissue sealant and the delivery of biological products such as growth factors needed for cardiac tissue repair. However, like most natural hydrogels, fibrin lacks strong mechanical properties. Fibrin has been shown to be effective in facilitating cardiac tissue repair. For example, Christman et al. reported that fibrin glue could be effectively employed as an injectable scaffold for increasing the retention and survival of transplanted skeletal myoblasts in rat models of myocardial infarction. Specifically, this team of investigators reported a reduction in infarct size, improved cardiac function, and increased blood flow to ischemic myocardium following the injection of fibrin glue hydrogels that acted as scaffolds for skeletal myoblasts. Providing evidence for the use of fibrin is a promising hydrogel for use in cardiac repair [17].

12.2.3.5 Alginate and biological delivery hydrogel systems

As the name suggests, alginate is derived from algae. Specifically, it is derived from the cell walls of brown algae. Alginate-based hydrogels are primarily used in drug delivery for wound healing due to their high biocompatibility, inexpensive cost, and simple gelation with Ca^{2+}, Mg^{2+}, Ba^{2+}, and Sr^{2+}. It also does not have a tendency to form blood clots, which makes it useful for cardiac tissue repair. However, alginate hydrogels are limited by their inadequate long-term stability due to the release of divalent cations into the surrounding tissues. In addition, they are also limited by their hydrophilicity. However, both of these issues can be overcome by various methods. Alginate has been shown to be effective for the treatment of myocardial infarction in various studies, and as a result, has become the first injectable biomaterial to enter into clinical trials for the treatment of myocardial infarction.

12.2.3.6 Keratin hydrogels

Keratin is a group of structural proteins commonly found in human hair, nails, and skin. It can self-assemble in fibrous scaffolds, making an ideal environment for cell proliferation. In addition, keratin hydrogels can be used to embed different cytokines and factors that promote cell growth and angiogenesis, making them useful for cardiac tissue repair and nerve repair. Also, keratin is widely available and highly biocompatible, making it a promising candidate for cardiac tissue repair.

12.2.3.7 Matrigel for the delivery of embryonic stem cells to treat myocardial infarction

Matrigel, though commonly used to improve cell adhesion, may also be used as an injectable hydrogel to aid in cardiac tissue repair. As a hydrogel used in this way, it prevents cell death and promotes angiogenesis to the damaged cardiac tissues. For instance, Kofidis et al. employed Matrigel for the delivery of embryonic stem cells as biologic cargo to infarcted myocardium, where the cells were encapsulated in the Matrigel injected in the infarcted left ventricle. Their study indicated that injectable Matrigel aids in stabilizing the heart's shape, geometry, and promotes its functionality after myocardial infarction. Overall, Matrigel shows promise as an agent to help repair cardiac tissue in ischemic heart disease [18].

12.2.4 The Use of Synthetic Hydrogels for Cardiac Tissue Repair Through the Delivery of Biological Cargo

Despite the many advantages of natural hydrogels, they suffer from challenges such as low mechanical strength and an increased risk of infection upon *in vivo* use, especially when intended for use as implantable devices for the long term and continuous delivery of biological cargo. Consequently, synthetic hydrogels are being produced in an attempt to try and overcome these problems. Some examples of synthetic hydrogels include poly(acrylic acid) derivatives (PAA), polyethylene glycol (PEG)/polyethylene oxide (PEO), polyvinyl alcohol (PVA), polyphosphazene, and polypeptides.

12.2.4.1 Examples of common synthetic polymers used for the design of biocompatible hydrogel systems for the delivery of biologic cargo for cardiovascular diseases

There are several derivatives of poly(acrylic acid) (PAA) that are used as hydrogels for cardiac tissue repair such as poly-2-hydroxyethyl methacrylate and poly-N-isopropyl-acrylamide. However, these hydrogels have several disadvantages. For one, neither are fully biodegradable, and secondly, poly-N-isopropyl-acrylamide is potentially toxic, teratogenic, or carcinogenic. Despite this, poly-N-isopropyl-acrylamide has been shown to be effective at improving cardiac tissue repair. However, future research and testing are warranted to fix these serious issues before they can be used in clinical practice.

PEG/PEO is FDA approved for several biomedical applications due to its biocompatibility and its ability to control the release of growth factors. In addition, it has low toxicity making it useful for injecting into the damaged cardiac tissues. Several studies have shown the PEG and its copolymers are promising hydrogels for use in cardiac tissue repair.

PVA is a biocompatible monomer produced through the hydrolysis of polyvinyl acetate. Contrary to natural hydrogels, PVA has strong mechanical properties, which make it ideal for cardiac tissue repair. In addition, PVA is high elastic, and therefore, it can be used as a matrix, making it ideal for tissue engineering and cardiac tissue repair. Despite these important advantages, PVA does have relatively poor adhesion properties, which can be improved by mixing it with various biological factors.

Several peptide hydrogels can be formed, including thermo-responsive hydrogels, ion-induced cross-linked peptide hydrogels, and pH-responsive hydrogels. Peptides are beneficial because they mimic the natural extracellular matrix and act as a scaffold for cell growth. Because of these properties, peptide hydrogels show great promise for use in cardiac tissue repair.

12.2.4.2 Composite hydrogels

Composite hydrogels are formed by combining natural and synthetic hydrogels. This is particularly useful because it exploits the beneficial properties and helps overcome the limitations of both types of hydrogels. The main method of achieving this is to develop nanocomposites by incorporating polymeric, metallic, ceramic, inorganic, or carbon-based nanoparticles. There are many examples of composite hydrogels that are currently being studied, including extracellular matrix-fibrin hydrogels, alginate-chitosan hydrogels, and ECM-PEG hydrogels. Composite and hybrid hydrogels are an ongoing area of research and they continue to show promise for cardiac tissue repair.

12.2.5 The Use of Hydrogels for the Treatment of Cardiac Rhythm Disorders: The Case of Arrhythmias

Arrhythmia is defined as abnormal heart rhythm and can be caused by a variety of factors including underlying cardiac diseases such as myocardial infarction and ischemia. There are several different types of arrhythmias, but for this section, we will focus on ventricular arrhythmias.

Ventricular Arrhythmias: During myocardial infarction/ischemia, there is massive cardiac tissue death, leading to a loss of cardiomyocytes. This causes the formation of scar tissue. Scar tissue has significantly decreased inotropy and chronotropy compared to cardiomyocytes. This can lead to ventricular arrhythmias such as ventricular fibrillation and ventricular tachycardia. Both of these can lead to sudden cardiac death (SCD). Current therapies for the treatment of arrhythmia are aimed at treating the causes of arrhythmia and restoring the normal rhythm. This can involve cardioversion and immediate defibrillation for the most serious cases. In addition, the patient can have an implanted cardioverter defibrillator placed that is designed to stop fibrillation. Finally, several pharmacotherapy options can be used to help maintain a normal sinus rhythm.

12.2.5.1 *Hydrogels, biologics, and the treatment of arrhythmias*

An ongoing area of research in the treatment of arrhythmias is the utilization of conductive biomaterials for creating hydrogels. It is imperative that these hydrogels are biocompatible, as they are being injected into the heart. One such example is the use of polypyrrole (PPy) and chitosan to create a polypyrrole-chitosan hydrogel.

Mihic et al. created a "conductive polymer by grafting pyrrole to the clinically tested biomaterial chitosan to create a polypyrrole (PPy)-chitosan hydrogel. Cyclic voltammetry showed that PPy-chitosan had semiconductive properties lacking in chitosan alone. PPy-chitosan did not reduce cell attachment, metabolism, or proliferation in vitro. Neonatal rat cardiomyocytes plated on PPy-chitosan showed enhanced Ca^{2+} signal conduction in comparison with chitosan alone. PPy-chitosan plating also improved electric coupling between skeletal muscles placed 25 mm apart in comparison with chitosan alone, demonstrating that PPy-chitosan can electrically connect contracting cells at a distance in which rats were injected with this hydrogel" [19]. The authors reported that the injection of the PPy-chitosan hydrogel decreased the QRS interval and increased the transverse activation velocity in comparison with saline or chitosan, suggesting improved electric conduction, and overall, better-maintained heart function after myocardial infarction [19]. This suggests that conductive hydrogels could be a promising therapy for managing arrhythmias and in the future, such conductive hydrogels systems may be further embedded with biological agents to circumvent conduction pathway blockades.

12.2.5.2 *Disadvantages of hydrogels usage in arrhythmias*

Despite the promise of hydrogels for treating cardiac arrhythmias, they also suffer from several limitations. First, biodegradability and biocompatibility are ongoing areas of concern and research and the lack of these attributes can lead to toxicity and further complications. Another area of concern is the possibility that these hydrogels can create a substrate for arrhythmias analogous to scar tissue formation, and in so doing, can cause more arrhythmias. There are very few studies on this. However, it has been shown that hydrogels with a low interstitial spread have the highest likelihood of being a substrate for arrhythmias. Finally, further research needs to be done to determine the optimal timing for injecting the hydrogels in order to minimize the inflammatory reaction and limit scar tissue formation. Despite these limitations, as discussed in previous sections, hydrogels and other biomaterials continue to show promise for the treatment of arrhythmias.

12.3 THE USE OF HYDROGELS FOR TISSUE ENGINEERING AND WOUND HEALING: NOVEL SUSTAINED CONTROLLED DELIVERY OF BIOLOGICALS

Many people each year suffer from tissue or organ death. While tissue and organ transplants are an acceptable option, they are with high cost and risk and are often limited by donor shortages. Tissue engineering, which is the practice of using engineered or synthetic materials to heal and regenerate damaged tissues or organs, is an ongoing area of research that attempts to solve this problem. Tissue engineering with hydrogels is a continual area of research aimed at improving this practice.

12.3.1 Hydrogels in Tissue Engineering

Hydrogels are used as scaffolds for biologics such as for providing a framework for cells and tissues to grow and can be used to generate various types of tissues, such as cartilage, bone, and smooth muscle. The most commonly used applications for hydrogel scaffolds in tissue engineering are space-filling, facilitating drug delivery, and organizing cells, and presenting stimuli to aid in tissue development. In addition to filling in spaces in damaged tissues, space-filling agents are also used to either prevent adhesion or facilitate adhesion between cells and tissues when needed. In addition, hydrogels can facilitate the delivery of drugs that can promote the formation of new blood vessels or encapsulate secretory cells. In order for hydrogels to be successful agents in tissue engineering, they must be sterile to prevent infection and they can't be immunogenic or trigger any other undesirable response.

12.3.2 The Use of Hydrogels for Delivery of Wound-Healing Biologics

A wound occurs when there is damage to the skin, which can result from trauma or some other medical conditions. Skin wounds are often classified as superficial, with partial thickness, or full-thickness, depending on the depth of the wound and layers of the skin and surrounding tissues that are involved. A wound is considered superficial if only the epidermis is damaged. A wound is classified as having partial thickness if the epidermis and dermis are damaged, but at least part of the dermis is still intact. Finally, a wound is considered a full-thickness wound if it penetrates through the epidermis and dermis into the subcutaneous fatty tissues. Hydrogels have been invented with wide capabilities for the treatment of several types of wounds and burns.

12.3.3 Characteristics of Hydrogels as Delivery Vehicles of Biologics for Wound Healing: General Considerations

Appropriate wound management products are imperative for proper wound healing. While gauze is still commonly used for wound care today, hydrogels are becoming more popular due to several chemical and physiological properties that make them ideal for wound healing, especially for necrotic or soughing wounds and burns. Physical properties of hydrogels that allow their use for wound healing include maintaining an appropriate moisture balance, absorbing wound secretions, allowing vapor and oxygen transmission at the site of the wound, and preventing further trauma to the wound upon removal of the dressing. Furthermore, hydrogels may be used for the continuous and slow, timed release of aqueous soluble wound healing biologics.

Traditionally, a dry environment was thought to be the ideal environment for wound healing. However, it has since been widely recognized that moist environments are the best choice for optimal wound healing. The moisture helps promote angiogenesis, which in turn will lead to the increased production of new epithelial tissues. A moist environment also helps to facilitate the debridement of the wound through the production and preservation of autolytic enzymes. This process helps prepare the wound for repair. In addition to enzymes, the moist environment also helps to facilitate the production of cytokines and other mediators and growth factors needed for wound healing and repair. This, in turn, leads to an inflammatory response and ultimately the migration of white blood cells to the wound through diapedesis. In addition, the moist environment helps prevent tissues from becoming dehydrated, which leads to less scarring and pain for the patient. Finally, the moist environment allows for the removal of the dressing without causing further trauma by preventing the dressing from adhering to the wound. Thus, hydrogels are being combined with other wound care products, which allow them to be used for many other types of wounds.

12.3.4 Disadvantages of Hydrogels in the Context of Wound-Healing

Despite the many advantageous properties of hydrogels, hydrogels also have several disadvantages. For one, the high-water content of hydrogels can potentially lead to the maceration of healthy tissues, which can lead to impaired wound healing and an increased risk for infection. Also, while hydrogels can absorb wound secretions, as discussed previously, this absorption process is usually minimal at best, especially when compared to other wound care products, such as hydrocolloids. This makes it harder to isolate bacteria, debris, and other contaminants. Thus, hydrogels are generally not suitable for infected wounds, bleeding wounds, and wounds with moderate to high levels of secretions. Finally, most hydrogels require a secondary dressing, as hydrogels are general not adhesive.

12.3.4.1 Nanocomposites/composite hydrogel products

Often, the following question is posed regarding hydrogels versus other wound healing products: are hydrogels the best product for wound healing? What distinct advantages may be offered by hydrogel systems for the delivery of biological wound healing products?

Initial studies that addressed these questions were phase I and phase II studies by Milne CT et al. [20], which compared hydrogels to collagenase for the treatment of pressure ulcers. The first of the two studies looked at wound debridement, while the second study evaluated maintenance debridement and wound closure.

These first two studies are phase I and phase II studies by Milne CT et al. [20] that compare hydrogels to collagenase for the treatment of pressure ulcers. The first of the two studies looked at wound debridement, while the second study looked at maintenance debridement and wound closure.

The first study sought to compare the time to complete wound debridement with collagenase compared to the hydrogel. There were 27 patients randomized to receive wound care with either collagenase or hydrogel. These patients were at least 18 years old and had at least 85% necrotic tissue on a pressure ulcer between 1 cm^2 and 64 cm^2. This study found that 11 out of 13 patients receiving collagenase achieved complete debridement of the wound by day 42, while only 4 out of 14 patients receiving hydrogel therapy achieved this goal. This was a statistically significant difference. In addition, there was a significantly greater weekly reduction of nonviable tissue in the collagenase group versus the hydrogel group and the wound size also decreased at a significantly faster rate in the collagenase group versus the hydrogel group. Overall, collagenase was found to be significantly more effective at fully debriding a pressure ulcer wound by day 42 than hydrogel. The phase II study is a continuation of this study and looks at maintenance debridement and wound closure.

The phase II study sought to evaluate maintenance debridement and wound closure using the same treatment groups as the phase I study. Patients were rolled over to the phase II study if complete debridement of all visible necrotic tissue was completed by day 42. There were 11 out of the original 13 patients in the collagenase group and 4 out of the original 14 patients in the hydrogel group that was included in this phase II study. This study found that there were no significant differences in maintenance debridement and wound closure between the two groups at day 84 for just phase II. However, collagenase was significantly favored when assessing closure rates from the onset of the pressure ulcer.

From these phases I and II trials, we see that hydrogels do show some benefit in healing pressure ulcers, especially when the wound has been completely debrided. However, it appears that collagenase is a better option for treating pressure from the onset of the wound as it was significantly better at completely debriding the wound by day 42 in these studies.

There are many hydrogel and hydrogel-containing products on the market indicated for different classifications of wounds. Some examples of currently available hydrogel products include Granugel®, Intrasite Gel®, Purilon Gel®, Aquaflo™, and Woundtab®. Granugel® is a combination of pectin, carboxymethylcellulose, pectin, and propylene glycol and is indicated for partial and full-thickness wounds, as well as dry cavity wounds. Intrasite Gel® is a combination of modified carboxymethylcellulose and propylene glycol and is indicated for shallow and deep open wounds. Purilon Gel® is a combination of sodium carboxymethylcellulose and water and is indicated for necrotic and sloughy wounds, as well as first and second-degree burns in combination with a secondary dressing. Aquaflo™ – used to fill shallow cavity wounds and to visualize wounds – is a combination of PEG and propylene glycol. Finally, Woundtab® is a combination of sulfonated copolymer, carboxymethylcellulose, glycerol, and water and is used as a secondary absorbent.

12.4 CONCLUSION

Hydrogels are an excellent new delivery medium for biological agents including monoclonal antibodies for the treatment of a variety of human diseases. Natural, synthetic, biodegradable, stimulus-sensitive polymers are being employed for the design of hydrogels which exploit the ability of the gel-sol transitions for imbibing biologics and biotherapeutic entities into hydrogels and for their long-term, tissue-specific release. The use of temperature-sensitive, pH sensitive, electrosensitive, and electroconductive polymers for the construction of hydrogels loaded with biologic drug cargo is beginning to revolutionize targeted and long-term drug delivery for complex human diseases such as cancers and specific cardiovascular disorders such as arrhythmias and tissue engineering and reconstruction options for myocardial infarction and for wound healing. However, despite rapid advancements, large-scale deployment of hydrogel-based biological drug delivery systems in humans requires much more research and development.

REFERENCES

1. Huang H, Qi X, Chen Y, Wu Z. Thermo-sensitive hydrogels for delivering biotherapeutic molecules: A review. *Saudi Pharm J.* 2019;27(7):990–999. doi:10.1016/j.jsps.2019.08.001.
2. Pan A, Wang Z, Chen B, et al. Localized co-delivery of collagenase and trastuzumab by thermosensitive hydrogels for enhanced antitumor efficacy in human breast xenograft. *Drug Deliv.* 2018;25(1):1495–1503. doi: 10.1080/10717544.2018.1474971.
3. Marcucci F, Bellone M, Rumio C, Corti A. Approaches to improve tumor accumulation and interactions between monoclonal antibodies and immune cells. *MAbs.* 2013 Jan–Feb; 5(1):34–46.
4. Rizwan M, Yahya R, Hassan A, et al. pH sensitive hydrogels in drug delivery: Brief history, properties, swelling, and release mechanism, material selection and applications [published correction appears in polymers (Basel). 2017 Jun 14; 9(6)]. *Polymers (Basel).* 2017;9(4):137. Published 2017 Apr 12. doi:10.3390/polym9040137.
5. Chang CH, Lin YH, Yeh CL, Chen YC, Chiou SF, Hsu YM, Chen YS, Wang CC. Nanoparticles incorporated in pH-sensitive hydrogels as amoxicillin delivery for eradication of Helicobacter pylori. *Biomacromolecules.* 2010 Jan 11;11(1):133–142. doi:10.1021/bm900985h.
6. Bellingeri RV, Picco NY, Alustiza FE, et al. pH-responsive hydrogels to protect IgY from gastric conditions: In vitro evaluation. *J Food Sci Technol.* 2015;52(5):3117–3122. doi:10.1007/s13197-014-1337-3.
7. Qiu Y, Park K. Environment-sensitive hydrogels for drug delivery. *Adv Drug Deliv Rev.* 2001 Dec 31; 53(3):321–339. doi:10.1016/s0169-409x(01)00203-4.
8. Zhao X, Kim J, Cezar CA, Huebsch N, Lee K, Bouhadir K. Active scaffolds for on-demand drug and cell delivery. *Proc Natl Acad Sci USA.* 2011;108:67–72.
9. Casolaro M, Casolaro I, Lamponi S. Stimuli-responsive hydrogels for controlled pilocarpine ocular delivery. *Eur J Pharm Biopharm.* 2012 Apr;80(3):553–561. doi:10.1016/j.ejpb.2011.11.013.
10. Kaur G, Adhikari R, Cass P, Bown M, Gunatillake P. Electrically conductive polymers and composites for biomedical applications. *RSC Advances.* 2015;5(47):37553–37567.
11. Marek SR, Peppas NA. Insulin release dynamics from poly(diethylaminoethyl methacrylate) hydrogel systems. *AIChE J.* 2013;59(10):3578–3585. doi:10.1002/aic.14108.
12. Pai CM, Bae YH, Mack EJ, Wilson DE, Kim SW. Concanavalin A microspheres for a self-regulating insulin delivery system. *J Pharm Sci.* 1992 Jun;81(6):532–536. doi:10.1002/jps.2600810612.
13. Dai W, Wold LE, Dow JS, Kloner RA. Thickening of the infarcted wall by collagen injection improves left ventricular function in rats: A novel approach to preserve cardiac function after myocardial infarction. *J Am Coll Cardiol.* 2005 Aug 16;46(4):714–719. doi:10.1016/j.jacc.2005.04.056.
14. Tabata Y, Ikada Y. Vascularization effect of basic fibroblast growth factor released from gelatin hydrogels with different biodegradabilities. *Biomaterials.* 1999 Nov;20(22):2169–2175. doi:10.1016/s0142-9612(99)00121-0.
15. Ifkovits JL, Tous E, Minakawa M, et al. Injectable hydrogel properties influence infarct expansion and extent of postinfarction left ventricular remodeling in an ovine model. *Proc Natl Acad Sci USA.* 2010;107(25):11507–11512. doi:10.1073/pnas.1004097107.

16. Abdalla S, Makhoul G, Duong M, Chiu RC, Cecere R. Hyaluronic acid-based hydrogel induces neovascularization and improves cardiac function in a rat model of myocardial infarction. *Interact Cardiovasc Thorac Surg*. 2013;17(5):767–772. doi:10.1093/icvts/ivt277.
17. Christman KL, Vardanian AJ, Fang Q, Sievers RE, Fok HH, Lee RJ. Injectable fibrin scaffold improves cell transplant survival, reduces infarct expansion, and induces neovasculature formation in ischemic myocardium. *J Am Coll Cardiol*. 2004 Aug 4;44(3):654–660. doi:10.1016/j.jacc.2004.04.040.
18. Kofidis T, Lebl DR, Martinez EC, Hoyt G, Tanaka M, Robbins RC. Novel injectable bioartificial tissue facilitates targeted, less invasive, large-scale tissue restoration on the beating heart after myocardial injury. *Circulation*. 2005 Aug 30;112(9 Suppl):I173–I177. doi:10.1161/CIRCULATIONAHA.104.526178.
19. Mihic A, Cui Z, Wu J, Vlacic G, Miyagi Y, Li SH, Lu S, Sung HW, Weisel RD, Li RK. A conductive polymer hydrogel supports cell electrical signaling and improves cardiac function after implantation into myocardial infarct. *Circulation*. 2015 Aug 25;132(8):772–784. doi:10.1161/CIRCULATIONAHA.114.014937.
20. Milne CT, Ciccarelli AO, Lassy M. A comparison of collagenase to hydrogel dressings in wound debridement. *Wounds*. 2010 Nov; 22(11):270–274.

13 Clinical Use of Biologics and Biosimilars in Oncology Settings

Rinda Devi Bachu, Shikha Kumari, Sai H.S. Boddu, R. Jayachandra Babu, and Amit K. Tiwari

Corresponding author: Amit K. Tiwari

Contents

13.1 THE NEED FOR BIOSIMILARS IN CANCER

Cancer is among the leading causes of death worldwide. Globally, cancer accounts for about one in every six deaths, which is more than HIV, tuberculosis, and malaria combined (1). In 2020, there were about 19.3 million estimated new cases and 10 million cancer-related deaths worldwide.

DOI: 10.1201/9780429485626-13

Among these deaths, one-quarter of the cases occur in low- and medium–Human Development Index countries, which lack resources and medical systems to address the disease burden (2). By 2040, the global cancer burden is expected to increase to an estimated 27.5 million cases and 16.3 million deaths based on the aging and growth of the population (3). The National Cancer Institute estimates the direct medical costs related to cancer treatment in the United States were $183 billion in 2015 and are expected to increase to $246 billion by 2030, a 34% raise (4). However, owing to the advances in personalized treatments and inflation, this increase is likely to be an underestimation (5, 6). Biologics have become an integral part of cancer treatment not only for their therapeutic outcomes, but also as supportive care agents (7). Biologics are composed of large and structurally complex molecular entities that require extensive immunogenic testing and pharmacovigilance strategies for monitoring any immune response that might be evoked in the body (8). Biologics account for about half of the pharmacological market in oncology; however, their high development and manufacturing costs pose a significant burden on the health care systems (9). With the expiration of biologics patents, biosimilars are being developed to reduce costs and increase patient access to novel cancer therapies (10, 11).

13.1.1 Biosimilar Market

Biosimilars are biological drugs that are designed to be highly similar to the existing marketed biologics (12). The high level of similarity to the original biologic is defined in terms of physicochemical characteristics, efficacy, and safety as outlined by the respective regulatory authorities (13, 14). However, biosimilars are not a generic version of biologics, as it is not possible to develop an identical biochemical entity. This is mainly due to the inherent complexity of the proteins and their associated manufacturing processes (15). Twenty-nine biosimilars for various indications have been approved in the United States (16), whereas 64 biosimilars were approved in Europe (17). The European Medical Agency (EMA) was the first to approve a biosimilar in 2006 and it provided guidance for biosimilar development and approval (18). However, the pathway for marketing biosimilars in the United States has been hindered by several obstacles (19). The FDA lacked a clear regulatory pathway for the approval of biosimilars until 2010. This was one of the main reasons for the slow adoption of biosimilars in the United States when compared to Europe (20). Moreover, the marketing of biosimilars in the United States is delayed by patent infringement lawsuits, exclusionary contracts, and anticompetitive tactics of brand name manufacturers (21). Even though the patent protection of several originator biologics was close to the expiration date, the market competition that was seen with chemical drugs through generics did not occur with biosimilars (22). Generics have been able to be marketed in the United States since 1984 due to the established abbreviated pathway through Hatch-Waxman Act (23).

13.1.2 Overview of Biosimilar Legislation and Regulation

Historically in the United States, biologics were regulated by the Public Health Hygienic Laboratory, a precursor of National Institutes of Health (NIH), which was then transferred to the Bureau of Biologics at the FDA in 1972 (24). A decade later, the Bureau of Drugs and Bureau of Biologics were merged into a single entity to form National Center for Drugs and Biologics (NCDB) (25). Owing to the high volume of new drug applications (NDAs) and to resolve other issues in drug and biologic evaluation, the Center for Drugs and Biologics was divided back into the Center for Drug Evaluation and Research (CDER) and the Center for Biologics Evaluation and Research (CBER) in 1987 (26). The jurisdictional responsibilities of the two centers were assigned through the Intercenter

Agreement issued by the FDA in 1991 (27). Traditional biologics, including vaccines, blood, blood products, allergenic extracts, certain devices, and test kits are regulated by CBER. The center also regulates gene and cellular therapy products as well as tissue transplants from human and non-human sources (28). Furthermore, CDER regulates prescription, over-the-counter and generic drugs (29), as well as most therapeutic biologics, including monoclonal antibodies, cytokines, growth factors, enzymes, and immunomodulators (29, 30).

Most biological products were approved under the Public Health Service Act (PHSA), while some of them are licensed as drugs under the Federal Food, Drug, and Cosmetic Act (FFDCA) (31). In 2010, Congress established an abbreviated licensure pathway for biological products that demonstrated to be biosimilar or interchangeable to a previously licensed biological product. This new regulatory authority for the FDA was accomplished through the Biologics Price Competition and Innovation Act (BPCIA) of 2009, which was enacted as Title VII of the Patient Protection and Affordable Care Act (ACA) (20). As a part of the implementation of BPCIA, three draft guidances on biosimilars development were released by the FDA in 2012 (32) and the final versions were released in 2015 (33). The BPCIA has also set periods of regulatory exclusivity for brand name biologics and biosimilars as well as laid procedures for resolving patent disputes (20). Biologics are offered 12 years of exclusivity, during which the FDA cannot approve any biosimilar or interchangeable product referencing the brand name biologic. However, a Biologics License Application (BLA) of a biosimilar or interchangeable product can be submitted after four years from the date on which the reference product was first licensed (34). Also, effective from March 23, 2020, biological products under BPCIA, which were previously approved as drugs under Section 505 of the FFDCA are transitioned to biological licenses under Section 351 of the PHSA (31).

Over the years, the FDA has released additional guidance on a variety of other areas related to biosimilars and these documents can be accessed through the FDA website (33). The agency's database "Purple Book" contains information about all the FDA-licensed biologics regulated by the CDER including their biosimilars and interchangeable products. In-depth information about the date on which the biological product was licensed, if the biological product has proven to be a biosimilar or interchangeable to an already licensed biological drug, and the expiration dates of applicable exclusivities of the reference biologics can be obtained. Also, the database provides information about licensed products regulated by the CBER (35).

13.1.3 European Regulatory Pathways for Biosimilars

Moreover, guidelines for the regulation of medicines in the European Union (EU) were well established. A dedicated pathway for the development and approval of biosimilars was introduced in 2004 (36). General guidelines on biosimilars were issued by EMA to introduce the concept and to provide biosimilar manufacturers with a user guide containing relevant scientific information (37). In the EU, biologics are offered eight years of exclusivity during which a biosimilar referencing the brand name biologic cannot be marketed (14). Biotechnology products including biosimilars are approved by the EMA through a marketing authorization application (MAA) following a centralized procedure (36). This procedure authorizes the manufacturers to market their products throughout the European Economic Area (EEA) with a single MAA. EEA includes all EU member states, and three countries of the European Free Trade Association (EFTA) – Iceland, Liechtenstein, and Norway (38). The MAA's for biosimilars are evaluated by EMA's scientific committees including, the Committee for Medical Products for Human Use (CHMP), Pharmacovigilance Risk Assessment Committee (PRAC) as well as EU experts and specialists on biological medicines (Biologics Working Party) and biosimilars (Biosimilar Working Party), respectively (36). The scientific opinion obtained after the EMA's evaluation is recommended to the European Commission, which ultimately decides if an EU-wide marketing authorization can be granted. Once approved, the decision of the commission is published in the Community Register of medicinal products for human use.

In addition, the EMA also publishes a European public assessment report (EPAR) for each application that has been granted/refused a marketing authorization (38). The complete list of centrally authorized biosimilars approved to date can be accessed from the EMA's website (17).

13.2 BIOSIMILARS IN ONCOLOGY

Despite the potential cost benefits, lack of knowledge and understanding of biosimilars and the possibility of immunogenicity have created an uncertain environment for healthcare professionals and patients (39). American Society of Clinical Oncology (ASCO) provides information and guidance to the oncology community on the use of biosimilars, their safety and efficacy, interchangeability, substitution, regulatory considerations, and the prescriber and patient education. CancerLinQ, an integrated real-time data resource, also provides valuable information on the use of biosimilars and their effectiveness (40). Currently, there are only a few biosimilars approved for cancer treatment and supportive care. Biosimilars are available for monoclonal antibodies (mAb), including Rituximab, Trastuzumab, and Bevacizumab, and supportive agents, including filgrastim, pegfilgrastim, epoetin-α (EPO-α), and epoetin-ζ (EPO-ζ) (10).

13.2.1 Filgrastim and Pegfilgrastim

The first-ever biosimilar product to be marketed in the United States was Zarxio® (filgrastim-sndz) and was approved by the FDA in March 2015 (41). Later in 2018, Nivestym® (filgrastim-aafi) was approved; both of these biosimilars can be used for the same indications as the reference drug, Neupogen® (filgrastim) (42). Filgrastim is a recombinant granulocyte colony-stimulating factor (G-CSF) that regulates neutrophil production from the bone marrow. Filgrastim is used to reduce febrile neutropenia in patients with non-myeloid malignancies receiving myelosuppressive anticancer agents or myeloablative chemotherapy followed by bone marrow transplantation. It is also used in patients with acute myeloid leukemia receiving induction or consolidation chemotherapy for reducing the time of neutrophil recovery and the duration of fever (43). In Europe, nine biosimilars of filgrastim are approved by the EMA, including Accofil® (44), Biograstim® (45), Filgrastim Hexal® (46), Filgrastim Ratiopharm® (47), Grastofil® (48), Nivestim® (49), Ratiograstim® (50), Tevegrastim® (51), and Zarzio® (52). However, the marketing of Biograstim® and Filgrastim Ratiopharm® was withdrawn by the EMA at the request of their respective marketing authorization holders (45, 47). Two other G-CSFs that are commonly used for treating chemotherapy-induced neutropenia (CIN) include pegfilgrastim and lenograstim. Filgrastim and lenograstim are short-acting G-CSFs that are injected daily during chemotherapy, while pegfilgrastim is a long-acting G-CSFs, administered once per chemotherapy cycle (53). Pegfilgrastim has an additional polyethylene glycol unit which causes an increase in the size of the molecule, thereby prolonging the half-life of the drug (9). Eight approved biosimilars for pegfilgrastim are available in Europe, including Pelgraz® (54), Udenyca® (55), Fulphila® (56), Pelmeg® (57), Ziextenzo® (58), Grasustek® (59), Cegfila® (60), and Nyvepria® (61). Whereas in the United States, for Neulasta® (pegfilgrastim) four biosimilars are approved: Fulphila® (pegfilgrastim-jmdb) (62), Udenyca® (pegfilgrastim-cbqv) (63), Ziextenzo® (pegfilgrastim-bmez) (64), and Nyvepria® (pegfilgrastim-apgf) (65). To, date no biosimilars for lenograstim are available.

13.2.2 Epoetins

Epoetins are used for treating chemotherapy-induced anemia (CIA), reducing the need for blood transfusions, thereby improving the quality of life. These are similar to erythropoietin hormone, secreted by the kidneys that stimulate red blood cell production (erythropoiesis) in the bone marrow and are also referred

TABLE 13.1 FDA-approved supportive care biosimilars in oncology

REFERENCE BIOLOGIC (ACTIVE SUBSTANCE)	*REFERENCE BIOLOGIC MANUFACTURER(S)*	*BIOSIMILAR (ACTIVE SUBSTANCE)*	*BIOSIMILAR MANUFACTURER*	*APPROVAL*
Neupogen® (filgrastim)	Amgen Inc.	Zarxio® (filgrastim-sndz)	Sandoz Inc.	2015
		Nivestym® (filgrastim-aafi)	Hospira Inc.	2018
Neulasta® (pegfilgrastim)	Amgen Inc.	Fulphila® (pegfilgrastim-jmdb)	Mylan GmbH	2018
		Udenyca® (pegfilgrastim-cbqv)	Coherus BioSciences, Inc.	2018
		Ziextenzo® (pegfilgrastim-bmez)	Sandoz Inc.	2019
		Nyvepria® (pegfilgrastim-apgf)	Hospira Inc.	2020
Epogen/Procrit (epoetin α)	Amgen Inc./Janssen Biotech Inc.	Retacrit® (epoetin alfa-epbx)	Hospira Inc.	2018

to as erythropoiesis-stimulating agents (ESAs) (66, 67). Five epoetin biosimilars are approved in Europe, including three of EPO-α biosimilars: Abseamed® (68), Binocrit® (69), epoetin-α hexal® (70), and two of EPO-ζ biosimilars: Retacrit® (71) and Silapo® (72). Whereas in the United States, only one ESA agent, Retacrit® (epoetin alfa-epbx) (73) has been approved for the reference drug, Epogen®/Procrit® (EPO-α). Both EPO-α and EPO-ζ have been approved for treating chemotherapy-induced and symptomatic anemia in patients with solid tumors, malignant lymphoma, or multiple myeloma (74, 75). Tables 13.1 and 13.2 list the biosimilar drugs approved for supportive cancer care by the FDA and EMA, respectively.

TABLE 13.2 EMA-approved supportive care biosimilars in oncology

REFERENCE BIOLOGIC (ACTIVE SUBSTANCE)	*REFERENCE BIOLOGIC MANUFACTURER*	*BIOSIMILAR*	*BIOSIMILAR MANUFACTURER*	*APPROVAL*
Neupogen® (filgrastim)	Amgen Inc.	Accofil®	Accord Healthcare S.L.U.	2014
		Filgrastim Hexal®	Hexal AG	2009
		Grastofil®	Accord Healthcare S.L.U.	2013
		Nivestim®	Pfizer Europe MA EEIG	2010
		Ratiograstim®	Ratiopharm GmbH	2008
		Tevegrastim®	Teva GmbH	2008
		Zarzio®	Sandoz GmbH	2009
Neulasta® (pegfilgrastim)	Amgen Inc.	Pelgraz®	Accord Healthcare S.L.U.	2018
		Fulphila®	Mylan S.A.S.	2018
		Udenyca®	ERA Consulting GmbH	2018
		Pelmeg®	Mundipharma Corporation (Ireland) Limited	2018
		Ziextenzo®	Sandoz GmbH	2018
		Grasustek®	Juta Pharma GmbH	2019
		Cegfila®	Mundipharma Corporation (Ireland) Limited	2019
		Nyvepria®	Pfizer Europe MA EEIG	2020
Eprex®/Erypo® (epoetin α)	Janssen-Cilag GmbH	Abseamed®	Medice Arzneimittel Pütter GmbH & Co. KG	2007
		Binocrit®	Sandoz GmbH	2007
		Epoetin-α hexal®	Hexal AG	2007
		Retacrit®	Pfizer Europe MA EEIG	2007
		Silapo®	Stada Arzneimittel AG	2007

13.3 MONOCLONAL ANTIBODIES

13.3.1 Bevacizumab

Bevacizumab is a recombinant humanized monoclonal antibody that targets vascular endothelial growth factor (VEGF-A) and inhibits the formation of new blood vessels (angiogenesis) and the growth of new tumors (76). Avastin® (bevacizumab) is used for various indications, including metastatic colorectal cancer (mCRC), non-squamous non–small cell lung cancer (NSCLC), glioblastoma, metastatic renal cell carcinoma (mRCC), and persistent, recurrent, or metastatic carcinoma of the cervix either as a single agent or in combination with chemotherapy/biologic response modifier (77). The patent of Avastin® in the United States expired in 2019, whereas in Europe, the patent will expire in 2022 (78). Currently, two biosimilars of Avastin® (bevacizumab) are available in the United States, including Mvasi® (bevacizumab-awwb) (79) and Zirabev® (bevacizumab-bvzr) (80). Both of these biosimilars (81, 82) along with a few others, including Aybintio® (83), Equidacent® (84), and Oyavas® (85), are approved in Europe. However, these biosimilars could face a delay in reaching the market until relevant patents and regulatory exclusivities expire (86). One more biosimilar product Alymsys® (87) is also currently under the EMA review. This biosimilar has already received a positive opinion from CHMP.

13.3.2 Rituximab

Rituxan® (rituximab) is a genetically engineered chimeric human monoclonal antibody that targets the CD20 antigen, found on the surface of B lymphocytes. Rituxan® is indicated for treating patients with non-Hodgkin's lymphoma (NHL) and chronic lymphocytic leukemia (CLL) (88). It is indicated for use as a single agent for relapsed/refractory, low-grade/follicular CD20-positive, B-cell NHL, and in patients with non-progressing, low-grade, CD20-positive, B-cell NHL after first-line cyclophosphamide, vincristine, and prednisolone (CVP) chemotherapy. It is also used in combination with chemotherapy in patients with previously untreated follicular and diffuse large B-cell, CD20 positive B-cell NHL, and as single-agent maintenance therapy in patients who achieved a complete/partial response to Rituxan®. In patients with CLL, Rituxan® is used in combination with chemotherapeutics: fludarabine and cyclophosphamide (FC) (89). The patent for Rituxan® (rituximab) in the United States expired in 2016 (90), which led to the development of biosimilars Truxima® (rituximab-abbs) (91), Ruxience® (rituximab-pvvr) (92), and recently Riabni® (rituximab-arrx) (93). Even in Europe, the patent for MabThera® (rituximab) expired in 2013 (90), and six biosimilars for rituximab were approved by EMA, including Blitzima® (94), Truxima® (95), Ruxience® (96), Riximyo® (97), Rixathon® (98), and Ritemvia® (99).

13.3.3 Trastuzumab

Herceptin® (trastuzumab) is a humanized monoclonal antibody that selectively binds to the extracellular domain of the human epidermal growth factor receptor 2 protein (HER2) (100). Herceptin® is indicated for patients with (a) metastatic HER2-overexpressing breast cancer either as a single agent or in combination with paclitaxel, (b) metastatic HER2-overexpressing gastric cancer in combination with cisplatin and capecitabine/5-fluorouracil, and (c) HER2-overexpressing breast cancer as an adjuvant treatment in combination with chemotherapeutics or as a single agent following multi-modality anthracycline-based treatment (101). The patent of Herceptin® (trastuzumab) in the United States expired in 2019, whereas in Europe, the patent expired in 2014 (102). Ogivri® (trastuzumab-dkst) was the first biosimilar to Herceptin®

TABLE 13.3 FDA-approved mAB biosimilars in oncology

REFERENCE BIOLOGIC (ACTIVE SUBSTANCE)	*REFERENCE BIOLOGIC MANUFACTURER*	*BIOSIMILAR (ACTIVE SUBSTANCE)*	*BIOSIMILAR MANUFACTURER*	*APPROVAL*
Avastin® (bevacizumab)	Genentech, Inc.	Mvasi® (bevacizumab-awwb)	Amgen Inc.	2017
		Zirabev® (bevacizumab-bvzr)	Pfizer Inc.	2019
Rituxan® (rituximab)	Genentech, Inc.	Truxima® (rituximab-abbs)	Celltrion, Inc.	2018
		Ruxience® (rituximab-pvvr)	Pfizer Ireland Pharmaceuticals	2019
		Riabni® (rituximab-arrx)	Amgen, Inc.	2020
Herceptin® (trastuzumab)	Genentech, Inc.	Ontruzant® (trastuzumab-dttb)	Samsung Bioepis Co., Ltd	2019
		Trazimera® (trastuzumab-qyyp)	Pfizer Inc.	2018
		Herzuma® (trastuzumab-pkrb)	Celltrion, Inc.	2018
		Kanjinti® (trastuzumab-anns)	Amgen Inc.	2019
		Ogivri® (trastuzumab-dkst)	Mylan GmbH	2017

(trastuzumab) to be approved by the FDA. Later in United States, four more biosimilars including Herzuma® (trastuzumab-pkrb) (103), Trazimera® (trastuzumab-qyyp) (104), Ontruzant® (trastuzumab-dttb) (105), and Kanjinti® (trastuzumab-anns) (106) were approved. Ontruzant® (107), Herzuma® (108), Trazimera® (109), Kanjinti® (110), and Ogivri® (111) are also approved in Europe. Recently, the EMA approved another biosimilar for trastuzumab, namely Zercepac® (112). Tables 13.3 and 13.4 list the biosimilar drugs approved for monoclonal antibodies – bevacizumab, rituximab, and trastuzumab by the FDA and EMA, respectively.

TABLE 13.4 EMA approved mAB biosimilars in oncology

REFERENCE BIOLOGIC (ACTIVE SUBSTANCE)	*REFERENCE BIOLOGIC MANUFACTURER*	*BIOSIMILAR*	*BIOSIMILAR MANUFACTURER*	*APPROVAL*
Avastin® (bevacizumab)	Roche Registration GmbH	Mvasi®	Amgen Technology (Ireland) UC	2018
		Zirabev®	Pfizer Europe MA EEIG	2019
		Aybintio®	Samsung Bioepis NL B.V.	2020
		Equidacent®	Centus Biotherapeutics Europe Limited	2020
		Oyavas®	STADA Arzneimittel AG	2021
MabThera® (rituximab)	Roche Registration GmbH	Blitzima®	Celltrion Healthcare Hungary Kft.	2017
		Truxima®	Celltrion Healthcare Hungary Kft.	2017
		Ruxience®	Pfizer Europe MA EEIG	2020
		Riximyo®	Sandoz GmbH	2017
		Rixathon®	Sandoz GmbH	2017
		Ritemvia®	Celltrion Healthcare Hungary Kft.	2017
Herceptin® (trastuzumab)	Roche Registration GmbH	Ontruzant®	Samsung Bioepis NL B.V.	2017
		Trazimera®	Pfizer Europe MA EEIG	2018
		Herzuma®	Celltrion Healthcare Hungary Kft.	2018
		Kanjinti®	Amgen Europe B.V.	2018
		Ogivri®	Mylan S.A.S.	2018
		Zercepac®	Accord Healthcare S.L.U.	2020

13.4 CONCLUSIONS AND FUTURE PERSPECTIVES

The use of biosimilars is rapidly evolving and will continue to play an important role in the future care of cancer patients (113). Many biosimilars are expected to be available in the coming years, and their use will largely depend on patient and provider acceptance, which is in turn based on an adequate understanding of the safety and efficacy of these agents in cancer treatment (40). Therefore, education of patients and provider on various aspects of biosimilars is necessary to increase confidence in biosimilars and for their successful incorporation in oncology practice. Furthermore, rigorous regulatory frameworks and close post-marketing monitoring of these drugs are required to ensure their safety and efficacy in a real-world setting (9).

REFERENCES

1. Roth GA, Abate D, Abate KH, Abay SM, Abbafati C, Abbasi N, et al. Global, regional, and national age-sex-specific mortality for 282 causes of death in 195 countries and territories, 1980–2017: a systematic analysis for the Global Burden of Disease Study 2017. The Lancet. 2018;392(10159):1736–88.
2. Sung H, Ferlay J, Siegel RL, Laversanne M, Soerjomataram I, Jemal A, et al. Global cancer statistics 2020: GLOBOCAN estimates of incidence and mortality worldwide for 36 cancers in 185 countries. CA: A Cancer Journal for Clinicians. 2021;71(3):209–49.
3. American Cancer Society. Global Cancer Facts & Figures 4th Edition. Atlanta: American Cancer Society; 2018.
4. Ofman JJ, Fendrick AM, Raza A. Novel multicancer early detection technology – potential value to employers and the workforce. The American Journal of Managed Care. 2020;26(10 Spec No.):SP363.
5. Bach PB. Limits on Medicare's ability to control rising spending on cancer drugs. The New England Journal of Medicine. 2009;360(6):626–33.
6. Memorial Sloan Kettering Cancer Center. Price and value of cancer drugs. [Available from: [mskcc.org/research-areas/programs-centers/health-policy-outcomes/cost-drugs].
7. Zelenetz AD, Ahmed I, Braud EL, Cross JD, Davenport-Ennis N, Dickinson BD, et al. NCCN Biosimilars White Paper: regulatory, scientific, and patient safety perspectives. Journal of the National Comprehensive Cancer Network. 2011;9(Suppl_4):S-1–S-22.
8. Weise M, Bielsky M-C, De Smet K, Ehmann F, Ekman N, Giezen TJ, et al. Biosimilars: what clinicians should know. Blood. 2012;120(26):5111–7.
9. Konstantinidou S, Papaspiliou A, Kokkotou E. Current and future roles of biosimilars in oncology practice. Oncology letters. 2020;19(1):45–51.
10. Patel KB, Arantes Jr LH, Tang WY, Fung S. The role of biosimilars in value-based oncology care. Cancer Management and Research. 2018;10:4591.
11. Chan JC, Chan AT. Biologics and biosimilars: what, why and how?. BMJ. 2017. [Available from https://doi.org/10.1136/esmoopen-2017-000180].
12. Gascón P, Tesch H, Verpoort K, Rosati MS, Salesi N, Agrawal S, et al. Clinical experience with Zarzio® in Europe: what have we learned? Supportive Care in Cancer. 2013;21(10):2925–32.
13. U.S. Food & Drug Administration. Guidance Document. Scientific Considerations in Demonstrating Biosimilarity to a Reference product. Accessed April 14, 2021. [Available from: https://www.fda.gov/regulatory-information/search-fda-guidance-documents/scientific-considerations-demonstrating-biosimilarity-reference-product].
14. European Medicines Agency. Biosimilar medicines: overview. Accessed April 14, 2021. [Available from: https://www.ema.europa.eu/en/human-regulatory/overview/biosimilar-medicines-overview].
15. Verrill M, Declerck P, Loibl S, Lee J, Cortes J. The rise of oncology biosimilars: from process to promise. Future Oncology. 2019;15(28):3255–65.
16. U.S. Food & Drug Administration (FDA). Biosimilar Product Information. Accessed April 19, 2021. [Available from: https://www.fda.gov/drugs/biosimilars/biosimilar-product-information].
17. European Medicines Agency. Medicines Type-Biosimilars. Accessed April 26, 2021. [Available from: https://www.ema.europa.eu/en/medicines/search_api_aggregation_ema_medicine_type s/field_ema_med_biosimilar].

18. Gherghescu I, Delgado-Charro MB. The biosimilar landscape: an overview of regulatory approvals by the EMA and FDA. Pharmaceutics. 2021;13(1):48.
19. Nabhan C, Valley A, Feinberg BA. Barriers to oncology biosimilars uptake in the United States. The Oncologist. 2018;23(11):1261.
20. Congressional Research Service. Biologics and Biosimilars: Background and Key Issues. Updated June 6, 2019. [Available from: https://crsreports.congress.gov/product/pdf/R/R44620].
21. Gupta R, Kesselheim AS, Downing N, Greene J, Ross JS. Generic drug approvals since the 1984 Hatch-Waxman Act. JAMA Internal Medicine. 2016;176(9):1391–3.
22. McCamish M, Woollett G, editors. Worldwide experience with biosimilar development. MAbs. 2011: Taylor & Francis.
23. Congressional Research Service. Biologics and Biosimilars: Background and Key Issues. Updated October 27, 2017. [Available from: https://crsreports.congress.gov/product/pdf/R/R44620/11].
24. Gruber MF, Marshall VB. Regulation and testing of vaccines. Plotkin's Vaccines. 2018;79:1547.
25. U.S. Food & Drug Administration. A Brief History of the Center for Drug Evaluation and Research. The Merger of Drug and Biologics. 2018. [Available from: https://www.fda.gov/about-fda/fda-history-exhibits/brief-history-center-drug-evaluation-and-research].
26. U.S. Food & Drug Administration. A Brief History of the Center for Drug Evaluation and Research. NCDB Divided Back into Drugs and Biologics Centers. [Available from: https://www.fda.gov/about-fda/fda-history-exhibits/brief-history-center-drug-evaluation-and-research].
27. U.S. Food & Drug Administration. Intercenter Agreement between the Center for Drug Evaluation and Research and the Center for Biologics Evaluation and Research. [Available from: https://www.fda.gov/combination-products/jurisdictional-information/intercenter-agreement-between-center-drug-evaluation-and-research-and-center-biologics-evaluation].
28. U.S. Food & Drug Administration. Regulated Products. CBER Regulated Products. Accessed April 25, 2021. [Available from: https://www.fda.gov/industry/regulated-products/cber-regulated-products].
29. U.S. Food & Drug Administration. FDA Organization. Center for Drug Evaluation and Research (CDER). Accessed April 25, 2021. [Available from: https://www.fda.gov/about-fda/fda-organization/center-drug-evaluation-and-research-cder].
30. U.S. Food & Drug Administration. Transfer of Therapeutic Products to the Center for Drug Evaluation and Research (CDER). Accessed April 25, 2021. [Available from: https://www.fda.gov/about-fda/center-biologics-evaluation-and-research-cber/transfer-therapeutic-products-center-drug-evaluation-and-research-cder].
31. U.S. Food & Drug Administration. Guidance, Compliance & Regulatory Information. "Deemed to be a License" Provision of the BPCI Act. Accessed April 25, 2021. [Available from: https://www.fda.gov/drugs/guidance-compliance-regulatory-information/deemed-be-license-provision-bpci-act].
32. Federal Register. A Notice by the Food and Drug Administration. Draft Guidances Relating to the Development of Biosimilar Products; Public Hearing; Request for Comments. [Available from: https://www.federalregister.gov/documents/2012/03/02/2012-5070/draft-guidances-relating-to-the-development-of-biosimilar-products-public-hearing-request-for].
33. U.S. Food & Drug Administration. Guidance, Compliance & Regulatory Information (Biologics). Biosimilars Guidelines. Accessed April 25, 2021. [Available from: https://www.fda.gov/vaccines-blood-biologics/general-biologics-guidances/biosimilars-guidances].
34. Competition BP. Title VII – Improving Access To Innovative Medical Therapies. 2010.
35. U.S. Food & Drug Administration (FDA). Purple Book. Accessed April 22, 2021 [Available from: https://purplebooksearch.fda.gov/].
36. European Medicines Agency. Biosimilars in the EU. Information Guide for Healthcare Professionals. Updated February, 2019. [Available from: https://www.ema.europa.eu/en/documents/leaflet/biosimilars-eu-information-guide-healthcare-professionals_en.pdf].
37. Wang J, Chow S-C. On the regulatory approval pathway of biosimilar products. Pharmaceuticals. 2012;5(4):353–68.
38. European Medicines Agency. Human Regulatory. Obtaining an EU Marketing Authorisation, Step by Step. Accessed April 27, 2021. [Available from: https://www.ema.europa.eu/en/human-regulatory/marketing-authorisation/obtaining-eu-marketing-authorisation-step-step].
39. Chopra R, Lopes G. Improving access to cancer treatments: the role of biosimilars. Journal of Global Oncology. 2017;3(5):596–610.
40. Lyman GH, Balaban E, Diaz M, Ferris A, Tsao A, Voest E, et al. American society of clinical oncology statement: biosimilars in oncology. Journal of Clinical Oncology. 2018;36(12):1260–5.

41. Awad M, Singh P, Hilas O. Zarxio (filgrastim-sndz): the first biosimilar approved by the FDA. Pharmacy and Therapeutics. 2017;42(1):19.
42. U.S. Food & Drug Administration (FDA). Biosimilar Product Information. Nivestym (filgrastim-aafi). Accessed January 30, 2021. [Available from: https://www.accessdata.fda.gov/drugsatfda_docs/label/2018/761080s000lbl.pdf].
43. Ludwig H, Bokemeyer C, Aapro M, Boccadoro M, Gascón P, Denhaerynck K, et al. Chemotherapy-induced neutropenia/febrile neutropenia prophylaxis with biosimilar filgrastim in solid tumors versus hematological malignancies: MONITOR-GCSF study. Future Oncology. 2019;15(8):897–907.
44. European Medicines Agency. Human medicine European public assessment report (EPAR): Accofil. Accessed January 31, 2021. [Available from: https://www.ema.europa.eu/en/medicines/human/EPAR/accofil].
45. European Medicines Agency. Human medicine European public assessment report (EPAR): Biograstim. Accessed January 31, 2021. [Available from: https://www.ema.europa.eu/en/medicines/human/EPAR/biograstim].
46. European Medicines Agency. Human medicine European public assessment report (EPAR): Filgrastim Hexal. Accessed January 31, 2021. [Available from: https://www.ema.europa.eu/en/medicines/human/EPAR/filgrastim-hexal].
47. European Medicines Agency. Human medicine European public assessment report (EPAR): Filgrastim ratiopharm. Accessed January 31, 2021. [Available from: https://www.ema.europa.eu/en/medicines/human/EPAR/filgrastim-ratiopharm].
48. European Medicines Agency. Human medicine European public assessment report (EPAR): Grastofil. Accessed January 31, 2021. [Available from: https://www.ema.europa.eu/en/medicines/human/EPAR/grastofil].
49. European Medicines Agency. Human medicine European public assessment report (EPAR): Nivestim. Accessed January 31, 2021. [Available from: https://www.ema.europa.eu/en/medicines/human/EPAR/nivestim].
50. European Medicines Agency. Human medicine European public assessment report (EPAR): Ratiograstim. Accessed January 31, 2021. [Available from: https://www.ema.europa.eu/en/medicines/human/EPAR/ratiograstim].
51. European Medicines Agency. Human medicine European public assessment report (EPAR): Tevagrastim. Accessed January 31, 2021. [Available from: https://www.ema.europa.eu/en/medicines/human/EPAR/tevagrastim].
52. European Medicines Agency. Human medicine European public assessment report (EPAR): Zarzio. Accessed January 31, 2021. [Available from: https://www.ema.europa.eu/en/medicines/human/EPAR/zarzio].
53. Cooper KL, Madan J, Whyte S, Stevenson MD, Akehurst RL. Granulocyte colony-stimulating factors for febrile neutropenia prophylaxis following chemotherapy: systematic review and meta-analysis. BMC Cancer. 2011;11(1):1–11.
54. European Medicines Agency. Human medicine European public assessment report (EPAR): Pelgraz. Accessed January 30, 2021. [Available from: https://www.ema.europa.eu/en/medicines/human/EPAR/pelgraz].
55. European Medicines Agency. Human medicine European public assessment report (EPAR): Udenyca. Accessed January 30, 2021. [Available from: https://www.ema.europa.eu/en/medicines/human/EPAR/udenyca].
56. European Medicines Agency. Human medicine European public assessment report (EPAR): Fulphila. Accessed January 30, 2021. [Available from: https://www.ema.europa.eu/en/medicines/human/EPAR/fulphila-0].
57. European Medicines Agency. Human medicine European public assessment report (EPAR): Pelmeg. Accessed January 30, 2021. [Available from: https://www.ema.europa.eu/en/medicines/human/EPAR/pelmeg].
58. European Medicines Agency. Human medicine European public assessment report (EPAR): Ziextenzo. Accessed January 30, 2021. [Available from: https://www.ema.europa.eu/en/medicines/human/EPAR/ziextenzo].
59. European Medicines Agency. Human medicine European public assessment report (EPAR): Grasustek. Accessed January 30, 2021. [Available from: https://www.ema.europa.eu/en/medicines/human/EPAR/grasustek].
60. European Medicines Agency. Human medicine European public assessment report (EPAR): Cegfila. Accessed January 30, 2021. [Available from: https://www.ema.europa.eu/en/medicines/human/EPAR/cegfila].
61. European Medicines Agency. Human medicine European public assessment report (EPAR): Nyvepria. Accessed January 30, 2021. [Available from: https://www.ema.europa.eu/en/medicines/human/EPAR/nyvepria].
62. U.S. Food & Drug Administration (FDA). Press Announcements. FDA approves first biosimilar to Neulasta to help reduce the risk of infection during cancer treatment. June 04, 2018. [Available from: https://www.fda.gov/news-events/press-announcements/fda-approves-first-biosimilar-neulasta-help-reduce-risk-infection-during-cancer-treatment].
63. U.S. Food & Drug Administration (FDA). Prescribing Information. Udenyca™ (pegfilgrastim-cbqv). Accessed January 30, 2021 [Available from: https://www.accessdata.fda.gov/drugsatfda_docs/label/2018/761039s000lbl.pdf].
64. U.S. Food & Drug Administration (FDA). Prescribing Information. Ziextenzo™ (pegfilgrastim-bmez). Accessed January 30, 2021 [Available from: https://www.accessdata.fda.gov/drugsatfda_docs/label/2019/761045lbl.pdf].
65. U.S. Food & Drug Administration (FDA). Prescribing Information. Nyvepria™ (pegfilgrastim-apgf). Accessed January 30, 2021 [Available from: https://www.accessdata.fda.gov/drugsatfda_docs/label/2020/761111lbl.pdf].

66. Tonelli M, Hemmelgarn B, Reiman T, Manns B, Reaume MN, Lloyd A, et al. Benefits and harms of erythropoiesis-stimulating agents for anemia related to cancer: a meta-analysis. CMAJ. 2009;180(11):E62–E71.
67. Jelkmann W. Physiology and pharmacology of erythropoietin. Transfusion Medicine and Hemotherapy. 2013;40(5):302–9.
68. European Medicines Agency. Human medicine European public assessment report (EPAR): Abseamed. Accessed February 8, 2021. [Available from: https://www.ema.europa.eu/en/medicines/human/EPAR/abseamed].
69. European Medicines Agency. Human medicine European public assessment report (EPAR): Binocrit. Accessed February 8, 2021. [Available from: https://www.ema.europa.eu/en/medicines/human/EPAR/binocrit].
70. European Medicines Agency. Human medicine European public assessment report (EPAR): Epoetin Alfa Hexal. Accessed February 8, 2021. [Available from: https://www.ema.europa.eu/en/medicines/human/EPAR/epoetin-alfa-hexal].
71. European Medicines Agency. Human medicine European public assessment report (EPAR): Retacrit. Accessed February 8, 2021. [Available from: https://www.ema.europa.eu/en/medicines/human/EPAR/retacrit].
72. European Medicines Agency. Human medicine European public assessment report (EPAR): Silapo. Accessed February 8, 2021. [Available from: https://www.ema.europa.eu/en/medicines/human/EPAR/silapo].
73. U.S. Food & Drug Administration (FDA). Prescribing Information. Retacrit® (epoetin alfa-epbx). Accessed February 8, 2021 [Available from: https://www.accessdata.fda.gov/drugsatfda_docs/label/2018/125545s000lbl.pdf].
74. Henry DH. The evolving role of epoetin alfa in cancer therapy. The Oncologist. 2004;9(1):97–107.
75. Losem C, Koenigsmann M, Rudolph C. Biosimilar Retacrit®(epoetin zeta) in the treatment of chemotherapy-induced symptomatic anemia in hematology and oncology in Germany (ORHEO)–non-interventional study. OncoTargets and Therapy. 2017;10:1295.
76. Moen MD. Bevacizumab. Drugs. 2010;70(2):181–9.
77. U.S. Food & Drug Administration (FDA). Prescribing Information. Avastin® (bevacizumab). Accessed February 2, 2021 [Available from: https://www.accessdata.fda.gov/drugsatfda_docs/label/2014/125085s301lbl.pdf].
78. Busse A, Lüftner D. What does the pipeline promise about upcoming biosimilar antibodies in oncology? Breast Care. 2019;14(1):10–6.
79. U.S. Food & Drug Administration (FDA). Prescribing Information. Mvasi™ (bevacizumab-awwb). Accessed February 4, 2021 [Available from: https://www.accessdata.fda.gov/drugsatfda_docs/label/2017/761028s000lbl.pdf].
80. U.S. Food & Drug Administration (FDA). Prescribing Information. Zirabev™ (bevacizumab-bvzr). Accessed February 4, 2021 [Available from: https://www.accessdata.fda.gov/drugsatfda_docs/label/2019/761099s000lbl.pdf].
81. European Medicines Agency. Human medicine European public assessment report (EPAR): Mvasi. Accessed February 4, 2021. [Available from: https://www.ema.europa.eu/en/medicines/human/EPAR/mvasi].
82. European Medicines Agency. Human medicine European public assessment report (EPAR): Zirabev. Accessed February 4, 2021.
83. European Medicines Agency. Human medicine European public assessment report (EPAR): Aybintio. Accessed February 4, 2021. [Available from: https://www.ema.europa.eu/en/medicines/human/EPAR/aybintio].
84. European Medicines Agency. Human medicine European public assessment report (EPAR): Equidacent. Accessed February 4, 2021. [Available from: https://www.ema.europa.eu/en/medicines/human/EPAR/equidacent].
85. European Medicines Agency. Human medicine European public assessment report (EPAR): Oyavas. Accessed February 4, 2021. [Available from: https://www.ema.europa.eu/en/medicines/human/summaries-opinion/oyavas].
86. Moorkens E, Vulto AG, Huys I, editors. An overview of patents on therapeutic monoclonal antibodies in Europe: are they a hurdle to biosimilar market entry? MAbs. 2020: Taylor & Francis.
87. European Medicines Agency. Human medicine European public assessment report (EPAR): Alymsys. Accessed February 4, 2021. [Available from: https://www.ema.europa.eu/en/medicines/human/summaries-opinion/alymsys].
88. Wang S, Weiner G. Rituximab: a review of its use in non-Hodgkin's lymphoma and chronic lymphocytic leukemia. Expert Opinion on Biological Therapy. 2008;8:759–68.
89. U.S. Food & Drug Administration (FDA). Prescribing Information. Rituxan® (rituximab). Accessed February 2, 2021 [Available from: https://www.accessdata.fda.gov/drugsatfda_docs/label/2012/103705s5367s5388lbl.pdf].
90. Greenwald M, Tesser J, Sewell KL. Biosimilars have arrived: rituximab. Arthritis. 2018;2018:1–6.
91. U.S. Food & Drug Administration (FDA). Prescribing Information. Truxima™ (rituximab-abbs). Accessed February 2, 2021 [Available from: https://www.accessdata.fda.gov/drugsatfda_docs/label/2018/761088s000lbl.pdf].
92. U.S. Food & Drug Administration (FDA). Prescribing Information. Ruxience™ (rituximab-pvvr). Accessed February 2, 2021 [Available from: https://www.accessdata.fda.gov/drugsatfda_docs/label/2019/761103s000lbl.pdf].
93. U.S. Food & Drug Administration (FDA). Prescribing Information. Riabni™ (rituximab-arrx). Accessed February 2, 2021 [Available from: https://www.accessdata.fda.gov/drugsatfda_docs/label/2020/761140s000lbl.pdf].
94. European Medicines Agency. Human medicine European public assessment report (EPAR): Blitzima. Accessed February 2, 2021. [Available from: https://www.ema.europa.eu/en/medicines/human/EPAR/blitzima].

95. European Medicines Agency. Human medicine European public assessment report (EPAR): Truxima. Accessed February 2, 2021. [Available from: https://www.ema.europa.eu/en/medicines/human/EPAR/truxima].
96. European Medicines Agency. Human medicine European public assessment report (EPAR): Ruxience. Accessed February 2, 2021. [Available from: https://www.ema.europa.eu/en/medicines/human/EPAR/ruxience].
97. European Medicines Agency. Human medicine European public assessment report (EPAR):Riximyo. Accessed February 2, 2021. [Available from: https://www.ema.europa.eu/en/medicines/human/EPAR/riximyo].
98. European Medicines Agency. Human medicine European public assessment report (EPAR): Rixathon. Accessed February 2, 2021. [Available from: https://www.ema.europa.eu/en/medicines/human/EPAR/rixathon].
99. European Medicines Agency. Human medicine European public assessment report (EPAR): Ritemvia. Accessed February 2, 2021. [Available from: https://www.ema.europa.eu/en/medicines/human/EPAR/ritemvia].
100. Yang J, Yu S, Yang Z, Yan Y, Chen Y, Zeng H, et al. Efficacy and safety of anti-cancer biosimilars compared to reference biologics in oncology: a systematic review and meta-analysis of randomized controlled trials. BioDrugs. 2019;33(4):357–71.
101. U.S. Food & Drug Administration (FDA). Prescribing Information. Herceptin® (trastuzumab). Accessed February 6, 2021 [Available from: https://www.accessdata.fda.gov/drugsatfda_docs/label/2010/103792s5250lbl.pdf].
102. Derbyshire M. Patent expiry dates for best-selling biologicals. Generics and Biosimilars Initiative Journal. 2015;4(4):178–80.
103. U.S. Food & Drug Administration (FDA). Prescribing Information. Herzuma® (trastuzumab-pkrb). Accessed February 6, 2021 [Available from: https://www.accessdata.fda.gov/drugsatfda_docs/label/2019/761091s001s002lbl.pdf].
104. U.S. Food & Drug Administration (FDA). Prescribing Information. Trazimera® (trastuzumab-qyyp). Accessed February 6, 2021 [Available from: https://www.accessdata.fda.gov/drugsatfda_docs/label/2019/761081s000lbl.pdf].
105. U.S. Food & Drug Administration (FDA). Prescribing Information. Ontruzant® (trastuzumab-dttb). Accessed February 6, 2021. [Available from: https://www.accessdata.fda.gov/drugsatfda_docs/nda/2019/761100Orig1s000Lbl.pdf].
106. U.S. Food & Drug Administration (FDA). Prescribing Information. Kanjinti® (trastuzumab-anns). Accessed February 6, 2021. [Available from: https://www.accessdata.fda.gov/drugsatfda_docs/label/2019/761073s000lbl.pdf].
107. European Medicines Agency. Human medicine European public assessment report (EPAR): Ontruzant. Accessed February 6, 2021. [Available from: https://www.ema.europa.eu/en/medicines/human/EPAR/ontruzant].
108. European Medicines Agency. Human medicine European public assessment report (EPAR): Herzuma. Accessed February 6, 2021. [Available from: https://www.ema.europa.eu/en/medicines/human/EPAR/herzuma].
109. European Medicines Agency. Human medicine European public assessment report (EPAR): Trazimera. Accessed February 6, 2021. [Available from: https://www.ema.europa.eu/en/medicines/human/EPAR/trazimera].
110. European Medicines Agency. Human medicine European public assessment report (EPAR): Kanjinti. Accessed February 6, 2021. [Available from: https://www.ema.europa.eu/en/medicines/human/EPAR/kanjinti].
111. European Medicines Agency. Human medicine European public assessment report (EPAR): Ogivri. Accessed February 6, 2021. [Available from: https://www.ema.europa.eu/en/medicines/human/EPAR/ogivri].
112. European Medicines Agency. Human medicine European public assessment report (EPAR): Zercepac. Accessed February 6, 2021. [Available from: https://www.ema.europa.eu/en/medicines/human/EPAR/zercepac].
113. Nixon N, Hannouf M, Verma S. The evolution of biosimilars in oncology, with a focus on trastuzumab. Current Oncology. 2018;25(Suppl 1):S171.

Monoclonal Antibodies in Cancer Therapeutics

14

Wenjia Wang and Yihui Shi

Corresponding author: Yihui Shi

Contents

14.1 INTRODUCTION

Biologic therapies, such as monoclonal antibody (mAb)-based immunotherapy, are rapidly advancing as a mainstay of cancer treatment, adding to the arsenal of traditionally prescribed treatment options like chemotherapy, radiation, and surgery. The mAbs are highly specific for their targets and possess multiple clinically significant mechanisms of action to suppress tumor growth, making them generally better tolerated and more effective than other available therapies (Baldo 2013; Baldo and Pham 2021; Zahavi and Weiner 2020). Immunotherapy can also be combined with additional therapies to achieve a greater effect with minimal drug–drug interactions (Baldo and Pham 2021; Zahavi and Weiner 2020). However, despite the progress in biological medicine development, continued high costs of mAb therapy add to the financial burden on healthcare budgets and represent a barrier to more widespread patient access to this treatment option (Cornes 2012; Taieb et al. 2021). With the expiration of exclusivity rights on several original biologic agents in recent years, biosimilar medicines have started to enter the market. Biosimilars undergo rigorous evaluation before approval and must demonstrate high

DOI: 10.1201/9780429485626-14

similarity to the original Food and Drug Administration (FDA)-approved reference product, with no clinically significant differences in safety, purity, and potency (FDA 2015). Therefore, the development of biosimilars could offer more affordable immunotherapy options and expand treatment access to patients. This chapter will give an overview of current oncological mAbs approved in the United States, focus on the available biosimilar products, and discuss future prospects in the development of biosimilars for cancer therapy.

14.2 THE mAbs

The mAbs used in cancer therapies are generated to interact with cell surface proteins involved in oncogenesis and cancer progression (Manis 2020; Weiner, Surana, and Wang 2010). Antibodies or immunoglobulins (Igs) are naturally produced by B lymphocytes as a part of the humoral adaptive immune response (Chaplin 2010). In 1975, Köhler and Milstein developed the hybridoma technology that allows for the cost-effective production of large quantities of relatively pure antibodies against a specific antigen (Köhler and Milstein 1975). Hybridoma cells are created by combining a short-lived plasma B cell with an immortal myeloma cell, allowing the generation of large amounts of one specific mAb (Parray et al. 2020). Antibodies produced by the normal immune response are usually polyclonal and bind different epitopes on an antigen or bind the same epitope with different affinities, while mAbs are produced by a single B-cell clone and bind a single epitope with high affinity and specificity (Manis 2020; Parray et al. 2020). This allows mAbs to achieve a more focused therapeutic result with fewer adverse side effects (Cersosimo 2003; Manis 2020).

Since the introduction of hybridoma technology and various advances in reducing immunogenicity, mAbs have risen to become one of the most successful cancer therapeutics in recent years (Doevendans and Schellekens 2019; Zahavi and Weiner 2020). Selecting tumor antigen targets for therapy requires careful analysis of antigen expression in both tumor cells and normal tissue, as well as knowledge of the role of the antigen in promoting tumor growth (Finn 2017; Scott, Allison, and Wolchok 2012). Ideally, tumor antigens should be overexpressed on tumor cells with low to no expression on normal cells, not shed into the circulation, and play a role in cancer cell progression and proliferation (Modjtahedi, Ali, and Essapen 2012). Different mAbs have been developed that utilize various mechanisms of action to fight cancer, such as direct destruction of tumor cells, indirect modulation of the tumor microenvironment, and enhanced activation of the immune system (Melero et al. 2007; Pento 2017; Vacchelli et al. 2013; Zahavi and Weiner 2020).

The mAbs possess a variety of mechanisms to induce tumor cell death (Figure 14.1). Some mAbs, like the CD20-targeting rituximab, can directly inhibit cell survival signaling and induce apoptosis. Others, like trastuzumab and rituximab, can mediate antibody-dependent cellular cytotoxicity (ADCC) and complement-dependent cytotoxicity (CDC). The mAbs that target growth factor receptors, like cetuximab (endothelial growth factor receptor [EGFR]) and trastuzumab (human epidermal growth factor receptor 2 [HER2]), and growth factors like bevacizumab (vascular endothelial growth factor [VEGF]), limit nutrient delivery to the cancer cells and inhibit proliferation (Coulson, Levy, and Gossell-Williams 2014; Modjtahedi, Ali, and Essapen 2012; Weiner, Surana, and Wang 2010). The immune checkpoint inhibitors target immunomodulatory molecules, like PD-L1, PD-1, and CTLA-4, to stimulate the immune response by inhibiting downregulation of T-cells (Coulson, Levy, and Gossell-Williams 2014; Modjtahedi, Ali, and Essapen 2012).

Antibody-drug conjugates (ADCs) are antibodies conjugated to a drug, toxin, or radioisotope. The high specificity of mAbs for their target antigen makes them useful for delivering potent effector molecules that can kill the tumor cell after the antibody binds and is internalized by receptor-mediated endocytosis (Shim 2020; Thomas, Teicher, and Hassan 2016). There are multiple ADCs available in the

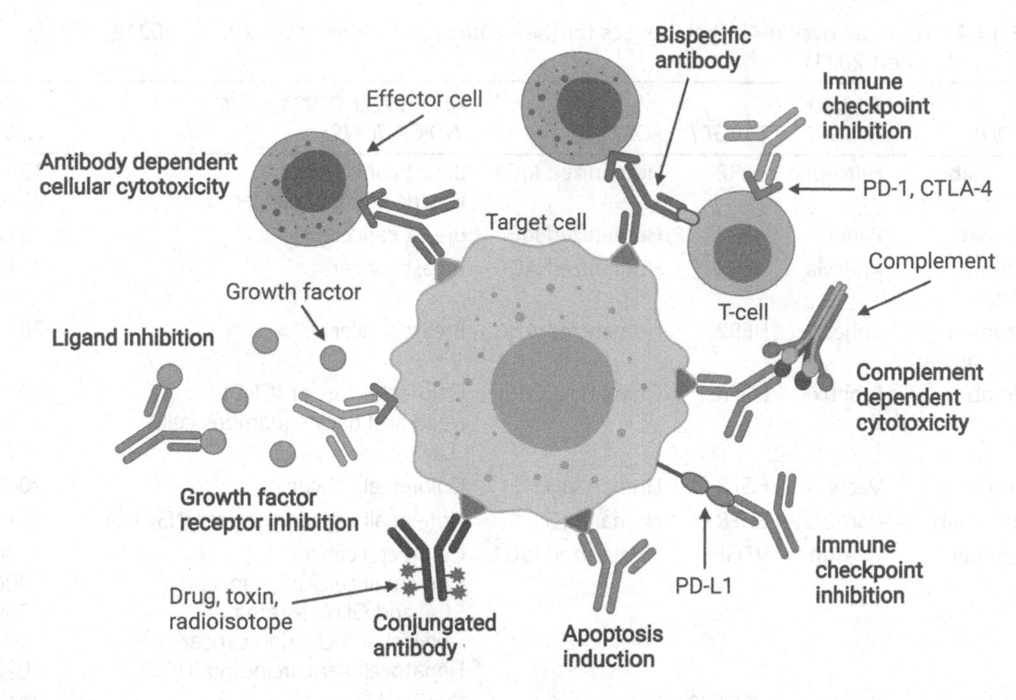

FIGURE 14.1 Mechanisms of action for mAbs in cancer therapy.

United States, with the most recently, Trodelvy (sacituzumab govitecan), approved in 2020 for the treatment of triple-negative breast cancer (Table 14.1).

Bispecific antibodies (bsAbs) can recognize two different epitopes or antigens simultaneously, and there is growing interest in applying their novel abilities in oncology. So far, only one bsAb has been approved in the United States for treating cancer, the bispecific T-cell engager (BiTE) blinatumomab for the treatment of Philadelphia chromosome-negative B-cell acute lymphoblastic leukemia (ALL) (Huang et al. 2020; Labrijn et al. 2019; Przepiorka et al. 2015). With two different binding specificities, blinatumomab can bind CD19 on a tumor cell and CD3 on a naïve T-cell, which engages the T-cell to induce direct cytotoxicity toward the tumor cell (Wu et al. 2015). In addition, the Fc region can also engage effector cells to mediate secondary immune functions, such as ADCC, antibody-dependent cellular phagocytosis (ADCP), or CDC (Heitmann et al. 2021).

There is a consistent flow of new biologics license applications (BLAs) submitted each year, which ensures a steady supply of novel antibody therapeutics entering the pipeline for approval (Kaplon et al. 2020; Kaplon and Reichert 2021). The low toxicity, high specificity, and multiple therapeutic effects of mAbs make them an effective treatment option for a variety of diseases. The development of mAbs to treat cancer has introduced innovative ways of harnessing the power of the immune system and significantly improved patient outcomes.

14.3 BIOSIMILARS

Biosimilars are biological products highly similar to an already licensed reference or originator product but may have slight differences in components that appear to have no clinical significance (FDA 2021d; Macdonald, Hartman, and Jacobs 2015). These are biological drugs that are similar but not identical to the

TABLE 14.1 FDA-approved mAb therapeutics for the treatment of cancer (FDA 2021c, 2021e, 2021f; Kaplon and Reichert 2021)

ANTIBODY	*BRAND NAME*	*TARGET*	*FORMAT*	*APPROVED ONCOLOGIC INDICATIONS*	*APPROVAL YEAR*
Trastuzumab	Herceptin	HER2	Humanized IgG1	Breast cancer (BC)	1998
				Gastric and GEJ cancer	2010
Pertuzumab	Perjeta	HER2	Humanized IgG1	Breast cancer	2012
Trastuzumab emtansine	Kadcyla	HER2	Humanized ADC	Breast cancer	2013
Trastuzumab deruxtecan	Enhertu	HER2	Humanized ADC	Breast cancer	2019
Cetuximab	Erbitux	EGFR	Chimeric IgG1	Colorectal cancer (CRC)	2004
				Head and neck squamous cell carcinoma	2006
Panitumumab	Vectibix	EGFR	Human IgG2	Colorectal cancer	2006
Necitumumab	Portrazza	EGFR	Human IgG1	Non-small cell lung cancer (NSCLC)	2015
Bevacizumab	Avastin	VEGF	Humanized IgG1	Colorectal cancer	2004
				Non-small cell lung cancer	2006
				RCC and Glioblastoma	2009
				Cervical and Ovarian cancer	2014
				Hepatocellular carcinoma (HCC)	2020
Ramucirumab	Cyramza	VEGFR2	Human IgG1	Gastric cancer	2014
Olaratumab	Lartruvo	PDGFRα	Human IgG1	Sarcoma	2016
Pembrolizumab	Keytruda	PD-1	Humanized IgG4	Melanoma	2014
				NSCLC	2015
				Head and neck squamous cell carcinoma	2016
				Solid tumors, cHL, Urothelial carcinoma, Gastric, and GEJ cancer	2017
				Cervical cancer, PMLBCL, HCC, and Merkel cell carcinoma	2018
				RCC, Endometrial carcinoma, SCLC, and Esophageal cancer	2019
				Bladder cancer, CRC, and TNBC	2020
Nivolumab	Opdivo	PD-1	Human IgG4	Melanoma	2014
				Lung cancer	2015
				Kidney cancer	2018
Cemiplimab	Libtayo	PD-1	Human IgG4	Cutaneous squamous-cell carcinoma	2018
Dostarlimab	Jemperli	PD-1	Humanized IgG4	Endometrial cancer	2021
Durvalumab	Imfinzi	PD-L1	Human IgG1	Bladder cancer	2017
Avelumab	Bavencio	PD-L1	Human IgG1	Urothelial carcinoma and Merkel cell carcinoma	2017
Atezolizumab	Tecentriq	PD-L1	Humanized IgG1	Bladder cancer and NSCLC	2016
				TNBC	2019
Mogamulizumab	Poteligeo	CCR4	Humanized IgG1	Cutaneous T-cell lymphoma	2018
Ipilimumab	Yervoy	CTLA-4	Human IgG1	Melanoma	2011
				RCC	2018
Blinatumomab	Blincyto	CD19, CD3	Mouse BiTE	Acute lymphoblastic leukemia	2014
Tafasitamab	Monjuvi	CD19	Humanized IgG1	Diffuse large B-cell lymphoma	2020

(Continued)

TABLE 14.1 FDA-approved mAb therapeutics for the treatment of cancer (FDA 2021c, 2021e, 2021f; Kaplon and Reichert 2021) (*Continued*)

ANTIBODY	*BRAND NAME*	*TARGET*	*FORMAT*	*APPROVED ONCOLOGIC INDICATIONS*	*APPROVAL YEAR*
Rituximab	Rituxan	CD20	Chimeric IgG1	Non-Hodgkin's lymphoma Chronic lymphocytic leukemia	1997 2010
Ibritumomab tiuxetan	Zevalin	CD20	Mouse IgG1-Y90 or In111	Non-Hodgkin's lymphoma	2002
Iodine (I131) tositumomab	Bexxar	CD20	Mouse IgG2-I131	Non-Hodgkin's lymphoma	2003
Obinutuzumab	Gazyva	CD20	Humanized IgG2	Chronic lymphocytic leukemia	2013
Ofatumumab	Arzerra	CD20	Human IgG1	Chronic lymphocytic leukemia	2014
Inotuzumab ozogamicin	Besponsa	CD22	Humanized ADC	Acute lymphoblastic leukemia	2017
Moxetumomab pasudotox	Lumoxiti	CD22	Mouse ADC	Hairy-cell leukemia	2018
Brentuximab vedotin	Adcetris	CD30	Chimeric ADC	Hodgkin's lymphoma and Anaplastic large-cell lymphoma	2011
Gemtuzumab ozogamicin	Mylotarg	CD33	Humanized ADC	Acute myeloid leukemia	2000
Daratumumab	Darzalex	CD38	Human IgG1	Multiple myeloma	2015
Isatuximab	Sarclisa	CD38	Chimeric IgG1	Multiple myeloma	2020
Elotuzumab	Empliciti	SLAMF7	Humanized IgG1	Multiple myeloma	2015
Belantamab mafodotin	Blenrep	BCMA	Humanized IgG1 ADC	Multiple myeloma	2020
Polatuzumab vedotin	Polivy	CD79B	Humanized ADC	B-Cell lymphoma	2019
Enfortumab vedotin	Padcev	Nectin-4	Human ADC	Bladder cancer	2019
Sacituzumab govitecan	Trodelvy	TROP-2	Humanized IgG1 ADC	TNBC	2020
Dinutuximab	Unituxin	GD2	Chimeric IgG1	Neuroblastoma	2015
Naxitamab	Danyelza	GD2	Humanized IgG1	Neuroblastoma	2020

Abbreviations: ADC, antibody-drug conjugate; BiTE, Bi-specific T-cell engager; HER2, human epidermal growth factor receptor 2; EGFR, endothelial growth factor receptor; VEGF, vascular endothelial growth factor; VEGFR2, vascular endothelial growth factor receptor 2; PDGFRα, platelet-derived growth factor receptor α; PD-1, programmed cell death 1; PD-L1, programmed cell death-ligand 1; CD, cluster of differentiation; CCR4, C-C chemokine receptor 4; CTLA-4, cytotoxic T-lymphocyte-associated protein 4; SLAMF7, signaling lymphocytic activation molecule family 7; BCMA, B cell maturation antigen; TROP-2, trophoblast cell-surface antigen 2; cHL, classic Hodgkin's lymphoma; GEJ, gastroesophageal junction; PMLBCL, primary mediastinal large B-cell lymphoma; RCC, renal cell carcinoma; SCLC, small cell lung cancer; NSCLC, non-small cell lung cancer; TNBC, triple-negative breast cancer.

reference drug in terms of characteristics like molecular structure (Figure 14.2), purity, potency, efficacy, safety, and immunogenicity. Biosimilars are rapidly being developed for different diseases and indications with the goal of increasing patient access to essential but expensive biological medications by providing treatment alternatives, but lack of confidence in biosimilar quality and efficacy due to unfamiliarity with the biosimilar development and approval process can hinder uptake. Increased awareness of the biosimilar pathway from development to approval could lead to greater use and acceptance of these effective and more affordable alternatives.

Although biosimilars represent cost savings compared to their originator drugs, the development of biosimilars is still a complicated and expensive process due to the intrinsic variability of biological

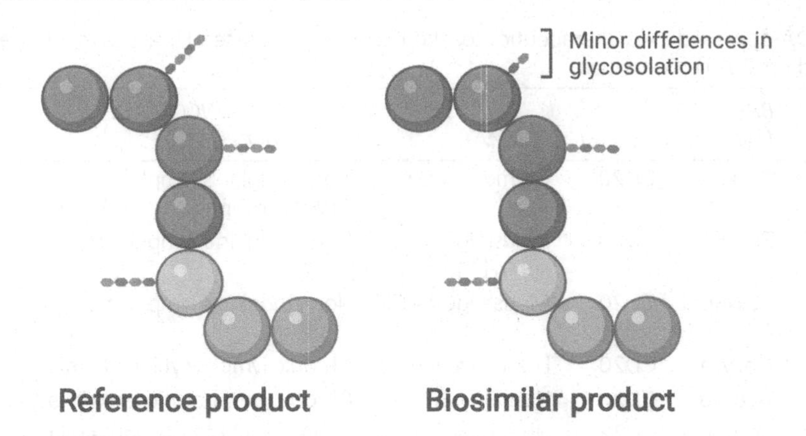

FIGURE 14.2 Schematic example of structural difference between biosimilar and reference biologic.

molecules in the manufacturing process (Cornes 2012; Ishii-Watabe and Kuwabara 2019). Biologics are large complex molecules produced in living cells, and it is difficult to make exact copies of these products, unlike small-molecule generic drugs where production can be more standardized (Rugo et al. 2016). Biosimilar manufacturers have to show that they can replicate the safety, purity, and efficacy profiles of the original biologic to the degree that can pass the rigorous standards for FDA approval (FDA 2018). mAbs are highly complex and can exert various mechanisms of action, so it is important to carefully determine the physicochemical properties, safety, efficacy, and potency of the biosimilars in comparison to the reference product through comparison studies and clinical trials.

14.4 APPROVAL PATHWAY

The Biologics Price Competition and Innovation (BPCI) Act enacted as part of the Affordable Care Act in 2010 established an abbreviated BLA pathway to approval for biosimilars, allowing extrapolation of indications, shortened time to approval, and decreased development costs (Chopra and Lopes 2017; FDA 2015). The approval process for biosimilars is based on a totality of evidence approach as opposed to the traditional process used in novel biosimilar drug development (FDA 2015; Thill 2019). If the same number of trials as a novel biologic drug was required for the biosimilar to be approved, the costs of developing the biosimilar could be just as high as or even higher than the cost to develop the original drug. Because safety and efficacy data have already been established in the various approved indications of the original biologics, it is possible for the FDA to extrapolate approval to all the same indications as the reference product without the need for additional clinical trials if the biosimilar demonstrated it is highly similar to the reference product.

To show biosimilarity to the originator product, biosimilars manufacturers must present compelling evidence in clinical and non-clinical studies that demonstrate comparable biophysicochemical properties as well as no clinical differences (FDA 2017; Taieb et al. 2021). Minor differences in formulation or structure that do not have a clinical effect are allowed. Clinically meaningful differences include differences in safety, purity, or potency, but minor differences in the rates of adverse events are not considered significant (FDA 2015). *In vitro* studies are initially performed to compare molecular structure and function, followed by *in vivo* studies to compare the activity, toxicity, and physiologic effects between the biosimilar and reference product. Finally, clinical studies are conducted to assess clinical efficacy, safety, immunogenicity, which includes a pharmacokinetic (PK) equivalence study and a confirmatory comparative clinical trial (FDA 2015; Taieb et al. 2021). Unlike the generic pathway for copies of small-molecule drugs, clinical trials are needed to demonstrate biosimilarity because biologic medicines are too large and complex to be characterized completely and copied identically (Doevendans and Schellekens 2019).

14.5 ECONOMIC CONSIDERATIONS

In many countries, cancer therapies are responsible for driving the steady increase of healthcare expenditures (Cornes 2012). Although cancer therapy is an area of exciting innovation for novel drug development, the process can be slow and expensive. Out of the thousands of compounds in preclinical trials, only a fraction reaches the clinical trial stage, and ultimately after a general development time of around 15 years, only 10%–20% are approved in the end (Schickedanz 2010). As the quality of biologics continues to improve, costs similarly continue to rise. Biologics are estimated to represent almost 40% of the total US prescription drug spending (FDA 2018). Even in wealthy countries, the high costs of novel biologic drug development can put a strain on healthcare budgets and limit patient access to these life-saving medicines (Cornes 2012; Saleem et al. 2020). Some of the most popular anti-cancer mAbs, such as bevacizumab (Avastin) which targets VEGF, trastuzumab (Herceptin) which targets HER2, and rituximab (Rituxan) which targets CD20, are among the most expensive oncologic drugs used in outpatient cancer centers (Saleem et al. 2020).

With the expiration of patents on original mAbs, biosimilars are anticipated to exert competitive pressure in the biologics market and mitigate rising healthcare costs (Nabhan and Feinberg 2017). According to one study, the use of biosimilars may result in savings of up to $54 billion from 2017 to 2026, which is about 3% of the total direct spending on biological drugs over that time period. The range of total savings is estimated to be from $24 to $150 billion, but actual savings will depend on the state of the biosimilars market and any changes in regulatory decisions or policy (Mulcahy, Hlavka, and Case 2018). Currently, biosimilars only make up a small portion of the biological medicines market, however, that is expected to change as exclusivity protections expire in coming years and more companies enter biosimilars development.

For biosimilars to achieve more widespread utilization and have an impact on decreasing healthcare costs, a better understanding of biosimilar development, safety, and efficacy among both patients and physicians is needed (Ishii-Watabe and Kuwabara 2019; Saleem et al. 2020). In one review, researchers found that oncologists and patients alike may not be completely familiar with the safety, efficacy, and guidelines around extrapolated indications pertaining to biosimilars and therefore have concerns regarding prescribing or receiving a biosimilar over a novel biologic drug (Nabhan and Feinberg 2017). More education, outreach, and studies demonstrating the reliability of available biosimilars could raise awareness and instill confidence in physicians to prescribe biosimilars in clinical practice, as well as promote patient acceptance of biosimilar medicines.

14.6 ONCOLOGIC BIOSIMILAR mAbs

As of May 2021, there are ten FDA-approved anti-cancer mAb biosimilars on the market, with more in varying stages of development (Busse and Lüftner 2019). Of those ten approved biosimilars, two for reference bevacizumab, three for reference rituximab, and five for reference trastuzumab (Table 14.2). A third promising bevacizumab biosimilar, SB8 (Samsung Bioepis and Merck & Co.), is pending FDA approval (Taieb et al. 2021). Several companies are in the process of developing biosimilars for the anti-EGFR mAb cetuximab (Erbitux), but most are still in preclinical development (Busse and Lüftner 2019).

TABLE 14.2 FDA-approved biosimilar mAb products for the treatment of cancer (FDA 2021c)

BIOSIMILAR	*BRAND NAME*	*APPROVAL DATE*	*REFERENCE PRODUCT*	*TARGET/FORMAT*
Rituximab-abbs	Truxima	November 2018	Rituxan (rituximab)	CD20/Chimeric IgG1
Rituximab-pvvr	Ruxience	July 2019		
Rituximab-arrx	Riabni	December 2020		
Bevacizumab-awwb	Mvasi	September 2017	Avastin (bevacizumab)	VEGF/Humanized IgG1
Bevacizumab-bvzr	Zirabev	June 2019		
Trastuzumab-dkst	Ogivri	December 2017	Herceptin (trastuzumab)	HER2/Humanized IgG1
Trastuzumab-pkrb	Herzuma	December 2018		
Trastuzumab-dttb	Ontruzant	January 2019		
Trastuzumab-qyyp	Trazimera	March 2019		
Trastuzumab-anns	Kanjinti	June 2019		

Abbreviations: HER2, human epidermal growth factor receptor 2; VEGF, vascular endothelial growth factor.

The naming convention for biosimilar mAbs, once they are approved, uses the name of the reference drug combined with a unique four-letter suffix that consists of meaningless lowercase letters (FDA 2015). This establishes a relationship between the biosimilar and reference product that allows proper tracking in pharmacovigilance programs and easy identification (Nabhan, Valley, and Feinberg 2018).

14.7 BEVACIZUMAB

Avastin (bevacizumab) is a humanized IgG1 mAb developed by Genentech Inc. that targets VEGF, which is overexpressed in many human solid tumors. Bevacizumab binds VEGF and prevents it from interacting with VEGFR-1 (Flt-1) and VEGFR-2 (KDR/Flk-1) receptors on the endothelial cell surface. The binding of VEGF to its receptors promotes angiogenesis, and therefore by blocking VEGF, bevacizumab inhibits the growth of new vasculature around the tumor and starves it of the nutrients needed for growth and proliferation (Baldo and Pham 2021; Genentech 2021c). Avastin was first approved by the FDA in 2004 for the treatment of metastatic colorectal cancer (mCRC) and has since gained additional indications for treating NSCLC, recurrent glioblastoma (rGBM), metastatic renal cell carcinoma (mRCC), persistent, recurrent, or metastatic cervical cancer, epithelial ovarian, fallopian tube, or primary peritoneal cancer, and most recently HCC in 2020 (Genentech 2004). It is also the first FDA-approved complex biologic drug indicated to treat gynecologic cancers (Monk et al. 2017a). The most common adverse reactions are epistaxis, headache, hypertension, rhinitis, proteinuria, taste alteration, dry skin, hemorrhage, lacrimation disorder, back pain, and exfoliative dermatitis (Genentech 2021c). Avastin combined with chemotherapy has been shown to increase patient survival in the indicated cancers, but patient access can be limited from high costs and lack of availability due to the complex manufacturing process of mAbs (Melosky et al. 2018; Monk et al. 2017b; Rosen, Jacobs, and Burkes 2017). Currently, there are two FDA-approved biosimilars of bevacizumab, with a third underway (Table 14.3).

In September 2017, the bevacizumab biosimilar mAb Mvasi (bevacizumab-awwb), developed by Amgen Inc., became the first oncological biosimilar drug approved in the United States based on a totality of evidence (FDA 2017; Lartigue 2018). The MAPLE trial (n = 642), a large phase III, randomized, double-blind, comparative clinical study, demonstrated the similarity of Mvasi (ABP 215; bevacizumab-awwb) to reference bevacizumab in patients with advanced non-squamous NSCLC. The primary efficacy endpoint was the risk ratio (RR) of the overall response rate (ORR). The ORR in patients taking ABP 215 (n = 328) compared with bevacizumab (n = 314) was 39.0% and 41.7%, respectively. The ORR RR was 0.93 (90% CI, 0.80–1.09), which was within the predefined equivalence margin of 0.67–1.50. Results showed similar safety and immunogenicity profiles between the biosimilar and reference product as well, with no

TABLE 14.3 Bevacizumab biosimilars: Comparing results of phase III trials (ClinicalTrials.gov 2021)

BIOSIMILAR	*EXTRAPOLATED INDICATIONS*	*PHASE III TRIAL DESIGN/ PRIMARY ENDPOINT*	*STUDY RESULTS (BIOSIMILAR VS REFERENCE PRODUCT)*	*SAFETY/ IMMUNOGENICITY (BIOSIMILAR VS REFERENCE PRODUCT)*
Mvasi (ABP 215; bevacizumab-awwb; Amgen)	mCRC, NSCLC, rGBM, mRCC, CC	MAPLE ($n = 642$) Adults ≥18 years Metastatic or recurrent NSCLC Combined with chemotherapy Risk ratio of ORR at 19 weeks (NCT01966003)	ORR 39.0% vs 41.7% Risk ratio of ORR 0.93 (90% CI, 0.80–1.09) Shows therapeutic equivalence Comparable PFS and OS	Serious TEAEs 26.2% vs. 23.0% ADA+ 1.4% vs. 2.5% Nab+ 0% vs. 0%
Zirabev (PF-06439535; bevacizumab-bvzr; Pfizer)	mCRC, NSCLC, rGBM, mRCC, CC, OC	($n = 719$) Adults ≥18 years Stage IIIb–IV NSCLC Combined with chemotherapy ORR at 19 weeks (NCT02364999)	ORR 45.3% vs 44.6% Shows therapeutic equivalence Comparable PFS and OS	Serious TEAEs 22.8% vs. 22.3% ADA+ 1.5% vs. 1.4% Nab+ 0% vs. 0.9%
(SB8; Samsung Bioepis and Merck & Co.)	Pending approval	($n = 763$) Adults ≥18 years Metastatic or recurrent NSCLC Combined with chemotherapy ORR at 24 weeks (NCT02754882)	ORR 47.6% vs 42.8% Within equivalence margins PFS 8.50 vs 7.90 months OS 14.9 vs 15.8 months	Serious TEAEs 19.8% vs. 21.3% ADA+ 16.1% vs. 11% Nab+ 7.1% vs. 8.1%

Abbreviations: mCRC, metastatic colorectal cancer; NSCLC, non-small cell lung cancer; rGBM, recurrent glioblastoma; mRCC, metastatic renal cell carcinoma; CC, cervical cancer; OC, ovarian cancer; ORR, overall response rate; TEAE, treatment-emergent adverse event; ADA, anti-drug antibody; nAb, neutralizing antibody; RP, reference product; PFS, progression-free survival; OS, overall survival; HTN, hypertension; PNA, pneumonia; PE, pulmonary embolism.

neutralizing antibodies detected in either arm of the study (Thatcher et al. 2019). Based on clinical PK and bioequivalence (BE) studies and non-clinical potency studies demonstrating biosimilarity to the reference bevacizumab, the FDA approved Mvasi for the treatment of metastatic CRC, NSCLC, rGBM, mRCC, and cervical cancer (Amgen 2021a; Markus et al. 2017; Thatcher et al. 2019). Since then, other clinical trials have confirmed biosimilarity and non-inferiority of Mvasi compared to Avastin (Apsangikar et al. 2017). It is not currently indicated for epithelial ovarian, fallopian tube, or primary peritoneal cancer, and HCC due to unexpired exclusivity protections held by the originator Avastin (FDA 2017).

The second bevacizumab biosimilar Zirabev (bevacizumab-bvzr), developed by Pfizer Inc., gained FDA approval in June 2019 (Pfizer 2021d). Results of phase III, randomized, double-blind study ($n = 719$) of Zirabev (PF-06439535; bevacizumab-bvzr) compared with reference bevacizumab-EU, each combined with paclitaxel and carboplatin, for the first-line treatment of advanced non-squamous NSCLC, showed similarity between the biosimilar and reference product. The primary endpoint of ORR by week 19 and confirmed by week 25 was 45.3% for the biosimilar group ($n = 358$) and 44.6% for the reference group ($n = 361$), with a 90% CI of 0.89–1.16. The RR 90% CI was within the prespecified equivalence margin, and efficacy and safety profiles were comparable between the two treatments (Pfizer 2021d; Reinmuth et al. 2019). Zirabev is indicated to treat mCRC, NSCLC, rGBM, mRCC, cervical cancer, and epithelial ovarian, fallopian tube, or primary peritoneal cancer, but not HCC as the exclusivity protection on this recently approved indication for Avastin does not expire until 2027 (FDA 2021d; Pfizer 2021c).

A third bevacizumab biosimilar, SB8, developed by Samsung Bioepis and marketed by Merck & Co. in the United States, is pending FDA approval (Taieb et al. 2021). Recent results from a phase III, double-blind study in patients with metastatic or recurrent non-squamous NSCLC showed an ORR of 47.6% for SB8 compared with 42.8% for reference bevacizumab, and the median progression-free survival (PFS) was 8.5 months and 7.9 months, respectively (Reck et al. 2020). Based on current data, SB8 shows promising similarity to the reference product in clinical efficacy, tolerability, immunogenicity, safety profile, as well as physicochemical and functional properties (Syed 2020).

14.8 TRASTUZUMAB

Trastuzumab, brand name Herceptin, was developed by Genentech Inc. and initially approved by the FDA in 1998. It is a humanized IgG1 kappa mAb that binds the extracellular domain of human epidermal growth factor receptor 2 (HER2; also known as HER2/neu, ErbB2, CD340, and p185) with high affinity. HER2 overexpression leads to increased activation of growth factor signaling pathways, and Herceptin has been shown to inhibit proliferation as well as preferentially induce ADCC in cancer cells that overexpress HER2 (Baldo and Pham 2021; Genentech 2021b). Many studies have shown that the use of trastuzumab with chemotherapy increases survival significantly in both early and metastatic disease, although Herceptin can cause serious side effects like cardiomyopathy, infusion reactions, embryo-fetal toxicity, and pulmonary toxicity (Genentech 2021b; Miller and Schwartzberg 2019; Nixon, Hannouf, and Verma 2018).

In addition to indications for HER2-overexpressing breast cancer, trastuzumab was also approved for HER2-overexpressing metastatic gastric or GEJ adenocarcinoma in 2010 (Genentech 2021b). Trastuzumab has been the standard of care for early and advanced HER2-positive BC since 2005 and was added to the WHO Essential Medicines List in 2015 (Lancet 2020). Just as with other biological therapies, trastuzumab comes at a high cost, with one news article estimating the price of a year-long course of treatment to be $76,700 (Grady 2018). In the past few years, five trastuzumab biosimilars have entered the market in the United States, bringing hope of more affordable and accessible treatment options without sacrificing quality (Table 14.4). After undergoing extensive preclinical testing and analysis, no clinically significant differences were found when the trastuzumab biosimilars were compared to reference Herceptin in the clinical trials that lead to FDA approval (Barbier et al. 2019; Miller and Schwartzberg 2019; Thill 2019).

The first trastuzumab biosimilar, Ogivri (trastuzumab-dkst, Mylan Inc.), gained FDA approval in December 2017. It is approved for all indications as the reference trastuzumab, Herceptin (Mylan 2021). Ogivri demonstrated BE in a phase I study and shows equivalence in safety and efficacy to reference trastuzumab based on the HERITAGE trial (n = 458) (Rugo et al. 2017; Waller et al. 2018). In phase III, multicenter, double-blind, randomized, parallel-group equivalence study of Ogivri (Myl 1401O; trastuzumab-dkst) compared to reference trastuzumab in patients with ERBB2-positive metastatic BC receiving taxane chemotherapy, the primary endpoint ORR at 24 weeks was 69.6% (95% CI, 63.62%–75.51%) for the biosimilar and 64.0% (95% CI, 57.81%–70.26%) for trastuzumab. The ORR ratio of 1.09 (90% CI, 0.974–1.211) was within equivalence boundaries of 0.81–1.24. There was no significant difference between Ogivri and Herceptin in any clinically important areas, including time to tumor progression, PFS, overall survival (OS), safety, and immunogenicity (Rugo et al. 2017).

Herzuma (trastuzumab-pkrb), developed by Celltrion Inc. and marketed by Teva Pharmaceuticals Inc. in the United States, was approved one year later in December 2018 for the treatment of HER2-overexpressing BC, and later gained approval for all current indications as Herceptin including HER2-overexpressing metastatic gastric or GEJ adenocarcinoma (Celltrion 2021a). In the randomized phase III clinical trial (n = 549) conducted, Herzuma (CT-P6; trastuzumab-pkrb) showed a similar therapeutic effect when compared to the reference trastuzumab in neoadjuvant therapy of HER2-positive early-stage breast cancer (EBC) (Stebbing et al. 2017). The primary endpoint was pathological complete response (pCR), defined as the absence of invasion tumor cells in the breast and axillary lymph nodes. The tissue specimens were obtained after surgical resection 3–6 weeks after the last treatment of the neoadjuvant period. An update to the study showed similar efficacy and safety of Herzuma compared to reference trastuzumab in adjuvant therapy following the initial neoadjuvant treatment and surgical resection (Esteva et al. 2019).

TABLE 14.4 Trastuzumab biosimilars: Comparing results of phase III trials (ClinicalTrials.gov 2021)

BIOSIMILAR	*EXTRAPOLATED INDICATIONS*	*PHASE III TRIAL DESIGN/ PRIMARY ENDPOINT*	*STUDY RESULTS (BIOSIMILAR VS. REFERENCE PRODUCT)*	*SAFETY/ IMMUNOGENICITY (BIOSIMILAR VS. REFERENCE PRODUCT)*
Ogivri (Myl 1401O; trastuzumab-dkst; Mylan)	HER2+ BC HER2+ metastatic GC or GEJ adenocarcinoma	HERITAGE (n = 458) Adults ≥18 years HER2+ recurrent or mBC First-line therapy combined with chemotherapy ORR at 24 weeks (NCT02472964)	ORR 69.6% vs. 64.0% Shows therapeutic equivalence Comparable PFS and OS	Serious TEAEs 38.1% vs. 36.2% ADA+ 2.4% vs. 2.8%
Herzuma (CT-P6; trastuzumab-pkrb; Celltrion and Teva)	HER2+ BC HER2+ metastatic GC or GEJ adenocarcinoma	(n = 549) Women ≥18 years Stage I–IIIa HER2+ EBC Neoadjuvant therapy combined with chemotherapy pCR 3–6 weeks after last dose (treated for 24 weeks) (NCT02162667)	pCR 46.8% vs. 50.4% Shows therapeutic equivalence Comparable in preventing progressive disease when used as adjuvant therapy	Serious TEAEs 7% vs. 8%
Ontruzant (SB3; trastuzumab-dttb; Samsung Bioepis and Merck & Co.)	HER2+ BC HER2+ metastatic GC or GEJ adenocarcinoma	(n = 800) Women 18–65 years Stage II–III HER2+ EBC Neoadjuvant therapy combined with chemotherapy Breast pCR at 24 weeks (NCT02149524)	Breast pCR 51.7% vs. 42.0% Total pCR 45.8% vs. 35.8% ORR 96.3% vs. 91.2% Shows therapeutic equivalence	Serious TEAEs 10.5% vs. 10.7% ADA+ 0.7% vs. 0.0%
Trazimera (PF-05280014; trastuzumab-qyyp; Pfizer)	HER2+ BC HER2+ metastatic GC or GEJ adenocarcinoma	REFLECTIONS B327-02 (n = 707) Women ≥18 years HER2+ mBC First-line therapy combined with chemotherapy ORR at 25 weeks (NCT01989676)	ORR 62.5% vs. 66.5% PFS 12.16 vs. 12.06 months Shows therapeutic equivalence Comparable PFS and OS	Serious TEAEs 15.2% vs. 15.9%
Kanjinti (ABP 980; trastuzumab-anns; Amgen)	HER2+ BC HER2+ metastatic GC or GEJ adenocarcinoma	LILAC (n = 696) Women ≥18 years HER2+ EBC Neoadjuvant therapy combined with chemotherapy pCR 3–7 weeks after last dose (NCT01901146)	pCR 48.0% vs. 40.5% Shows therapeutic equivalence Comparable results in adjuvant therapy and in switching from RP	Serious TEAEs 4.95% vs. 1.39%

Abbreviations: HER2, human epidermal growth factor receptor 2; BC, breast cancer; EBC, early breast cancer; mBC, metastatic breast cancer; GC, gastric cancer; GEJ, gastroesophageal junction; pCR, pathologic complete response; ORR, overall response rate; TEAE, treatment-emergent adverse event; ADA, anti-drug antibody; nAb, neutralizing antibody; RP, reference product; PFS, progression-free survival; OS, overall survival.

Shortly after, three trastuzumab biosimilars gained FDA approval within months of each other in 2019, all sharing the same indications with Herceptin. Ontruzant (trastuzumab-dttb) was approved in January 2019 (Merck 2021), Trazimera (trastuzumab-qyyp) in March 2019 (Pfizer 2021b), and Kanjinti (trastuzumab-anns) in June 2019 (Amgen 2021b).

Ontruzant (trastuzumab-dttb, Samsung Bioepis and Merck & Co.) showed PK equivalence to trastuzumab in a Phase I study in 109 healthy male subjects (Pivot et al. 2016). Results of the subsequent phase III randomized, double-blind, multicenter study comparing Ontruzant (SB3, trastuzumab-dttb) to trastuzumab for the neoadjuvant treatment of newly diagnosed HER2-positive early or locally advanced BC demonstrated similar efficacy, safety, and immunogenicity (Pivot et al. 2018). The study analyzed 800 patients who completed the treatment according to protocol, with 402 randomized to receive the biosimilar and 398 to receive the reference product, and the primary endpoint was based on pCR rate of the primary breast tumor by week 24. The pCR rates in the breast at the conclusion of treatment were 51.7% in Ontruzant and 42.0% in trastuzumab. The adjusted ratio of breast pCR was within equivalence margins at 1.259 (95% CI, 1.085–1.460). The adjusted difference was 10.79% (95% CI, 4.12%–17.26%), with the lower limit within and upper limit outside the equivalence margin. Both products showed no significant differences in safety, tolerability, and immunogenicity profiles. The results were published in 2018, and Ontruzant gained FDA approval for all extrapolated indications as Herceptin in January 2019.

Trazimera (trastuzumab-qyyp, Pfizer Inc.) demonstrated biosimilarity to trastuzumab in the REFLECTIONS B327-02 trial when given as first-line treatment for HER2-positive metastatic breast cancer (Pegram et al. 2019). The primary endpoint was ORR at 25 weeks. Out of the 707 patients who received treatment, 352 were randomized to receive Trazimera (PF-05280014, trastuzumab-qyyp) plus paclitaxel and 355 to receive trastuzumab-EU plus paclitaxel. The ORR was 62.5% vs. 66.5% in the biosimilar and reference groups, respectively. At about one year follow-up after starting treatment, there were no significant differences between groups in PFS or OS. The safety and immunogenicity profiles were also similar between the two groups.

The most recently approved trastuzumab biosimilar, Kanjinti (trastuzumab-anns, Amgen Inc.), was shown to be comparable to trastuzumab based on results of studies including the LILAC trial (Kolberg et al. 2019; von Minckwitz et al. 2018). The randomized, double-blind, phase III study compared the clinical efficacy and safety of Kanjinti (ABP 980, trastuzumab-anns) and trastuzumab in 725 women with HER2-positive EBC (von Minckwitz et al. 2018). Around 364 patients received Kanjinti, and 361 received trastuzumab, but the primary endpoint of pCR was assessable in 358 patients who received the biosimilar and 338 who received trastuzumab. Specimens obtained after surgical resection of breast and lymph node tissue 3–7 weeks after initial neoadjuvant treatment showed pCR in 48% (95% CI, 43–53) and 41% (95% CI, 35–46) of patients receiving Kanjinti compared with trastuzumab, respectively. After the initial neoadjuvant treatment, adjuvant therapy was continued, and a switching study was performed as well. The results showed equivalence between Kanjinti and trastuzumab in regard to clinical efficacy, safety, and immunogenicity in both the initial neoadjuvant and adjuvant settings. The incidence of treatment-emergent adverse events was slightly higher in the biosimilar group, but it is not considered clinically significant.

14.9 RITUXIMAB

Initially approved by the FDA in 1997, Rituxan (rituximab) developed by Genentech Inc., is the earliest therapeutic monoclonal antibody on the market in the United States for treating cancer (Genentech 2021a; Grillo-Lopez 2000). It is a cytolytic chimeric mouse/human monoclonal IgG1 kappa mAb that targets the CD20 antigen expressed on the surface of all B-cells before they differentiate into antibody-secreting plasma cells (Genentech 2021a; Smith 2003). CD20 is an ideal target for mAb therapy because it is expressed at high levels on B-cells but not in other tissues, not downregulated after antibody binding, and not shed into the circulation, so sufficient amounts of antibody can reach the intended target (Baldo and Pham 2021; Smith 2003). Although the exact role of CD20 in B-cell development is not known, it may be involved in transmembrane signaling controlling intracellular calcium, cell cycle progression, and apoptosis (Tedder and Engel 1994). Upon rituximab binding to CD20, possible mechanisms of cell lysis include direct signaling of apoptosis, CDC, and ADCC (Smith 2003; Maloney et al. 1997). Rituximab is indicated for the treatment of B-cell cancers in adult patients, including NHL and CLL, as well as several autoimmune conditions, and the most common serious adverse reactions are infusion-related reactions (Genentech 2021a).

The three rituximab biosimilars currently on the market in the United States are FDA-approved for all the same oncological indications as their reference product, Rituxan (Table 14.5). Truxima (rituximab-abbs) was the first to be approved in November 2018 (Celltrion 2021b), followed by Ruxience (rituximab-pvvr) in July 2019 (Pfizer 2021a), and Riabni (rituximab-arrx) in December 2020 (Amgen 2021c). With no clinically meaningful differences from the originator rituximab, these biosimilars will offer much-needed cost-effective treatment options for patients and prescribers.

Truxima (rituximab-abbs) was developed by Celltrion Inc. and marketed by Teva Pharmaceuticals Inc. in the United States. It is the first FDA-approved biosimilar of rituximab and became available in November 2018. A phase III, randomized, double-blind, parallel-group, active-controlled study in adult patients (≥18 years) with previously untreated CD20-positive stage II–IV low-tumor-burden follicular lymphoma (LTB-FL) showed similar efficacy, safety, immunogenicity, and therapeutic equivalence between Truxima (CT-P10; rituximab-abbs) and the reference Rituxan (rituximab). The 258 patients enrolled in the trial were randomly assigned to either Truxima or Rituxan. Around 83% of patients assigned to Truxima (n = 130) and 81% of patients assigned to Rituxan (n = 128) achieved the primary efficacy endpoint of ORR by 28 weeks, with a treatment difference estimate of 1.8%. Therapeutic equivalence was shown by a 2-sided 90% CI of –6.43% to 10.20%, which was within the prespecified equivalence range of ±17%. Around 5% of patients who received Truxima and 2% who received Rituxan experienced at least one treatment-emergent serious adverse event (Ogura et al. 2018).

TABLE 14.5 Rituximab biosimilars: Comparing results of phase III trials (ClinicalTrials.gov 2021)

BIOSIMILAR	*EXTRAPOLATED INDICATIONS*	*PHASE III TRIAL DESIGN/ PRIMARY ENDPOINT*	*STUDY RESULTS (BIOSIMILAR VS. REFERENCE PRODUCT)*	*SAFETY/ IMMUNOGENICITY (BIOSIMILAR VS. REFERENCE PRODUCT)*
Truxima (CT-P10; rituximab-abbs; Celltrion/Teva)	NHL, CLL	(n = 258) Adults ≥18 years Stage II–IV LTB-FL ORR at 28 weeks (NCT02260804)	ORR 83% vs. 81% Shows therapeutic equivalence	Serious TEAEs 10.77% vs. 10.94%
Ruxience (PF-05280586; rituximab-pvvr; Pfizer)	NHL, CLL	REFLECTIONS B328-06 (n = 394) Adults ≥18 years Stage II–IV LTB-FL ORR at 26 weeks (NCT02213263)	ORR 75.5% vs. 75.7% Shows therapeutic equivalence	Serious TEAEs 8.67% vs. 7.61%
Riabni (ABP 798; rituximab-arrx; Amgen)	NHL, CLL	JASMINE (n = 256) Adults ≥18 years Grade 1–3a/Stage II–IV LTB-FL RD of ORR at 28 weeks (NCT02747043)	ORR 78.0% vs. 70.2% RD of ORR 7.7% Shows therapeutic equivalence	Serious TEAEs 3.91% vs. 3.97%

Abbreviations: NHL, non-Hodgkin's lymphoma; CLL, chronic lymphocytic leukemia; LTB-FL, low-tumor-burden follicular lymphoma; ORR, overall response rate; RD, risk difference.

The second FDA-approved rituximab biosimilar Ruxience (rituximab-pvvr) was developed by Pfizer Inc. and entered the market in July 2019. It was shown to be similar to reference rituximab in the 52-week REFLECTIONS B328-06 study (n = 394). The phase III, randomized, double-blind, parallel-group comparative study evaluated the efficacy, safety, immunogenicity, PK, and pharmacodynamics (PD) of Ruxience (PF-05280586; rituximab-pvvr) vs. the reference rituximab, MabThera (rituximab-EU) in adult patients with previously untreated CD20-positive stage II–IV LTB-FL. 75.5% of patients who received the biosimilar (n = 196) and 70.7% of patients who received the reference rituximab (n = 198) achieved the primary endpoint ORR at week 26, with a treatment difference estimate of 4.66%. The two-sided 95% CI of –4.16% to 13.47% was within the prespecified equivalence margin of ±16%. The results showed similar safety, efficacy, immunogenicity, PK, and PD profiles between PF-05280586 and rituximab-EU (Sharman et al. 2020).

The most recent rituximab biosimilar to enter the market is Riabni (rituximab-arrx), developed by Amgen Inc. and FDA-approved in December 2020. It was shown to be similar with no clinical differences to reference rituximab in the JASMINE study (n = 256), a phase III, randomized, double-blind, active-controlled, multiple-dose, clinical similarity study to evaluate the safety, efficacy, PK, PD, immunogenicity, and tolerability of Riabni (ABP 798; rituximab-arrx) compared to rituximab in adult patients with CD20-positive grade 1, 2, or 3a or stage II–IV LTB-FL. The patients treated with the biosimilar ABP 798 (n = 128) and rituximab reference product (n = 126) had an ORR of 78.0% and 70.2%, respectively, by week 28. The primary endpoint was the risk difference (RD) of ORR at week 28, which showed a point estimate of 7.7%. The 1-sided 95% CI was –1.4% and 16.8% for the lower and upper confidence limits of RD in ORR, which was within the prespecified equivalence margin of–15% to 35.5% (Niederwieser et al. 2020).

14.10 SUMMARY

Since their conception, monoclonal antibodies have climbed to the top of recommended cancer treatments in terms of both efficacy and costs. Novel mAbs are continuing to be developed in addition to biosimilar products, and manufacturers of originator products have been trying to identify new disease targets and clinical applications for their existing mAbs to combat the growing competition of new biosimilars entering the market (Grilo and Mantalaris 2019). Biosimilars will likely lower costs and promote a healthy competitive environment in drug development and innovation as more products gain approval.

With the continued improvement of biomanufacturing processes and introduction of new technologies to facilitate faster and more scalable production systems, biosimilars could one day reach the status of interchangeable drugs, which the FDA defines as a biosimilar product expected to produce the same clinical result as the reference product in any given patient, and show no additional risk or decreased efficacy if a patient switches back and forth between the interchangeable product and reference product vs. using the reference product without switching (FDA 2017; Grilo and Mantalaris 2019). An interchangeable product can be substituted for the reference product without intervention from the prescribing physician just like how generic drugs are substituted for brand name drugs at the pharmacy (Alvarez et al. 2020; FDA 2021b). Since not much clinical data is available yet on switching between reference mAbs and biosimilars used for cancer therapy, additional carefully designed switching studies need to be conducted to pave the way for interchangeability (Declerck et al. 2018). Continued pharmacovigilance programs and physician reporting of adverse events are important to establish long-term outcomes (Daller 2016).

As of now, there are no FDA-approved interchangeable biosimilar medications, but with more refined techniques and experience, continued development of biosimilars could eventually lead to the approval of interchangeable biosimilars and expand cost-effective treatment options for patients (FDA 2021b; Grilo and Mantalaris 2019).

ACKNOWLEDGMENT

Figures 14.1 and 14.2 were created with Biorender.com.

REFERENCES

Alvarez, D. F., G. Wolbink, C. Cronenberger, J. Orazem, and J. Kay. 2020. Interchangeability of biosimilars: What level of clinical evidence is needed to support the interchangeability designation in the United States? *BioDrugs* 34 (6):723–732.

Amgen, Inc. 2021a. *MVASI (Bevacizumab-awwb) Prescribing Information*. U.S. Food and Drug Administration [cited May 5 2021]. Available from https://www.pi.amgen.com/~/media/amgen/repositorysites/pi-amgen-com/mvasi/mvasi_pi_hcp_english.pdf.

———. 2021b. *KANJINTI (trastuzumab-anns) Prescribing Information*. U.S. Food and Drug Administration [cited May 3 2021]. Available from https://www.pi.amgen.com/~/media/amgen/repositorysites/pi-amgen-com/kanjinti/kanjinti_pi.ashx.

———. 2021c. *RIABNI (rituximab-arrx) Prescribing Information*. Amgen Inc. [cited May 8 2021]. Available from https://www.pi.amgen.com/~/media/amgen/repositorysites/pi-amgen-com/riabni/riabni_pi_english.ashx.

Apsangikar, P. D., S. R. Chaudhry, M. M. Naik, S. B. Deoghare, and J. Joseph. 2017. Comparative pharmacokinetics, efficacy, and safety of bevacizumab biosimilar to reference bevacizumab in patients with metastatic colorectal cancer. *Indian J Cancer* 54 (3):535–538.

Baldo, B. 2013. Adverse events to monoclonal antibodies used for cancer therapy: Focus on hypersensitivity responses. *OncoImmunology* 2 (10):e26333.

Baldo, B. A., and N. H. Pham. 2021. Targeted drugs for cancer therapy: Small molecules and monoclonal antibodies. In Drug Allergy: Springer International Publishing.

Barbier, L., P. Declerck, S. Simoens, P. Neven, A. G. Vulto, and I. Huys. 2019. The arrival of biosimilar monoclonal antibodies in oncology: Clinical studies for trastuzumab biosimilars. *Br J Cancer* 121 (3):199–210.

Busse, A., and D. Lüftner. 2019. What does the pipeline promise about upcoming biosimilar antibodies in oncology? *Breast Care* 14 (1):10–16.

Celltrion, Inc. 2021a. *HERZUMA (Trastuzumab-pkrb) Prescribing Information*. U.S. Food and Drug Administration [cited May 3 2021]. Available from https://www.herzuma.com/globalassets/herzuma-consumer/pdfs/herzuma-prescribing-information.pdf.

———. 2021b. *TRUXIMA (Rituximab-abbs) Prescribing Information*. Teva Pharmaceuticals USA, Inc. [cited May 8 2021]. Available from https://www.truxima.com/globalassets/truxima-dtc/pdfs/truxima-prescribing-information.pdf.

Cersosimo, R. J. 2003. Monoclonal antibodies in the treatment of cancer, Part 1. *Am J Health Syst Pharm* 60 (15):1531–1548.

Chaplin, D. D. 2010. Overview of the immune response. *Journal of Allergy and Clinical Immunology* 125 (2):S3–S23.

Chopra, R., and G. Lopes. 2017. Improving access to cancer treatments: The role of biosimilars. *J Glob Oncol* 3 (5):596–610.

ClinicalTrials.gov. 2021. U.S. National Library of Medicine. Available from https://clinicaltrials.gov/.

Cornes, P. 2012. The economic pressures for biosimilar drug use in cancer medicine. *Targeted Oncology* 7 (S1):57–67.

Coulson, A., A. Levy, and M. Gossell-Williams. 2014. Monoclonal antibodies in cancer therapy: Mechanisms, successes and limitations. *The West Indian Med J* 63 (6):650–654.

Daller, J. 2016. Biosimilars: A consideration of the regulations in the United States and European Union. *Regul Toxicol Pharmacol* 76:199–208.

Declerck, P., G. Bakalos, E. Zintzaras, B. Barton, and T. Schreitmüller. 2018. Monoclonal antibody biosimilars in oncology: Critical appraisal of available data on switching. *Clin Therapeut* 40 (5):798–809.e2.

Doevendans, E., and H. Schellekens. 2019. Immunogenicity of innovative and biosimilar monoclonal antibodies. *Antibodies (Basel)* 8 (1):21–31.

Esteva, F. J., Y. V. Baranau, V. Baryash, et al. 2019. Efficacy and safety of CT-P6 versus reference trastuzumab in HER2-positive early breast cancer: Updated results of a randomised phase 3 trial. *Cancer Chemother Pharmacol* 84 (4):839–847.

FDA. 2015. *Scientific Considerations in Demonstrating Biosimilarity to a Reference Product Guidance for Industry.* U.S. Food and Drug Administration. Available from https://www.fda.gov/media/82647/download

———. 2017. *FDA Approves First Biosimilar for the Treatment of Cancer.* U.S. Food and Drug Administration. Available from https://www.fda.gov/news-events/press-announcements/fda-approves-first-biosimilar-treatment-cancer

———. 2018. *Prescribing Biosimilar and Interchangeable Products.* U.S. Food and Drug Administration. Available from https://www.fda.gov/drugs/biosimilars/prescribing-biosimilar-and-interchangeable-products.

———. 2021a. Biosimilars Action Plan: Balancing Innovation and Competition. U.S. Food and Drug Administration [cited May 10 2021]. Available from https://www.fda.gov/media/114574/download.

———. 2021b. *Biosimilar and Interchangeable Biologics: More Treatment Choices.* U.S. Food and Drug Administration [cited May 8 2021]. Available from https://www.fda.gov/consumers/consumer-updates/biosimilar-and-interchangeable-biologics-more-treatment-choices#:~:text=Interchangeable%20Biosimilar%20Medications, no%20FDA%2Dapproved%20interchangeable%20medications.

———. 2021c. *Biosimilar Product Information.* U.S. Food and Drug Administration [cited April 25 2021]. Available from https://www.fda.gov/drugs/biosimilars/biosimilar-product-information.

———. 2021d. *Orphan Drug Designations and Approvals Search.* U.S. Food and Drug Administration. Available from https://www.accessdata.fda.gov/scripts/opdlisting/oopd/detailedIndex.cfm?cfgridkey=619317.

———. 2021e. Drugs@FDA: FDA-Approved Drugs. U.S. Food and Drug Administration [cited April 25 2021]. Available from https://www.accessdata.fda.gov/scripts/cder/daf/.

———. 2021f. *Novel Drug Approvals for 2021.* U.S. Food and Drug Administration [cited April 25 2021]. Available from https://www.fda.gov/drugs/new-drugs-fda-cders-new-molecular-entities-and-new-therapeutic-biological-products/novel-drug-approvals-2021.

Finn, O. J. 2017. Human tumor antigens yesterday, today, and tomorrow. *Cancer Immunol Res* 5 (5):347–354.

Genentech, Inc. 2021a. *RITUXAN (Rituximab) Prescribing Information.* U.S. Food and Drug Administration [cited May 8 2021]. Available from https://www.gene.com/download/pdf/rituxan_prescribing.pdf.

———. 2021b. *HERCEPTIN (Trastuzumab) Prescribing Information.* Genentech, Inc. [cited May 5 2021]. Available from https://www.gene.com/download/pdf/herceptin_prescribing.pdf.

———. 2021c. *AVASTIN (Bevacizumab) Prescribing Information.* U.S. Food and Drug Administration [cited May 5 2021]. Available from https://www.accessdata.fda.gov/drugsatfda_docs/label/2020/125085s332lbl.pdf.

Grady, Denise. 2018. Good news on early breast cancer: Herceptin treatment can be shortened. The New York Times. Available from https://www.nytimes.com/2018/05/16/health/breast-cancer-herceptin-genentech.html

Grillo-Lopez, A. J. 2000. Rituximab: An insider's historical perspective. *Semin Oncol* 27 (6 Suppl 12):9–16.

Grilo, António L., and A. Mantalaris. 2019. The increasingly human and profitable monoclonal antibody market. *Trends in Biotechnology* 37 (1):9–16.

Heitmann, J. S., M. Pfluegler, G. Jung, and H. R. Salih. 2021. Bispecific antibodies in prostate cancer therapy: Current status and perspectives. *Cancers (Basel)* 13 (3):549–566.

Huang, S., S. M. J. van Duijnhoven, Ajam Sijts, and A. van Elsas. 2020. Bispecific antibodies targeting dual tumor-associated antigens in cancer therapy. *J Cancer Res Clin Oncol* 146 (12):3111–3122.

Ishii-Watabe, A., and T. Kuwabara. 2019. Biosimilarity assessment of biosimilar therapeutic monoclonal antibodies. *Drug Metab Pharmacokinet* 34 (1):64–70.

Kaplon, H., M. Muralidharan, Z. Schneider, and J. M. Reichert. 2020. Antibodies to watch in 2020. *mAbs* 12 (1):1703531.

Kaplon, H., and J. M. Reichert. 2021. Antibodies to watch in 2021. *mAbs* 13 (1):1860476.

Köhler, G., and C. Milstein. 1975. Continuous cultures of fused cells secreting antibody of predefined specificity. *Nature* 256 (5517):495–497.

Kolberg, H.-C., M. Colleoni, P. Santi, et al. 2019. Totality of scientific evidence in the development of ABP 980, a biosimilar to trastuzumab. *Target Oncol* 14 (6):647–656.

Labrijn, A. F., M. L. Janmaat, J. M. Reichert, and P. W. H. I. Parren. 2019. Bispecific antibodies: A mechanistic review of the pipeline. *Nat Rev Drug Discov* 18 (8):585–608.

Lancet. 2020. Biosimilars: A new era in access to breast cancer treatment. *Lancet* 395 (10217):2.

Lartigue, J. de 2018. Bevacizumab-AWWB becomes first biosimilar approved for cancer treatment. *JCSO* 16 (2):e60–e62.

Macdonald, J. C., H. Hartman, and I. A. Jacobs. 2015. Regulatory considerations in oncologic biosimilar drug development. *mAbs* 7 (4):653–661.

Maloney, D. G., A. J. Grillo-LóPez, C. A. White, et al. 1997. IDEC-C2B8 (Rituximab) anti-CD20 monoclonal antibody therapy in patients with relapsed low-grade non-Hodgkin's lymphoma. *Blood* 90 (6):2188–2195.

Manis, J. P. 2020. Overview of therapeutic monoclonal antibodies. In *UpToDate*, edited by D. E. Furst. Waltham, MA: UpToDate, Inc.

Markus, R., V. Chow, Z. Pan, and V. Hanes. 2017. A phase I, randomized, single-dose study evaluating the pharmacokinetic equivalence of biosimilar ABP 215 and bevacizumab in healthy adult men. *Cancer Chemother Pharmacol* 80 (4):755–763.

Melero, I., S. Hervas-Stubbs, M. Glennie, D. M. Pardoll, and L. Chen. 2007. Immunostimulatory monoclonal antibodies for cancer therapy. *Nat Rev Cancer* 7 (2):95–106.

Melosky, B., D. A. Reardon, A. B. Nixon, J. Subramanian, A. H. Bair, and I. Jacobs. 2018. Bevacizumab biosimilars: Scientific justification for extrapolation of indications. *Future Oncol* 14 (24):2507–2520.

Merck, Inc. 2021. *ONTRUZANT (trastuzumab-dttb) Prescribing Information*. U.S. Food and Drug Administration [cited May 3 2021]. Available from https://www.merck.com/product/usa/pi_circulars/o/ontruzant/ontruzant_pi.pdf.

Miller, E. M., and L. S. Schwartzberg. 2019. Biosimilars for breast cancer: A review of HER2-targeted antibodies in the United States. *Ther Adv Med Oncol* 11:1758835919887044.

Modjtahedi, H., S. Ali, and S. Essapen. 2012. Therapeutic application of monoclonal antibodies in cancer: Advances and challenges. *Br Med Bull* 104:41–59.

Monk, B. J., W. K. Huh, J. A. Rosenberg, and I. Jacobs. 2017a. Will bevacizumab biosimilars impact the value of systemic therapy in gynecologic cancers? *Gynecol Oncol Res Pract* 4 (7):7–14.

Monk, B., P. Lammers, T. Cartwright, and I. Jacobs. 2017b. Barriers to the access of bevacizumab in patients with solid tumors and the potential impact of biosimilars: A physician survey. *Pharmaceuticals* 10 (4):19.

Mulcahy, A. W., J. P. Hlavka, and S. R. Case. 2018. Biosimilar cost savings in the United States: Initial experience and future potential. *Rand Health Quart* 7 (4):3.

Mylan, Inc. 2021. *OGIVRI (trastuzumab) Prescribing Information*. U.S. Food and Drug Administration [cited May 2 2021]. Available from https://dailymed.nlm.nih.gov/dailymed/fda/fdaDrugXsl.cfm?type=display&setid=6b7938e6-14c7-4a65-9605-967542ecfb8f.

Nabhan, C., and B. A. Feinberg. 2017. Behavioral economics and the future of biosimilars. *J Natl Compreh Cancer Network* 15 (12):1449–1451.

Nabhan, C. A. Valley, and B. A. Feinberg. 2018. Barriers to oncology biosimilars uptake in the United States. *The Oncol.* 23 (11):1261–1265.

NCT01901146. *Efficacy and Safety Study of ABP 980 Compared with Trastuzumab in Women with HER2-positive Early Breast Cancer.* Available from https://ClinicalTrials.gov/show/NCT01901146.

NCT01966003. *Efficacy and Safety Study of ABP 215 Compared with Bevacizumab in Patients with Advanced Non-Small Cell Lung Cancer.* Available from https://ClinicalTrials.gov/show/NCT01966003.

NCT01989676. *A Study Of PF-05280014 Trastuzumab-Pfizer or Herceptin® Trastuzumab-EU Plus Paclitaxel in HER2 Positive First Line Metastatic Breast Cancer Treatment (REFLECTIONS B327-02).* Available from https://ClinicalTrials.gov/show/NCT01989676.

NCT02149524. *A Study to Compare the Effect of SB3 and Herceptin® in Women with HER2 Positive Breast Cancer.* Available from https://ClinicalTrials.gov/show/NCT02149524.

NCT02162667. *Efficacy and Safety Evaluating Study of CT-P6 in Her2 Positive Early Breast Cancer.* Available from https://ClinicalTrials.gov/show/NCT02162667.

NCT02213263. *A Study of PF-05280586 (Rituximab-Pfizer) or MabThera® (Rituximab-EU) for the First-Line Treatment of Patients with CD20-Positive, Low Tumor Burden, Follicular Lymphoma (REFLECTIONS B328-06).* Available from https://ClinicalTrials.gov/show/NCT02213263.

NCT02260804. *To Compare Efficacy and Safety Between CT-P10 and Rituxan in Patients With Low Tumour Burden Follicular Lymphoma.* Available from https://ClinicalTrials.gov/show/NCT02260804.

NCT02364999. *A Comparative Study Of PF-06439535 Plus Paclitaxel-Carboplatin and Bevacizumab Plus Paclitaxel-Carboplatin Patients with Advanced Non-Squamous NSCLC.* Available from https://ClinicalTrials.gov/show/NCT02364999.

NCT02472964. *Study of Efficacy and Safety of Myl1401O + Taxane vs Herceptin©+ Taxane for 1st Line, Met. Br. Ca.* Available from https://ClinicalTrials.gov/show/NCT02472964.

NCT02747043. *Study to Assess if ABP798 is Safe & Effective in Treating Non-Hodgkin Lymphoma Compared to Rituximab.* Available from https://ClinicalTrials.gov/show/NCT02747043.

NCT02754882. *A Study Comparing SB8 and Avastin® in Patients with Advanced Non-squamous Non-small Cell Lung Cancer.* Available from https://ClinicalTrials.gov/show/NCT02754882.

Niederwieser, D., C. Hamm, P. Cobb, et al. 2020. Efficacy and safety of ABP 798: Results from the JASMINE trial in patients with follicular lymphoma in comparison with rituximab reference product. *Target Oncol* 15 (5):599–611.

Nixon, N.A., M.B. Hannouf, and S. Verma. 2018. The evolution of biosimilars in oncology, with a focus on trastuzumab. *Curr Oncol* 25 (11):171–179.

Ogura, M., J. M. Sancho, S. G. Cho, et al. 2018. Efficacy, pharmacokinetics, and safety of the biosimilar CT-P10 in comparison with rituximab in patients with previously untreated low-tumour-burden follicular lymphoma: A randomised, double-blind, parallel-group, phase 3 trial. *Lancet Haematol* 5 (11):e543–e553.

Parray, H. A., S. Shukla, S. Samal, et al. 2020. Hybridoma technology a versatile method for isolation of monoclonal antibodies, its applicability across species, limitations, advancement and future perspectives. *Intl Immunopharmacol* 85:106639.

Pegram, M. D., I. Bondarenko, M. M. C. Zorzetto, et al. 2019. PF-05280014 (a trastuzumab biosimilar) plus paclitaxel compared with reference trastuzumab plus paclitaxel for HER2-positive metastatic breast cancer: A randomised, double-blind study. *Brit J Cancer* 120 (2):172–182.

Pento, J. T. 2017. Monoclonal antibodies for the treatment of cancer. *Anticancer Res* 37 (11):5935–5939.

Pfizer, Inc. 2021a. *RUXIENCE (rituximab-pvvr) Prescribing Information*. Pfizer Laboratories Div Pfizer Inc. [cited May 8 2021]. Available from http://labeling.pfizer.com/ShowLabeling.aspx?id=12090.

———. 2021b. TRAZIMERA (trastuzumab-qyyp) Prescribing Information. U.S. Food and Drug Administration [cited May 3 2021]. Available from http://labeling.pfizer.com/ShowLabeling.aspx?id=12725.

———. 2021c. *ZIRABEV (bevacizumab-bvzr) Prescribing Information*. Pfizer Laboratories Div [cited May 3 2021]. Available from http://labeling.pfizer.com/ShowLabeling.aspx?id=11860.

———. 2021d. Zirabev Digital Product Monograph. Pfizer USA [cited May 15 2021]. Available from https://www.pfizerpro.com/sites/default/files/pp-zir-usa-0164_zirabev_digital_product_monograph_fr12_mg.pdf?advert=advert.

Pivot, X., I. Bondarenko, Z. Nowecki, et al. 2018. Phase III, randomized, double-blind study comparing the efficacy, safety, and immunogenicity of SB3 (trastuzumab biosimilar) and reference trastuzumab in patients treated with neoadjuvant therapy for human epidermal growth factor receptor 2–Positive early. *J Clin Oncol* 36 (10):968–974.

Pivot, X., E. Curtit, Y. J. Lee, et al. 2016. A randomized phase I pharmacokinetic study comparing biosimilar candidate SB3 and trastuzumab in healthy male subjects. *Clin Ther* 38 (7):1665–1673.e3.

Przepiorka, D., C. W. Ko, A. Deisseroth, et al. 2015. FDA approval: Blinatumomab. *Clin Cancer Res* 21 (18):4035–4039.

Reck, M., A. Luft, I. Bondarenko, et al. 2020. A phase III, randomized, double-blind, multicenter study to compare the efficacy, safety, pharmacokinetics, and immunogenicity between SB8 (proposed bevacizumab biosimilar) and reference bevacizumab in patients with metastatic or recurrent nonsquamous. *Lung Cancer* 146:12–18.

Reinmuth, N., M. Bryl, I. Bondarenko, et al. 2019. PF-06439535 (a bevacizumab biosimilar) compared with reference bevacizumab (Avastin®), both plus paclitaxel and carboplatin, as first-line treatment for advanced non-squamous non-small-cell lung cancer: A randomized, double-blind study. *BioDrugs* 33 (5):555–570.

Rosen, L. S., I. A. Jacobs, and R. L. Burkes. 2017. Bevacizumab in colorectal cancer: Current role in treatment and the potential of biosimilars. *Target Oncol* 12 (5):599–610.

Rugo, H. S., A. Barve, C. F. Waller, et al. 2017. Effect of a proposed trastuzumab biosimilar compared with trastuzumab on overall response rate in patients with ERBB2 (HER2)–Positive metastatic breast cancer. *JAMA* 317 (1):37.

Rugo, H. S., K. M. Linton, P. Cervi, J. A. Rosenberg, and I. Jacobs. 2016. A clinician's guide to biosimilars in oncology. *Cancer Treat Rev* 46:73–79.

Saleem, T., H. Qurashi, M. Jamali, J. Chan Gomez, and T. Kanderi. 2020. Biosimilars as a future, promising solution for financial toxicity: A review with emphasis on bevacizumab. *Cureus* 12 (7):e9300.

Schickedanz, A. 2010. Of value: A discussion of cost, communication, and evidence to improve cancer care. *Oncologist* 15 Suppl 1:73–79.

Scott, A. M., J. P. Allison, and J. D. Wolchok. 2012. Monoclonal antibodies in cancer therapy. *Cancer Immun* 12:14.

Sharman, J. P., A. M. Liberati, K. Ishizawa, et al. 2020. A randomized, double-blind, efficacy and safety study of PF-05280586 (a rituximab biosimilar) compared with rituximab reference product (MabThera®) in subjects with previously untreated CD20-positive, low-tumor-burden follicular lymphoma (LTB-FL). *BioDrugs* 34 (2):171–181.

Shim, H. 2020. Bispecific antibodies and antibody-drug conjugates for cancer therapy: Technological considerations. *Biomolecules* 10 (3):360–391.

Smith, M. R. 2003. Rituximab (monoclonal anti-CD20 antibody): Mechanisms of action and resistance. *Oncogene* 22 (47):7359–7368.

Stebbing, J., Y. Baranau, V. Baryash, et al. 2017. CT-P6 compared with reference trastuzumab for HER2-positive breast cancer: a randomised, double-blind, active-controlled, phase 3 equivalence trial. *Lancet Oncol* 18 (7):917–928.

Syed, Y. Y. 2020. SB8: A bevacizumab biosimilar. *Target Oncol* 15 (6):787–790.

Taieb, J., E. Aranda, S. Raouf, H. Dunn, and D. Arnold. 2021. Clinical and regulatory considerations for the use of bevacizumab biosimilars in metastatic colorectal cancer. *Clin Colorectal Cancer* 20 (1):42–51.e3.

Tedder, T. F., and P. Engel. 1994. CD20: A regulator of cell-cycle progression of B lymphocytes. *Immunol Today* 15 (9):450–454.

Thatcher, N., J. H. Goldschmidt, M. Thomas, et al. 2019. Efficacy and safety of the biosimilar ABP 215 compared with bevacizumab in patients with advanced nonsquamous non-small cell lung cancer (MAPLE): A randomized, double-blind, phase III study. *Clin Cancer Res* 25 (7):2088–2095.

Thill, M. 2019. Biosimilar trastuzumab in clinical trials: Differences or not? *Breast Care* 14 (1):17–22.

Thomas, A., B. A. Teicher, and R. Hassan. 2016. Antibody-drug conjugates for cancer therapy. *The Lancet Oncol* 17 (6):e254–e262.

Vacchelli, E., A. Eggermont, J. Galon, et al. 2013. Trial watch: Monoclonal antibodies in cancer therapy. *Oncoimmunology* 2 (1):e22789.

von Minckwitz, G., M. Colleoni, H. C. Kolberg, et al. 2018. Efficacy and safety of ABP 980 compared with reference trastuzumab in women with HER2-positive early breast cancer (LILAC study): A randomised, double-blind, phase 3 trial. *Lancet Oncol* 19 (7):987–998.

Waller, C. F., A. Vutikullird, T. E. Lawrence, et al. 2018. A pharmacokinetics phase 1 bioequivalence study of the trastuzumab biosimilar MYL-1401O vs. EU-trastuzumab and US-trastuzumab. *Brit J Clin Pharmacol* 84 (10):2336–2343.

Weiner, L. M., R. Surana, and S. Wang. 2010. Monoclonal antibodies: Versatile platforms for cancer immunotherapy. *Nat Rev Immunol* 10 (5):317–327.

Wu, J., J. Fu, M. Zhang, and D. Liu. 2015. Blinatumomab: A bispecific T cell engager (BiTE) antibody against CD19/CD3 for refractory acute lymphoid leukemia. *J Hematol Oncol* 8:104–111.

Zahavi, D., and L. Weiner. 2020. Monoclonal antibodies in cancer therapy. *Antibodies (Basel)* 9 (3):34–54.

15

Molecule-Targeted Drugs That Treat Colorectal Cancer

Qian-Ming Du

Corresponding author: Qian-Ming Du

Contents

DOI: 10.1201/9780429485626-15

15.1 INTRODUCTION

Colorectal cancer (CRC) is the third most commonly diagnosed malignant disease worldwide, next to lung cancer and breast cancer [1, 2]. Due to its occult clinical manifestations, an estimated 50%–60% of cases are diagnosed with advanced-stage, and 80%–90% have unresectable liver metastases, facing more worse overall survival (OS) time, with a 5-year OS rate being less than 20% [3].

The major therapeutic approaches used for CRC are surgery, chemotherapy, and irradiation. Indeed, when localized at the bowel, CRC is highly treatable and often curable. Surgery is the primary treatment modality and has a higher cure rate. Nevertheless, recurrence and metastasis after surgery are major problems because they lead to a poorer prognosis, with a median of OS being approximately 24 months [4]. For patients with metastatic CRC (mCRC), resection is not an option, palliative systemic therapy remains the treatment of choice. The aim of therapy of these patients is to extend their OS time by delaying disease progression as well as preserving physical mobility and the quality of life [5]. The standard treatments comprise conventional cytotoxic chemotherapy and new biologic agents, of which the former is the most important [6, 7]. However, chemotherapy does not specifically target tumor cells and is blunted due to insufficient drug exposure in tumor tissues, presence of drug-resistant tumor cells, and systemic toxicity.

In the past decades, the major advances in clinical management of mCRC are attributable to the introduction of novel biologics and biosimilars to clinical settings [8]. Most biologic agents are used to target the signaling cascades responsible for the regulation of tumor growth, drug resistance, metastasis, angiogenesis, and apoptosis [9]. The effectiveness of biologic agents is the result of their specific binding to certain therapeutic targets of cancer to be treated. The clinical use of targeted agents has increased the median OS of patients with mCRC to more than 30 months [10].

This chapter introduces the concept of immunotherapy and gene therapy, summarizes the molecular-targeted drugs that treat mCRC (currently in use or under development) as well as their mechanisms of action, and provides information on the leading clinical trials for CRC.

15.2 IMMUNOTHERAPY

Tumor immunotherapy has considerable potential for treating CRC. Indeed, various modalities, such as monoclonal antibodies (mAbs), immune-checkpoint inhibitors, cancer vaccines, adoptive cell therapy (ACT), complement inhibitors, and cytokines, are investigated in clinical trials, most of which are in phase I and II clinical trials, and some of which have yielded promising results. To date, more than 24 immunotherapy-based clinical trials have been completed in patients with CRC, and approximately 40 are recruiting or about to recruit patients [11].

15.2.1 mAb Therapy

An mAb therapy involves the use of humanized antibodies against mCRC. For example, bevacizumab, aflibercept, regorafenib, and ramucirumab all inhibit the growth of new blood vessels, whereas cetuximab

TABLE 15.1 Targeted drugs approved by the FDA for mCRC [10]

GROUP	DRUG	THERAPEUTIC TARGET	APPROVAL YEAR
Angiogenesis inhibitors	Bevacizumab	VEGF-A	2004
	Aflibercept	VEGF-A/B, PIGF	2012
	Regorafenib	VEGFR1–3, TIE2	2012
	Ramucirumab	VEGFR-2	2015
EGFR inhibitors	Panitumumab	EGFR	2006
	Cetuximab	EGFR	2009
BRAF inhibitors	Encorafenib	BRAF	2020

and panitumumab selectively target the epidermal growth factor receptor (EGFR). Several mAb therapies for CRC are in clinical trials. For instance, adecatumumab is used to target epithelial cell adhesion molecule, labetuzumab targets carcinoembryonic antigen (CEA), and pemtumomab targets mucins [12].

Cetuximab was the first targeted drug approved by the United States Food and Drug Administration (FDA) in 2009 for patients with mCRC who harbor wild-type KRAS [13]. Bevacizumab, a humanized mAb, was approved (in 2004). Subsequently, targeted drug development has been a focus of research, resulting in the approval of five other drugs by the FDA, which can be categorized into three types according to the signaling cascade targeted [14] (also see Table 15.1). Patients with mCRC are not always beneficial from the combination of vascular endothelial growth factor (VEGF) and EGFR inhibitors, but they could benefit more from the combination of these biologics with chemotherapy [15].

15.2.1.1 Inhibitors of angiogenesis

Inhibitors of angiogenesis suppress the generation of new blood vessels, representing a new approach for cancer therapy. Angiogenesis is a complex and highly modulated process in which new blood vessels are formed from pre-existing vessels. It plays an essential role in normal growth and in the development of diverse diseases [16]. Angiogenesis is tightly regulated under physiological conditions. A variety of growth factors, chemokines, and angiogenic enzymes [17]. However, the presence of excess of proangiogenic factors may induce tumor progression. In the early 1970s, Folkman explored the potential of inhibitors of angiogenesis for preventing the formation of new blood vessels [18]. This assumption was subsequently confirmed, and the growth of new vessels is now regarded as a hallmark of cancer [19].

The VEGF signaling pathway is implicated in the growth of new blood vessels. VEGF binds to VEGF tyrosine kinase receptors on the surface of endothelial cells and activates intracellular tyrosine kinases, resulting in angiogenesis. Overexpressed VEGF in cancer cells facilitates, in turn, angiogenesis, tumor growth, and metastasis upon binding to its receptors [20]. The VEGF family consists of VEGF-A, -B, -C, and -D, placental growth factor (PlGF), and several VEGF tyrosine kinase receptors, including VEGFR-1 (Flt-1), VEGFR-2 (Flk-2 or KDR), and VEGFR-3 (Flt-4) [21]. Neuropilins (NRP-1 and -2) also modulate angiogenesis by interacting with VEGF(R), class 3 semaphorins, and EGFR. VEGF-A (formerly VEGF) is a potent proangiogenic substance under both physiologic and pathophysiologic circumstances. Within CRC tissues, VEGF-A binds to VEGFR-1 and VEGFR-2, promoting angiogenesis [20]. Approximately 50% of CRC cases express VEGF, which is validated as a biomarker to predict a poorer prognosis in CRC [22]. Because VEGFs play an important role in angiogenesis, inhibition of the VEGF pathway represents a treatment strategy for mCRC.

Large molecules that target VEGF-A, such as mAbs, have been developed. Also, several small-molecule tyrosine kinase inhibitors, which inhibit VEGFR by binding to its ATP-binding site, have been synthesized. Other methods for inhibiting VEGF include the use of soluble decoy receptors. Finally, bevacizumab, aflibercept, regorafenib, and ramucirumab are FDA-approved inhibitors of angiogenesis that suppress tumor vascularization by targeting VEGF or its receptor [23].

15.2.1.1.1 Bevacizumab

Bevacizumab, a humanized mAb against VEGF, prevents interaction of VEGF with the VEGFR [20] as the first anti-angiogenic agent approved for clinical use in combination with chemotherapy [24]. It has been approved for the treatment of mCRC in 2004. Bevacizumab alone is ineffective against mCRC but demonstrates significantly greater efficacy when combined with chemotherapy than the use of chemotherapy alone [25]. For example, progression-free survival (PFS) and OS were significantly improved by bevacizumab in combination with standard chemotherapy as the first- or second-line regimen [26, 27]. Bevacizumab promotes the regression of existing microvessels, normalization of surviving mature vasculature, and inhibition of neovascularization and vessel growth, which enhances the delivery of the co-administered chemotherapy drugs into the tumor [28]. The most frequent side effects of bevacizumab are class-related and are therefore also seen with other VEGF inhibitors, including hypertension and hemorrhage. These adverse events are often asymptomatic and reversible after ceasing the drug or with other appropriate therapy [15].

15.2.1.1.2 Aflibercept

Aflibercept (VEGF-trap, Zaltrap) is a recombinant fusion protein composed of VEGFR-1 and VEGFR-2 ligand-binding components fused to the Fc portion of human IgG1 [29]. It functions as a decoy receptor that prevents binding of the ligands (VEGF-A, and -B, as well as PlGF) to their receptors. Aflibercept was approved in both the United States and Europe for use in combination with FOLFIRI in patients with mCRC who showed disease progression following an oxaliplatin-based regimen. This approval was based on the VELOUR trial [30], in which 1,226 patients manifesting disease progression during or within 6 months of receiving oxaliplatin-based chemotherapy were treated with FOLFIRI and either aflibercept or placebo. The combination regimen resulted in a significant improvement in OS, PFS, and response rate (RR). Although the magnitude of the effect on OS, PFS, and RR was similar to that of bevacizumab beyond progression in the TML study [31], the level of toxicity was slightly higher. Also, the VEGF-related adverse events were consistent with other anti-VEGF agents (e.g., hypertension, bleeding, and proteinuria), but aflibercept increased the incidence of the chemotherapy-associated adverse events, such as diarrhea, stomatitis, fatigue, and neutropenia. The benefit of aflibercept as a second-line agent was similar in bevacizumab-pretreated and -naïve patients. When combined with chemotherapy, aflibercept is an important option for mCRC where oxaliplatin failed, particularly in patients suffering rapid progression during or after treatment with bevacizumab-containing regimens or those with symptoms. Therefore, anti-EGFR antibodies should be used as third-line agents in view of their comparable activity as second- and third-line therapies for mCRC.

15.2.1.1.3 Regorafenib

Regorafenib (BAY 73–4506) is a tyrosine kinase inhibitor targeting VEGFR-1, -2, and -3, as well as PDGFRβ, Tie-2, c-KIT, FGFR-1, RET, and B-RAF [32]. Therefore, regorafenib exhibits anti-proliferative and anti-angiogenic properties. Regorafenib is the first multi-kinase inhibitor with proof of efficacy against mCRC. The CORRECT study investigated the use of regorafenib (160 mg orally once daily for the first 3 weeks of each 4-week cycle) or placebo in 760 patients with chemotherapy-resistant mCRC [33], demonstrating a modest but significant difference in OS (6.4 vs. 5 months). Because only 1% of patients in the regorafenib arm showed responses to the drug, regorafenib seemed to slow tumor progression rather than to induce a therapeutic response. The most frequent grade 3/4 toxicities were hand–foot skin reactions, fatigue, hypertension, diarrhea, and skin rash. Based on the CORRECT study, regorafenib received approval from the FDA in October 2012 for the treatment of patients with chemorefractory mCRC. Among the unanswered questions on regorafenib in mCRC are whether the mechanisms of action are direct blockade of a specific tyrosine kinase, rather than inhibition of angiogenesis, as well as whether there are biomarkers predictive of the activity of the drug, how to manage optimally the adverse events, and what are other indications of regorafenib in mCRC.

Biomarkers for anti-angiogenic agents

Despite long searching, validated biomarkers of anti-angiogenic therapy are not yet available for routine clinical use. A number of studies have evaluated systemic measurements (blood pressure), genotypes

(VEGF or IL-8 polymorphisms), circulating biomarkers (such as plasma VEGF), tissue markers (tumor microvessel density), and imaging parameters (K_{trans}) [34]. However, escape from anti-angiogenic agents can be mediated by both tumor cells and by cells from the microenvironment. Although anti-angiogenic agents are targeted, their mechanisms of action are largely unknown, hampering the selection of patients who would benefit from such drugs. Identification of a suitable biomarker would not only prevent unnecessary toxicity but also reduce the treatment cost. Blood and tumor samples from trials, such as the TML and the VELOUR study, might facilitate the identification of patients who would benefit from anti-angiogenic agents.

In a phase II trial involving 43 patients with mCRC treated with FOLFIRI in combination with bevacizumab, the levels of FGF-2, hepatocyte growth factor (HGF), PlGF, SDF-1, and macrophage chemoattractant protein-3 increased significantly, which may represent mechanisms of resistance [35]. Also, an increased baseline IL-8 (promoter of angiogenesis) level was correlated with decreased PFS [36].

Because variation in response to the drugs is mediated both by differences among patients or tumors, single nucleotide polymorphisms (SNPs) have the potential as candidate biomarkers linked to varied responses to the anti-angiogenic agents. To date, several SNPs in the VEGF pathway have been found to predict response to bevacizumab in non-CRC studies. However, no VEGF polymorphism has been validated as a marker of a response to anti-angiogenic agents in patients with mCRC [36, 37]. This is, in part, because of the lack of large prospective studies with a suitably powered biomarker component.

15.2.1.2 EGFR inhibitors

EGFR is a cell-surface receptor expressed by cancer cells and is associated with a poorer prognosis in a number of types of cancers [38]. EGFR is implicated in the regulation of tumor-cell division, repair, survival, and metastasis. The binding of specific ligands to EGFR activates the receptor, triggering a signal transduction cascade and cell proliferation.

EGFR, also known as ErbB1/HER1, is a transmembrane glycoprotein that belongs to the human epidermal growth factor receptor (HER)-ErbB family of receptor tyrosine kinases. Other members of the ErbB family of cell-membrane receptors include ErbB2/HER-2 or neu, ErbB3/HER-3, and ErbB4/HER-4 [39]. All these receptors are transmembrane glycoproteins consisting of an extracellular ligand-binding domain, a hydrophobic transmembrane domain, and, with the exception of ErbB3, a cytoplasmic (i.e., intracellular) tyrosine kinase domain that has a carboxy terminal region with tyrosine autophosphorylation sites [40] (ErbB3 lacks intrinsic tyrosine kinase activity). The binding of specific ligands to these ErbB receptors triggers the autophosphorylation of its intracellular domain [41], followed by activation of downstream signaling. The two major ligands that activate EGFR are epidermal growth factor (EGF) and transforming growth factor-α. Heregulins (growth factors) stimulate ErbB3 and ErbB4 [42]. There are currently unknown ligands for ErbB2, but it is an important partner for receptor dimerization and the resultant activation of the intrinsic tyrosine kinase activity of the receptor. The most commonly activated downstream signaling pathways are those centered on Ras/Raf mitogen-activated protein kinase (MAPK) and phosphatidylinositol-3-kinase/protein kinase B (PI3K/AKT), both of which are responsible for cancer cell proliferation, survival, invasion, metastasis, and neoangiogenesis [43].

EGFR plays a role in CRC, and EGFR may function in intestinal homeostasis [44, 45] and in stem-cell maintenance, proliferation, and differentiation [45, 46]. EGFR is variably expressed on nearly all normal cells, particularly those of epithelial origin, such as liver, skin, and gastrointestinal tract cells, but not on hematopoietic cells. Although its function in normal intestinal development and crypt homeostasis is poorly understood, there is evidence that EGFR hyperactivation is involved in multiple cancer types. Several mechanisms have been reported to contribute to this phenomenon, such as mutations in the kinase domain of EGFR, overexpression of EGFR and its ligands, and changes in gene copy number [15]. These findings, coupled with the druggable properties of EGFR, led to the development of anti-EGFR agents.

Inhibition of EGFR may suppress the growth and progression of EGFR-expressing tumors [47]. The anti-EGFR mAbs, such as cetuximab and panitumumab, inhibit its dimerization and subsequent phosphorylation and signal transduction by targeting the extracellular domain. They have been approved by the FDA for the treatment of CRC, and many other relevant mAbs are under development. Despite first being introduced into clinical practice for the treatment of non-selected mCRC patients, panitumumab and cetuximab are now restricted to patients with wild-type KRAS tumors.

15.2.1.2.1 Cetuximab

Cetuximab is a chimeric mAb targeting the extracellular domain of EGFR. By binding the extracellular component of the receptor, it blocks the ligand binding-induced receptor dimerization and activation of a tyrosine kinase. As an mAb, it also elicits an antibody-dependent cellular cytotoxicity against cancer cells [48, 49].

As the first anti-EGFR compound approved for the treatment of mCRC, cetuximab was approved in 2009 for the treatment of irinotecan-refractory mCRC. Although the rate of the response to cetuximab alone is only 10% [12], it can improve the OS of patients with chemorefractory mCRC when administered as the last-line treatment [50]. Moreover, several retrospective studies have reported that activating mutations in the KRAS oncogene are associated with intrinsic resistance to cetuximab [47]. This led the United States and European health authorities to restrict the use of cetuximab to patients with KRAS wild-type tumors. The most common side effects are skin rash and hypomagnesemia [47], with the severity of skin rash being positively correlated with the treatment response. Amphiregulin and epiregulin expression is also predictive biomarker in KRAS wild-type patients.

15.2.1.2.2 Panitumumab

Panitumumab is a fully-humanized IgG2 mAb with lacking murine proteins that targets the extracellular domain of EGFR with high affinity. Like cetuximab, by targeting the extracellular domain of EGFR, panitumumab inhibits EGF-dependent tumor cell activation and proliferation mediated by its natural ligands [51]. In particular, panitumumab inhibits proliferation by suppressing the EGFR signaling pathway, inducing cycle arrest, inhibiting angiogenesis, downregulating EGFR expression, and stimulating receptor internalization [52]. The most common side effects are skin rash and hypomagnesemia. A skin rash is a biomarker of the response to treatment with panitumumab.

15.2.1.2.3 Biomarkers for EGFR response

Anti-EGFR inhibitors play an important role in the treatment of mCRC. In particular, cetuximab and panitumumab, each alone and in combination with chemotherapy, and as first- and next-line agents, improve the outcomes of patients with mCRC. Nevertheless, the first clinical trials involved a non-selected population of patients with chemorefractory mCRC and explained the low RR [47]. It was always clear that only a specific subset of patients with mCRC would benefit from anti-EGFR mAbs, and that primary resistance could play a role. This can be explained, in part, by the hypothesis that only a limited number of tumors are EGFR positive and would be affected by anti-EGFR mAbs after the failure of the standard cytotoxic chemotherapy. The populations of patients who would benefit from these compounds are currently unknown, but the discovery of biomarkers has improved the therapeutic index of cetuximab and panitumumab. In this context, both negative and positive predictors of the response have been identified.

In contrast to HER2/Neu in breast cancer, in which the protein level is correlated with the response to trastuzumab, detection of EGFR expression in primary CRC tumors by immunohistochemistry is not considered a marker of the response to anti-EGFR mAbs because it is not correlated with the clinical response or clinical benefit [47]. The incidence of EGFR kinase domain mutations in CRC tumors is very low, and so it is highly unlikely that they drive the response to targeted anti-EGFR agents [53]. Also, little evidence supports the ability of EGFR amplification and EGFR gene copy-number variation to predict the treatment response [47]. In contrast, activating alterations in the KRAS oncogene, which are present in 30%–40% of cases with CRC tumors [54], are associated with the lack of response to anti-EGFR inhibitors. Previous studies [47] confirm that the presence of KRAS-activating mutations is correlated with primary resistance to anti-EGFR mAbs. This led the authorities to restrict the use of cetuximab and panitumumab to patients with KRAS wild-type mCRC. Although KRAS is a predictive biomarker of the lack of response to anti-EGFR mAbs, it cannot be used to identify mCRC patients who would benefit from cetuximab and panitumumab. This is, in part, due to its low sensitivity (around 47%), to the fact that only 20%–40% of wild-type patients respond well to anti-EGFR compounds, and that the effects of point mutations in KRAS vary considerably. Indeed, a mutation in codon G13D, which accounts for ~20% of cases, is associated with the response to cetuximab; this is in contrast to codon 12 (70% of cases), which is a marker of resistance [47]. These findings have yet to be confirmed in prospective trials.

Other genes downstream of EGFR signaling, such as BRAF, NRAS, and PIK3CA, have also been investigated. Mutations in NRAS are associated with the lack of response to cetuximab [55], while the roles of mutations in PIK3CA and BRAF are controversial [56].

15.2.1.3 Promising targeted drugs

Although agents targeting EGFR or the VEGF pathway for the treatment of advanced CRC are under development, their efficacy can be limited by mutations that lead to activation of downstream signaling pathways. In particular, 41.6% of patients with CRC harbor BRAF or KRAS mutations, and anti-EGFR therapies are unsuitable for such patients. Therefore, novel therapeutic approaches are needed for patients with BRAF or KRAS mutations [57].

There have been many studies investigating EGFR and VEGF signaling pathways over the last decade. The clinical utility of traditional chemotherapy is reduced by resistance, toxicity, and cost [58]. To overcome these limitations, drugs targeting these two signaling cascades have been investigated in clinical trials, which makes it possible by our improved understanding of signaling cascades [59]. The MEK, PI3K, MET, and programmed death (PD)-receptor signaling pathways are targeted in mCRC treatment. Summarized in Table 15.2 and Figure 15.1 are the targeted drugs and immune-checkpoint inhibitors for mCRC currently undergoing clinical trials or on the verge of entering the market.

15.2.1.3.1 VEGF signaling cascade

The anti-VEGF agents, such as sorafenib, cediranib, famitinib, vanucizumab, fruquintinib, and nintedanib are at present under evaluation in clinical trials for the treatment of cancer [60]. Of them, amitinib and fruquintinib, developed by the Chinese pharmaceutical industry, are in Phase III trials and have different mechanisms of action. If demonstrated to be safe and efficacious, both will be marketed in China. This will benefit patients because amitinib and fruquintinib are far less costly than brand-name drugs.

15.2.1.3.2 EGFR signaling cascade

HER-1 (also known as EGFR), HER-2, and HER-3 are [39]. In the EGFR signaling cascade, EGF binds to the extracellular domain of EGFR, leading to phosphorylation of the cytoplasmic tyrosine-kinase domain, followed by activation of the Raf/MEK/ERK and PI3K/Akt pathways, which promote cell proliferation [47].

TABLE 15.2 Biological drugs demonstrating the potential for the treatment of mCRC [10]

TYPE	*THERAPEUTIC TARGET*	*DRUG (MARKED OR UNDERWAY)*
VEGF signaling cascade	VEGF inhibitor	Sorafenib, cediranib, famitinib, vanucizuma, apatinib, fruquintinib, nintedanib, Donafenib
EGFR/Raf/MEK signaling cascade	EGFR inhibitor	Sym004, AMG510, BVD-523
	BRAF inhibitor	LGX818, vemurafenib (Zelboraf)
	MEK inhibitor	Trametinib, MEK-162, selumetinib (AZD6244)
	P13K inhibitor	BKM120
Miscellaneous	C-MET inhibitor	Tivantinib (ARQ 197), rilotumumab, INC280
	PD-1 inhibitor	Nivolumab, pembrolizumab (MK-3475)

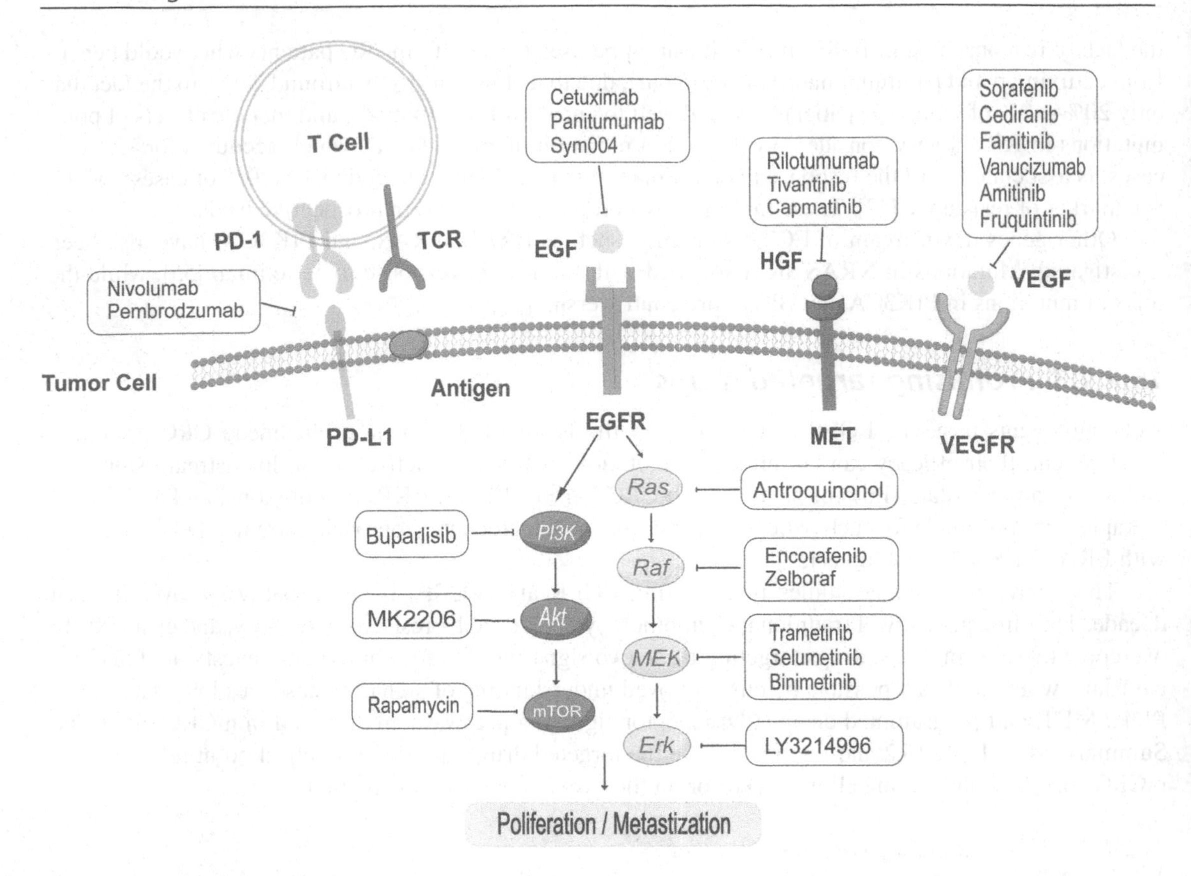

FIGURE 15.1 A perspective on targeted therapy for mCRC [10].

Notes: PD-1/PD-L1: Programmed cell death protein 1/Programmed cell death 1 ligand 1; TCR: T cell receptor; EGF/EGFR: epidermal growth factor/epidermal growth factor receptor; HGF: hepatocyte growth factor; MET: mesenchymal–epithelial transition factor; VEGF/VEGFR: vascular endothelial growth factor/vascular endothelial growth factor receptor; PI3k: phosphatidylinositol 3 kinase; Akt: protein kinase B; mTOR: mammalian Target of Rapamycin; Ras: rat sarcoma; Raf: rapidly accelerated fibrosarcoma; MEK: mitogen-activated protein kinase kinase; Erk: extracellular signal-regulated kinase.

15.2.1.3.3 Raf/MEK/Erk pathway

The MAPK pathway is frequently activated in CRC. Phosphorylation of RAS activates RAF and ERK, leading to the expression of a number of genes involved in cell proliferation, propagation, and survival. BRAF and its downstream target, MEK, are kinases of the MAPK pathway and play an important role in cell proliferation.

LGX818 and vemurafenib are BRAF inhibitors [47]. Most metastatic melanomas harbor one or more mutations in BRAFV600. This finding provided insight into drug design and cancer treatment. Targeting BRAFV600 has gradually become a novel therapeutic strategy [61]. Vemurafenib is a highly selective BRAFV600 inhibitor initially identified in a large-scale drug screening and has been approved by the FDA for the treatment of melanoma. Four phase I/II trials of vemurafenib for the treatment of mCRC are underway.

Trametinib, selumetinib, and MEK-162 are inhibitors of MEK [47]. Trametinib is indicated for the treatment of patients with unresectable or metastatic melanoma. However, most trials of trametinib for the treatment of mCRC are phase I or II, and so phase III trials are needed before this agent can be used clinically.

15.2.1.3.4 PI3K/Akt pathway

PI3Ks are a family of intracellular lipid kinases that convert phosphatidylinositol-4,5-biphosphate (PIP2) to phosphatidylinositol-3,4,5-trisphosphate (PIP3). PIP3 is an important second messenger that activates AKT through phosphorylation. Phospho-AKT phosphorylates more than 100 proteins, including mTOR [47]. The PI3K pathway was discovered over 20 years ago and is now understood to play a central role in various cellular functions. Emerging data suggest the PI3K/AKT/mTOR cascade to be involved in the development of CRC.

Several drugs that interfere with components of the PI3K axis are under development. For example, BKM120, also named buparlisib, is a PI3K inhibitor with a favorable pharmacokinetic profile, a high level of specificity, preliminary anti-tumor activity, and considerable tolerability in a phase I clinical trial [62]. In this trial, the most frequent adverse events were rash (37%), anorexia (37%), hyperglycemia (37%), diarrhea (37%), mood alteration (37%), nausea (31%), and fatigue (26%). Moreover, BKM120 is rapidly absorbed, has a half-life of 40 h, and around 3-fold between-subject variation in the steady-state accumulation. Two trials of BKM120 in patients with CRC are underway, one is in combination with panitumumab and the other with irinotecan.

15.2.1.3.5 Others

HGF is the ligand for the cellular mesenchymal/epithelial transition (c-MET, also called MET). C-MET is expressed in a variety of cell types, such as epithelial, endothelial, and hematopoietic progenitor cells [63]. The HGF/c-MET signaling cascade is involved in embryogenesis, organogenesis, tissue regeneration, carcinogenesis, and other biological processes. The role of the HGF/c-MET axis in solid and hematological malignancies has been confirmed in mice by conditional knockout of MET [64].

All Amgen-sponsored clinical studies of rilotumumab in the treatment of advanced gastric cancer were terminated by Amgen in 2014, and no further trials have been initiated. The c-MET inhibitor, tivantinib [65], has completed a phase I trial for mCRC, and tivantinib plus cetuximab is now under investigation in a phase II trial for the treatment of CRC resistant to anti-EGFR agents. INC280, a new drug from Novartis, is undergoing phase I/II trials in patients with MET-positive mCRC and head-and-neck squamous cell carcinoma (HNSCC) who showed disease progression after cetuximab or panitumumab therapy.

15.2.2 Immune-Checkpoint Inhibitors

The development of immune-checkpoint inhibitors of melanomas and lung tumors is an important therapeutic innovation. The immune system plays a crucial role in modulating the response to mAb therapy. Therefore, combinations of novel cytotoxic agents with immune-checkpoint inhibitors have been investigated. Tumors can evade destruction by the immune system. The results of studies of immune-checkpoint inhibitors, such as the anti-CTLA-4 antibodies, ipilimumab and tremelimumab, and anti-PD-1/PD-L1 antibodies, have demonstrated that these agents enhance the local immune response. Antibodies against CTLA-4 promote the activation of T-cells by antigen-presenting cells in lymphatic tissue. By contrast, anti-PD-1 antibodies inhibit signaling pathways in the tumor microenvironment to improve the activity of effector T cells and thus augment the tumor-mediated immune blockade.

Under physiological conditions, PD-1, an immune checkpoint or coinhibitory molecule that prevents autoimmunity, is expressed on activated T cells. The binding of PD-1 to its ligand PD-L1 or PD-L2 suppresses the function of cytotoxic T cells. T cell-induced anti-tumor responses can be evaded by solid tumors via the PD-1/PD-L1 pathway. Inhibition of the PD-1/PD-L1 pathway by immune checkpoint inhibitors prevents the interaction of PD-1 with its ligands and removes inhibitory signals in T cells, leading to recognition and attack of the tumor by cytotoxic T cells [47]. The recent development of immune-checkpoint inhibitors represents a major advance [66].

The anti-PD-1 checkpoint inhibitors nivolumab and pembrolizumab and other novel anti-PD-1 or -PD-L1 antibodies, such as MPDL3280A, avelumab, MEDI-4736, and AMP-224, are used alone or in combination with other agents for the treatment of gastrointestinal tumors [67].

Anti-PD-1or -PD-L1 mAbs, such as Nivolumab and Pembrolizumab, have considerable potential for the treatment of mCRC. Nivolumab [47] (Opdivo, BMS) was the first immune-checkpoint inhibitor approved by the US FDA to treat progressive advanced-stage squamous or non-squamous non-small-cell lung cancer (NSCLC) after platinum-based chemotherapy. Nivolumab and pembrolizumab are fully human high-affinity IgG4 mAbs against PD-1, which prevent PD-L1/2 from binding to its receptor [68]. Their clinical activity against multiple types of cancers (i.e., melanoma and NSCLC) has been demonstrated. In the future, the clinical use of an immune-checkpoint inhibitor is expected to increase rapidly.

There is a lack of targeted and validated molecular markers for personalized immunotherapy. At the 2015 ASCO annual meeting, the results of a study on the correlation of immunotherapy with specific genetic markers for DNA mismatch repair (MMR) were presented. The MMR defect is a rare form of genetic instability characterized by loss of function of the MMR system. Tumors with genetic defects in MMR harbor much more mutations than those of the same type without such repair defects. Tumors with MMR defects are therefore particularly susceptible to the immune-checkpoint blockade. An investigator-initiated phase II trial examined the efficacy of pembrolizumab in 48 evaluable, heavily pretreated patients with advanced mCRC and other solid tumors. The MMR-deficient (MMR^-) mCRC group achieved an objective response rate (ORR) of 62% (n = 8/13) and a disease control rate (DCR) of 92%, while the MMR proficient (MMR^+) mCRC group showed no ORR (n = 0/25) and a DCR of only 16%. At the time of analysis, the median PFS and the OS of the MMR^- mCRC group was >9 months, while the MMR^+ mCRC group showed a PFS of 2.3 months and an OS of 7.6 months. In patients with other MMR tumors, the ORR was 60% and the DCR was 70%. The adverse reactions (>10%) of pembrolizumab were skin rash (17%), pancreatitis (15%), and thyroiditis/hypothyroidism (10%) [69].

However, some patients with MMR deficiency did not show an objective response to pembrolizumab. The absence of objective response in patients with MMR-proficient tumors suggests that the major value of this study lies in identifying patients with mCRC who should not receive anti-PD-1 antibody therapy. An estimated 85% of CRC patients are MMR-proficient and probably should not be treated with anti-PD-1 antibodies. The enhanced immune response to MMR tumors is explained by their production of more immunologic proteins due to faulty DNA reading frames that re-stimulate as "non-self" recognizable antigens. MMR-deficient tumors are found in about 15%–20% of stage II, 10% stage III, and ~5% stage IV CRC. An MMR defect is particularly common in patients with hereditary non-polyposis CRC, known as Lynch syndrome [47]. Immune-checkpoint inhibitors represent a new therapeutic option for MMR-deficient mCRC.

15.2.3 Cancer Vaccines

Cancer vaccines induce antigen-specific T- or B-cell activity by supplying antigens to APC-like dendritic cells (DCs). Such vaccines include components that activate antigen-pulsed DCs and guide them to local lymph nodes, e.g., DC vaccines and OncoVAX [70].

15.2.3.1 DC vaccines

Because the majority of CRC cases express CEA, which is a tumor-associated antigen DC, they can be pulsed with CEA mRNA or CEA peptides. Most of the CRC patients who were administered with the DC vaccine evoked CEA-specific T cell immune activities [70].

15.2.3.2 Oncovax

Oncovax comprises cancer cells and an immune-stimulating adjuvant to evoke an anti-tumor immune response to prevent relapse of CRC after surgery. The combination of specific immunotherapy with surgery markedly improves survival [71].

15.2.4 Adoptive T-Cell Therapy

Although immune-checkpoint inhibitors are effective against many solid tumors, they rely on boosting a pre-existing population of tumor-reactive T cells in the patient. Thus, in poorly immunogenic types of cancers, immune-checkpoint therapy alone likely fails [72]. The administration of tumor-recognizing T cells via ACT shows promise for poorly immunogenic cancers and may augment the response to tumors that responded to immune-checkpoint therapy [73].

ACT enhances anti-tumor immunity and increases vaccine efficacy. Recent studies have focused on endowing effector T cells with desired antigen receptors, such as chimeric antigen receptor (CAR) T cells. Indeed, ex vivo-expanded human Vδ1 γδ T cells displayed remarkable therapeutic efficacy in a mouse model of human CRC [74].

15.2.5 Complement Inhibitors

Complement, as part of the innate and adaptive immune system, comprises more than 30 proteins and fragments and is involved in the immunosurveillance response to cancer upon activation. Its pro-inflammatory role and contribution to the development of inflammatory diseases suggest involvement in carcinogenesis [75]. Its stimulation is essential for the immunosurveillance response to CRC. Other mechanisms, including increasing the levels of tumorigenic growth factors and cytokines, preventing apoptosis, enhancing angiogenesis, and promoting immunosuppression, also contribute to complement-associated tumorigenesis [76]. Further, the interaction between the tumor and complement may protect against complement-mediated lysis by inducing the expression of complement-regulatory proteins, resulting in a sublytic level of the membrane attack complex and thus preventing cell lysis as a result [77].

Complement can be inhibited by antibodies or other proteins. Various protein inhibitors of complement, such as cobra venom factor, humanized cobra venom factor, and recombinant staphylococcus aureus super antigen-like protein 7, have been assessed in murine models of CRC. Complement depletion suppresses CRC progression by enhancing the host immune response and reducing the immunosuppressive effect of the tumor microenvironment. Therefore, the complement could be used as a component of combination immunotherapies [77].

15.2.6 Cytokine Therapy

Cytokines are essential for tumor immunology, particularly for CRC, the growth of which is determined by the inflammatory process and the immune response. Cytokines, such as tumor necrosis factor (TNF) and interleukin (IL)-6, trigger the synthesis of the central oncogenic factors nuclear factor-κB and inducer of signal transducer and activator of transcription 3 (STAT3) in intestinal cells, all of which enhance cell proliferation and resistance to apoptosis in CRC [78].

TNF is a key proinflammatory cytokine produced by both hematopoietic and non-hematopoietic cells. TNF binds as a homotrimer to two distinct receptors on the cell surface: TNF receptor 1 (TNFR1; also known as p55 receptor) and TNFR2 (also known as p75 receptor) [79]. Although TNF was first identified as a tumor-suppressive cytokine based on its ability to induce apoptosis of certain cell types, it also exerts tumor-promoting effects and links inflammation to cancer. Intestinal epithelial cells are highly sensitive to TNF due to their high level of expression of TNFR1, which activates NF-κB-dependent oncogenic pathways [78]. Recent data demonstrate a novel protective role for TNF. Mice deficient for both TNF and IL-10 develop colitis and cancer, demonstrating greater progression than mice lacking IL-10 alone, and they show high serum levels of IL-6, IFN-γ, and IL-17A [80]. Furthermore, the development of spontaneous colitis in these mice is prevented by antibiotic treatment, suggesting that inappropriate bacterial handling due to TNF deficiency can promote colitis and colitis-associated cancer (CAC) [80]. By contrast, hematopoietic cell-specific deficiency of TNF dramatically attenuates spontaneous adenoma formation in the non-colitis-based Apc$^{\Delta 468}$ model of CRC (in which transgenic mice express a mutant form of *Apc*, which encodes adenomatous polyposis coli (APC) protein) [81]. Similarly, TNF is required for the tumor-promoting effect of obesity in both the dextran sodium sulfate (DSS)/azoxymethane (AOM) model of CAC and in human CRC xenografts [82].

IL-6, produced by diverse cell types, is a crucial mediator of inflammation and immunity. The IL-6 receptor transduces signals via gp130 (also known as IL-6Rβ), which is the shared receptor chain of the IL-6 family and a strong activator of STAT3. IL-6 plays several important roles in cancer progression by driving cell proliferation, migration, and angiogenesis [83]. Mouse and human studies have emphasized its role in both CAC and sporadic CRC. IL-6-dependent STAT3 signaling is a critical promoter of cancer-cell proliferation and survival in the DSS/AOM model [84]. Intriguingly, IL-6 production in this experimental system is dependent on sphingosine-1-phosphate (S1P) signaling through the S1P receptor, and tumorigenesis is strongly attenuated when mice are treated with a prodrug, that is, an inhibitor of the S1P pathway (FTY720; also known as fingolimod) [85].

Intestinal cytokine networks are critical mediators of tissue homeostasis, inflammation, and tumorigenesis. In both CAC and sporadic CRC, established cytokines, such as IL-6 and TNF, are being joined by a growing catalog of novel players with similar biochemical functions, such as IL-17A and IL-22. Collectively, these cytokines promote several key hallmarks of cancer, including resistance to apoptosis, aberrant growth and proliferation, induction of genetic instability, angiogenesis, and invasiveness and metastasis. The multitude of pro-tumorigenic cytokines in CRC implies that combinatorial or broad-spectrum anti-cytokine therapies may be advantageous. Both understanding of how the molecular features of tumors dictate aberrant cytokine responses and exploitation of these links for patient stratification will be essential for the successful application of immunomodulatory therapies for CRC [78].

15.3 GENE THERAPY

Gene therapy offers a novel approach to treating rare and sometimes life-threatening genetic diseases and may require new responsibilities for pharmacy practice. The American Society of Gene and Cell Therapy (ASGCT) defines gene therapy as the introduction or removal of genetic material or modification of gene expression to alter the biological function of an individual's genetic code with the objective of achieving a therapeutic benefit [86]. These approaches include replacing a non-functional gene with a functioning healthy gene, inactivating a disease-causing gene, or introducing a new or modified gene into the body.

The types of gene therapy include (1) gene-replacement therapy, in which a fully functioning gene is introduced to replace a mutated one, (2) addition of genes for complex cancerous and infectious diseases, in which a new gene is introduced into the body to help fight a certain disease, (3) gene inhibition or "knockdown" to inactivate a mutated gene that is overproducing its product by targeting RNA, and (4) gene editing, targeted changes to a gene sequence [47]. All involve the introduction of genetic components (typically nucleic acids: DNA or RNA) for treating various diseases, including cancer. Gene therapy can also be utilized to induce an immune response.

The progression of CRC is mediated by genetic mutations and aberration. Modification or correction of defective genes and prevention of gene overexpression can prevent the development of CRC. Alterations of multiple genes are implicated in colon carcinogenesis. Point mutations, formation of oncogenes, de-regulation or deletion of proto-oncogenes, and lack of function of suppressor-oncogenes may all lead to cancer. Until November 2017, around 2,600 clinical trials had been conducted in 38 countries, with over half undergoing phase I trials [87]. Among 1,309 trials of gene therapy worldwide, only 45 reached phase III. A total of 11 gene therapies for CRC are undergoing trials in the United Kingdom [88]. Humans have 50,000 to 100,000 genes, of which only a few are related to the cell cycle. It has been reported that at least 30% of CRCs are due to defective genes. Few genes, then, are associated with familial CRC. Gene therapy involves transfer of genes to tumor cells to suppress the abnormal function of mutated gene(s), inhibiting tumor progression [89].

15.3.1 Methods of Transferring Genes into Mammalian Cells

In general, there are two ways of transferring genes into mammalian cells, one is to transfer an exogenous gene into cells by loco regional or systemic administration *in vivo*, and the other is to culture cells from the patient, transfect them with the required gene *in vitro* (*ex vivo*), and re-introduce them into the patient. Transfer of genes into a cell requires a vector, which may be viral or non-viral. A suitable vector is important for efficient gene transfer.

15.3.1.1 Viral vectors

Viral vectors can be used to introduce DNA into eukaryotic cells and are typically more efficient than non-viral transfection systems. A variety of viral vectors have been employed to transfer genes into cells to induce transient (e.g., adenovirus and vaccinia virus) or permanent (e.g., retrovirus and adeno-associated virus) transgene expression, with each approach having advantages and disadvantages [90]. The serious complications of viral-mediated gene therapy include virus dissemination that causes toxic shock and leukemia [91], an immune reaction to the vector that causes failure of gene therapy, and inappropriate insertion of vectors and transgenes that may cause mutations that lead to the development of cancer.

15.3.1.2 Retrovirus

The first human gene therapy trial using a retrovirus was performed by Rosenberg in 1989 [92, 93]. He introduced genes encoding resistance to neomycin into human lymphocytes and infused them into five patients with advanced melanoma. Only one patient achieved remission. The major drawbacks of retroviruses are the lack of tissue specificity, low titers, and the need for cell replication for infectivity. A recent animal study showed a significant reduction in the number of tumor foci in the liver when the gene encoding cytosine deaminase (CD) was transferred using a retrovirus and the prodrug 5-fluorocytosine (5-FC) was infused into the portal circulation [94].

15.3.1.3 Adenovirus

Adenovirus is responsible for the common cold. Replication-deficient recombinant adenoviral vectors are predominantly used for gene therapy of CRC because they can be produced at a high titer and they readily infect a number of cell types [95]. They also have tropism for the liver and can infect both dividing and quiescent cells [96]. Therefore, these vectors are also useful for gene therapy for patients with liver metastases.

15.3.1.4 Other viruses

Herpes simplex virus, which causes cold sores, can infect a wide range of dividing cells [97]. Its advantage is that unwanted side effects of herpes simplex infection can be treated with antiviral agents, such as acyclovir. Other viruses with the potential as vector include vaccinia virus, lentivirus, and Japanese hemagglutinin virus.

15.3.1.5 Non-viral vectors

There are at least five non-viral methods that introduce DNA into mammalian cells: (1) calcium phosphate transfection [98], (2) Diethylaminoethyl (DEAE)-dextran transfection, (3) electroporation [99], (4) liposome-mediated transfection [100], and (5) plasmid-mediated gene transfer [101]. Of them, the first two procedures involve the creation of a chemical environment that promotes the attachment of DNA to the cell surface, followed by its endocytosis by unknown pathways.

Electroporation involves the generation of an electric field to open up pores in the cell membrane, through which DNA diffuses into the cell. Therefore, this technique is suitable for any cell type. Electrogene therapy is an in vivo application of electroporation in which naked plasmid DNA is injected, and electric pulses are delivered directly to the tissue [99]. However, it is less efficient *in vivo* than virally mediated gene transfer. Electrically-assisted gene transfer is clinically advantageous because of its low immunogenicity, the ease of preparing large quantities of plasmid DNA, reproducibility, and the availability of electro-pulsators for human use [99].

In liposome-mediated transfection, liposomes containing cationic and neutral lipids mediate the transfection of DNA. The advantages of this method include ease to prepare, ability to inject large lipid/DNA complexes, and low immunogenicity [100]. The mechanism(s) by which genes are transferred using this method is poorly understood.

Plasmids can also be used to transfer DNA. Plasmids are self-replicating extrachromosomal DNA molecules. *E. coli* carries plasmids that encode resistance to antibiotics, heavy metals, and bacteriophages. The replication of these plasmids may or may not require plasmid-coded proteins and may or may not be synchronized with the cell cycle. Some plasmids can be freely transferred from one bacterium to another. Artificial plasmids were first constructed in 1970 using fragments of DNA and naturally occurring plasmids. All plasmids have a replicator, a selectable marker, and a cloning site. A replicator is a stretch of DNA that contains the site at which DNA replication begins. A selectable marker is a gene that encodes resistance to an antibiotic. The cloning site is a restriction endonuclease cleavage site into which foreign DNA can be introduced without interfering with the ability of plasmids to replicate. Uptake of plasmid DNA can be enhanced by using a hand-held Swiss jet injector, which uses pressurized air to force small volumes (3–10 μL) of naked DNA into tissues [102]. Plasmids are introduced into *E. coli* by transformation and constitute an important tool in recombinant DNA technology. The addition of new genes to a recipient cell introduces a heritable modification in phenotype. Plasmids targeting the liver can be injected into veins [103].

15.3.2 Basis of Gene Therapy and Its Application in CRC

The gene therapy approaches in current use can be classified into five major categories: (1) enzyme/prodrug systems (suicide gene therapy), (2) gene correction (tumor suppressor gene replacement therapy, or oncogene inactivation), (3) immune-gene therapy, (4) drug-resistance gene therapy, and (5) miscellaneous.

15.3.2.1 Virus-directed enzyme-prodrug therapy (VDEPT)

VDEPT is the transfer of genes encoding bacterial, viral, or fungal enzymes into tumor cells, which convert an inactive prodrug into short-lived toxic metabolites, limiting the toxic effect to tumor cells [104, 105]. This type of therapy is also known as suicide-gene therapy. Examples include: (1) thymidine kinase and ganciclovir [106], (2) CD and 5-FC [47], (3) thymidine phosphorylase and 5-fluorouracil (5-FU) [107], and (4) nitroreductase and the prodrug CB1954 [108]. One of the major advantages of this type of therapy is the so-called "bystander effect" [109]. This is a phenomenon by which small molecules, such as an active drug metabolite, are able to pass between cells via gap junctions so that untransfected cells are also affected. Thymidine kinase of herpes simplex phosphorylates ganciclovir, inhibiting DNA polymerase and leading to cell death. In an animal study, the thymidine kinase/ganciclovir combination showed profound bystander effects on CRC [106]. Surprisingly, herpes simplex virus thymidine kinase gene therapy without ganciclovir increased tumor growth and Cox-2 expression [110]. In vivo, monocyte chemoattractant protein gene therapy plus thymidine kinase/ganciclovir exhibited an enhanced anti-tumor effect [111].

In double enzyme/prodrug therapy, genes encoding two enzymes, such as CD and uracil phosphoribosyl transferase, are used in combination with 5-FC. This form of suicide gene therapy is more effective than CD/5-FC regimens (a single enzyme) [112]. In a novel technique, cytokines are combined with enzyme/prodrug therapy. For example, granulocyte-macrophage colony-stimulating factor (GM-CSF) gene transfer plus thymidine kinase/ganciclovir resulted in complete inhibition of liver metastasis [113]. Some gene therapies can be administered intraperitoneally (i.p.); indeed, i.p. administration of a yeast CD gene increased the level of the enzyme in liver metastases [114].

15.3.2.2 Gene correction

Cancer may result from an imbalance between proto-oncogenes and tumor suppressor genes. Gene correction therapy is aimed at reversing some genetic abnormalities by introducing a tumor suppressor gene or inactivating a proto-oncogene by an anti-sense method [109]. The tumor suppressor gene *TP53* is the most commonly mutated gene in human cancers [115] and is present in 20%–69% of CRC cases. Its product, p53, regulates the cell cycle and repairs abnormal DNA. If DNA cannot be repaired by p53, growth arrest or apoptosis occurs [116]. Loss of p53 leads to uncontrolled and aggressive cellular growth. In a phase I controlled trial, an adenovirus encoding wild-type *TP53* gene was delivered by hepatic artery infusion to 16 patients with *TP53*-mutated liver metastases of CRC [117]. The side effects were fever and transient derangement of liver function. Although the gene was expressed in subsequently resected tumors, there was no significant change in their radiological appearance. This suggests that a low level of p53 results in cell-cycle arrest, whereas a high level triggers apoptosis [118].

Kras, which encodes p21, a protein implicated in signal induction and control of proliferation, is frequently mutated in CRC [119]. The upregulation of p21 by p53 results in cellular arrest, but if p53 does not upregulate p21, apoptosis is triggered [120]. Reduced p21 expression is correlated with metastatic colon cancer. *Kras* mutations can be detected in DNA purified from stool [121]. In vitro, transfection of *Kras*-overexpressing CRC cells with apoptotic genes (bax, caspase-8, and PKG) promoted apoptosis [119]. The function of mutated genes can be inhibited by anti-sense oligonucleotides, which prevent the translocation of mRNAs. Anti-sense Bcl-2 was reported to potentiate apoptosis of lymphoma cells and anti-sense *Kras* suppressed the growth of CRC cells [88].

15.3.2.3 Immune-gene therapy

Immunological mechanisms are important for eliminating cancer. Individuals with immune deficiency, such as those infected with human immunodeficiency virus (HIV), are at high risk for cancer. Cancer cells are recognized and destroyed by $CD8^+$ cytotoxic T cells and natural killer (NK) cells. However, many cancer cells escape such immune-mediated destruction by not expressing HLA class I antigens. A phase I trial of immunotherapy for colorectal metastases using intralesional injection of HLA-B7 cDNA in liposomes to 15 HLA-B7-negative patients resulted in production of HLA-B7 protein in around 50% of the patients but showed no therapeutic benefit [122]. This suggests the complexity of the molecular mechanisms of cancer. CEA is typically expressed by CRC. Immunization with DCs transfected with RNA encoding tumor antigens stimulates tumor antigen-specific immune responses in vitro and in animal models. In a phase II study [123] involving patients with resected hepatic metastases of CRC, the safety and feasibility of administering DCs transfected with CEA mRNA were assessed. There were no major side effects, and 9 out of 13 patients showed relapse for a median of 122 days. An immunologic response was detected in biopsies from the sites of DC injection and in peripheral blood samples.

Activation of cytotoxic lymphocytes and NK cells by directly transferring cytokines (such as IL-2 and IL-12) into tumor cells could result in an anti-tumor effect [124]. The systemic administration of ILs is limited by their side effects. In vivo, intratumoral IL-2 gene therapy induced tumor regression in 76% of mice harboring cancer [125]. In a phase I/II clinical trial involving patients with unresectable CRC who received intratumoral injection of adenovirus-IL-2 at the time of surgery, one patient's tumor showed increased expression of membrane-bound IL-2 receptor and another patient showed necrosis of the tumor mass. In a study involving mice with experimentally induced colorectal tumors, adenovirus containing the mouse IL-12 gene was injected into the tumor, resulting in tumor regression and extended survival time [126]. CD40 is a costimulatory protein expressed by APCs that binds to its ligand CD40L on T cells to activate them. In vivo gene transfer of CD40L into CRC cells caused tumor regression mediated by

$CD8^+$ T lymphocytes. A mouse vaccinated with IL-12 transfected CT26 colon cancer cells develops long-lasting memory so that it can reject all tumors expressing the same epitope [127].

15.3.2.4 Drug resistance gene therapy

The main factor limiting chemotherapy is bone-marrow toxicity. Multiple drug resistance gene (usually *ABCB1*) may confer bone marrow resistance to vinca alkaloids, anthracyclines, and paclitaxel [128], but is still at the experimental stage.

15.3.2.5 Chemo-gene therapy

The susceptibility of cancer cells to chemotherapy is enhanced when concomitant use of gene therapy. The concept "chemo-gene therapy" involves the administration of a chemotherapeutic agent, such as 5-FU in combination with a gene, such as Flt3L, exerting a synergistic effect on cancer [129]. Flt3L is a hematopoietic factor that stimulates immune cells to kill cancer cells. Another promising such combination is wild-type p53 and 5-FU/cisplatin [130]. The synergistic effect of such combinations of gene therapy and chemotherapy is capable of eradicating tumors. The overexpression of interferon (IFN) alpha in CRC cells greatly enhances their susceptibility to 5-FU [131]. Some cancer cells are resistant to chemotherapeutic agents, which can be overcome by the combined use of gene therapy and chemotherapy. In vivo, suicide gene therapy using thymidine kinase/ganciclovir plus thymidylate synthase inhibitors increases the survival of mice with CRC xenografts [132]. Administration of the TRAIL gene and actinomycin D suppressed metastatic liver tumors [133]. Inhibition of IGF receptor greatly enhances the apoptosis of cancer cells upon exposure to chemotherapy [134]. Adenovirus-mediated transfer of caspase-8, an enzyme involved in apoptosis, increases the sensitivity of CRC cells to 5-FU [135]. Finally, excessive production of Bax protein reportedly potentiates the effect of chemotherapeutic agents [136].

15.3.3 Obstacles That Gene Therapy Would Face

Several hurdles must be overcome for more widespread use of gene therapy. Cancer cells can develop further mutations after correction of an existing genetic defect [137]. Viral amount uptaked by tumor cells may not be sufficient to cause the desired effect. Prior radiotherapy has been shown to increase viral uptake by various cancer cells [138]. Also, replication-competent virus reportedly improves gene transfer. Systemic effects of viruses and the short-lived effect of gene therapy may be problematic [139]. Systemic dissemination of virus following intratumoral administration may be reduced by the inclusion of alginate, which also prolongs the effect of gene therapy. Fibroblasts transfected with the IL-12 gene enclosed in alginate microspheres showed sustained release of IL-12, which suppressed adjacent CRC [140]. Occasionally, death can result from intra-arterial infusion of viral vectors [141]. Injection of viral vectors into the hepatic artery can cause fever, rigor, and fatigue. Viral dissemination can be prevented by using vectors that respond to antiviral therapy, such as herpes simplex. Direct injection of viral-mediated gene therapy may result in a poor distribution of the gene within the tumor. Leukemia and death due to viral dissemination must be overcome before gene therapy can be applied on a routine basis [88].

15.3.4 New Molecular Targets for Gene Therapy

Genes related to apoptosis show promise for gene therapy. Bcl-XL, a protein involved in apoptosis, is overexpressed by CRC cells. In vivo, adenovirus-mediated RNA interference therapy targeting Bcl-XL has shown efficacy against CRC [142]. BiK belongs to the proapoptotic Bcl-2 family and mutants in the BiK gene prolong the survival of mice with peritoneal carcinomatosis [143]. Inhibition of signal-regulating kinase-1, a regulator of apoptosis, causes tumor regression [144], as does Bax gene therapy [145].

Growth factors are important targets for gene therapy. NK4 is an antagonist of HGF and may be effective against disseminated peritoneal cancer [146]. Suppression of EGFR induces apoptosis, and an in vitro study using LoVo cells showed suppression of EGPR to have the potential as a target for gene therapy [147]. Overexpression of inhibitory binding protein, such as insulin-like growth factor binding protein-4, promotes the apoptosis of CRC cells [148]. Cellular adhesion molecules are also targets of gene therapy. For example, disruption of CD-44 promoted the regression of cancer [149].

The heat-shock protein (Hsp) is a potential target for the treatment of CRC. For instance, Hsp90beta is phosphorylated upon exposure to 5-FU, but its clinical utility requires further evaluation [150]. Tissue plasminogen activator (tPA) influences liver metastases, and gene therapy based on tPA improved the survival of mice with liver metastases [151]. Chemokines, such as fractalkine, can suppress the growth of cancer cells [152]. Mutations in the gene encoding gap junction connexin 43 are frequently found in CRC, which warrants further investigation [153]. Cyclin-dependent kinase 4 (CDK4) is involved in regulation of the cell cycle, and anti-sense inhibition of CDK4 in cell culture caused increased apoptosis and decreased proliferation [154].

TNF-related apoptosis-inducing ligand (TRIAL) is a member of the TNF superfamily that induces apoptosis of CRC cells, but not normal cells [155]. Inhibition of PI3K/Akt by RNA interference sensitizes resistant colon cancer cells to TRAIL-induced cell death by inducing the expression of TRAIL receptors [156]. Cancer cells can reportedly acquire resistance to TRAIL gene therapy [157].

Short interfering RNA (siRNA) targeting VEGF has been reported to reduce the proliferation of CRC cells [158]. In vivo, angiopoietin-1 overexpression in CRC cells suppresses tumor growth [159]. Endostatin inhibits endothelial proliferation, and gene therapy using endostatin suppresses angiogenesis and exerts a direct inhibitory effect on some cancer cells [160].

Gene therapy is useful for cancers with altered gene expression, the number of which is increasing as knowledge advances. The mechanisms by which gene therapy prevents and eliminates CRC are largely unclear. Chemo-gene therapy will in future play an important role in the treatment of advanced cancer. Several gene therapy trials are underway worldwide; based on the results of these trials, gene therapy may be used either alone or as an adjunct to other treatment modalities.

15.4 SUMMARY AND FUTURE PERSPECTIVES

Despite surgery, radiotherapy, chemotherapy, targeted therapy, and other treatment modalities, CRC remains an important cause of cancer-related death in China and the rest of the world. How to eliminate cancer while minimizing side effects is challenging. 5-FU-based chemotherapy, which has been used for several decades, is the gold standard of treatment for mCRC. Several signaling pathways play an essential role in the development and progression of CRC, and thus targeting these signaling cascades has therapeutic potential. The identification of genes and biomarkers has improved the efficacy of conventional therapies by specifically targeting tumor cells. Gene therapy and immunotherapies (such as cytokines, cancer vaccines, ACT, and mAbs) were recently introduced for the treatment of CRC and are more effective than conventional modalities.

ACKNOWLEDGMENTS

This work was supported by the National Natural Science Foundation of China (No. 82172558) and Distinguished Young Scholars of Nanjing (JQX20008).

REFERENCES

1. Padma VV, An overview of targeted cancer therapy, BioMedicine 5(4) (2015) 19.
2. Siegel R, Desantis C, Jemal A, Colorectal cancer statistics, 2014, CA: A Cancer Journal for Clinicians 64(2) (2014) 104–17.
3. Chibaudel B, Tournigand C, Bonnetain F, Richa H, Benetkiewicz M, André T, de Gramont A, Therapeutic strategy in unresectable metastatic colorectal cancer: an updated review, Therapeutic Advances in Medical Oncology 7(3) (2015) 153–69.
4. Poston GJ, Figueras J, Giuliante F, Nuzzo G, Sobrero AF, Gigot JF, Nordlinger B, Adam R, Gruenberger T, Choti MA, Bilchik AJ, Van Cutsem EJ, Chiang JM, D'Angelica MI, Urgent need for a new staging system in advanced colorectal cancer, Journal of Clinical Oncology: Official Journal of the American Society of Clinical Oncology 26(29) (2008) 4828–33.
5. Eisterer W, Prager G, Chemotherapy, still an option in the twenty-first century in metastatic colorectal cancer?, Cardiovascular and Interventional Radiology 42(9) (2019) 1213–20.
6. Chibaudel B, Tournigand C, André T, de Gramont A, Therapeutic strategy in unresectable metastatic colorectal cancer, Therapeutic Advances in Medical Oncology 4(2) (2012) 75–89.
7. Bang HJ, Littrup PJ, Currier BP, Goodrich DJ, Choi M, Heilbrun LK, Goodman AC, Percutaneous cryoablation of metastatic lesions from colorectal cancer: efficacy and feasibility with survival and cost-effectiveness observations, ISRN Minimally Invasive Surgery 2012 (2012) 942364.
8. Nordlinger B, Adam R, Arnold D, Zalcberg JR, Gruenberger T, The role of biological agents in the resection of colorectal liver metastases, Clinical Oncology 24(6) (2012) 432–42.
9. Gharwan H, Groninger H, Kinase inhibitors and monoclonal antibodies in oncology: clinical implications, Nature Reviews Clinical Oncology 13(4) (2016) 209–27.
10. Geng F, Wang Z, Yin H, Yu J, Cao B, Molecular targeted drugs and treatment of colorectal cancer: recent progress and future perspectives, Cancer Biotherapy & Radiopharmaceuticals 32(5) (2017) 149–60.
11. Nappi A, Berretta M, Romano C, Tafuto S, Cassata A, Casaretti R, Silvestro L, Divitiis C, Alessandrini L, Fiorica F, Ottaiano A, Nasti G, Metastatic colorectal cancer: role of target therapies and future perspectives, Current Cancer Drug Targets 18(5) (2018) 421–9.
12. Cunningham D, Humblet Y, Siena S, Khayat D, Bleiberg H, Santoro A, Bets D, Mueser M, Harstrick A, Verslype C, Chau I, Van Cutsem E, Cetuximab monotherapy and cetuximab plus irinotecan in irinotecan-refractory metastatic colorectal cancer, New England Journal of Medicine 351(4) (2004) 337–45.
13. Brodowicz T, Ciuleanu TE, Radosavljevic D, Shacham-Shmueli E, Vrbanec D, Plate S, Mrsic-Krmpotic Z, Dank M, Purkalne G, Messinger D, Zielinski CC, FOLFOX4 plus cetuximab administered weekly or every second week in the first-line treatment of patients with KRAS wild-type metastatic colorectal cancer: a randomized phase II CECOG study, Annals of Oncology: Official Journal of the European Society for Medical Oncology 24(7) (2013) 1769–77.
14. Lee JJ, Sun W, Options for second-line treatment in metastatic colorectal cancer, Clinical Advances in Hematology & Oncology 14(1) (2016) 46–54.
15. Prenen H, Vecchione L, Van Cutsem E, Role of targeted agents in metastatic colorectal cancer, Targeted Oncology 8(2) (2013) 83–96.
16. Chou T, Finn RS, Brivanib: a review of development, Future Oncology 8(9) (2012) 1083–90.
17. Kerbel R, Folkman J, Clinical translation of angiogenesis inhibitors, Nature Reviews Cancer 2(10) (2002) 727–39.
18. Folkman J, Tumor angiogenesis: therapeutic implications, New England Journal of Medicine 285(21) (1971) 1182–6.
19. Cartwright TH, Treatment decisions after diagnosis of metastatic colorectal cancer, Clinical Colorectal Cancer 11(3) (2012) 155–66.
20. Bergsland EK, Vascular endothelial growth factor as a therapeutic target in cancer, American Journal of Health-System Pharmacy: Official Journal of the American Society of Health-System Pharmacists 61 (2004) S4–11.
21. Brenner H, Kloor M, Pox CP, Colorectal cancer, Lancet 383(9927) (2014) 1490–502.
22. Lee JC, Chow NH, Wang ST, Huang SM, Prognostic value of vascular endothelial growth factor expression in colorectal cancer patients, European Journal of Cancer 36(6) (2000) 748–53.
23. Custodio A, Barriuso J, de Castro J, Martínez-Marín V, Moreno V, Rodríguez-Salas N, Feliu J, Molecular markers to predict outcome to antiangiogenic therapies in colorectal cancer: current evidence and future perspectives, Cancer Treatment Reviews 39(8) (2013) 908–24.

24. Ellis LM, Hicklin DJ, VEGF-targeted therapy: mechanisms of anti-tumour activity, Nature Reviews. Cancer 8(8) (2008) 579–91.
25. Gordon MS, Margolin K, Talpaz M, Sledge GW, Holmgren E, Benjamin R, Stalter S, Shak S, Adelman D, Phase I safety and pharmacokinetic study of recombinant human anti-vascular endothelial growth factor in patients with advanced cancer, Journal of Clinical Oncology: Official Journal of the American Society of Clinical Oncology 19(3) (2001) 843–50.
26. Bendell JC, Bekaii-Saab TS, Cohn AL, Hurwitz HI, Kozloff M, Tezcan H, Roach N, Mun Y, Fish S, Flick ED, Dalal D, Grothey A, Treatment patterns and clinical outcomes in patients with metastatic colorectal cancer initially treated with FOLFOX-bevacizumab or FOLFIRI-bevacizumab: results from ARIES, a bevacizumab observational cohort study, The Oncologist 17(12) (2012) 1486–95.
27. Shankaran V, Mummy D, Koepl L, Blough D, Yim YM, Yu E, Ramsey S, Adverse events associated with bevacizumab and chemotherapy in older patients with metastatic colorectal cancer, Clinical Colorectal Cancer 12(3) (2013) 204–13.e1.
28. Tejpar S, Prenen H, Mazzone M, Overcoming resistance to antiangiogenic therapies, The Oncologist 17(8) (2012) 1039–50.
29. Chu QS, Aflibercept (AVE0005): an alternative strategy for inhibiting tumour angiogenesis by vascular endothelial growth factors, Expert Opinion on Biological Therapy 9(2) (2009) 263–71.
30. Van Cutsem E, Tabernero J, Lakomy R, Prenen H, Prausová J, Macarulla T, Ruff P, van Hazel GA, Moiseyenko V, Ferry D, McKendrick J, Polikoff J, Tellier A, Castan R, Allegra C, Addition of aflibercept to fluorouracil, leucovorin, and irinotecan improves survival in a phase III randomized trial in patients with metastatic colorectal cancer previously treated with an oxaliplatin-based regimen, Journal of Clinical Oncology: Official Journal of the American Society of Clinical Oncology 30(28) (2012) 3499–506.
31. Bennouna J, Sastre J, Arnold D, Österlund P, Greil R, Van Cutsem E, von Moos R, Viéitez JM, Bouché O, Borg C, Steffens CC, Alonso-Orduña V, Schlichting C, Reyes-Rivera I, Bendahmane B, André T, Kubicka S, Continuation of bevacizumab after first progression in metastatic colorectal cancer (ML18147): a randomised phase 3 trial, The Lancet. Oncology 14(1) (2013) 29–37.
32. Strumberg D, Schultheis B, Regorafenib for cancer, Expert Opinion on Investigational Drugs 21(6) (2012) 879–89.
33. Grothey A, Van Cutsem E, Sobrero A, Siena S, Falcone A, Ychou M, Humblet Y, Bouché O, Mineur L, Barone C, Adenis A, Tabernero J, Yoshino T, Lenz HJ, Goldberg RM, Sargent DJ, Cihon F, Cupit L, Wagner A, Laurent D, Regorafenib monotherapy for previously treated metastatic colorectal cancer (CORRECT): an international, multicentre, randomised, placebo-controlled, phase 3 trial, Lancet 381(9863) (2013) 303–12.
34. Jain RK, Duda DG, Willett CG, Sahani DV, Zhu AX, Loeffler JS, Batchelor TT, Sorensen AG, Biomarkers of response and resistance to antiangiogenic therapy, Nature Reviews. Clinical Oncology 6(6) (2009) 327–38.
35. Kopetz S, Hoff PM, Morris JS, Wolff RA, Eng C, Glover KY, Adinin R, Overman MJ, Valero V, Wen S, Lieu C, Yan S, Tran HT, Ellis LM, Abbruzzese JL, Heymach JV, Phase II trial of infusional fluorouracil, irinotecan, and bevacizumab for metastatic colorectal cancer: efficacy and circulating angiogenic biomarkers associated with therapeutic resistance, Journal of Clinical Oncology 28(3) (2010) 453–9.
36. Lambrechts D, Claes B, Delmar P, Reumers J, Mazzone M, Yesilyurt BT, Devlieger R, Verslype C, Tejpar S, Wildiers H, de Haas S, Carmeliet P, Scherer SJ, Van Cutsem E, VEGF pathway genetic variants as biomarkers of treatment outcome with bevacizumab: an analysis of data from the AViTA and AVOREN randomised trials, The Lancet. Oncology 13(7) (2012) 724–33.
37. Schneider BP, Shen F, Miller KD, Pharmacogenetic biomarkers for the prediction of response to antiangiogenic treatment, The Lancet. Oncology 13(10) (2012) e427–36.
38. Herbst RS, Shin DM, Monoclonal antibodies to target epidermal growth factor receptor-positive tumors: a new paradigm for cancer therapy, Cancer 94(5) (2002) 1593–611.
39. Yarden Y, The EGFR family and its ligands in human cancer. Signalling mechanisms and therapeutic opportunities, European Journal of Cancer 37(4) (2001) S3–8.
40. Yarden Y, Sliwkowski MX, Untangling the ErbB signalling network, Nature Reviews. Molecular Cell Biology 2(2) (2001) 127–37.
41. Ciardiello F, Tortora G, EGFR antagonists in cancer treatment, New England Journal of Medicine 358(11) (2008) 1160–74.
42. Breuleux M, Schoumacher F, Rehn D, Küng W, Mueller H, Eppenberger U, Heregulins implicated in cellular functions other than receptor activation, Molecular Cancer Research: MCR 4(1) (2006) 27–37.
43. Fiske WH, Threadgill D, Coffey RJ, ERBBs in the gastrointestinal tract: recent progress and new perspectives, Experimental Cell Research 315(4) (2009) 583–601.

44. Erwin CR, Helmrath MA, Shin CE, Falcone RA, Stern LE, Warner BW, Intestinal overexpression of EGF in transgenic mice enhances adaptation after small bowel resection, The American Journal of Physiology 277(3) (1999) G533–40.
45. Sato T, Vries RG, Snippert HJ, van de Wetering M, Barker N, Stange DE, van Es JH, Abo A, Kujala P, Peters PJ, Clevers H, Single Lgr5 stem cells build crypt-villus structures in vitro without a mesenchymal niche, Nature 459(7244) (2009) 262–5.
46. Sato T, van Es JH, Snippert HJ, Stange DE, Vries RG, van den Born M, Barker N, Shroyer NF, van de Wetering M, Clevers H, Paneth cells constitute the niche for Lgr5 stem cells in intestinal crypts, Nature 469(7330) (2011) 415–8.
47. Tran KA, Cheng MY, Mitra A, Ogawa H, Shi VY, Olney LP, Kloxin AM, Maverakis E, MEK inhibitors and their potential in the treatment of advanced melanoma: the advantages of combination therapy, Drug Design, Development and Theory 10 (2016) 43–52.
48. Kimura H, Sakai K, Arao T, Shimoyama T, Tamura T, Nishio K, Antibody-dependent cellular cytotoxicity of cetuximab against tumor cells with wild-type or mutant epidermal growth factor receptor, Cancer Science 98(8) (2007) 1275–80.
49. Kawaguchi Y, Kono K, Mimura K, Sugai H, Akaike H, Fujii H, Cetuximab induce antibody-dependent cellular cytotoxicity against EGFR-expressing esophageal squamous cell carcinoma, International Journal of Cancer 120(4) (2007) 781–7.
50. Jonker DJ, O'Callaghan CJ, Karapetis CS, Zalcberg JR, Tu D, Au HJ, Berry SR, Krahn M, Price T, Simes RJ, Tebbutt NC, van Hazel G, Wierzbicki R, Langer C, Moore MJ, Cetuximab for the treatment of colorectal cancer, The New England Journal of Medicine 357(20) (2007) 2040–8.
51. Yang XD, Jia XC, Corvalan JR, Wang P, Davis CG, Development of ABX-EGF, a fully human anti-EGF receptor monoclonal antibody, for cancer therapy, Critical Reviews in Oncology/Hematology 38(1) (2001) 17–23.
52. Foon KA, Yang XD, Weiner LM, Belldegrun AS, Figlin RA, Crawford J, Rowinsky EK, Dutcher JP, Vogelzang NJ, Gollub J, Thompson JA, Schwartz G, Bukowski RM, Roskos LK, Schwab GM, Preclinical and clinical evaluations of ABX-EGF, a fully human anti-epidermal growth factor receptor antibody, International Journal of Radiation Oncology, Biology, Physics 58(3) (2004) 984–90.
53. Tsuchihashi Z, Khambata-Ford S, Responsiveness to cetuximab without mutations in EGFR, New England Journal of Medicine 353(2) (2005) 208–9.
54. Samowitz WS, Curtin K, Schaffer D, Robertson M, Leppert M, Slattery ML, Relationship of Ki-ras mutations in colon cancers to tumor location, stage, and survival: a population-based study, Cancer Epidemiology, Biomarkers & Prevention: A publication of the American Association for Cancer Research, Cosponsored by theAmerican Society of Preventive Oncology 9(11) (2000) 1193–7.
55. De Roock W, Claes B, Bernasconi D, De Schutter J, Biesmans B, Fountzilas G, Kalogeras KT, Kotoula V, Papamichael D, Laurent-Puig P, Penault-Llorca F, Rougier P, Vincenzi B, Santini D, Tonini G, Cappuzzo F, Frattini M, Molinari F, Saletti P, De Dosso S, Martini M, Bardelli A, Siena S, Sartore-Bianchi A, Tabernero J, Macarulla T, Di Fiore F, Gangloff AO, Ciardiello F, Pfeiffer P, Qvortrup C, Hansen TP, Van Cutsem E, Piessevaux H, Lambrechts D, Delorenzi M, Tejpar S, Effects of KRAS, BRAF, NRAS, and PIK3CA mutations on the efficacy of cetuximab plus chemotherapy in chemotherapy-refractory metastatic colorectal cancer: a retrospective consortium analysis, The Lancet. Oncology 11(8) (2010) 753–62.
56. Saridaki Z, Tzardi M, Papadaki C, Sfakianaki M, Pega F, Kalikaki A, Tsakalaki E, Trypaki M, Messaritakis I, Stathopoulos E, Mavroudis D, Georgoulias V, Souglakos J, Impact of KRAS, BRAF, PIK3CA mutations, PTEN, AREG, EREG expression and skin rash in ≥ 2 line cetuximab-based therapy of colorectal cancer patients, PloS One 6(1) (2011) e15980.
57. Vatandoost N, Ghanbari J, Mojaver M, Avan A, Ghayour-Mobarhan M, Nedaeinia R, Salehi R, Early detection of colorectal cancer: from conventional methods to novel biomarkers, Journal of Cancer Research and Clinical Oncology 142(2) (2016) 341–51.
58. Kumar B, Singh S, Skvortsova I, Kumar V, Promising targets in anti-cancer drug development: recent updates, Current Medicinal Chemistry 24(42) (2017) 4729–52.
59. Peeters M, Price T, Biologic therapies in the metastatic colorectal cancer treatment continuum–applying current evidence to clinical practice, Cancer Treatment Reviews 38(5) (2012) 397–406.
60. Cidon EU, Alonso P, Masters B, Markers of response to antiangiogenic therapies in colorectal cancer: where are we now and what should be next?, Clinical Medicine Insights. Oncology 10 (2016) 41–55.
61. Schreuer M, Jansen Y, Planken S, Chevolet I, Seremet T, Kruse V, Neyns B, Combination of dabrafenib plus trametinib for BRAF and MEK inhibitor pretreated patients with advanced BRAF V600 -mutant melanoma: an open-label, single arm, dual-centre, phase 2 clinical trial, The Lancet Oncology 18(4) (2017) 464–72.

62. Rodon J, Braña I, Siu LL, De Jonge MJ, Homji N, Mills D, Di Tomaso E, Sarr C, Trandafir L, Massacesi C, Eskens F, Bendell JC, Phase I dose-escalation and -expansion study of buparlisib (BKM120), an oral pan-Class I PI3K inhibitor, in patients with advanced solid tumors, Investigational New Drugs 32(4) (2014) 670–81.
63. Jiang Y, Zhang K, Gao S, Wang G, Huang J, Wang J, Chen L, Discovery of potent c-MET inhibitors with new scaffold having different quinazoline, pyridine and tetrahydro-pyridothienopyrimidine headgroups, Molecules 21(5) (2016) 612.
64. Parikh RA, Wang P, Beumer JH, Chu E, Appleman LJ, The potential roles of hepatocyte growth factor (HGF)-MET pathway inhibitors in cancer treatment, OncoTargets and Therapy 7 (2014) 969–83.
65. Rimassa L, Assenat E, Peck-Radosavljevic M, Pracht M, Zagonel V, Mathurin P, Rota Caremoli E, Porta C, Daniele B, Bolondi L, Mazzaferro V, Harris W, Damjanov N, Pastorelli D, Reig M, Knox J, Negri F, Trojan J, López López C, Personeni N, Decaens T, Dupuy M, Sieghart W, Abbadessa G, Schwartz B, Lamar M, Goldberg T, Shuster D, Santoro A, Bruix J, Tivantinib for second-line treatment of MET-high, advanced hepatocellular carcinoma (METIV-HCC): a final analysis of a phase 3, randomised, placebo-controlled study, The Lancet. Oncology 19(5) (2018) 682–93.
66. Trapani JA, Darcy PK, Immunotherapy of cancer, Australian Family Physician 46(4) (2017) 194–9.
67. Wu X, Gu Z, Chen Y, Chen B, Chen W, Weng L, Liu X, Application of PD-1 blockade in cancer immunotherapy, Computational and Structural Biotechnology Journal 17 (2019) 661–74.
68. Soo RA, Shedding light on the molecular determinants of response to anti-PD-1 therapy, Translational Lung Cancer Research 4(6) (2015) 816–9.
69. Le DT, Uram JN, Wang H, Bartlett BR, Kemberling H, Eyring AD, Skora AD, Luber BS, Azad NS, Laheru D, Biedrzycki B, Donehower RC, Zaheer A, Fisher GA, Crocenzi TS, Lee JJ, Duffy SM, Goldberg RM, de la Chapelle A, Koshiji M, Bhaijee F, Huebner T, Hruban RH, Wood LD, Cuka N, Pardoll DM, Papadopoulos N, Kinzler KW, Zhou S, Cornish TC, Taube JM, Anders RA, Eshleman JR, Vogelstein B, Diaz LA, PD-1 blockade in tumors with mismatch-repair deficiency, New England Journal of Medicine 372(26) (2015) 2509–20.
70. Tiwari A, Saraf S, Verma A, Panda PK, Jain SK, Novel targeting approaches and signaling pathways of colorectal cancer: an insight, World Journal of Gastroenterology 24(39) (2018) 4428–35.
71. Fong L, Hou Y, Rivas A, Benike C, Yuen A, Fisher GA, Davis MM, Engleman EG, Altered peptide ligand vaccination with Flt3 ligand expanded dendritic cells for tumor immunotherapy, Proceedings of the National Academy of Sciences of the United States of America 98(15) (2001) 8809–14.
72. Sharma P, Allison JP, The future of immune checkpoint therapy, Science 348(6230) (2015) 56–61.
73. Met Ö, Jensen KM, Chamberlain CA, Donia M, Svane IM, Principles of adoptive T cell therapy in cancer, Seminars in Immunopathology 41(1) (2018) 49–58.
74. Wu D, Wu P, Wu X, Ye J, Wang Z, Zhao S, Ni C, Hu G, Xu J, Han Y, Zhang T, Qiu F, Yan J, Huang J, Ex vivo expanded human circulating Vδ1 γδT cells exhibit favorable therapeutic potential for colon cancer, Oncoimmunology 4(3) (2015) e992749.
75. Markiewski MM, Lambris JD, Is complement good or bad for cancer patients? A new perspective on an old dilemma, Trends in Immunology 30(6) (2009) 286–92.
76. Rutkowski MJ, Sughrue ME, Kane AJ, Mills SA, Parsa AT, Cancer and the complement cascade, Molecular Cancer Research: MCR 8(11) (2010) 1453–65.
77. Downs-Canner S, Magge D, Ravindranathan R, O'Malley ME, Francis L, Liu Z, Sheng Guo Z, Obermajer N, Bartlett DL, Complement inhibition: a novel form of immunotherapy for colon cancer, Annals of Surgical Oncology 23(2) (2016) 655–62.
78. West NR, McCuaig S, Franchini F, Powrie F, Emerging cytokine networks in colorectal cancer, Nature Reviews. Immunology 15(10) (2015) 615–29.
79. Locksley RM, Killeen N, Lenardo MJ, The TNF and TNF receptor review superfamilies: integrating mammalian biology, Cell 104(4) (2001) 487–501.
80. Hale LP, Greer PK, A novel murine model of inflammatory bowel disease and inflammation-associated colon cancer with ulcerative colitis-like features, PloS One 7(7) (2012) e41797.
81. Blatner NR, Mulcahy MF, Dennis KL, Scholtens D, Bentrem DJ, Phillips JD, Ham S, Sandall BP, Khan MW, Mahvi DM, Halverson AL, Stryker SJ, Boller AM, Singal A, Sneed RK, Sarraj B, Ansari MJ, Oft M, Iwakura Y, Zhou L, Bonertz A, Beckhove P, Gounari F, Khazaie K, Expression of RORγt marks a pathogenic regulatory T cell subset in human colon cancer, Science Translational Medicine 4(164) (2012) 164ra159.
82. Flores MBS, Rocha GZ, Damas-Souza DM, Osório-Costa F, Dias MM, Ropelle ER, Camargo JA, de Carvalho RB, Carvalho HF, Saad MJA, Carvalheira JBC, RETRACTED: obesity-induced increase in tumor necrosis factor-α leads to development of colon cancer in mice, Gastroenterology 143(3) (2012) 741–53.e4.
83. Taniguchi K, Karin M, IL-6 and related cytokines as the critical lynchpins between inflammation and cancer, Seminars in Immunology 26(1) (2014) 54–74.

84. Grivennikov S, Karin E, Terzic J, Mucida D, Yu GY, Vallabhapurapu S, Scheller J, Rose-John S, Cheroutre H, Eckmann L, Karin M, IL-6 and Stat3 are required for survival of intestinal epithelial cells and development of colitis-associated cancer, Cancer Cell 15(2) (2009) 103–13.
85. Liang J, Nagahashi M, Kim EY, Harikumar KB, Yamada A, Huang WC, Hait NC, Allegood JC, Price MM, Avni D, Takabe K, Kordula T, Milstien S, Spiegel S, Sphingosine-1-phosphate links persistent STAT3 activation, chronic intestinal inflammation, and development of colitis-associated cancer, Cancer Cell 23(1) (2013) 107–20.
86. Petrich J, Marchese D, Jenkins C, Storey M, Blind J, Gene replacement therapy: a primer for the health-system pharmacist, Journal of Pharmacy Practice 33(6) (2020) 846–55.
87. Ginn SL, Amaya AK, Alexander IE, Edelstein M, Abedi MR, Gene therapy clinical trials worldwide to 2017: an update, The Journal of Gene Medicine 20(5) (2018) e3015.
88. Durai R, Yang SY, Seifalian AM, Winslet MC, Principles and applications of gene therapy in colon cancer, Journal of Gastrointestinal Liver Disease 17(1) (2008) 59–67.
89. Armaghany T, Wilson JD, Chu Q, Mills G, Genetic alterations in colorectal cancer, Gastrointestinal Cancer Research 5(1) (2012) 19–27.
90. Young LS, Searle PF, Onion D, Mautner V, Viral gene therapy strategies: from basic science to clinical application, The Journal of Pathology 208(2) (2006) 299–318.
91. Puck JM, Malech HL, Gene therapy for immune disorders: good news tempered by bad news, The Journal of Allergy and Clinical Immunology 117(4) (2006) 865–9.
92. Aebersold P, Kasid A, Rosenberg SA, Selection of gene-marked tumor infiltrating lymphocytes from post-treatment biopsies: a case study, Human Gene Therapy 1(4) (1990) 373–84.
93. Rosenberg SA, Aebersold P, Cornetta K, Kasid A, Morgan RA, Moen R, Karson EM, Lotze MT, Yang JC, Topalian SL, Gene transfer into humans–immunotherapy of patients with advanced melanoma, using tumor-infiltrating lymphocytes modified by retroviral gene transduction, New England Journal of Medicine 323(9) (1990) 570–8.
94. Hiraoka K, Kimura T, Logg CR, Tai CK, Haga K, Lawson GW, Kasahara N, Therapeutic efficacy of replication-competent retrovirus vector-mediated suicide gene therapy in a multifocal colorectal cancer metastasis model, Cancer Research 67(11) (2007) 5345–53.
95. Zwacka RM, Dunlop MG, Gene therapy for colon cancer, Hematology/Oncology Clinics of North America 12(3) (1998) 595–615.
96. Li HJ, Everts M, Pereboeva L, Komarova S, Idan A, Curiel DT, Herschman HR, Adenovirus tumor targeting and hepatic untargeting by a coxsackie/adenovirus receptor ectodomain anti-carcinoembryonic antigen bispecific adapter, Cancer Research 67(11) (2007) 5354–61.
97. Nakano K, Todo T, Zhao G, Yamaguchi K, Kuroki S, Cohen JB, Glorioso JC, Tanaka M, Enhanced efficacy of conditionally replicating herpes simplex virus (G207) combined with 5-fluorouracil and surgical resection in peritoneal cancer dissemination models, The Journal of Gene Medicine 7(5) (2005) 638–48.
98. Maitra A, Calcium phosphate nanoparticles: second-generation nonviral vectors in gene therapy, Expert Review of Molecular Diagnostics 5(6) (2005) 893–905.
99. Cemazar M, Golzio M, Sersa G, Rols MP, Teissié J, Electrically-assisted nucleic acids delivery to tissues in vivo: where do we stand?, Current Pharmaceutical Design 12(29) (2006) 3817–25.
100. Karmali PP, Chaudhuri A, Cationic liposomes as non-viral carriers of gene medicines: resolved issues, open questions, and future promises, Medicinal Research Reviews 27(5) (2007) 696–722.
101. Li S, Huang L, Nonviral gene therapy: promises and challenges, Gene Therapy 7(1) (2000) 31–4.
102. Walther W, Stein U, Fichtner I, Schlag PM, Low-volume jet injection for efficient nonviral in vivo gene transfer, Molecular Biotechnology 28(2) (2004) 121–8.
103. Liu F, Song Y, Liu D, Hydrodynamics-based transfection in animals by systemic administration of plasmid DNA, Gene Therapy 6(7) (1999) 1258–66.
104. Kerr DJ, Young LS, Searle PF, McNeish IA, Gene directed enzyme prodrug therapy for cancer, Advanced Drug Delivery Reviews 26 (1997) 173–84.
105. Walther W, Stein U, Fichtner I, Kobelt D, Aumann J, Arlt F, Schlag PM, Nonviral jet-injection gene transfer for efficient in vivo cytosine deaminase suicide gene therapy of colon carcinoma, Molecular Therapy: The Journal of the American Society of Gene Therapy 12(6) (2005) 1176–84.
106. Link CJ, Levy JP, McCann LZ, Moorman DW, Gene therapy for colon cancer with the herpes simplex thymidine kinase gene, Journal of Surgical Oncology 64(4) (1997) 289–94.
107. Kanyama H, Tomita N, Yamano T, Miyoshi Y, Ohue M, Fujiwara Y, Sekimoto M, Sakita I, Tamaki Y, Monden M, Enhancement of the anti-tumor effect of 5'-deoxy-5-fluorouridine by transfection of thymidine phosphorylase gene into human colon cancer cells, Japanese Journal of Cancer Research: GANN 90(4) (1999) 454–9.

108. Chung-Faye G, Palmer D, Anderson D, Clark J, Downes M, Baddeley J, Hussain S, Murray PI, Searle P, Seymour L, Harris PA, Ferry D, Kerr DJ, Virus-directed, enzyme prodrug therapy with nitroimidazole reductase: a phase I and pharmacokinetic study of its prodrug, CB1954, Clinical Cancer Research: An Official Journal of the American Association forCancer Research 7(9) (2001) 2662–8.
109. Chung-Faye GA, Kerr DJ, Young LS, Searle PF, Gene therapy strategies for colon cancer, Molecular Medicine Today 6(2) (2000) 82–7.
110. Konson A, Ben-Kasus T, Mahajna JA, Danon A, Rimon G, Agbaria R, Herpes simplex virus thymidine kinase gene transduction enhances tumor growth rate and cyclooxygenase-2 expression in murine colon cancer cells, Cancer Gene Therapy 11(12) (2004) 830–40.
111. Kagaya T, Nakamoto Y, Sakai Y, Tsuchiyama T, Yagita H, Mukaida N, Kaneko S, Monocyte chemoattractant protein-1 gene delivery enhances antitumor effects of herpes simplex virus thymidine kinase/ganciclovir system in a model of colon cancer, Cancer Gene Therapy 13(4) (2006) 357–66.
112. Koyama F, Sawada H, Hirao T, Fujii H, Hamada H, Nakano H, Combined suicide gene therapy for human colon cancer cells using adenovirus-mediated transfer of Escherichia coli cytosine deaminase gene and Escherichia coli uracil phosphoribosyltransferase gene with 5-fluorocytosine, Cancer Gene Therapy 7(7) (2000) 1015–22.
113. Hayashi S, Emi N, Yokoyama I, Namii Y, Uchida K, Takagi H, Inhibition of establishment of hepatic metastasis in mice by combination gene therapy using both herpes simplex virus-thymidine kinase and granulocyte macrophage-colony stimulating factor genes in murine colon cancer, Cancer Gene Therapy 4(6) (1997) 339–44.
114. Nyati MK, Symon Z, Kievit E, Dornfeld KJ, Rynkiewicz SD, Ross BD, Rehemtulla A, Lawrence TS, The potential of 5-fluorocytosine/cytosine deaminase enzyme prodrug gene therapy in an intrahepatic colon cancer model, Gene Therapy 9(13) (2002) 844–9.
115. Linderholm B, Norberg T, Bergh J, Sequencing of the tumor suppressor gene TP 53, Methods in Molecular Medicine 120 (2006) 389–401.
116. Liu MC, Gelmann EP, P53 gene mutations: case study of a clinical marker for solid tumors, Seminars in Oncology 29(3) (2002) 246–57.
117. Chung-Faye GA, Kerr DJ, ABC of colorectal cancer: innovative treatment for colon cancer, The BMJ 321(7273) (2000) 1397–9.
118. Chen X, Ko LJ, Jayaraman L, Prives C, p53 levels, functional domains, and DNA damage determine the extent of the apoptotic response of tumor cells, Genes & Development 10(19) (1996) 2438–51.
119. Dvory-Sobol H, Kazanov D, Arber N, Gene targeting approach to selectively kill colon cancer cells, with hyperactive K-Ras pathway, Biomedicine & Pharmacotherapy (2005) S370–4.
120. Bukholm IK, Nesland JM, Protein expression of p53, p21 (WAF1/CIP1), bcl-2, Bax, cyclin D1 and pRb in human colon carcinomas, Virchows Archiv: An International Journal of Pathology 436(3) (2000) 224–8.
121. Sidransky D, Tokino T, Hamilton SR, Kinzler KW, Levin B, Frost P, Vogelstein B, Identification of ras oncogene mutations in the stool of patients with curable colorectal tumors, Science 256(5053) (1992) 102–5.
122. Rubin J, Galanis E, Pitot HC, Richardson RL, Burch PA, Charboneau JW, Reading CC, Lewis BD, Stahl S, Akporiaye ET, Harris DT, Phase I study of immunotherapy of hepatic metastases of colorectal carcinoma by direct gene transfer of an allogeneic histocompatibility antigen, HLA-B7, Gene Therapy 4(5) (1997) 419–25.
123. Morse MA, Nair SK, Mosca PJ, Hobeika AC, Clay TM, Deng Y, Boczkowski D, Proia A, Neidzwiecki D, Clavien PA, Hurwitz HI, Schlom J, Gilboa E, Lyerly HK, Immunotherapy with autologous, human dendritic cells transfected with carcinoembryonic antigen mRNA, Cancer Investigation 21(3) (2003) 341–9.
124. Alves A, Vibert E, Trajcevski S, Solly S, Fabre M, Soubrane O, Qian C, Prieto J, Klatzmann D, Panis Y, Adjuvant interleukin-12 gene therapy for the management of colorectal liver metastases, Cancer Gene Therapy 11(12) (2004) 782–9.
125. Mazzolini G, Qian C, Xie X, Sun Y, Lasarte JJ, Drozdzik M, Prieto J, Regression of colon cancer and induction of antitumor immunity by intratumoral injection of adenovirus expressing interleukin-12, Cancer Gene Therapy 6(6) (1999) 514–22.
126. Caruso M, Pham-Nguyen K, Kwong YL, Xu B, Kosai KI, Finegold M, Woo SL, Chen SH, Adenovirus-mediated interleukin-12 gene therapy for metastatic colon carcinoma, Proceedings of the National Academy of Sciences of the United States of America 93(21) (1996) 11302–6.
127. Adris S, Chuluyan E, Bravo A, Berenstein M, Klein S, Jasnis M, Carbone C, Chernajovsky Y, Podhajcer OL, Mice vaccination with interleukin 12-transduced colon cancer cells potentiates rejection of syngeneic non-organ-related tumor cells, Cancer Research 60(23) (2000) 6696–703.
128. Ueda K, Cornwell MM, Gottesman MM, Pastan I, Roninson IB, Ling V, Riordan JR, The mdrl gene, responsible for multidrug-resistance, codes for P-glycoprotein, Biochemical and Biophysical Research Communications 141(3) (1986) 956–62.

129. Hou S, Kou G, Fan X, Wang H, Qian W, Zhang D, Li B, Dai J, Zhao J, Ma J, Li J, Lin B, Wu M, Guo Y, Eradication of hepatoma and colon cancer in mice with Flt3L gene therapy in combination with 5-FU, Cancer Immunology, Immunotherapy 56(10) (2007) 1605–13.
130. Ogawa N, Fujiwara T, Kagawa S, Nishizaki M, Morimoto Y, Tanida T, Hizuta A, Yasuda T, Roth JA, Tanaka N, Novel combination therapy for human colon cancer with adenovirus-mediated wild-type p53 gene transfer and DNA-damaging chemotherapeutic agent, International Journal of Cancer 73(3) (1997) 367–70.
131. Sabaawy HM, Ikehara S, Adachi Y, Quan S, Feldman E, Kancherla R, Abraham NG, Ahmed T, Enhancement of 5-fluorouracil cytotoxicity on human colon cancer cells by retrovirus-mediated interferon-alpha gene transfer, International Journal of Oncology 14(6) (1999) 1143–51.
132. Wildner O, Blaese RM, Candotti F, Enzyme prodrug gene therapy: synergistic use of the herpes simplex virus-cellular thymidine kinase/ganciclovir system and thymidylate synthase inhibitors for the treatment of colon cancer, Cancer Research 59(20) (1999) 5233–8.
133. Ishii M, Iwai M, Harada Y, Kishida T, Asada H, Shin-Ya M, Itoh Y, Imanishi J, Okanoue T, Mazda O, Soluble TRAIL gene and actinomycin D synergistically suppressed multiple metastasis of TRAIL-resistant colon cancer in the liver, Cancer Letters 245 (2007) 134–43.
134. Adachi Y, Lee CT, Coffee K, Yamagata N, Ohm JE, Park KH, Dikov MM, Nadaf SR, Arteaga CL, Carbone DP, Effects of genetic blockade of the insulin-like growth factor receptor in human colon cancer cell lines, Gastroenterology 123(4) (2002) 1191–204.
135. Uchida H, Shinoura N, Kitayama J, Watanabe T, Nagawa H, Hamada H, 5-Fluorouracil efficiently enhanced apoptosis induced by adenovirus-mediated transfer of caspase-8 in DLD-1 colon cancer cells, The Journal of Gene Medicine 5(4) (2003) 287–99.
136. Kobayashi T, Sawa H, Morikawa J, Zhang W, Shiku H, Bax induction activates apoptotic cascade via mitochondrial cytochrome c release and Bax overexpression enhances apoptosis induced by chemotherapeutic agents in DLD-1 colon cancer cells, Japanese Journal of Cancer Research: GANN 91(12) (2000) 1264–8.
137. Weiss MB, Vitolo MI, Baerenfaller K, Marra G, Park BH, Bachman KE, Persistent mismatch repair deficiency following targeted correction of hMLH1, Cancer Gene Therapy 14(1) (2007) 98–104.
138. Zhang M, Li S, Li J, Ensminger WD, Lawrence TS, Ionizing radiation increases adenovirus uptake and improves transgene expression in intrahepatic colon cancer xenografts, Molecular Therapy: The Journal of the American Society of Gene Therapy 8(1) (2003) 21–8.
139. Zamir G, Zeira E, Gelman AE, Shaked A, Olthoff KM, Eid A, Galun E, Replication-deficient adenovirus induces host topoisomerase I activity: implications for adenovirus-mediated gene expression, Molecular Therapy: The Journal of the American Society of Gene Therapy 15(4) (2007) 772–81.
140. Zheng S, Xiao ZX, Pan YL, Han MY, Dong Q, Continuous release of interleukin 12 from microencapsulated engineered cells for colon cancer therapy, World Journal of Gastroenterology 9(5) (2003) 951–5.
141. Reid T, Galanis E, Abbruzzese J, Sze D, Andrews J, Romel L, Hatfield M, Rubin J, Kirn D, Intra-arterial administration of a replication-selective adenovirus (dl1520) in patients with colorectal carcinoma metastatic to the liver: a phase I trial, Gene Therapy 8(21) (2001) 1618–26.
142. Zhu H, Zhu Y, Hu J, Hu W, Liao Y, Zhang J, Wang D, Huang X, Fang B, He C, Adenovirus-mediated small hairpin RNA targeting Bcl-XL as therapy for colon cancer, International Journal of Cancer 121(6) (2007) 1366–72.
143. Lan KL, Yen SH, Liu RS, Shih HL, Tseng FW, Lan KH, Mutant Bik gene transferred by cationic liposome inhibits peritoneal disseminated murine colon cancer, Clinical & Experimental Metastasis 24(6) (2007) 461–70.
144. Kuwamura H, Tominaga K, Shiota M, Ashida R, Nakao T, Sasaki E, Watanabe T, Fujiwara Y, Oshitani N, Higuchi K, Ichijo H, Arakawa T, Iwao H, Growth inhibition of colon cancer cells by transfection of dominant-negative apoptosis signal-regulating kinase-1, Oncology Reports 17(4) (2007) 781–6.
145. Lemoine NR, McNeish IA, Gene transfer: Bax to the future for cancer therapy, Gut 53(4) (2004) 478–9.
146. Jie JZ, Wang JW, Qu JG, Hung T, Suppression of human colon tumor growth by adenoviral vector-mediated NK4 expression in an athymic mouse model, World Journal of Gastroenterology 13(13) (2007) 1938–46.
147. Wu X, Deng Y, Wang G, Tao K, Combining siRNAs at two different sites in the EGFR to suppress its expression, induce apoptosis, and enhance 5-fluorouracil sensitivity of colon cancer cells, TheJournal of Surgical Research 138(1) (2007) 56–63.
148. Durai R, Yang SY, Sales KM, Seifalian AM, Goldspink G, Winslet MC, Increased apoptosis and decreased proliferation of colorectal cancer cells using insulin-like growth factor binding protein-4 gene delivered locally by gene transfer, Colorectal Disease: The Official Journal of the Association of Coloproctology of Great Britain and Ireland 9(7) (2007) 625–31.
149. Subramaniam V, Vincent IR, Gilakjan M, Jothy S, Suppression of human colon cancer tumors in nude mice by siRNA CD44 gene therapy, Experimental and Molecular Pathology 83(3) (2007) 332–40.

150. Negroni L, Samson M, Guigonis JM, Rossi B, Pierrefite-Carle V, Baudoin C, Treatment of colon cancer cells using the cytosine deaminase/5-fluorocytosine suicide system induces apoptosis, modulation of the proteome, and Hsp90beta phosphorylation, Molecular Cancer Therapeutics 6(10) (2007) 2747–56.
151. Hayashi S, Yokoyama I, Namii Y, Emi N, Uchida K, Takagi H, Inhibitory effect on the establishment of hepatic metastasis by transduction of the tissue plasminogen activator gene to murine colon cancer, Cancer Gene Therapy 6(4) (1999) 380–4.
152. Vitale S, Cambien B, Karimdjee BF, Barthel R, Staccini P, Luci C, Breittmayer V, Anjuère F, Schmid-Alliana A, Schmid-Antomarchi H, Tissue-specific differential antitumour effect of molecular forms of fractalkine in a mouse model of metastatic colon cancer, Gut 56(3) (2007) 365–72.
153. Iaitskiĭ NA, Dubina MV, Vasil'ev SV, Popov DE, Krutovskikh VA, [The scientific and clinical value of molecular genetic changes in the intercellular gap junctions observed in colon cancer], Vestnik Rossiiskoi akademii meditsinskikh nauk undefined 10 (2003) 24–9.
154. Ye YJ, Zhu XG, Wang S, Wang YC, Sang JL, Antisense to CDK4 inhibits the growth of human colon cancer cells HT29, Zhonghua yi xue za zhi 86(12) (2006) 846–9.
155. Jacob D, Bahra M, Schumacher G, Neuhaus P, Fang B, Gene therapy in colon cancer cells with a fiber-modified adenovector expressing the TRAIL gene driven by the hTERT promoter, Anticancer Research 24 (2004) 3075–9.
156. Rychahou PG, Murillo CA, Evers BM, Targeted RNA interference of PI3K pathway components sensitizes colon cancer cells to TNF-related apoptosis-inducing ligand (TRAIL), Surgery 138(2) (2005) 391–7.
157. Zhang L, Gu J, Lin T, Huang X, Roth JA, Fang B, Mechanisms involved in development of resistance to adenovirus-mediated proapoptotic gene therapy in DLD1 human colon cancer cell line, Gene Therapy 9(18) (2002) 1262–70.
158. Mulkeen AL, Silva T, Yoo PS, Schmitz JC, Uchio E, Chu E, Cha C, Short interfering RNA-mediated gene silencing of vascular endothelial growth factor: effects on cellular proliferation in colon cancer cells, Archives of Surgery 141(4) (2006) 367–74; discussion 374.
159. Stoeltzing O, Ahmad SA, Liu W, McCarty MF, Wey JS, Parikh AA, Fan F, Reinmuth N, Kawaguchi M, Bucana CD, Ellis LM, Angiopoietin-1 inhibits vascular permeability, angiogenesis, and growth of hepatic colon cancer tumors, Cancer Research 63(12) (2003) 3370–7.
160. Dkhissi F, Lu H, Soria C, Opolon P, Griscelli F, Liu H, Khattar P, Mishal Z, Perricaudet M, Li H, Endostatin exhibits a direct antitumor effect in addition to its antiangiogenic activity in colon cancer cells, Human Gene Therapy 14(10) (2003) 997–1008.

16

Rising from the Ashes *The Curious Case of the Development of Biologics for the Treatment of Neuroblastoma*

Mandeep Rajpal, Anitha K. Shenoy, and Ashim Malhotra

Corresponding author: Ashim Malhotra

Contents

DOI: 10.1201/9780429485626-16

16.1 INTRODUCTION: NEUROBLASTOMA, DISCOVERY, AND TIMELINE

Neuroblastoma is a rare cancer that occurs in children originating from primitive neural cells in the abdomen, chest, or other regions of the body. Following its initial discovery in 1864 by Rudolph Virchow [1], its etiology and pathology remained challenging to dissect, understand, and replicate in the laboratory. Three physicians played a quintessential role in demystifying and describing for the first time what came to be known as neuroblastoma. Pepper (1901) and Hutchison (1907) separately published a series of papers describing neuroblastomas in children, which was consolidated and contextualized as referring to the same disease through the work of Homer Wright (1910). It was Homer Wright who first attributed these tumors to primitive neural cells such as those found during the embryological development of the human nervous system, in particular, the human neural crest, which is a part of the nascent central nervous system [1].

Although it was initially a challenge to understand the molecular basis for neuroblastoma, significant progress has been made over recent times in understanding the diverse manifestations, genetic and biochemical heterogeneity, and clinical challenges associated with the treatment of neuroblastoma. It is the tireless efforts in foundational science discovery research and the advent of modern molecular biology that have made possible clinical strategies for the treatment of neuroblastoma such as the ability to risk-stratify a patient at the time of diagnosis. However, despite substantial augmentation of our understanding of the underlying genetic and biochemical abrogation originating neuroblastoma, it remains a challenging disease, necessitating the development of novel therapeutic modalities. As our understanding of the pathobiology of neuroblastoma evolves, new tumor targets, novel antibody- and cell-mediated immunotherapy agents, and the deployment of "big data" strategies such as molecular and genetic tumor profiling have led to a large number of clinical trials for neuroblastoma. Combining these developing therapies with the current treatment regimens is needed to improve outcomes for high-risk and relapse patients [2].

Apart from surgery, chemotherapy, and radiotherapy as the primary treatment for neuroblastoma, immunotherapy with biologics has shown promise in recent times. However, the increasing burden of costs of biologics has effectively undermined their role as preeminent treatment options. Nonetheless, as biologics lose patent protection, pharmaceutical companies have seized this opportunity to develop highly similar versions of these biologics as breakthrough alternatives, inherently changing the narrative of immuno-oncotherapy.

Biosimilars are similar versions of the biologic prototype. They act similarly in efficacy, safety, and immunogenicity but are different from the generic versions of a biologic. Unlike generics, biosimilars are not exact copies of the originator product. As a more affordable and alternative treatment option, biosimilars are increasingly gaining widespread use for oncologic diseases. However, the promise of these therapies in cancer treatment is limited by sparse and sporadic availability because biologics have only a few versions of biosimilars indicated for use in particular diseases, especially in pediatric cancers.

This relatively narrow scope of use spawned the current debate in the scientific and medical community, who wrestle with the concept of extrapolation. Extrapolation refers to the process in which the totality of the evidence, including a clinical evaluation therein, can be approved for use in an indication other than that approved in a phase III clinical trial of biosimilars [3]. The extension of the use of biosimilars in other indications can be accepted and justified if scientific data is available and can be approved for all indications of the original biologic. Only a few biosimilars are available for the treatment of neuroblastoma. The concept of extrapolation seems a viable recourse for clinicians as they consider alternate treatment for tumor regression. Here, we provide an overview of neuroblastoma biology and drug discovery.

16.2 NEUROBLASTOMA: INCIDENCE AND BACKGROUND

Neuroblastoma is a relatively rare pediatric cancer that occurs in about one in 100,000 children and is slightly more common in boys. Originating from immature nerve cells, among childhood and infant cancers, neuroblastoma is a leading cause of infant and child mortality with 700–800 new patient cases reported in the United States each year according to the American Cancer Society [4]. Neuroblastomas account for more than 7% of all malignancies in patients younger than 15 and around 15% of all pediatric oncology-related deaths [3]. It is the most common extracranial solid tumor in childhood and the most frequently diagnosed neoplasm during infancy [3].

A brief timeline of a century's worth of research in its elucidation, diagnosis, and treatment is provided in Figure 16.1 to underscore the challenges in unraveling the complexities related to this often-devastating disease.

Neuroblastomas originate from the sympathoadrenal lineage of the embryonic neural crest cells and can therefore engender tumors anywhere in the sympathetic nervous system, from the neck to the pelvis, but most frequently arise in the adrenal gland and the abdomen but are also commonly encountered in the chest and the nerve tissue near the spine [5].

Clinical outcomes remain unpredictable and vary from the much-desired spontaneous regression of the tumor to the opposed and often fatal scenario of metastatic disease with a poor prognosis. Thus, from the perspective of clinical outcomes, neuroblastoma is a heterogeneous disease and one that appears to affect different areas of the body in an unpredictable manner. The likelihood of resolution is multifactorial and depends on the age at diagnosis, the extent of disease, and the specific tumor biology driving progression in the affected individual. Accordingly, clinicians need to stratify patients at the time of diagnosis to enable categorization based on the genetic and biochemical basis of tumor drivers in the particular patient coupled with clinical factors that will sculpt outcomes. This scheme of classification allows incorporation of the knowledge of specific risk categories and helps determine patient-specific treatment [6].

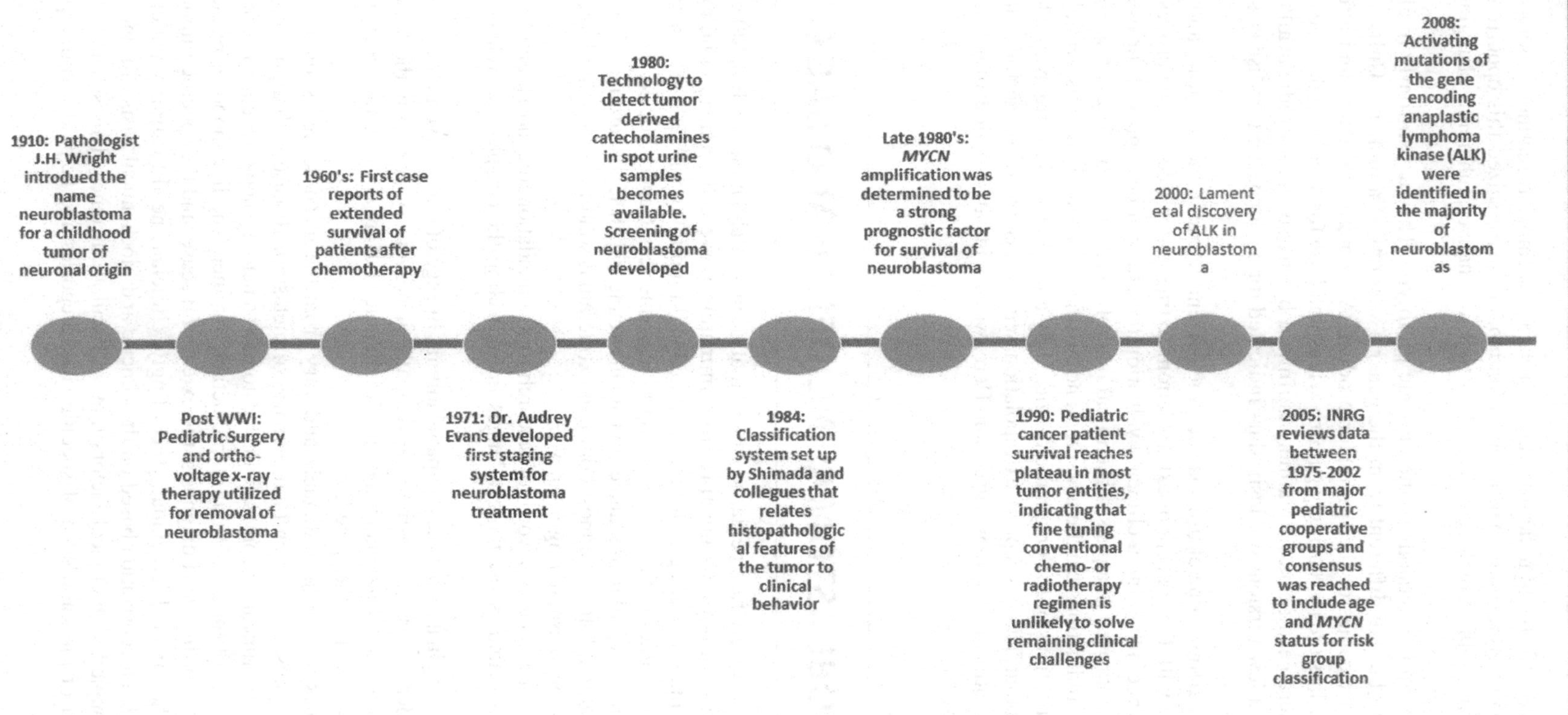

FIGURE 16.1 An exploration of the discovery, description, clinical characteristics, and staging of neuroblastoma.

16.3 DELINEATING THE MOLECULAR PATHOGENESIS OF NEUROBLASTOMA

16.3.1 Discovery and Initial Insights into the Molecular Basis of a Seemingly Strange Disease

Biedler et al.'s seminal 1973 paper described for the first time that patient-derived neuroblastoma cell lines, grown *ex vivo* as primary cultures, manifested as cells with very different phenotypes [7]. These cell types were named "N-type" (for neuroblast-like) and "S-type" (for substrate-adherent) cells. This group established the primary cell line SK-N-SH from the bone marrow of a four-year-old female patient suffering from neuroblastoma and, using these cells, reported that these morphologically distinct cells were able to interconvert spontaneously upon growth in culture, suggesting some degree of cross and inter-differentiation. This was an important observation since the N-type cells were small, had neurite processes, closely resembled neurons, and grew as focal aggregates, while the S-type cells grew as flat, anchorage-dependent cells.

As both N-type and S-type cells were originally derived from neuroblastomas of a single patient, the masking and unmasking of phenotypic and morphological diversity suggested complex genetic and epigenetic programming, at least *in vitro*. This fascinating observation presented many scientific hurdles in characterizing the molecular basis for this phenotypic heterogeneity, formidable challenges which would remain largely unsolved for over 40 years till the availability of novel molecular technologies and capabilities resulted in the publication of work from two groups in 2017.

Since the time of the publication of Biedler's work, other groups have helped establish the universality of the heterogeneity of phenotypes of cells grown from patient-derived neuroblastoma primary cultures. Later groups reported startling differences in the genetic programming between the N-type and S-type cells grown in cultures derived from cell lines including SK-N-SH and LAN-1. For example, Rettig et al. reported that the "N-type cells presented activity for enolase and expressed chromogranin, MC25, and AO10 antigens and the NGF receptor, whereas S-type cells did not. Nine antigens were found to be strongly expressed and specific to the S-type, for example, mel-CSPG, β2-microglobulin, and HLA class I antigens [8]".

However, despite such advances, the field of neuroblastoma research has remained muddled due to the unprecedented nature of morphological and phenotypic variability in cells grown in culture from patient-derived neuroblastomas. For instance, while some patient-derived neuroblastoma N-type cells show commonalities in phenotypic and biochemical properties, the S-type cell phenotypes have remained challenging to reproduce from among different patient sources. The main problem has remained the proper classification of the type of cells that the S-phenotype represents; there is debate on whether these cells more closely resemble the Schwann cells in the nervous system or glial cells or neuronal or mesenchymal cells or fibroblasts based on the proteins and enzymes they express [9, 10]. This conflict was further compounded by the discovery in 1989 of a third cell state, called the Intermediate or "I-type". The I-type cells were shown to be moderately adherent *in vitro* with sparse cytoplasm and neurites while expressing the vimentin protein, and their discovery heralded a new idea in the field of neuroblastoma etiology.

16.3.2 Compounding Liner Discovery: The Case of Stem-Cell Like Cells and Their Role in Neuroblastoma

The discovery of the intermediate-morphology cells suggested the presence of progenitor cells, which some argued, could be the originating precursor cell population for subsequent N-type and S-type differentiation; a sort-of developmental stage that could differentiate into the N-type or S-type cells. It was argued that these ancestral, rudimentary-differentiated I-type cells more closely resembled cancer stem cells since they retained the phenotypic and biochemical expression patterns of both the more predominant cultured cell types, the N-type and S-type cells. In this context, Ross et al. published important work in 1995 showing that the I-type cells not only retained the ability to be induced into either the N-type phenotype upon exposure to retinoic acid or into the S-type phenotype upon exposure to 5-Bromo-2′-deoxyuridine but also retained a capacity for self-renewal. This ability for self-renewal indicated stability and the lack of spontaneous differentiation, hinting that these aggressive cell populations represented a multi-potent precursor cell type that was similar to cancer stem cells and perhaps seeded the continuous growth and progression of neuroblastomas in the patient [11].

Terminologies have since been updated, with the N-type cells now called NOR cells to call attention to their dependence on sympathetic noradrenergic and adrenergic programming while the so-called S-type cells are now more appropriately rechristened as MES cells since they more closely resemble mesenchymal cells in culture.

16.3.3 The Genetic and Biochemical Basis for NOR and MES Heterogeneity in Neuroblastoma

NOR cells have been shown to have an intricate genetic and epigenetic program that facilitates the observed cellular transdifferentiation. However, even with the advancements of the past two decades in the tools of molecular biology, it has remained challenging to identify the various molecular, biochemical, and genetic clues, and in particular, their chronological and temporal activation and inactivation that results in the formation and differentiation axis of neuroblastomas. Scientific interest and effort have concentrated on the identification of main genes and transcription factors that regulate the overall programming of the observed phenotypic and genetic cellular heterogeneity (Figure 16.2).

Specifically, for example, some of the first experiments to close the knowledge gap about the genetic basis of neuroblastomas were conducted using the primary cell line established from metastatic stage-4

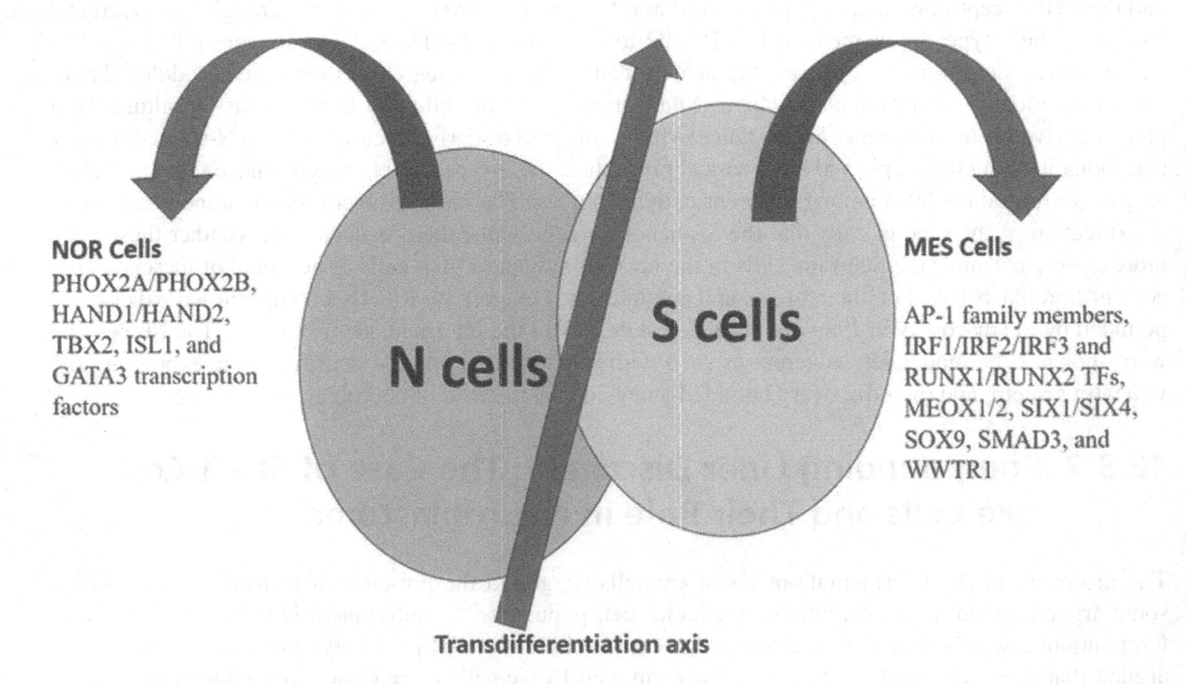

FIGURE 16.2 Neuroblastomas from patients originate cells with starkly different phenotypic characteristics in culture. N cells (now called NOR cells) and S cells (rechristened as MES cells).

neuroblastoma called NBL-W. Conducted by Foley et al. and reported in 1991, these experiments established the differential expression of the proto-oncogenic DNA-binding base-helix-loop-helix protein called N-myc in the phenotypically distinct N-type and S-type subsets of cultured cells [12]. This was a landmark discovery of the role of specific transcription factors in driving neuroblastomas and heralded a new era of scientific discoveries. The N-myc protein is the product of the NMYC gene, which is expressed in high copy numbers in fetal brain cells and hence is known as a high-amplification gene. The work conducted by Foley's group was significant because it showed for the first time how despite a high copy number, N-myc expression was differentially regulated. In other words, the downregulation of the N-myc protein seemed to form the basis that allowed the differentiation of the patient-derived neuroblastoma cells grown in culture into the S-type phenotype.

Since the time of this first discovery, a panoply of transcription factors has been discovered that seem to drive the observed differentiation in cell types derived from patient neuroblastomas. For instance, two teams of scientists, Bovea et al. [13] and van Groningen et al. [14], published findings in 2017 identifying what is known as "core regulatory circuitries (CRCs)" of transcription factors that can regulate the interconversion and transdifferentiation observed for the NOR and MES cell types of neuroblastoma-derived cells. As a result, an "analysis of the SEs characterized in the NOR identity first suggested that the NOR CRC contained the PHOX2A/PHOX2B, HAND1/HAND2, TBX2, ISL1, and GATA3 TFs, among others", where SEs signify super-enhancer sequences that aid with transcription. These CRCs were ultimately found to be regulated by the MYC proto-oncogene in neuroblastoma cells, though the exact mechanism of MYC regulation and control remains unelucidated.

For the MES phenotype, Gautier et al. [15] summarized the findings in the research field by commenting that "on the one hand, AP-1 family members, IRF1/IRF2/IRF3 and RUNX1/RUNX2 TFs, among others, were identified as CRC members in the MES cell lines in one study [Boevua et al.], whereas on the other hand, MEOX1/2, SIX1/SIX4, SOX9, SMAD3, and WWTR1, among others, were highlighted in the second study [van Grogenin et al.]".

16.3.4 Summary of the Genetics and Molecular Biology of Neuroblastoma

Detection of neuroblastoma *in utero* or early life indicates that the tumor results from an early impairment of the normal developmental and embryological processes. As neuroblastomas originate from a noradrenergic lineage, insight into its multi-faceted molecular regulation is anticipated to lead to the identification of novel targets. Cancer stem cells are being explored in their role in propagating malignant disease and are foreseen to prove helpful in developing insight and treating neuroblastoma. Many of the molecularly guided therapies for neuroblastoma that have been discovered and are continuing to be developed emerge from an increased understanding of biomarkers such as *MYCN* and *ALK*. Amplification of these biomarkers serves as validation and are strong prognostic factors of survival of neuroblastoma. However, directly targeting each of these markers has been challenging. The most effective approach, as it appears in numerous clinical research studies, is to indirectly target these markers to control their regulation. Researchers undertake studies to find additional prognostic markers.

16.4 STAGING NEUROBLASTOMAS: CHALLENGES IN CLINICAL PRACTICE

The International Neuroblastoma Staging System (INSS) Committee provides guidelines for a standardized staging system for neuroblastoma [5] (Table 16.1). Staging is essential for pharmacotherapeutic planning beginning with identifying the type of cancer, its location, spread, and whether it is affecting other parts of the body (metastasis). The use of diagnostic tools further substantiates and confirms the staging

TABLE 16.1 Classifying neuroblastoma using the International Neuroblastoma Staging System

Stage 1: The tumor can be removed completely during surgery. Lymph nodes attached to the tumor removed during surgery may or may not contain cancer but other lymph nodes near the tumor do not.

Stage 2A: The tumor is located only in the area it started and cannot be completely removed during surgery. Nearby lymph nodes do not contain cancer.

Stage 2B: The tumor is located only in the area where it started and may or may not be completely removed during surgery but nearby lymph nodes do contain cancer.

Stage 3: The tumor cannot be removed with surgery. It has spread to regional lymph nodes (lymph nodes near the tumor) or other areas near the tumor but not to other parts of the body.

Stage 4: The original tumor has spread to distant lymph nodes (lymph nodes in other parts of the body), bones, bone marrow, liver, skin, and/or other organs, except for those listed in stage 4S below.

Stage 4S: The original tumor is located only where it started (as in stage 1, 2A, or 2B), and it has spread only to the skin, liver, and/or bone marrow in infants younger than one year of age. The spread to the bone marrow is minimal, and usually, less than 10% of cells examined show cancer.

Source: https://www.cancer.net/cancer-types/neuroblastoma-childhood/stages-and-groups

process. Accordingly, the INSS and its subsidiary, the International Neuroblastoma Risk Group Staging System, developed a staging protocol based on imaging criteria, specifically, imaging-defined risk factors identified from images taken before surgery (Table 16.2). This risk factor classification system employs a combination of clinical, pathological, and genetic markers to predict the clinical behavior of the tumor and how it might respond to treatment. Consequently, these markers define how patients are stratified into four different categories: (i) very low-risk, (ii) low-risk, (iii) intermediate-risk, or (iv) high-risk [4]. Understanding the pathological and developmental stage of neuroblastoma helps clinicians in developing individualized and personalized therapy.

TABLE 16.2 The classification of neuroblastomas using the International Neuroblastoma Risk Group Staging System

KNOWLEDGE REGARDING THE PRESENCE OR ABSENCE OF IMAGE DEFINED RISK FACTORS (IDRFs) IS REQUIRED FOR THIS STAGING SYSTEM

Stage L1: The tumor is located only in the area where it started; no image-identified risk factors (IDRFs) are found on imaging scans, such as CT or MRI.

Stage L2: The tumor has not spread beyond the area where it started and the nearby tissue; IDRFs are found on imaging scans, such as CT or MRI.

Stage M: The tumor has spread to other parts of the body (except stage MS, see below).

Stage MS: The tumor has spread to only the skin, liver, and/or bone marrow (less than 10% bone marrow involvement) in patients younger than 18 months.

Source: https://www.cancer.net/cancer-types/neuroblastoma-childhood/stages-and-groups

16.5 FORMULATING AN APPROACH FOR THE CLINICAL MANAGEMENT OF NEUROBLASTOMAS: WHAT HAVE WE LEARNED FROM ITS GENETICS AND MOLECULAR BIOLOGY?

Treatment of neuroblastoma is highly dependent on an assessment of patient risk factors. The risk stratification system of the Children's Oncology Group incorporates patient-specific features such as age, INSS stage, tumor histopathology, DNA index, and MYCN gene status to assign patients to risk categories of low, intermediate, and high. Current management involves the combinatorial approach of observation, surgery, chemotherapy, radiotherapy, and the use of biological agents. These initial approaches may be followed with more intense treatments based on whether any residual tumor remains after surgical resection. Therefore, to ensure optimal clinical outcomes, the timely and efficient integration of biological and clinical data is necessary to determine the course of individual treatment for specific patients.

16.5.1 The Clinical Evaluation of Neuroblastoma Therapy: Assessment of Response to Treatment

Although neuroblastomas originate as primary tumors of the sympathetic nervous system, they commonly metastasize to the bone and bone marrow [6, 7]. These sites by nature are highly diffuse and consequently, an accurate evaluation of metastatic sites and response to therapy is challenging. Thus, it follows that ideally, a specific and consistent biomarker would greatly assist in evaluating patient responses to therapy.

16.6 THE USE OF BIOLOGICS AND BIOSIMILARS FOR THE TREATMENT OF NEUROBLASTOMA

16.6.1 What Are Biologics and Biosimilars?

Unlike traditional pharmaceuticals that are small, chemically synthesized molecules, biologic medications such as monoclonal antibodies, cytokines, and cytokine antagonists, interleukins, interferons, proteins, blood, and others are derived from cells and tissues of living organisms. Another category of biological products is recombinant proteins made by cells that have been transformed to contain genetic information encoding a particular protein. Structurally, biologics are three-dimensional and are typically protein-based. They are sometimes called biologic response modifiers because they act to alter cellular responses. These medications hold particular promise in improving survival and quality of life in cancer care.

Interestingly, the market share for biologics has been steadily enlarging worldwide, suggesting global expansion in their development and deployment for the treatment of a wide variety of human diseases. For instance, in 2015, biologics were the top globally selling product for cancer treatment [16]. Despite the considerable advances these medications have prompted, their biosimilars have shown ever more promise as newer and more affordable alternatives. Globally, there exist varying definitions of what exactly a biosimilar is.

16.6.2 The Use of Biologicals and Biosimilars in Neuroblastoma: The New Hope!

While advances in research provide further insight regarding the pathogenesis of neuroblastoma and have allowed for the precise risk stratification of children, improving clinical outcomes, many patients, unfortunately, suffer relapses. Therefore, low- and intermediate-risk relapse patients may benefit from further conventional treatments including a second surgery and intensive chemotherapy. However, patients with a high-risk phenotype continue to endure frequent relapse with little prospect for long-term survival. Recurrence of high-risk neuroblastoma is difficult to treat and there exists no broadly effective cure. In light of these challenges, the idea of exploiting the body's natural immune system to treat cancer, known generally as immunotherapy, has opened avenues for promise in advanced oncologic disease states such as neuroblastoma. This has begun to partially address the vast gaps in drug availability for rare diseases such as neuroblastoma, where the economics of drug discovery and development does not lie in favor.

Clinical trials involving advanced therapies targeting specific protein markers and pathways have demonstrated promise toward long-term survival and tumor regression. Effective immunotherapies with biologics and biosimilars have come to constitute a part of the multi-modal therapeutic strategies for treating neuroblastoma patients. Often, children suffering from neuroblastoma-associated tumors receive a combination of chemotherapy, radiation therapy, immunotherapy, and treatment with new-age drugs such as monoclonal antibodies.

16.6.3 Regulatory and Manufacturing Considerations for Biologics and Biosimilars

Due to the complexities and proprietary nature of manufacturing biosimilars concerning their originator counterparts, it is not possible to develop an exact duplicate. Hence, a degree of heterogeneity exists between a biosimilar and its originator product.

The Patient Protection and Affordable Care Act, also known as the Biologics Price Competition and Innovation Act, was signed into law in 2010 [17]. This law transformed the economic landscape of biologics, rendering the Food and Drug Administration (FDA) approval process easier to navigate and complete for biosimilars from the bench to the bedside. For example, the abbreviated FDA drug application for biosimilars encompasses the statutory provision of creating a shortened approval pathway for biologic products that are evidenced to be highly similar to an FDA-approved biologic product. It seems intuitive that

Given the impact of biologics and biosimilars in oncologic treatment, their development and use for the treatment of neuroblastoma are of paramount importance. Biologics and their biosimilars are the leading medications in the treatment of high-risk pediatric patients who have had little to no solid tumor regression with advanced surgery, chemotherapy, and radiation.

Biologics for treating neuroblastoma including granulocyte-macrophage colony-stimulating factor (GM-CSF), interleukin-2 (IL-2), immunocytokines, which are monoclonal antibodies with cytokines linked to their Fc (tail) ends, and monoclonal antibodies, a few of the products that have been used as induction and maintenance immunotherapies, respectively. In particular, monoclonal antibodies are the mainstay medications that have garnered promise as highly effective immunotherapies in cancer treatment. Immune cell-based therapies such as the use of T cells, natural killer cells, armed T cells, and chimeric antigen receptor T cells are newer experimental biological approaches being explored for combatting neuroblastoma [18].

16.7 MONOCLONAL ANTIBODIES AS THE MAINSTAY FOR TREATING NEUROBLASTOMA: PRE-CLINICAL AND CLINICAL TRIALS

Several therapeutic options utilizing monoclonal antibodies have proven effective against neuroblastoma antigens.

16.7.1 Disialoganglioside (GD2) as a Neuroblastoma Target: The Case for the Anti-GD2 Monoclonal Antibodies

Gangliosides are carbohydrate-containing sphingolipids. Chemically, they are composed of a ceramide attached to N-acetylmuramic acid or another sialic acid. They are named starting with the letter G, followed by the letters M, D, etc., to indicate the number of sialogangliosides. Thus, GD2 is a disialogangloside and is very interesting in its properties. Although most gangliosides are commonly expressed on many cells and tissues in the human body, making them extremely unsuitable for targeting in cancer therapy, GD2 is unique in that it is expressed only by tumor cells and in very low expression by normal cells in the body.

16.7.2 The Clinical and Therapeutic Advantages of GD2

Before birth, GD2 is expressed on neural and mesenchymal stem cells, and following birth is found attached to the cell membranes of peripheral neurons. Significantly, it is found in very large quantities on tumor cell surfaces such as for neuroblastoma cells, retinoblastoma, rhabdomyosarcoma, and osteosarcoma. Importantly, it is found in relatively large concentrations on the cell membranes of neuroblastoma cells – to the extent of 5–10 million molecules/cell. This gives it the distinct advantage of targeting neuroblastoma cells with a high degree of specificity. Thus, monoclonal antibodies against human GD2 are being designed and tested in clinical trials for the treatment of neuroblastomas of a variety of origins.

16.7.3 The Development of Anti-GD2 Antibodies

The story of the development of monoclonal antibodies against GD2 showcases a part of the history of the development of biological agents and drugs. For example, the first experimental antibody drugs against GD2, which could be used for neuroblastoma, were anti-human antibodies raised in the mouse, and with the advent of newer technologies, these antibodies were improved, primarily to reduce immunogenicity and minimize adverse events. This is reflected in the development of murine monoclonals, followed by chimeric, and then humanized antibodies. m3F8, heat-modified m3F8, and 14.G2a are important examples of murine anti-human GD2 antibodies that have been tested in clinical trials through phase I (m3F8, heat-modified m3F8, and 14.G2a) and phase II (m3F8 and m3F8 combinations with other biologics).

Dinutuximab is an important example of a chimeric antibody against GD2, which has gone through phase I and randomized phase II/III trials as the sole drug and in combination with other biologics, through phase II (dinutuximab + GMCSF + temsirolimus) and phase III (dinutuximab + IL-2 + GMCSF) [19]. Interestingly, it should be pointed out that the phase trial, which established the initial toxicity profile of dinutuximab occurred as early as 1995 and was followed by others.

16.7.4 Dinutuximab versus Dinutuximab Beta

Dinutuximab is manufactured in the laboratory using the murine myeloma cell line called SP2/0 in the United States, while its counterpart in Europe is manufactured using the Chinese hamster ovary cells (CHO cells). Both are monoclonal antibodies in which the variable and heavy and light chains are murine in amino acid sequence, while the Fc region (the antibody tail) is composed of a human amino acid sequence. However, due to the use of the human CHO cell line for its production, dinutuximab-beta has a glycation pattern that more closely resembles glycation of antibodies in humans, thereby reducing its ability to clicit a hypersensitivity response. Interestingly, the murine cell line SP2/0 is commonly reported to be infected with murine viruses. Production of dinutuximab-beta in the human CHO cells is also helpful in avoiding such viral contamination of the final drug product [20].

16.7.5 Dinutuximab and the US FDA Approval Process: Clinical Trials for Unituxin

Currently, in the United States dinutuximab is sold under the brand name of Unituxin. Unituxin is an orphan drug and went through the priority drug review process at the US FDA. Orphan drugs are those that treat rare diseases and the priority review process shortens the drug review process by about four months, allowing high-impact drugs that can have significant clinical impact in areas where limited treatment options are available to be reviewed.

16.7.6 The History and Development of Unituxin

Unituxin has an important and interesting developmental history. It was developed by the National Cancer Institute in collaboration with Yu et al. at the University of California at San Diego and underwent extensive investigation and analysis for over 20 years by the Children's Oncology Group. As of the time of writing, it has been tried in approximately 2,000 children with neuroblastoma. It was initially manufactured by the National Cancer Institute under compassionate use license and eventually was contracted by them to a company called United Therapeutics (Silver Spring, Maryland).

Approved by the FDA in 2015, Unituxin was the first therapy approved by the FDA specifically for patients suffering from neuroblastoma. Unituxin is a biologic that is available as an intravenous injection of dinutuximab and is administered in combination with GMCSF, IL-2, and 13-cis-retinoic acid for the second-line treatment of children with high risk for neuroblastoma. According to the FDA, Unituxin needs to be included in a multi-modal regimen that includes surgery, radiation therapy, and chemotherapy regimens, including the use of Unituxin.

Yu et al. conducted a randomized, open-label, multi-center clinical trial, the results of which were published in 2010 [21]. It was the outcome of this trial, along with one other, that formed the basis for the FDA application of Unituxin. The main trial objective was to ascertain whether the addition of Unituxin, GM-CSF, and IL-2 combination would improve outcomes such as overall patient survival and tumor incidence for children suffering from neuroblastoma who had already received intensive multi-modal therapy as compared to those children who were only received retinoic acid. A total of 226 children suffering from neuroblastoma were included in the study, with a median enrollment age of 3.8 years old with 87% of the children from the non-Hispanic or Latino background, 10% from Hispanic or Latino ethnicity, and 4% with unknown ethnicity. Around 82% identified as White, 7% Black or African American, 3% Asian, 1% Asian Hawaiian or Pacific Islander, 1% multiple races, and 6% from an unknown race. Around 60% of the enrolled patients were male. Patients were randomly assigned to receive either Unituxin in combination with the other biologicals listed above or retinoic acid alone.

Three years after the assignment, the study found that 63% of the participants were alive and tumor-free in the Unituxin group as compared to 46% of the patients in the group that received only retinoic acid, following multi-modal therapy regimens. The study also found that there was a 43% reduction in the risk of an event (tumor) compared to retinoic acid treatment alone (HR = 0.57 [0.37–0.89]; $p = 0.01$).

16.7.7 The Adverse Events Profile of Dinutuximab and Other Disadvantages

Despite this finding, important patient-centered challenges remain with the use of Unituxin therapy. A major concern is the reporting of pain by 85% of the study group patients who received the Unituxin/retinoic acid regimen compared to 16% who reported pain in the group that received retinoic acid alone. This is because Unituxin irritates nerve cells and produces severe pain that requires treatment with narcotic analgesics with treatment before, during, and up to two hours after Unituxin injection. The FDA has included this information as a Black Box Warning for Unituxin and noted that it "may also cause other serious side effects including infections, eye problems, electrolyte abnormalities, and bone marrow suppression" [22].

Other common side effects of Unituxin include "severe pain, fever, low platelet counts, infusion reactions, low blood pressure, low levels of salt in the blood (hyponatremia), elevated liver enzymes, anemia, vomiting, diarrhea, low potassium levels in the blood, capillary leak syndrome (which is characterized by a massive leakage of plasma and other blood components from blood vessels into neighboring body cavities and muscles), low numbers of infection-fighting white blood cells (neutropenia and lymphopenia), hives, and low blood calcium levels" [22].

16.7.8 Humanized Anti-GD2 Monoclonal Antibody Drugs

On the other hand, because dinutuximab is a chimeric antibody and is generated using murine cells lines such as SP2/0, as explained above, it may engender hypersensitivity reactions in some patients. Not only is this hypersensitivity generated from differences in glycation patterns for the engendered antibody, but it also is a consequence of the production of human antibodies against proteins of murine origin, a phenomenon that is well documented for other murine or chimeric antibodies. From the patient's perspective, such hypersensitivity reactions may result in life-threatening adverse events and may also increase the already high degree of pain associated with this drug due to its irritant effects on neurons. Thus, an important goal in the drug development process for dinutuximab and related anti-GD2 monoclonal antibody drugs was to consider "humanizing" the monoclonal antibody to ameliorate adverse responses.

To address these concerns, humanized antibodies may be used instead. hu14.18K322A and Hu3F8 are examples of humanized monoclonal antibodies that may be used to treat neuroblastomas. For instance, as a humanized antibody, the basis of the design of Hu3F8 is the combination of amino acid sites that render it active by allowing immunorecognition specific epitopes on GD2 while restructuring other parts of the monoclonal antibody such as the Fc (tail) portion to bestow structural similarities with human antibodies. This technology retains the ability of the chimeric antibody to recognize the epitopes similar to the design of the murine antibody, while replacement of the amino acid sequences in the Fc portion ameliorates hypersensitivity and immunogenicity, considerably enhancing its usability in diverse patient populations. Both Hu3F8 and hu14.18K322A have undergone phase I studies, with the latter showing a toxicity profile similar to dinutuximab though it also showed marked improvement of tolerability at higher doses [23].

16.7.9 The Humanized Forms of Anti-GD2 Monoclonal Antibodies: Information from Comparative Analysis Studies

Cheung et al. developed a chimeric (ch3F8) and humanized (hu3F8) anti-GD2 monoclonal antibodies and tested their characteristics in the laboratory setting. The parameters that were compared included (1) measuring the binding and dissociation affinities of the chimeric and humanized anti-GD2 monoclonal antibodies to the receptor, (2) the ability of the drugs to effectuate tumor regression *in vitro* and *in vivo*, and (3) the ability of the drugs to engender the antibody-dependent cellular toxicity responses (ADCC) through the activation of polymorphonuclear cells (PMNCs, neutrophils).

The authors reported that "in GD2 binding studies by SPR, ch3F8 and hu3F8 maintained K(D) comparable to m3F8. Unlike other anti-GD2 antibodies, m3F8, ch3F8, and hu3F8 had substantially slower k(off.). Similar to m3F8, both ch3F8 and hu3F8 inhibited tumor cell growth *in vitro*, while cross-reactivity with other gangliosides was comparable to that of m3F8. Both peripheral blood mononuclear cell (PBMC)-ADCC and polymorphonuclear leukocytes (PMN)-ADCC of ch3F8 and hu3F8-IgG1 were more potent than m3F8. This superiority was consistently observed in ADCC assays, irrespective of donors or NK-92MI-transfected human CD16 or CD32, whereas complement-mediated cytotoxicity (CMC) was reduced. As expected, hu3F8-IgG4 had near absent PBMC-ADCC and CMC. Hu3F8 and m3F8 had similar tumor-to-non tumor ratios in biodistribution studies. The anti-tumor effect against neuroblastoma xenografts was better with hu3F8-IgG1 than m3F8. In conclusion, humanizing m3F8 produced next-generation anti-GD2 antibodies with substantially more potent ADCC *in vitro* and anti-tumor activity *in vivo*" [24].

The study above served as an important milestone in the development of humanized anti-GD2 monoclonal antibodies that could be used for the treatment of neuroblastoma in humans. For example, following the initial reporting of potential success in laboratory findings such as those above, hu3F8 testing moved into the clinical trials stage, with five clinical trials at one point in time. Similar to the strategy employed for dinutuximab, in addition to trials employing hu3F8 alone, combinatorial approaches were also tried and implemented. For example, a comparative analysis of two clinical trials, one of the humanized anti-GD2 antibody version developed by the Memorial Sloan Kettering Cancer Center in collaboration with

Y-mAbs Therapeutics, namely clinical trial NCT01419834 [25], and the other as reported by Kushner et al. in 2016 showed that combination of hu3F8 with GM-CSF was a promising approach for the treatment of neuroblastomas [26]. These and other similar trials laid the groundwork for the November 2020 FDA approval of the first humanized anti-GD2 antibody for the treatment of neuroblastomas.

16.7.10 Danyelza, Humanized Anti-GD2 Antibody, Hu3F8: FDA Approval for the Treatment of Relapsed Neuroblastoma

Extending the information regarding hu3F8 discussed above, a landmark drug development event occurred in 2020 with the November 25, 2020 announcement by the US FDA, which approved the use of naxitamab (hu3F8), marketed under the brand name of Danyelza for use in combination with GM-CSF "for pediatric patients one year of age and older and adult patients with relapsed or refractory high-risk neuroblastoma in the bone or bone marrow demonstrating a partial response, minor response, or stable disease to prior therapy [25]".

The FDA approved the use of naxitamab under its priority review, breakthrough therapy, and orphan drug designations. According to the FDA website, this approval was based on two clinical trials (Trial 1/NCT03363373 and Trial 2/NCT01757626) conducted across four centers in Spain and in the United States, which together included 97 patients with high-risk neuroblastoma in the bone and the bone marrow.

These trials included 61 males and 36 females, of whom the majority (90 patients) were under 12 years old, 5 patients from between the ages of 2 and 17, and two patients who were 18 and older. A total of 67 patients were White, 17 Asian, 3 Black or African American, 2 American Indian or Alaskan Native, while 8 others did not report their race. Furthermore, 6 patients self-reported as being of Hispanic or Latino origin, while 84 were not Hispanic or Latino, and 7 did not report their ethnicity.

Both the trials were designed for patients who had previously been treated for neuroblastoma but had either stopped responding or had relapsed. Patients received naxitamab by intravenous injection on days 1, 3, and 5 of each treatment cycle and also received subcutaneously administered GM-CSF daily, starting 5 days before the start of treatment with naxitamab and continuing with the GM-CSF treatment until the last day of the naxitamab dose for each cycle of the treatment regimen.

In its published report, FDA included that in one of the clinical trials, "approximately 45 percent of patients receiving Danyelza with GM-CSF experienced a complete or partial shrinkage of their cancer; for 30% of those patients, shrinkage lasted 6 months or longer. In another trial, 34% of patients receiving Danyelza with GM-CSF experienced a complete or partial shrinkage of their cancer; for 23% of those patients, shrinkage lasted 6 months or longer" [27].

In summary, anti-GD2 ganglioside antibodies such as the chimeric dinutuximab or the humanized naxitamab, both in combination with GM-CSF and both approved under orphan or priority use by the FDA for high-risk neuroblastoma patients offer promising therapies for what has been a difficult-to-treat disease. Thus, biologics such as monoclonal antibodies truly have the potential to alter the field of drug development, particularly regarding lesser-known, difficult to treat or recalcitrant or relapsing cancers, justifying the expenses related to the cost of research and development and the ultimate production of these important drugs to alleviate human suffering. Unfortunately, it will be more years for the development and approval of the more affordable biosimilars for the treatment of patients with neuroblastoma.

16.7.11 Targeting Less-Specific Proteins in Neuroblastoma: The Theoretical Consideration for the Use of Bevacizumab

The discovery and drug development process for the treatment of challenging cancers such as neuroblastomas is by and large a slow-moving field that needs to be driven first by breakthrough scientific discoveries, followed by months, if not years, of careful research and drug development and animal toxicity studies and careful and controlled human trials in patients. Thus, the drug pipeline and approval process to get neuroblastoma-based target-specific drugs is excruciatingly and exasperatingly slow, resulting in many delays, which is unspeakably tragic for the many children who ultimately succumb to this unfortunate and aggressive cancer.

The reality of the concerns mentioned above necessitates the use of combinatorial approaches that not only include multi-modal therapy such as surgery, radiation therapy, and chemotherapy but perhaps also the borrowing of approaches from the treatment of other cancers that may not be specific to neuroblastomas but instead target the underlying biology of tumor growth. One such drug target in neuroblastoma that has been studied is the vascular endothelial growth factor (VEGF) which may contribute to the growth of neuroblastomas [28].

Generally, growing tumors are often starved for oxygen and nutrient supplies, which triggers angiogenesis, driving the growth of new blood vessels that allow the promiscuous proliferation of the growing tumors. Serving primarily a regulatory role in angiogenesis, VEGF tyrosine kinase receptors on the surface of endothelial cells promote proliferation, migration, and survival and act as potent microvascular enhancers [28]. Thus, VEGF is implicated in tumor (1) angiogenesis, (2) vascularization, and (3) metastases.

The monoclonal antibody, bevacizumab, is an anti-VEGF that has been approved in the United States for treating a range of cancers. Concerning neuroblastoma, *in vivo* studies of bevacizumab in minimally inflicted mice demonstrated disease suppression [29]. These studies elucidated VEGF involvement in calreticulin neuronal differentiation in neuroblastoma. These advances in our understanding provide insight into the possibility of extrapolating these findings to fit that for clinical studies. As such, bevacizumab is approved for the treatment of adult patients with glioblastoma, a condition not very dissimilar to neuroblastoma that derives from nervous system cells.

As the patent of the originator Bevacizumab product has reached expiration, its biosimilar, Mvasi©, has been approved by the US FDA for all eligible indications authorized for its originator counterpart [9]. Thus, it is plausible to reflect upon the potential usefulness of Mvasi in the possible future treatment of neuroblastomas.

16.8 CONCLUSION

Although neuroblastoma presents many challenges that hinder precise molecular dissection of its etiology and causation on the one hand and treatment modalities on the other, recent advances in technology-enabled the production of target-specific monoclonal antibodies, many against the GD2 ganglioside found on neurons. Anti-GD2 antibodies run the gamut from murine to chimeric to humanized, with clinical trials providing evidence for combinatorial approaches such as multi-modal therapy including radiation therapy, chemotherapy, and the combination of anti-GD2 chimeric or humanized antibodies with biologics such as GM-CSF, which show tumor regression in children in clinical trials. However, the rarity of this orphan disease skews the economics of large-scale investment into research and drug development into an unfavorable position, necessitating the development of academia-government-industry partnerships as evidence for the development of the humanized anti-GD2 naxitimab, which became the latest drug in 2020 to be approved by the US FDA for use in the treatment of neuroblastoma.

REFERENCES

1. Rothenberg AB, Berdon WE, D'Angio GJ, Yamashiro DJ, Cowles RA. Neuroblastoma-remembering the three physicians who described it a century ago: James Homer Wright, William Pepper, and Robert Hutchison. *Pediatr Radiol.* 2009 Feb;39(2):155–60. doi: 10.1007/s00247-008-1062-z. Epub 2008 Nov 26. PMID: 19034443.

2. Implementation of the Biologics Price Competition and Innovation Act of 2009 | FDA [Internet]. [cited 2021 Apr 17]. Available from: https://www.fda.gov/drugs/guidance-compliance-regulatory-information/implementation-biologics-price-competition-and-innovation-act-2009
3. Thill M, Thatcher N, Hanes V, Lyman GH. Biosimilars: what the oncologist should know. *Future Oncol.* 2019 Apr;15(10):1147–65.
4. American Cancer Society. Last accessed on August 21, 2021. https://www.cancer.org/cancer/neuroblastoma/about/key-statistics.html
5. Van Arendonk KJ, Chung DH. Neuroblastoma: tumor biology and its implications for staging and treatment. *Children (Basel).* 2019 Jan 17;6(1):12. doi:10.3390/children6010012
6. Sokol E, Desai AV. The evolution of risk classification for neuroblastoma. *Children (Basel).* 2019 Feb 11;6(2):27. doi:10.3390/children6020027
7. Biedler JL, Helson L, Spengler BA. Morphology and growth, tumorigenicity, and cytogenetics of human neuroblastoma cells in continuous culture. *Cancer Res.* 1973 Nov; 33(11):2643–52
8. Rettig WJ, Spengler BA, Chesa PG, Old LJ, Biedler JL. Coordinate changes in neuronal phenotype and surface antigen expression in human neuroblastoma cell variants. *Cancer Res.* 1987 Mar 1;47(5):1383–9.
9. Ross RA, Biedler JL. Presence and regulation of tyrosinase activity in human neuroblastoma cell variants in vitro. *Cancer Res.* 1985;45:1628–32.
10. DeClerck YA, Bomann ET, Spengler BA, Biedler JL. Differential collagen biosynthesis by human neuroblastoma cell variants. *Cancer Res.* 1987;47:6505–10.
11. Ross RA, Spengler BA, Domenech C, Porubcin M, Rettig WJ, Biedler JL. Human neuroblastoma I-type cells are malignant neural crest stem cells. *Cell Growth Differ.* 1995;6:449.
12. Foley J, Cohn SL, Salwen HR, Chagnovich D, Cowan J, Mason KL, Parysek LM. Differential expression of N-myc in phenotypically distinct subclones of a human neuroblastoma cell line. *Cancer Res.* 1991 Dec 1;51(23 Pt 1):6338–45.
13. Boeva V, Louis-Brennetot C, Peltier A, Durand S, Pierre-Eugène C, Raynal V, Etchevers HC, Thomas S, Lermine A, Daudigeos-Dubus E, Geoerger B, Orth MF, Grünewald TGP, Diaz E, Ducos B, Surdez D, Carcaboso AM, Medvedeva I, Deller T, Combaret V, Lapouble E, Pierron G, Grossetête-Lalami S, Baulande S, Schleiermacher G, Barillot E, Rohrer H, Delattre O, Janoueix-Lerosey I. Heterogeneity of neuroblastoma cell identity defined by transcriptional circuitries. *Nat Genet.* 2017 Sep;49(9):1408–13.
14. van Groningen T, Koster J, Valentijn LJ, Zwijnenburg DA, Akogul N, Hasselt NE, Broekmans M, Haneveld F, Nowakowska NE, Bras J, van Noesel CJM, Jongejan A, van Kampen AH, Koster L, Baas F, van Dijk-Kerkhoven L, Huizer-Smit M, Lecca MC, Chan A, Lakeman A, Molenaar P, Volckmann R, Westerhout EM, Hamdi M, van Sluis PG, Ebus ME, Molenaar JJ, Tytgat GA, Westerman BA, van Nes J, Versteeg R. Neuroblastoma is composed of two super-enhancer-associated differentiation states. *Nat Genet.* 2017 Aug;49(8):1261–6.
15. Gautier M, Thirant C, Delattre O, Janoueix-Lerosey I. Plasticity in neuroblastoma cell identity defines a noradrenergic-to-mesenchymal transition (NMT). Cancers (Basel) 2021 Jun;13(12):2904. doi:10.3390/cancers13122904
16. Immunotherapy for Cancer – National Cancer Institute [Internet]. [cited 2021 Apr 19]. Available from: https://www.cancer.gov/about-cancer/treatment/types/immunotherapy
17. U.S. Food and Drug Administration. Implementation of the biologics price competition and innovation act of 2009. Last accessed on August 21, 2021. https://www.fda.gov/drugs/guidance-compliance-regulatory-information/implementation-biologics-price-competition-and-innovation-act-2009
18. Sait S, Modak S. Anti-GD2 immunotherapy for neuroblastoma. *Expert Rev Anticancer Ther.* 2017;17(10):889–904. doi:10.1080/14737140.2017.1364995
19. Handgretinger R, Anderson K, Lang P et al. A phase I study of human/mouse chimeric antiganglioside GD2 antibody ch 14.18 in patients with neuroblastoma. *Eur J Cancer.* 1995;31:261–7.
20. Ladenstein R, Weixler S, Baykan B et al. Ch14.18 antibody produced in CHO cells in relapsed or refractory Stage 4 neuroblastoma patients: a SIOPEN Phase 1 study. *mAbs.* 2013;5(5):801–9.
21. Yu A, Gilman A, Ozkaynak M et al. Anti-GD2 antibody with GM-CSF, interleukin-2, and isotretinoin for neuroblastoma. *N Engl J Med.* 2010;363(14):1324–34.
22. U.S. Food and Drug Administration. FDA approves first therapy for high-risk neuroblastoma. Last accessed on August 21, 2021. http://wayback.archive-it.org/7993/20170112023806/http://www.fda.gov/NewsEvents/Newsroom/PressAnnouncements/ucm437460.htm.
23. Osenga KL, Hank JA, Albertini MR et al. A phase I clinical trial of the hu14.18-IL2 (EMD 273063) as a treatment for children with refractory or recurrent neuroblastoma and melanoma: a study of the Children's Oncology Group. *Clin Cancer Res: An Off J Am Assoc Cancer Res.* 2006;12(6):1750–9.

24. Cheung NK, Guo H, Hu J, Tassev DV, Cheung IY. Humanizing murine IgG3 anti-GD2 antibody m3F8 substantially improves antibody-dependent cell-mediated cytotoxicity while retaining targeting in vivo. *Oncoimmunology.* 2012 Jul 1;1(4):477–86. doi: 10.4161/onci.19864. PMID: 22754766; PMCID: PMC3382886.
25. Clinical Trials.gov. Humanized 3F8 monoclonal antibody (Hu3F8) in patients with high-risk neuroblastoma and GD2-positive tumors. Last accessed on August 21, 2021. https://clinicaltrials.gov/ct2/show/NCT01419834.
26. Kushner B, Cheung IY, Basu E et al. Phase I study of anti-GD2 humanized 3F8 plus GM-CSF: High dosing and major responses in patients with resistant high-risk neuroblastoma. *Adv Neuroblastoma Res.* 2016;Abs 57;2016.
27. U.S. Food and Drug Administration. FDA grants accelerated approval to naxitamab for high-risk neuroblastoma in bone or bone marrow. Last accessed on August 21, 2021. https://www.fda.gov/drugs/resources-information-approved-drugs/fda-grants-accelerated-approval-naxitamab-high-risk-neuroblastoma-bone-or-bone-marrow
28. Roy Choudhury S, Karmakar S, Banik NL, Ray SK. Targeting angiogenesis for controlling neuroblastoma. *J Oncol.* 2012; 2012:782020.
29. Melosky B, Reardon DA, Nixon AB, Subramanian J, Bair AH, Jacobs I. Bevacizumab biosimilars: scientific justification for extrapolation of indications. *Future Oncol.* 2018 Oct;14(24):2507–20.

17

Paradigm Shift in Cancer Therapy

Chimeric Antigen Receptor T Cell Therapy

Anh Nguyen and Yihui Shi

Corresponding author: Yihui Shi

Contents

17.1 INTRODUCTION

Cytotoxic T lymphocytes play a crucial role in controlling and killing cancer cells. Chimeric antigen receptor (CAR) T cells are autologous T cells that have been changed to produce a tumor antigen-specific CAR against tumor-associated antigens (TAAs) following *ex vivo* expansion. T cell-mediated immunotherapies have been known to enhance the efficacy of cancer treatments. Approaches to how to effectively use CAR-T cell therapy for treatments are still undergoing research. The therapeutic goal is to increase anti-tumor efficacy and safety through target selection and target specificity. There are several side effects of CAR-T cell treatment, including cytokine release syndrome (CRS), systemic inflammatory response upon the release of cytokines, on-target off-tissues, and neurotoxicity (Brentjens et al. 2011; Hay et al. 2017). A more recent challenge of CAR-T cell therapy is translating into solid tumors. Targeted antigens for hematologic malignancies are not always transferable to solid tumors, limiting treatment intensity and effectiveness. CAR-T cell therapy also exhibits

DOI: 10.1201/9780429485626-17

impressive results in curbing autoimmunity. To this end, we review the challenges faced and prospects of novel CAR designs in adoptive CAR-T therapy in treatments of cancer and autoimmune diseases.

17.2 GENERATIONS OF CAR-T CELLS

Researchers have created several generations of CAR-T cells to improve the efficacy and persistence of CAR-T therapy (Figure 17.1). CAR-T cells have three parts: a single-chain variable domain of an antibody (scFv), a transmembrane domain, and a signal transduction domain of the T-cell receptor (TCR) (Brudno and Kochenderfer 2018). The first-generation CAR-T cells use CD3ζ as an intracellular signaling domain to enhance the therapeutic effect (Eshhar et al. 1993). The second-generation CAR-T cells express a costimulatory molecule (CD27, CD28, OX40, and 4-1BB) to prolong the anti-tumor effects (Song et al. 2012). The third-generation CAR-T cells express two costimulatory domains (CD28 and 4-1BB) (Till et al. 2012). Some modified third-generation CAR-T cells can secret cytokines and pro-inflammatory ligands to enhance the efficacy and persistence of CAR-T cells (Hombach et al. 2012). More recently, the fourth-generation CAR-T cells (TRUCKs) are comprised of an intracellular domain with two costimulatory molecules (CD28 and 4-1BB), a transmembrane domain, an extracellular domain, and cytokine genes activated upon recognition by a specific antigen (Chmielewski and Abken 2015).

Upon discovering T cells in 1961, scientists achieved early success in bone marrow transplantation to treat cancer as the first known immunotherapy in 1973. That laid the path for developing four generations of CAR-T cell therapy in the past three decades, starting with developing first-generation CARs in 1993 (Table 17.1).

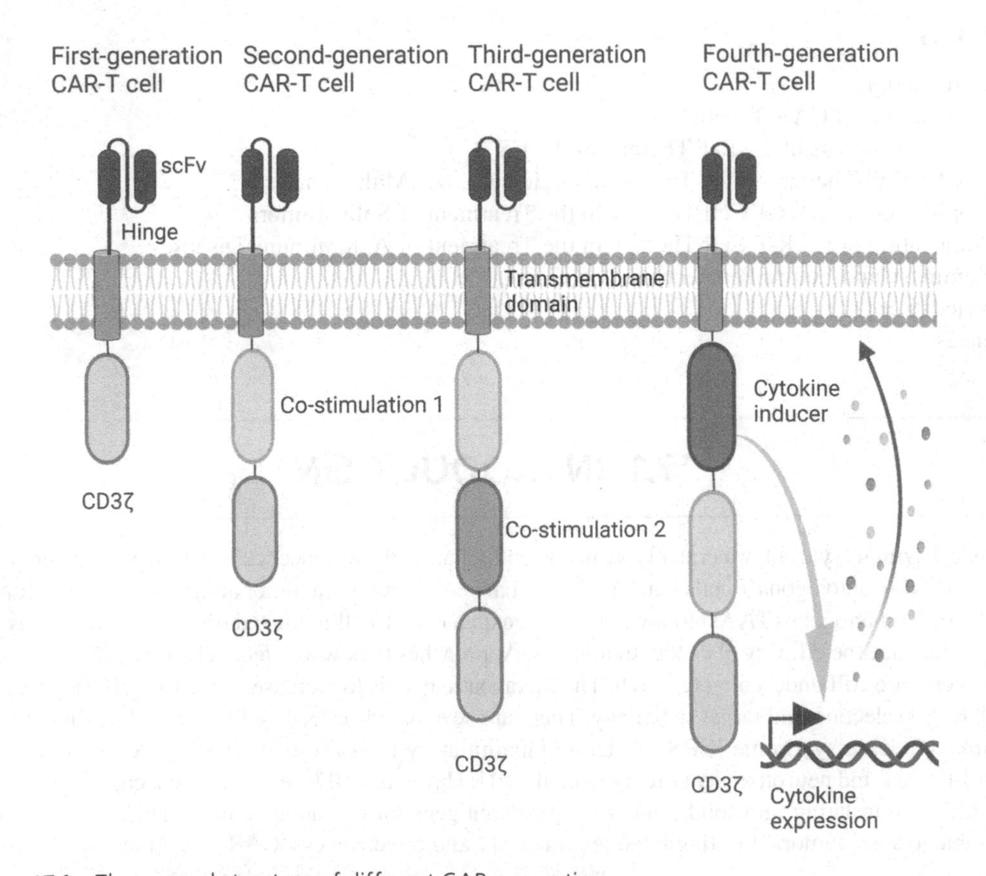

FIGURE 17.1 The general structure of different CAR generations.

TABLE 17.1 Progress in CAR-T cell therapy

TIME	*GENERATION*	*RESEARCHER*	*INSTITUTE*	*REFERENCES*
1993	First-generation CARs	Zelig Eshhar	NCI, Rosenberg lab	Han et al. (2021)
2002	Second-generation CARs (CD28-based)	Michael Sadelain	Memorial Sloan Kettering Cancer (MSKCC)	Han et al. (2021)
2004	Second-generation CARs (4-1BB-based)	Dario Campana	National University of Singapore (NUS)	Han et al. (2021)
2010	Third-generation CARs	Michael Sadelain	Memorial Sloan Kettering Cancer (MSKCC)	Han et al. (2021)
2013	Fourth-generation CARs (TRUCK)	Hinrich Abken	University of Cologne	Han et al. (2021)

17.3 CURRENT PROGRESS OF CAR-T THERAPY IN THE US

CAR-T cells have exhibited rousing success and efficacy in the treatment of hematologic malignancies. In 2017, the first CAR-T cell therapy, tisagenlecleucel (Kymriah®; Novartis), crossed the regulatory finish line and received approval from the US Food and Drug Administration (FDA) for treatment of relapsed/refractory hematological malignancies. The second FDA-approved CAR-T cell is axicabtagene ciloleucel (Yescarta®; Kite Pharma). The CAR-T cells that have been broadly applied in the clinic have been second-generation CARs. The FDA continues to approve brexucabtagene autoleucel (Tecartus®) in 2020 for relapsed/refractory mantle cell lymphoma (MCL) (Liu et al. 2017; AlDallal 2020; Reagan and Friedberg 2021). In 2021, lisocabtagene maraleucel (Breyanzi®) and idecabtagene vicleucel (Abecma®) were approved to treat adult patients with relapsed/refractory B-cell non-Hodgkin lymphomas and multiple myeloma (MM), respectively (Table 17.2).

TABLE 17.2 FDA-approved CAR-T therapy

DRUG	*TARGET*	*INDICATION*
Tisagenlecleucel (Kymriah®)[a]	CD19	Pediatric or young patients up to age 25 years with relapsed or refractory acute lymphoblastic leukemia (ALL) Patients with relapsed or refractory large B-cell lymphoma (LBCL) after two or more lines of therapy (diffuse large B-cell lymphoma (DLBCL), high-grade B-cell lymphoma (BCL), DLBCL that results from follicular lymphoma (FL))
Axicabtagene ciloleucel (Yescarta®)[b]	CD19	Patients with relapsed large B-cell lymphoma (LBCL) following two or more lines of systemic therapy (DLBCL, primary mediastinal B-cell lymphoma (PMBCL), high-grade BCL, DLBCL that results from FL) Patients with relapsed or refractory FL after two or more lines of systemic therapy
Brexucabtagene autoleucel (Tecartus®)[b]	CD19	Patients with relapsed or refractory mantle cell lymphoma (MCL) with five or more previous lines of therapy
Lisocabtagene maraleucel (Breyanzi®)[c]	CD19	Patients with relapsed or refractory large B-cell lymphoma (LBCL) after two or more lines of systemic therapy (DLBCL, high-grade BCL, DLBCL transformed from indolent lymphoma, PMBCL, FL grade 3B)
Idecabtagene vicleucel (Abecma®)[d]	BCMA	Patients with relapsed or refractory multiple myeloma (MM) after four or more prior lines of therapy

Notes:

[a] Novartis.
[b] Kite Pharma/Gilead.
[c] Juno Therapeutics (Bristol-Myers Squibb).
[d] Celgene (Bristol-Myers Squibb).

Concurrently, pre-clinical and clinical trials on both hematologic and solid tumors are in progress using combined CAR-T cell therapies for a necessary boost in therapeutic effects, specificity, and persistence. Most current trials are focusing on B-cell malignancies, including relapsed or refractory large B-cell lymphoma (LBCL), acute myeloid leukemia (AML), and multiple myeloma (MM).

17.4 CAR-T CELL THERAPY IN THE TREATMENT OF HEMATOLOGIC MALIGNANCIES

The B cell biomarker, CD19, is expressed at a high level on the surface of malignant B cells, including B cell acute lymphoblastic leukemia (B-ALL) (Lee et al. 2015; Davila and Brentjens 2016; Turtle et al. 2016), chronic lymphocytic leukemia (CLL) (Porter et al. 2011, 2015), and non-Hodgkin lymphoma (Kochenderfer et al. 2013, 2015). Targeted therapeutic CAR-T cells against CD19 have been extensively studied to treat B cell malignancies. Besides CD19, other targeted antigens for CAR-T immunotherapy include CD20, BCMA, C-type lectin-like molecule 1 (CLL1), mesothelin (MSLN), and PD-1/PD-L1 (Jensen et al. 2010; Wei et al. 2019; Lyu et al. 2020).

Transmembrane receptor, BCMA, exclusively expressing on differentiated plasma cells, is a confirmed target for multiple myeloma (MM). Recent research comparing BCMA/CD3 bispecific T cells with traditional second-generation anti-BCMA CAR-T cells has shown similar *in vitro* and *in vivo* tumor-killing efficacy. However, anti-BCMA CAR-T cells did not demonstrate superior anti-tumor dynamics. In fact, BCMA/CD3 bispecific T cells eradicated tumors *in vivo* more rapidly than anti-BCMA CAR-T cells which require a specific time to expand in the tumor microenvironment and to become fully effective after infusion (DiLillo et al. 2021). Concurrently, Dann et al. targeted BCMA antigen through the first pre-clinical and clinical study of BCMA-targeted fourth-generation CAR-T cells expressing IL-7 and CCL19 (BCMA-7 × 19 CAR-T cells) to improve tumor-killing activity in relapsed or refractory MM patients. The research has demonstrated that BCMA-7 × 19 CAR-T cells exhibited more robust expansion, differentiation, and cytotoxicity than second-generation anti-BCMA CAR-T cells. Investigators have been conducting a human clinical trial with promising safety and efficacy, and the preliminary results implicate clinical application of BCMA-targeted fourth-generation CAR-T cells for malignancies with BCMA overexpression (Duan et al. 2021).

Studies reported the remission rate of CD19 CAR-T cell therapy in B-ALL is up to 90%. For more control of CAR-T cell reactivity, Roybal et al. modified anti-CD19 CAR-T cells with a synthetic Notch (synNotch) receptor, creating CD19 synNotch CAR-T cells that possess a customized gene expression program and regulate cytokine releases and other therapeutic molecules independent of TCR or costimulatory bindings (Morsut et al. 2016; Roybal et al. 2016). In another study, investigators designed CAR-T cells that co-target CD19 and MSLN, a TAA broadly overexpressed on various malignant tumor cells (Beatty et al. 2014). Blankenstein et al. studied CD19/MSLN CAR-T cells that recognize the first antigen through anti-CD19 CAR and the other through anti-MSLN CAR. When both antigens are simultaneously engaged, CD19/MSLN CAR-T cells can be fully activated. However, this strategy has a limited effect due to transient effectiveness or partial activation of T cells once CAR expression of one antigen is absent (Blankenstein 2016). Additionally, the limitations exist when dual antigens, CD19 and MSLN, fail to coexpress in one tumor cell or when both antigens are coexpressed in non-cancerous cells, resulting in B cell depletion by anti-CD19 CAR-T cells (Kochenderfer et al. 2010, 2012).

There remain the challenges associated with CAR-T cell therapy. Emerging CD19-negative tumor cells can surpass anti-CD19 CAR-T detection. The loss of target means tumor cells can escape the killing activity of CAR-T cells, suggesting that the stability of antigen expression is crucial for CAR-T cell efficacy (Gardner et al. 2016; Maude 2018; Maude et al. 2018). The loss of antigen occurs most commonly in B-ALL (14%), CLL, and PMLBCL. A typical loss of target associated with CD19 antigen is different truncated forms of CD19 generated due to gene mutation, causing CD19-bearing cells undetected to CAR-T cells (Wei et al. 2019). The selective pressure treated with blinatumomab against CD19/CD3 has increased efficacy in B cell malignancies. Also, antigen escape is not limited to CD19. An AML patient has shown reduced expression of CD33 when treated with anti-CD33 CAR-T cell therapy. Recently, Perera et al. took a different approach by targeting cytokine receptor CCR4 that is extensively expressed in a range of T-cell malignancies. The CCR4 CAR-T cells exert potent cytotoxicity against tumor tissues in a broader spectrum of T-cell malignancies, including T-cell leukemia (ATL), cutaneous T-cell lymphoma (CTCL), anaplastic large cell lymphoma (ALCL) (Perera et al. 2017).

In a study on patients with persistent renal and PMLBCL infused with anti-CD19 CAR-T cells, CD19 translocated from cell surface to cytosol, thus losing CAR-T cell efficacy. Also, investigators faced a similar challenge studying the effectiveness of CD22 CAR-T cells when CD22 is undetected due to altered signaling pathways and epigenetic modification, which occur more often compared to gene mutation. CD22 is also known to be more likely subject to antigen loss than CD19 (Wei et al. 2019). The challenges suggest a multiple-antigen approach and endogenous anti-tumor activity. An example is the novel bi-specific humanized CD37/CD19 CAR-T cells that specifically target either CD37-positive or CD37 and CD19-positive cells to surmount the loss of CD19 antigen inhibiting xenograft tumor growth with impressive efficacy (Golubovskaya et al. 2021).

Recently, researchers have designed CD7-targeted CAR-T cells, which are restricted to T cells and natural killer (NK) cells. CD7 is highly expressed on the surface of T-ALL and different hematologic malignancies. Scientists can target CD7-positive hematologic malignancies that lack targeted therapies using CAR-T cells. Cooper et al. generated novel CD7-targeted CAR-T cells through CRISPR/Cas9 gene editing to delete CD7 and T-cell receptor α chain (TRAC) in allogeneic T cells to create CD7-deleted and TRAC-deleted CAR-T cells targeting CD7 (UCART7). UCART7 knocking out CD7 prevents fratricide through the TCR-mediated immune response without compromising their potency. This strategy has directed CAR-T cells to kill T-ALL cells *in vitro* at 95% efficiency compared to CD19-CAR-T cells as a control and target T-ALL *in vivo* without graft vs. host disease. Additionally, the researchers found that UCART7 potentially kills recipient alloreactive T cells and NK cells, thus inhibiting the rejection of host T cells and NK cells while maintaining UCART7 vigorous activity (Cooper and DiPersio 2019).

17.5 APPLICATION OF CAR-T CELL THERAPY IN THE TREATMENT OF SOLID TUMORS

Solid tumors present barriers to the development of efficient CAR-T cell therapies. The most significant obstacles include antigen heterogeneity, antigen escape, inefficient trafficking and infiltration, and an immunosuppressive and hostile tumor microenvironment, including hypoxia, suppressive immune cells, immunosuppressive cytokines, immune checkpoint molecules, and outgrowing tumor vascularization (Heyman and Yang 2019). Unlike B cell malignancies that specifically express CD19 on the surface, it is challenging to determine the ideal surface target of TAAs or tumor-specific antigens (TSAs) universally expressed in the spectrum of solid cancer tissues.

Therefore, dual-targeting is superior to single-antigen targeting to differentiate between target and off-target cells and improve the limited therapeutic effects. Although the challenges of adoptive CAR-T cell therapy in solid tumors are multi-factorial, researchers have made significant progress in applying CAR-T cell therapy in solid tumors. To date, the efficacy of CAR-T cell therapy is generally limited to cell surface antigens. Given the shortage of TSAs and ideal TAAs for a spectrum of solid tumor cell lines, several TAAs (such as MSLN, MUC1, CD171, ICAM1, EGFR) have been studied to overcome this challenge.

MSLN, glycoprotein anchored in the plasma membrane, is highly expressed in many solid tumors, including metastatic pleural mesothelioma, pancreatic, ovarian, and lung adenocarcinoma comparison to normal tissues (Chang and Pastan 1996; Hassan and Ho 2008; Pastan and Hassan 2014). MSLN can act as a promising target to overcome the antigen loss in CD19 CAR-T cells. The over-expression of MSLN makes it an attractive target for CAR-T cell immunotherapy (Hassan and Ho 2008). Ye et al. investigated the anti-tumor reactivity of MSLN CAR-T cells in a mice xenograft model of non-small cell lung cancer (NSCLC) and mesothelioma. CAR-T cells targeting MSLN kill tumor tissue significantly higher than control T cells (Ye et al. 2019). Second-generation MSLN CAR-T cells constructed with costimulatory domain (CD28, 4-1BB, or OX40) along with CD3ζ induced higher T cell activation, persistence, and cytotoxicity and allowed dose-dependent CAR-T therapeutics (Lanitis et al. 2012; Moon et al. 2014).

Investigators further engineered third-generation MSLN CAR-T cells with two costimulatory domains, including CD28, 4-1BB, OX40, TLR2, and DAP10, to enhance CAR-T cell efficacy. Anti-MSLN CD3ζ/CD28/TLR2 CARs exhibited strong tumor-killing ability and released an efficiently higher level of IL-2, IFN-γ, and GM-CSF, compared to normal second-generation CAR-T cells without TLR2 costimulation (Lai et al. 2018). Furthermore, researchers engineered T cells that simultaneously express CARs targeting CD19 and MSLN to direct CAR-T cells to MSLN-positive tumor cells. MSLN CAR-T cells improved anti-tumor activity *in vivo,* and clinical trials are in progress in lung cancer, ovarian cancer, malignant pleural mesothelioma (MPM), pancreatic ductal adenocarcinoma (PDAC), triple-negative breast cancer (TNBC) (Sotoudeh et al. 2019; Ye et al. 2019; Klampatsa, Dimou, and Albelda 2020).

CAR-T cell therapy directed against MSLN remains a promising therapeutic strategy. However, clinical trials do not guarantee the efficacy and persistence of MSLN CAR-T cells after infusion to patients with different types of MSLN-positive tumors. Insufficient tumor infiltration can limit CAR-T cell potency, or transient mRNA expression of CAR temporarily redirects T cell activity and increases toxic side effects (Zhang et al. 2016). Researchers at the University of Pennsylvania induced CAR expression by designing MSLN-specific mRNA CAR-T therapy that showed anti-tumor activity. However, patients require frequent infusion due to the transient expression of a transgene. In another study, Yang et al. discovered a novel approach to CAR-T cell therapy by harvesting exosomes secreted from MSLN-targeted CAR-T cells. The study demonstrated that exosomes facilitated the anti-tumor effect through targeting MSLN-positive TNBC cells and producing perforin and granzyme B for tumor killing with low toxicity *in vivo* (Yang et al. 2020).

Furthermore, MUC1 is the second targetable tumor antigen (Qi et al. 2015). MUC1 is overexpressed by most adenocarcinomas. Studies have shown effectiveness in pancreatic cancer using CAR-T cells targeting MUC1 (Tarp et al. 2007; Posey et al. 2016). The overexpression of glycosylated MUC1 in over 90% of breast cancer cases and 95% of TNBC cases causes MUC1 a relevant antigen for targeted immunotherapy (Roy et al. 2017). Recently, Zhou et al. generated novel tMUC1-specific CAR-T cells comprising scFv of tMUC antibody, TAB004, to reduce the tumor growth in TNBC. This novel CAR construct attacks only tMUC1-expressing breast epithelial cells (Zhou et al. 2020). MUC1 targeted CAR-T cells engineered with IL-4 also enhance anti-tumor efficacy (Wilkie et al. 2010; Ramachandran, Dimberg, and Essand 2017). Currently, MUC1-CAR-T cell therapy is the most studied in clinical trials. Researchers have evaluated the therapeutic efficacy safety of MUC1-CAR-T cells in patients with relapsed or refractory TNBC in a phase I/II study (NCT02587689). Besides MUC1, tissue factors (TF) are abundantly expressed on the surface of tumor cells in lung cancer, melanoma, and other cancers. TF-targeted CAR-T cells have efficiently killed TF-positive tumor cells *in vitro* (Zhang et al. 2017). In a different effort, scientists replaced the scFv domain in CARs with a modified avidin motif to create CAR-T cells that could recognize a range of targets with biotinylated tumor-specific molecules. Subsequently, there are continued studies involved in developing universal CAR, including anti-5B9-tag scFv, anti-FITC scFv, and anti-PNE (peptide neoepitope) scFv (Rodgers et al. 2016; Cho, Collins, and Wong 2018; Cartellieri et al. 2016; Ma et al. 2016).

Investigators noted that neural tumor cells like glioblastoma and neuroblastoma exclusively express adhesion molecule glycosylated-CD171 instead of normal cells. CAR-T cells targeting glycosylated

CD171 have specifically encountered tumor cells, reducing "on-target off-tissue" toxicity and enhancing safety in a pre-clinical trial of neuroblastoma (Kunkele et al. 2017; Wachowiak et al. 2018). In a pre-clinical *in vitro* test, Andersch et al. constructed CD171- and GD2-specific CAR-T cells effectively kill retinoblastoma cells upon encountering either antigens. The study suggested that both antigens are highly effective targets for CAR-T therapy against retinoblastoma. However, the administration of sequential CAR-T treatment, CD171-specific CAR-T before GD2-specific CAR-T cells followed by subsequent CD171-specific CAR-T, increased tumor-killing efficacy and overcame the tumor antigen escape (Andersch et al. 2019). Clinical trials have currently been studied on the effectiveness of CD171-directed CAR-T cells on patients with recurrent or refractory neuroblastoma (NCT02311621) (Kunkele et al. 2017).

Recently, intercellular adhesion molecule-1 (ICAM1) has been selected as a therapeutic target for TNBC. TNBC does not respond to anti-HER2-targeted therapies or hormone therapy because of the absence of estrogen receptor (ER), progesterone receptor (PR), and human epidermal growth factor receptor 2 (HER2). However, TNBC is reported to overexpress ICAM1 required for cell adhesion and migration of immune cells to sites of inflammation. Elevated levels of ICAMs in tumor tissue are associated with metastasis and poor prognosis. Wei et al. have created CAR-T cells expressing extracellular scFv structure against ICAM1 to recognize ICAM1-expressing TNBC cells preferentially. In *in vitro* and *in vivo* studies, anti-ICAM1 CAR-T cells significantly reduced tumor growth and improved individual survival, suggesting immunotherapy against ICAM1-positive TNBC tumors an excellent anti-tumor response (Wei et al. 2020). Recently, Dong et al. delineate that IFNγ upregulates ICAM1 expression in tumor cells and overcomes the inhibitory function of PD-L1/PD-1 on CAR-T activity, priming ICAM1-overexpressing tumor cells for enhanced CAR-T cell cytotoxicity, which is abolished in ICAM1 knockout cells (Dong et al. 2021).

Among tumor antigens, EGFR variant III (EGFRvIII) is abundantly expressed in many cancer types and associated with survival, invasion, angiogenesis, and resistance to radiation and chemotherapy. In a study of EGFR-amplified glioblastoma patients, 57% of 106 primary patients and 87.5% of 40 recurring EGFR-amplified glioblastoma patients had EGFRvIII (Felsberg et al. 2017). Therefore, EGFRvIII is an ideal therapeutic target for CAR-T cell therapy in many cancers as EGFRvIII-specific CAR-T cells can target different tumors with EGFRvIII overexpression, including glioblastoma and lung cancer (Miao et al. 2014; Sahin et al. 2018). EGFRvIII is recognized by a biotinylated monoclonal antibody (biotin-4G1). A study designed biotin-4G1 specific avidin-CAR-T cells that targeted only TSA EGFRvIII in glioblastoma (Learn et al. 2004; Liu et al. 2015; Johnson et al. 2015). Zhang et al., for the first time, found that EGFRvIII-targeted CAR-T cells efficiently killed metastatic EGFRvIII-positive lung cancer in both *in vivo* and *in vitro* and prolonged animal survival (Zhang et al. 2019). To increase CAR-T cell specificity on tumor cells and reduce "on-target off-tumor" cytotoxicity, Jiang et al. engineered the M27-derived CAR construct co-targeting EGFR and EGFRvIII, entirely restricted to human cancer. M27-scFv structure designed to recognize with high specificity both EGFR- and EGFRvIII-positive glioblastoma but not EGFR expressed in normal cells. This CAR is also a promising strategy to treat other cancer types with EGFR and EGFRvIII overexpression (Jiang et al. 2018). Li et al. improved the efficacy of anti-tumor therapeutics through combinatory targeting of EGFRvIII and trapping TGF-β. The investigators engineered anti-EGFRvIII CAR-T cells with TGFRII ectodomain to generate T cells resistant to TGF-β. These CAR-T cells have controlled tumor growth, enhanced anti-tumor ability, and increased animal survival in a glioma mouse model (Li et al. 2020). Investigators also combined therapy targeting EGFRvIII and PD-1 blockage by PD-1 siRNA or CRISPR/Cas9 model, and this strategy showed enhanced efficacy in both second-and third-generation anti-EGFRvIII CAR-T cells. The anti-EGFRvIII CAR-T cells with disruption of genomic PD-1 inhibited tumor growth and offered a more potent therapeutic effect on EGFRvIII- and PD-L1-positive glioblastoma cells (Choi et al. 2019; Nakazawa et al. 2020; Zhu et al. 2020). These findings suggest a consistent effect to one reported by a study of PD-1 disrupted MSLN-targeted CAR-T cells (Hu et al. 2019). Despite promising results on animal models, a recent clinical trial of patients with recurrent glioblastoma did not exhibit clinically comparable effects in tumor suppression or individual survival (Goff et al. 2019).

17.6 APPLICATION OF CAR-T CELL THERAPY IN THE TREATMENT OF AUTOIMMUNE DISEASES

Regulatory T cells (Tregs) are a subset of T cells that can suppress immune responses and control autoimmunity. The mechanisms that Tregs suppress immune responses include the production of the immunosuppressive cytokines (IL-10, TGF-β) and reduced proliferation, differentiation, and activation of effector T cells. They may also directly suppress B cell activation and inhibit the proliferation and differentiation of NK cells (Chen et al. 2003). Identification of Tregs potent immunosuppression enables researchers to develop a therapeutic approach, CAR-Tregs, to treat autoimmune disease. Tregs modified with CAR construct to induce suppressive function toward antigens of interest. The costimulatory domain CD28 on CAR-Tregs does not interfere with the expression of FoxP3 for Treg differentiation and development but inhibits cytokine production and suppresses immune responses (Boroughs et al. 2019).

Multiple sclerosis is an autoimmune disease of the central nervous system (CNS). Fransson et al. developed CAR/FoxP3-engineered Tregs targeting myelin oligodendrocyte glycoprotein (MOG) and expressing FoxP3 that is important for the development and function of Tregs. CARαMOG-expressing Tregs can bind to MOG-expressing cells and prevent immune attacks on MOG-positive oligodendrocytes in the CNS. At the same time, transcription factor FoxP3 maintains the function of the Treg phenotype. CNS-targeting Tregs have shown an efficient suppression of inflammation and improvement of disease symptoms (Fransson et al. 2012). The therapeutic approach exhibited effectiveness in other autoimmune diseases, like ulcerative colitis and type 1 diabetes. To suppress the inflammatory reaction in ulcerative colitis, anti-CEA CAR-Tregs are generated to target carcinoembryonic antigen (CEA) on the cell surface and alleviate the disease symptoms (Blat et al. 2014). Likewise, Tenspolde et al. engineered Tregs with novel insulin-specific CAR to treat NOD/TtJ female mice with type 1 diabetes. Converted insulin CAR-Tregs have shown a successful suppressive capacity in diabetic mice. The promising results from CAR-Treg treatment in multiple sclerosis, colitis, and type 1 diabetes suggest a possibility of adoptive immunotherapy such as CAR-Tregs in general autoimmunity (Tenspolde et al. 2019). In another approach, adoptive immunotherapy using CAR-T cells targeting CD19-positive B cells has been proven effective in sustaining B cell depletion and decreasing disease manifestations in systemic lupus erythematosus (Kansal et al. 2019).

17.7 SUMMARY AND FUTURE PERSPECTIVES

CAR-T immunotherapy is a field of cancer treatment and autoimmune diseases with great potential. Thus far, CD19 is the most studied target, while exploring multiple targets and universal CAR-T cells have remained a research focus. The evolution of strategies as treatments for hematological malignancies has improved the initial efficacy and persistence of CAR-T cells. Several generations of CAR-T cells have been designed to enhance the tumor-killing activity, endurance, infiltration to solid tumor tissues, and regulation of the immune microenvironment. Combined therapies have been developed as an emerging strategy to recognize tumor tissues and minimize "on-target off-tumor" effects. Future studies could be focused on further identifying ideal target selection and targeting multiple TAAs. Equally important, optimal timing in treatment, the fitness, and exhaustion of CAR-T cells have been further addressed for prolonged survival and expansion within the tumor immunosuppression microenvironment. With the introduction of checkpoint inhibitors and novel approaches in the regional delivery of CAR-T cells, cancer immunotherapy may become more potent for treating solid tumors. Finally, researchers continue to target gene identifying and editing to enhance the persistence, expansion, and function of CAR-T cells.

ACKNOWLEDGMENT

Figure 17.1 was created with Biorender.com

REFERENCES

AlDallal, S. M. 2020. Yescarta: a new era for non-Hodgkin lymphoma patients. *Cureus* 12 (11):e11504.

Andersch, L., J. Radke, A. Klaus, et al. 2019. CD171- and GD2-specific CAR-T cells potently target retinoblastoma cells in preclinical in vitro testing. *BMC Cancer* 19 (1):895.

Beatty, G. L., A. R. Haas, M. V. Maus, et al. 2014. Mesothelin-specific chimeric antigen receptor mRNA-engineered T cells induce anti-tumor activity in solid malignancies. *Cancer Immunol Res* 2 (2):112–20.

Blankenstein, T. 2016. Receptor combinations hone T-cell therapy. *Nat Biotechnol* 34 (4):389–91.

Blat, D., E. Zigmond, Z. Alteber, T. Waks, and Z. Eshhar. 2014. Suppression of murine colitis and its associated cancer by carcinoembryonic antigen-specific regulatory T cells. *Mol Ther* 22 (5):1018–28.

Boroughs, A. C., R. C. Larson, B. D. Choi, et al. 2019. Chimeric antigen receptor costimulation domains modulate human regulatory T cell function. *JCI Insight* 1–19.

Brentjens, R. J., I. Riviere, J. H. Park, et al. 2011. Safety and persistence of adoptively transferred autologous CD19-targeted T cells in patients with relapsed or chemotherapy refractory B-cell leukemias. *Blood* 118 (18):4817–28.

Brudno, J. N., and J. N. Kochenderfer. 2018. Chimeric antigen receptor T-cell therapies for lymphoma. *Nat Rev Clin Oncol* 15 (1):31–46.

Cartellieri, M., A. Feldmann, S. Koristka, et al. 2016. Switching CAR T cells on and off: a novel modular platform for retargeting of T cells to AML blasts. *Blood Cancer J* 6 (8):e458.

Chang, K., and I. Pastan. 1996. Molecular cloning of mesothelin, a differentiation antigen present on mesothelium, mesotheliomas, and ovarian cancers. *Proc Natl Acad Sci USA* 93 (1):136–40.

Chen, Z. M., M. J. O'Shaughnessy, I. Gramaglia, et al. 2003. IL-10 and TGF-beta induce alloreactive CD4+CD25- T cells to acquire regulatory cell function. *Blood* 101 (12):5076–83.

Chmielewski, M., and H. Abken. 2015. TRUCKs: the fourth generation of CARs. *Expert Opin Biol Ther* 15 (8):1145–54.

Cho, J. H., J. J. Collins, and W. W. Wong. 2018. Universal chimeric antigen receptors for multiplexed and logical control of T cell responses. *Cell* 173 (6):1426–38.e11.

Choi, B. D., X. Yu, A. P. Castano, et al. 2019. CRISPR-Cas9 disruption of PD-1 enhances activity of universal EGFRvIII CAR T cells in a preclinical model of human glioblastoma. *J Immunother Cancer* 7 (1):304.

Cooper, M. L., and J. F. DiPersio. 2019. Chimeric antigen receptor T cells (CAR-T) for the treatment of T-cell malignancies. *Best Pract Res Clin Haematol* 32 (4):101097.

Davila, M. L., and R. J. Brentjens. 2016. CD19-targeted CAR T cells as novel cancer immunotherapy for relapsed or refractory B-cell acute lymphoblastic leukemia. *Clin Adv Hematol Oncol* 14 (10):802–8.

DiLillo, D. J., K. Olson, K. Mohrs, et al. 2021. A BCMAxCD3 bispecific T cell-engaging antibody demonstrates robust anti-tumor efficacy similar to that of anti-BCMA CAR T cells. *Blood Adv* 5 (5):1291–304.

Dong, E., X. Z. Yue, L. Shui, et al. 2021. IFN-gamma surmounts PD-L1/PD1 inhibition to CAR-T cell therapy by upregulating ICAM-1 on tumor cells. *Signal Transduct Target Ther* 6 (1):20.

Duan, D., K. Wang, C. Wei, et al. 2021. The BCMA-targeted fourth-generation CAR-T cells secreting IL-7 and CCL19 for therapy of refractory/recurrent multiple myeloma. *Front Immunol* 12:609421.

Eshhar, Z., T. Waks, G. Gross, and D. G. Schindler. 1993. Specific activation and targeting of cytotoxic lymphocytes through chimeric single chains consisting of antibody-binding domains and the gamma or zeta subunits of the immunoglobulin and T-cell receptors. *Proc Natl Acad Sci USA* 90 (2):720–4.

Felsberg, J., B. Hentschel, K. Kaulich, et al. 2017. Epidermal growth factor receptor variant III (EGFRvIII) positivity in EGFR-amplified glioblastomas: prognostic role and comparison between primary and recurrent tumors. *Clin Cancer Res* 23 (22):6846–55.

Fransson, M., E. Piras, J. Burman, et al. 2012. CAR/FoxP3-engineered T regulatory cells target the CNS and suppress EAE upon intranasal delivery. *J Neuroinflammation* 9:112.

Gardner, R., D. Wu, S. Cherian, et al. 2016. Acquisition of a CD19-negative myeloid phenotype allows immune escape of MLL-rearranged B-ALL from CD19 CAR-T-cell therapy. *Blood* 127 (20):2406–10.

Goff, S. L., R. A. Morgan, J. C. Yang, et al. 2019. Pilot trial of adoptive transfer of chimeric antigen receptor-transduced T cells targeting EGFRvIII in patients with glioblastoma. *J Immunother* 42 (4):126–35.

Golubovskaya, V., H. Zhou, F. Li, et al. 2021. Novel CD37, humanized CD37 and bi-specific humanized CD37-CD19 CAR-T cells specifically target lymphoma. *Cancers (Basel)* 13:1–13.

Han, D., Z. Xu, Y. Zhuang, Z. Ye, and Q. Qian. 2021. Current progress in CAR-T cell therapy for hematological malignancies. *J Cancer* 12 (2):326–34.

Hassan, R., and M. Ho. 2008. Mesothelin targeted cancer immunotherapy. *Eur J Cancer* 44 (1):46–53.

Hay, K. A., L. A. Hanafi, D. Li, et al. 2017. Kinetics and biomarkers of severe cytokine release syndrome after CD19 chimeric antigen receptor-modified T-cell therapy. *Blood* 130 (21):2295–306.

Heyman, B., and Y. Yang. 2019. Chimeric antigen receptor T cell therapy for solid tumors: current status, obstacles and future strategies. *Cancers (Basel)* 11:1–21.

Hombach, A. A., J. Heiders, M. Foppe, M. Chmielewski, and H. Abken. 2012. OX40 costimulation by a chimeric antigen receptor abrogates CD28 and IL-2 induced IL-10 secretion by redirected CD4(+) *T cells. Oncoimmunology* 1 (4):458–66.

Hu, W., Z. Zi, Y. Jin, et al. 2019. CRISPR/Cas9-mediated PD-1 disruption enhances human mesothelin-targeted CAR T cell effector functions. *Cancer Immunol Immunother* 68 (3):365–77.

Jensen, M. C., L. Popplewell, L. J. Cooper, et al. 2010. Antitransgene rejection responses contribute to attenuated persistence of adoptively transferred CD20/CD19-specific chimeric antigen receptor redirected T cells in humans. *Biol Blood Marrow Transplant* 16 (9):1245–56.

Jiang, H., H. Gao, J. Kong, et al. 2018. Selective targeting of glioblastoma with EGFRvIII/EGFR bitargeted chimeric antigen receptor T cell. *Cancer Immunol Res* 6 (11):1314–26.

Johnson, L. A., J. Scholler, T. Ohkuri, et al. 2015. Rational development and characterization of humanized anti-EGFR variant III chimeric antigen receptor T cells for glioblastoma. *Sci Transl Med* 7 (275):275ra22.

Kansal, R., N. Richardson, I. Neeli, et al. 2019. Sustained B cell depletion by CD19-targeted CAR T cells is a highly effective treatment for murine lupus. *Sci Transl Med* 11:1–13.

Klampatsa, A., V. Dimou, and S. M. Albelda. 2020. Mesothelin-targeted CAR-T cell therapy for solid tumors. *Expert Opin Biol Ther* 21 (4):473–86.

Kochenderfer, J. N., M. E. Dudley, R. O. Carpenter, et al. 2013. Donor-derived CD19-targeted T cells cause regression of malignancy persisting after allogeneic hematopoietic stem cell transplantation. *Blood* 122 (25):4129–39.

Kochenderfer, J. N., M. E. Dudley, S. A. Feldman, et al. 2012. B-cell depletion and remissions of malignancy along with cytokine-associated toxicity in a clinical trial of anti-CD19 chimeric-antigen-receptor-transduced T cells. *Blood* 119 (12):2709–20.

Kochenderfer, J. N., M. E. Dudley, S. H. Kassim, et al. 2015. Chemotherapy-refractory diffuse large B-cell lymphoma and indolent B-cell malignancies can be effectively treated with autologous T cells expressing an anti-CD19 chimeric antigen receptor. *J Clin Oncol* 33 (6):540–9.

Kochenderfer, J. N., W. H. Wilson, J. E. Janik, et al. 2010. Eradication of B-lineage cells and regression of lymphoma in a patient treated with autologous T cells genetically engineered to recognize CD19. *Blood* 116 (20):4099–102.

Kunkele, A., A. Taraseviciute, L. S. Finn, et al. 2017. Preclinical assessment of CD171-directed CAR T-cell adoptive therapy for childhood neuroblastoma: CE7 epitope target safety and product manufacturing feasibility. *Clin Cancer Res* 23 (2):466–77.

Lai, Y., J. Weng, X. Wei, et al. 2018. Toll-like receptor 2 costimulation potentiates the anti-tumor efficacy of CAR T Cells. *Leukemia* 32 (3):801–8.

Lanitis, E., M. Poussin, I. S. Hagemann, et al. 2012. Redirected anti-tumor activity of primary human lymphocytes transduced with a fully human anti-mesothelin chimeric receptor. *Mol Ther* 20 (3):633–43.

Learn, C. A., T. L. Hartzell, C. J. Wikstrand, et al. 2004. Resistance to tyrosine kinase inhibition by mutant epidermal growth factor receptor variant III contributes to the neoplastic phenotype of glioblastoma multiforme. *Clin Cancer Res* 10 (9):3216–24.

Lee, D. W., J. N. Kochenderfer, M. Stetler-Stevenson, et al. 2015. T cells expressing CD19 chimeric antigen receptors for acute lymphoblastic leukaemia in children and young adults: a phase 1 dose-escalation trial. *Lancet* 385 (9967):517–28.

Li, Y., H. Wu, G. Chen, et al. 2020. Arming anti-EGFRvIII CAR-T with TGFbeta trap improves antitumor efficacy in glioma mouse models. *Front Oncol* 10:1117.

Liu, Y., X. Chen, W. Han, and Y. Zhang. 2017. Tisagenlecleucel, an approved anti-CD19 chimeric antigen receptor T-cell therapy for the treatment of leukemia. *Drugs Today (Barc)* 53 (11):597–608.

Liu, X., K. Liu, J. Qin, et al. 2015. C/EBPbeta promotes angiogenesis through secretion of IL-6, which is inhibited by genistein, in EGFRvIII-positive glioblastoma. *Int J Cancer* 136 (11):2524–34.

Lyu, L., Y. Feng, X. Chen, and Y. Hu. 2020. The global chimeric antigen receptor T (CAR-T) cell therapy patent landscape. *Nat Biotechnol* 38 (12):1387–94.

Ma, J. S., J. Y. Kim, S. A. Kazane, et al. 2016. Versatile strategy for controlling the specificity and activity of engineered T cells. *Proc Natl Acad Sci U S A* 113 (4):E450–8.

Maude, S. L. 2018. Tisagenlecleucel in pediatric patients with acute lymphoblastic leukemia. *Clin Adv Hematol Oncol* 16 (10):664–6.

Maude, S. L., T. W. Laetsch, J. Buechner, et al. 2018. Tisagenlecleucel in children and young adults with B-cell lymphoblastic leukemia. *N Engl J Med* 378 (5):439–48.

Miao, H., B. D. Choi, C. M. Suryadevara, et al. 2014. EGFRvIII-specific chimeric antigen receptor T cells migrate to and kill tumor deposits infiltrating the brain parenchyma in an invasive xenograft model of glioblastoma. *PLoS One* 9 (4):e94281.

Moon, E. K., L. C. Wang, D. V. Dolfi, et al. 2014. Multifactorial T-cell hypofunction that is reversible can limit the efficacy of chimeric antigen receptor-transduced human T cells in solid tumors. *Clin Cancer Res* 20 (16):4262–73.

Morsut, L., K. T. Roybal, X. Xiong, et al. 2016. Engineering customized cell sensing and response behaviors using synthetic notch receptors. *Cell* 164 (4):780–91.

Nakazawa, T., A. Natsume, F. Nishimura, et al. 2020. Effect of CRISPR/Cas9-mediated PD-1-disrupted primary human third-generation CAR-T cells targeting EGFRvIII on in vitro human glioblastoma cell growth. *Cells* 9:1–20.

Pastan, I., and R. Hassan. 2014. Discovery of mesothelin and exploiting it as a target for immunotherapy. *Cancer Res* 74 (11):2907–12.

Perera, L. P., M. Zhang, M. Nakagawa, et al. 2017. Chimeric antigen receptor modified T cells that target chemokine receptor CCR4 as a therapeutic modality for T-cell malignancies. *Am J Hematol* 92 (9):892–901.

Porter, D. L., W. T. Hwang, N. V. Frey, et al. 2015. Chimeric antigen receptor T cells persist and induce sustained remissions in relapsed refractory chronic lymphocytic leukemia. *Sci Transl Med* 7 (303):303ra139.

Porter, D. L., B. L. Levine, M. Kalos, A. Bagg, and C. H. June. 2011. Chimeric antigen receptor-modified T cells in chronic lymphoid leukemia. *N Engl J Med* 365 (8):725–33.

Posey, A. D., Jr., R. D. Schwab, A. C. Boesteanu, et al. 2016. Engineered CAR T cells targeting the cancer-associated Tn-glycoform of the membrane mucin MUC1 control adenocarcinoma. *Immunity* 44 (6):1444–54.

Qi, X. W., F. Zhang, H. Wu, et al. 2015. Wilms' tumor 1 (WT1) expression and prognosis in solid cancer patients: a systematic review and meta-analysis. *Sci Rep* 5:8924.

Ramachandran, M., A. Dimberg, and M. Essand. 2017. The cancer-immunity cycle as rational design for synthetic cancer drugs: novel DC vaccines and CAR T-cells. *Semin Cancer Biol* 45:23–35.

Reagan, P. M., and J. W. Friedberg. 2021. Axicabtagene ciloleucel and brexucabtagene autoleucel in relapsed and refractory diffuse large B-cell and mantle cell lymphomas. *Future Oncol* 17 (11):1269–83.

Rodgers, D. T., M. Mazagova, E. N. Hampton, et al. 2016. Switch-mediated activation and retargeting of CAR-T cells for B-cell malignancies. *Proc Natl Acad Sci USA* 113 (4):E459–68.

Roy, L. D., L. M. Dillon, R. Zhou, et al. 2017. A tumor specific antibody to aid breast cancer screening in women with dense breast tissue. *Genes Cancer* 8 (3–4):536–49.

Roybal, K. T., J. Z. Williams, L. Morsut, et al. 2016. Engineering T cells with customized therapeutic response programs using synthetic notch receptors. *Cell* 167 (2):419–32.e16.

Sahin, A., C. Sanchez, S. Bullain, P. Waterman, R. Weissleder, and B. S. Carter. 2018. Development of third generation anti-EGFRvIII chimeric T cells and EGFRvIII-expressing artificial antigen presenting cells for adoptive cell therapy for glioma. *PLoS One* 13 (7):e0199414.

Song, D. G., Q. Ye, M. Poussin, G. M. Harms, M. Figini, and D. J. Powell, Jr. 2012. CD27 costimulation augments the survival and anti-tumor activity of redirected human T cells in vivo. *Blood* 119 (3):696–706.

Sotoudeh, M., S. I. Shirvani, S. Merat, N. Ahmadbeigi, and M. Naderi. 2019. MSLN (mesothelin), ANTXR1 (TEM8), and MUC3A are the potent antigenic targets for CAR T cell therapy of gastric adenocarcinoma. *J Cell Biochem* 120 (4):5010–7.

Tarp, M. A., A. L. Sorensen, U. Mandel, et al. 2007. Identification of a novel cancer-specific immunodominant glycopeptide epitope in the MUC1 tandem repeat. *Glycobiology* 17 (2):197–209.

Tenspolde, M., K. Zimmermann, L. C. Weber, et al. 2019. Regulatory T cells engineered with a novel insulin-specific chimeric antigen receptor as a candidate immunotherapy for type 1 diabetes. *J Autoimmun* 103:102289.

Till, B. G., M. C. Jensen, J. Wang, et al. 2012. CD20-specific adoptive immunotherapy for lymphoma using a chimeric antigen receptor with both CD28 and 4-1BB domains: pilot clinical trial results. *Blood* 119 (17):3940–50.

Turtle, C. J., L. A. Hanafi, C. Berger, et al. 2016. CD19 CAR-T cells of defined CD4+:CD8+ composition in adult B cell ALL patients. *J Clin Invest* 126 (6):2123–38.
Wachowiak, R., M. Krause, S. Mayer, et al. 2018. Increased L1CAM (CD171) levels are associated with glioblastoma and metastatic brain tumors. *Medicine (Baltimore)* 97 (38):e12396.
Wei, J., X. Han, J. Bo, and W. Han. 2019. Target selection for CAR-T therapy. *J Hematol Oncol* 12 (1):62.
Wei, H., Z. Wang, Y. Kuang, et al. 2020. Intercellular adhesion molecule-1 as target for CAR-T-cell therapy of triple-negative breast cancer. *Front Immunol* 11:573823.
Wilkie, S., S. E. Burbridge, L. Chiapero-Stanke, et al. 2010. Selective expansion of chimeric antigen receptor-targeted T-cells with potent effector function using interleukin-4. *J Biol Chem* 285 (33):25538–44.
Yang, P., X. Cao, H. Cai, et al. 2020. The exosomes derived from CAR-T cell efficiently target mesothelin and reduce triple-negative breast cancer growth. *Cell Immunol* 360:104262.
Ye, L., Y. Lou, L. Lu, and X. Fan. 2019. Mesothelin-targeted second generation CAR-T cells inhibit growth of mesothelin-expressing tumors in vivo. *Exp Ther Med* 17 (1):739–47.
Zhang, B. L., D. Y. Qin, Z. M. Mo, et al. 2016. Hurdles of CAR-T cell-based cancer immunotherapy directed against solid tumors. *Sci China Life Sci* 59 (4):340–8.
Zhang, Z., J. Jiang, X. Wu, et al. 2019. Chimeric antigen receptor T cell targeting EGFRvIII for metastatic lung cancer therapy. *Front Med* 13 (1):57–68.
Zhang, Q., H. Wang, H. Li, et al. 2017. Chimeric antigen receptor-modified T Cells inhibit the growth and metastases of established tissue factor-positive tumors in NOG mice. *Oncotarget* 8 (6):9488–99.
Zhou, R., M. Yazdanifar, L. D. Roy, et al. 2020. Corrigendum: CAR T cells targeting the tumor MUC1 glycoprotein reduce triple-negative breast cancer growth. *Front Immunol* 11:628776.
Zhu, H., Y. You, Z. Shen, and L. Shi. 2020. EGFRvIII-CAR-T cells with PD-1 knockout have improved anti-glioma activity. *Pathol Oncol Res* 26 (4):2135–41.

Biologics and Biosimilars Used for Diabetes

18

Angela Penney, Bahaar Kaur Muhar, and Simeon Kotchoni
Corresponding author: Simeon Kotchoni

Contents

DOI: 10.1201/9780429485626-18

18.1 INTRODUCTION

Diabetes mellitus is the fifth-leading cause of mortality in most developed countries and continues to impact the lives of many people in the United States each day.[1] In this chapter, we will discuss the foundations of diabetes mellitus – including the definition, diagnostic criteria, prevalence/incidence, different types, symptoms, and complications of the disease. The main focus of this chapter will be devoted to the description and use of biologics and biosimilars to treat diabetes disease.

The main category of biologic utilized for the treatment of diabetes is exogenous insulin, and the various types of insulin preparations will be further detailed. Since it is crucial for blood glucose levels to be meticulously controlled around the clock, insulin treatment typically requires at least once daily administration, if not more. Therefore, the costs can quickly add up. In an attempt to make these vital biologics more accessible and affordable to all, manufacturers opened the door to biosimilar research and development.

A biosimilar is a biologic product that is highly similar to a reference biologic product, which already exists on the market with approval from the Food and Drug Administration (FDA). However, the realm of biosimilar insulin remains a novel field, with few already securing FDA approval. The biosimilar treatment options specified in this chapter include those in any stage of development – comprising of those that are at their final stage of clinical trials but still in the process of gaining FDA approval, as well as those that are currently progressing through clinical trials, or even yet to commence with clinical trials.

18.2 DIABETES MELLITUS

18.2.1 Definition and Diagnosis

Diabetes mellitus (diabetes meaning "passing of urine" and mellitus meaning "sweet") is clinically defined as a chronic, metabolic disease characterized by elevated levels of blood glucose (or blood sugar), which causes excess glucose to diffuse into the urine and over time leads to serious damage to the heart, blood vessels, eyes, kidneys, and nerves.[2] The diagnostic criteria for this condition evaluate an individual's hemoglobin A1C (HbA1c), fasting plasma glucose (FPG), oral glucose tolerance test (OGTT), and random plasma glucose (RPG) test (Table 18.1). An individual's A1C is indicative of the average blood glucose levels over the previous two to three months as it measures the percentage of glycated hemoglobin circulating in the blood. Measuring this component in the blood is significant because when glucose binds to hemoglobin, it forms glycated hemoglobin (hemoglobin A1C), and the glucose will not dissociate from the hemoglobin in the lifetime of the red blood cell.

TABLE 18.1 Various factors involved in the diagnostic criteria of diabetes and prediabetes

DIAGNOSIS	*A1C (PERCENT)*	*FASTING PLASMA GLUCOSE (FPG)*	*ORAL GLUCOSE TOLERANCE TEST (OGTT)*	*RANDOM PLASMA GLUCOSE TEST (RPG)*
Normal	Below 5.7	99 or below	139 or below	
Prediabetes	5.7–6.4	100–125	140–199	
Diabetes	6.5 or above	126 or above	200 or above	200 or above

Thus, explaining why the measurement of hemoglobin A1C is a direct reflection of how much glucose has been circulating through the blood over the last two to three months. As for the FPG, OGTT, and RPG, these are all measurements of glucose levels in the blood and offer insight into how effectively the body is processing glucose.

18.2.2 Prevalence/Incidence

As reported by the CDC in 2020, 34.2 million (10.5% of the US population) adults in the United States (US) have diabetes, and 1 in 5 of them do not know they have it.[3] Moreover, 88 million people (34.5% of the US population) aged 18 years or older already have prediabetes, and 24.2 million of them are 65 years or older. Prevalence, defined as the proportion of individuals who *have* a condition at or during a particular time period, was recorded from 2013 to 2019 on diabetes in the US.[3] According to these statistics, the most prevalent race/ethnicity group with diabetes was non-Hispanic Black (16.4%), followed by non-Hispanic Asian (14.9%), Hispanic (14.7%), and non-Hispanic White (11.9%).[3] On the other hand, incidence, referring to the proportion or rate of individuals who *develop* a condition during a particular time period, was reported at 1.5 million new cases of diabetes in 2018.[4] Additionally, non-Hispanic blacks and people of Hispanic origin had a higher incidence in comparison to non-Hispanic whites.

18.2.3 Various Classifications of Diabetes

There are multiple ways for this condition to arise, leading to multiple types of diabetes – mainly type I and type II (Table 18.2). Additionally, there is also prediabetes, which is essentially a precursor to diagnosable type II diabetes, as well as gestational diabetes and Maturity Onset Diabetes of Youth (MODY).

TABLE 18.2 Comparison between type I and type II diabetes in various categories

	TYPE I	*TYPE II*
Etiology	Autoimmune destruction of pancreatic β cells	Begins as insulin resistance and leads to inadequate β cell compensation
Age of onset	Under 30 years of age but typically under 15 years of age – usually appears in juveniles to early adolescents	Usually over 40 years of age – known as adult-onset diabetes
Genetic predisposition	Has been tied specifically with different variants of the human leukocyte antigen (HLA) gene	No clear genetic component but is known to be a combination of genetic and lifestyle factors
Prevalence	Approximately 6% of diabetes cases in the US	Approximately 91% of cases in the US
Insulin Resistance	Not typically seen in type I	The mainstay of type II
Insulin levels	Absent or very negligible number	Higher than normal as the β cells are trying to compensate for the insulin resistance
Insulin action	Absent or negligible action	Dramatically diminished
Acute complications	• Hypoglycemia • Diabetic ketoacidosis (DKA)	• Hyperglycemia • Hyperosmolar hyperglycemic syndrome (HHS)
Pharmacological interventions	Insulin is the common treatment	Insulin as well as various other drug classes to target the insulin resistance

18.2.3.1 Type I diabetes

Type I diabetes is an autoimmune disease that causes the destruction of β cells in the islets of Langerhans of the pancreas. These pancreatic beta cells are responsible for the production and secretion of insulin in the body, and insulin is a hormone that is required in order for the body to properly intake glucose into the cells for use as energy. Without this essential hormone, glucose from food cannot enter the cells and begins to accumulate in the bloodstream instead, resulting in an elevated blood glucose concentration. This leads to a cascade of adverse health compilations later on, especially when left untreated.

18.2.3.2 Type II diabetes

Type II diabetes is more of a condition that develops over time where it begins with insulin resistance. This means that the cells in the muscle, fat, and liver become resistant to insulin and thus do not intake enough glucose into the cells for energy. Subsequently, as the pancreas tries to compensate for the consistently elevated level of blood glucose, it wears down the β cells and ends with inadequate β cell function. This renders the pancreas unable to produce enough insulin to manage glucose in the bloodstream and results in the diagnosis of diabetes. In the case of prediabetes, the individual has blood glucose levels that are higher than normal but not yet high enough to be diagnosed as type II diabetes.

18.2.3.3 Gestational diabetes

Gestational diabetes is another type of diabetes that occurs exclusively in pregnant women who did not previously have diabetes but get diagnosed with diabetes for the first time in the second or third trimester of pregnancy. Hormone changes during pregnancy lead to insulin resistance and result in the pancreatic β cells not being able to adequately compensate with increased insulin secretion in order to maintain normal blood glucose levels. Normally, the individual's blood sugar levels will return to normal after the baby is born. However, 50% of women with gestational diabetes will advance on to develop type II diabetes later in life if careful preventative measures are not taken.[5]

18.2.3.4 MODY

MODY is a category for various inherited forms of diabetes mellitus that are more likely to develop during adolescence and young adulthood. It got its name because the symptoms resemble that of adult-type diabetes (type II) but was found in younger people.[6] MODY is a genetic autosomal dominant disorder that is characterized by an onset of hyperglycemia at an early age, along with impaired insulin secretion with no insulin resistance.

18.2.4 Symptoms and Complications

The three hallmark symptoms of diabetes mellitus are polydipsia, polyuria, and polyphagia – increased thirst, urination, and hunger, respectively. Diabetes is characterized by an inadequate uptake of glucose into the cells for utilization as energy, which results in an increased hunger sensation due to the depleted energy levels. Additionally, since the body is not able to uptake all of the excess glucose, it gets filtered into the urine and excreted. However, the high osmotic gradient caused by the elevated urine glucose concentration results in increased water following into the urine for excretion, resulting in polyuria. This increase in urination then triggers the body to attempt to replace the water loss by signaling increased thirst. Additional symptoms include fatigue, unexplained weight loss, blurred vision, numbness and/or tingling in the hands or feet, and sores that do not heal.

Some short-term complications that are commonly experienced in diabetes patients include acute hyperglycemia and hypoglycemia from untreated and overtreated diabetes, respectively. In type I diabetes, it is possible to experience diabetic ketoacidosis (DKA), which is characterized by a high concentration of ketone bodies due to the absence of insulin– resulting in metabolic acidosis (blood pH of 6.8–7.3) combined with hyperglycemia. This condition is more common among type I diabetics as there is a complete absence of insulin while type II diabetics typically still produce insulin. On the other hand, hyperglycemic hyperosmolar syndrome (HHS) is more common among type II diabetics and is characterized by extreme hyperglycemia and severe dehydration but without the presence of ketone bodies and ketoacidosis.

Long-term complications of both types of diabetes include chronic issues to the microvascular and macrovascular systems due to the stress and inflammation of the body's vasculature. The microvascular complications may include retinopathy, nephropathy, and neuropathy; meanwhile, the major macrovascular complications include stroke, heart diseases (i.e. coronary artery disease (CAD)), and peripheral vascular diseases (PVDs).

18.2.5 Treatment Options

The initial recommended non-pharmacological treatment options, as well as prevention options for diabetes, are lifestyle changes. This entails diet and nutrition modifications along with weight loss measures, such as increases in exercise and physical activity. Additionally, smoking cessation is recommended.

As for pharmacological treatment, there are various different classes of drugs and biologics that are utilized to treat diabetes that will be further discussed later on.

18.3 BIOLOGICS AND BIOSIMILARS AS TREATMENT FOR DIABETES

A biologic is a compound that is manufactured with living organisms, commonly used in vaccines, monoclonal antibodies, and insulin products. These biological agents differ from "drugs" in that they are generally larger, more complex molecules and drugs are typically manufactured through a chemical synthesis process while biologics utilize living systems as the main component. Since the methods of pharmacological treatment of diabetes typically utilize insulin or insulin analogs, a biologic is a more likely candidate to be used, as opposed to a drug.

18.3.1 Biologics for Diabetes – Exogenous Insulin

18.3.1.1 Structure and mechanism

Human insulin is a small globular protein comprised of 51 amino acids in two polypeptide chains (Chains A and B) that are linked with two inter-chain disulfide bridges as well as a disulfide loop within just Chain A (Figure 18.1).[8] This polypeptide hormone is synthesized in the β-cells of the pancreas and is secreted in response to rising levels of glucose in the blood. Insulin promotes the uptake of glucose, amino acids, and fatty acids from the blood into the liver, adipose tissue, and skeletal muscle cells for storage as glycogen, protein, and triglycerides, respectively.[7] Therefore, without this essential hormone, the body is not able to adequately acquire and utilize energy for normal functioning.

18.3.1.2 Insulin types

Insulin replacement therapy is the mainstay treatment for type I diabetes and adjunct treatment for type II when diet and other therapies are not providing sufficient glycemic control. The four major types of exogenous insulin are detailed in the following sections (Figure 18.2).

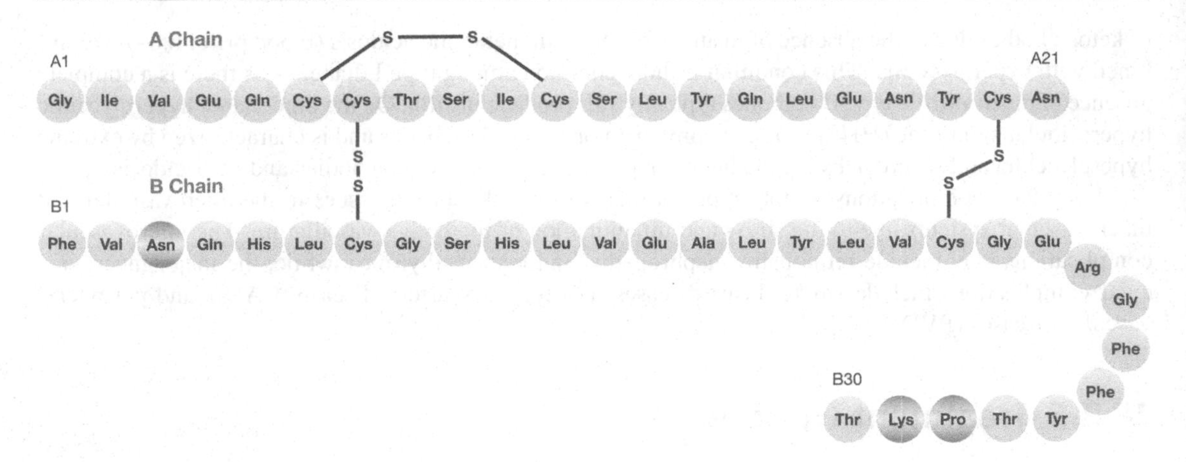

FIGURE 18.1 Structure of human insulin. A 21-amino acid A chain is linked by two disulfide bonds to a 30-amino acid B chain to form the functional insulin monomer. Purple-colored residues represent ones that are commonly mutated in fast-acting insulin analogs in order to speed up absorption upon subcutaneous injection.[8]

18.3.1.3 Regular insulin (short acting)

Regular insulin is a short-acting human insulin that is used to cover glycemic peaks during mealtimes. Examples of regular insulin include Humulin R and Novolin R. They have an onset of action around 0.5–1 hour with peak action from 2 to 3 hours and a duration of 6–8 hours. In order to achieve optimum insulin activity for carbohydrate metabolism, this type of insulin should be injected approximately 20–30 minutes prior to meal consumption.[1]

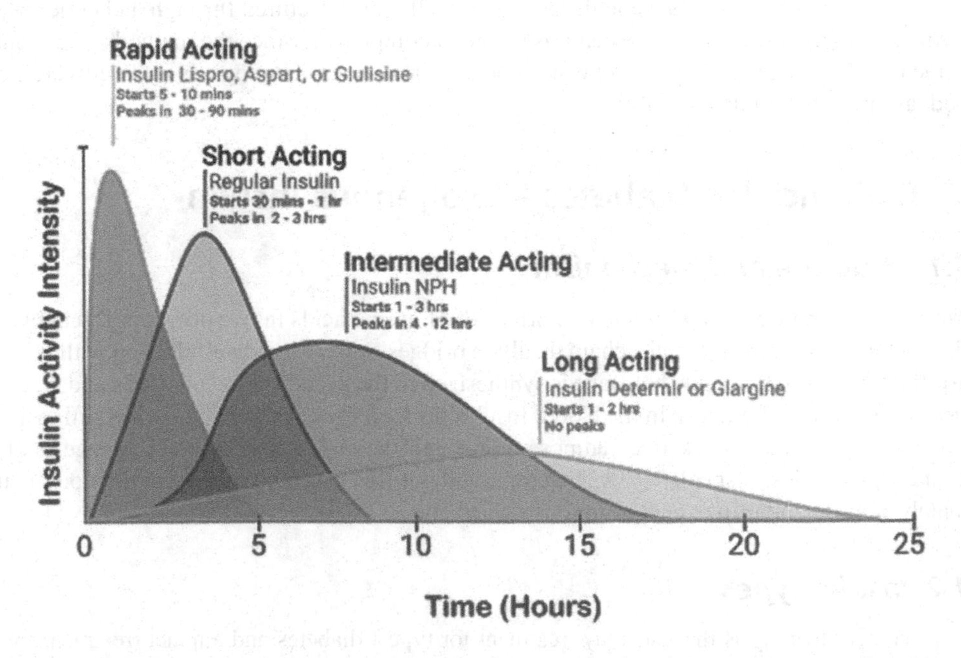

FIGURE 18.2 Different types of exogenous insulin and their intensity of action at various durations of time. (Courtesy of InsuLearn.)

18.3.1.4 NPH insulin (intermediate acting)

NPH, Neutral Protamine Hagedorn, is an intermediate-acting human insulin designed to mimic the baseline insulin secretion and is used for basal insulin therapy. Along with regular insulin, this is another form of human insulin that is used to mimic a natural physiological state. Humulin N and Novolin N are common examples of NPH insulins. These intermediate-acting insulins have an onset of action around 1–3 hours with peak action from 4 to 12 hours and a duration of 12+ hours.

18.3.1.5 Rapid-acting insulin

Rapid-acting insulins are insulin analogs, which are similar to human insulin, but they are altered along their amino acid sequences in order to produce different chemical properties. By a single or two amino acid alterations in the polypeptide chain, the ability of insulin to associate into hexamers is reduced such that they are readily absorbed, however, these modifications do not change the biological properties of these analogs.[9,10] Examples of rapid-acting insulins include Novolog, Humalog, and Apidra. Novolog, or Aspart, has an aspartic acid substituted for the B28 proline. Similarly, in Humalog, or lispro, the positions of proline at position 28 on the B chain and lysine at position 29 on the B chain have been reversed. Apidra, or glulisine, contains two amino acid substitutions: the B3 asparagine is replaced by a lysine residue, and the B29 lysine is swapped with glutamic acid. Since these are fast-acting insulins that have been bioengineered to get absorbed into the bloodstream quickly, they display a quick onset of action – at a mere 5–10 minutes. These have a peak of action from 30 to 90 minutes and a duration of 4–6 hours (Tables 18.3 and 18.4).

18.3.1.6 Long-acting insulin

Basal analogs are similar to NPH in activity, but they are modified to exhibit glycemic control over prolonged periods of time. One way to achieve this desired effect is to design analogs with more positively charged amino acids on the polypeptide chain so as to raise the isoelectric point of insulin to near neutral pH, decreasing the solubility of insulin at neutral pH after injection and reducing the speed of absorption into the blood.[11] Common examples of long-acting insulins are Levemir, Lantus, and Tresiba. Levemir (Detemir) is altered via removal of the B30 threonine and attachment of a 14-C fatty acid (i.e. myristic acid) to the B29 lysine.[1] The addition of the fatty acid chain aids in insulin hexamer formation and facilitates the binding of insulin to plasma albumin, delaying free insulin release and perpetuating the activity of insulin. The key difference in this type is that basal analogs retain a more stable profile and are associated with less weight gain and nocturnal hypoglycemia (Table 18.4).

18.3.1.7 Pre-mixed insulin

Lastly, pre-mixed insulin does not utilize a novel mechanism of action or formulation but is rather a pre-formulated combination of both fast-acting and basal insulin, which is typically used in basal-bolus therapy with the advantage of fewer total injections while providing extended periods of glycemic control (Table 18.4).[12,13]

TABLE 18.3 Different types of exogenous insulin and their corresponding onset, peak, and duration

	TYPE	*ONSET*	*PEAK*	*DURATION*
Human insulin	Regular insulin (short acting)	½–1 hour	2–3 hours	6–8 hours
	NPH (intermediate acting)	1–3 hours	4–12 hours	12+ hours
Insulin analogs	Rapid acting	5–10 mins	30–90 mins	4–6 hours
	Long acting	1–2 hours	–	12–24 hours

TABLE 18.4 Classification of insulin biologics currently approved by the FDA and published in the Purple Book (Database of Licensed Biological Products)[13]

	BRAND NAME	*GENERIC/PROPER NAME*	*MANUFACTURER*
Rapid acting	Humalog	Insulin lispro	Eli Lilly
	Novolog	Insulin aspart	Novo Nordisk
	Fiasp	Insulin aspart	Novo Nordisk
	Apidra	Insulin glulisine	Sanofi
	Afrezza	Insulin human	Mannkind and Sanofi
Short acting	Humulin R U-100	Insulin human	Eli Lilly
	Humulin R U-500	Insulin human	Eli Lilly
	Novolin R	Insulin human	Novo Nordisk
	Myxredlin	Insulin human	Baxter
Intermediate acting	Humulin N	Insulin isophane human	Eli Lilly
	Novolin N	Insulin isophane human	Novo Nordisk
Long acting	Lantus	Insulin glargine	Sanofi
	Toujeo	Insulin glargine	Sanofi
	Levemir	Insulin detemir	Novo Nordisk
	Tresiba	Insulin degludec	Novo Nordisk
Pre-mixed insulin	Humalog Mix 50/50	Insulin lispro protamine and insulin lispro	Eli Lilly
	Humalog Mix 75/25	Insulin lispro protamine and insulin lispro	Eli Lilly
	Humulin 70/30	Insulin isophane human and insulin human	Eli Lilly
	Novolog Mix 50/50	Insulin aspart protamine and insulin aspart	Novo Nordisk
	Novolog Mix 70/30	Insulin aspart protamine and insulin aspart	Novo Nordisk
	Novolin 70/30	Insulin isophane human and insulin human	Novo Nordisk
	Ryzodeg 70/30	Insulin degludec and insulin aspart	Novo Nordisk
	Xultophy 100/3.6	Insulin degludec and liraglutide	Novo Nordisk
	Soliqua 100/33	Insulin glargine and lixisenatide	Sanofi

18.3.1.8 Novel insulin delivery systems

Apart from the standard subcutaneous injectable insulins, there is also a type of inhaled insulin that exists as a powder form. This treatment gets inhaled to quickly pass through the lungs and enter the bloodstream rapidly within a minute. An example of this novel type of insulin is Afrezza. Structurally, this is human insulin, but it gets characterized as a rapid-acting insulin due to its kinetic properties – exhibiting a peak of action at 12–15 minutes and a duration of approximately 160 minutes. However, the disadvantage to this insulin type is the lower bioavailability, at about 20%–25%, in comparison to 70% in subcutaneous injection forms.

Additionally, there is a transbuccal insulin currently progressing through clinical trials – a liquid spray of human insulin marketed as Oral-Lyn.[8] This treatment is administered to the inner lining of the cheek and contains surfactants to facilitate the transport of insulin across the mucosa for entry into the bloodstream.

18.3.2 Biosimilars for Diabetes Treatment

Similar to the use of generic drugs in place of a reference drug, a biosimilar can be used in place of a reference biologic. The United States enacted the Biologics Price Competition and Innovation Act (BPCIA) in 2010 to establish a pathway for the FDA to approve biologic products as biosimilar to already-approved biologics, allowing access to more affordable versions of these critical medicines for patients. A biosimilar, as defined by the FDA, is a "biological product that is highly similar to and has no clinically

TABLE 18.5 Various insulin biosimilars that are either approved or undergoing clinical trials and their corresponding manufacturers and reference products[15]

MANUFACTURER	*BIOSIMILAR NAME*	*REFERENCE PRODUCT*
Merk	Lusduna	Lantus
Sanofi	Admelog	Humalog
Eli Lilly	Lyumjev	Humalog
Biocon-Mylan	Mylan insulin glargine	Lantus
Julphar Gulf Pharmaceutical Industries	Julphar insulin R	Humulin R
	Julphar insulin N	Huminsulin Basal
	Julphar insulin 30–70	Huminsulin Profil III
Gan and Lee	Basalin	Gan and Lee insulin glargine
Wockhardt	Glaritus	Lantus
Boehringer Ingelheim and Eli Lilly (BI-Lilly)	Basaglar[16]	Lantus
Mylan and Biocon	Semglee[17]	Lantus

meaningful differences from an existing FDA-approved reference product".[14] Therefore, it is common to view biosimilars as analogous to generic drugs.

Biosimilars mirror the structure and function of its reference biologic, so they will have the same amino acid sequence, but there may be differences in the more subtle molecular characteristics and clinical profiles.[12] Since biosimilars are an up-and-coming aspect of pharmaceutical development, many are still undergoing clinical trials, but there are some biosimilars for a few of the aforementioned types of exogenous insulin (Table 18.3). For example, Admelog (Table 18.5) was the first biosimilar mealtime insulin with the reference biologic being the rapid-acting insulin Humalog (insulin lispro).[15] They are both short-acting insulins to control mealtime glycemic spikes and are used to treat both type I and type II diabetes, giving diabetic patients a more cost-effective alternative to insulin lispro. Another example of a biosimilar but for a more long-acting insulin is Basaglar – a biosimilar basal insulin that mimics the action of insulin glargine, or Lantus. Most importantly, Basaglar markets for approximately 15% cheaper than Lantus, indicating a significant distinction.

18.3.3 Advantages of Biosimilars

With diabetes being one of the most prevalent diseases in the country, a huge proportion of the population is subject to the financial burden of pharmacological treatment. The main advantage to utilizing biosimilars in place of biologics for the treatment of diabetes is simply the cost aspect. Biosimilars possess a significant financial advantage over biologics as the biosimilars are not required to undergo the entire research and development stage along with full-scale clinical trials like the reference biologic did – being more time efficient and saving substantial financial resources.

Furthermore, the introduction of more biosimilars into the market can facilitate market expansion and create competition. This not only expands on the number of insulin brands available to those with diabetes but also offers advantages to the public population as enabling more market competition can potentially bring down the prices for consumers. Due to this potential for increased accessibility and affordability when it comes to diabetes treatments, biosimilars are essentially increasing the availability of pharmacological treatment, especially to the less fortunate subsets of the population, such as the uninsured and underinsured groups.

Lastly, the introduction of a greater variety of biosimilars into the market facilitates the potential for enhanced individualized patient care.[18] Physiologically, certain individuals may respond differently to various formulations of insulins, and the abundance of options gives the prescriber expanded flexibility in treatment strategies.

18.3.4 Disadvantages of Biosimilars

Although the consumer advantages of biosimilar production are plentiful, that is not really the case for the development side. A major disadvantage is that biosimilar manufacturing is much more complicated and time intensive than generic drug production due to the intricacy of working with larger, more complex molecules in living systems. Problems arise as there are no clear metrics to verify that two biologics expressed in different cell lines in different factories subjected to different purification methods are, in fact, therapeutically identical. Thus, approval for a biosimilar requires exhaustive safety, efficacy, and immunogenicity studies, as well as independent clinical trials.[8] Therefore, these extensive processes come at the expense of the manufacturer and fundamentally limits how much the final retail price can essentially be reduced. Small molecule generic drugs can afford to offer discounts as high as 90% off of the reference product price, but, unfortunately, this substantial discount is simply not feasible with biosimilars without eliminating the financial incentive for companies to produce them.

18.4 CONCLUSION

It has been estimated that by 2025 there will be nearly 380 million people suffering from diabetes mellitus if this issue is not controlled.[1] If left untreated, this disease can lead to detrimental consequences, including hypoglycemic shock and even death. Therefore, it is imperative that individuals who are diagnosed with any type of diabetes are given proper access to the necessary pharmacological treatments in order to manage it – indicating the significance of biologics and biosimilars. While insulin biologics are available for patients to obtain, they can be extremely costly, especially for individuals lacking adequate insurance coverage. Although this does not occur without underlying manufacturing roadblocks, biosimilars were developed to offer the same treatment at a more affordable price point and upon the introduction of biosimilars to the market, many individuals who previously could not afford to pay for the reference biologic, or those that did not possess insurance, could now gain access to the critical, life-saving medication that they need.

REFERENCES

1. Ahmad, K., *Insulin sources and types: a review of insulin in terms of its mode on diabetes mellitus.* J Tradit Chin Med, 2014. **34**(2): p. 234–7.
2. Prevention, C.f.D.C.a. *What is diabetes*? June 11, 2020 [cited 2021 March 30]; Available from: https://www.cdc.gov/diabetes/basics/diabetes.html.
3. *Prevention, C.f.D.C.a. National Diabetes Statistics Report*, 2020. August 28, 2020 [cited 2021 March 31]; Available from: https://www.cdc.gov/diabetes/data/statistics-report/index.html.
4. Prevention, C.f.D.C.a. *Principles of Epidemiology in Public Health Practice: An Introduction to Applied Epidemiology and Biostatistics.* May 18, 2012 [cited 2021 March 30]: 3rd Edition; Available from: https://www.cdc.gov/csels/dsepd/ss1978/lesson3/section2.html.
5. *Prevention, C.f.D.C.a. Gestational Diabetes.* May 30, 2019 [cited 2021 March 31]; Available from: https://www.cdc.gov/diabetes/basics/gestational.html.
6. *Children's, N. Diabetes: MODY* [cited 2021 March 31]; Available from: https://www.nationwidechildrens.org/conditions/diabetes-mody.
7. Mohamed, I., *Diabetes Mellitus I.* November 4, 2019.
8. Gilroy, C.A., K.M. Luginbuhl, and A. Chilkoti, *Controlled release of biologics for the treatment of type 2 diabetes.* J Control Release, 2016. **240**: p. 151–64.

9. Howey, D.C., et al., *[Lys(B28), Pro(B29)]-human insulin. A rapidly absorbed analogue of human insulin.* Diabetes, 1994. **43**(3): p. 396–402.
10. Mudaliar, S.R., et al., *Insulin aspart (B28 asp-insulin): a fast-acting analog of human insulin: absorption kinetics and action profile compared with regular human insulin in healthy nondiabetic subjects.* Diabetes Care, 1999. **22**(9): p. 1501–6.
11. Rosskamp, R.H. and G. Park, *Long-acting insulin analogs.* Diabetes Care, 1999. **22**(Suppl 2): p. B109–13.
12. Lisa S., B. Rotenstein, B.A. Nina Ran, P. Joseph, B.A. Shivers, M.D. Mark Yarchoan and L.C. Kelly, *Opportunities and challenges for biosimilars: what's on the horizon in the global insulin market?* Clin Diabetes, October 2012. **30**(4): 138–50.
13. Administration, U.F.a.D., *Purple Book: Database of Licensed Biological Products.* 2021, FDA: Purple Book.
14. Administration, U.S.F.a.D., *Drugs @ FDA Glossary.* Available from: https://www.accessdata.fda.gov/scripts/cder/daf/index.cfm?event=glossary.page.
15. White, J. and J. Goldman, *Biosimilar and follow-on insulin: the ins, outs, and interchangeability.* J Pharm Technol, September 28, 2018. **35**: p. 25–35.
16. Learn, d. *FDA Approves New Insulin Glargine Basaglar – The First "Biosimilar" Insulin in the US.* January 11, 2016 [cited 2021 April 1]; Available from: https://diatribe.org/fda-approves-new-insulin-glargine-basaglar-first-biosimilar-insulin-us.
17. Jeremias, S., *Biosimilar FDA Roundup: June 2020.* July 1, 2020 [cited 2021 April 2]; Available from: https://www.centerforbiosimilars.com/view/biosimilar-fda-roundup-june-2020.
18. Kim, A.P. and R.J. Bindler, *The future of biosimilar insulins.* Diabetes Spectr, 2016. **29**(3): p. 161–6.

Clinical Use of Biologics and Biosimilars for Asthma 19

Gewedy Berhe, Dilip Shah, Sayeed Ahmad, and Mohd Shahid

Corresponding author: Mohd Shahid

Contents

19.1 DEFINITION

According to the US Food and Drug Administration (FDA) "biological products include a wide range of products such as vaccines, blood and blood components, allergenics, somatic cells, gene therapy, tissues, and recombinant therapeutic proteins. Biologics can be composed of sugars, proteins, or nucleic acids or complex combinations of these substances, or may be living entities such as cells and tissues. They are isolated from a variety of natural sources – human, animal, or microorganism – and may be produced by biotechnology methods and other cutting-edge technologies. Gene-based and cellular biologics, for example, often are at the forefront of biomedical research and may be used to treat a variety of medical

DOI: 10.1201/9780429485626-19

conditions for which no other treatments are available".[1] Bases on their mechanism of action, biologics can be broadly divided into monoclonal antibodies (neutralize a specific protein), receptor modulators (agonist/antagonist) or enzyme inhibitors/activators.

Pharmaceutical biologics are often manufactured by using the biological system or extracted from it. They are, in general, large molecules with molecular weight ranges in thousands of kDa, unlike small molecules, which typically have a molecular weight less than 1 kDa and are chemically synthesized with highly predictable structures and functions. The biological system required for manufacturing biologics can vary from simple microbial agents, complex mammalian and plant cell lines, and fungi. These systems can be directed to synthesize a biologic of interest by using complex techniques such as recombinant DNA technology, gene cloning, and product isolation and purification. Recent advancements in molecular and genetic technologies have dramatically improved the manufacturing of biologics that ensures the consistency and quality of the final product.[2] On the other hand, a biosimilar is a biological product that is highly similar to an already approved biologic.[3] According to FDA, "a biosimilar is a biological product that is highly similar to and has no clinically meaningful differences from an existing FDA-approved reference product". European Union European Medicines Agency (EU-EMA) defines a biosimilar "as a biological medicine highly similar to another already approved biological medicine (the 'reference medicine')". Biosimilars approval follows the same standards of pharmaceutical quality, safety, and efficacy that apply to all biological medicines.[4] Whereas, World Health Organization (WHO) defines biosimilar as "a biotherapeutic product that is similar in terms of quality, safety and efficacy to an already licensed reference biotherapeutic product".[5]

To qualify as a biosimilar a biological product needs to have structural and functional similarities with comparable pharmacokinetic and pharmacodynamic properties of the reference compounds, for example, pharmacological effects such as cytokines levels, blood glucose, etc. The manufacturer also must produce evidence regarding the quality control of the processes of synthesis and composition of the formulation. Moreover, state-of-the-art technology should be used to compare the purity, chemical identity, and bioactivity characteristics of biosimilar.

While a biosimilar should be highly similar to the reference product, minor differences in clinically inactive components are acceptable. For example, minor differences in the properties of stabilizer or buffer between biosimilar and reference product. These differences, however, need to be carefully evaluated and approved by FDA. In addition to being highly similar to the reference product, a biosimilar must have no clinically meaningful difference from the originator. A biosimilar must demonstrate high clinical similarity to the reference product in terms of safety, purity, and potency (efficacy). As mentioned above, this is accomplished through comparable pharmacokinetic and pharmacodynamic human studies. If needed, further clinical studies such as assessment of clinical immunogenicity are also performed.[2]

19.2 CURRENT STATUS OF BIOLOGICS IN HEALTHCARE

Biologics and biosimilars have a significant role in healthcare. The use of biologics and biosimilars can be a very useful tool after the failure of therapies in diseases with limited treatment options as well disease states that have a plethora of treatment options that have been exhausted as well. In 1982, human insulin-first DNA recombinant protein was approved by US-FDA.[6] The health care industry has seen a rapid growth in the number of biologics ever since, especially in the field of oncology and immunology. For instance, adalimumab, an anti-inflammatory biologic launched by AbbVie has dramatically changed the treatment and quality of life of patients affected by rheumatoid arthritis, plaque psoriasis, Crohn's disease, or ulcerative colitis. Rituximab (Rituxan, Roche) is a monoclonal antibody used to treat non-Hodgkin's lymphoma, chronic lymphocytic leukemia, and rheumatoid arthritis. Etanercept (Enbrel, Pfizer/Amgen) is a Tumor necrosis factor-alpha (TNF-α) inhibitor used to treat a myriad of diseases such as non-Hodgkin's lymphoma, chronic lymphocytic leukemia, rheumatoid arthritis, and plaque psoriasis. The discovery of bevacizumab made it possible to develop the first anti-angiogenic therapy for the treatment of patients with an advanced cancer. Likewise, omalizumab was the first FDA-approved biologic for the treatment of severe asthma.[7]

Biologics are now at the forefront in terms of total revenue generation in pharmaceutical industry. In addition, FDA provides an accelerated approval process for the development of biosimilar due to the short patent life of the original compounds.[2] Approximately $194 billion will be lost in revenues between the years 2017 and 2022 if the biosimilar is not produced for the several biologics that are due expiration.[8] Thus, pharmaceutical companies are being encouraged to move toward biologics/biosimilar areas.

Although biologics and biosimilars are highly similar structurally and functionally, there are minor differences that may exist between the two. However, these minor differences are acceptable and do not negatively affect product performance. Biologics take a significantly longer time to develop in comparison to biosimilars by at least half a decade. The cost of developing these products is another major difference between biosimilars and biologics. Costs to develop a biologic product can be upward into the billions of dollars while biosimilars costs hover in the hundreds of millions. Biologics cost significantly more than biosimilars seeing as biosimilars are created to have a cheaper option for patients similar to brand and generics in non-biologic medications.

The approval process for a new biologic follows the same standards as any other new drug. To determine the dosage and safety profile of a new biologic an Investigational New Drug Application (IND) must be submitted to the FDA to start Phase 1. Next, the biologic advances to Phase 2 to determine the efficacy and side effects in a smaller population size. This is followed by Phase 3 to further assess the efficacy and side effects in larger populations. The typical approval process for a new biologic takes about 10–12 years. In contrast, biosimilars undergo a shorter regulatory approval process.[9] Since biosimilars are copies of the reference products that have already completed the approval process, initial investigation and efficacy stages (Phases 1 and 2) are not required. Marketed product needed to undergo formulation development and dosage estimation. Biosimilars being a "highly similar" to the marketed product do not need to undergo these studies either, accelerating and shortening the process to 7–8 years. As a result, a biosimilar biologic can be developed at a 10%–20% of the cost required for a new biologic.[10] However, the manufacturer must demonstrate that the biosimilar is highly similar to the reference product in terms of pharmacokinetics and clinical efficacy, and as such a minimum of two clinical studies are required for the approval of a biosimilar.

19.3 ASTHMA

19.3.1 Epidemiology

According to National Asthma Education and Prevention Program, asthma is defined as "a common chronic disorder of the airways that is complex and characterized by variable and recurring symptoms, airflow obstruction, bronchial hyperresponsiveness, and an underlying inflammation. The interaction of these features of asthma determines the clinical manifestations and severity of asthma and the response to treatment".[11] According to the Centers for Disease Control and Prevention (CDC), 1 in 13 Americans (approximately 25 millions) have asthma. 2018 National Health Interview Survey indicated that currently, 24.7 million people are affected by asthma. While females represent 60% (14.9 million) of those affected, males making up 40% (9.78 million). At 13.4% of the national population, African Americans have the highest rates of asthma, more common in children than adults. Within children, boys have higher rates of asthma than girls.[12]

While it is well known that asthma incidence and prevalence are higher in children, asthma-related morbidity and mortality are higher in adults. Gender also plays a role in asthma pathophysiology. Until the fifth decade of their lifespan, women have higher asthma incidences than men. Similarly, boys in the adolescent and prepubertal stages have a higher asthma burden than their female counterparts (Table 19.1).[13]

TABLE 19.1 Type of asthma and their most common allergens

TYPE OF ASTHMA	*ALLERGENS*	*FEATURES*	*GROUP*
Allergy	Airborne		
	House dust mite	More prevalent in children and a common trigger throughout the whole year	Children>adolescents>adults
	Animal hair/dander	A frequent trigger of asthma	Children>adolescents>adults
	Pollen	A frequent but seasonal cause of severe asthma attacks	A common trigger and can lead to hospitalization
	Mold	Can exacerbate asthma control	Children>adolescents>adults
	Occupational allergens (e.g., paints, resins, latex, cleaning agent)	More common in adults. Caused by breathing in chemical fumes, gases, dust, or other substances. Can trigger asthma attack and exacerbate pre-existing asthma	Adults>adolescents>children
	Food • Egg • Cow's milk • Shellfish • Nuts	Can trigger life-threatening asthma attack but is rare	More common in children
Non-Allergy	Non-Allergic Triggers		
	Respiratory viral infections plus exercise or cold air	Changes in the flow and content of inhaled air trigger attack	Attacks in sensitive individuals mostly during exertion
	Air pollution including auto exhaust	More related to exacerbation of asthma attack but may also trigger an attack	A major cause of increase in new cases in children and adults
	Indoor air pollution	Detergents, cleaning agents, house dust, cooking smell, and indoor smoking	Children
	Occupational agents, e.g., bleach containing chlorine and ammonia	Inhaling fumes from an irritant, such as chlorine, can trigger immediate asthma symptoms in the absence of allergy	Adolescents and adults
	Smoking:		
	First-hand	Exposure increases risk in childhood or sensitive individuals	Adolescents and adults
	Second-hand		Exposure during pregnancy increases risk for childhood asthma

Source: Modified from Dharmage et al.[14]

19.4 ASTHMA PATHOPHYSIOLOGY

Asthma is an obstructive pulmonary disorder with exacerbations characterized by recurrent episodes of wheezing, breathlessness, chest tightness, and cough at night and in the early morning. The 1991, 1997, and 2007 National Institute of Health Guidelines on Asthma (NIH Guidelines) define asthma as follows: "Asthma is a chronic inflammatory disorder of the airways in which many cells and cellular elements play a role: in particular, mast cells, eosinophils, T lymphocytes, macrophages, neutrophils, and epithelial cells". Even with the vast advancement in the field, the cause of asthma remains unknown. However, epidemiological studies have identified several important environmental and genetic risk factors that are associated with asthma. Mechanistically, the central component of allergic asthma is the development of a Th2/IgE (immunoglobulin E) immune response to allergens that activates and recruits several inflammatory cells such as leukocytes and mast cells.

Airway inflammation and damage are central to asthma pathophysiology. It is characterized by airway dysfunction due to the release of cytotoxic cytokines that directly damage the epithelium and airway remodeling. In the late stage, asthma progresses to a severe stage characterized by the presence of mucus gland hypertrophy, airway smooth muscle proliferation, and fibrosis. Based on the pathophysiology and actors involved an asthma attack can be divided into two broad stages: the early/immediate phase and the late phase (Figure 19.1). The early phase is manifested as bronchoconstriction, vasodilation (increased vascular permeability), and eosinophilia and can occur within minutes of exposure to allergen. In contrast, late phase is dominated by mucus hypersecretion, airway hypertrophy and fibrosis, airway epithelial damage, and severe inflammation concomitant with the early phase markers. The late phase can take several hours or days to develop after initial exposure to allergen. The early phase begins with the production of IgE antibodies and their interaction with mast cells and basophils. Upon activation mast, cells release cytokines and asthma mediators, including histamine, prostaglandins, and leukotrienes (LTs) that essentially mediate the early phase of an asthma attack.[15] It is followed by massive recruitment of inflammatory leukocytes and hypertrophy of airway smooth muscles and mucus gland that progress the asthma attack into severe stage (late phase).[15,16]

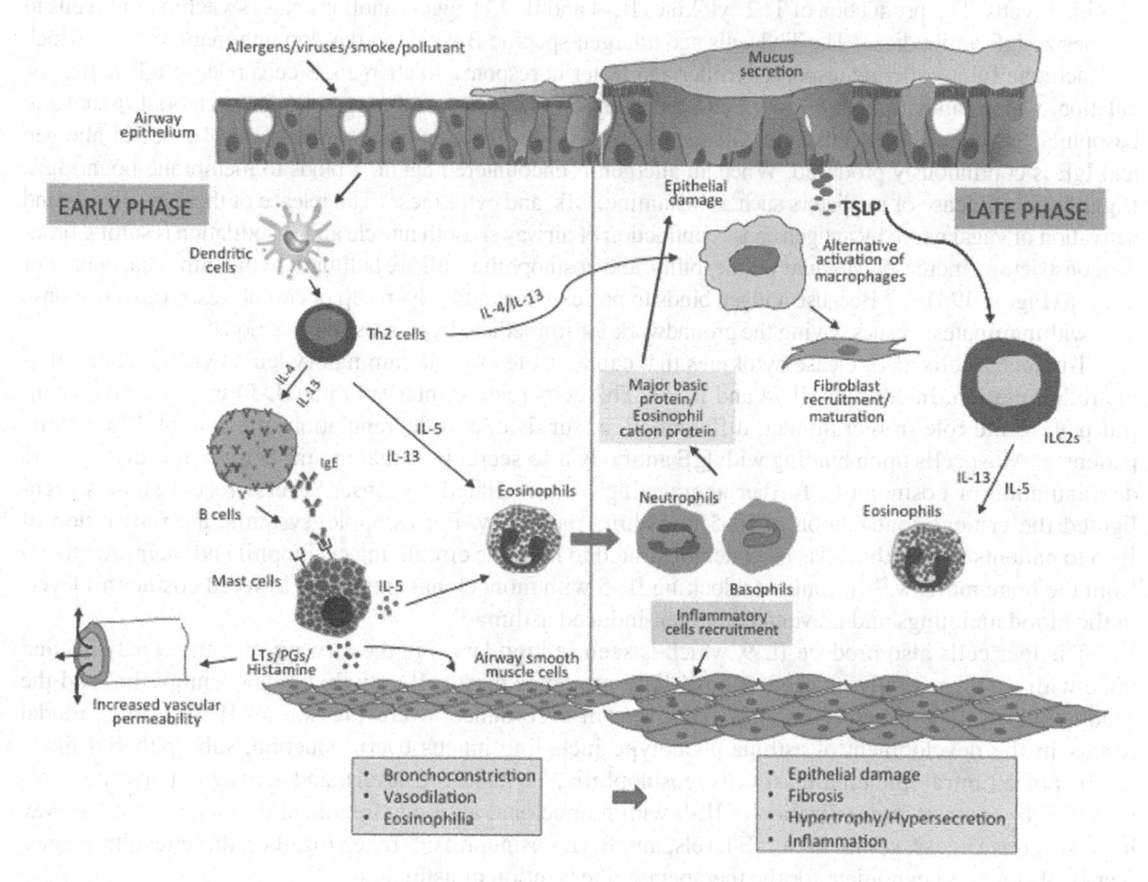

FIGURE 19.1 Exposure to allergens elicits a Th2 response via dendritic cells. It increases the production of IL-4, IL-5, and IL-13 type 2 cytokines, activating the production of IgE by B cells. Mast cells become activated by interaction with IgE antibody and release various mediators such as LTs, PGs, and histamine. On re-exposure to allergen, the immediate reaction is triggered by stimulation of mast cells via Ige-Fc receptors binding. Activated mast cells release various large amounts of LTs, PGs, and histamine driving bronchoconstriction, increased vascular permeability, and eosinophilia. This is followed by massive leukocytes recruitment including basophils and neutrophils. These cells secret cytotoxic metabolites such as major basic protein and eosinophil cationic protein that cause damage to airway epithelium to initiate the late phase of asthma. Factors released by inflammatory cells further induce hypertrophy of airway smooth muscles and goblet cells, mucus hypersecretion, recruitment of fibroblasts, and inflammation. Allergens directly activate epithelium to release TSLP and recruitment of ILC2 cells. ILC2 – innate lymphoid cell type 2; TSLP – thymic stromal lymphopoietin; LTs – leukotrienes; PGs – prostaglandins.

19.4.1 Immediate Phase

Antigen-presenting cells (APCs), which are mainly dendritic cells in the airways that monitor the environment for infections, detect and endocytose inhaled allergens. Dendritic cells present antigens on the cell surface through the major histocompatibility complex (MHC) proteins to the CD4+ naive T cells that contain receptors specific to the antigen. Dendritic cells play an important role in the formation of an immune response and function as the immune system's gatekeepers. Dendritic cells have the ability to induce tolerance or elicit an immunological response by producing a variety of cells mediators and cell surface molecules.[17] It regulates the nature of the immune response by promoting differentiation of CD4+ T cells into various types of T-helper (Th) cells with precise functions[17,18] (Figure 19.1).

T cells, upon interaction with dendritic cells, differentiate into T-helper 2 (Th2)-specific cells, which are important in allergy diathesis.[17,18] T-helper 2 lymphocytes are responsible for the generation of allergic reactions and the stimulation of inflammatory cells by secreting cytokines IL-4 and IL-13 that mediate the allergic illness.[19] Several T-cell receptors and co-stimulatory molecules on the surfaces of Th2 cells engage allergen-specific B cells. The production of Th2 cytokines IL-4 and IL-13 triggers antibody class switching in B cells to synthesize IgE antibodies.[19] The Th2 cells and allergen-specific B cells can develop into memory cells which can facilitate future allergic responses earlier and faster in response to allergen. B cells release IgE in the circulation, which binds to high-affinity FcR1 receptors on the surfaces of mast cells (in interstitial tissue) and basophils,[19] major effector cells in an allergic reaction.[20] This reaction occurs even in the absence of allergen and IgE is continuously produced. When an allergen is encountered again, it binds to membrane-bound IgE, triggering the release of mediators such as histamine, LTs, and cytokines.[19] The release of these mediators and activation of vagal nerve by antigen cause contraction of airway smooth muscle and vasodilation resulting bronchoconstriction, increased vascular permeability, and eosinophilia – all are hallmark of the immediate phase of asthma (Figure 19.1).[21–23] Because antigen binds to pre-existing antibody–receptor complexes, a rapid response (i.e., within minutes) ensues, laying the groundwork for immediate hypersensitive reactions.

T-helper 2 cells also release cytokines that can activate other inflammatory leukocytes to elicit allergic inflammation. In addition, IL-4 and IL-13, Th2 cells release Interleukin-5 (IL-5) which exerts a central pathogenic role in recruitment, differentiation, survival, and degranulation of eosinophil in asthma patients.[24] Mast cells upon binding with IgE antibody also secret IL-5 that in turn helps in recruitment and degranulation of eosinophils, further aggravating Th2-mediated response. Several recent studies highlighted the critical contribution of IL-5 to asthma pathology. For example, systemic administration of IL-5 to patients with asthma has been demonstrated to increase circulating eosinophil and their precursors from the bone marrow.[25] In contrast, blocking IL-5 with monoclonal antibodies lowered eosinophil levels in the blood and lungs and prevented allergen-induced asthma.[25]

T helper cells also produce IL-9, which is seen in high levels in the airways of asthma patients and mice with experimentally induced asthma. IL-9 promotes mast cell growth, tissue eosinophilia, and the production of other T helper 2 (Th2) cytokines. In accordance, overexpression of IL-9 in mice model results in the development of asthma phenotype including mucus overproduction, sub-epithelial fibrosis, increased intra-epithelial mast cells, eosinophilia, elevated IgE levels, and airway hyperresponsiveness.[26–28] In contrast, neutralization of IL-9 with monoclonal antibody treatment in mice reduced airway hyperresponsiveness, serum total IgE levels, and BAL eosinophils.[29] Taken together, these results suggest that IL-9 is a good candidate for the therapeutic intervention of asthma.

Mast cells are the principal makers of cysteinyl LTs, arachidonic acid-derived lipid mediators[21–23] that play a major role in asthma pathophysiology. LTs bind to their G protein-coupled receptors on the cell surfaces of structural airway cells. They cause smooth muscle contractions, increase small blood vessel permeability, increase mucus secretion, and attract leukocytes to the airway. Thus, blocking LTs using LT receptor antagonists or lipoxygenase inhibitor (inhibit LTs synthesis) is an effective way to treat asthma patients.[30] Mast cells also release inflammatory cytokines, chemokines, and proteases that contribute to airway inflammation.[30] These mediators cause an inflammatory reaction in the epithelium, cause direct injury to the airway, and draw in more leukocytes.

19.4.2 Late Phase

The late phase of asthma begins within the next several hours of antigen exposure. In this phase, mediators released by mast cells and eosinophils further drive massive recruitment of eosinophils, basophils, neutrophils, and helper and memory T-cells to the site of injury. Basophils and eosinophils secret major basic protein and eosinophil cationic protein, causing damage to the airway epithelium. This further aggravates bronchoconstriction and inflammation that occurs during the early phase.[31] The late-phase reaction is dominated by the recruitment of leukocytes, notably eosinophils, neutrophils, and more T cells. Factors released by inflammatory cells induce hypertrophy of airway smooth muscles and goblet cells, increase mucus secretion, recruitment of fibroblasts, and inflammation, resulting in airway hypertrophy and fibrosis (Figure 19.1). In addition, Th2 cytokines promote alternative activation of macrophages which drives fibroblast recruitment to cause airway fibrosis. Another important feature of late-phase asthma is goblet cell hyperplasia and mucus overproduction. IL-13 produced by Th2 cells induces goblet cell hyperplasia and mucus hypersecretion from bronchial submucosal glands. In addition, it induces airway hyperresponsiveness and promotes subepithelial fibrosis and airway smooth muscle proliferation.[32] It induces inflammation by acting primarily on the airway epithelium and recruits eosinophils by upregulating multiple chemokines expression, including CCL11,[32] CCL2, CCL3, and CCL6.[33] Furthermore, allergens can also directly stimulate the release of thymic stromal lymphopoietin (TSLP) from the activated epithelium. TSLP drives recruitment of innate lymphoid type 2 cells (ILC2) cells and release of cytokine IL-13 and IL-5 to further cause eosinophilia. TSLP also promotes the differentiation of Th2 lymphocytes via dendritic cells, increasing secretion of inflammatory cytokines IL-4, IL-5, IL-9, and IL-13. These cytokines then drive increase IgE production from B cells, airway hypertrophy, and release of cytotoxic cytokines. TSLP also promotes the activation of neutrophils that in turn activate T helper 17 (Th17) lymphocytes to cause neutrophilic asthma. In contrast, TSLP directly stimulates ILC2s in non-allergic asthma driving recruitment and activation of eosinophils via releasing IL-5 and IL-13 (Figure 19.1).

19.4.3 Role of IL-17

Besides Th2 cells, Th17 cells also contribute to neutrophilic asthma by recruiting neutrophils via IL-17 secretion. Th17 cell-derived cytokines and immunological components that drive neutrophilic infiltration to the airways are mechanistically linked to many cases of severe asthma. In mouse models, allergen sensitization leads Th17 cells to migrate to the lungs, where they increase neutrophil infiltration and eosinophilic inflammation driven by Th2.[34] T-helper 17 cells produce IL-17, which has been found in high concentrations in the airways and blood of asthmatics.[34] The IL-17 family has six members (A-F), all of which appear to be implicated in a variety of other inflammatory disorders, including rheumatoid arthritis, multiple sclerosis, inflammatory bowel disease, and psoriasis, according to a new pre-clinical study.[34] IL-17 upregulates the expression of a wide range of cytokines, chemokines, adhesion molecules, and growth factors. Its precise involvement in asthma and allergy disease, as well as its interactions with Th2 and other leukocyte pathways, are ongoing research topics with implications for disease causation and treatment.

19.5 CURRENTLY AVAILABLE BIOLOGICS FOR ASTHMA

The majority of asthma patients positively respond to standard controller therapy, achieving their target of controlling the disease. However, a significant percentage of patients (~5%) responds inadequately to standard treatment with an inhaled corticosteroid (ICS) and a long-acting bronchodilator (e.g., β2 agonists). As a result, disease in these patients remains uncontrolled, leading to severe asthma and frequent hospitalization, and is associated with high morbidity, mortality, and poor quality of life.[11,35] Research investigation over the past decade has significantly improved our understanding of asthma pathophysiology. It is now evident that asthma is a multi-factorial and a heterogeneous disease that involves multiple pathophysiological pathways, which in turn manifests in different phenotypes/endotypes. This led to the discovery of promising targeted therapies that have proven effective in the treatment of uncontrolled severe asthma. Specifically, the emergence of novel biologic therapies has provided personalized treatment strategies for these patients. Currently, there are five approved biologics for asthma treatment: omalizumab, mepolizumab, reslizumab, benralizumab, and dupilumab – with several others currently in development (Tables 19.2 and 19.3).[36] These drugs target specific mediator or inflammatory modulators that play a crucial role in the pathogenesis of asthma. For example, omalizumab targets and binds with IgE, whereas mepolizumab, reslizumab, and benralizumab all target IL-5, thereby interfering with eosinophils activation. By blocking these specific pathways, these drugs alleviate the degree of release of inflammatory mediators of the allergic response, reduce eosinophil levels, and inhibit IL-4 and IL-13 cytokine-induced responses, including the release of proinflammatory cytokines, chemokines, and IgE (Figure 19.2).[36]

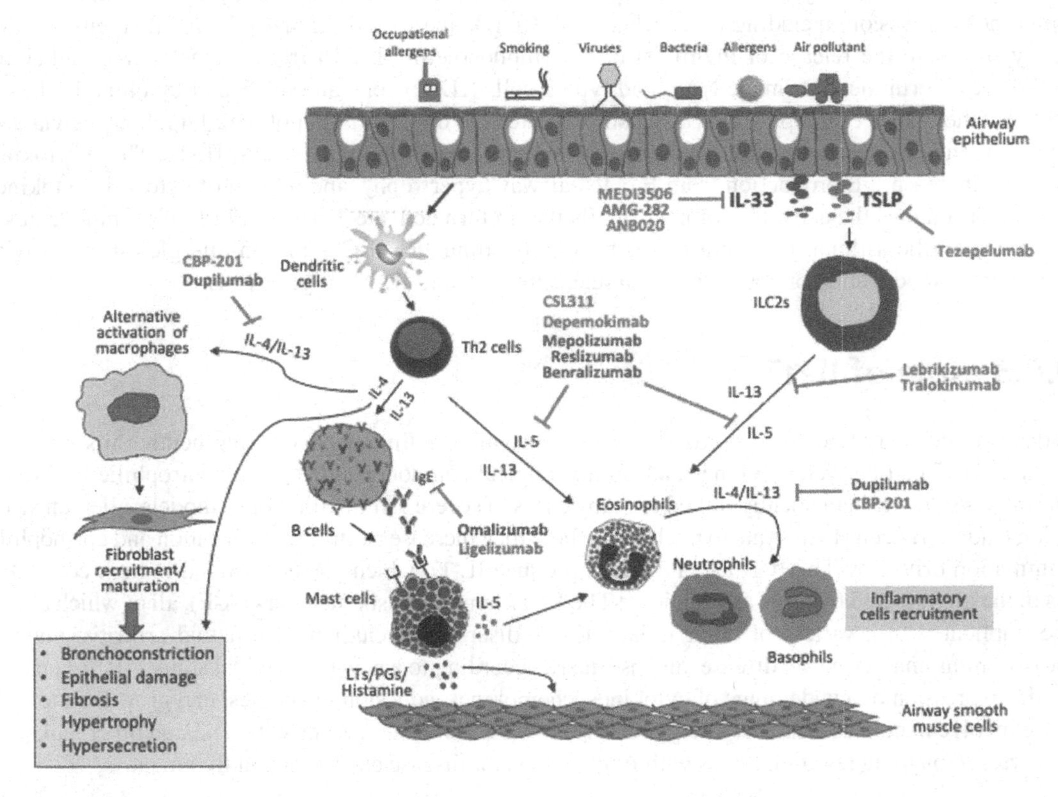

FIGURE 19.2 Schematic representation of asthma immunopathophysiology. Sites of action of FDA approved (blue bold font) and the investigational (red bold font) biologics are indicated. In asthma, the interaction of genetic susceptibility and environmental exposures – such as with allergens, viruses, pollutants, and irritants – creates airway inflammation. Dendritic cells present allergens to naive T-helper (Th0) cells, resulting in their activation into Th2 cells. Th2 cells release type 2 cytokines IL-4, IL-5, and IL-13 that stimulate IgE production from B cells and activate mast cells. These cytokines, along with mast cells, recruit eosinophils, basophils, and neutrophils. Moreover, environmental allergens directly activate the airway epithelium, increasing the secretion of cytokines such as IL-33 and TSLP to drive the recruitment of ILC2 cells. IL-4 and IL-13 promote alternative activation of macrophages that cause fibroblast recruitment and maturation. Together, this complex process results in airway hyperresponsiveness, epithelial damage, fibrosis, hypertrophy and hypersecretion, and massive inflammation. ILC2 – innate lymphoid cell type 2; LTs – leukotrienes; PGs – prostaglandins.

TABLE 19.2 Pharmacokinetic and pharmacodynamic profile of asthma biologics

BIOLOGICS	*OMALIZUMAB (XOLAIR)*	*MEPOLIZUMAB (NUCALA)*	*RESLIZUMAB (CINQAIR)*	*BENRALIZUMAB (FASENRA)*	*DUPILUMAB (DUPIXENT)*
Mechanism of action (Pharmacology)	Anti-IgE antibody	Anti-IL-5 monoclonal antibody (IgG1 kappa)	Anti-IL-5 monoclonal antibody (IgG4, kappa)	Anti-IL-5 monoclonal antibody (IgG1, kappa)	Anti-IL-4R monoclonal antibody. Inhibits both IL-4 and IL-13 signaling
Pharmacokinetic	Absorption				
	Absolute bioavailability of ~62% with SC administration	Mepolizumab bioavailability was estimated ~80%	Peak serum concentrations of reslizumab are often observed at the end of the infusion. The serum concentration declines in a biphasic manner	The absorption half-life was ~3.6 days, absolute bioavailability ~58%, SC administration	Need further studies
	Distribution				
	Vd, SC administration was 78 ± 32 mL/kg	Vd, SC administration 3.6 L for a 70-kg individual	Vd 5L, suggesting minimal distribution to the extravascular tissues	Central and peripheral Vd of benralizumab was 3.2 L and 2.5 L, respectively, for a 70-kg individual	The estimated total volume of distribution was approximately 4.8 ± 1.3 L
	Metabolism				
	IgE complexes are metabolized by reticuloendothelial system and endothelial cells in the liver	Undergoes proteolytic degradation via enzymes that are widely distributed in the body and not restricted to hepatic tissue	Degraded by enzymatic proteolysis into small peptides and amino acids	Degraded by proteolytic enzymes widely distributed in the body and not restricted to hepatic tissue	The metabolic pathway of dupilumab has not been characterized
	Elimination				
	Elimination half-life 26 days, with apparent clearance averaging 2.4 ± 1.1 mL/kg/day	Half-life 16 to 22 days. Systemic clearance of mepolizumab in 0.28 L/day for a 70-kg individual	Reslizumab clearance was approximately 7 mL/hour. Reslizumab has a half-life of about 24 days	Clearance 0.29 L/d for a 70-kg person Elimination half-life was ~15 days[37]	Expected to be degraded into small peptides and amino acids via catabolic pathways (not well documented)

(*Continued*)

TABLE 19.2 Pharmacokinetic and pharmacodynamic profile of asthma biologics (*Continued*)

BIOLOGICS	*OMALIZUMAB (XOLAIR)*	*MEPOLIZUMAB (NUCALA)*	*RESLIZUMAB (CINQAIR)*	*BENRALIZUMAB (FASENRA)*	*DUPILUMAB (DUPIXENT)*
Onset of action	Omalizumab was absorbed slowly, reaching peak serum concentrations after an average of 7–8 days[38]	Decreases in blood eosinophil count observed on the third day following SC administration[39]	Decreased blood eosinophil count observed as early as 2–3 days following IV administration[37]	Decreased blood eosinophil count observed 24 hours after IV administration[40]	Nonlinear pharmacokinetics over SC dose range of 75–600 mg[41]
Duration of action	Total IgE does not return to pre-treatment concentrations for ≤1 year after drug discontinuance in patients with asthma[42]	Following drug discontinuance, blood and sputum eosinophil counts increase within three months[43]	Eosinophils returned toward baseline 120 days after last dose[37]	Decreased blood eosinophil count reported to last for at least 8–12 weeks[40]	Not well defined
Indication	Moderate to severe persistent asthma in patients six years of age and older with a positive skin test or *in vitro* reactivity to a perennial aeroallergen and symptoms that are inadequately controlled with inhaled corticosteroids	Add-on maintenance treatment of patients with severe asthma aged 12 years and older, and with an eosinophilic phenotype	Severe asthma aged 18 years and older, and with an eosinophilic phenotype	Add-on maintenance treatment of patients with severe asthma aged 12 years and older and with an eosinophilic phenotype	Add-on maintenance treatment in patients with moderate-to-severe asthma aged 12 years and older with an eosinophilic phenotype or with oral corticosteroid dependent asthma
Adverse effects	Arthralgia, general pain, leg pain, fatigue, dizziness, fracture, arm pain, pruritus, dermatitis, earache[42]	Opportunistic Infections: Herpes Zoster Injection site reactions (e.g., pain, erythema, swelling, itching, burning sensation) Hypersensitivity reactions (e.g., rash, pruritus, headache, and myalgia) headache (19%), injection-site reaction (8%), back pain (5%), and fatigue (5%)[44]	Anaphylaxis Malignancy (6/1028) patients receiving 3 mg/kg CINQAIR had at least one malignant neoplasm reported compared to 2/730 (0.3%) patients in the placebo group[45]	Headache, pharyngitis, pyrexia, hypersensitivity Reactions[46-48]	Injection site reactions (e.g., erythema, swelling, warmth) Ocular effects (e.g., conjunctivitis, blepharitis, keratitis, pruritus, dry eye) Herpes simplex virus infection[41,49,50]

Abbreviations: IL-4R, interleukin 4 receptor; Vd, volume of distribution; IgE, immunoglobulin E; IgG, immunoglobulin G; SC, subcutaneous; IV, intravenous.

TABLE 19.3 Summary of FDA-approved asthma biologics

BIOLOGIC	*DOSING*	*EFFICACY*	*CONSIDERATION*
Omalizumab	Divide doses of more than 150 mg among more than one injection site to limit injections to not more than 150 mg per site. Asthma: 75–375 mg SC every 2 or 4 weeks. Determine dose (mg) and dosing frequency by serum total IgE level (IU/mL), measured before the start of treatment, and body weight (kg). See the dose determination charts	Reduces asthma exacerbation by 25%	For SC administration only
Mepolizumab	100 mg SC every 4 weeks (bodyweight ≥ 40 kg); 40 mg SC every 4 weeks (bodyweight < 40 kg)	Reduces asthma exacerbation by 50%	Patients with eosinophilic phenotype
Reslizumab	3 mg/kg IV every 4 weeks	Reduces asthma exacerbation by 50%–60%	Only IV biologic approved
Benralizumab	30 mg SC every 4 weeks × 3 doses, then every 8 weeks	Reduces asthma exacerbation by 25%–60%	Longer duration of action of 8 weeks
Dupilumab	400 or 600 mg (two injections) SC followed by 200 or 300 mg every other week	Reduces asthma exacerbation by 50%–70%	Can be self-administered

Abbreviations: SC, subcutaneous; IV, intravenous; IgE, immunoglobulin E.

A detailed description of their pharmacokinetic and pharmacodynamic profile of FDA-approved asthma biologics is given in Tables 19.2 and 19.3.

Two specific endotypes, type 2 (T2) high and low, are taken into consideration when considering a patient for candidacy of biologic therapies. These phenotypes are based on the levels of IL-4, IL-5, and IL-13 cytokines that play a critical role in asthma pathogenesis. In T2 high asthma, exposure to allergens leads to a massive secretion of IL-4, IL-5, and IL-13 cytokines, resulting in activation of inflammatory cells including mast cells, eosinophils, and basophils, and pulmonary cells including alveolar cells and smooth muscle cells. Secretion of vasoactive and hypertrophic factors and activation of vagus nerve eventually lead to bronchoconstriction, mucus production, and airway remodeling.[51,52] Several biologics have been currently approved for the treatment of severe asthma with T2-high endotype (Figure 19.2). The American Thoracic Society (ATS) and European Respiratory Society (ERS) Task Force defines severe asthma as "asthma which requires treatment with high dose inhaled corticosteroids (ICS) (see Table 19.2 for doses in adults and children) plus a second controller (and/or systemic CS) to prevent it from becoming 'uncontrolled' or which remains 'uncontrolled' despite this therapy".[53] In contrast, T2-low asthma is mostly caused by neutrophil-mediated inflammation. This involves both Th1 and Th17 immune response and is characterized by a high level of IL-17. Patients with T2-low asthma are generally less allergic to common allergens and respond poorly to steroids. No biologic is approved currently for T2-low asthma.[54]

19.6 LIMITATIONS OF ASTHMA BIOLOGICS

1. Asthma pathophysiology is complex with the involvement of multiple mediators and pathways. To contribute to asthma pathophysiology, these pathways overlap and exhibit redundancy in action. This may result in incomplete success in the therapy outcome with biologics as they target one specific mediator in the complex pathophysiology. As a result, biologics usually have a limited efficacy spectrum than standard therapies.[55]

2. Due to the high cost involved with the usage of biologics therapy, pharmacoeconomic studies are necessary to identify the ideal target patient population, which further adds to the cost and time required for their approval.[56]
3. Many biologics require reconstitution before administration. This may warrant the patient to be present at the clinic before the formulation of the drug, increasing visit time and the cost involved. Additional step in formulation may also increase the risk for errors and contamination.[57]
4. Conditions required for transport and storage also add to limitations associated with their therapeutic usage. In general, since biologics/biosimilars are biological products, they are highly unstable and easy to degrade. Thus, they require low temperature and special handling during storage and transport. For example, benralizumab is stored in the prefilled syringe at 2°C–8°C and protected from light. After removal from the storage temperature, these products need to be used in a short period of time of 1–2 weeks.
5. In addition, monitoring of patients after administration of biologic agents is highly recommended to observe for prophylactic reactions, if any.[41,42]

19.7 EMERGING BIOLOGICS/BIOSIMILAR FOR ASTHMA

Major efforts in recent times have further unraveled the underlying mechanisms of severe asthma. Improved understanding and classification of different patterns of asthma have played a crucial in the development of precision medicines with biologic therapies for the treatment of severe asthma. Degree and type of specific cytokines activation is one of the bases of the broad classification of asthma. Type 2 asthma inflammation is caused by exposure to allergens that results in the activation of Th2 cells and ILC2, increasing T2 cytokines (IL-4, IL-5, and IL-13) release to activate eosinophils and mast cells. In contrast, low T2 asthma is mostly caused by the exposure to viruses, bacteria, and irritants that activate non-type 2 cells, including Th17 cells, resulting in activation of neutrophils through IL-6 and IL-17. This new understanding has dramatically improved our understanding of the immunopathogenesis of asthma and led to the discovery of a wide range of novel drug targets, such as IL-33, thymic stromal lymphopoietin (TSLP), the transcription factor GATA-3, IL-25, IL-4Rα, and the prostaglandin DP2 (CRTH2) receptor (Figures 19.2 and 19.3). The delivery of these new targets has paved the way for the development of biologic therapeutic agents. While the currently available biologics target downstream mediators of T2 inflammation, emerging new therapeutic agents are being proposed and studied for their ability to target various upstream targets of asthma inflammation. Targeting upstream mediators of T2 or non-T2 inflammation in asthma will potentially provide more personalized and additional treatment options for patients with uncontrolled asthma (Figure 19.2). Among these, the most promising target which has produced promising results in clinical trials is TSLP. In a recently completed Phase 3 randomized controlled trial of patients with moderate asthma, tezepelumab, a monoclonal antibody against TSLP, reduced asthma exacerbations unrelated to baseline AEC and decreased markers of Th2 inflammation, IgE, and nitric oxide levels.[58,59]

Thymic stromal lymphopoietin TSLP is a member of the IL-2 family of cytokines and shares homology with IL-7. It also belongs to the family of alarmins, which instantly release upon exposure to injury. TSLP is released by airway epithelial cells when airway tissue is exposed to an injury stimulus such as allergens, cigarette smoke, airborne pollutants, viruses, bacteria, and chemical irritants (Figure 19.3).[59,60] To produce its action, TSLP interacts with its receptor TSLPR expressed on the target cells and affects their functions including dendritic cells, ILC2, eosinophils, fibroblasts, and mast cells (Figure 19.3). TSLP promotes the differentiation of Th2 lymphocytes via dendritic cells increasing the secretion of inflammatory cytokines IL-4, IL-5, IL-9, and IL-13. These cytokines, in turn, activate

B cells, eosinophils, mast cells, and airway smooth muscle cells, respectively, to increase IgE production, release of cytotoxic cytokines, airway hypertrophy. TSLP also promotes the activation of neutrophils that in turn activate Th17 lymphocytes to cause neutrophilic asthma. In contrast, TSLP directly stimulates ILC2s in non-allergic asthma driving recruitment and activation of eosinophils via releasing IL-5 and IL-13 (Figure 19.3).[61–63]

Tezepelumab is the first-in-class anti-TSLP monoclonal antibodies that bind with human TSLP, preventing its interaction with its receptor TSLPR. By blocking the action of TSLP, Tezepelumab

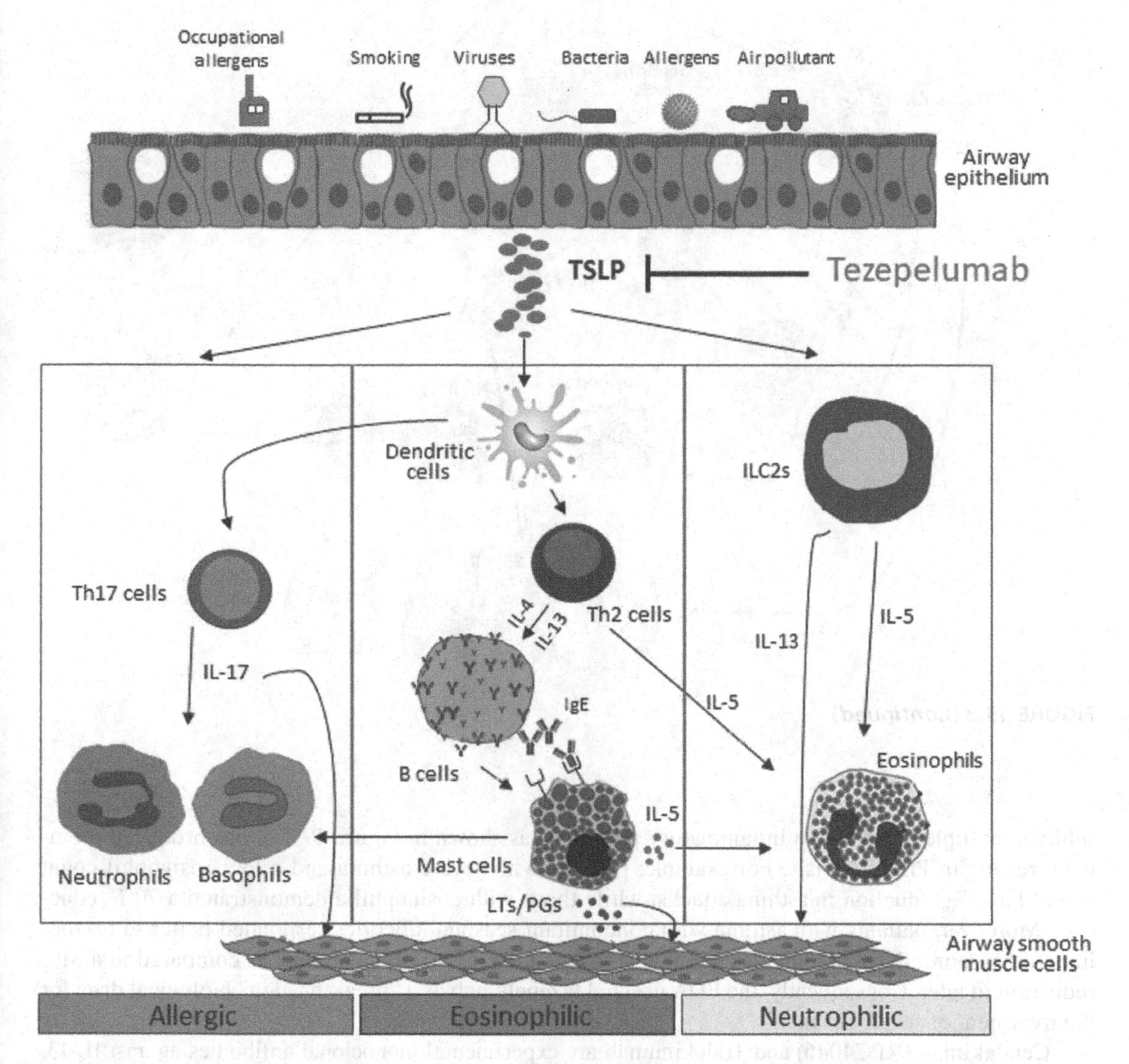

FIGURE 19.3 Thymic stromal lymphopoietin (TSLP), an upstream mediator of asthma immunopathobiology plays a crucial role in different types of asthma. In allergic asthma, TSLP activate a Th2 response via dendritic cells to release type 2 cytokines such as IL-4, IL-5, and IL-13. This results in the production of IgE and activation of mast cells. In eosinophilic asthma, TSLP recruits ILC2 to release IL-5 and IL-13, followed by infiltration and activation of eosinophils. TSLP can also recruit Th17 cells to increase neutrophil recruitment via secreting IL-17. Tezepelumab works by binding with TSLP and preventing its interaction with its receptor, thereby blocking downstream cellular signaling. ILC2 – innate lymphoid cell type 2; LTs – leukotrienes; PGs – prostaglandins; Th – T helper. Modified from Pelaia C, 2021.[63] (*Continued*)

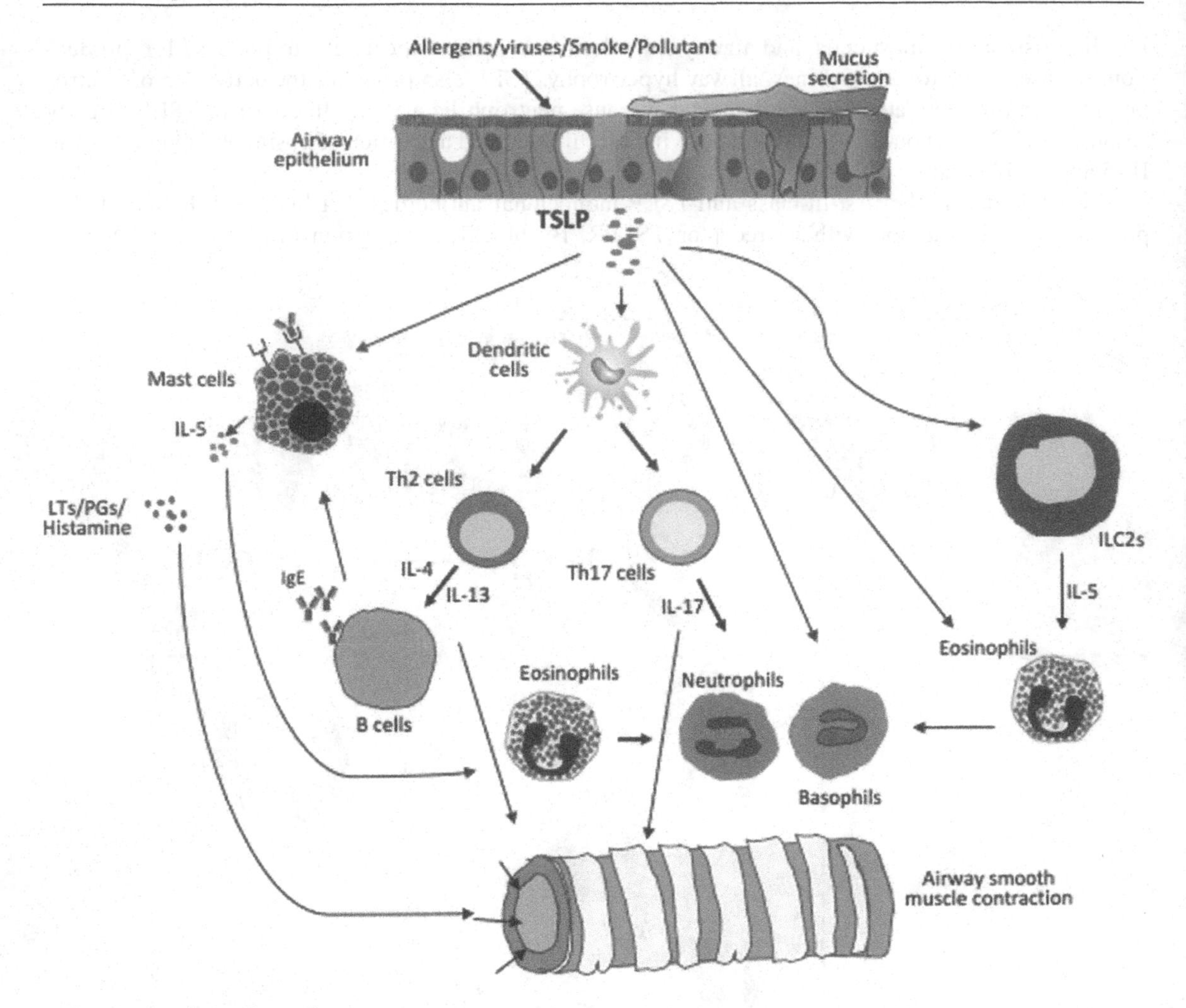

FIGURE 19.3 (*Continued*)

inhibits multiple downstream inflammatory pathways, as shown in Figure 19.3. It has produced promising results in Phase 2 trials. For example, patients with severe asthma and a low eosinophil count showed a 41% reduction in asthma attacks, while those with eosinophilia demonstrated a 70% reduction. Moreover, patients with asthma with concomitant seasonal allergies responded better to tezepelumab than non-allergic asthmatics and had a 58% reduction in asthma attacks as compared to a 51% reduction in later. Consequently, the FDA granted tezepelumab as a "breakthrough" biological drug for the treatment of severe asthma.[58,64]

Cendakimab (RPC4046) and Tralokinumab are experimental monoclonal antibodies against IL-13, preventing its interaction with its receptor, generating a unique transcriptional profile at the site of inflammation that may, in turn, result in improved efficacy in Th2-type allergic diseases such as asthma and Eosinophilic oesophagitis.[65,66]

Ligelizumab (QGE031) is in a Phase 2 trial for allergic asthma, atopic dermatitis, and bullous pemphigoid. It is a fully-humanized IgG1 monoclonal antibody directed against human IgE, which binds to the Cε3 domain of IgE.[67] It has demonstrated greater efficacy than omalizumab on inhaled and skin allergen responses in patients with mild allergic asthma.[67] A list of emerging biologics being evaluated in different clinical trials for asthma is given in Table 19.4.

TABLE 19.4 Emerging biologic under development for the treatment of Asthma (also refer to Figure 19.2)

BIOLOGIC	*MECHANISM OF ACTION*	*INDICATIONS*	*CURRENT CLINICAL PHASE*	*TRIALS NUMBER*	*SPONSOR*
CBP-201	Anti-IL-4Rα monoclonal antibody	Moderate to severe persistent asthma with type 2 inflammation	Phase 2	NCT04773678	Connect Biopharmaceuticals
Tezepelumab	Anti-thymic stromal lymphopoietin (TSLP) monoclonal antibody	Children aged ≥5–11 years with asthma Inadequately controlled asthma	Phase 1 Phase 3	NCT04673630 NCT03347279	AstraZeneca and Amgen
MEDI3506	Anti-IL-33 monoclonal antibody	Adults with uncontrolled moderate to severe asthma	Phase 2	NCT04570657	AstraZeneca
CSL311	Antagonist of the combined activities of IL-3, GM-CSF, and IL-5 (Anti-βc receptor monoclonal antibody)	Patients with mild asthma	Phase 1	NCT04082754	CSL Behring
GSK3511294 (Depemokimab)	Anti-IL-5 receptor monoclonal antibody	Patients with severe uncontrolled asthma with an eosinophilic phenotype	Phase 3	NCT04718103	GlaxoSmithKline
Lebrikizumab	Anti-IL-13 receptor monoclonal antibody	Uncontrolled severe asthma	Phase 2	NCT02340234	Dermira, Inc., a subsidiary of Eli Lilly
Tralokinumab	Anti-IL-13 receptor monoclonal antibody	Uncontrolled severe asthma	Phase 3	NCT03131648	AstraZeneca
Ligelizumab (QGE031)	Anti-IgE monoclonal antibody	Allergic asthma, atopic dermatitis, and bullous pemphigoid	Phase 2	NCT01716754	Novartis
AMG-282	Anti-IL-33 monoclonal antibody	Mild atopic asthma and rhinosinusitis	Phase 2	NCT01928368, NCT02170337	Amgen
ANB020	Anti-IL-33 monoclonal antibody	Severe asthma	Phase 2	NCT03469934	AnaptysBio, Inc.

Abbreviations: TSLP, thymic stromal lymphopoietin; GM-CSF, granulocytes-macrophage colony-stimulating factor.

19.8 CONSIDERATION WHEN AND WHICH BIOLOGIC TO START

Considering the high cost and limited data available for the biologics, a clear and objective approach should be incorporated before considering a patient the candidate for a biologic. Most asthma patients who comply and adhere with their controller therapy do not need a biologic. Biologic therapy is usually considered when disease symptoms remain uncontrolled or the risk of asthma attack exacerbation is high despite adherence to controller therapy (or Global Initiative for Asthma, GINA, step 4 therapy).[68] GINA step 4 therapy includes medium dose ICS and LABA both in adults and children. Choosing the most appropriate biologic for a patient is critical and requires a thorough understanding of asthma endotype and classification.

The GINA 2019 guidelines provide basic guidance regarding when and which biologic to initiate in an asthma patient.[68] Patients on maintenance oral steroids for asthma should be considered for an appropriate biologic therapy. Uncontrolled/severe asthma should be assessed for T2 inflammation before initiating a biologic. Blood eosinophil ≥ 150/μL, and/or fractional exhaled nitric oxide (FeNO) ≥ 20 ppb, and/or sputum eosinophils ≥ 2%, and/or asthma is clinically allergen-driven, and/or need for maintenance oral corticosteroids (OCS) will confirm T2 asthma. T2-targeted biologic should be considered in patients with exacerbations or poor symptoms control on high dose OCS-LABA and have eosinophilic or allergic biomarker and need maintenance OCS. Adherence with step 4 therapy, assessment of full pulmonary function tests, and oral steroid use status needs to be determined before considering a biologic.

Which biologic: Is the patient eligible for one of the following treatments? If yes, begin the treatment and assess the response for at least 4 months, and the trial should be extended for 12 months if good response.[68]

Anti-IgE: Sensitization on skin prick testing, total serum IgE and weight within dosage range, exacerbation in last year.

Anti-IL-4/Anti-IL-5R: Exacerbation in last year and blood eosinophils ≥ 300/μL.

Anti-IL-4R: Exacerbation in last year, blood eosinophils ≥ 150/μL, FeNO ≥ 25 ppb, or need for maintenance OCS.

Another important point to consider while choosing a specific biologic for a patient is that if a patient requires maintenance OCS to control asthma symptoms, usefulness of anti-IgE therapy is not established. Allergic asthma with high IgE levels – in which anti-IgE therapy is likely to be effective – may not necessitate OCS therapy. Additionally, more studies are warranted to develop the efficacy of biologics in T2-low asthma or non-eosinophilic patients. This group has been distinctively missing from the biologic trials. There is a need to explore novel targets in T2-low or non-eosinophilic asthma in order to develop new personalized biologics for this group of asthma patients

19.9 CONCLUSION

The discovery of biologics/biosimilars has paved the way to treat multiple disorders more effectively, which was not achievable otherwise. For instance, the availability of several biologics drugs such as adalimumab, etanercept, and rituximab has dramatically changed the treatment options and improved the overall quality of life of patients affected by rheumatoid arthritis, plaque psoriasis, Crohn's disease, or ulcerative colitis. Likewise, omalizumab – an anti-IgE monoclonal antibody – has reduced the disease burden in patients with severe asthma. A considerable number of asthma patients respond inadequately to standard available therapies, including ICS, resulting in exacerbation of asthma symptoms and poor quality of life. There are currently five FDA-approved biologics available for the treatment of asthma which has provided promising and effective targeted therapies for these patients. These drugs target specific inflammatory mediators such as IgE, IL-5, and IL-4R. These drugs have been shown to significantly reduce asthma exacerbation in patients with uncontrolled asthma. However, due to the high cost involved with their manufacturing and special handling requirements, as well as the narrow mechanism of action use of biologics is limited to cases where standard therapies have failed. There is a need for the development of novel and cost-effective biologics with a broader mechanism of action. With the advancement of our understanding of asthma pathophysiology, several biologics targeting upstream inflammatory mediators of asthma biology have been recently discovered and are currently in clinical trials. These emerging biologics target upstream mediators of asthma, the blockade of which should supposedly provide better control of asthma symptoms and exacerbation. These targets include IL-33, TSLP, GATA-3, and IL-25. Targeting upstream mediators of asthma inflammation should potentially provide more personalized and additional treatment options for patients with uncontrolled asthma. One of the most promising targets which have shown effectiveness in clinical trials is TSLP (blocked by tezepelumab). Another limitation of current biologics is that they are effective only in eosinophilic asthma. There is a need to develop new biologics that will be more geared toward patients with non-eosinophilic or T2 low asthma. A better understanding of this type of asthma and endotyping (Type 2 high or low) may potentially help us develop more specific and effective targets. This will allow us to develop novel biologics and identify the specific target patient population to provide personalized medicine for different asthma groups.

SOURCES OF FUNDING

Dr. Shahid's current research activities at Chicago State University are supported by National Institutes of Health grant 1SC3 GM135023-01 and American Heart Association grant 20AIREA35150000.

REFERENCES

1. US FDA. *What Are "Biologics" Questions and Answers.* 2018; Accessed on Jan 30, 2022.
2. ATC CJaC. Biologics and biosimilars: What, why and how? *ESMO Open.* 2017;2:e000180.
3. Agbogbo FK, Ecker DM, Farrand A, Han K, Khoury A, Martin A, McCool J, Rasche U, Rau TD, Schmidt D, Sha M and Treuheit N. Current perspectives on biosimilars. *Journal of Industrial Microbiology & Biotechnology.* 2019;46:1297–1311.
4. Agency EM. *Biosimilar Medicines: Overview [Internet].* London: European Medicines Agency. 2018, 2021.
5. Kang HN and Knezevic I. Regulatory evaluation of biosimilars throughout their product life-cycle. *Bulletin of the World Health Organization.* 2018;96:281–285.
6. Alexander GC, Ogasawara K, Wiegand D, Lin D and Breder CD. Clinical development of biologics approved by the US Food and Drug Administration, 2003–2016. *Therapeutic Innovation & Regulatory Science.* 2019;53:752–758.
7. Busse W, Corren J, Lanier BQ, McAlary M, Fowler-Taylor A, Cioppa GD, van As A and Gupta N. Omalizumab, anti-IgE recombinant humanized monoclonal antibody, for the treatment of severe allergic asthma. *The Journal of Allergy and Clinical Immunology.* 2001;108:184–190.
8. Kabir ER, Moreino SS and Sharif Siam MK. The breakthrough of biosimilars: A twist in the narrative of biological therapy. *Biomolecules.* 2019;9:410.
9. US FDA. *Biosimilar Development, Review, and Approval.* 2017; Accessed on Jan 30, 2022.
10. DiMasi JA, Grabowski HG and Hansen RW. Innovation in the pharmaceutical industry: New estimates of R&D costs. *Journal of Health Economics.* 2016; 47:20–33.
11. Moore WC, Bleecker ER, Curran-Everett D, Erzurum SC, Ameredes BT, Bacharier L, Calhoun WJ, Castro M, Chung KF, Clark MP, Dweik RA, Fitzpatrick AM, Gaston B, Hew M, Hussain I, Jarjour NN, Israel E, Levy BD, Murphy JR, Peters SP, Teague WG, Meyers DA, Busse WW and Wenzel SE. Characterization of the severe asthma phenotype by the National Heart, Lung, and Blood Institute's Severe Asthma Research Program. *The Journal of Allergy and Clinical Immunology.* 2007; 119:405–413.
12. Pate C, Zahran HS, Qin X, Johnson C, Hummelman E, Malilay J. Asthma Surveillance _ United States, 2006-2018. *MMWR Surveill Summ* 2021; 70 (No. SS-5):1–32
13. Fuhlbrigge AL JB, Wright R. Gender and asthma. *Immunology and Allergy Clinics of North America.* 2002;22:753–789.
14. Dharmage SC, Perret JL and Custovic A. Epidemiology of asthma in children and adults. *Frontiers in Pediatrics.* 2019;7:246.

15. Liu MC, Hubbard WC, Proud D, Stealey BA, Galli SJ, Kagey-Sobotka A, Bleecker ER and Lichtenstein LM. Immediate and late inflammatory responses to ragweed antigen challenge of the peripheral airways in allergic asthmatics. Cellular, mediator, and permeability changes. *The American Review of Respiratory Disease*. 1991;144:51–58.
16. Zhu Z, Homer RJ, Wang Z, Chen Q, Geba GP, Wang J, Zhang Y and Elias JA. Pulmonary expression of interleukin-13 causes inflammation, mucus hypersecretion, subepithelial fibrosis, physiologic abnormalities, and eotaxin production. *The Journal of Clinical Investigation*. 1999;103:779–788.
17. Lambrecht BN and Hammad H. The role of dendritic and epithelial cells as master regulators of allergic airway inflammation. *Lancet*. 2010;376:835–843.
18. Liu YJ. Thymic stromal lymphopoietin and OX40 ligand pathway in the initiation of dendritic cell-mediated allergic inflammation. *The Journal of Allergy and Clinical Immunology*. 2007;120:238–244; quiz 245–246.
19. Levine SJ and Wenzel SE. Narrative review: The role of Th2 immune pathway modulation in the treatment of severe asthma and its phenotypes. *Annals of Internal Medicine*. 2010;152:232–237.
20. Brightling CE, Bradding P, Symon FA, Holgate ST, Wardlaw AJ and Pavord ID. Mast-cell infiltration of airway smooth muscle in asthma. *New England Journal of Medicine*. 2002;346:1699–1705.
21. Boyce JA. Mast cells: Beyond IgE. *The Journal of Allergy and Clinical Immunology*. 2003;111:24–32; quiz 33.
22. Galli SJ, Kalesnikoff J, Grimbaldeston MA, Piliponsky AM, Williams CM and Tsai M. Mast cells as "tunable" effector and immunoregulatory cells: Recent advances. *Annual Review of Immunology*. 2005;23:749–786.
23. Robinson DS. The role of the mast cell in asthma: Induction of airway hyperresponsiveness by interaction with smooth muscle? *The Journal of Allergy and Clinical Immunology*. 2004;114:58–65.
24. Pelaia C, Paoletti G, Puggioni F, Racca F, Pelaia G, Canonica GW and Heffler E. Interleukin-5 in the pathophysiology of severe asthma. *Frontiers in Physiology*. 2019;10:1514.
25. Blanchard C and Rothenberg ME. Biology of the eosinophil. *Advances in Immunology*. 2009;101:81–121.
26. McLane MP, Haczku A, van de Rijn M, Weiss C, Ferrante V, MacDonald D, Renauld JC, Nicolaides NC, Holroyd KJ and Levitt RC. Interleukin-9 promotes allergen-induced eosinophilic inflammation and airway hyperresponsiveness in transgenic mice. *American Journal of Respiratory Cell and Molecular Biology*. 1998;19:713–720.
27. Temann UA, Geba GP, Rankin JA and Flavell RA. Expression of interleukin 9 in the lungs of transgenic mice causes airway inflammation, mast cell hyperplasia, and bronchial hyperresponsiveness. *The Journal of Experimental Medicine*. 1998;188:1307–1320.
28. Levitt RC, McLane MP, MacDonald D, Ferrante V, Weiss C, Zhou T, Holroyd KJ and Nicolaides NC. IL-9 pathway in asthma: New therapeutic targets for allergic inflammatory disorders. *The Journal of Allergy and Clinical Immunology*. 1999;103:S485–S491.
29. Kung TT, Luo B, Crawley Y, Garlisi CG, Devito K, Minnicozzi M, Egan RW, Kreutner W and Chapman RW. Effect of anti-mIL-9 antibody on the development of pulmonary inflammation and airway hyperresponsiveness in allergic mice. *American Journal of Respiratory Cell and Molecular Biology*. 2001;25:600–605.
30. Montuschi P and Peters-Golden ML. Leukotriene modifiers for asthma treatment. *Clinical and Experimental Allergy: Journal of the British Society for Allergy and Clinical Immunology*. 2010;40:1732–1741.
31. Stewart AG, Tomlinson PR, Fernandes DJ, Wilson JW and Harris T. Tumor necrosis factor alpha modulates mitogenic responses of human cultured airway smooth muscle. *American Journal of Respiratory Cell and Molecular Biology*. 1995;12:110–119.
32. Rael EL and Lockey RF. Interleukin-13 signaling and its role in asthma. *The World Allergy Organization Journal*. 2011;4:54–64.
33. Barnes PJ. The cytokine network in asthma and chronic obstructive pulmonary disease. *The Journal of Clinical Investigation*. 2008;118:3546–3556.
34. Park SJ and Lee YC. Interleukin-17 regulation: An attractive therapeutic approach for asthma. *Respiratory Research*. 2010;11:78.
35. Antonicelli L, Bucca C, Neri M, De Benedetto F, Sabbatani P, Bonifazi F, Eichler HG, Zhang Q and Yin DD. Asthma severity and medical resource utilisation. *The European Respiratory Journal*. 2004;23:723–729.
36. McGregor MC, Krings JG, Nair P and Castro M. Role of biologics in asthma. *American Journal of Respiratory and Critical Care Medicine*. 2019;199:433–445.
37. Hom S and Pisano M. Reslizumab (Cinqair): An interleukin-5 antagonist for severe asthma of the eosinophilic phenotype. *P&T: A Peer-reviewed Journal for Formulary Management*. 2017;42:564–568.
38. Thomson NC and Chaudhuri R. Omalizumab: Clinical use for the management of asthma. *Clinical Medicine Insights Circulatory, Respiratory and Pulmonary Medicine*. 2012;6:27–40.
39. Pouliquen IJ, Kornmann O, Barton SV, Price JA and Ortega HG. Characterization of the relationship between dose and blood eosinophil response following subcutaneous administration of mepolizumab. *International Journal of Clinical Pharmacology and Therapeutics*. 2015;53:1015–1027.

40. Busse WW, Katial R, Gossage D, Sari S, Wang B, Kolbeck R, Coyle AJ, Koike M, Spitalny GL, Kiener PA, Geba GP and Molfino NA. Safety profile, pharmacokinetics, and biologic activity of MEDI-563, an anti-IL-5 receptor alpha antibody, in a phase I study of subjects with mild asthma. *The Journal of Allergy and Clinical Immunology.* 2010;125:1237–1244.e2.
41. Administration UFaD. DUPIXENT® (dupilumab) injection, for subcutaneous use Initial U.S. Approval. 2017.
42. Administration UFaD. XOLAIR® (omalizumab) for injection, for subcutaneous use Initial U.S. Approval: 2. 2003.
43. Haldar P, Brightling CE, Singapuri A, Hargadon B, Gupta S, Monteiro W, Bradding P, Green RH, Wardlaw AJ, Ortega H and Pavord ID. Outcomes after cessation of mepolizumab therapy in severe eosinophilic asthma: A 12-month follow-up analysis. *The Journal of Allergy and Clinical Immunology.* 2014;133:921–923.
44. Fala L. Nucala (mepolizumab): First IL-5 antagonist monoclonal antibody FDA approved for maintenance treatment of patients with severe asthma. *American Health & Drug Benefits.* 2016;9:106–110.
45. Padilla Galo A, Labor M, Tiotiu A, Baiardini I, Scichilone N and Braido F. Impact of reslizumab on outcomes of severe asthmatic patients: Current perspectives. *Patient Related Outcome Measures.* 2018;9:267–273.
46. Administration UFaD. FASENRA (benralizumab) injection, for subcutaneous use Initial U.S. Approval. 2017.
47. Bleecker ER, FitzGerald JM, Chanez P, Papi A, Weinstein SF, Barker P, Sproule S, Gilmartin G, Aurivillius M, Werkström V and Goldman M. Efficacy and safety of benralizumab for patients with severe asthma uncontrolled with high-dosage inhaled corticosteroids and long-acting β(2)-agonists (SIROCCO): a randomised, multicentre, placebo-controlled phase 3 trial. *Lancet.* 2016;388:2115–2127.
48. FitzGerald JM, Bleecker ER, Nair P, Korn S, Ohta K, Lommatzsch M, Ferguson GT, Busse WW, Barker P, Sproule S, Gilmartin G, Werkström V, Aurivillius M and Goldman M. Benralizumab, an anti-interleukin-5 receptor α monoclonal antibody, as add-on treatment for patients with severe, uncontrolled, eosinophilic asthma (CALIMA): A randomised, double-blind, placebo-controlled phase 3 trial. *Lancet.* 2016;388:2128–2141.
49. Simpson EL, Bieber T, Guttman-Yassky E, Beck LA, Blauvelt A, Cork MJ, Silverberg JI, Deleuran M, Kataoka Y, Lacour JP, Kingo K, Worm M, Poulin Y, Wollenberg A, Soo Y, Graham NM, Pirozzi G, Akinlade B, Staudinger H, Mastey V, Eckert L, Gadkari A, Stahl N, Yancopoulos GD and Ardeleanu M. Two phase 3 trials of dupilumab versus placebo in atopic dermatitis. *The New England Journal of Medicine.* 2016;375:2335–2348.
50. Blauvelt A, de Bruin-Weller M, Gooderham M, Cather JC, Weisman J, Pariser D, Simpson EL, Papp KA, Hong HC, Rubel D, Foley P, Prens E, Griffiths CEM, Etoh T, Pinto PH, Pujol RM, Szepietowski JC, Ettler K, Keméňy L, Zhu X, Akinlade B, Hultsch T, Mastey V, Gadkari A, Eckert L, Amin N, Graham NMH, Pirozzi G, Stahl N, Yancopoulos GD and Shumel B. Long-term management of moderate-to-severe atopic dermatitis with dupilumab and concomitant topical corticosteroids (LIBERTY AD CHRONOS): A 1-year, randomised, double-blinded, placebo-controlled, phase 3 trial. *Lancet.* 2017;389:2287–2303.
51. Fahy JV. Type 2 inflammation in asthma – Present in most, absent in many. *Nature Reviews Immunology.* 2015;15:57–65.
52. Israel E and Reddel HK. Severe and difficult-to-treat asthma in adults. *The New England Journal of Medicine.* 2017;377:965–976.
53. Chung KF, Wenzel SE, Brozek JL, Bush A, Castro M, Sterk PJ, Adcock IM, Bateman ED, Bel EH, Bleecker ER, Boulet LP, Brightling C, Chanez P, Dahlen SE, Djukanovic R, Frey U, Gaga M, Gibson P, Hamid Q, Jajour NN, Mauad T, Sorkness RL and Teague WG. International ERS/ATS guidelines on definition, evaluation and treatment of severe asthma. *The European Respiratory Journal.* 2014;43:343–373.
54. Bullens DM, Truyen E, Coteur L, Dilissen E, Hellings PW, Dupont LJ and Ceuppens JL. IL-17 mRNA in sputum of asthmatic patients: Linking T cell driven inflammation and granulocytic influx? *Respiratory Research.* 2006;7:135.
55. McCracken JL, Tripple JW and Calhoun WJ. Biologic therapy in the management of asthma. *Current Opinion in Allergy and Clinical Immunology.* 2016;16:375–382.
56. Quirce S, Phillips-Angles E, Domínguez-Ortega J and Barranco P. Biologics in the treatment of severe asthma. *Allergologia et immunopathologia.* 2017;45(Suppl 1):45–49.
57. Gelhorn HL, Balantac Z, Ambrose CS, Chung YN and Stone B. Patient and physician preferences for attributes of biologic medications for severe asthma. *Patient Preference and Adherence.* 2019;13:1253–1268.
58. Corren J, Parnes JR, Wang L, Mo M, Roseti SL, Griffiths JM and van der Merwe R. Tezepelumab in adults with uncontrolled asthma. *The New England Journal of Medicine.* 2017;377:936–946.
59. Corren J and Ziegler SF. TSLP: From allergy to cancer. *Nature Immunology.* 2019;20:1603–1609.
60. Mitchell PD and O'Byrne PM. Epithelial-derived cytokines in asthma. *Chest.* 2017;151:1338–1344.
61. Gauvreau GM, Sehmi R, Ambrose CS and Griffiths JM. Thymic stromal lymphopoietin: Its role and potential as a therapeutic target in asthma. *Expert Opinion on Therapeutic Targets.* 2020;24:777–792.
62. Camelo A, Rosignoli G, Ohne Y, Stewart RA, Overed-Sayer C, Sleeman MA and May RD. IL-33, IL-25, and TSLP induce a distinct phenotypic and activation profile in human type 2 innate lymphoid cells. *Blood Advances.* 2017;1:577–589.

63. Pelaia C, Pelaia G, Crimi C, Maglio A, Gallelli L, Terracciano R and Vatrella A. Tezepelumab: A potential new biological therapy for severe refractory asthma. *International Journal of Molecular Sciences*. 2021;22:4369.
64. Gauvreau GM, O'Byrne PM, Boulet LP, Wang Y, Cockcroft D, Bigler J, FitzGerald JM, Boedigheimer M, Davis BE, Dias C, Gorski KS, Smith L, Bautista E, Comeau MR, Leigh R and Parnes JR. Effects of an anti-TSLP antibody on allergen-induced asthmatic responses. *The New England Journal of Medicine*. 2014;370:2102–2110.
65. Tripp CS, Cuff C, Campbell AL, Hendrickson BA, Voss J, Melim T, Wu C, Cherniack AD and Kim K. RPC4046, a novel anti-interleukin-13 antibody, blocks IL-13 binding to IL-13 $\alpha 1$ and $\alpha 2$ receptors: A randomized, double-blind, placebo-controlled, dose-escalation first-in-human study. *Advances in Therapy*. 2017;34:1364–1381.
66. Panettieri RA, Jr., Sjöbring U, Péterffy A, Wessman P, Bowen K, Piper E, Colice G and Brightling CE. Tralokinumab for severe, uncontrolled asthma (STRATOS 1 and STRATOS 2): two randomised, double-blind, placebo-controlled, phase 3 clinical trials. *The Lancet Respiratory Medicine*. 2018;6:511–525.
67. Gauvreau GM, Arm JP, Boulet LP, Leigh R, Cockcroft DW, Davis BE, Mayers I, FitzGerald JM, Dahlen B, Killian KJ, Laviolette M, Carlsten C, Lazarinis N, Watson RM, Milot J, Swystun V, Bowen M, Hui L, Lantz AS, Meiser K, Maahs S, Lowe PJ, Skerjanec A, Drollmann A and O'Byrne PM. Efficacy and safety of multiple doses of QGE031 (ligelizumab) versus omalizumab and placebo in inhibiting allergen-induced early asthmatic responses. *The Journal of Allergy and Clinical Immunology*. 2016;138:1051–1059.
68. Global Initiative for Asthma. *Global Strategy for Asthma Management and Prevention 2021*. Available from: www.ginaasthma.org. 2019.

Biologics and Biosimilars

Potential Therapeutics for Autoimmune Renal Diseases

20

Shaofei Wang, Yubin Li, Mengyao Liu, Hongbin Wang, and Jiajun Fan

Corresponding authors: Hongbin Wang & Jiajun Fan

Contents

DOI: 10.1201/9780429485626-20

ABBREVIATIONS

ANCA	Anti-neutrophil cytoplasmic autoantibody
C3Gs	Complement 3 glomerulopathies
cGN	Crescentic glomerulonephritis
DDD	Dense deposit disease
GBM	Glomerular basement membrane
HSP	Henoch-Schonlein purpura
HSPN	Henoch-Schonlein purpura nephritis
IgAN	IgA nephropathy
IgA	Immunoglobulin A
LN	Lupus nephritis
MN	Membranous nephropathy
MPO	Myeloperoxidase
NETs	Neutrophil extracellular traps
RPGN	Rapidly progressive glomerulonephritis
SLE	Systemic lupus erythematosus

20.1 INTRODUCTION

Autoimmune renal diseases are a group of chronic renal disorders that arise from a dysfunction of the host immune system with similar immunopathological characteristics of occurrence of autoantibodies against cellular or molecular components and intrarenal deposition of immune complexes as well as inflammatory lesions and subsequent functional loss of renal cells (1, 2). Recent data have shown that the number of patients with renal diseases that results from autoimmune injury including IgA nephropathy, Henoch-Schonlein purpura nephritis (HSPN), lupus nephritis (LN), complement C3 nephropathy, anti-neutrophil cytoplasmic antibody-associated glomerulonephritis, membranous nephropathy (MN), anti-glomerular basement membrane (GBM) glomerulonephritis, and crescentic glomerulonephritis (cGN) is continuously increasing (3). According to clinical statistics, 95% of renal diseases are caused by immune system disorders, while most autoimmune renal diseases are uncommon among all autoimmune diseases where the diagnosis and treatment strategies are limited (4). Simply suppressing hypersensitivity reactions and inflammatory responses could only temporarily relieve the symptoms. Thus, understanding the pathogenesis of specific autoimmune renal diseases, establishing standard diagnostic procedures in the clinic, and developing novel therapies including new drugs and innovative technologies are in urgent need to reverse the passive situation (5). Herein, we will enumerate representative examples of autoimmune renal diseases (Figure 20.1), introduce their cellular pathogenesis, summarize biologics and biosimilars for these autoimmune renal diseases, and discuss their challenges and perspectives.

20.1.1 IgA Nephropathy (IgAN)

IgA nephropathy (IgAN), characterized by predominant immunoglobulin A (IgA) deposition in the glomerular mesangium, was first described by Berger and Hinglais in 1968 and thus also known as Berger's disease (6). IgAN, as one of the most common causes of glomerulonephritis worldwide, is a leading cause of chronic kidney diseases and end-stage renal failure. Even with timely diagnosis and proper management, up to 40% of IgAN patients will progress to end-stage renal failure (7). The susceptibility to IgAN and the risk of disease progression are governed by genetic and environmental factors (8). Currently, the diagnosis of IgAN relies on a histopathologic evaluation of kidney biopsy (9). The abnormal glycosylated IgA1 accumulation together with nephritogenic galactose-deficient-IgA1-containing immune complexes formed in the circulating system has been considered as the origin of the glomerular immune deposits (10). Besides, the O-glycosylation pathway, anti-glycan immune response, the mucosal defense system, antigen processing and presentation, and the alternative pathway of complement system also contributed to the pathogenesis of IgAN (11). Due to increased international clinical and genetic studies collaboration, new genetic susceptibility loci, multi-hit pathogenesis model, Oxford pathology scoring system, and IgAN treatment guidelines have been developed in most recent years (12–14).

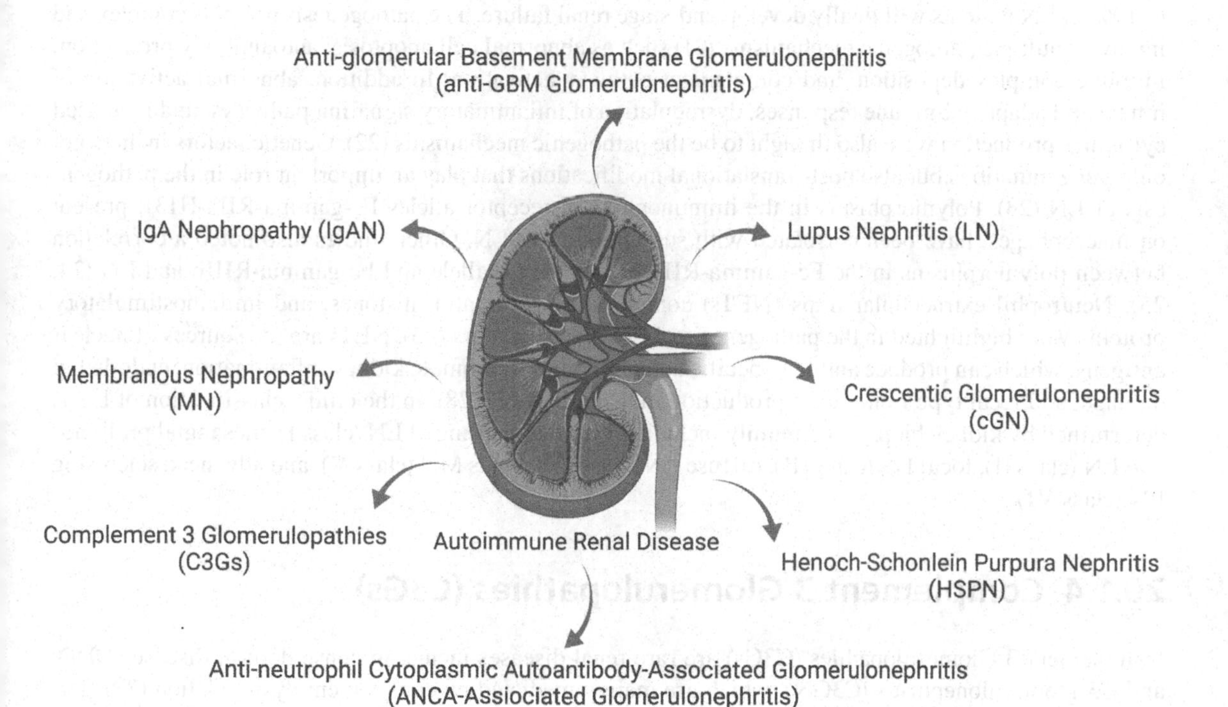

FIGURE 20.1 The major categories of autoimmune renal diseases.

20.1.2 Henoch-Schonlein Purpura Nephritis (HSPN)

Henoch-Schonlein purpura (HSP) is a systemic vasculitis featured by pathological changes such as small vasculitis. HSP accompanied with renal injury is recognized as HSPN (15, 16). HSPN occurs in 30%–50% of HSP patients. Among them, almost 20% of patients with HSPN devolve into nephritic and nephrotic syndromes (17). Because HSPN is a self-limited disease, it is rare to develop into chronic kidney injury and end-stage renal failure; while renal tubular disease and renal interstitial fibrosis could affect the prognosis of HSPN (15). The glomerular deposition and accumulation of IgA1-containing immune complexes in the mesangium, the subepithelial and the subendothelial space contribute to the pathogenesis of HSPN (17). Although HSPN and IgAN are considered to be related diseases, there is a lot of different features between HSPN and IgAN: HSPN is mostly diagnosed in children with early ages while IgAN has a wide peak age range between fifteen and thirty years old. Different from IgAN, nephritic syndromes and/or nephrotic syndromes are usually presented in HSPN, and these syndromes are correlated with hypersensitivity. Besides, it is common to see that endocapillary and extracapillary inflammations together with fibrin deposits in the glomerulus in HSPN, and leukocytes infiltration in the tissues is a major feature of HSPN vasculitis (18, 19).

20.1.3 Lupus Nephritis (LN)

Systemic lupus erythematosus (SLE) is an autoimmune disorder that results from the recognition of self-antigen such as DNA and proteins released from cell debris. It is an example of a type III hypersensitivity reaction associated with a wide range of clinical and immunologic manifestations. Among them, as a frequent complication of SLE, LN has become the most common cause of morbidity and mortality (20). Clinical evidence showed that more than 50% of SLE patients present kidney diseases in the clinic, and up to 10% of LN patients will finally develop end-stage renal failure. The pathogenesis of LN is complex and involves multiple pathogenic mechanisms (21) such as abnormal cell apoptosis, autoantibody production, immune complex deposition, and complement pathway activation. In addition, abnormal activation of innate and adaptive immune responses, dysregulation of inflammatory signaling pathways, and increased cytokines production were also thought to be the pathogenic mechanisms (22). Genetic factors include not only gene mutations but also post-translational modifications that play an important role in the pathogenesis of LN (23). Polymorphisms in the immunoglobulin receptor alleles Fc-gamma-RIIa-H131, present on macrophages, have been correlated with susceptibility to LN. Other studies also noted a correlation between polymorphisms in the Fc-gamma-RIIa-F158 receptor allele and Fc-gamma-RIIIb and LN (24, 25). Neutrophil extracellular traps (NETs) composed of chromatin, histones, and immunostimulatory proteins were highlighted in the pathogenesis of LN in recent years (26). NETs are the sources of nuclear antigens, which can produce antigen-specific autoantibodies to promote kidney inflammation, endothelial damage, and local type I interferon production in the kidney (27, 28). In the clinic, classification of LN is determined by kidney biopsy, and mainly including minimal mesangial LN (class I), mesangial proliferative LN (class II), focal LN (class III), diffuse LN (class IV), lupus MN (class V), and advanced sclerosing LN (class VI).

20.1.4 Complement 3 Glomerulopathies (C3Gs)

Complement 3 Glomerulopathies (C3Gs) are rare renal diseases including dense deposit disease (DDD) and C3 glomerulonephritis (C3GN), which are mainly mediated by complement dysregulation (29). The quintessential feature of these disorders with glomerular pathology is associated with C3 cleavage product deposition and defective complement action and regulation (30). Glomerular deposits of C3 alone, without immunoglobulin, are the hallmark of alternative pathway dysregulation of complement systems through inherited or acquired defects (31). C3Gs are classified into DDD and C3GN based on electron microscopy images, specific genetic types, and autoantibodies. Berger and Galle first named DDD in 1962 based on the electron-dense materials noted while analyzing electron micrographs of the lamina densa of the GBM (32). Immunofluorescence showed predominant C3 deposition along with its breakdown products in the GBM, while less or no immunoglobulin was stained together (33). DDD is usually seen in both younger individuals and adults. The main presenting characteristics are proteinuria, hematuria, hypertension, and renal failure (33). Distinguishable from DDD, C3GN features prominent C3 deposition in the mesangium and capillary wall, while intramembranous deposition is discontinuous. But the pathophysiology for alternate pathway activation in C3GN is similar to DDD with fluid-phase dysregulation because of varying mutations or autoantibodies, thus C3GN and DDD shared a similar disease course in a large US cohort of patients with C3Gs (34).

20.1.5 Membranous Nephropathy (MN)

Membranous nephropathy (MN) is an autoimmune disease of the glomerulus and one of the leading causes of nephrotic syndrome (35, 36). Approximately 30% of MN cases progressed to end-stage renal disease. About 80% of cases are renal limited (primary MN), and 20% are associated with other systemic diseases or exposures (secondary MN) (36). Since 2009, autoantibodies to the phospholipase A2 receptor and thrombospondin domain-containing 7A have been identified to be associated with primary MN and are used for MN diagnosis (37). Antibody levels correlate with disease severity and possess significant biomarker values in monitoring disease progression and treatment response (38).

20.1.6 Anti-Glomerular Basement Membrane (GBM) Glomerulonephritis

Anti-GBM glomerulonephritis is a rare autoimmune renal disease induced by IgG autoantibodies attacking against the GBM (39). And these autoantibodies directly bind against the non-collagenous 1 domain of type IV collagen in the GBM. In the kidney, binding of these autoantibodies with the GBM results in the activation of the complement signaling pathway and can lead to quickly progressive glomerulonephritis (40). Besides, anti-GBM antibodies could also bind with the alveolar basement membrane. Glomerulonephritis is usually accompanied by pulmonary hemorrhage in anti-GBM diseases (41). In some circumstances, anti-GBM glomerulonephritis also occurs with other diseases at the same time, such as anti-neutrophil cytoplasmic autoantibody (ANCA)-associated vasculitis; while it is rarely occurring with MN (42).

20.1.7 Anti-Neutrophil Cytoplasmic Autoantibody-Associated Glomerulonephritis (ANCA-Associated Glomerulonephritis)

ANCA-associated vasculitis, especially Wegener's granulomatosis and microscopic polyangiitis, is the most common cause of rapidly progressive glomerulonephritis (RPGN) worldwide, and the morphologic changes in the renal biopsy is the gold standard for establishing the diagnosis (43). The infiltration of white blood cells (including T cells) and the proliferation of resident glomerular cells can lead to glomerular crescents formation and the glomerulus anatomy destruction, which finally leading to loss of renal function. Current treatment options are unspecific and hampered by toxic side effects that worsen the patients' prognosis.

ANCA-associated glomerulonephritis is characterized by immunofluorescence microscopy by little or no glomerular staining for Igs or complement, the so-called "pauci-immune" staining pattern (44). RPGN, also called cGN, is the most aggressive form of autoimmune kidney disease. It can destroy renal structure and function within days or weeks, leading to end-stage renal failure, accompanying severe morbidity and mortality. In the vast majority of cGN patients, both anti-GBM antibodies and ANCAs specific for myeloperoxidase (MPO-ANCA) have been detected (45).

20.2 CELLULAR PATHOGENESIS OF AUTOIMMUNE RENAL DISEASES

Many factors may contribute to the pathology of autoimmune renal injury, including genetic, epigenetic, hormonal, environmental, and immunological factors. Immune complexes, autoantibodies, pathogenic T cells, and inflammatory cytokines have been documented as key effectors for pathogenesis of renal inflammation.

20.2.1 T Cells Dysfunction in Autoimmune Renal Diseases

Under physiological conditions, T cells are crucial for regulating immune responses, controlling inflammation, maintaining immune homeostasis, and executing tolerance against self-antigens. In contrast, under pathological conditions, T cells drive immune response to fight pathogens and can support B cells to produce antibodies and increase self-reactive B-cell survival and differentiation. Moreover, T cells can trigger cytokine secretion and enhance inflammation and tissue injury, which can affect the kidney, leading to the development of different types of nephritis. While naive $CD8^+$ T cells differentiate into double negative (DN) and cytotoxic T lymphocytes (CTL), naive $CD4^+$ T cells differentiate into Th1, Th2, Th17, Treg, and T follicular helper (TFH) cell subgroups. Different subsets of T cells play various roles in the development of nephritis.

Th1 and Th2 cells: In mouse models of anti-GBM glomerulonephritis, SLE, and ANCA-associated vasculitis, Th1 cells and their secreted cytokine IFN-γ contribute considerably to the development of cGN and diffuse proliferative LN via the recruitment of macrophages (46–48). Despite the accumulation of Th1 cells in the kidneys of patients with diffuse proliferative LN or ANCA-associated cGN, the roles of Th1 cells in human renal autoimmune disease have not yet been well established. Th2 cells and their signature cytokine IL-4 play important roles in the development of membranous LN in MRL/lpr mice (48–50). There is a subset of T cells that can produce cytokine IL-17, including Th17 cells, γδT cells, and DN T cells, by releasing cytokines such as IL-23, leading to the release of IL-17 and the recruitment of natural killer (NK) cells and neutrophils (51).

Th17 cells: A subset of αβT cells are identified as Th17 cells with the expression of the transcription factor RORγt. In humans and mice, Th17 cells can produce IFN-γ, IL-17, IL-21, IL-22, and granulocyte-macrophage colony-stimulating factor (GM-CSF) to promote autoimmune pathology (52, 53). In different studies of autoimmune disease, Th17 cells have been recognized to be involved in the development and progression of chronic nephritis. The interaction between C-C chemokine receptor type 6 (CCR6) expressed on Th17 cells and the C-C motif chemokine ligand 20 (CCL20) that is expressed by kidney mesangial cells contributes to the recruitment of Th17 cells to the kidney (54). Th17 cells are also observed to be accumulated in the kidneys of patients with ANCA-associated cGN and in different mouse models of glomerulonephritis indicating that Th17 cells might be therapeutic targets for autoimmune renal diseases (55–58).

γδT cells: γδT cells, a minor population of T cells with innate-like and adaptive features, are associated with mucosal immunity (59). γδT cells play important roles in the secretion of inflammatory cytokines, antigen presentation, the inhibition of T_{reg} cells, and the promotion of antibody production by B cells in the absence of antigen and have been associated with the development of several autoimmune diseases. IL-17 produced by γδT cells that are stimulated by IL-23 from dendritic cells can drive kidney inflammation and the infiltration of proinflammatory cells in tubular areas and lead to glomerulonephritis (60–62). Depletion of γδT cells and inhibition of IL-17 production from γδT cells lead to limited renal injury and reduced neutrophil recruitment in nephrotoxic serum-induced glomerulonephritis (60). γδT cells are presented as their active phenotype Vδ2 T cells in kidney biopsy samples, which express IL-21 and CD40 ligand, suggesting the involvement of B-cell stimulation and accumulation of macrophages, neutrophils, and other T cells. The study revealed that Vδ2 T cells can directly cause the damages of kidney epithelial cells (63).

Double-negative T cells: Double-negative (DN) T cells originated from continuously stimulated $CD8^+$ T cells and are characterized as $TCR\alpha\beta^+CD4^-CD8^-$. In autoimmune diseases, the number and frequency of DN T cells are increased. In SLE patients, DN T cells accumulate in kidney and produce IL-17 (64).

T follicular helper (TFH) cells: TFH cells are defined as a subtype of $CD4^+$ T cells that express the C-X-C chemokine receptor type 5 (CXCR5) and transcription factor BCL6. In response to C-X-C motif chemokine ligand 13 (CXCL13), TFH cells migrate into germinal centers. PD-1, CD40L, and inducible T-cell co-stimulator are expressed on the surface of THF cells. THF cells also secrete IL-21 to stimulate B-cell activation, immunoglobulin class switching, and the differentiation of B cells into memory B cells and plasmablasts (46, 65, 66). The pathogenic role of TFH cells has been established in lupus-prone mice (65). TFH cells are increased in patients with SLE, rheumatoid arthritis, multiple sclerosis (MS), and autoimmune thyroid disease. Anti-double stranded DNA autoantibody and numbers of plasmablast B cells correlate with extrafollicular Th cells in SLE patients. In patients with LN, these Th cells probably promote B-cell activation and plasmablast generation within the kidney (67).

T regulatory cells (T_{reg}): The deficiency of T_{reg} cells has been documented in multiple autoimmune disorders (68, 69). T_{reg} cells develop in the thymus and are characterized as $TCR\alpha\beta^+Foxp3^+CD4^+$ phenotypic T cells. They confer suppressive activity and control inflammation and immune responses to self-antigen to prevent autoimmune diseases through multiple mechanisms. Any autoimmune or inflammatory conditions may indicate an insufficiency of T_{reg} cells (70). T_{reg} cells have been confirmed to contribute to tissue repair through uncharacterized mechanisms (71, 72). It was demonstrated that T_{reg} cells can repair lung tissue injury caused by infection, which is dependent on amphiregulin with growth factor activity (73), speculating that T_{reg} cells may play a role in the repairment of kidney injury caused by autoimmune mechanisms.

$CD8^+$ T cells: $CD8^+$ cytotoxic T cells play important roles in renal pathology. In mouse models, $CD8^+$ T cells are involved in a glomerular injury resulting from ANCA-associated vasculitis, and the depletion of $CD8^+$ cells mitigated anti-GBM nephritis (74). $CD8^+$ cells were also found in the kidneys of ANCA-positive patients and the presence of $CD8^+$ T cells correlates with poor prognostic outcomes (75). It has been revealed that high levels of $CD8^+$ T cells are a biomarker in both lupus- and ANCA-associated nephritis (76).

20.2.2 B Cells Dysfunction in Autoimmune Renal Diseases

The effector mechanisms that B lymphocytes-mediated immune disorders are through the production of antibodies and antigen presentation that results in T lymphocytes activation and secretion of various cytokines. The success of B cell depletion therapeutics in diseases such as cryoglobulinemic glomerulonephritis and ANCA-associated vasculitis has improved our understanding of the contribution of B cells to the etiology and pathogenesis of immune diseases. Studies from mouse models and humans suggest that the production of pro-inflammatory cytokines from B cells is relevant to the pathology of autoimmune diseases. IL-6 levels in mice and humans are correlated to the prognostics of MS in a mouse model and human (91, 92). Whether IL-6 is associated with the pathology of autoimmune diseases is still questionable. Other inflammatory cytokines, such as TNF and IFN-λ have been demonstrated to be linked with kidney disease, such as glomerulonephritis (93, 94). B cells were found in the renal interstitium in patients with interstitial nephritis and the disease will go into remission as the B cell number decreases, indicating B cells found within the kidney might contribute to immune pathological function through secretion of various cytokines (95).

B cells are key players in humoral immunity and regulate T cell activation and tolerance through antigen presentation and cytokines production (77). Mounting evidence reveals the existence of an immunoglobin-independent contribution of B cells to immune reactions by the experiments with B cells depletion or B cells that are unable to secrete antibody. B cell depletion did not affect T cell activation status in naïve mice, however, can reduce T cell proliferation and reduce memory reaction in response to antigen or pathogens (78). B cells can function as antigen-presenting cells to T cells due to their spatial proximity to T cells in lymphoid tissues (79). In contrast to the non-specific antigen captured by macrophages or dendritic cells, B cells uptake a specific antigen by immunoglobulin receptor and then bring it to MHC class-II rich compartments and present antigens either endocytosed or associated with endocytosed antigens. It appears that B cells are particularly essential in the presentation of antigen with low abundance, such as autoantigen. Depletion of B cells can cause decreased T cell activation and effector functions owing to reduced effective antigen presentation, which demonstrates the effectiveness of treatment in various autoimmune diseases (80). Animal studies revealed that in response to immune reactions, B cells produce pro-inflammatory and anti-inflammatory cytokines that can modulate T cell responses (81). B cells can be activated *in vitro* in the presence of Th1 cells and Th2 cells to generate effector cytokines (82). IFN-γ and IL-4 can prime and polarize B cells to be B effector 1 (Be1) cells and Be2 cells, respectively. While Be1 cells predominantly produce type 1 cytokines, such as IFN-γ and IL-12, Be2 cells can generate type 2 cytokine IL-14 and high levels of IL-2, IL-6, and IL-10. Be1 cells and Be2 cells can be differentially identified with distinctive types of bacterial infections (83). While Be1 cells were initially considered to have a role in autoimmunity, Be2 cells were thought to contribute to allergy.

Recent studies identified a group of cells, termed as B_{reg} cells, which can suppress pathogenic T cell response and thus assist immune tolerance. B_{reg} cells have been described in both humans and mice. No unique cell marker for B_{regs} has been identified. The most common definition for B_{reg} cells is that the B cells can produce IL-10 and inhibit inflammation (84). In mice, B_{reg} cells development is dependent on cytokines IL-1, IL-6, and IL-21, which are generated from T cells. Interestingly, Th17 cells are also polarized by those cytokines, suggesting B_{reg} cells arise to support immune homeostasis in response to immune reactions (85). It was found that while low levels of IFN-γ can drive B_{reg} cells differentiation, the high levels of IFN-γ cause the differentiation of plasmablasts that can no longer generate IL-10 (85). Interestingly, Moderate levels of inflammation promote induction and maintenance of B_{reg} cells that can restore homeostasis. Experiments in mouse models have demonstrated that BCR is required for engagement in B_{reg} cell development (86). CD40 has also been revealed to be critical in the generation and development of B_{reg} cells (87–89). CD40 stimulation enhances IL-10 production by murine B cells and mice with the deficiency of CD40 were unable to recover from various autoimmune diseases (such as lupus, CIA, and EAE) due to lack of IL-10 generation (88, 90). Moreover, toll-like receptor (TLR) activation by agonists can induce IL-10 generation by mouse and human B cells in vitro (91). Mice with the specific deletion of both *Tlr2* and *Tlr4* or of *Myd88* in B cells will develop chronic inflammatory diseases that are linked to a failure of control of T cell-mediated inflammation (92). In summary, specific antigen recognition, CD40 ligation, TLRs activation, and acute inflammation are essential inducers of B_{reg} cells and IL-10 generations. IL-10 producing B cells can ameliorate arthritis in mouse models by decreasing Th1 and Th17 cell responses (93, 94). B_{reg} cells were also found to enhance the proliferation of T_{reg} cells, which is consistent with the observation of decreased T_{reg} cell numbers in B cell deficiency mice (95, 96). Other than IL-10, cytokine IL-35 can also be produced from plasmablasts, which can contribute to the suppression of inflammation in mice. In addition, IL-35 can promote IL-10 generation from human B cells, contributing to the immune suppressive pathway (97, 98). It is now clear that deficiency of B_{reg} cell number and suppressive function has been documented in various autoimmune disorders, such as psoriasis, MS, RA, and SLE. Reduction of B_{reg} cell numbers and/or dysfunction have been identified in various diseases that could affect the kidney. B_{reg} cells might be used as a possible biomarker of disease activity and the prediction of prognostic and future relapses. As we noticed, many diseases causing renal dysfunction are associated with autoantibodies present in patients. However, some kidney diseases, such as focal segmental glomerulosclerosis or minimal change diseases, do not have a clear correlation with autoantibodies and appear to respond well to immunomodulatory treatment, indicating immune dysregulation leads to their pathogenesis and disease progression (99). Treatment in autoimmune diseases usually starts with the removal of antibodies using plasmapheresis, the interference with their actions by immunoglobulins, or the suppression of B cell or T cell activities. The current nuanced understanding of regulatory functions of T_{reg} cells and B_{reg} cells would lead to therapies that might have short-term and long-term effects on the pathogenesis and progression of autoimmune diseases.

20.3 BIOLOGICS AND BIOSIMILARS VERSUS SMALL MOLECULES FOR THE TREATMENT OF AUTOIMMUNE RENAL DISEASES

Biologics and biosimilars originated from living organisms, cells, microorganisms, or even cell-free biosynthetic systems (100) through highly complex manufacturing processes, include a wide range of biological products such as therapeutic peptides and proteins, vaccines, allergenics, blood components, extracellular vesicles (101), gene therapy products, cellular therapy products, and tissue-based products. In the past decades, biologics and biosimilars, also termed biopharmaceuticals, have emerged as a unique category of pharmaceuticals with remarkable therapeutic effects for the management of cancer, inflammatory diseases, and autoinflammatory disorders by targeting the underlying immunopathogenic molecular events and signaling pathways. Given the immunopathogenic similarities of autoimmune renal diseases and non-renal autoinflammatory diseases, biologics and biosimilars developed as potential therapeutics of autoimmune diseases by targeting critical molecular mechanisms in the injurious autoimmune processes are probably applicable for the management of autoimmune renal diseases.

The existing conventional therapies targeting the overreactive immune system by immunosuppressive small molecules can be somewhat effective for autoimmune renal diseases. However, to sufficiently control the progression of diseases, it is almost inevitable to repeatedly apply high doses of these systemic immunosuppressive agents for various autoimmune diseases of distinct pathogenesis. Therefore, the conventional immunosuppressive therapeutics are usually non-specific and non-curative for autoinflammatory disorders, with a high risk of life-threatening side effects such as infectious and cancerous complications. Biologics and biosimilars possess some clear advantages over the conventional therapeutic small molecules for the treatment of autoimmune renal diseases, such as good safety profiles and minimal toxicity concerns, and most importantly, well-defined molecular mechanisms and high targeting specificity.

The most prominent theoretical advantage of biologics and biosimilars over conventional immunosuppressive small molecules is that they can target specific mediators and signal transduction pathways contributing to the initiation and progression of autoimmune renal diseases. In general, biopharmaceuticals are specifically designed to target immunopathogenic processes of autoinflammatory diseases, such as autoantibodies production, cytokine abnormalities, the activation of autoreactive T and B cells, and the deposition of immune complexes. Due to their high targeting specificity, biological products generally lead to less systemic adverse effects as compared to conventional therapies. However, it should be noted that biopharmaceuticals that selectively target exact cell subsets, cytokines, receptors, or pathways within the immune system may be still not specific enough due to the unexpected biological interactions with the host immune system. Another major advantage of biologics and biosimilars over conventional small molecules is that they are designed and developed with well-defined molecular mechanisms for the treatment of autoimmune diseases. The autoantigens or injurious inflammatory pathways responsible for disease initiation and progression are well-defined in a variety of autoimmune renal diseases. And biologics and biosimilars, such as monoclonal antibodies or receptor ligands, which are specifically designed to act as competitive inhibitors of autoantigens or injurious inflammatory pathways, are theoretically effective for the management of these disorders. Collectively, due to these noble advantages, biologics and biosimilars will probably become the mainstream therapeutics for the management of autoimmune renal diseases in the future. The major differences between biologics and biosimilars and small molecules are summarized in Table 20.1.

TABLE 20.1 The major differences between biologics and biosimilars and small molecules

	BIOLOGICS AND BIOSIMILARS	*SMALL MOLECULES*
Categories and examples	Therapeutic peptides and proteins (e.g., antibodies, hormones, cytokines, growth factors, enzymes, immunomodulators); vaccines; allergenics; blood components; extracellular vesicles; gene therapy products; cellular therapy products (e.g., engineered cells); tissue-based products (e.g., allogeneic transplants)	Chemicals
Origin	Living organisms, cells, microorganisms, or cell-free biosynthesis systems	Nonliving vessels
Production methods	Biological synthesis (recombinant DNA technology; protein engineering; antibody technology)	Chemical synthesis (organic and inorganic chemical reactions)
Components	Proteins, carbohydrates, nucleic acids, extracellular vesicles, cells, tissues, or mixtures of these substances	Compounds
Batch consistency	Highly dependent on starting materials and manufacturing processes	Largely independent on starting materials and manufacturing processes
Molecular weight	High (typically >1 kDa)	Low (mostly <1 kDa)
Homogeneity	Heterogeneous	Homogeneous
Structural complexity	Complicated	Relatively simple
Characterization	Unlikely to be completely characterized	Can be well-characterized by standardized analytical techniques
Stability	Commonly unstable under room temperature (heat, shear, and pH sensitive)	Usually stable under room temperature
Immunogenicity	Immunogenic	Nonimmunogenic
Shipment and storage condition	Refrigeration	Room temperature
Administration route	Systemic routes (e.g., injection or infusion)	Various routes (e.g., oral administration)
	Usually administered under close clinical supervision in the hospital	Availability in pharmacies
Costs	Expensive	Less expensive
Specificity	High	Low
Toxicity	Limited toxicity	Systemic toxicity

20.4 BIOLOGICS AND BIOSIMILARS FOR AUTOIMMUNE RENAL DISEASE

Up to now, a series of biologics that target the various pathological stages during the progression of autoimmune diseases have been developed for autoimmune renal disease therapy. These stages contain the activation of autoimmune T cells and B cells and the upregulation of autoimmune response and effector pathways afterward, such as the production of inflammatory cytokines and dysregulation of complements (Figure 20.2).

20.4.1 Biologic Therapeutics Against B Cells

As was reported previously, loss of self-tolerance and then generation of autoantibodies is regarded as one of the most important reasons that resulted in multiple autoimmune renal diseases including LN, MN, and ANCA glomerulonephritis (1). B cells, as well as their-derived plasma cells, as the source of these autoantibodies, usually play critical roles in the development of LN, which are therefore considered as promising therapeutic targets for autoimmune renal disease therapies (77).

Due to the key function of B cells in the development of autoimmune renal diseases, specifically targeting B-cell subsets by monoclonal antibodies are probable approaches for the treatment of autoimmune

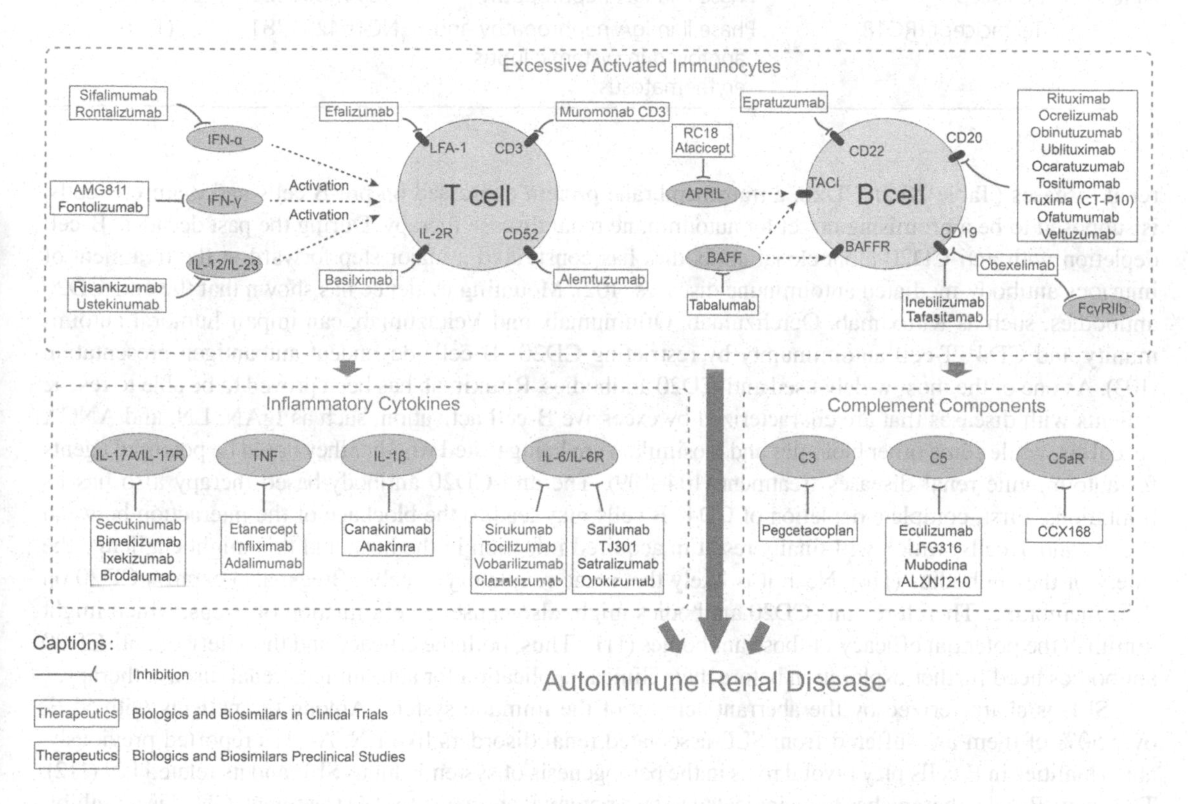

FIGURE 20.2 Biologic therapeutics target various pathological stages in autoimmune renal disease. The pathological stages including activation of overreacted lymphocytes, production of inflammatory cytokine, and dysregulation of complement are illustrated in this figure. Based on these stages, the related biological agents either in clinical trials (in red) or preclinical studies (in blue), are discussed.

TABLE 20.2 Biologics or biosimilars targets B cells in autoimmune renal disease

TARGET	*BIOLOGIC OR BIOSIMILAR*	*CLINICAL STATUS*	*RELATED CLINICAL TRIALS*	*REFERENCES*
CD20	Rituximab	1. Phase III in idiopathic membranous nephropathy 2. Phase IV in IgA nephropathy 3. Phase III in lupus nephritis 4. Phase III in ANCA vasculitis and end-stage renal disease	1. NCT01508468 2. NCT02571842 3. NCT01765842 4. NCT03323476	(107, 108, 125–128)
	Ocrelizumab	Phase III in lupus nephritis	NCT00626197	(129)
	Obinutuzumab	1. Phase III in membranous nephropathy 2. Phase II in lupus nephritis	1. NCT04629248 2. NCT05039619	(130–132)
CD19	Obexelimab (bifunctional protein that binds CD19 and FcγRIIb)	Phase II in systemic lupus erythematosus and its-related lupus nephritis	NCT02725515	(133)
CD22	Epratuzumab	Phase III in systemic lupus erythematosus and its-related lupus nephritis	NCT00111306 NCT00383214	(134, 135)
BAFF	Tabalumab	Basic research in autoimmune renal disease and Phase III in systemic lupus erythematosus	NCT01205438	(136, 137)
APRIL	Atacicept	Phase II in IgA nephropathy	NCT02808429	(138)
	Telitacicept (RC18)	Phase II in IgA nephropathy and approved in systemic lupus erythematosus	NCT04291781	(139)

renal diseases (Table 20.2). CD20, a transmembrane protein expressed on pre-B cells and mature B cells, is supposed to be a promising target for autoimmune renal disease therapy. During the past decades, B-cell depletion with anti-CD20 monoclonal antibodies has constituted a major step forward in the treatment of injurious antibody-mediated autoimmune diseases (102). Mounting evidence has shown that the anti-CD20 antibodies, such as Rituximab, Ocrelizumab, Oftumumab, and Veltuzumab, can impair humoral autoimmunity and CD4$^+$ T-cell autoimmunity by restricting CD20$^+$ B cells-dependent autoantigen presentation (103). As one of the most widely used anti-CD20 antibodies, Rituximab has been proved to be able to rescue patients with diseases that are characterized by excessive B-cell activation, such as IgAN, LN, and ANCA vasculitis, while some other biologics and biosimilars are being tested whether they could be potential agents for autoimmune renal diseases treatment (104–109). The anti-CD20 antibody-based therapy also has its limitations. First, complete depletion of CD4$^+$ B cells may lead to the blockage of the interaction between B cells and T cells, which will finally result in acquired infection in the host, and this might challenge the safety of the antibodies (110). Next, it is likely that some regulatory B cells (Bregs) also express CD20 on their membrane. Therefore, anti-CD20 antibodies might also cause the elimination of Bregs, which might diminish the potential efficacy of those antibodies (111). Thus, both the efficacy and the safety of anti-CD20 antibodies need further exploration before their clinical application for autoimmune renal disease therapy.

SLE is characterized by the aberrant activity of the immune system. Among the patients with SLE, over 50% of them are suffered from SLE-associated renal disorders like LN. As was reported previously, abnormalities in B cells play pivotal roles in the pathogenesis of systemic lupus SLE and its related LN (112). Thus, anti-B-cells therapy has been indicated to be a potential approach for LN treatment. CD22 is an inhibitory co-receptor of the B cell receptor (BCR), which is constitutively expressed on the membrane of IgD$^+$ and IgM$^+$ mature B cells (113). Through interacting with its cis-ligands like CD45 and IgM, CD22 will be internalized into cells and then phosphorylated, which will finally result in the removal of BCR complexes from the cell surface and B cell inhibition (114, 115). Epratuzumab is a humanized IgG1 anti-CD22 monoclonal antibody, which is designed for the treatment of SLE patients. It can simulate cis-signaling-activation of CD22 and promote the internalization of the molecule, which will cause a reduction in both B cell activity and the number of B cells in peripheral blood (116). Although the phase III clinical trial of epratuzumab has been terminated due to interruption in drug supply, the analysis of the clinical data still exhibited lower BILAG scores in epratuzumab-treated patients when compared to those treated with placebo (117, 118). Therefore, the data of the trials have indicated the appropriate efficacy and safety of epratuzumab for SLE treatment, which will support its future research and development for SLE clinical use.

CD19, the member of the immunoglobulin super-family, is critically important for the proliferation, differentiation, and maturation of B cells. CD19 is reported to be expressed on mature B cells, B-cell precursors, and almost all plasma cells, whereas CD20 can only be found on B cells (119). Inebilizumab (also known as MEDI-551), a humanized and afucosylated IgG1 monoclonal antibody that binds to CD19, has been established for the treatment of a variety of autoimmune diseases associated with excessive B cell activity. Recently, FDA approved Inebilizumab for its treatment of neuromyelitis optica spectrum disorder (NMOSD), and the potential of the antibody in autoimmune renal disease therapy has also been investigated (120). Evidence illustrated the ability of Inebilizumab to deplete B cells and IgG-specific antibody-secreting cells in mice with SLE (121). The antibody treatment caused a remarkable reduction of autoantibodies but had minimal influence on total serum immunoglobulins, which suggested its potential for SLE and SLE-related LN therapy.

Obexelimab (also known as XmAb5871) is a bifunctional protein that targets CD19, with a high affinity to FcγRIIb (122). A phase II study of 104 SLE patients using Obexelimab was carried out, and low disease activity was observed in patients with Obexelimab treatment (123, 124). During the period of Obexelimab treatment, the researchers found that Obexelimab could suppress B cell activation and inhibit B cell proliferation induced by cytokines and cellular factors such as IL-4, BAFF, and lipopolysaccharide (LPS). Since the Fc domain of Obexelimab targets only FcγRIIb on B cells instead of that on natural killer (NK) cells, B cell depletion through CD19 blockade would not likely to suppress the function of innate immunity of the host. Therefore, treatment with Obexelimab might be a potential approach for clinical SLE care.

20.4.2 Biologic Therapeutics Against T Cells

As one of the most important components in host acquired immunity, T cells have deservedly become the key effectors and regulators in the process of autoimmune renal diseases. Although T cells usually promote the host immune responses to fight pathogens through inflammation and cytotoxicity, they can also attenuate immune responses to control inflammation during effector immune responses, maintain immune homeostasis under physiologic conditions and enforce tolerance against self-antigens. Actually, impairment and destruction of T-cell tolerance will result in the generation of autoantibodies, inflammation, and tissue infiltration, which can cause the occurrence of various types of autoimmune renal injury (140). Therefore, inhibition of T-cell-driven autoimmunity through immunosuppressive agents is likely to be an effective approach to ameliorate autoimmune balance in the host. A variety of biologics or biosimilars have been established to attenuate the T-cell activity and autoimmune renal diseases (Table 20.3).

TABLE 20.3 Biologics or biosimilars targets T cells in autoimmune renal disease

TARGET	*BIOLOGIC OR BIOSIMILAR*	*CLINICAL STATUS*	*RELATED CLINICAL TRIALS*	*REFERENCE*
CD52	Alemtuzumab	Phase I/II in type 1 diabetes mellitus and end-stage renal disease	NCT00692562	(148)
IL-2R	Basiliximab	Phase IV in glomerulonephritis and IgA nephropathy	NCT02523768	(150)
LFA-1	Efalizumab	Phase I/II in autoimmune renal transplantation rejection	NCT00777400	(151)

Muromonab CD3 (also known as OKT3), a murine IgG2a antibody, was the only anti-CD3 monoclonal antibody approved by FDA to rescue patients with renal transplantation-induced autoimmune injury (141–143). Although it is a murine-originated antibody with severe immunogenicity, Muromonab CD3 was also considered to be a safe and effective agent for the prophylaxis and first-line treatment of acute rejection of solid organ transplantations in the past few years (144). As was reported previously, Muromonab CD3 could directly bind the CD3 antigen on peripheral human T cells, promote ADCC and CDC effect against T cells, resulting in the almost complete depletion of peripheral T cells. Clinical data showed that the antibody seemed to be better than high-dose steroids as first-line treatment for acute renal allograft rejection (141, 145). Muromonab CD3 has been withdrawn from the US market due to its high immunogenicity and potential safety problems. However, the data of another clinical research have still illustrated that $CD3^+$ T cells are infiltrated in kidney tissues and are associated with the development of LN (146). Therefore, these results still suggest a potential approach for LN treatment by using anti-CD3 antibodies.

CD52, a glycoprotein of 12 amino acids anchored to glycosylphosphatidylinositol, can be monitored in peripheral blood or on the cell surface of immune cells, such as mature lymphocytes, NK cells, eosinophils, neutrophils, B cells, and T cells. Previous studies indicated that CD52 can promote T-cell activation either by its intracellular signal pathways or by its interaction with the Siglec-10 on T cells (147), and the finding also suggested that targeting CD52 may be an effective way to reduce T cell activity. The anti-CD52 monoclonal antibody, Alemtuzumab, has been approved to use in clinics for attenuating the T cell responses in patients with organ transplantations. Interestingly, the efficacy of Alemtuzumab has also been evaluated in patients with type 1 diabetes mellitus and end-stage renal disease. Some clinical data have shown no significant acute as well as chronic T-cell response in patients after Alemtuzumab treatment, as indicated by the normal kidney function (148), indicating that anti-CD52 therapeutics is probably to be a potential way for ameliorating excessive T-cell response. Although CD52 is a hallmark expressed on B cells, eliminating B cells by using only anti-CD52 antibodies still has the risk to trigger the secondary autoimmunities after the incomplete B cells depletion by the antibodies. In patients with MS, Alemtuzumab could deplete over 85% of $CD19^+$ B cells, whereas a 180% increase of immature B cells occurred with conversion to mature B cells after 12-month antibody treatment. Secondary autoimmunities occur after Alemtuzumab therapy, such as Goodpasture syndrome, Graves' disease, and idiopathic thrombocytopenic purpura (149). Fortunately, the related secondary autoimmunities could be stopped by CD20 antibodies. Thus, therapeutics targeting both CD52 and CD20 is probably to be a potential approach for the treatment of patients with autoimmune renal disease.

20.4.3 Biologic Therapeutics Against Inflammatory Cytokines

As the key effectors of activated T cells and B cells, which participate in the regulation of both innate and adaptive immune responses, cytokines always play major roles in the promotion and regulation of inflammatory responses in the kidney through their ability to modulate the signaling between infiltrating immunocytes and renal cells (152). Therefore, cytokines have been regarded as the potential therapeutic targets for autoimmune renal disease treatment and various biologic agents have been thus generated (Table 20.4).

Interleukin 6 (IL-6) is a multi-functional cytokine that can not only regulate the immune and inflammatory response but also modulate renal metabolism and development (153). Recent studies have shown that the depletion of IL-6 or dysfunction of IL-6/IL-6 receptor (IL-6R) signaling is likely beneficial to the therapy of various autoimmune renal injuries (154). Thus, a series of biologic drugs against IL-6 or IL-6R, such as Sirukumab, Tocilizumab, Clazakizumab, and Vobarilizumab have been tested for their potential to treat autoimmune renal disease in the clinic (155). Among those four biologics, Sirukumab and Clazakizumab target IL-6, and the others bind IL-6R (156).

On the one hand, the depletion of IL-6 may contribute to the controlling progression of autoimmune kidney diseases. The results of clinical trials have shown the beneficial effect of Clazakizumab on late antibody-mediated renal rejection, while a phase II clinical study exhibit a dispiriting outcome of

TABLE 20.4 Biologics or biosimilars targets inflammatory cytokines in autoimmune renal disease

TARGET	*BIOLOGIC OR BIOSIMILAR*	*CLINICAL STATUS*	*RELATED CLINICAL TRIALS*	*REFERENCES*
IL-6/IL-6R	Sirukumab	Phase II in lupus nephritis	NCT01273389	(158)
	Tocilizumab	Phase I in systemic lupus erythematosus and its-related lupus nephritis	NCT00046774	(161, 175)
	Vobarilizumab (ALX-0061)	Phase II in systemic lupus erythematosus and its related lupus nephritis	NCT02437890	(176)
	Clazakizumab	Phase II in transplant glomerulopathy, kidney transplant failure and rejection, and antibody-mediated rejection	NCT03380962 NCT03380377 NCT03444103	(159)
TNF	Infliximab	1. Phase II/III in systemic lupus erythematosus and its related lupus nephritis 2. Phase II in renal transplant rejection 3. Phase II in ANCA associated vasculitis	1. NCT00368264 2. NCT04114188 3. NCT00753103	(170, 171, 177, 178)
	Humira (adalimumab)	Phase II in focal segmental glomerulosclerosis	NCT00814255 NCT04009668	(179, 180)
TWEAK	BIIB023	Phase II in lupus nephritis	NCT01930890 NCT01499355	(181, 182)
IL-12/IL-23	Ustekinumab	Phase III in systemic lupus erythematosus and its related lupus nephritis	NCT03517722 NCT04060888	(183)
IFN-α	Sifalimumab	Phase II in systemic lupus erythematosus and its related lupus nephritis	NCT00979654 NCT00657189	(184)
	Rontalizumab	Phase II in systemic lupus erythematosus and its related lupus nephritis	NCT00962832	(185)
IFN-γ	AMG 811	1. Phase I in systemic lupus erythematosus and its-related lupus nephritis 2. Phase I in nephritis	1. NCT01164917 2. NCT00818948	(186)
IL-17A/IL-17R	Secukinumab	Phase III in lupus nephritis	NCT04181762	(187, 188)

Sirukumab in LN treatment (157–159). Importantly, the failure of Sirukumab in LN therapy is not due to the unsatisfactory efficacy of the antibody. This is mainly because of the severe side effects that occurred during its clinical use. When compared to those treated with a placebo, patients treated with Sirukumab are more likely to gain cardiovascular events, tuberculosis, chronic hepatic injuries, opportunistic infections, and malignant tumors (160). On the other hand, binding to IL-6R is also a promising approach for autoimmune renal disease therapy. Tocilizumab, the first inhibitor of IL-6R, has been demonstrated to trigger dosage-related decreases in the absolute neutrophil count, with a median decrease of 38% and 56% in SLE patients treated with 4 mg/kg or 8 mg/kg of the antibody, respectively (161). Of course, there are a lot of anti-IL-6/IL-6R antibodies or fusion proteins, such as Olamkicept (TJ301), Satralizumab, and Olokizumab, that has not been evaluated for their treatment of the autoimmune renal injury. The positive effects of those biologics on other autoimmune diseases including neuromyelitis optica spectrum disorder, psoriasis, and rheumatoid arthritis have been well observed in the clinic (162–164). Therefore, anti-IL6/IL-6R biologics are suggested to be effective approaches for autoimmune renal disease therapy.

Tumor necrosis factor (TNF) is a pro-inflammatory cytokine that has important roles in maintaining the balance of host immunity and cellular homeostasis. Over the past decades, TNF has been regarded as one of the most intensively studied cytokines of the immune system. The binding of TNF to its receptor can upregulate the activity of NF-κB-based inflammatory response, resulting in the progression of many inflammatory and autoimmune disorders, including rheumatoid arthritis, inflammatory bowel disease, and various autoimmune kidney diseases (165, 166). Thus, mounting studies have been carried out to determine whether TNF-targeting therapies are effective therapeutic strategies in these diseases. Infliximab, a chimeric antibody against TNF-α, can bind both the soluble subunit and transmembrane precursor of TNF molecules (167).

Due to its high specificity and affinity to TNF-α, Infliximab can efficiently neutralize TNFs in peripheral blood, followed by its interference with the initiation and progression of TNF-related autoimmune diseases (168, 169). The effect of Infliximab on ameliorating excessive immune response has been investigated in three subtypes of autoimmune renal diseases. Clinical data have shown the prolonged improvement of the renal function in patients with SLE and LN during their treatment by Infliximab (170). Another clinical trial of infliximab against ANCA-associated vasculitis has proved that the antibody is effective for triggering remission in 88% of patients with ANCA-associated vasculitis in renal tissues (171). Similar results can also be found in the related trials of Etanercept and Adalimumab (172, 173). Although evidence supports anti-TNF therapies to be promising ways for autoimmune renal diseases treatment, the safety of those biologics should be further evaluated. The crucial adverse effects of anti-TNF therapies are to induce infection, tuberculosis, and acquired autoimmune diseases. In a study containing 233 patients with rheumatoid arthritis, Crohn's disease, ankylosing spondylitis, psoriatic arthritis, and other autoimmune diseases, 92 of patients were found to gain lupus after treatment with anti-TNF agents (infliximab in 40 cases, etanercept in 37, and adalimumab in 15) (174). Therefore, the use of anti-TNF antibodies for autoimmune renal disease in the clinic is still nascent and should be further evaluated.

20.4.4 Biologic Therapeutics Against Complements

The complement system, as one part of innate immunological components, is helpful in both fighting against the invasion of pathogens and regulation of the immune balance. In autoimmune renal diseases, complements are overacted due to the continuous generation of self-autoantibodies, which can result in tissue injury and trigger local inflammation in kidney (189, 190). Thus, complement inhibitory biologics are supposed to be beneficial for the treatment of autoimmune renal diseases (Table 20.5).

Among various autoimmune kidney failures, C3 glomerulonephritis is the disease mainly in relation to the dysregulation of components, which is characterized by the C3 component of complement accumulated in kidney tissues. Eculizumab, a humanized monoclonal antibody against C5, has the ability to block the interaction between C3 complex and C5, and therefore prevent the formation of C5-dependent membrane attack complex (191, 192). It has been approved for the treatment of paroxysmal nocturnal hemoglobinuria as well as atypical hemolytic uremic syndrome, and its potential for C3 glomerulonephritis therapy has been evaluated as well. Clinical data have shown that eculizumab triggers a therapeutic response in some but not all patients with C3 glomerulonephritis (193). This is possible because of the complicated pathophysiological basis of C3 glomerulonephritis (194). Therefore, treatment of eculizumab should be based on evaluating the gene and function of the complement system in C3 glomerulonephritis patients.

TABLE 20.5 Biologics or biosimilars targets complements in autoimmune renal disease

TARGET	*BIOLOGIC OR BIOSIMILAR*	*CLINICAL STATUS*	*RELATED CLINICAL TRIALS*	*REFERENCES*
C5	Eculizumab	1. Phase I in C3 glomerulonephritis 2. Phase II in membranoproliferative glomerulonephritis	1. NCT01221181 2. NCT02093533	(193, 195)
	Tesidolumab (LFG316)	Phase I in pre-sensitized end-stage renal disease	NCT02878616	(196)
C5aR	CCX168	Phase II in IgA nephropathy	NCT02384317	(197)

20.5 CHALLENGES AND PERSPECTIVES

Despite the prominent efficacy, high specificity, and good safety profiles of biological therapy, a variety of technical and regulatory issues should be properly addressed before the clinical application of a biologic or biosimilar for the management of a certain subtype of autoimmune renal disease. In recent years, numerous efforts have been devoted to the development of biopharmaceuticals as therapeutics for inflammatory diseases and autoinflammatory disorders. However, due to the lack of suitable animal models, low incidence and insufficient subtyping of diseases, and high variation of individual difference in disease severity and progression, current evidence in favor of the clinical applications of biologics and biosimilars for autoimmune renal diseases is still limited. In order to speed up the translation of basic discoveries to clinical applications, more efforts should be devoted to the generation of highly relevant and clearly characterized animal models of autoimmune renal diseases. More importantly, characterizing the molecular signatures of individual patients will be necessary to precisely define the molecular subsets of autoimmune renal diseases. Due to structurally complexity of biological products, slight alterations in the starting materials, manufacturing processes, purification protocols may result in undesirable adverse effects or even compromised therapeutic effects, thus raising safety or efficiency concerns for the clinical application of biological drugs. Therefore, it is also necessary to adopt a relatively strict regulatory policy to enhance quality control of biologics and biosimilars. Besides, health care professionals must spare no efforts to take their responsibility of routinely reporting therapeutic performances and adverse effects of biologics and biosimilars during clinical trials and practice.

Although biopharmaceuticals potentially exhibit excellent effectiveness in the management of autoimmune renal diseases, there is an uncertainty concerning the safety profile of these biological products. Immunogenicity of biologics and biosimilars is a dominant safety challenge to be addressed during the development of biomolecules as therapeutics of autoimmune diseases. Generally, the immunogenicity of a biologic or biosimilar is highly dependent on product-intrinsic factors (e.g., size, structure, composition, impurities, formulation, species, degree of foreignness, post-translational modifications, and other properties contributing to immunogenic potential of biological products), product-extrinsic factors (e.g., route of administration, dosing regimen, duration of treatment, pharmacokinetics, and existence of endogenous equivalents), and patient-specific factors (e.g., age, sex, genetic profile, prior exposure, history of medication, health condition and immune status) (198). Due to structural differences in comparison with native or endogenous molecules, biotherapeutics can probably elicit activation of B cells and induce the production of antibodies against themselves after being perceived or recognized by the immune system as foreign substances. The antibodies produced against the biomolecules can not only bind to the therapeutics administered to the patients and thus, to some extent, compromise or even completely eliminate their beneficial effects, but also may lead to mild or severe adverse effects due to their potential to interact with autologous molecules. Therefore, further understanding of contributing factors of immunogenicity is urgently needed to provide potential strategies for mitigating the incidence of immunogenic responses to biologics.

Since the immunogenicity is a complex dynamic process that evolves and changes over time, it is a necessity to have a systematic and dynamic evaluation of immunogenicity before approval of therapeutic biologics and biosimilars. The risk of undesirable immunogenicity of biologics or biosimilars can be minimized during the process of drug screening and development by reasonable immunogenicity prediction and evaluation via various techniques, such as *in vitro* and in *vivo* immunogenicity assays and predictive mathematical modeling and simulation. And it is highly recommended that the overall assessment of immunogenicity of biopharmaceuticals should take both the probability of occurrence and the severity of consequence into account. However, it should be noted that it is currently impossible to predict the likelihood of a certain patient developing serious immunogenic responses after treatment by biological pharmaceuticals, thus making it difficult for health care professionals to prescribe the optimal biologics or biosimilars. Nowadays, various strategies are available to diminish the immunogenicity of biologics, especially for therapeutic proteins products. For instance, surface modification of therapeutic proteins via

PEGylating can effectivity mask the immunogenic epitopes and reduce their recognition by the immune system, thereby minimizing the immunogenicity of therapeutic proteins. In parallel, the removal of immunogenic epitopes via site-specific mutagenesis can also reduce the immunogenicity of the biologics. Besides, with the advances of nanotechnology, novel drug delivery systems are intensively investigated to realize controlled and targeted delivery of unmodified and potentially immunogenic protein therapeutics, thus offering an alternative strategy of overcoming the immunogenicity of biomacromolecules and avoiding detrimental humoral immune responses. An in-depth understanding of molecular and cellular mechanism underlying immunogenic responses to biopharmaceuticals will facilitate the development of strategies to reduce the incidence of immunogenicity and improve the safety profile of the biological products.

Difficulty in identification of the population of patients that will respond to a biologic or biosimilar drug is also a major challenge for the clinical application of these biological therapeutics. Due to distinct molecular subtypes, medical histories, clinical or pathological stages, and immune status, patients with autoinflammatory renal diseases may probably respond differently to a specific biological therapy. For instance, in some cases, the patients fail to have adequate responses to biologics from the beginning (primary failure) partly due to different pathotypes of diseases, while in others, the patients may initially respond sufficiently but gradually lose responses over time (secondary failure) most likely due to the production of antibodies against the biologics or alteration of pharmacokinetics and pharmacodynamics. Besides, there is still a proportion of patients that have to optimize dosing regimen or terminate the biological therapy due to the emergence of severe or even life-threatening adverse events. However, in the present time, it is not technically feasible to discriminate among different subgroups of patients (e.g., super responders, responders, partial responders, and non-responders) before receiving biologics therapy, largely owing to the lack of specific biomarkers as an indicator for treatment responses to certain biopharmaceuticals. Thus, to ensure a more appropriate and specific targeting therapy for individuals, molecular subtypes and immunological characteristics should be determined to guide the selection of biologics or biosimilars for the management of autoimmune renal disorders. Moreover, it is crucial to discover valuable biomarkers for predicting therapeutic responses and outcomes as well as responsive rates in individual patients prior to initiation of biological therapies. During the process of biological therapy, additional biomarkers should be used for real-time evaluation of therapeutic efficacy of biologics and guide the adjustment of the therapy regimes. It is still challenging to decide the initiation, alteration, and termination of biological therapy in autoimmune renal diseases, indicating the need to better recognize the clinical relevance of phenotypes and biomarkers.

Since the immune system is a highly complex network comprised of a series of interacting cellular and molecular components, simple manipulation of a single pathogenic molecule by biopharmaceuticals may exert unpredictable effects on patients with autoinflammatory diseases. Besides, the genetic, epigenetic, and environmental factors add an extra layer of complexity to the network of the human immune system. All these factors make the interaction between biologics and the immune system complicated than expected. The biologics or biosimilars specially developed for targeting crucial mediators or molecular pathways may induce redundant interactions between the biopharmaceuticals and immune systems, thus constituting a limitation of these biological therapeutics for the management of autoimmune renal diseases. Therefore, to recognize the limitations of the targeted therapy by biologics before their clinical application, we need to gradually improve the understanding of their interactions with the human immune system to avoid potential risks of undesirable interactions. And, predictably, we will be able to develop novel targeted biological therapeutics and select suitable ones for the corresponding phenotypes based on a more comprehensive knowledge on the human immune system. Novel techniques, including high-throughput single-cell sequencing, transcriptomic and proteomic analysis, will promote the overall understanding of autoimmune renal diseases and accelerate the discovery of therapeutic targets (199, 200). Recent advances in high-throughput single-cell sequencing enable the identification and characterization of distinct subsets of immunocytes within the immune system, thus providing a potential strategy for global characterization of the immune status of patients with autoimmune renal diseases. In parallel, the transcriptional signatures of subpopulations of immunocytes can largely facilitate the identification of novel immunopathogenic molecules and biomarkers (201). Moreover, transcriptomic and proteomic analysis of renal biopsy specimen further promote the identification of the dominant immunopathogenic mediators and signaling pathways contributing to the progression of autoinflammatory renal diseases. The further identification of disease-initiating autoantigens and corresponding autoantibodies for autoimmune renal diseases not only facilitates the characterization and monitorization of autoimmunity during the disease progression but also provides potential diagnostic biomarkers and therapeutic targets for the management of autoinflammatory diseases. With the noble advances of these techniques, a growing number of biologics and biosimilars has been developed to target different pathological stages during the initiation and progression of autoimmune diseases, including the occurrence of autoantibodies, cytokine abnormalities, the activation of autoreactive T and B cells, and the formation of immune complexes.

Deciphering precise molecular mechanisms contributing to autoimmune inflammatory responses will further facilitate the identification of novel therapeutic targets of autoimmune renal diseases by biologics and biosimilars. In recent years, a broad range of biopharmaceuticals targeting dominant mediators and mechanisms of autoimmune diseases are being researched and developed for the management of these disorders. Autoimmune renal diseases do share some molecular similarities in the pathogenesis, thereby enabling the application of existing biologics developed for non-renal autoinflammatory diseases with proven efficacy for application in the management of autoimmune renal diseases. However, it is not advisable to simply incorporating biologics and biosimilars initially developed for other non-renal autoimmune diseases into the therapy regimens of autoimmune renal diseases since the biologics or biosimilars effective for non-renal autoimmune diseases may not work for the treatment of autoimmune renal diseases. Nowadays, our understandings of the molecular basis of pathogenesis of autoinflammatory diseases are still limited, with the major pathogenesis of autoimmune renal diseases varying during the disease progression. Due to the heterogeneity of the pathogenesis of a single category of autoimmune renal diseases, a biologic or biosimilar targeting a specific pathogenic pathway responsible for the pathogenesis of the disease may not effective for the other patients with distinct pathogenesis.

In parallel, owning to the complexity of the pathogenesis of autoimmune renal diseases, targeting a single pathogenic pathway may not be enough effective for the management of these disorders. The recent advances of integrated proteomics and transcriptomics of renal biopsy samples offer a potential opportunity to better define the critical molecular events and signaling pathways that contribute to injurious renal inflammatory responses. Collectively, an in-depth understanding of the pathogenesis of autoimmune renal diseases will eventually allow for a more specific and efficient approach for the management of disease progression. It should be noted that targeting a particular autoinflammatory molecule by biologics or biosimilars may also lead to unprecedented therapeutic benefits and durable clinical responses in certain subtypes or specific cohorts of patients with autoimmune renal diseases. And there is still a risk of initiating unexpected autoimmune diseases by the continuous utility of these biological pharmaceuticals. For instance, it has been reported that a broad range of autoimmune disorders, including psoriasis, lupus, inflammatory bowel disease, and autoimmune hepatitis are correlated with exposure to biologics (202). Besides, the quantity and diversity of biologics-induced autoimmune diseases will probably increase exponentially with the prevalence of clinical uses of biopharmaceuticals. On this occasion, it is imperative to identify subsets of patients with a high risk of developing other inflammatory disorders after biologics therapies by means of certain cellular or molecular markers. Furthermore, it will be a necessity for health care professionals to have a thorough and updated knowledge on potential autoimmune adverse events induced by biologics for a more specialized and individualized management of autoimmune renal diseases.

ACKNOWLEDGMENTS

This work was sponsored by the Shanghai Sailing Program (20YF1402800), the National Natural Science Foundation of China (82002915, 81901232, 81803529, 32070935), and the Shanghai Rising-Star Program (21QB1401800).

REFERENCES

1. Tecklenborg J, Clayton D, Siebert S, Coley SM. The role of the immune system in kidney disease. Clin Exp Immunol. 2018;192(2):142–50.
2. Imig JD, Ryan MJ. Immune and inflammatory role in renal disease. Compr Physiol. 2013;3(2):957–76.
3. Holdsworth SR, Gan PY, Kitching AR. Biologics for the treatment of autoimmune renal diseases. Nat Rev Nephrol. 2016;12(4):217–31.
4. Kronbichler A, Mayer G. Renal involvement in autoimmune connective tissue diseases. BMC Med. 2013;11:95.
5. Foster MH, Ord JR. Emerging immunotherapies for autoimmune kidney disease. Hum Vaccin Immunother. 2019;15(4):876–90.
6. Oka K, Nishimura K, Kishikawa H, Ichikawa Y. IgA1 dominant subclass of latent IgA mesangial deposition in donated kidney. Int J Nephrol Renovasc Dis. 2016;9:313–7.
7. Rodrigues JC, Haas M, Reich HN. IgA nephropathy. Clin J Am Soc Nephrol. 2017;12(4):677–86.
8. Li M, Yu X. Genetic study of immunoglobulin A nephropathy: From research to clinical application. Nephrology (Carlton). 2018;23(Suppl 4):26–31.
9. Wyatt RJ, Julian BA. IgA nephropathy. N Engl J Med. 2013;368(25):2402–14.
10. Reily C, Ueda H, Huang ZQ, Mestecky J, Julian BA, Willey CD, et al. Cellular signaling and production of galactose-deficient IgA1 in IgA nephropathy, an autoimmune disease. J Immunol Res. 2014; 2014:197548.
11. Kiryluk K, Novak J. The genetics and immunobiology of IgA nephropathy. J Clin Invest. 2014;124(6):2325–32.
12. Working Group of the International Ig ANN, the Renal Pathology S, Roberts IS, Cook HT, Troyanov S, Alpers CE, et al. The Oxford classification of IgA nephropathy: Pathology definitions, correlations, and reproducibility. Kidney Int. 2009;76(5):546–56.
13. Suzuki H, Kiryluk K, Novak J, Moldoveanu Z, Herr AB, Renfrow MB, et al. The pathophysiology of IgA nephropathy. J Am Soc Nephrol. 2011;22(10):1795–803.
14. Magistroni R, D'Agati VD, Appel GB, Kiryluk K. New developments in the genetics, pathogenesis, and therapy of IgA nephropathy. Kidney Int. 2015; 88(5):974–89.
15. Liu F, Wang C, Wang R, Wang W, Li M. Henoch-Schonlein purpura nephritis with renal interstitial lesions. Open Med (Wars). 2018; 13:597–604.
16. Rai A, Nast C, Adler S. Henoch-Schonlein purpura nephritis. J Am Soc Nephrol. 1999;10(12):2637–44.
17. Pohl M. Henoch-Schonlein purpura nephritis. Pediatr Nephrol. 2015;30(2):245–52.
18. Davin JC, Ten Berge IJ, Weening JJ. What is the difference between IgA nephropathy and Henoch-Schonlein purpura nephritis? Kidney Int. 2001;59(3):823–34.
19. Sanders JT, Wyatt RJ. IgA nephropathy and Henoch-Schonlein purpura nephritis. Curr Opin Pediatr. 2008;20(2):163–70.
20. Yu F, Haas M, Glassock R, Zhao MH. Redefining lupus nephritis: Clinical implications of pathophysiologic subtypes. Nat Rev Nephrol. 2017;13(8):483–95.
21. Lech M, Anders HJ. The pathogenesis of lupus nephritis. J Am Soc Nephrol. 2013;24(9):1357–66.
22. Yung S, Yap DY, Chan TM. A review of advances in the understanding of lupus nephritis pathogenesis as a basis for emerging therapies. F1000Res. 2020;9. F1000 Faculty Rev-905
23. Iwamoto T, Niewold TB. Genetics of human lupus nephritis. Clin Immunol. 2017;185:32–9.
24. Salmon JE, Millard S, Schachter LA, Arnett FC, Ginzler EM, Gourley MF, et al. Fc gamma RIIA alleles are heritable risk factors for lupus nephritis in African Americans. J Clin Invest. 1996;97(5):1348–54.
25. Karassa FB, Trikalinos TA, Ioannidis JP, Fc gamma R-SLEm-ai. The Fc gamma RIIIA-F158 allele is a risk factor for the development of lupus nephritis: A meta-analysis. Kidney Int. 2003;63(4):1475–82.
26. Villanueva E, Yalavarthi S, Berthier CC, Hodgin JB, Khandpur R, Lin AM, et al. Netting neutrophils induce endothelial damage, infiltrate tissues, and expose immunostimulatory molecules in systemic lupus erythematosus. J Immunol. 2011;187(1):538–52.
27. Hakkim A, Furnrohr BG, Amann K, Laube B, Abed UA, Brinkmann V, et al. Impairment of neutrophil extracellular trap degradation is associated with lupus nephritis. Proc Natl Acad Sci USA. 2010;107(21):9813–8.
28. Li Y, Li M, Weigel B, Mall M, Werth VP, Liu ML. Nuclear envelope rupture and NET formation is driven by PKCalpha-mediated lamin B disassembly. EMBO Rep. 2020;21(8):e48779.
29. Fakhouri F, Fremeaux-Bacchi V, Noel LH, Cook HT, Pickering MC. C3 glomerulopathy: A new classification. Nat. Rev. Nephrol. 2010;6(8):494–9.
30. Zipfel PF, Skerka C, Chen Q, Wiech T, Goodship T, Johnson S, et al. The role of complement in C3 glomerulopathy. Mol. Immunol. 2015;67(1):21–30.

31. D'Agati VD, Bomback AS. C3 glomerulopathy: What's in a name? Kidney Int. 2012;82(4):379–81.
32. Berger J, Galle P. Unusual change of the basal membranes of the kidney. Journal d'urologie et de nephrologie. 1962;68:116–22.
33. Master Sankar Raj V, Gordillo R, Chand DH. Overview of C3 Glomerulopathy. Front. Pediatr. 2016;4:45.
34. Bomback AS, Santoriello D, Avasare RS, Regunathan-Shenk R, Canetta PA, Ahn W, et al. C3 glomerulonephritis and dense deposit disease share a similar disease course in a large United States cohort of patients with C3 glomerulopathy. Kidney Int. 2018;93(4):977–85.
35. Makker SP, Tramontano A. Idiopathic membranous nephropathy: An autoimmune disease. Semin Nephrol. 2011;31(4):333–40.
36. Couser WG. Primary membranous nephropathy. Clin J Am Soc Nephrol. 2017;12(6):983–97.
37. Beck LH, Jr. PLA2R and THSD7A: Disparate paths to the same disease? J Am Soc Nephrol. 2017;28(9): 2579–89.
38. Gu Y, Xu H, Tang D. Mechanisms of primary membranous nephropathy. Biomolecules. 2021;11(4):513.
39. Fischer EG, Lager DJ. Anti-glomerular basement membrane glomerulonephritis: A morphologic study of 80 cases. Am J Clin Pathol. 2006;125(3):445–50.
40. Borza DB. Autoepitopes and alloepitopes of type IV collagen: Role in the molecular pathogenesis of anti-GBM antibody glomerulonephritis. Nephron Exp Nephrol. 2007;106(2):e37–43.
41. McAdoo SP, Pusey CD. Anti-glomerular basement membrane disease. Clin J Am Soc Nephrol. 2017;12(7):1162–72.
42. Ahmad SB, Santoriello D, Canetta P, Bomback AS, D'Agati VD, Markowitz G, et al. Concurrent anti-glomerular basement antibody disease and membranous nephropathy: A case series. Am J Kidney Dis. 2021;78(2):219–225.el.
43. Rutgers A, Sanders JS, Stegeman CA, Kallenberg CG. Pauci-immune necrotizing glomerulonephritis. Rheum Dis Clin North Am. 2010;36(3):559–72.
44. Scaglioni V, Scolnik M, Catoggio LJ, Christiansen SB, Varela CF, Greloni G, et al. ANCA-associated pauci-immune glomerulonephritis: Always pauci-immune? Clin Exp Rheumatol. 2017;35 Suppl 103(1):55–8.
45. Rutgers A, Slot M, van Paassen P, van Breda Vriesman P, Heeringa P, Tervaert JW. Coexistence of anti-glomerular basement membrane antibodies and myeloperoxidase-ANCAs in crescentic glomerulonephritis. Am J Kidney Dis. 2005;46(2):253–62.
46. Craft JE. Follicular helper T cells in immunity and systemic autoimmunity. Nat Rev Rheumatol. 2012;8(6):337–47.
47. Steinmetz OM, Turner JE, Paust HJ, Lindner M, Peters A, Heiss K, et al. CXCR3 mediates renal Th1 and Th17 immune response in murine lupus nephritis. J Immunol. 2009;183(7):4693–704.
48. Paust HJ, Riedel JH, Krebs CF, Turner JE, Brix SR, Krohn S, et al. CXCR3+ Regulatory T Cells Control TH1 responses in crescentic GN. J Am Soc Nephrol. 2016;27(7):1933–42.
49. Masutani K, Akahoshi M, Tsuruya K, Tokumoto M, Ninomiya T, Kohsaka T, et al. Predominance of Th1 immune response in diffuse proliferative lupus nephritis. Arthritis Rheum. 2001;44(9):2097–106.
50. Yap DY, Lai KN. Pathogenesis of renal disease in systemic lupus erythematosus–the role of autoantibodies and lymphocytes subset abnormalities. Int J Mol Sci. 2015;16(4):7917–31.
51. Suarez-Fueyo A, Bradley SJ, Klatzmann D, Tsokos GC. T cells and autoimmune kidney disease. Nat Rev Nephrol. 2017;13(6):329–43.
52. Yamagata T, Skepner J, Yang J. Targeting Th17 effector cytokines for the treatment of autoimmune diseases. Arch Immunol Ther Exp (Warsz). 2015;63(6):405–14.
53. Gaffen SL, Jain R, Garg AV, Cua DJ. The IL-23-IL-17 immune axis: From mechanisms to therapeutic testing. Nat Rev Immunol. 2014;14(9):585–600.
54. Turner JE, Paust HJ, Steinmetz OM, Panzer U. The Th17 immune response in renal inflammation. Kidney Int. 2010;77(12):1070–5.
55. Krebs CF, Paust HJ, Krohn S, Koyro T, Brix SR, Riedel JH, et al. Autoimmune renal disease is exacerbated by S1P-receptor-1-dependent intestinal Th17 cell migration to the kidney. Immunity. 2016;45(5):1078–92.
56. Velden J, Paust HJ, Hoxha E, Turner JE, Steinmetz OM, Wolf G, et al. Renal IL-17 expression in human ANCA-associated glomerulonephritis. Am J Physiol Renal Physiol. 2012; 302(12):F1663–73.
57. Koga T, Ichinose K, Tsokos GC. T cells and IL-17 in lupus nephritis. Clin Immunol. 2017;185:95–9.
58. Krebs CF, Turner JE, Paust HJ, Kapffer S, Koyro T, Krohn S, et al. Plasticity of Th17 cells in autoimmune kidney diseases. J Immunol. 2016;197(2):449–57.
59. Lalor SJ, McLoughlin RM. Memory γδ T cells-newly appreciated protagonists in infection and immunity. Trends Immunol. 2016;37(10):690–702.
60. Turner JE, Krebs C, Tittel AP, Paust HJ, Meyer-Schwesinger C, Bennstein SB, et al. IL-17A production by renal γδ T cells promotes kidney injury in crescentic GN. J Am Soc Nephrol. 2012;23(9):1486–95.

61. Peng X, Xiao Z, Zhang J, Li Y, Dong Y, Du J. IL-17A produced by both γδ T and Th17 cells promotes renal fibrosis via RANTES-mediated leukocyte infiltration after renal obstruction. J Pathol. 2015;235(1):79–89.
62. Yin S, Mao Y, Li X, Yue C, Zhou C, Huang L, et al. Hyperactivation and in situ recruitment of inflammatory Vdelta2 T cells contributes to disease pathogenesis in systemic lupus erythematosus. Sci Rep. 2015;5:14432.
63. Chen H, You H, Wang L, Zhang X, Zhang J, He W. Chaperonin-containing T-complex protein 1 subunit zeta serves as an autoantigen recognized by human Vdelta2 γδ T cells in autoimmune diseases. J Biol Chem. 2016;291(38):19985–93.
64. Crispin JC, Oukka M, Bayliss G, Cohen RA, Van Beek CA, Stillman IE, et al. Expanded double negative T cells in patients with systemic lupus erythematosus produce IL-17 and infiltrate the kidneys. J Immunol. 2008;181(12):8761–6.
65. Ueno H. T follicular helper cells in human autoimmunity. Curr Opin Immunol. 2016;43:24–31.
66. Morita R, Schmitt N, Bentebibel SE, Ranganathan R, Bourdery L, Zurawski G, et al. Human blood CXCR5(+) CD4(+) T cells are counterparts of T follicular cells and contain specific subsets that differentially support antibody secretion. Immunity. 2011;34(1):108–21.
67. Liarski VM, Kaverina N, Chang A, Brandt D, Yanez D, Talasnik L, et al. Cell distance mapping identifies functional T follicular helper cells in inflamed human renal tissue. Sci Transl Med. 2014;6(230):230ra46.
68. Ferretti C, La Cava A. Adaptive immune regulation in autoimmune diabetes. Autoimmun Rev. 2016;15(3):236–41.
69. Ghali JR, Wang YM, Holdsworth SR, Kitching AR. Regulatory T cells in immune-mediated renal disease. Nephrology (Carlton). 2016;21(2):86–96.
70. Klatzmann D, Abbas AK. The promise of low-dose interleukin-2 therapy for autoimmune and inflammatory diseases. Nat Rev Immunol. 2015;15(5):283–94.
71. Kasper IR, Apostolidis SA, Sharabi A, Tsokos GC. Empowering regulatory T cells in autoimmunity. Trends Mol Med. 2016;22(9):784–97.
72. Arpaia N, Green JA, Moltedo B, Arvey A, Hemmers S, Yuan S, et al. A distinct function of regulatory T cells in tissue protection. Cell. 2015;162(5):1078–89.
73. Comte D, Karampetsou MP, Kis-Toth K, Yoshida N, Bradley SJ, Mizui M, et al. Engagement of SLAMF3 enhances CD4+ T-cell sensitivity to IL-2 and favors regulatory T-cell polarization in systemic lupus erythematosus. Proc Natl Acad Sci U S A. 2016;113(33):9321–6.
74. Chang J, Eggenhuizen P, O'Sullivan KM, Alikhan MA, Holdsworth SR, Ooi JD, et al. CD8+ T cells effect glomerular injury in experimental anti-myeloperoxidase GN. J Am Soc Nephrol. 2017;28(1):47–55.
75. McKinney EF, Lyons PA, Carr EJ, Hollis JL, Jayne DR, Willcocks LC, et al. A CD8+ T cell transcription signature predicts prognosis in autoimmune disease. Nat Med. 2010;16(5):586–91, 1p following 91.
76. Kopetschke K, Klocke J, Griessbach AS, Humrich JY, Biesen R, Dragun D, et al. The cellular signature of urinary immune cells in Lupus nephritis: New insights into potential biomarkers. Arthritis Res Ther. 2015;17:94.
77. Oleinika K, Mauri C, Salama AD. Effector and regulatory B cells in immune-mediated kidney disease. Nat Rev Nephrol. 2019;15(1):11–26.
78. Bouaziz JD, Yanaba K, Venturi GM, Wang Y, Tisch RM, Poe JC, et al. Therapeutic B cell depletion impairs adaptive and autoreactive CD4+ T cell activation in mice. Proc Natl Acad Sci USA. 2007;104(52):20878–83.
79. Rodriguez-Pinto D. B cells as antigen presenting cells. Cell Immunol. 2005;238(2):67–75.
80. Liossis SN, Sfikakis PP. Rituximab-induced B cell depletion in autoimmune diseases: Potential effects on T cells. Clin Immunol. 2008;127(3):280–5.
81. Shen P, Fillatreau S. Antibody-independent functions of B cells: A focus on cytokines. Nat Rev Immunol. 2015;15(7):441–51.
82. Harris DP, Haynes L, Sayles PC, Duso DK, Eaton SM, Lepak NM, et al. Reciprocal regulation of polarized cytokine production by effector B and T cells. Nat Immunol. 2000;1(6):475–82.
83. Wojciechowski W, Harris DP, Sprague F, Mousseau B, Makris M, Kusser K, et al. Cytokine-producing effector B cells regulate type 2 immunity to H. polygyrus. Immunity. 2009; 30(3):421–33.
84. Mauri C, Bosma A. Immune regulatory function of B cells. Annu Rev Immunol. 2012;30:221–41.
85. Burkett PR, Meyer zu Horste G, Kuchroo VK. Pouring fuel on the fire: Th17 cells, the environment, and autoimmunity. J Clin Invest. 2015;125(6):2211–9.
86. Matsumoto M, Oshiumi H, Seya T. Antiviral responses induced by the TLR3 pathway. Rev Med Virol. 2011;21(2):67–77.
87. Fillatreau S, Sweenie CH, McGeachy MJ, Gray D, Anderton SM. B cells regulate autoimmunity by provision of IL-10. Nat Immunol. 2002;3(10):944–50.
88. Mauri C, Gray D, Mushtaq N, Londei M. Prevention of arthritis by interleukin 10-producing B cells. J Exp Med. 2003;197(4):489–501.

89. Mizoguchi A, Mizoguchi E, Takedatsu H, Blumberg RS, Bhan AK. Chronic intestinal inflammatory condition generates IL-10-producing regulatory B cell subset characterized by CD1d upregulation. Immunity. 2002;16(2):219–30.
90. Blair PA, Chavez-Rueda KA, Evans JG, Shlomchik MJ, Eddaoudi A, Isenberg DA, et al. Selective targeting of B cells with agonistic anti-CD40 is an efficacious strategy for the generation of induced regulatory T2-like B cells and for the suppression of lupus in MRL/lpr mice. J Immunol. 2009;182(6):3492–502.
91. Liu BS, Cao Y, Huizinga TW, Hafler DA, Toes RE. TLR-mediated STAT3 and ERK activation controls IL-10 secretion by human B cells. Eur J Immunol. 2014;44(7):2121–9.
92. Lampropoulou V, Hoehlig K, Roch T, Neves P, Calderon Gomez E, Sweenie CH, et al. TLR-activated B cells suppress T cell-mediated autoimmunity. J Immunol. 2008;180(7):4763–73.
93. Carter NA, Vasconcellos R, Rosser EC, Tulone C, Munoz-Suano A, Kamanaka M, et al. Mice lacking endogenous IL-10-producing regulatory B cells develop exacerbated disease and present with an increased frequency of Th1/Th17 but a decrease in regulatory T cells. J Immunol. 2011;186(10):5569–79.
94. Carter NA, Rosser EC, Mauri C. Interleukin-10 produced by B cells is crucial for the suppression of Th17/Th1 responses, induction of T regulatory type 1 cells and reduction of collagen-induced arthritis. Arthritis Res Ther. 2012;14(1):R32.
95. Sun JB, Flach CF, Czerkinsky C, Holmgren J. B lymphocytes promote expansion of regulatory T cells in oral tolerance: Powerful induction by antigen coupled to cholera toxin B subunit. J Immunol. 2008;181(12):8278–87.
96. Tadmor T, Zhang Y, Cho HM, Podack ER, Rosenblatt JD. The absence of B lymphocytes reduces the number and function of T-regulatory cells and enhances the anti-tumor response in a murine tumor model. Cancer Immunol Immunother. 2011;60(5):609–19.
97. Shen P, Roch T, Lampropoulou V, O'Connor RA, Stervbo U, Hilgenberg E, et al. IL-35-producing B cells are critical regulators of immunity during autoimmune and infectious diseases. Nature. 2014;507(7492):366–70.
98. Wang RX, Yu CR, Dambuza IM, Mahdi RM, Dolinska MB, Sergeev YV, et al. Interleukin-35 induces regulatory B cells that suppress autoimmune disease. Nat Med. 2014;20(6):633–41.
99. Vivarelli M, Massella L, Ruggiero B, Emma F. Minimal change disease. Clin J Am Soc Nephrol. 2017;12(2):332–45.
100. Carlson ED, Gan R, Hodgman CE, Jewett MC. Cell-free protein synthesis: Applications come of age. Biotechnol Adv. 2012;30(5):1185–94.
101. S ELA, Mager I, Breakefield XO, Wood MJ. Extracellular vesicles: Biology and emerging therapeutic opportunities. Nat Rev Drug Discov. 2013;12(5):347–57.
102. Arzoo K, Sadeghi S, Liebman HA. Treatment of refractory antibody mediated autoimmune disorders with an anti-CD20 monoclonal antibody (rituximab). Ann Rheum Dis. 2002;61(10):922–4.
103. Crickx E, Weill JC, Reynaud CA, Mahevas M. Anti-CD20-mediated B-cell depletion in autoimmune diseases: Successes, failures and future perspectives. Kidney Int. 2020;97(5):885–93.
104. Fanouriakis A, Kostopoulou M, Alunno A, Aringer M, Bajema I, Boletis JN, et al. 2019 update of the EULAR recommendations for the management of systemic lupus erythematosus. Ann Rheum Dis. 2019;78(6):736–45.
105. Mok CC. Current role of rituximab in systemic lupus erythematosus. Int J Rheum Dis. 2015;18(2):154–63.
106. Cancarevic I, Malik BH. Use of rituximab in management of rapidly progressive glomerulonephritis. Cureus. 2020;12(1):e6820.
107. Ruggenenti P, Cravedi P, Chianca A, Perna A, Ruggiero B, Gaspari F, et al. Rituximab in idiopathic membranous nephropathy. J Am Soc Nephrol. 2012;23(8):1416–25.
108. Lafayette RA, Canetta PA, Rovin BH, Appel GB, Novak J, Nath KA, et al. A randomized, controlled trial of rituximab in IgA nephropathy with proteinuria and renal dysfunction. J Am Soc Nephrol. 2017;28(4):1306–13.
109. Rovin BH, Furie R, Latinis K, Looney RJ, Fervenza FC, Sanchez-Guerrero J, et al. Efficacy and safety of rituximab in patients with active proliferative lupus nephritis: The lupus nephritis assessment with rituximab study. Arthritis Rheum. 2012;64(4):1215–26.
110. Md Yusof MY, Vital EM, McElvenny DM, Hensor EMA, Das S, Dass S, et al. Predicting severe infection and effects of hypogammaglobulinemia during therapy with rituximab in rheumatic and musculoskeletal diseases. Arthritis Rheumatol. 2019;71(11):1812–23.
111. Musette P, Bouaziz JD. B cell modulation strategies in autoimmune diseases: New concepts. Front Immunol. 2018;9:622.
112. Yap DYH, Chan TM. B cell abnormalities in systemic lupus erythematosus and lupus nephritis-role in pathogenesis and effect of immunosuppressive treatments. Int J Mol Sci. 2019;20(24):6231.
113. Silver K, Cornall RJ. Isotype control of B cell signaling. Sci STKE. 2003;2003(184):pe21.
114. Cesano A, Gayko U. CD22 as a target of passive immunotherapy. Semin Oncol. 2003;30(2):253–7.

115. Clark EA, Giltiay NV. CD22: A Regulator of Innate and Adaptive B Cell Responses and Autoimmunity. Front Immunol. 2018;9:2235.
116. Ereno-Orbea J, Sicard T, Cui H, Mazhab-Jafari MT, Benlekbir S, Guarne A, et al. Molecular basis of human CD22 function and therapeutic targeting. Nat Commun. 2017;8(1):764.
117. Gottenberg JE, Dorner T, Bootsma H, Devauchelle-Pensec V, Bowman SJ, Mariette X, et al. Efficacy of epratuzumab, an anti-CD22 monoclonal IgG antibody, in systemic lupus erythematosus patients with associated Sjogren's syndrome: Post hoc analyses from the EMBODY Trials. Arthritis Rheumatol. 2018;70(5):763–73.
118. Li J, Wei MM, Song Q, Guo XH, Shao L, Liu Y. Anti-CD22 epratuzumab for systemic lupus erythematosus: A systematic review and meta-analysis of randomized controlled trials. Exp Ther Med. 2019;18(2):1500–6.
119. Kehrl JH, Riva A, Wilson GL, Thevenin C. Molecular mechanisms regulating CD19, CD20 and CD22 gene expression. Immunol Today. 1994;15(9):432–6.
120. Cree BAC, Bennett JL, Kim HJ, Weinshenker BG, Pittock SJ, Wingerchuk DM, et al. Inebilizumab for the treatment of neuromyelitis optica spectrum disorder (N-MOmentum): A double-blind, randomised placebo-controlled phase 2/3 trial. Lancet. 2019;394(10206):1352–63.
121. Gallagher S, Yusuf I, McCaughtry TM, Turman S, Sun H, Kolbeck R, et al. MEDI-551 treatment effectively depletes B cells and reduces serum titers of autoantibodies in mice transgenic for Sle1 and human CD19. Arthritis Rheumatol. 2016;68(4):965–76.
122. Horton HM, Chu SY, Ortiz EC, Pong E, Cemerski S, Leung IW, et al. Antibody-mediated coengagement of Fc gamma RIIb and B cell receptor complex suppresses humoral immunity in systemic lupus erythematosus. J Immunol. 2011;186(7):4223–33.
123. Zhao Q. Bispecific antibodies for autoimmune and inflammatory diseases: Clinical progress to date. BioDrugs. 2020;34(2):111–9.
124. Szili D, Cserhalmi M, Banko Z, Nagy G, Szymkowski DE, Sarmay G. Suppression of innate and adaptive B cell activation pathways by antibody coengagement of Fc gamma RIIb and CD19. MAbs. 2014;6(4):991–9.
125. Fervenza FC, Appel GB, Barbour SJ, Rovin BH, Lafayette RA, Aslam N, et al. Rituximab or cyclosporine in the treatment of membranous nephropathy. N Engl J Med. 2019;381(1):36–46.
126. Atisha-Fregoso Y, Malkiel S, Harris KM, Byron M, Ding L, Kanaparthi S, et al. Phase II randomized trial of rituximab plus cyclophosphamide followed by belimumab for the treatment of lupus nephritis. Arthritis Rheumatol. 2021;73(1):121–31.
127. Stone JH, Merkel PA, Spiera R, Seo P, Langford CA, Hoffman GS, et al. Rituximab versus cyclophosphamide for ANCA-associated vasculitis. N Engl J Med. 2010;363(3):221–32.
128. Jones RB, Furuta S, Tervaert JW, Hauser T, Luqmani R, Morgan MD, et al. Rituximab versus cyclophosphamide in ANCA-associated renal vasculitis: 2-year results of a randomised trial. Ann Rheum Dis. 2015;74(6):1178–82.
129. Mysler EF, Spindler AJ, Guzman R, Bijl M, Jayne D, Furie RA, et al. Efficacy and safety of ocrelizumab in active proliferative lupus nephritis: Results from a randomized, double-blind, phase III study. Arthritis Rheum. 2013;65(9):2368–79.
130. Marinov AD, Wang H, Bastacky SI, van Puijenbroek E, Schindler T, Speziale D, et al. The type II anti-CD20 antibody obinutuzumab (GA101) is more effective than rituximab at depleting B cells and treating disease in a murine lupus model. Arthritis Rheumatol. 2021;73(5):826–36.
131. Klomjit N, Fervenza FC, Zand L. Successful treatment of patients with refractory PLA2R-associated membranous nephropathy with obinutuzumab: A report of 3 cases. Am J Kidney Dis. 2020;76(6):883–8.
132. Sethi S, Kumar S, Lim K, Jordan SC. Obinutuzumab is effective for the treatment of refractory membranous nephropathy. Kidney Int Rep. 2020;5(9):1515–8.
133. Bag-Ozbek A, Hui-Yuen JS. Emerging B-cell therapies in systemic lupus erythematosus. Ther Clin Risk Manag. 2021; 17:39–54.
134. Clowse ME, Wallace DJ, Furie RA, Petri MA, Pike MC, Leszczynski P, et al. Efficacy and safety of epratuzumab in moderately to severely active systemic lupus erythematosus: Results from two phase III randomized, double-blind, placebo-controlled trials. Arthritis Rheumatol. 2017;69(2):362–75.
135. Tsuru T, Tanaka Y, Kishimoto M, Saito K, Yoshizawa S, Takasaki Y, et al. Safety, pharmacokinetics, and pharmacodynamics of epratuzumab in Japanese patients with moderate-to-severe systemic lupus erythematosus: Results from a phase 1/2 randomized study. Mod Rheumatol. 2016;26(1):87–93.
136. Oon S, Huq M, Godfrey T, Nikpour M. Systematic review, and meta-analysis of steroid-sparing effect, of biologic agents in randomized, placebo-controlled phase 3 trials for systemic lupus erythematosus. Semin Arthritis Rheum. 2018;48(2):221–39.
137. Rovin BH, Dooley MA, Radhakrishnan J, Ginzler EM, Forrester TD, Anderson PW. The impact of tabalumab on the kidney in systemic lupus erythematosus: Results from two phase 3 randomized, clinical trials. Lupus. 2016;25(14):1597–601.

138. Samy E, Wax S, Huard B, Hess H, Schneider P. Targeting BAFF and APRIL in systemic lupus erythematosus and other antibody-associated diseases. Int Rev Immunol. 2017;36(1):3–19.
139. Dhillon S. Telitacicept: First approval. Drugs. 2021;81(14):1671–1675.
140. Rayner F, Isaacs JD. Therapeutic tolerance in autoimmune disease. Semin Arthritis Rheum. 2018;48(3):558–62.
141. Parlevliet KJ, Schellekens PT. Monoclonal antibodies in renal transplantation: A review. Transpl Int. 1992;5(4):234–46.
142. Chow FY, Polkinghorne K, Saunder A, Kerr PG, Atkins RC, Chadban SJ. Historical controlled trial of OKT3 versus basiliximab induction therapy in simultaneous pancreas-renal transplantation. Nephrology (Carlton). 2003;8(4):212–6.
143. Bock HA, Gallati H, Zurcher RM, Bachofen M, Mihatsch MJ, Landmann J, et al. A randomized prospective trial of prophylactic immunosuppression with ATG-fresenius versus OKT3 after renal transplantation. Transplantation. 1995;59(6):830–40.
144. Jones-Hughes T, Snowsill T, Haasova M, Coelho H, Crathorne L, Cooper C, et al. Immunosuppressive therapy for kidney transplantation in adults: A systematic review and economic model. Health Technol Assess. 2016;20(62):1–594.
145. Sevmis S, Emiroglu R, Karakayali F, Yagmurdur MC, Dalgic A, Moray G, et al. OKT3 treatment for steroid-resistant acute rejection in kidney transplantation. Transplant Proc. 2005;37(7):3016–8.
146. Zabinska M, Krajewska M, Koscielska-Kasprzak K, Klinger M. CD3(+)CD8(+)CD28(-) T lymphocytes in patients with lupus nephritis. J Immunol Res. 2016;2016:1058165.
147. Zhao Y, Su H, Shen X, Du J, Zhang X, Zhao Y. The immunological function of CD52 and its targeting in organ transplantation. Inflamm Res. 2017;66(7):571–8.
148. Tan J, Yang S, Cai J, Guo J, Huang L, Wu Z, et al. Simultaneous islet and kidney transplantation in seven patients with type 1 diabetes and end-stage renal disease using a glucocorticoid-free immunosuppressive regimen with alemtuzumab induction. Diabetes. 2008;57(10):2666–71.
149. Baker D, Herrod SS, Alvarez-Gonzalez C, Giovannoni G, Schmierer K. Interpreting lymphocyte reconstitution data from the pivotal phase 3 trials of alemtuzumab. JAMA Neurol. 2017;74(8):961–9.
150. Machhi R, Mandelbrot DA, Al-Qaoud T, Astor BC, Parajuli S. Characteristics and graft survival of kidney transplant recipients with renal cell carcinoma. Am J Nephrol. 2020;51(10):777–85.
151. Vincenti F, Mendez R, Pescovitz M, Rajagopalan PR, Wilkinson AH, Butt K, et al. A phase I/II randomized open-label multicenter trial of efalizumab, a humanized anti-CD11a, anti-LFA-1 in renal transplantation. Am J Transplant. 2007;7(7):1770–7.
152. Tipping PG, Holdsworth SR. Cytokines in glomerulonephritis. Semin Nephrol. 2007;27(3):275–85.
153. Su H, Lei CT, Zhang C. Interleukin-6 signaling pathway and its role in kidney disease: An update. Front Immunol. 2017; 8:405.
154. Cash H, Relle M, Menke J, Brochhausen C, Jones SA, Topley N, et al. Interleukin 6 (IL-6) deficiency delays lupus nephritis in MRL-Faslpr mice: The IL-6 pathway as a new therapeutic target in treatment of autoimmune kidney disease in systemic lupus erythematosus. J Rheumatol. 2010;37(1):60–70.
155. Yao X, Huang J, Zhong H, Shen N, Faggioni R, Fung M, et al. Targeting interleukin-6 in inflammatory autoimmune diseases and cancers. Pharmacol Ther. 2014;141(2):125–39.
156. Jones SA, Fraser DJ, Fielding CA, Jones GW. Interleukin-6 in renal disease and therapy. Nephrol Dial Transplant. 2015;30(4):564–74.
157. Thanarajasingam U, Niewold TB. Sirukumab: A novel therapy for lupus nephritis? Expert Opin Investig Drugs. 2014;23(10):1449–55.
158. Rovin BH, van Vollenhoven RF, Aranow C, Wagner C, Gordon R, Zhuang Y, et al. A Multicenter, randomized, double-blind, placebo-controlled study to evaluate the efficacy and safety of treatment with sirukumab (CNTO 136) in patients with active lupus nephritis. Arthritis Rheumatol. 2016;68(9):2174–83.
159. Doberer K, Duerr M, Halloran PF, Eskandary F, Budde K, Regele H, et al. A randomized clinical trial of anti-IL-6 antibody clazakizumab in late antibody-mediated kidney transplant rejection. J Am Soc Nephrol. 2021;32(3):708–22.
160. Lin P. Targeting interleukin-6 for noninfectious uveitis. Clin Ophthalmol. 2015;9:1697–702.
161. Illei GG, Shirota Y, Yarboro CH, Daruwalla J, Tackey E, Takada K, et al. Tocilizumab in systemic lupus erythematosus: Data on safety, preliminary efficacy, and impact on circulating plasma cells from an open-label phase I dosage-escalation study. Arthritis Rheum. 2010;62(2):542–52.
162. Yamamura T, Kleiter I, Fujihara K, Palace J, Greenberg B, Zakrzewska-Pniewska B, et al. Trial of satralizumab in neuromyelitis optica spectrum disorder. N Engl J Med. 2019;381(22):2114–24.
163. Kerschbaumer A, Sepriano A, Smolen JS, van der Heijde D, Dougados M, van Vollenhoven R, et al. Efficacy of pharmacological treatment in rheumatoid arthritis: A systematic literature research informing the 2019 update of the EULAR recommendations for management of rheumatoid arthritis. Ann Rheum Dis. 2020;79(6):744–59.

164. Raimondo MG, Biggioggero M, Crotti C, Becciolini A, Favalli EG. Profile of sarilumab and its potential in the treatment of rheumatoid arthritis. Drug Des Devel Ther. 2017;11:1593–603.
165. Aggarwal BB. Signalling pathways of the TNF superfamily: A double-edged sword. Nat Rev Immunol. 2003;3(9):745–56.
166. Brenner D, Blaser H, Mak TW. Regulation of tumour necrosis factor signalling: Live or let die. Nat Rev Immunol. 2015;15(6):362–74.
167. Kerdel FA, Strober BE. Tumor necrosis factor inhibitors in psoriasis: An update. Semin Cutan Med Surg. 2014;33(2 Suppl 2):S31–6.
168. Hemperly A, Vande Casteele N. Clinical pharmacokinetics and pharmacodynamics of infliximab in the treatment of inflammatory bowel disease. Clin Pharmacokinet. 2018;57(8):929–42.
169. Subedi S, Gong Y, Chen Y, Shi Y. Infliximab and biosimilar infliximab in psoriasis: Efficacy, loss of efficacy, and adverse events. Drug Des Devel Ther. 2019;13:2491–502.
170. Malaise O, von Frenckell C, Andre B, Malaise MG. The management of systemic lupus erythematosus with biological therapies. Rev Med Suisse. 2013;9(395):1507–11.
171. Booth A, Harper L, Hammad T, Bacon P, Griffith M, Levy J, et al. Prospective study of TNF alpha blockade with infliximab in anti-neutrophil cytoplasmic antibody-associated systemic vasculitis. J Am Soc Nephrol. 2004;15(3):717–21.
172. Micheloud D, Nuno L, Rodriguez-Mahou M, Sanchez-Ramon S, Ortega MC, Aguaron A, et al. Efficacy and safety of etanercept, high-dose intravenous gammaglobulin and plasmapheresis combined therapy for lupus diffuse proliferative nephritis complicating pregnancy. Lupus. 2006;15(12):881–5.
173. Liu Y, Shi Y, Ren R, Xie J, Wang W, Chen N. Advanced therapeutics in focal and segmental glomerulosclerosis. Nephrology (Carlton). 2018;23(Suppl 4):57–61.
174. Ramos-Casals M, Brito-Zeron P, Munoz S, Soria N, Galiana D, Bertolaccini L, et al. Autoimmune diseases induced by TNF-targeted therapies: Analysis of 233 cases. Medicine (Baltimore). 2007;86(4):242–51.
175. Shirota Y, Yarboro C, Fischer R, Pham TH, Lipsky P, Illei GG. Impact of anti-interleukin-6 receptor blockade on circulating T and B cell subsets in patients with systemic lupus erythematosus. Ann Rheum Dis. 2013;72(1):118–28.
176. Karkhur S, Hasanreisoglu M, Vigil E, Halim MS, Hassan M, Plaza C, et al. Interleukin-6 inhibition in the management of non-infectious uveitis and beyond. J Ophthalmic Inflamm Infect. 2019;9(1):17.
177. Furie R, Toder K, Zapantis E. Lessons learned from the clinical trials of novel biologics and small molecules in lupus nephritis. Semin Nephrol. 2015;35(5):509–20.
178. Hayat SJ, Uppal SS, Narayanan Nampoory MR, Johny KV, Gupta R, Al-Oun M. Safety and efficacy of infliximab in a patient with active WHO class IV lupus nephritis. Clin Rheumatol. 2007;26(6):973–5.
179. Joy MS, Gipson DS, Powell L, MacHardy J, Jennette JC, Vento S, et al. Phase 1 trial of adalimumab in focal segmental glomerulosclerosis (FSGS): II. Report of the FONT (novel therapies for resistant FSGS) study group. Am J Kidney Dis. 2010;55(1):50–60.
180. Trachtman H, Vento S, Herreshoff E, Radeva M, Gassman J, Stein DT, et al. Efficacy of galactose and adalimumab in patients with resistant focal segmental glomerulosclerosis: Report of the font clinical trial group. BMC Nephrol. 2015;16:111.
181. Michaelson JS, Wisniacki N, Burkly LC, Putterman C. Role of TWEAK in lupus nephritis: A bench-tobedside review. J Autoimmun. 2012;39(3):130–42.
182. Wisniacki N, Amaravadi L, Galluppi GR, Zheng TS, Zhang R, Kong J, et al. Safety, tolerability, pharmacokinetics, and pharmacodynamics of anti-TWEAK monoclonal antibody in patients with rheumatoid arthritis. Clin Ther. 2013;35(8):1137–49.
183. van Vollenhoven RF, Hahn BH, Tsokos GC, Wagner CL, Lipsky P, Touma Z, et al. Efficacy and safety of ustekinumab, an IL-12 and IL-23 inhibitor, in patients with active systemic lupus erythematosus: Results of a multicentre, double-blind, phase 2, randomised, controlled study. Lancet. 2018;392(10155):1330–9.
184. Khamashta M, Merrill JT, Werth VP, Furie R, Kalunian K, Illei GG, et al. Sifalimumab, an anti-interferon-alpha monoclonal antibody, in moderate to severe systemic lupus erythematosus: A randomised, double-blind, placebo-controlled study. Ann Rheum Dis. 2016;75(11):1909–16.
185. Kalunian KC, Merrill JT, Maciuca R, McBride JM, Townsend MJ, Wei X, et al. A Phase II study of the efficacy and safety of rontalizumab (rhuMAb interferon-alpha) in patients with systemic lupus erythematosus (ROSE). Ann Rheum Dis. 2016;75(1):196–202.
186. Boedigheimer MJ, Martin DA, Amoura Z, Sanchez-Guerrero J, Romero-Diaz J, Kivitz A, et al. Safety, pharmacokinetics and pharmacodynamics of AMG 811, an anti-interferon-gamma monoclonal antibody, in SLE subjects without or with lupus nephritis. Lupus Sci Med. 2017;4(1):e000226.
187. Costa R, Antunes P, Salvador P, Oliveira P, Marinho A. Secukinumab on refractory lupus nephritis. Cureus. 2021;13(8):e17198.

188. Satoh Y, Nakano K, Yoshinari H, Nakayamada S, Iwata S, Kubo S, et al. A case of refractory lupus nephritis complicated by psoriasis vulgaris that was controlled with secukinumab. Lupus. 2018;27(7):1202–6.
189. Thurman JM, Yapa R. Complement therapeutics in autoimmune disease. Front Immunol. 2019;10:672.
190. Vignesh P, Rawat A, Sharma M, Singh S. Complement in autoimmune diseases. Clin Chim Acta. 2017;465:123–30.
191. Akaishi T, Nakashima I. Efficiency of antibody therapy in demyelinating diseases. Int Immunol. 2017;29(7):327–35.
192. Alabbad S, AlGaeed M, Sikorski P, Kaminski HJ. Monoclonal antibody-based therapies for myasthenia gravis. BioDrugs. 2020;34(5):557–66.
193. Bomback AS, Smith RJ, Barile GR, Zhang Y, Heher EC, Herlitz L, et al. Eculizumab for dense deposit disease and C3 glomerulonephritis. Clin J Am Soc Nephrol. 2012;7(5):748–56.
194. Hohenstein B, Amann K, Menne J. Membranoproliferative glomerulonephritis and C3 glomerulopathy. Internist (Berl). 2019;60(5):458–67.
195. Herlitz LC, Bomback AS, Markowitz GS, Stokes MB, Smith RN, Colvin RB, et al. Pathology after eculizumab in dense deposit disease and C3 GN. J Am Soc Nephrol. 2012;23(7):1229–37.
196. Jordan SC, Kucher K, Bagger M, Hockey HU, Wagner K, Ammerman N, et al. Intravenous immunoglobulin significantly reduces exposure of concomitantly administered anti-C5 monoclonal antibody tesidolumab. Am J Transplant. 2020;20(9):2581–8.
197. Reddy YN, Siedlecki AM, Francis JM. Breaking down the complement system: A review and update on novel therapies. Curr Opin Nephrol Hypertens. 2017;26(2):123–8.
198. Sethu S, Govindappa K, Alhaidari M, Pirmohamed M, Park K, Sathish J. Immunogenicity to biologics: Mechanisms, prediction and reduction. Arch Immunol Ther Exp (Warsz). 2012;60(5):331–44.
199. Rovin BH, Klein JB. Proteomics and autoimmune kidney disease. Clin Immunol. 2015;161(1):23–30.
200. Stewart BJ, Ferdinand JR, Clatworthy MR. Using single-cell technologies to map the human immune system – implications for nephrology. Nat Rev Nephrol. 2020;16(2):112–28.
201. Brodin P, Davis MM. Human immune system variation. Nat Rev Immunol. 2017;17(1):21–9.
202. Perez-De-Lis M, Retamozo S, Flores-Chavez A, Kostov B, Perez-Alvarez R, Brito-Zeron P, et al. Autoimmune diseases induced by biological agents. A review of 12,731 cases (BIOGEAS registry). Expert Opin Drug Saf. 2017;16(11):1255–71.

21 Therapeutics of Recombinant Human Clotting Factor VIII (rhFVIII) for Hemophilia A

Bang-Shun He and Meng-Qiu Xiong

Corresponding author: Bang-Shun He

Contents

21.1 INTRODUCTION

Hemophilia is an inherited disease characterized by a bleeding disorder that is caused by the deficiency of coagulation factors VIII (FVIII) and IX (FIX). In clinical settings, hemophilia is classified as two subtypes, one is hemophilia A, which results from the absence or ineffectiveness of FVIII, and the other is hemophilia B, which is due to inadequate or ineffective FIX. Genetically, both hemophilia A and B are X-linked, and therefore, these two inherited diseases are almost manifested in males relative to females, who are rarely severely affected. Actually, an estimated one-third of the newly diagnosed cases of the X-linked disorder with no family history demonstrated the presence of hemophilia due to the occurrence of spontaneous or

DOI: 10.1201/9780429485626-21

sporadic mutations (Mirchandani et al. 2011; Myrin-Westesson, Baghaei, and Friberg 2013). Hemophilia A affects about 1 in 5,000–10,000 males, while hemophilia B affects about 1 in 40,000 males at birth. Thus, the prevalence of hemophilia A is about three-fold more frequent than that of hemophilia B.

Signs and symptoms of hemophilia vary with severity, depending on the circulating levels of clotting factors. In clinical settings, patients with less than 1% active factors are classified as severe hemophilia, those with 1%–5% active factor demonstrate moderate hemophilia, and those with between 5% and 40% of normal levels of active clotting factor are defined as mild hemophilia. Cases with mild hemophilia may experience bleeding only after trauma or surgery happens. By contrast, patients with moderate hemophilia occasionally have spontaneous hemorrhages those with severe hemophilia suffer spontaneous hemorrhages. Characteristic symptoms of patients with mild to moderate hemophilia may show internal or external bleeding episodes, such as unexplained and excessive bleeding from cuts or injuries, many large or deep bruises, unusual bleeding after vaccinations, pain, swelling or tightness in joints, blood in urine or stool, frequent nosebleeds, and more.

Aside from hemophilia caused by genetic changes, there is also a bleeding disorder named acquired hemophilia (AH), which is not inherited but develops suddenly at some times, such as autoimmune disorders, malignant disease, use of certain drugs, and pregnancy. However, approximately half of cases have no known causes. For the potential risk factors, most cases have immune system disorder with producing antibodies that mistakenly attack clotting factor. Therefore, AH is an autoimmune disorder, and approximately half of the patients having coexisting disorders or conditions, including autoimmune disorders, such as rheumatoid arthritis and lupus. Different from hemophilia caused by genetic change, AH occurs in only 1–1.5 per 1 million people per year (Collins et al. 2004), affecting both men and women, in particular the elderly whose median age of onset is approximately 76.1 ± 7.2 years old (Godaert et al. 2018).

The majority of patients with hemophilia have a known family history. However, about one-third of cases occur without a known family history. Hemophilia is diagnosed with blood tests to determine the level of clotting factors. Note that before choosing an appropriate screening test, the detailed demographic and clinical characteristics of patients should be collected, including sex, age at the first time of hemorrhagic episode, the type of current and past bleeding episodes, history of bleeding in the family, and consanguineous marriage status of the parents. The screening tests are recommended as follows: complete blood count (CBC), activated partial thromboplastin time (APTT), prothrombin time (PT), and fibrinogen testing. To further determine the type and severity of hemophilia, determination of the type and the levels of clotting factors is required for patient care. Genetic testing is currently available for the clotting factor genes to determine the causes of hemophilia. Disease-causing mutations are identified in up to 98% of cases who suffer hemophilia. Thus, prenatal diagnosis is aimed to determine whether a mother is a carrier of all known mutations present in the *FVIII* or *FIX* gene. In some cases, it is used to diagnose individuals who have mild symptoms of hemophilia.

21.2 THE HISTORY OF HEMOPHILIA TREATMENT

Until the early 1960s, severe hemophilia was a life-threatening disease or resulted in chronic disability due to its serious complications, including intracranial hemorrhage, repeated musculoskeletal bleeding (Srivastava et al. 2013), and the treatments were limited to whole blood transfusions. In the mid-1960s, the treatment was improved with the preparation of cryoprecipitated plasma containing factor VIII from fresh frozen plasma, which was a great advancement for allowing intravenous administration of more factor VIII in a smaller volume. Subsequently, the freeze-dried (lyophilized) factor VIII or factor IX concentrates were developed by the methods of separating factor VIII and IX from pooled plasma.

The availability of plasma-derived components (fresh frozen plasma, cryoprecipitated, and lyophilized) made the replacement therapy of hemorrhage more specific and effective. Thus, the concentrated clotting factor rapidly became the preferred treatment option for acute bleeding episodes in hemophilia patients, and the availability of these concentrates paves the way for home treatment, which greatly improves the quality of life of hemophilia patients. Further, in clinical practice, the comorbidity of hemophilia, such as the chronic arthropathy of major joints, was proved to be prevented by early prophylaxis.

Consequently, the on-demand treatment was replaced gradually by prophylaxis for reducing joint bleedings (Manco-Johnson et al. 2007; Gringeri et al. 2011).

Unfortunately, in the 1980–1990s, the replace treatment of hemophilia with plasma-derived concentrates resulted in dramatically increased infections, such as human immunodeficiency virus (HIV), and hepatitis A (HAV), B (HBV), or C (HCV), all of which were transmitted by factor concentrates manufactured from pooled plasma from donors, with thousands of hemophilia patients died of acquired infections, especially HIV (Lee 2009). Subsequently, to avoid blood-borne transmission of hepatitis viruses or HIV, screening of viruses in blood to be donated (i.e., NAT, nucleic acid amplification testing) were applied. Therefore, viral inactivation techniques have been used to improve the safety of plasma-derived products when plasma-derived factor concentrates were manufactured. Since then, no blood-borne transmission of hepatitis viruses or HIV has occurred for the past two decades or so (Mannucci 2008; Franchini and Mannucci 2012).

With the development of recombinant DNA technology, recombinant human clotting factors have been developed as a safer alternative to plasma-derived products since the *FVIII* and *FIX*-coding genes were cloned in 1982 and 1984, respectively, which paved the way to industrial production of recombinant coagulation factors by mammalian cell cultures (Pipe 2008; Franchini and Lippi 2010; Monahan and Di Paola 2010). In 1989, the clinical efficacy of recombinant FVIII was firstly reported in two patients with hemophilia A, and recombinant FIX concentrate was commercially available in 1998 (Franchini et al. 2013). The production of recombinant coagulation made replacement therapy for hemophilia safe and more widely available without concerns of product safety.

21.3 CHALLENGES OF HEMOPHILIA TREATMENT: THE DEVELOPMENT OF INHIBITOR DURING THE PROCESS OF THE TREATMENT

Regardless of the use of on-demanded therapy or home prophylaxis, the formation of neutralizing antibodies and the occurrence of inhibitors induced by replacement treatment with plasma derived or recombinant coagulation factors make factor concentrate replacement ineffective for the treatment or prevention of bleeds. In clinical settings, inhibitor remains a challenging complication of hemophilia treatment. Actually, the development of inhibitor is a common problem in hemophilia patients, more frequently occurring in hemophilia A (severe hemophilia A cases: overall lifelong risk of 25%–40% (Iorio, Fischer, and Makris 2017); mild to moderate hemophilia A: 5%–15%) than in hemophilia B (Dimichele 2002, 2007; Eckhardt et al. 2013).

Inhibitors are most often induced during the first 20 exposure days (EDs) to factor replacement, and subsequently, the risk decreases dramatically, particularly from 20 to 50 ED. Thereafter (50–150 ED), the risk at the steady state is approximately 2–5 per 1,000 patients per year (Xuan et al. 2014). In hemophilia A patients, FVIII inhibitors are mainly IgG, predominantly IgG1, and IgG4 subclasses. IgG4 is mainly in patients with high-titer inhibitors (HTI, >5 BU/mL), whereas IgG1 is more abundant in patients with low-titer inhibitors (LTI, <5 BU/mL). Moreover, inhibitors to FVIII may also be found in patients with acquired hemophilia as auto-antibodies.

21.4 EMERGING DRUGS FOR THE TREATMENT OF HEMOPHILIA

Although the availability of the plasma derived or recombinant components make the on-demand treatment feasible, greatly extending the lives of patients with hemophilia, the replacement treatment for hemophilia is not always practicable worldwide because of its high cost especially in developing countries. Actually,

TABLE 21.1 Biological products that are indicated for hemophilia A

PRODUCT	*FACTOR*	*TECHNOLOGY USED*	*REFERENCES*
Eloctate®, Biogen Idec	FVIII	Fusion protein with the Fc fragment of IgG1 (rFVIII-Fc)	Powell et al. (2012); Mahlangu et al. (2014)
Adynovate®, Baxalta	FVIII	Random PEGylation (BAX 855; 20 kDa)	Mullins et al. (2017); Brand et al. (2016)
NN7088, Novo Nordisk	FVIII	Site-specific glycoPEGylation (N8-GP; 40 kDa)	Tiede et al. (2013); Giangrande et al. (2017)
BAY 94-9027, Bayer	FVIII	Site-specific PEGylation (K1804C PEGylation; 60 kDa)	Coyle et al. (2014); Santagostino et al. (2020)
Afstyla®, CSL Behring	FVIII	Single chain rFVIII (CSL627)	Mahlangu et al. (2016)
Kowaltry®, Bayer Iblias®, CSL-Behring	FVIII	Full-length rFVIII product (BAY 81-873)	Kavakli et al. (2015)
Nuwiq®, Octapharma	FVIII	B-domain deleted (Simoctocogalfa)	Lissitchkov et al. (2016); Klukowska et al. (2016)
Alprolix®, Biogen Idec	FIX	Fusion protein with the Fc fragment of IgG1	Fischer et al. (2017); Pasi et al. (2017)
delvion®, CSL-Behring	FIX	Fusion protein with albumin	Kenet et al. (2016); Negrier et al. (2016)
N9-GP, Novo Nordisk	FIX	Site-specific glycoPGEylation	Collins et al. (2014); Carcao et al. (2016)
IXinity®, Cangene	FIX	comparable to the Thr148 allelic form of FIX	Srivastava et al. (2013); Srivastava et al. (2013)
Rixubis®, Baxalta	FIX	Recombinant DNA technology	Rixubis – FDA prescribing information, side effects, and uses (Drugs.com 2020)

about 70%–80% of patients with hemophilia worldwide are undertreated or not treated at all (Alzoebie et al. 2013). Currently, several commercial coagulation factor products have been available for clinical use. Plasma-derived products are wildly used in clinical settings (Mannucci 2008). On the other hand, with the rapid progress made in recombinant DNA technology, the recombinant human concentrates have been developed as a safer alternative means than plasma-derived products since the FVIII-coding gene was cloned in 1984 (Gitschier et al. 1984; Toole et al. 1984; Vehar et al. 1984) and the first clinical research study about the efficacy of the first-generation full-length rhFVIII product was reported (White et al. 1989). The first- and second-generation products were prepared in a mammalian cell system, some of which contained human plasma proteins due to the use of human- or animal-derived materials, and therefore the risk of transmission of pathogenic agents still exists, in particular previously unknown pathogens (Meeks and Josephson 2006). In contrast, the third-generation products do not use any additional animal or human proteins in the process of manufacturing, almost eliminating all potential risks of transmission of blood-borne pathogens (Franchini and Lippi 2010). Accordingly, recombinant concentrates have been proved to be safer alternatives to plasma-derived products. Types of hemophilia drugs were developed for clinical treatment (see Table 21.1).

21.5 THE THIRD-GENERATION RECOMBINANT ADVATE

The plasma- and albumin-free, octocogalfa antihemophilic factor (trade name: Advate, rAHF-PFM) is a typical third-generation recombinant human full-length coagulation FVIII. Here, Advate is described as an example of hemophilia drugs. rAHF-PFM is a product without any human or animal blood derived additives throughout cell culture and formulation. rAHF-PFM, a dimeric glycoprotein that constitutes 2,332 amino acid residues with a molecular mass of approximately 280 kDa and has a similar amino acid sequence to that of human plasma derived FVIII. It is secreted by the CHO cells transfected with the *FVIII* gene and is purified from the culture medium by preparation chromatography (Ding et al. 2014). Further studies have documented that lyophilized rAHF-PFM is stable under the different conditions (e.g., 92% residual factor VIII activity at 5°C for 30 months, 80% residual activity at room temperature for 18 months, and 84% residual activity at 40°C for 3 months) (Parti et al. 2005) and is also stable during the period of continuous infusion (Fernandez et al. 2006). In patient care, lyophilized rAHF-PFM is extensively used for the prevention and control of hemorrhagic episodes and perioperative management in patients with hemophilia A.

The pharmacokinetic profile of rAHF-PFM is similar to that of recombinant (a first-generation rFVIII antihemophilic factor) (Shapiro 2007), as estimated with the area under the plasma concentration versus time curve (AUC), adjusted recovery, plasma half-life ($t_{½}$), peak drug plasma concentrations (C_{max}), mean residence time (MRT), and volume distribution at the steady-state (Vss) (Ding et al. 2014; Tarantino et al. 2004b). Further clinical studies have revealed that AUC, incremental recovery, and terminal phase $t_{½}$ are lower and clearance is higher in infants and children than in adolescents and adults, as summarized in Table 21.2 (Tarantino et al. 2004b; Blanchette et al. 2008; Collins et al. 2009; Bjorkman et al. 2010), and that the intra-patient variability of rAHF-PFM pharmacokinetics is uniformly less than the inter-patient variability by comparison with short- and long-term exposure (Bjorkman et al. 2010). Patients receiving the same dosage of rAHF-PFM may have varying levels of FVIII. For infants or young children, the elimination rate of rFVIII ($t_{½}$ or clearance) is associated with the annual incidence of all joint bleeds (Collins et al. 2009); moreover, $t_{½}$ and the intervals of dosing had larger effects on FVIII trough concentrations (C_{min}) and time per week with FVIII < 1 IU/dL than *in vivo* recoveries (IVR) and infused dose per kg body weight. In addition, the relationship between the pharmacokinetics of rAHF-PFM and body weight is thought to be useful for the dose adjustments of rAHF-PFM to achieve adequate rAHF-PFM target levels (Bjorkman et al. 2012). Recently, Valentino et al. reported that BAX 855, a PEGylated form of unmodified rAHF-PFM, provided a longer duration of protection from hemarthrosis than did pretreatment with unmodified rAHF-PFM by improving the pharmacokinetic and pharmacodynamics properties (Valentino et al. 2015).

As for the application of rAHF-PFM for the prophylaxis and treatment of bleeding episodes in patients with moderate or severe hemophilia A, accumulated data have suggested the efficacy of rAHF-PFM. Five prospective clinical research studies on the efficacy of rAHF-PFM were pooled to show that 88% of 1,724 bleeding episodes were categorized as "excellent/good" and that 90% of bleeding episodes were managed with one or two infusions. In addition, patients who received prophylactic therapy and complied with at least the minimum prescribed doses and infusion intervals had a significantly lower frequency of bleeding episodes than those who were non-compliant to the therapy (Shapiro et al. 2009); moreover, a post-analysis has also revealed that compared with on-demand therapy, both standard factor VIIII prophylaxis and pharmacokinetic-tailored prophylaxis with FVIII significantly reduced annualized joint bleeding rate (Valentino et al. 2014). In addition, a meta-analysis based on 120 patients revealed overall median annualized bleeding rate (ABR) was 2.0 (Romanov et al. 2015). All the released studies indicated that rAHF-PFM has acceptable effectiveness for on-demanded or prophylaxis treatment for hemophilia A in adults and children.

The safety of rAHF-PFM is also concerned in its clinical applications. Adverse events (AEs) and inhibitors are the two major aspects of its safety profiling. An updated study integrated analyses of 12 clinical interventional studies of rAHF-PFM used for hemophilia A and showed that 93 AEs were reported to be related to the use of rAHF-PFM in 45 of 418 patients (10.8%), of which most AEs were reported as FVIII inhibitors, pyrexia, and headache, and that 81.7% of the AEs were considered mild or moderated (Shapiro et al. 2015). Moreover, another study based on a post-authorization safety study (PASS) global program has also demonstrated that 5 of 83 treatment-related AEs were reported to be serious AEs (with the exception of FVIII inhibitor development) in five patients (5/1188) (Iorio et al. 2014). These integrated studies showed the rarity of AEs after the use of rAHF-PFM.

For patients with hemophilia A, the development of inhibitors remains the most serious complications that are frequently triggered by the treatment with FVIII replacement therapy. As reported, inhibitors

TABLE 21.2 Pharmacokinetic parameters of Advate in patients after receiving a single infusion of 50 IU/kg

AGE (YRS)	*N*	*AUC(0–48 h) (IU/h/dL)*	*RECOVERY (IU/dL/kg)*	*HALF-LIFE (h)*	C_{max} *(IU/dL)*	*MRT (h)*	*Vss (dL/kg)*	*CL (dL/h/kg)*	
0.08–2	8	1217 (312)	2.11 (0.51)	8.59 (1.34)	105 (26)	–	0.43 (0.09)	0.043 (0.009)	Shapiroet al. (2009)
2–12	55	1254 (468)	1.96 (0.48)	9.95 (1.91)	99 (27)	–	0.53 (0.12)	0.045 (0.015)	
12–16	28	1410 (527)	2.29 (0.57)	12.27 (3.61)	115 (29)	–	0.55 (0.11)	0.040 (0.014)	
>16	76	1717 (497)	2.54 (0.60)	12.16 (3.02)	128 (30)	–	0.48 (0.10)	0.032 (0.010)	
<6 (3.1 ± 1.5)	52	1208 (391)	1.88 (0.42)	9.71 (1.89)	95.0 (22.4)	12.2 (3.1)	51.4 (12.3)	0.044 (0.014)	Blanchette et al. (2008)
	47	1236 (401)	1.90 (0.43)	9.88 (1.89)	95.6 (23.3)	12.5 (3.1)	51.4 (12.9)	0.043 (0.014)	
35.8 (19–72)	17	1380 (420)[a]	110.4 (25.3)	13.6 (3.8)	118 (25)	16.3 (5.0)	0.617 (0.018)	0.040 (0.014)	Di Paola et al. (2007)
18 (10–65)	30	1534(436)	2.4 (0.5)	12.0 (4.3)	120 (26)	15.69 (6.21)	0.47 (0.10)	0.03 (0.01)	Tarantino et al. (2004a)
—	30	1534(436)	2.41 (0.50)	11.98 (4.28)	120 (26)	15.68(6.21)	0.47 (0.10)	0.03 (0.01)	Product information[b]

Abbreviations: AUC: area under the plasma drug concentration-time curve; yrs: years; Advate: PI 08022013; CL: clearance; MRT: mean residence time; Vss: volume of distribution at steady state; C_{max}: peak plasma drug concentration occur in approximately 30% of individuals with severe hemophilia and increase the risk of uncontrollable bleeding and disability, making the treatment of bleeding episodes more difficult (Ingerslev et al. 1996; Lusher 2000; Leissinger 2004). Regardless of the presence of little or no amount of any animal- or human-derived proteins during the process of production, FVIII inhibitors would be produced in patients with hemophilia A treated with recombinant FVIII (Franchini et al. 2012). For rAHF-PFM, Tarantino et al. reported that approximately 1 of 108 previously treated patients was inhibitor-positive after 26-day exposure (Tarantino et al. 2004b). In contrast, Guenter et al. reported that the inhibitor was developed in 16 of 55 (29.1%) patients who received at least one infusion of rAHF-PFM, and that non-Caucasian ethnicity and high-intensity treatment were associated with a high risk of inhibitor development (Auerswald et al. 2012). Recently, an integrated analysis of safety data from 12 clinical interventional studies of rAHF-PFM for hemophilia A was reported, documenting that 4.06% (17/418) patients developed inhibitor (Shapiro et al. 2015), consistent with the results reported as 2% (21/1188) derived from a meta-analysis of Advate-PASS studies. In addition, none of 219 patients with low-titer inhibitors or inhibitor history were found positive for a high-titer inhibitor during the period of study (an average of 196 days) (Romanov et al. 2015). Furthermore, in another study of 12 patients with high-titer inhibitors who were enrolled for received immune tolerance induction (ITI) by rAHF-PFM, a high incidence (75%) of tolerance for inhibitors was achieved, including 7 of 10 patients (70%) with high-titer inhibitors (Valentino et al. 2009). As summarized in Table 21.3, the accumulated published data shown that rAHF-PFM is safe FVIII for hemophilia A in patient care.

TABLE 21.3 The occurrence of inhibitors in patients receiving rAHF-PFM

CHARACTERISTICS OF PATIENTS	*EXPOSURE DAYS (MEDIAN)*	*NO. OF INHIBITOR TESTS (MEDIAN)*	*REFERENCES*
418 patients (median aged 18.7 yrs) from 12 interventional studies with FVIII levels <=2% of normal, including 55 PUPs/MTPs from all rAHF-PFM phase I–IV studies.	97.0 (1–709)	PTP: 1/127 (0.36%) PUP/MTP: 16/55 (29.1%)	Shapiro et al. (2015)[a]
152 patients, 69 % had severe HA, 116 (76.3%) patients aged ≧16 yrs.	116 (1–642)	1/144 (0.69 %)	Pollmann et al. (2013)
66 patients from 24 international sites, 55 (18 PUPs and 37 MTPs) received at least one infusion of rAHF-PFM.	498 (82–1360)	16/55 (29.1%)	Auerswald et al. (2012)
58 patients, aged 7–53 (mean 24); weighing 47–75 (mean: 59.1) kg. Severe: 8 (13.8%), moderately severe: 31 (53.5%), moderate: 14 (21.14%), HA: 5 (8.62%)	6 mo	1/54 (1.85%)	Zhang et al. (2011)
113 patients (>18 yrs:76; <18 yrs: 37; mild: 8, moderate: 4, severe: 101; prophylaxis:71, OD: 42).	Most (>85%) >100 EDs	1/113 (0.88%)	Bacon et al. (2011)
234 patients (FVIII: C ≤ 2%) (median aged 14.7, range: 0.02–72.7 yrs).	178 (10–598)	1/198 (0.51%)	Shapiro et al. (2009)[b]
58 patients with age ≥5 years, baseline FVIII:C ≤ 2%, and ≥150 prior FVIII Eds.	2 wk	0/65 (0)	Negrier et al. (2008)
53 patients, aged 3.1 ± 1.5 yrs (<3 yrs: 24; 3–5 yrs: 29) and 50 prior EDs; patients for OD: 5 (9.4%), prophylaxis: 39 (73.6%), both: 9 (17.0%).	156 (14–384)	0/53 (0)	Blanchette et al. (2008)
Median aged 18 yrs, 96% of patients had baseline factor VIII <1%, and 108 received rAHF-PFM.	117	1/108 (0.93%)	Tarantino et al. (2004a)

Abbreviations: HA, hemophilia A; PUP, previously untreated patient; IU, international unit; MTP, minimally treated patient; OD, on-demand; BI, bolus infusion; yr, years; mo, months; wk, weeks

[a] Part of data were from Auerswald *et al.* (2012), Shapiro *et al.* (2009), Blanchette *et al.* (2008), and Tarantino *et al.* (2004b).

[b] Part of data were from Tarantino *et al.* (2004), Blanchette *et al.* (2008), and Negrier *et al.* (2008).

21.6 CURRENT STATUS OF THE USE OF ADVATE IN CHINA

Hemophilia care in China has made rapid progress with the fast development of the national economy and further improvement of healthcare conditions in the recent two decades. However, fundamental changes began in 1993 with the involvement of WFH to help set up a foundation of comprehensive care in China. In 2004, under the help of WFH, the Hemophilia Treatment Centers Collaborative Network of China (HTCCNC), composed of six centers (Beijing, Tianjin, Jinan, Hefei, Shanghai, and Guangzhou), was found to develop special strategies for improving hemophilia care, such as registration in Tianjin, nursing and prophylaxis in Guangzhou, laboratory diagnosis in Shanghai, and physiotherapy in Beijing. To date, the health care for hemophilia A in China has made greater achievements in the education for hemophilia A care (Chen et al. 2014; Sun et al. 2014), laboratory diagnosis (Hua et al. 2010; Dai et al. 2012; Ding et al. 2012; He et al. 2013), and prenatal diagnosis (Dai et al. 2012), second aryprophylaxis (Wu et al. 2011), and comprehensive therapy (Xuan et al. 2013).

Nowadays, like most other countries, replacement therapy of plasma-derived and rFVIII concentrates has been widely applied for patients with hemophilia A in China. However, plasma-derived FVIII concentrates produced by China's companies possessed the major share of the marketed products, and patients with hemophilia A may be at increased risk of blood-borne infections. A retrospective study of Chinese patient cohort with hemophilia revealed that 12.62% of 926 patients were infected with HCV, and 0.22% HIV-positive (Xuan et al. 2013), 4.5%–10.4% HBsAg-positive, and 39.6%–45.5% HCV-Ab-positive (Shi et al. 2007; Zhang et al. 2011), rFVIII with lower risk of transmission of blood-borne infections has been accepted as the best choice for replacement therapy. Up to date, the rFVIII products that will be used in clinical settings in China are largely imported after official approval of the three categories of rFVIII (Kogenate FS®, in 2007; Advate® and Xyntha® in 2012). Currently, there are two available clinical research studies on rAHF-PFM in Chinese patients (Zhang et al. 2011; Xuan et al. 2013) due to the short history of use of rAHF-PFM. A multicenter prospective clinical study was reported, evaluating the efficacy, safety, and immunogenicity of rAHF-PFM in patients with hemophilia A (Zhang et al. 2011), in which 58 patients (8 severe, 45 moderate, and 5 mild) were enrolled for treatment of 6 months, and the response to the first rAHF-PFM treatment was grouped as either "excellent" (82.8%) or "improved" (17.2%) in all subjects, with an inhibitor of four units measured in one patient at clinic visit on day 180 after discharge. A retrospective analysis of 1,226 Chinese patients with hemophilia reported that, of 102 patients treated with the third-generation rFVIII (Xyntha® or Advate®), inhibitors were measured in three patients (two from Advate® and one from both Advate® and Xyntha®) after rAHF-PFM treatment. In contrast, after 110 patients were treated with both plasma-derived FVIII and rFVIII, 8 were inhibitor-positive, but only one was derived from rAHF-PFM. The incidence of the inhibitors was ranked in an ascending order as follows: 2.9% for the third-generation FVIII (Advate® and Xyntha®), 10.5% for the first-generation FVIII, 11.0% for the second-generation FVIII, and 14.3% for plasma-derived FVIII. In general, the use of rAHF-PFM in Chinese patients with hemophilia A is well tolerated for the excellent safety and efficacy and low rate of inhibitor positive (Xuan et al. 2014).

21.7 THE FUTURE DIRECTIONS OF HEMOPHILIA A TREATMENT

Hemophilia A is a genetic deficiency in clotting factor VIII. Currently, the alternative treatment with FVIII is the major treatment of choice for hemophilia A. Gene therapy could be an ideal treatment for this disease ultimately. In the past decades, clinical trials based on gene therapy have achieved a goal in the production of the active factor VIII mediated by the virus vector (Nathwani, Davidoff, and Tuddenham 2017). In the future, the development of genome editing technology (Ormond et al. 2017), in particular the improvement of precise genome editing, will provide a useful method for hemophilia A gene therapy (Barbieri et al. 2017), which will make a new era for hemophilia A patients.

In conclusion, rAHF-PFM has been shown to be safe and effective for the prevention and treatment of bleeding episodes and perioperative management inpatient with hemophilia A. As a prophylactic treatment regimen, rAHF-PFM is more effective in preventing bleeding episodes than on-demand therapy. In China, rAHF-PFM has been officially approved for patients with hemophilia A for two years, according to its clinical safety and effectiveness in patient care.

FINANCIAL DISCLOSURE

This work was supported in part by grants from the National Natural Science Foundation of China (81200401), Jiangsu 333 Talent training program, and Jiangsu Medical Young Investigator training program (QNRC2016066).

REFERENCES

Alzoebie, A., M. Belhani, P. Eshghi, et al. 2013. Establishing a harmonized haemophilia registry for countries with developing health care systems. *Haemophilia* 19 (5):668–73.

Auerswald, G., A. A. Thompson, M. Recht, et al. 2012. Experience of Advate rAHF-PFM in previously untreated patients and minimally treated patients with haemophilia A. *Thromb Haemost* 107 (6):1072–82.

Bacon, C. L., E. Singleton, B. Brady, et al. 2011. Low risk of inhibitor formation in haemophilia A patients following en masse switch in treatment to a third generation full length plasma and albumin-free recombinant factor VIII product (ADVATE®). *Haemophilia* 17 (3):407–11.

Barbieri, E. M., P. Muir, B. O. Akhuetie-Oni, C. M. Yellman, and F. J. Isaacs. 2017. Precise editing at DNA replication forks enables multiplex genome engineering in eukaryotes. *Cell* S0092-8674 (17):31256–4.

Bjorkman, S., V. S. Blanchette, K. Fischer, et al. 2010. Comparative pharmacokinetics of plasma- and albumin-free recombinant factor VIII in children and adults: the influence of blood sampling schedule on observed age-related differences and implications for dose tailoring. *J Thromb Haemost* 8 (4):730–6.

Bjorkman, S., M. Oh, G. Spotts, et al. 2012. Population pharmacokinetics of recombinant factor VIII: the relationships of pharmacokinetics to age and body weight. *Blood* 119 (2):612–8.

Blanchette, V. S., A. D. Shapiro, R. J. Liesner, et al. 2008. Plasma and albumin-free recombinant factor VIII: pharmacokinetics, efficacy and safety in previously treated pediatric patients. *J Thromb Haemost* 6 (8):1319–26.

Brand, B., R. Gruppo, T. T. Wynn, et al. 2016. Efficacy and safety of pegylated full-length recombinant factor VIII with extended half-life for perioperative haemostasis in haemophilia A patients. *Haemophilia* 22 (4):e251–8.

Carcao, M., M. Zak, F. Abdul Karim, et al. 2016. Nonacog beta pegol in previously treated children with hemophilia B: results from an international open-label phase 3 trial. *J Thromb Haemost* 14 (8):1521–9.

Chen, L., J. Sun, P. Hilliard, et al. 2014. "Train-the-Trainer": an effective and successful model to accelerate training and improve physiotherapy services for persons with haemophilia in China. *Haemophilia* 20 (3):441–5.

Collins, P. W., V. S. Blanchette, K. Fischer, et al. 2009. Break-through bleeding in relation to predicted factor VIII levels in patients receiving prophylactic treatment for severe hemophilia A. *J Thromb Haemost* 7 (3):413–20.

Collins, P., N. Macartney, R. Davies, S. Lees, J. Giddings, and R. Majer. 2004. A population based, unselected, consecutive cohort of patients with acquired haemophilia A. *Br J Haematol* 124 (1):86–90.

Collins, P. W., G. Young, K. Knobe, et al. 2014. Recombinant long-acting glycoPEGylated factor IX in hemophilia B: a multinational randomized phase 3 trial. *Blood* 124 (26):3880–6.

Coyle, T. E., M. T. Reding, J. C. Lin, L. A. Michaels, A. Shah, and J. Powell. 2014. Phase I study of BAY 94-9027, a PEGylated B-domain-deleted recombinant factor VIII with an extended half-life, in subjects with hemophilia A. *J Thromb Haemost* 12 (4):488–96.

Dai, J., Y. Lu, Q. Ding, H. Wang, X. Xi, and X. Wang. 2012. The status of carrier and prenatal diagnosis of haemophilia in China. *Haemophilia* 18 (2):235–40.

Di Paola, J., M. P. Smith, R. Klamroth, et al. 2007. ReFacto (R)(1) and Advate (R)(2): a single-dose, randomized, two-period crossover pharmacokinetics study in subjects with haemophilia A. *Haemophilia* 13 (2):124–30.

Dimichele, D. 2002. Inhibitors: resolving diagnostic and therapeutic dilemmas. *Haemophilia* 8 (3):280–7.

Dimichele, D. 2007. Immune tolerance therapy for factor VIII inhibitors: moving from empiricism to an evidence-based approach. *J Thromb Haemost* Suppl 1:143–50.

Ding, Q. L., Y. L. Lu, J. Dai, X. D. Xi, X. F. Wang, and H. L. Wang. 2012. Characterisation and validation of a novel panel of the six short tandem repeats for genetic counselling in Chinese haemophilia A pedigrees. *Haemophilia* 18 (4):621–5.

Ding, P., Y. Yang, L. Cheng, et al. 2014. The relationship between seven common polymorphisms from five DNA repair genes and the risk for breast cancer in northern Chinese women. *PLoS One* 9 (3):e92083.

Donna D. 2007. Inhibitor development in haemophilia B: an orphan disease in need of attention. *Br J Haematol* 138 (3):305–15.

Drugs.com. Rixubis. https://www.drugs.com/pro/rixubis.html#s-34093-5. Published 2020. Accessed February 2, 2022.

Eckhardt, C. L., A. S. van Velzen, M. Peters, et al. 2013. Factor VIII gene (F8) mutation and risk of inhibitor development in nonsevere hemophilia A. *Blood* 122 (11):1954–62.

Fernandez, M., T. Yu, E. Bjornson, H. Luu, and G. Spotts. 2006. Stability of ADVATE, antihemophilic factor (recombinant) plasma/albumin-free method, during simulated continuous infusion. *Blood Coagul Fibrinolysis* 17 (3):165–71.

Fischer, K., R. Kulkarni, B. Nolan, et al. 2017. Recombinant factor IX Fc fusion protein in children with haemophilia B (Kids B-LONG): results from a multicentre, non-randomised phase 3 study. *Lancet Haematol* 4 (2):e75–e82.

Franchini, M., F. Frattini, S. Crestani, C. Sissa, and C. Bonfanti. 2013. Treatment of hemophilia B: focus on recombinant factor IX. *Biologics: Targets Ther* 7:33–8.

Franchini, M., and G. Lippi. 2010. Recombinant factor VIII concentrates. *Semin Thromb Hemost* 36 (5):493–7.

Franchini, M., and P. M. Mannucci. 2012. Past, present and future of hemophilia: a narrative review. *Orphanet J Rare Dis* 7:24.

Franchini, M., A. Tagliaferri, C. Mengoli, and M. Cruciani. 2012. Cumulative inhibitor incidence in previously untreated patients with severe hemophilia A treated with plasma-derived versus recombinant factor VIII concentrates: a critical systematic review. *Crit Rev Oncol Hematol* 81 (1):82–93.

Giangrande, P., T. Andreeva, P. Chowdary, et al. 2017. Clinical evaluation of glycoPEGylated recombinant FVIII: efficacy and safety in severe haemophilia A. *Thromb Haemost* 117 (2):252–61.

Gitschier, J., W. I. Wood, T. M. Goralka, et al. 1984. Characterization of the human factor VIII gene. *Nature* 312 (5992):326–30.

Godaert, L., S. Bartholet, S. Colas, L. Kanagaratnam, J. L. Fanon, and M. Drame. 2018. Acquired hemophilia A in aged people: a systematic review of case reports and case series. *Semin Hematol* 55 (4):197–201.

Gringeri, A., B. Lundin, S. von Mackensen, L. Mantovani, P. M. Mannucci, and Esprit Study Group. 2011. A randomized clinical trial of prophylaxis in children with hemophilia A (the ESPRIT study). *J Thromb Haemost* 9 (4):700–10.

He, Z., J. Chen, S. Xu, et al. 2013. A strategy for the molecular diagnosis in hemophilia a in Chinese population. *Cell Biochem Biophys* 65 (3):463–72.

Hua, B. L., Z. Y. Yan, Y. Liang, et al. 2010. Identification of seven novel mutations in the factor VIII gene in 18 unrelated Chinese patients with hemophilia A. *Chin Med J (Engl)* 123 (3):305–10.

Ingerslev, J., D. Freidman, D. Gastineau, et al. 1996. Major surgery in haemophilic patients with inhibitors using recombinant factor VIIa. *Haemostasis* 26 (Suppl 1):118–23.

Iorio, A., K. Fischer, and M. Makris. 2017. Large scale studies assessing anti-factor VIII antibody development in previously untreated haemophilia A: what has been learned, what to believe and how to learn more. *Br J Haematol* 178 (1):20–31.

Iorio, A., M. Marcucci, J. Cheng, et al. 2014. Patient data meta-analysis of post-authorization safety surveillance (PASS) studies of haemophilia A patients treated with rAHF-PFM. *Haemophilia* 20 (6):777–83.

Kavakli, K., R. Yang, L. Rusen, et al. 2015. Prophylaxis vs. on-demand treatment with BAY 81-8973, a full-length plasma protein-free recombinant factor VIII product: results from a randomized trial (LEOPOLD II). *J Thromb Haemost* 13 (3):360–9.

Kenet, G., H. Chambost, C. Male, et al. 2016. Long-acting recombinant fusion protein linking coagulation factor IX with albumin (rIX-FP) in children. Results of a phase 3 trial. *Thromb Haemost* 116 (4):659–68.

Klukowska, A., T. Szczepański, V. Vdovin, S. Knaub, M. Jansen, and R. Liesner. 2016. Novel, human cell line-derived recombinant factor VIII (Human-cl rhFVIII, Nuwiq®) in children with severe haemophilia A: efficacy, safety and pharmacokinetics. *Haemophilia* 22 (2):232–9.

Lee, C. A. 2009. The best of times, the worst of times: a story of haemophilia. *Clin Med* 9 (5):453–8.

Leissinger, C. A. 2004. Prevention of bleeds in hemophilia patients with inhibitors: emerging data and clinical direction. *Am J Hematol* 77 (2):187–93.

Lissitchkov, T., K. Hampton, M. von Depka, et al. 2016. Novel, human cell line-derived recombinant factor VIII (human-cl rhFVIII; Nuwiq®) in adults with severe haemophilia A: efficacy and safety. *Haemophilia* 22 (2):225–31.

Lusher, J. M. 2000. Inhibitor antibodies to factor VIII and factor IX: management. *Semin Thromb Hemost* 26 (2):179–88.

Mahlangu, J., K. Kuliczkowski, F. A. Karim, et al. 2016. Efficacy and safety of rVIII-SingleChain: results of a phase 1/3 multicenter clinical trial in severe hemophilia A. *Blood* 128 (5):630–7.

Mahlangu, J., J. S. Powell, M. V. Ragni, et al. 2014. Phase 3 study of recombinant factor VIII Fc fusion protein in severe hemophilia A. *Blood* 123 (3):317–25.

Manco-Johnson, M. J., T. C. Abshire, A. D. Shapiro, et al. 2007. Prophylaxis versus episodic treatment to prevent joint disease in boys with severe hemophilia. *N Engl J Med* 357 (6):535–44.

Mannucci, P. M. 2008. Back to the future: a recent history of haemophilia treatment. *Haemophilia* 14 (Suppl 3):10–8.

Meeks, S. L., and C. D. Josephson. 2006. Should hemophilia treaters switch to albumin-free recombinant factor VIII concentrates. *Curr Opin Hematol* 13 (6):457–61.

Mirchandani, G. G., J. H. Drake, S. L. Cook, B. C. Castrucci, H. S. Brown, and C. P. Labaj. 2011. Surveillance of bleeding disorders, Texas, 2007. *Am J Prev Med* 41 (6 Suppl 4):S354–9.

Monahan, P. E., and J. Di Paola. 2010. Recombinant factor IX for clinical and research use. *Semin Thromb Hemost* 36 (5):498–509.

Mullins, E. S., O. Stasyshyn, M. T. Alvarez-Roman, et al. 2017. Extended half-life pegylated, full-length recombinant factor VIII for prophylaxis in children with severe haemophilia A. *Haemophilia* 23 (2):238–46.

Myrin-Westesson, L., F. Baghaei, and F. Friberg. 2013. The experience of being a female carrier of haemophilia and the mother of a haemophilic child. *Haemophilia* 19 (2):219–24.

Nathwani, A. C., A. M. Davidoff, and E. G. D. Tuddenham. 2017. Advances in gene therapy for hemophilia. *Hum Gene Ther* 28 (11):1004–12.

Negrier, C., F. Abdul Karim, L. M. Lepatan, et al. 2016. Efficacy and safety of long-acting recombinant fusion protein linking factor IX with albumin in haemophilia B patients undergoing surgery. *Haemophilia* 22 (4):e259–66.

Negrier, C., A. Shapiro, E. Berntorp, et al. 2008. Surgical evaluation of a recombinant factor VIII prepared using a plasma/albumin-free method: efficacy and safety of Advate in previously treated patients. *Thromb Haemost* 100 (2):217–23.

Ormond, K. E., D. P. Mortlock, D. T. Scholes, et al. 2017. Human germline genome editing. *Am J Hum Genet* 101 (2):167–76.

Parti, R., A. Schoppmann, H. Lee, and L. Yang. 2005. Stability of lyophilized and reconstituted plasma/albumin-free recombinant human factor VIII (ADVATE rAHF-PFM). *Haemophilia* 11 (5):492–6.

Pasi, K. J., K. Fischer, M. Ragni, et al. 2017. Long-term safety and efficacy of extended-interval prophylaxis with recombinant factor IX Fc fusion protein (rFIXFc) in subjects with haemophilia B. *Thromb Haemost* 117 (3):508–18.

Pipe, S. W. 2008. Recombinant clotting factors. *Thromb Haemost* 99 (5):840–50.

Pollmann, H., R. Klamroth, N. Vidovic, et al. 2013. Prophylaxis and quality of life in patients with hemophilia A during routine treatment with ADVATE [antihemophilic factor (recombinant), plasma/albumin-free method] in Germany: a subgroup analysis of the ADVATE PASS post-approval, non-interventional study. *Ann Hematol* 92 (5):689–98.

Powell, J. S., N. C. Josephson, D. Quon, et al. 2012. Safety and prolonged activity of recombinant factor VIII Fc fusion protein in hemophilia A patients. *Blood* 119 (13):3031–7.

Romanov, V., M. Marcucci, J. Cheng, L. Thabane, and A. Iorio. 2015. Evaluation of safety and effectiveness of factor VIII treatment in haemophilia A patients with low titre inhibitors or a personal history of inhibitor. Patient data meta-analysis of rAFH-PFM post-authorization safety studies. *Thromb Haemost* 114 (1):56–64.

Santagostino, E., G. Kenet, K. Fischer, T. Biss, S. Ahuja, and M. Steele. 2020. PROTECT VIII Kids: BAY 94-9027 (PEGylated recombinant factor VIII) safety and efficacy in previously treated children with severe haemophilia A. *Haemophilia* 26 (3):e55–65.

Shapiro, A. D. 2007. Anti-hemophilic factor (recombinant), plasma/albumin-free method (octocog-alpha; Advate) in the management of hemophilia A. *Vasc Health Risk Manag* 3 (5):555–65.

Shapiro, A., R. Gruppo, I. Pabinger, et al. 2009. Integrated analysis of safety and efficacy of a plasma- and albumin-free recombinant factor VIII (rAHF-PFM) from six clinical studies in patients with hemophilia A. *Exp Opin Biol Ther* 9 (3):273–83.

Shapiro, A. D., C. Schoenig-Diesing, L. Silvati-Fidell, W. Y. Wong, and V. Romanov. 2015. Integrated analysis of safety data from 12 clinical interventional studies of plasma- and albumin-free recombinant factor VIII (rAHF-PFM) in haemophilia A. *Haemophilia* 21 (6):791–8.

Shi, J., Y. Zhao, J. Wu, J. Sun, L. Wang, and R. Yang. 2007. Safety and efficacy of a sucrose-formulated recombinant factor VIII product for the treatment of previously treated patients with haemophilia A in China. *Haemophilia* 13 (4):351–6.

Srivastava, A., A. K. Brewer, E. P. Mauser-Bunschoten, et al. 2013. Guidelines for the management of hemophilia. *Haemophilia* 19 (1):e1–47.

Sun, J., P. E. Hilliard, B. M. Feldman, et al. 2014. Chinese hemophilia joint health score 2.1 reliability study. *Haemophilia* 20 (3):435–40.

Tarantino, M., T. Abshire, A. Shapiro, et al. 2004a. Global clinical evaluation of an rFVIII prepared using a plasma/albumin-free method, ADVATE rAHF-PFM. *Pediatric Research* 55 (4):286a–286a.

Tarantino, M. D., P. W. Collins, C. R. Hay, et al. 2004b. Clinical evaluation of an advanced category antihaemophilic factor prepared using a plasma/albumin-free method: pharmacokinetics, efficacy, and safety in previously treated patients with haemophilia A. *Haemophilia* 10 (5):428–37.

Tiede, A., B. Brand, R. Fischer, et al. 2013. Enhancing the pharmacokinetic properties of recombinant factor VIII: first-in-human trial of glycoPEGylated recombinant factor VIII in patients with hemophilia A. *J Thromb Haemost* 11 (4):670–8.

Toole, J. J., J. L. Knopf, J. M. Wozney, et al. 1984. Molecular cloning of a cDNA encoding human antihaemophilic factor. *Nature* 312 (5992):342–7.

Valentino, L. A., L. Cong, C. Enockson, et al. 2015. The biological efficacy profile of BAX 855, a PEGylated recombinant factor VIII molecule. *Haemophilia* 21 (1):58–63.

Valentino, L. A., M. Recht, J. Dipaola, et al. 2009. Experience with a third generation recombinant factor VIII concentrate (Advate) for immune tolerance induction in patients with haemophilia A. *Haemophilia* 15 (3):718–26.

Valentino, L. A., C. M. Reyes, B. Ewenstein, et al. 2014. ADVATE prophylaxis: post hoc analysis of joint bleeding episodes. *Haemophilia* 20 (5):630–8.

Vehar, G. A., B. Keyt, D. Eaton, et al. 1984. Structure of human factor VIII. *Nature* 312 (5992):337–42.

White, G. C. 2nd, C. W. McMillan, H. S. Kingdon, and C. B. Shoemaker. 1989. Use of recombinant antihemophilic factor in the treatment of two patients with classic hemophilia. *N Engl J Med* 320 (3):166–70.

Wu, R., K. H. Luke, M. C. Poon, et al. 2011. Low dose secondary prophylaxis reduces joint bleeding in severe and moderate haemophilic children: a pilot study in China. *Haemophilia* 17 (1):70–4.

Xuan, M., F. Xue, R. Fu, et al. 2013. Retrospective analysis of 1,226 Chinese patients with haemophilia in a single medical centre. *J Thromb Thrombolysis* 38 (1):92–7.

———. 2014. Retrospective analysis of 1,226 Chinese patients with haemophilia in a single medical centre. *J Thromb Thrombolysis* 38 (1):92–7.

Zhang, L., Y. Zhao, J. Sun, et al. 2011. Six-month clinical observation on safety and efficacy of a full-length recombinant factor VIII for on-demand treatment of Chinese patients with haemophilia A. *Haemophilia* 17 (3):538–41.

Zhang, L., Y. Zhao, J. Sun, X. Wang, M. Yu, and R. Yang. 2011. Clinical observation on safety and efficacy of a plasma- and albumin-free recombinant factor VIII for on-demand treatment of Chinese patients with haemophilia A. *Haemophilia* 17 (2):191–5.

22 Complement as New Immunotherapy Target

Past, Present, and Future

Yubin Li and Hongbin Wang

Corresponding author: Hongbin Wang

Contents

DOI: 10.1201/9780429485626-22

22.1 INTRODUCTION

Complement-like proteins and their activators have a long evolutionary history and have been found in ancient animals as primitive as Cephalochordata, Urochordata, Echinochordata, and Cnardarian anthozoans (Pinto et al., 2003, Nonaka and Kimura, 2006, Miller et al., 2007, Pinto et al., 2007). The complement system is a very important component of host defense and was originally thought to have an auxiliary role in enhancing the functions of antibodies or phagocytes. As a part of the human innate immune system, the complement system comprises of a set of more than 50 soluble proteins in the blood and cell surface receptors and regulators that act in a highly coordinated way to kill microbes and expedite the removal of apoptotic cells without damaging the healthy host cells.

The main premise underlining complement activation is that the activation of upstream complement components through a series of proteolytic cleavages will cleave proteins downstream of the complement cascade. The complement system can be activated through three major pathways: the classical pathway (CP), the mannose-binding lectin pathway (LP), and the alternative pathway (AP), respectively. The activation of CP and LP pathways generates C3 convertase complex C4bC2a (Muller-Eberhard et al., 1967, Fujita, 2002, Endo et al., 2015). In contrast to these two pathways, the AP pathway generates another C3 convertase complex (C3bBb) (Kawasaki et al., 1983, Lachmann, 2009). The terminal pathway starts with the formation of C5 convertase (C3bC4bC2a; C3bBbC3b). The inclusion of C3b into the C3 convertase leads to the formation of C5 convertase that will cleave C5 into C5a and C5b. The final event of complement activation is the formation of the membrane attack complex (MAC, or C5b-9), which is able to kill invading pathogens by forming a lytic pore in the membrane. Complement activation fragmented peptides (C3a, C4a, and C5a) are involved in inflammation, opsonization, and anaphylatoxic effects. The response of complement activation is a balance between positive and negative regulators that allow very fine control to avoid a potentially misdirected or excessive activation of the system. However, this defense system can also be improperly activated and attack host cells, contributing to a broad spectrum of immune, inflammatory, and age-related diseases. When the dysfunction occurs in the regulatory proteins (i.e. factor H) in the complement system, the powerful cell-killing property of complement can be turned against "self". Dysregulated or excessive complement activation is now recognized as a key pathogenic driver in a wide spectrum of immune-mediated and inflammatory diseases, ranging from hematological and aging-related ocular pathologies to cancer, autoimmunity, oral dysbiotic diseases and neuroinflammatory, and neurodegenerative disorders. The list of diseases in which complement has a role as either primary or secondary to other triggers is growing. Currently, eculizumab, humanized anti-C5 monoclonal antibody (mAb) and complement protein C3 inhibitor pegcetacoplan, a pegylated cyclic peptide are available complement therapeutic drugs in the clinic for the treatment of complement-related diseases, such as atypical hemolytic uremic syndrome (aHUS) and paroxysmal nocturnal hemoglobinuria (PNH). Nevertheless, many drug candidates, including peptides, antibodies, and small molecules, are in the preclinical and clinical phases of drug development.

Below we first discuss the current state of discovery related to the human complement system, including an exploration of its molecular regulation and activation, followed by a discussion of genetic and biochemical delinquencies on human immunity.

22.2 ACTIVATION OF COMPLEMENT SYSTEM

CP pathway activation is initiated by the association of C1q with immune complexes (IgG or IgM) to form a multimolecular complex with C1r and C1s serine proteases. This complex cleaves C4 then associates with the product C4b into a complex, which cleaves C2 (into C2a and C2b) to form the C3 convertase

(C4bC2a). On the other hand, the LP pathway activation starts by binding of mannose-binding lectin (MBL)/ficolins/collectin 11 to the mannose residues on microbial surfaces. MBL-associated serine proteases (MASP-1, MASP-2, and MASP-3) recruited by bound MBL, functioning similar to C1r and C1s, cleaves C4 and C2 to form the same C3 convertase as the CP. The AP of complement activation is distinctive as it is activated without pattern recognition molecules. Instead, a "tick over" mechanism involves the spontaneous hydrolysis of C3 into $C3_{(H_2O)}$, which acts similar to C3b and binds factor B (FB). The AP C3 convertase (C3bBb) is generated through a chain of reactions involving factor D (FD) and properdin. C3 convertase cleaves C3 into C3a and C3b. C3b then binds with the C3 convertase to form C5 convertases (C4aC2aC3b; C3bBbC3b), which cleaves C5 into C5a and C5b. C5b binds C6, C7, C8, and C9 molecules to form the MAC, the lytic machinery of complement (Figure 22.1). Two models have been proposed for the role of properdin in the AP activation: (1) properdin binds to a surface and provides a platform for C3 convertase assembly by acting as a pattern recognition molecule, and (2) properdin binds to C3b attached to a surface, which in turn promotes AP convertase formation (Spitzer et al., 2007, Lesher et al., 2013). A recent study by Harboe *et al.* challenged the view of properdin as a pattern recognition molecule, and argued that the experimental conditions used to test this hypothesis should be carefully considered, with emphasis on controlling initial C3 activation under physiological conditions (Harboe et al., 2017). Complement can also be activated through other mechanisms. For instance, some coagulation pathway proteases can directly cleave C5 and C3 (Amara et al., 2008). MASPs are reported to cleave C3 to C3b, thereby triggering the AP (Matsushita and Fujita, 1995). MASP-1, without the requirement of MBL, by cleaving factor D from the pro- to mature form, can lead to AP activation (Takahashi et al., 2010). C3 can be cleaved through MBL but a C2 by-pass

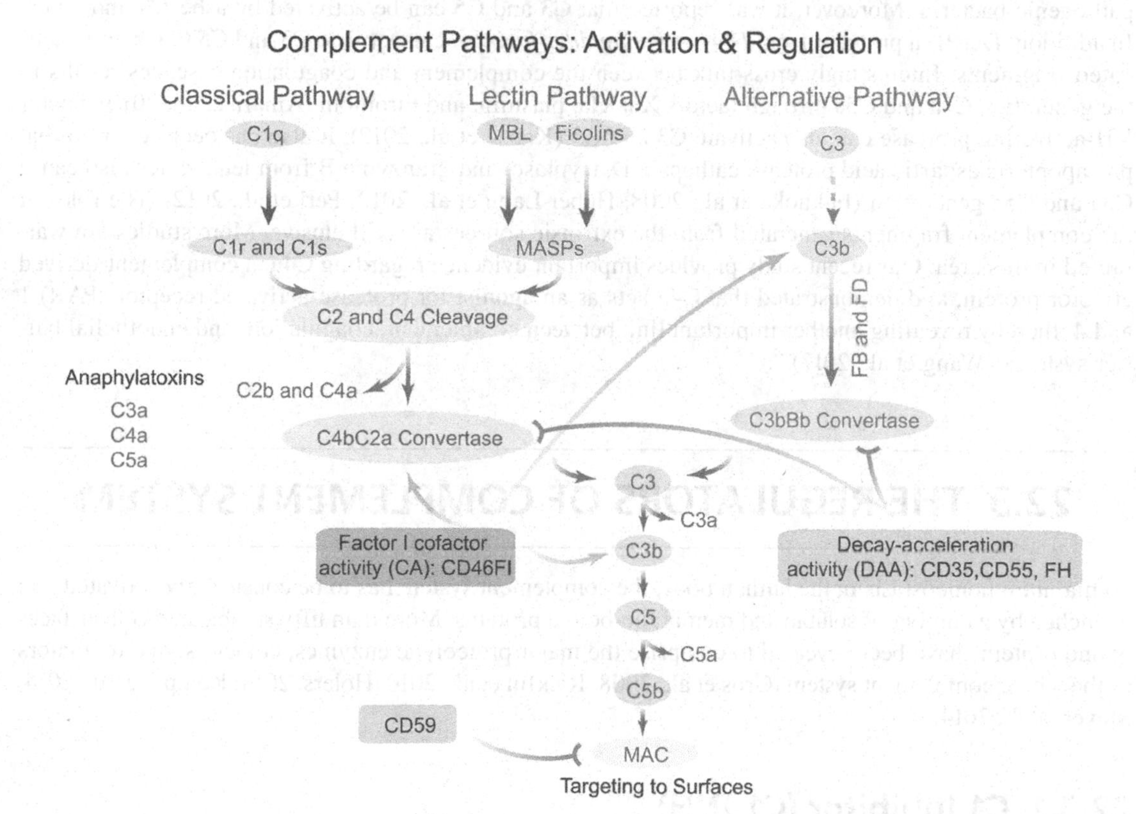

FIGURE 22.1 Complement pathway.

activation mechanism (Selander et al., 2006). Now AP pathway may be considered to be responsible for complement amplification that was initiated by other pathways, but rather as an entirely independent pathway (Lutz and Jelezarova, 2006, Harboe and Mollnes, 2008).

22.2.1 Anaphylatoxin

Upon complement activation, a group of biologically active small protein fragments (74–77 amino acid residues) are generated. C3a and C5a are from all three pathways and C4a is from CP and LP pathways. They were originally called anaphylatoxins that can act as potent chemoattractants and secretagogues on a wide variety of cell types. These complement fragments are rapidly metabolized by carboxypeptidases, becoming des-arginated forms (C3a desArg, C4a desArg, and C5a desArg). Research works have attempted to elucidate the relationship between the levels of anaphylatoxins in different disease states. This could be problematic due to assay techniques and the nature of the biological samples. Plasma collected with EDTA is more reliable since EDTA could effectively block all three major routes of complement activation. It has been reported that the levels of complement peptides are clearly elevated in inflammatory diseases. More research needs to be done to explore the anaphylatoxin as a clinical marker for different diseases.

Complement C3 and C5 can also be cleaved through extrinsic pathways other than canonic complement activation pathways. For example, C5 was the first reported to be cleaved by trypsin (Cochrane and Muller-Eberhard, 1968). Gingipain-1, from porphyromonas gingivalis, cleaves C3 and C5 to generate the fragments with higher molecular weights than C3a and C5a, which can produce leukocyte chemoattractant activities (Wingrove et al., 1992). Complement C5 and C3 also can be activated by many pathogenic bacteria. Moreover, it was reported that C3 and C5 can be activated by asbestos and silica. In addition, DerP1, a protease from *Dermatophagoides farinae*, can activate C3 and C5 to generate activated fragments. Interestingly cross-talk between the complement and coagulation cascades results in the generation C3a and C5a through factors Xa/XIa, plasmin, and thrombin (Amara et al., 2010). Factor VII-activating protease can also activate C3 and C5 (Kanse et al., 2012). It also has been reported that pro-apoptotic aspartic acid protease cathepsin D, tryptase, and granzyme B from leukocytes also cause C3a and C5a generation (Fukuoka et al., 2008, Huber-Lang et al., 2012, Perl et al., 2012). The roles of the complement fragments generated from the extrinsic sources are still elusive. More studies are warranted in this area. Our recent study provides important evidence regarding C4a, a complement-derived effector protein, and demonstrated that C4a acts as an agonist for protease-activated receptor (PAR) 1 and 4, thereby revealing another important link between complement, coagulation, and endothelial barrier systems (Wang et al., 2017).

22.3 THE REGULATORS OF COMPLEMENT SYSTEM

To maintain homeostasis in the human body, the complement system has to be consistently activated and quenched by a number of soluble and membrane-bound proteins. More than fifty soluble and cell surface-bound proteins have been revealed to comprise the major proteolytic enzymes, cofactors, and regulators of the whole complement system (Gros et al., 2008, Ricklin et al., 2010, Holers, 2014, Kemper et al., 2014, Meyer et al., 2014).

22.3.1 C1 Inhibitor (C1-INH)

C1-INH is a member of the serine protease inhibitor (serpin) family, which can regulate the activation of CP and LP pathways. C1-INH can irreversibly inhibit C1r and C1s in the C1 complex. It also can block

MASP-1 and -2 activity (Degn et al., 2013). It can cause the disassociation of C1r:C1s from C1q and consequently limits the time during which the active C1s is able to cleave C4 and C2. It can also inhibit the activity of factor XIIa and kallikrein. C1-INH deficiency and mutations have been associated with hereditary angioedema (HAE) and autoimmune diseases, such as systemic lupus erythematosus (SLE) (Davis, 2005). Functional C1-INH deficiency can lead to abnormal bradykinin production, which causes vascular permeability and angioedema (Cicardi and Zuraw, 2018).

22.3.2 C4b-Binding Protein (Also Known as C4bp)

C4bp is a fluid-phase regulator of the CP and LP that can accelerate the decay of C3 convertase (C4bC2a) by displacing irreversibly C2a from C4b. C4bp is a large glycoprotein that consists of seven alpha- and one beta-chain, both made of complement control protein modules, and with a spider-like appearance (Blom et al., 2004, Hofmeyer et al., 2013). It also acts as a cofactor for plasma protease factor I. Although no C4bp deficiency has been identified yet, non-synonymous polymorphism (R240H) has been found to be associated with atypical hemolytic-uremic syndrome (aHUS), a disease with excessive complement activation (Blom et al., 2008).

22.3.3 Complement Factor H (CFH)

CFH plays a crucial role in regulating APs. When CFH is dysfunctional or deficient, AP activation in plasma becomes vigorous and will lead to the deficiency of secondary complement via consumption of C3 and other complement components (Nilsson et al., 2011b). Despite its essential role in the regulation of AP, it also has been revealed to contribute to a decay-accelerating activity of C3 convertase of the CP pathway. Moreover, acting as a cofactor for factor I, it facilitates the fragmentation of C3b and C4b into their inactive products iC3b and C4c/C4d, respectively (Meyer et al., 2014). Several mutations of CFH have been described leading to either impaired functional activity or more expression on cell membrane but not in plasma (Hakobyan et al., 2008). Anti-CFH autoantibodies in the plasma hamper the interaction with C3b on the cell membrane (Hakobyan et al., 2010). Decreased protection of host cells from complement attack by dysfunctional CFH usually causes aHUS. Interestingly, cancer cells were reported to over-express CFH to evade complement-mediated cancer killing (Junnikkala et al., 2000). A clinical study showed that the event of free survival after rituximab treatment in follicular lymphoma patients is associated with a CFH polymorphism, although the study had relatively low power and this finding should be confirmed by further studies (Charbonneau et al., 2012).

22.3.4 Complement Factor I (CFI)

CFI, a protease, together with C4-binding protein, serves as a cofactor to degrade the C4b component in C3 convertase (C4bC2a) for CP and LP. It also can break down various components in AP. In addition, CFI together with CFH or factor H-like protein 1, serving as a cofactor, can permanently inactivate C3b to iC3b by proteolysis. iC3b then is further fragmented to C3dg and C3c by CFI (Crossley and Porter, 1980, Nilsson et al., 2011a).

22.3.5 Complement Receptor Type 1 (CR1; CD35)

CR1, a membrane-bound complement regulator, is expressed on both polymorphonuclear leukocytes and macrophages. The association of C3b or C4b with CR1 causes a decay-accelerating activity toward the C3 or C5 convertase. Despite its function as a receptor, CR1 is a major cofactor for CFI for the inactivation of C3b and C4b. Moreover, iC3b interacts with CR1 to be cleaved into inactive form C3c and C3dg to prevent further direct cell lysis (Liu and Niu, 2009).

22.3.6 Membrane Cofactor Protein (MCP, CD46)

MCP, a ubiquitously expressed membrane glycoprotein (absent only on erythrocytes), serves as a cofactor for the cleavage of C3b and C4b (Andrews et al., 1985). MCP blocks the formation of C3 convertase in CP and AP. Several functional polymorphisms were identified to be associated with aHUS (Richards et al., 2003). MCP KO mice show no phenotype related to complement functions due to its expression limited to the testis in mice.

22.3.7 Decay-Accelerating Factor (DAF; CD55)

DAF, a glycosyl phosphatidylinositol (GPI)-linked glycoprotein widely expressed on the surface of blood cells, placenta, vascular endothelial cells (ECs), and epithelial cells. A soluble form of MCP is found in body fluids such as plasma, tears, and urine. Through its decay-accelerating activity, DAF can destabilize C3 and C5 convertase in CP and AP (Medof et al., 1984). Similar to CFH, DAF accelerates the disassociation of AP C3 convertase, C3bBb (Liszewski et al., 1996, Walport, 2001). DAF also can decrease the stability of the C3 convertase of CP and LP, C4bC2a, by accelerating its disassociation to C4b and C2a (Kim and Song, 2006). In contrast to CR1, DAF acts in a *cis*-fashion by inhibiting complement activation on the same cell as it is expressed (Meyer et al., 2014). A mutation in a gene called phosphatidylinositol glycan class A (PIGA), which is involved in the first step of GPI biosynthesis, leads to the loss of DAF and protectin (CD59) that results in PNH characterized by, among others, complement-mediated hemolytic anemia (Brodsky, 2014).

22.3.8 Protectin

Protectin is a key regulator of the terminal pathway that inhibits the C9 association with C5b-8 to stop MAC C5b-9 formation. Protectin is expressed on circulating blood cells, spermatozoa, ECs, and epithelial cells. Like many other proteins, DAF and protectin are both linked to the cell membrane by a phosphatidylinositol glycolipid (PIG) tail (Noris and Remuzzi, 2013). As mentioned in the previous section, PIGA is encoded on chromosome X. Both DAF and protectin fail to function properly in people with a somatic mutation in PIGA (Botto et al., 2009). This will cause PNH, which is manifested by episodes of intravascular red blood cell lysis by complement. Red blood cells lacking protectin only are susceptible to destruction due to spontaneous activation of the complement cascade (Risitano and Rotoli, 2008).

22.4 COMPLEMENT IN INNATE AND ADAPTIVE IMMUNITY

Innate and adaptive immunity are two interconnected sub-immune systems that play critical roles in host defense. The innate immunity system consists of immunological effectors that provide nonspecific and instant responses to keep viruses, bacteria, parasites, and other foreign particles out of the body. The innate immune system executes immune reactions through complement, inflammation, and non-specific cellular responses. There are many types of white blood cells in the innate immune system, including phagocytic cells, macrophages, mast cells, neutrophils, eosinophils, basophils, natural killer cells (NK cells), and dendritic cells. In contrast, the adaptive immune system is composed of T and B lymphocytes, which have various antigen-specific receptors that can generate defined identification and pathogens elimination. In addition, adaptive immunity can mold immune responses and generate long-lived immunological memory against re-infections.

The complement system undertakes its function through three key effector pathways, including opsonization/phagocytosis, lysis, and inflammation. Complement was originally discovered as a heat-labile component of normal plasma that could enhance the opsonization of bacteria by antibodies and allow antibodies to kill bacteria. It is one of the major mechanisms by which pathogen recognition is converted into an effective host defense against initial infection. Three complement pathways (CP, LP, and AP) can be initiated independently of the antibody as part of innate immunity. One of the important effects of complement system activation is the formation of a terminal MAC. The MAC complex formation leads to a pore in the lipid bilayer membrane and destroys pathogen membrane integrity. Consequently, this process will kill the pathogen by destroying the proton gradient across the pathogen cellular membrane (Papadimitriou et al., 1991, Esser, 1994, Cragg et al., 2000). C3a, C4a, and C5a, highly related proinflammatory molecules, can elicit chemoattractant effects, recruit phagocytes to sites of infection, and activate them by association with their specific G protein-coupled receptors (C3aR and C5aR1/C5aL2). Those complement fragments can increase in vascular permeability, leukocyte recruitment, smooth muscle contraction, other cell functions (e.g. phagocytosis, chemotaxis, and migration), and release of inflammatory mediators (Wetsel, 1995, Haas and van Strijp, 2007). The binding of C3b molecules and their inactive fragments (iC3b/C3c/C3dg) to the pathogen is the fundamental event in complement activation. Bound C3b and its inactive fragments are recognized by specific complement receptors (CR1, CR2, CR3, CR4, and CRIg) on phagocytic cells that will engulf pathogens opsonized by C3b and its inactive fragments (Ross et al., 1985, van Lookeren Campagne et al., 2007). CR1 is mainly expressed in peripheral blood cells. Association of CR1 with ligands serves to clear immune complex as well as enhance proinflammatory molecules secretion, like IL-1 and prostaglandins (Bacle et al., 1990, Krych-Goldberg and Atkinson, 2001).

As early as in the 1970s, the question of whether the complement system was involved in adaptive immune responses beyond innate immune functions was raised (Nussenzweig, 1971). Subsequent studies discovered that human C3 deficiency results in impaired humoral immune responses, which provided direct evidence that an intact complement system is critical in maintaining efficient adaptive responses (Pepys, 1972). Using complement deficiency animals, components in the complement CP were demonstrated to function to trap and present antigens to lymphocytes for reactions (Papamichail et al., 1975, Ochs et al., 1986, O'Neil et al., 1988). Complement effectors have been reported to engage in humoral immunity at numerous stages of B cell differentiation and functions (Carroll, 2004, Carroll, 2008). Complement can enhance B cell immunity primarily through complement receptors (CRs). The complement opsonins (e.g. C3d) bind to CR1 (CD35) and CR2 (CD21), expressed on B lymphocytes and follicular dendritic cells, to reduce B cell activation threshold and enhance B-cell immunity (Kinoshita et al., 1990, van Lookeren Campagne et al., 2007, Carroll, 2008). Emerging evidence also reveals that complement acts as a "natural adjuvant" and an instructor of the humoral immune response. CR2, CD19, and the tetraspan protein CD81 can form the B-cell coreceptor complex (CD21-CD19-CD81) that supports enhanced signal via the B-cell receptor (BCR). CR2-deficient mice were observed to have decreased the repertoire of natural antibodies and a striking decrease in ischemia/reperfusion-induced injury despite a normal level of IgM (Fleming et al., 2002, Reid et al., 2002). Moreover, these mice showed impaired antibody production and had a reduced number of B1a cells (Ahearn et al., 1996). B cells without CR1/CR2 receptors fail to survive within a germinal center (GC) when put in competition with wild-type B cells, suggesting that co-receptor signaling is essential for B cells clonal selection (Fischer et al., 1998). In addition to the complement receptors, anaphylatoxins were also demonstrated to modulate B cells' functions. C3a and C3adesArg have been demonstrated to adversely regulate the polyclonal immune response and decrease IL-6 and TNF-α secretion (Morgan et al., 1982, Fischer and Hugli, 1997). However, C5a was reported to induce trafficking and migration of different B cells (Morgan et al., 1983, Kupp et al., 1991). In summary, all the evidence indicates that complement plays an important role in the generation of the B cells antibody response.

Since many of the complement- and CR-deficient animal models developed normal T-cell functions whereas B cell humoral responses were impaired (Gustavsson et al., 1995, Da Costa et al., 1999), the complement system was considered to be more critical in regulating B cell *vs* T cell functions. However, during the pulmonary influenza challenge, the priming of both CD4 and CD8 T-cells was significantly reduced in C3-KO mice, indicating a more general regulatory role in adaptive immunity of complement (Kopf et al., 2002). The involvement of complement in T-cell immune responses to viral and alloantigens has now been demonstrated in many other studies (Kim et al., 2004, Fang et al., 2007, Li et al., 2008, Peng et al., 2008). However, the mechanisms of regulation of complement on T-cell functions were not as well elucidated as those related to B-cell functions. Thus, this area represents a critical area of study to understand the role of complement in regulating T-cell adaptive immune responses. The DAF-KO mice model has facilitated the characterization of the potential role of complement in T-cell immunity. It was demonstrated that following antigen re-stimulation, DAF-KO mice have an augmented Th1 response with enhanced secretion of interferon-γ and IL-2 and decreased IL-10 secretion (Heeger et al., 2005, Liu et al., 2005). DAF-KO mice displayed enhanced T-cell immunity by revealing the pathology of multiple sclerosis and autoimmune encephalomyelitis in the mouse model (Liu et al., 2005, Liu et al., 2008). Enhanced T-cell immune responses were also displayed in lymphocytic choriomeningitis virus (LCMV) infection in DAF-KO mice (Fang et al., 2007). C5aR has been demonstrated to be important for modulating T-cell immunity in different models. For example, C5aR antagonist treatments generate fewer antigen-specific CD8 T lymphocytes after infection of influenza type A. A stronger inflammatory response was observed as a result of the synergism of both C5aR and Toll-like receptor (TLR)-4 receptors than either functioning alone (Zhang et al., 2007). C3a/C3aR- and C5a/C5aR-triggered signaling in antigen-presenting cells (APCs) are critical in both costimulatory and survival signals that are needed for effector T cell responses (Lalli et al., 2008, Liu et al., 2008, Strainic et al., 2008). Further studies found that absent C3a and C5a receptor signaling into $CD4^+$ T cells enables autoinductive TGF-β1 signaling and induction of $Foxp3^+$ T regulatory cells (Kwan et al., 2013, Strainic et al., 2013). Complement modulating proteins, CR1, MCP, and DAF, have also been associated with the modulation of APC and T-cell functions. Studies showed that cross-linking of MCP, DAF, or CR1 on macrophages, APCs, or T cells significantly impaired the cytokines (i.e. IL-12, IL-2) production from those immune cells (Karp et al., 1996, Cattaneo, 2004, Wagner et al., 2006). The complement system clearly serves to play a functional role in regulating T cells function, supporting the link between increased T-cell immune responses and complement activation. However, ongoing studies need to be done to elucidate the precise mechanisms by which complement regulates T-cell function.

22.5 COMPLEMENT AND COAGULATION

The host defense relies on a strong interconnection between the complement system, the contact and coagulation cascades, and cellular components, including platelets and ECs. Apart from its important role in fighting infection, the complement system is also involved in tissue regeneration (Clark et al., 2008), clearance of debris (Pickering and Walport, 2000), and pathogenesis of multiple diseases (Alchi and Jayne, 2010, Sweigard et al., 2014). Similarly, the coagulation cascade plays a crucial role in the pathology of multiple diseases and fighting infections as well, despite its major function in the maintenance of hemostasis (Muller and Renne, 2008). Interestingly, the complement and coagulation systems contribute to a number of general common features. Activation of the complement system after bacterial infections results in enhanced activation of the intrinsic and extrinsic coagulation cascades and a decreased fibrinolytic activity. Moreover, severe shock and acute blood loss are not only related to disseminated intravascular coagulopathy (DIC) but also with rapid and massive complement activation. The potent anaphylatoxin C3a and C5a can in turn exaggerate coagulation cascades (Amara et al., 2008). Furthermore, activation of both systems directs to the formation of zymogens and proteolytic complexes that are typically serine proteases with high-substrate specificity.

Substantial experimental evidence demonstrates that, in addition to modulating the individual complement pathway activation, complement regulator components can also modulate the coagulation system. For example, complement C1 inhibitor (C1-INH) plays a role in the inhibition of FXIIa and kallikrein of the contact system/intrinsic coagulation pathway. Notably, the lack of the inhibition of these enzymes by C1-INH links with inappropriate bradykinin production and causes angioedema with enhanced vascular permeability. C3 and MAC components combinations were demonstrated to stimulate the thrombin-mediated platelet accumulation and serotonin secretion (Polley and Nachman, 1978). Complement C5b-9 aggregation on cell surfaces can cause changes in the membrane-linked components, affecting the activation of platelets during the initiation phase of coagulation (Sims and Wiedmer, 1991). Activation of platelets through sublytic concentrations of the MAC components can cause temporary membrane depolarization (Wiedmer and Sims, 1985), granule secretion (Ando et al., 1988), and initiation of platelet-catalyzed thrombin production and clotting (Wiedmer et al., 1986). In addition, the generation of MAC provokes the release of membrane microparticles from platelets and ECs (Sims et al., 1988, Hamilton et al., 1990), which can express binding sites for FVa and are responsible for the proteolytic generation of thrombin from its proenzyme in the presence of prothrombinase complex. It has also been reported that the platelet surface possesses a C1q receptor. The binding of C1q to the respective receptor on platelet surfaces induces the accumulation of platelets through a P-selectin-dependent pathway (Peerschke et al., 1993, Skoglund et al., 2010). Furthermore, platelet activation and aggregation are directly induced by C3 activation products C3a and its derivative C3adesArg (Polley and Nachman, 1983). Recent studies support these aggregative properties of C3a/C3adesArg by demonstrating the delayed response in thrombosis following vessel wall injury in C3 KO mice. Moreover, the abnormal platelet hyper-reactivity was observed in mice that are deficient in the negative regulator of MAC (CD59b, a potential mouse variant of CD59) (Qin et al., 2003, Gushiken et al., 2009). C5a is a potent anaphylatoxin that activates procoagulant function by a number of mechanisms on cells. C5a stimulates the expression of TF by ECs and neutrophils (Ikeda et al., 1997, Ritis et al., 2006). In addition, C5a induces the exchange of mast cell and basophil effects from profibrinolytic to prothrombotic through the overexpression of plasminogen activator inhibitor-1 (PAI-1) (Wojta et al., 2002). Furthermore, interaction within the two membrane receptors, TF and C5a receptor (C5aR), warrants further investigation between the two systems (Ritis et al., 2006). MASP-2, an essential enzyme of LP activation, is capable of generating thrombin by direct cleavage of prothrombin (Krarup et al., 2007). In the same way, the terminal complement complex (TCC) C5b-9 shows identical activity to prothrombin even in the lack of FV (Wiedmer et al., 1986). Moreover, both the sub-lytic concentration of MAC and the cytolytically inactive TCC demonstrate procoagulant activity mediated through the induction of TF expression via ECs (Tedesco et al., 1997). Wang *et al.* recently provided important evidence that C4a, a complement activation-derived effector protein, acts as an agonist for PAR1 and PAR4, thereby revealing another important link between the complement, coagulation, and endothelial barrier systems (Wang et al., 2017).

On the other hand, activation of coagulation cascades also induces the activation of complement pathways. A study revealed the contribution of thrombin and plasmin in the activation of complement cascade during liver regeneration while lacking C4 and AP activity (Dunkelberger and Song, 2010). In addition, thrombin can initiate the DAF expression in a PAR1-dependent manner (Lidington et al., 2000). Furthermore, in vitro study revealed that this induction results in significantly decreased C3 deposition and the complement-mediated ECs lysis. Similarly, thrombin-activatable fibrinolysis inhibitor (TAFI), also recognized as carboxypeptidase R or plasma carboxypeptidase B, produced by a thrombomodulin-thrombin complex, can play a dual role in the downregulation of plasmin-mediated fibrinolysis and the inactivation of complement anaphylatoxin C3a and C5a (Campbell et al., 2001, Walker and Bajzar, 2004). Additionally, FXIIa has been explored to activate the CP of complement cascade directly through the activation of the C1qrs complex (Ghebrehiwet, 1981). In contrast to the established complement activation pathways, coagulation cascades can activate complement C3 and C5 independent of generating thrombin (Ward and Gao, 2009). Amara *et al.* also demonstrated that FIXa, FXa, FXIa, and plasmin mediated cleavage of C3 and C5, with the generation of C3a and C5a, respectively, independent of the recognized activation of complement pathways (Amara et al., 2008). Activated platelets that are vital contributors in the coagulation cascade can also activate both the CP and the AP (Del Conde et al., 2005, Peerschke et al., 2006, Saggu et al., 2013). However, the pathophysiological influence of this activation is unknown, although complement activation components are recognized to activate platelets (Del Conde et al., 2005), which may direct to a positive or negative feedback loop. The linkage of complement activation with coagulation system via the regulation of cytokine network has also been demonstrated in many studies, which may evaluate the equilibrium between anticoagulant and procoagulant pathways (Levi et al., 2010). For instance, the anaphylatoxins C3a and C5a are demonstrated to control the generation and secretion of TNFα and IL-6 from Kupffer's liver cells (Markiewski and Lambris, 2007, Markiewski et al., 2009). TNF-α and IL-6 may subsequently promote TF expression from ECs and blood cells, whereas IL-6 may also affect the platelets propensity for clotting (Burstein et al., 1996, Shebuski and Kilgore, 2002). Additionally, the C1q-dependent generation of IL-6, IL-8, MCP-1, and macrophage inflammatory protein 1β (MIP-1β) from ligament fibroblasts of gingivitis and periodontitis has also been explored (Verardi et al., 2007). The inflammatory cytokines can also control the antithrombotic proteins expressions, such as thrombomodulin (Levi and van der Poll, 2010).

22.6 COMPLEMENT IN DISEASES AND COMPLEMENT-BASED THERAPY

Complement system-related diseases mainly result from deficiency of complement system component, abnormal regulation of complement system, or stimulation of complement system by abnormal stimuli (Markiewski and Lambris, 2007, Noris and Remuzzi, 2013). The role of complement in various disorders and diseases is complex, and complement-based therapy is multifaceted. Here we overviewed several kinds of diseases and disorders tightly associated with complement (Figure 22.2) and the development of their therapeutic strategies based on complement (Figure 22.3).

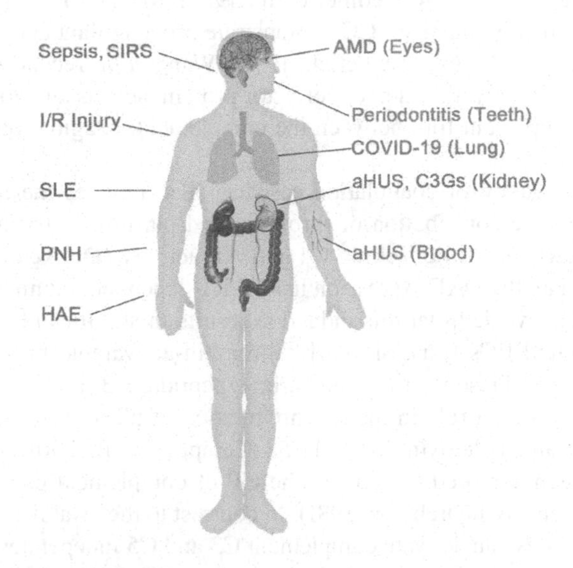

FIGURE 22.2 Complement-related diseases. SIRS: Systemic inflammatory response syndrome; I/R injury: Ischemia/reperfusion injury; SLE: Systemic lupus erythematosus; PNH: Paroxysmal nocturnal hemoglobinuria; HAE: Hereditary angioedema; AMD: Age-related macular degeneration; COVID-19: Coronavirus disease 2019; aHUS: Atypical hemolytic uremic syndrome; C3Gs: C3 glomerulopathies. Purple labels systemic diseases associated with complement; Red labels local diseases associated with complement.

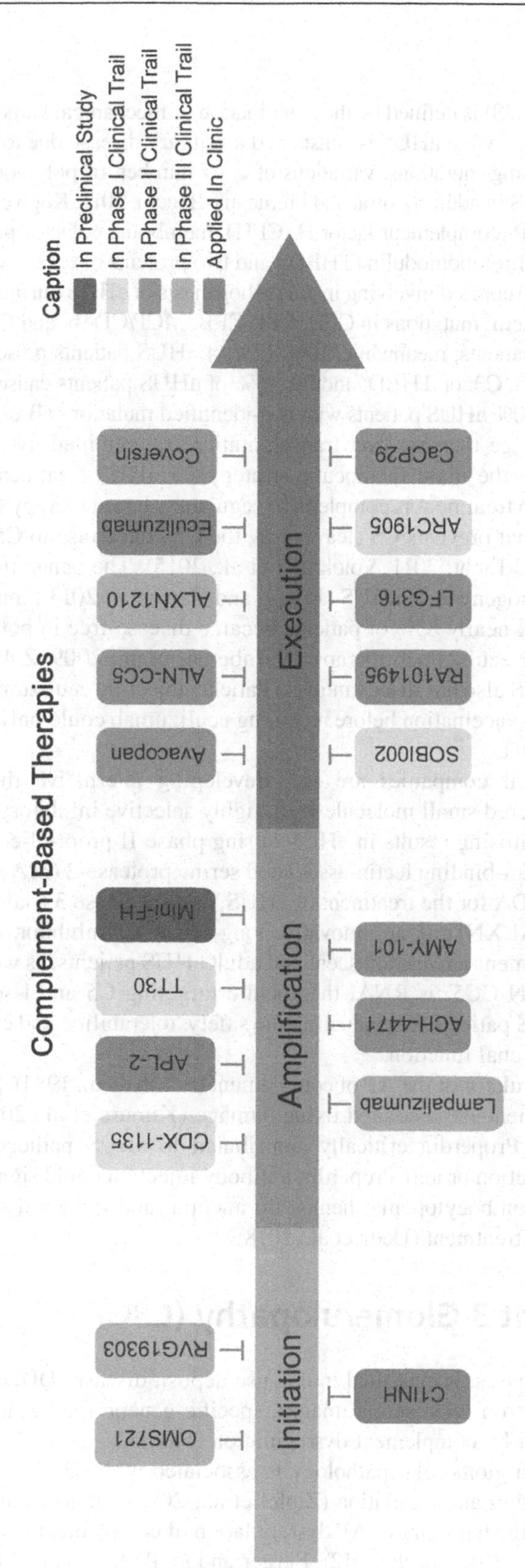

FIGURE 22.3 Complement-based therapy. Complement activation process includes initiation, amplification and execution. New drugs development targeting complement mainly focuses on initiation prevention, amplification prevention and attenuation and execution suppression. Although few complement-based drugs are now applied in clinic (green), there are still a series of drugs targeting different stages of complement activation being developed.

22.6.1 aHUS

Hemolytic uremic syndrome (HUS) is defined by the combination of mechanical hemolytic anemia, thrombocytopenia, and renal impairment, while aHUS is considered a primary disease due to abnormal regulation of complement AP with predisposing mutations, variations of copy number, or polymorphisms in complement genes, and is the majority of HUS in adults (Loirat and Fremeaux-Bacchi, 2011, Kopp et al., 2012). Four regulatory proteins of complement AP (complement factor H (CFH), membrane cofactor protein (MCP or CD46), complement factor I (CFI) and thrombomodulin (THBD)) and two proteins of C3 convertase (C3 and complement factor B (CFB)) have been reported involving in the pathogenesis of aHUS during recent decades (Joseph and Gattineni, 2013). Among them, mutations in CFH, CFI, CFB, MCP/CD46, and C3 comprise nearly 50% of known mutations in aHUS patients; meanwhile, 10%–12% of aHUS patients present combined mutations of CFH, CFI, CFB, MCP/CD46, C3, or THBD; another 10% of aHUS patients caused by complement CFH autoantibodies; the remaining 30% aHUS patients with non-identified mutations (Bresin et al., 2013).

Plasma infusion or exchange therapy, liver transplantation or combined liver-renal transplantation, and eculizumab treatment are the main therapeutic strategy for aHUS treatment (Zuber et al., 2012). Among them, only eculizumab treatment is complement-regulatory based therapy (Rathbone et al., 2013). Eculizumab is a mAb to C5 that prevents C5 cleavage by the C5 convertase to C5b, thus preventing the formation of TCC (Waters and Licht, 2011, Volokhina et al., 2015). The generation of TCC is the most classical character for the pathogenesis of aHUS (Joseph and Gattineni, 2013); thus, eculizumab showed good results in aHUS patients: nearly 85% of patients became disease-free in both plasma-resistant and plasma-dependent aHUS after eculizumab therapy (Nurnberger et al., 2009, Zuber et al., 2011). While eculizumab treatment for aHUS also has disadvantages: Patients injecting eculizumab increased Neisseria meningitis infection risk, even vaccination before receiving eculizumab could only show insufficient protective effects (Bouts et al., 2011).

Nowadays, pharmaceutical companies are still developing alternative therapeutics for aHUS. Avacopan, an orally administered small molecule with highly selective inhibitory effects on C5a receptor 1 (C5aR1), has shown promising results in aHUS during phase II proof-of-concept study. OMS721, the antibodies targeting mannan-binding lectin-associated serine protease-2 (MASP-2), has received Fast Track Designation from the FDA for the treatment of aHUS, and the phase 3 trial evaluating OMS721 in aHUS patients is underway. ALXN1210, an innovative long-acting C5 inhibitor, is also being evaluated for complement inhibitor treatment-naïve adolescent and adult aHUS patients, as well as pediatric patients with aHUS. Cemdisiran (ALN-CC5) is RNAi therapeutic targeting C5 and is now being initiated in phase 2 clinical trials on aHUS patients for evaluating the safety, tolerability, and effect of C5 knockdown on hematologic response and renal function.

Properdin is a positive regulator of the AP of complement (Smith et al., 1984). Inhibition of Properdin could ameliorate the complement AP-mediated tissue damage (Kimura et al., 2010, Miwa et al., 2013). Ueda Y et al. reported that Properdin critically contributed to aHUS pathogenesis. Suppression of Properdin through genetic deletion or anti-Properdin antibody injection could significantly inhibit aHUS induced premature death, thrombocytopenia, hemolytic anemia, and kidney disease, and might be of therapeutic benefits for aHUS treatment (Ueda et al., 2018).

22.6.2 Complement 3 Glomerulopathy (C3G)

C3G, a class of rare renal diseases, is classified into dense deposit disease (DDD) and C3 glomerulonephritis (C3GN) based on electron microscopy images, specific genetic types, and autoantibodies, both of which are mainly mediated by complement dysregulation (Fakhouri et al., 2010). The quintessential feature of these disorders with glomerular pathology is associated with C3 cleavage product deposition and defective complement action and regulation (Zipfel et al., 2015). Glomerular deposits of C3 alone, without immunoglobulin, are the hallmark of AP dysregulation of complement systems through inherited or acquired defects (D'Agati and Bomback, 2012). Berger and Galle first named DDD in 1962 from the electron-dense materials noted by electron microscopy in the lamina densa of the glomerular basement membrane (GBM) (Berger and Galle, 1962). Immunofluorescence showed predominant C3 deposition along with its breakdown products in the GBM, while less or no immunoglobulin was stained together (Master Sankar Raj et al., 2016). DDD is usually diagnosed in youth and adults with DDD also exist. The mainly presenting characters are proteinuria, hematuria, hypertension, and renal failure (Master Sankar Raj et al., 2016). In difference with DDD, C3GN is featured by C3 dominant deposition in the mesangium and capillary wall while intramembranous deposition is discontinuous. But the pathophysiology for alternate pathway activation in C3GN is similar to DDD with fluid-phase dysregulation because of varying mutations or autoantibodies; thus C3GN and DDD shared a similar disease course in a large United States cohort of patients with C3Gs (Bomback et al., 2018).

Up to now, there is no specific therapeutics for C3G approved. Supportive therapies such as angiotensin-converting enzyme inhibitors and angiotensin II receptor blockers controlling proteinuria and renal hemodynamics, lipid-lowering drugs including statins delaying the renal disorders are usually used for C3G treatment (Zoja et al., 2002, Appel et al., 2005, Christensen et al., 2006). Besides, plasma infusions or exchange might also work for offending autoantibodies remove and essential complement factors replenish (McGinley et al., 1985, Berthe-Aucejo et al., 2014). Welte et al. reported seven patients with C3G (5 C3GN and 2 DDD) were treated with C5 inhibitor eculizumab. Four patients presented with significantly improved or stable renal function and urinary protein; one patient showed positive response initially but relapsed and non-response after discontinuation and re-initiation of treatment. Two patients had impaired renal function and increased urinary protein with eculizumab treatment (Welte et al., 2018). Although the eculizumab therapeutic response is heterogeneous to patients, eculizumab might also be a therapeutic option for C3G and should still be deeply evaluated.

Avacopan, a selective and potent complement 5a receptor inhibitor, is now involved in a randomized, double-blind, placebo-controlled phase 2 study to evaluate the safety and efficacy in patients with C3G. While ACH-5228, a next-generation oral small molecule inhibitor of complement factor D; and ACH-4471, a first-in-class, oral inhibitor of complement factor D, has received positive opinion for orphan drug designation for C3G treatment in the European Union. AMY-101, a novel peptidic complement inhibitor developed based on the third-generation compstatin analog Cp40, could bind with C3 and inhibit C3 cleavage. The phase I clinical study of AMY-101 has been completed and now phase Ib/II clinical trials have been initiated including C3G indications.

Soluble CR1 can regulate C3 convertase and is also the only cofactor of factor I to promote cleavage of inactive C3b and inactive C4b into fragments. Zhang et al. found that soluble CR1 could reestablish regulation of the alternative complement pathway, CR1 prevented dysregulation of C3 convertase, even in the presence of C3 nephritic factors. *In vivo* experiments confirmed that soluble CR1 therapy stopped AP activation, resulting in the normalization of serum C3 levels and clearance of iC3b from GBM, suggesting soluble CR1 therapy improves complement regulation in C3G (Zhang et al., 2013). CDX-1135 (TP10), a soluble CR1, is now under phase I trial for DDD treatment.

22.6.3 Paroxysmal Nocturnal Hemoglobinuria (PNH)

PNH is a rare, clonal, acquired, life-threatening hematopoietic disease characterized by the destruction of red blood cells mediated by the complement system, especially by the AP of complement systems (Mastellos et al., 2014). Somatic mutations of phosphatidylinositol glycan class A caused glycosylphosphatidylinositol (GPI) anchor proteins deficiency, while the absence of two kinds of GPI anchor proteins, CD55, which regulating formation and stability of the C3 and C5 convertases, and CD59, which blocks MAC formation, leads to abnormal activation of the complement system, thus triggered PNH (Rollins and Sims, 1990, Rosse, 1997). In clinical settings, PNH patients mainly present chronic hemolytic anemia, bone marrow failure, thrombophilia, and peripheral blood cytopenias. (Parker et al., 2005).

Intravascular hemolysis via the MAC is the clinical hallmark of the disease, and the C5 blockade is currently the only approved treatment for PNH. Eculizumab, a first-in-class mAb that inhibits terminal complement, is the treatment of choice for patients with a severe manifestation of PNH and is the only antibody therapy approved for PNH by US FDA (Brodsky, 2014). Eculizumab treatment could decrease intravascular hemolysis, reduce thrombosis risk, and improve the quality of PNH patients' life (Brodsky et al., 2008). Bone marrow transplantation remains the only cure for PNH but should be reserved for patients with a suboptimal response to eculizumab as it is still a therapy associated with high morbidity and mortality (Varela and Brodsky, 2013).

Apart from eculizumab, some other candidates targeting C5 through inhibition of C5 activation by C5 convertases are still under clinical trials. ALXN1210, a long-acting C5 inhibitor, has been enrolled in a phase 3 trial for complement inhibitor treatment-naïve patients with PNH and showed positive top-line results in March 2018. Coversin, a recombinant small protein, is a second-generation and potentially best-in-class complement inhibitor acting on the prevention of C5a release and MAC formation. Coversin was initiated in phase II trial in February 2016 for patients resistant to eculizumab and now has completed phase II COBALT trial in patients with PNH in February 2018. SOBI 002 is also a small-molecular biologic protein targeting C5 under clinical trials (Berglund and Stromberg, 2016). RA101495, a synthetic macrocyclic peptide binding to C5, is now under clinical phase 2 trials for patients with PNH. Besides, based on RNAi technology for C5 knockdown, Cemdisiran (ALN-CC5) is now under phase I/II for PNH indication.

Although targeting C5 has reached a breakthrough in PNH treatment, recent studies have confirmed that residual anemia may persist in some patients regardless of sustained fluid-phase terminal complement inhibition (Risitano, 2013), and polymorphic variations in complement genes like C5 and CR1 might be responsible for the variable response to anti-C5 therapy (Nishimura et al., 2014, Rondelli et al., 2014). Thus, targeting upstream components of the complement system might bring more benefits for PNH therapy. APL-2 is a C3 inhibitor and has been approved by US FDA for the treatment of adult patients with PNH in 2021 and showed rapid and durable improvements in LDH and hemoglobin levels in PNH patients. AMY-101, a novel peptidic C3 inhibitor developed based on the third-generation compstatin analog Cp40, has been designated as an orphan drug for the treatment of PNH by the FDA and the European Medicines Agency (EMA). TT30 is a recombinant human fusion protein consisting of the iC3b/C3d-binding region of complement receptor 2 and the inhibitory domain of the complement AP regulator factor H (Fridkis-Hareli et al., 2011), and was involved in a phase I trial for untreated PNH patients. Mini-CFH is the second analogous agent that results in selective inhibition of activation and amplification of the complement AP and was found to be more effective than TT30. All these studies based on interfering upstream components of the complement system could give more alternative options for PNH treatment.

22.6.4 Age-Related Macular Degeneration (AMD)

AMD is an inflammatory disease that causes visual impairment and blindness (Lim et al., 2012). Characterized by retinal pigment epithelium, Bruch's membrane, and choriocapillaris tissue complex damage, AMD has become the leading cause of severe vision loss in elderly adults (McHarg et al., 2015). There have two phenotypes of AMD including neovascular AMD (wet type) and atrophic AMD (dry type). Wet AMD usually develops due to choroidal neovascularization, while dry AMD, which is characterized by geographic atrophy (GA), develops due to the death of retinal pigment epithelium. (Kuno and Fujii, 2011) The wet type of AMD affects approximately 10%–15% of patients with AMD, while about 85%–90% of AMD patients are in dry type (Gehrs et al., 2006).

The cause and development of AMD are complex and influenced by aging factors, environmental factors, nutritional factors, and genetic variants (Fritsche et al., 2014). In 2005, researchers reported that *complement factor H* (*CFH*), located on chromosome 1q, is a major AMD susceptibility gene. *CFH* gene was the first of the complement genes found to be associated with AMD risk. (Edwards et al., 2005, Haines et al., 2005, Klein et al., 2005). After that, three other complement genes, CFI, C3, and C9, were found existing non-synonymous rare variants related to AMD based on targeted genomic re-sequencing (Seddon et al., 2013). As variants in complement system genes alter an individual's risk of developing AMD, these findings highlighted the role of the complement system in the development of AMD (Bradley et al., 2011). Local and systemic inflammations in AMD are mediated by the deregulated action of the AP of the complement system, and the products of complement activation in AMD can be detected in the retina and blood. The treatment of neovascular AMD is currently based on monoclonal antibodies against vascular endothelial growth factor (anti-VEGF), which aims to prevent fragile new blood vessels growth.

Although the proteins of the complement system are central to the development of this disease, and many pharmaceutical scientists are developing candidates targeting complement components for AMD therapy, up to now, there is no AMD therapeutic strategy based on the complement system on the market. Lampalizumab is an antigen-binding fragment of complement factor D binding humanized mAb and was involved in clinical trials for patients with AMD, but its failure to reduce AMD-related GA lesion area, and failed in phase 3 trials in 2017. LFG316 is a full-human mAb with high-affinity binding to human and cynomolgus C5 to prevent cleavage by the C5 convertase to C5a and C5b. LFG316 is now being tested as a therapy for the treatment of GA under phase 2 trials. A prospective, double-blinded, randomized clinical trial was conducted to evaluate the effect of eculizumab on the growth of GA in AMD patients but failed to decrease the growth rate of GA significantly (Yehoshua et al., 2014). ARC1905 is a PEGylated RNA aptamer targeting complement C5 and showed promise for enhanced efficacy in AMD. A phase 1 trial testing ARC1905 in patients with exudative AMD in combination with ranibizumab has been completed, and the study evaluating ARC1905 in patients with nonexudative AMD is still in progress. POT-4 is an acyclic peptide binding to complement C3 and C3b, the phase 2 study evaluating 12 monthly doses on the progression of GA has been successfully finished using a reformulated version of POT-4 (APL-2), APL-2 could significantly slow the disease over 12 months (Tolentino et al., 2015).

22.6.5 Periodontitis

Periodontitis is a prevalent oral infection-derived inflammatory disease in tooth-supporting structures (Pihlstrom et al., 2005). Tooth-related anaerobic bacteria growing in a biofilm and infecting tooth surfaces including gingiva, bone, and periodontal ligament mainly induce periodontitis (Kinane et al., 2017). With the loss of connective tissue and bone support, Periodontitis has become the major cause of teeth loss in adults (Delima et al., 2002). Besides, environmental factors, behavioral factors, and genetic mutations also contribute to the occurrence of Periodontitis (Kinane et al., 2005). Meanwhile, periodontitis is also tightly associated with increased inflammation in the body like diabetes, pneumonia, atherosclerosis, and rheumatoid arthritis (Koromantzos et al., 2011, Bansal et al., 2013, Bartova et al., 2014, Mikuls et al., 2014).

The periodontal pathogen *Porphyromonas gingivalis* was considered the main bacteria associated with periodontitis (Hajishengallis, 2009). It could suppress antimicrobial responses, stimulate local inflammation, and cause collateral tissue damage (Hajishengallis et al., 2008). Besides, *Porphyromonas gingivalis* could subvert complement receptor 3 and C5a anaphylatoxin receptor signals. And more evidence has confirmed that complement is involved in periodontal *Porphyromonas gingivalis* transformation and in the inflammatory process leading to periodontal bone destruction (Hajishengallis et al., 2013). *Treponema denticola*, another periodontal pathogen, is a Gram-negative, anaerobic spirochete. Researchers have reported that *Treponema denticola* could activate the complement system (Schenkein and Berry, 1991), and *Treponema denticola* could also evade complement-mediated killing through binding factor H to its surface via the FhbB protein (Miller et al., 2014). Components of the complement system are present in the periodontal tissues and the system is activated and contributes to periodontal pathogenesis (Hajishengallis, 2010, Damgaard et al., 2015). Thus, targeting complement for periodontitis treatment might be an effective strategy.

The receptor C3R blockade promoted IL-12-mediated clearance of *Porphyromonas gingivalis* and negated its virulence *in vivo*, suggesting that complement receptor 3 blockade may represent a promising immunomodulatory approach for controlling human periodontitis and possibly associated systemic diseases (Hajishengallis et al., 2007). Researchers found that C3 was required for maximal periodontal inflammation and bone loss and for the sustenance of the dysbiotic microbiota. C3-targeted intervention will be a therapeutic strategy for human periodontitis. And the peptidic inhibitor of the central complement component

C3, AMY-101 (Cp40), appeared to be a promising candidate drug for the adjunctive treatment of human periodontitis (Maekawa et al., 2014, Kajikawa et al., 2017). Besides, mice deficient with C5aR were resistant to periodontitis, and local treatment with a C5aR antagonist inhibited periodontal inflammation through down-regulation of proinflammatory cytokines and further protected against bone loss. Thus, local targeting of C5aR would also be a powerful candidate for the treatment of human periodontitis (Abe et al., 2012).

22.6.6 Hereditary Angioedema (HAE)

HAE is a very rare but potentially life-threatening genetic disease (Lumry, 2018). Characterized by acute, recurrent, and self-limited edematous episodes of the face, extremities, trunk, genitals, upper airways, or the intestinal tract, HAE occurs in around 1:10000 to 1:50000 individuals and affects nearly 200,000 individuals all over the world (Cicardi and Agostoni, 1996, Bork, 2010). HAE is mainly caused by functional C1 esterase inhibitor deficiency (Banerji, 2010). C1 esterase inhibitor is a soluble, highly glycosylated serine protease inhibitor controlling various proteases involved in the complement, coagulation, and other systems and is also a key regulator of the complement system (Feussner et al., 2014). Besides, a study on the complete kinetic follow-up of symptoms and complement parameters during HAE attack indicated that C4a might be a useful biomarker for the prediction of edematous attacks as the level of C4a increased seven hours before HAE attack occurred while there's no increase on C1 esterase inhibitor's level even after resolution of the attack (Veszeli et al., 2018), while another study suggested that the high copy number of the C4B could be a protective factor against disease severity in HAE (Blasko et al., 2007).

There have three types of HAE according to different underlying causes and different serum C1 esterase inhibitor levels (Carugati et al., 2001). Type I and type II HAE exist mutations in the C1 esterase inhibitor gene located on chromosome 11 and are major forms of HAE: In type I HAE, levels of both functional and antigenic C1 esterase inhibitor are lower, while in type II HAE, only functional C1 esterase inhibitor level is reduced, level of antigenic C1 esterase inhibitor is normal or increased. Type III HAE does not present C1 esterase inhibitor deficiency but may be related to coagulation factor XII gene mutations (Agostoni et al., 2004, Bork, 2009). And all these three types of HAE have similar symptoms and signs (Csuka et al., 2017).

Although C1 esterase inhibitor was available for HAE treatment since the 1980s, the FDA approved the first human plasma-derived C1 esterase inhibitor in October 2008 for HAE prevention (Lunn et al., 2010). After that, nano-filtered, pasteurized human plasma-derived C1 esterase inhibitor and recombinant C1 esterase inhibitor were followed approved for HAE therapy. Besides, HAE patients with hyperreactivity to kallikrein-kinin cascade resulted in massive release of bradykinin, which is the mediator of pain and swelling. Plasma kallikrein inhibitor, ecallantide, and selective bradykinin-B2 receptor antagonist, icatibant, were also approved for HAE treatment. The optimal treatment for HAE attacks still needs to be defined, and the treatment options are still expanding.

22.6.7 SLE

SLE is a chronic autoimmune disease in which the body's immune system mistakenly attacks healthy tissues, including skin, joints, kidneys, and other organs. The cause of SLE is complex and has not been clearly studied. SLE is presumably caused by a genetic susceptibility coupled with an environmental trigger that results in defects in the immune system.

The complement system plays a major role in the initiation and development of SLE. However, the role of complement in SLE is complex since it may both prevent and exacerbate the disease. C1q deficiency is the strongest genetic risk factor for SLE (Walport et al., 1998). And the other genetic associations also include mutations in the complement receptors 1 and 2 (Asokan et al., 2013, Kim et al., 2016), as the expression of complement receptors 1 and 2 is decreased by nearly 50% on B cells of SLE patients and complement receptors 1 and 2 deficiency changed IgG3 autoantibody production and IgA glomerular deposition in SLE models (Boackle et al., 2004). Apart from C1q, CR1, and CR2, researchers also found that complement C4 inhibited systemic autoimmunity through complement receptors CR1- and CR2-independent mechanisms (Chen et al., 2000). Besides, SLE generated numerous disease manifestations involving contributions from complement such as glomerulonephritis and the increased risk of thrombosis. Most of the complement system is present in plasma; complement is very accessible and may be suitable as a biomarker for diagnosis or monitoring of disease activity (Leffler et al., 2014).

A clinical trial of eculizumab in SLE patients, especially those with active nephritis, was needed to assess its usefulness in the management of SLE, and the role of the mAb in SLE was investigated in a phase I single-dose study to explore the safety and collect pharmacodynamics and pharmacokinetic data of intravenously administered eculizumab in patients with SLE (Robak and Robak, 2009).

22.6.8 Sepsis

Sepsis is a potentially life-threatening complication of an infection. Sepsis occurs when chemicals are released into the bloodstream to fight the infection trigger inflammatory responses throughout the body. This inflammation can trigger a cascade of changes that can damage multiple organ systems, causing them to fail.

Sepsis is a potent activator of the hemostatic and complement systems (Lupu et al., 2014). Basic and clinical studies suggested that activation of complements, such as C3 and C5, was involved in the development of sepsis (Haeney, 1998). And all three upstream complement pathways were activated in septic shock. Measurement of complement activity is useful for evaluating the severity of sepsis (Nakae et al., 1994). The AP may be activated earlier than the CP pathway, and both pathways were essential to remove endotoxin, while the LP pathway also showed a protective role against sepsis (Charchaflieh et al., 2012). However, despite its role in protection, complements can also contribute to the development of severe complications that significantly worsen the prognosis of septic patients (Markiewski et al., 2008).

Severe sepsis leads to massive activation of coagulation and complement cascades that could contribute to multiple organ failure and death (Silasi-Mansat et al., 2010). Complement-coagulation interplay contributed to the progression of severe sepsis and blocking the harmful effects of complement activation products, especially during the organ failure stage of severe sepsis, is a potentially important therapeutic strategy (Amara et al., 2008). As coagulation proteases may act as natural C3 and C5 convertases, generating biologically active anaphylatoxins, linking both cascades via multiple direct interactions in terms of a complex serine protease system, inhibition of factor XII in septic baboons attenuated the activation of complement and fibrinolytic systems and reduced the release of IL-6 and neutrophil elastase (Jansen et al., 1996, Amara et al., 2010). While a recent study showed that only *Escherichia coli* challenge activated the complement system but not of thrombin-antithrombin (coagulation) and plasmin-antiplasmin (fibrinolysis). Despite inducing a strong burst of thrombin and plasmin, factor Xa/phospholipids infusion did not produce measurable levels of complement activation *in vivo*, suggesting that *in vivo* generated thrombin and plasmin do not directly activate the complement in nonhuman primates (Keshari et al., 2017).

Sepsis is often associated with hemostatic changes ranging from subclinical activation of blood coagulation, which may contribute to localized venous thromboembolism, to acute disseminated intravascular coagulation, characterized by widespread microvascular thrombosis and subsequent consumption of platelets and coagulation protein, eventually causing bleeding manifestations (Semeraro et al., 2012, Semeraro et al., 2015). Thrombus formation leading to vaso-occlusive events is a major cause of death and involves complex interactions between coagulation, fibrinolytic, and innate immune systems. Researchers found that plasmin bridged thrombosis and the immune response by liberating C5a and inducing MAC assembly. Complement inhibition could effectively attenuate collagen deposition and fibrotic responses in the lung after severe sepsis; inhibition complement could prove an attractive strategy for preventing sepsis-induced fibrosis of the lung (Silasi-Mansat et al., 2015). CaCP 29, a novel humanized mAb targeting complement C5a, was developed for the treatment of inflammation, particularly in septic organ dysfunction and systemic inflammatory response syndrome, is now involved in clinical trials (Ricklin et al., 2017).

22.6.9 Ischemia-Reperfusion (I-R) Injury

Cellular damage after reperfusion of ischemic tissues is defined as ischemia-reperfusion (I-R) injury. I-R injury is characterized by an initial deprivation of blood flow to tissues/organs followed by restoration of blood flow (Gorsuch et al., 2012). I-R injury could induce oxidant production, complement activation, leucocyte-EC adhesion, platelet-leucocyte aggregation, increased microvascular permeability, and decreased endothelium-dependent relaxation and could lead to multiorgan dysfunction and death (Eltzschig and Collard, 2004).

C3 cleavage associated with chemotaxis and leukocyte activation in the myocardial ischemia model was first studied on complement-mediated ischemic injury (Hill and Ward, 1971). Since then, the role of complement activation during I-R injury was widely investigated in multiple organs of the I-R injury model including the central neuron system, gastrointestinal system, lung, kidney, and retina (D'Ambrosio et al., 2001, Wada et al., 2001, Kuehn et al., 2008, Tuboly et al., 2016). Activation of CP, AP, and LP pathways and the generation of anaphylatoxins C3a and C5a contribute to I-R injury-mediated polymorphonuclear leukocytes recruitment, ROS generation, and cytokines release (Banz and Rieben, 2012). Attenuation of complement-mediated tissue injury was further demonstrated by complement components suppression or depletion (Diepenhorst et al., 2009). Using animal models, researchers found that soluble complement receptor type 1 (sCR1) could ameliorate local and remote organ injury after intestinal I-R injury (Hill et al., 1992), sCR1 also presented complement inhibitory and anti-inflammatory activities in model of reperfusion injury of ischemic myocardium and reduced myocardial infarction (Weisman et al., 1990). Human C1 esterase inhibitors could also suppress murine mesenteric I-R induced local organ injury.

22.6.10 Coronavirus Disease 2019 (COVID-19)

Coronavirus disease 2019 (COVID-19), caused by severe acute respiratory syndrome-coronavirus 2 (SARS-CoV-2), has become a global pandemic since December 2019 (Li, 2020). According to COVID-19 Dashboard contributed by the Center for Systems Science and Engineering (CSSE) at John Hopkins University, as of May 3, 2021, COVID-19 has caused 153,036,350 infections and 3,205,632 deaths all over the world (Dong et al., 2020). The activation of complement and contact system play a critical role during coronavirus pathogenesis and targeting complement activating pathway might open new strategies for COVID-19, especially in severe cases treatment (Holter et al., 2020).

C3 inhibitor AMY-101 was the first C3 inhibitor successfully used for a COVID-19 patient who had severe pneumonia and systemic hyper-inflammation in Italy. Amyndas Pharmaceuticals S.A. has launched Phase II clinical trials for patients with acute respiratory distress syndrome (ARDS) mediated by SARS-CoV-2 (Mastaglio et al., 2020). As the only approved complement system inhibitor for FDA, the anti-C5 mAb eculizumab was also showed potent effects for severe COVID-19 related ARDS and is currently in ongoing clinical trials (Diurno et al., 2020). Besides, a long-acting version of Eculizumab, Ravulizumab, was also launched for clinical trials in patients with COVID-19 severe pneumonia, acute lung injury, and ARDS in multi-countries (Smith et al., 2020). Anti-C5a monoclonal antibodies, BDB-001 developed by Staidson Biopharmaceuticals and IFX-1 developed by InflaRx GmbH, are also enrolling COVID-19 patients with severe ARDS for clinical trials (Polycarpou et al., 2020, Li and Chen, 2021).

22.7 SUMMARY AND FUTURE PERSPECTIVES

Collectively, complement, an ancient system, plays a central role in innate immunity and tightly linked innate immune response with adaptive response, is involved in a growing number of diseases. Although very few complement-based drugs have been approved in clinic, the clarity of complement systems and their roles in disease mechanisms will bring benefits for the complement-based drug development landscape. The chapter overviewed the classification of the complement pathway, the process of complement activation and regulation, the functions of complement in innate and adaptive immunity, and highlighted the role of complement in coagulation. In addition, this chapter further focused on diseases and disorders tightly associated with complement and prospected the development of their therapeutic strategies based on complement.

ACKNOWLEDGMENT

The authors would like to thank Dr. Jiajun Fan (Fudan University) for his help during figure preparation.

REFERENCES

Abe, T., Hosur, K. B., Hajishengallis, E., Reis, E. S., Ricklin, D., Lambris, J. D. & Hajishengallis, G. 2012. Local complement-targeted intervention in periodontitis: proof-of-concept using a C5a receptor (CD88) antagonist. *J Immunol*, 189, 5442–8.

Agostoni, A., Aygoren-Pursun, E., Binkley, K. E., Blanch, A., Bork, K., Bouillet, L., Bucher, C., Castaldo, A. J., Cicardi, M., Davis, A. E., De Carolis, C., Drouet, C., Duponchel, C., Farkas, H., Fay, K., Fekete, B., Fischer, B., Fontana, L., Fust, G., Giacomelli, R., Groner, A., Hack, C. E., Harmat, G., Jakenfelds, J., Juers, M., Kalmar, L., Kaposi, P. N., Karadi, I., Kitzinger, A., Kollar, T., Kreuz, W., Lakatos, P., Longhurst, H. J., Lopez-Trascasa, M., Martinez-Saguer, I., Monnier, N., Nagy, I., Nemeth, E., Nielsen, E. W., Nuijens, J. H., O'grady, C., Pappalardo, E., Penna, V., Perricone, C., Perricone, R., Rauch, U., Roche, O., Rusicke, E., Spath, P. J., Szendei, G., Takacs, E., Tordai, A., Truedsson, L., Varga, L., Visy, B., Williams, K., Zanichelli, A. & Zingale, L. 2004. Hereditary and acquired angioedema: problems and progress: proceedings of the third C1 esterase inhibitor deficiency workshop and beyond. *J Allergy Clin Immunol*, 114, S51–131.

Ahearn, J. M., Fischer, M. B., Croix, D., Goerg, S., Ma, M., Xia, J., Zhou, X., Howard, R. G., Rothstein, T. L. & Carroll, M. C. 1996. Disruption of the Cr2 locus results in a reduction in B-1a cells and in an impaired B cell response to T-dependent antigen. *Immunity*, 4, 251–62.

Alchi, B. & Jayne, D. 2010. Membranoproliferative glomerulonephritis. *Pediatr Nephrol*, 25, 1409–18.

Amara, U., Flierl, M. A., Rittirsch, D., Klos, A., Chen, H., Acker, B., Bruckner, U. B., Nilsson, B., Gebhard, F., Lambris, J. D. & Huber-Lang, M. 2010. Molecular intercommunication between the complement and coagulation systems. *J Immunol*, 185, 5628–36.

Amara, U., Rittirsch, D., Flierl, M., Bruckner, U., Klos, A., Gebhard, F., Lambris, J. D. & Huber-Lang, M. 2008. Interaction between the coagulation and complement system. *Adv Exp Med Biol*, 632, 71–9.

Ando, B., Wiedmer, T., Hamilton, K. K. & Sims, P. J. 1988. Complement proteins C5b-9 initiate secretion of platelet storage granules without increased binding of fibrinogen or von Willebrand factor to newly expressed cell surface GPIIb-IIIa. *J Biol Chem*, 263, 11907–14.

Andrews, P. W., Knowles, B. B., Parkar, M., Pym, B., Stanley, K. & Goodfellow, P. N. 1985. A human cell-surface antigen defined by a monoclonal antibody and controlled by a gene on human chromosome 1. *Ann Hum Genet*, 49, 31–9.

Appel, G. B., Cook, H. T., Hageman, G., Jennette, J. C., Kashgarian, M., Kirschfink, M., Lambris, J. D., Lanning, L., Lutz, H. U., Meri, S., Rose, N. R., Salant, D. J., Sethi, S., Smith, R. J., Smoyer, W., Tully, H. F., Tully, S. P., Walker, P., Welsh, M., Wurzner, R. & Zipfel, P. F. 2005. Membranoproliferative glomerulonephritis type II (dense deposit disease): an update. *J Am Soc Nephrol*, 16, 1392–403.

Asokan, R., Banda, N. K., Szakonyi, G., Chen, X. S. & Holers, V. M. 2013. Human complement receptor 2 (CR2/CD21) as a receptor for DNA: implications for its roles in the immune response and the pathogenesis of systemic lupus erythematosus (SLE). *Mol Immunol*, 53, 99–110.

Bacle, F., Haeffner-Cavaillon, N., Laude, M., Couturier, C. & Kazatchkine, M. D. 1990. Induction of IL-1 release through stimulation of the C3b/C4b complement receptor type one (CR1, CD35) on human monocytes. *J Immunol*, 144, 147–52.

Banerji, A. 2010. Current treatment of hereditary angioedema: An update on clinical studies. *Allergy Asthma Proc*, 31, 398–406.

Bansal, M., Khatri, M. & Taneja, V. 2013. Potential role of periodontal infection in respiratory diseases – a review. *J Med Life*, 6, 244–8.

Banz, Y. & Rieben, R. 2012. Role of complement and perspectives for intervention in ischemia-reperfusion damage. *Ann Med*, 44, 205–17.

Bartova, J., Sommerova, P., Lyuya-MI, Y., Mysak, J., Prochazkova, J., Duskova, J., Janatova, T. & Podzimek, S. 2014. Periodontitis as a risk factor of atherosclerosis. *J Immunol Res*, 2014, 636893.

Berger, J. & Galle, P. 1962. Unusual change of the basal membranes of the kidney. *J Urol Nephrol (Paris)*, 68, 116–22.

Berglund, M. M. & Stromberg, P. 2016. The clinical potential of Affibody-based inhibitors of C5 for therapeutic complement disruption. *Expert Rev Proteomics*, 13, 241–3.

Berthe-Aucejo, A., Sacquepee, M., Fila, M., Peuchmaur, M., Perrier-Cornet, E., Fremeaux-Bacchi, V. & Deschenes, G. 2014. Blockade of alternative complement pathway in dense deposit disease. *Case Rep Nephrol*, 2014, 201568.

Blasko, B., Szeplaki, G., Varga, L., Ronai, Z., Prohaszka, Z., Sasvari-Szekely, M., Visy, B., Farkas, H. & Fust, G. 2007. Relationship between copy number of genes (C4A, C4B) encoding the fourth component of complement and the clinical course of hereditary angioedema (HAE). *Mol Immunol*, 44, 2667–74.

Blom, A. M., Bergstrom, F., Edey, M., Diaz-Torres, M., Kavanagh, D., Lampe, A., Goodship, J. A., Strain, L., Moghal, N., Mchugh, M., Inward, C., Tomson, C., Fremeaux-Bacchi, V., Villoutreix, B. O. & Goodship, T. H. 2008. A novel non-synonymous polymorphism (p.Arg240His) in C4b-binding protein is associated with atypical hemolytic uremic syndrome and leads to impaired alternative pathway cofactor activity. *J Immunol*, 180, 6385–91.

Blom, A. M., Villoutreix, B. O. & Dahlback, B. 2004. Functions of human complement inhibitor C4b-binding protein in relation to its structure. *Arch Immunol Ther Exp (Warsz)*, 52, 83–95.

Boackle, S. A., Culhane, K. K., Brown, J. M., Haas, M., Bao, L., Quigg, R. J. & Holers, V. M. 2004. CR1/CR2 deficiency alters IgG3 autoantibody production and IgA glomerular deposition in the MRL/lpr model of SLE. *Autoimmunity*, 37, 111–23.

Bomback, A. S., Santoriello, D., Avasare, R. S., Regunathan-Shenk, R., Canetta, P. A., Ahn, W., Radhakrishnan, J., Marasa, M., Rosenstiel, P. E., Herlitz, L. C., Markowitz, G. S., D'agati, V. D. & APPEL, G. B. 2018. C3 glomerulonephritis and dense deposit disease share a similar disease course in a large United States cohort of patients with C3 glomerulopathy. *Kidney Int*, 93, 977–985.

Bork, K. 2009. Hereditary angioedema with normal cl inhibition. *Curr Allergy Asthma Rep*, 9, 280–5.

Bork, K. 2010. Diagnosis and treatment of hereditary angioedema with normal C1 inhibitor. *Allergy Asthma Clin Immunol*, 6, 15.

Botto, M., Kirschfink, M., Macor, P., Pickering, M. C., Wurzner, R. & Tedesco, F. 2009. Complement in human diseases: lessons from complement deficiencies. *Mol Immunol*, 46, 2774–83.

Bouts, A., Monnens, L., Davin, J. C., Struijk, G. & Spanjaard, L. 2011. Insufficient protection by Neisseria meningitidis vaccination alone during eculizumab therapy. *Pediatr Nephrol*, 26, 1919–20.

Bradley, D. T., Zipfel, P. F. & Hughes, A. E. 2011. Complement in age-related macular degeneration: a focus on function. *Eye (Lond)*, 25, 683–93.

Bresin, E., Rurali, E., Caprioli, J., Sanchez-Corral, P., Fremeaux-Bacchi, V., Rodriguez De Cordoba, S., Pinto, S., Goodship, T. H., Alberti, M., Ribes, D., Valoti, E., Remuzzi, G., Noris, M. & European working party on complement genetics in renal diseases. 2013. Combined complement gene mutations in atypical hemolytic uremic syndrome influence clinical phenotype. *J Am Soc Nephrol*, 24, 475–86.

Brodsky, R. A. 2014. Paroxysmal nocturnal hemoglobinuria. *Blood*, 124, 2804–11.

Brodsky, R. A., Young, N. S., Antonioli, E., Risitano, A. M., Schrezenmeier, H., Schubert, J., Gaya, A., Coyle, L., De Castro, C., Fu, C. L., Maciejewski, J. P., Bessler, M., Kroon, H. A., Rother, R. P. & Hillmen, P. 2008. Multicenter phase 3 study of the complement inhibitor eculizumab for the treatment of patients with paroxysmal nocturnal hemoglobinuria. *Blood*, 111, 1840–7.

Burstein, S. A., Peng, J., Friese, P., Wolf, R. F., Harrison, P., Downs, T., Hamilton, K., Comp, P. & Dale, G. L. 1996. Cytokine-induced alteration of platelet and hemostatic function. *Stem Cells*, 14 Suppl 1, 154–62.

Campbell, W., Okada, N. & Okada, H. 2001. Carboxypeptidase R is an inactivator of complement-derived inflammatory peptides and an inhibitor of fibrinolysis. *Immunol Rev*, 180, 162–7.

Carroll, M. C. 2004. The complement system in B cell regulation. *Mol Immunol*, 41, 141–6.

Carroll, M. C. 2008. Complement and humoral immunity. *Vaccine, 26 Suppl 8*, I28–33.

Carugati, A., Pappalardo, E., Zingale, L. C. & Cicardi, M. 2001. C1-inhibitor deficiency and angioedema. *Mol Immunol*, 38, 161–73.

Cattaneo, R. 2004. Four viruses, two bacteria, and one receptor: membrane cofactor protein (CD46) as pathogens' magnet. *J Virol*, 78, 4385–8.

Charbonneau, B., Maurer, M. J., Fredericksen, Z. S., Zent, C. S., Link, B. K., Novak, A. J., Ansell, S. M., Weiner, G. J., Wang, A. H., Witzig, T. E., Dogan, A., Slager, S. L., Habermann, T. M. & Cerhan, J. R. 2012. Germline variation in complement genes and event-free survival in follicular and diffuse large B-cell lymphoma. *Am J Hematol*, 87, 880–5.

Charchaflieh, J., Wei, J., Labaze, G., Hou, Y. J., Babarsh, B., Stutz, H., Lee, H., Worah, S. & Zhang, M. 2012. The role of complement system in septic shock. *Clin Dev Immunol*, 2012, 407324.

Chen, Z., Koralov, S. B. & Kelsoe, G. 2000. Complement C4 inhibits systemic autoimmunity through a mechanism independent of complement receptors CR1 and CR2. *J Exp Med*, 192, 1339–52.

Christensen, M., Su, A. W., Snyder, R. W., Greco, A., Lipschutz, J. H. & Madaio, M. P. 2006. Simvastatin protection against acute immune-mediated glomerulonephritis in mice. *Kidney Int*, 69, 457–63.

Cicardi, M. & Agostoni, A. 1996. Hereditary angioedema. *N Engl J Med*, 334, 1666–7.

Cicardi, M. & Zuraw, B. L. 2018. Angioedema due to bradykinin dysregulation. *J Allergy Clin Immunol Pract*, 6, 1132–41.

Clark, A., Weymann, A., Hartman, E., Turmelle, Y., Carroll, M., Thurman, J. M., Holers, V. M., Hourcade, D. E. & Rudnick, D. A. 2008. Evidence for non-traditional activation of complement factor C3 during murine liver regeneration. *Mol Immunol*, 45, 3125–32.

Cochrane, C. G. & Muller-Eberhard, H. J. 1968. The derivation of two distinct anaphylatoxin activities from the third and fifth components of human complement. *J Exp Med*, 127, 371–86.

Cragg, M. S., Howatt, W. J., Bloodworth, L., Anderson, V. A., Morgan, B. P. & Glennie, M. J. 2000. Complement mediated cell death is associated with DNA fragmentation. *Cell Death Differ*, 7, 48–58.

Crossley, L. G. & Porter, R. R. 1980. Purification of the human complement control protein C3b inactivator. *Biochem J*, 191, 173–82.

Csuka, D., Veszeli, N., Varga, L., Prohaszka, Z. & Farkas, H. 2017. The role of the complement system in hereditary angioedema. *Mol Immunol*, 89, 59–68.

D'agati, V. D. & Bomback, A. S. 2012. C3 glomerulopathy: what's in a name? *Kidney Int*, 82, 379–81.

D'ambrosio, A. L., Pinsky, D. J. & Connolly, E. S. 2001. The role of the complement cascade in ischemia/reperfusion injury: implications for neuroprotection. *Mol Med*, 7, 367–82.

Da Costa, X. J., Brockman, M. A., Alicot, E., Ma, M., Fischer, M. B., Zhou, X., Knipe, D. M. & Carroll, M. C. 1999. Humoral response to herpes simplex virus is complement-dependent. *Proc Natl Acad Sci U S A*, 96, 12708–12.

Damgaard, C., Holmstrup, P., Van Dyke, T. E. & Nielsen, C. H. 2015. The complement system and its role in the pathogenesis of periodontitis: current concepts. *J Periodontal Res*, 50, 283–93.

Davis, A. E., 3rd. 2005. The pathophysiology of hereditary angioedema. *Clin Immunol*, 114, 3–9.

Degn, S. E., Thiel, S. & Jensenius, J. C. 2013. Recombinant expression of the autocatalytic complement protease MASP-1 is crucially dependent on co-expression with its inhibitor, C1 inhibitor. *Protein Expr Purif*, 88, 173–82.

Del Conde, I., Cruz, M. A., Zhang, H., Lopez, J. A. & Afshar-Kharghan, V. 2005. Platelet activation leads to activation and propagation of the complement system. *J Exp Med*, 201, 871–9.

Delima, A. J., Karatzas, S., Amar, S. & Graves, D. T. 2002. Inflammation and tissue loss caused by periodontal pathogens is reduced by interleukin-1 antagonists. *J Infect Dis*, 186, 511–6.

Diepenhorst, G. M., Van Gulik, T. M. & Hack, C. E. 2009. Complement-mediated ischemia-reperfusion injury: lessons learned from animal and clinical studies. *Ann Surg*, 249, 889–99.

Diurno, F., Numis, F. G., Porta, G., Cirillo, F., Maddaluno, S., Ragozzino, A., De Negri, P., Di Gennaro, C., Pagano, A., Allegorico, E., Bressy, L., Bosso, G., Ferrara, A., Serra, C., Montisci, A., D'amico, M., Schiano Lo Morello, S., Di Costanzo, G., Tucci, A. G., Marchetti, P., Di Vincenzo, U., Sorrentino, I., Casciotta, A., Fusco, M., Buonerba, C., Berretta, M., Ceccarelli, M., Nunnari, G., Diessa, Y., Cicala, S. & Facchini, G. 2020. Eculizumab treatment in patients with COVID-19: preliminary results from real life ASL Napoli 2 Nord experience. *Eur Rev Med Pharmacol Sci*, 24, 4040–7.

Dong, E., Du, H. & Gardner, L. 2020. An interactive web-based dashboard to track COVID-19 in real time. *Lancet Infect Dis*, 20, 533–4.

Dunkelberger, J. R. & Song, W. C. 2010. Complement and its role in innate and adaptive immune responses. *Cell Res*, 20, 34–50.

Edwards, A. O., Ritter, R., 3rd, Abel, K. J., Manning, A., Panhuysen, C. & Farrer, L. A. 2005. Complement factor H polymorphism and age-related macular degeneration. *Science*, 308, 421–4.

Eltzschig, H. K. & Collard, C. D. 2004. Vascular ischaemia and reperfusion injury. *Br Med Bull*, 70, 71–86.

Endo, Y., Matsushita, M. & Fujita, T. 2015. New insights into the role of ficolins in the lectin pathway of innate immunity. *Int Rev Cell Mol Biol*, 316, 49–110.

Esser, A. F. 1994. The membrane attack complex of complement. assembly, structure and cytotoxic activity. *Toxicology*, 87, 229–47.

Fakhouri, F., Fremeaux-Bacchi, V., Noel, L. H., Cook, H. T. & Pickering, M. C. 2010. C3 glomerulopathy: a new classification. *Nat Rev Nephrol*, 6, 494–9.

Fang, C., Miwa, T., Shen, H. & Song, W. C. 2007. Complement-dependent enhancement of CD8+ T cell immunity to lymphocytic choriomeningitis virus infection in decay-accelerating factor-deficient mice. *J Immunol*, 179, 3178–86.

Feussner, A., Kalina, U., Hofmann, P., Machnig, T. & Henkel, G. 2014. Biochemical comparison of four commercially available C1 esterase inhibitor concentrates for treatment of hereditary angioedema. *Transfusion*, 54, 2566–73.

Fischer, M. B., Goerg, S., Shen, L., Prodeus, A. P., Goodnow, C. C., Kelsoe, G. & Carroll, M. C. 1998. Dependence of germinal center B cells on expression of CD21/CD35 for survival. *Science*, 280, 582–5.

Fischer, W. H. & Hugli, T. E. 1997. Regulation of B cell functions by C3a and C3a(desArg): suppression of TNF-alpha, IL-6, and the polyclonal immune response. *J Immunol*, 159, 4279–86.

Fleming, S. D., Shea-Donohue, T., Guthridge, J. M., Kulik, L., Waldschmidt, T. J., Gipson, M. G., Tsokos, G. C. & Holers, V. M. 2002. Mice deficient in complement receptors 1 and 2 lack a tissue injury-inducing subset of the natural antibody repertoire. *J Immunol*, 169, 2126–33.

Fridkis-Hareli, M., Storek, M., Mazsaroff, I., Risitano, A. M., Lundberg, A. S., Horvath, C. J. & Holers, V. M. 2011. Design and development of TT30, a novel C3d-targeted C3/C5 convertase inhibitor for treatment of human complement alternative pathway-mediated diseases. *Blood*, 118, 4705–13.

Fritsche, L. G., Fariss, R. N., Stambolian, D., Abecasis, G. R., Curcio, C. A. & Swaroop, A. 2014. Age-related macular degeneration: genetics and biology coming together. *Annu Rev Genomics Hum Genet*, 15, 151–71.

Fujita, T. 2002. Evolution of the lectin-complement pathway and its role in innate immunity. *Nat Rev Immunol*, 2, 346–53.

Fukuoka, Y., Xia, H. Z., Sanchez-Munoz, L. B., Dellinger, A. L., Escribano, L. & Schwartz, L. B. 2008. Generation of anaphylatoxins by human beta-tryptase from C3, C4, and C5. *J Immunol*, 180, 6307–16.

Gehrs, K. M., Anderson, D. H., Johnson, L. V. & Hageman, G. S. 2006. Age-related macular degeneration – emerging pathogenetic and therapeutic concepts. *Ann Med*, 38, 450–71.

Ghebrehiwet, B. 1981. C1q inhibitor (C1qINH): functional properties and possible relationship to a lymphocyte membrane-associated C1q precipitin. *J Immunol*, 126, 1837–42.

Gorsuch, W. B., Chrysanthou, E., Schwaeble, W. J. & Stahl, G. L. 2012. The complement system in ischemia-reperfusion injuries. *Immunobiology*, 217, 1026–33.

Gros, P., Milder, F. J. & Janssen, B. J. 2008. Complement driven by conformational changes. *Nat Rev Immunol*, 8, 48–58.

Gushiken, F. C., Han, H., Li, J., Rumbaut, R. E. & Afshar-Kharghan, V. 2009. Abnormal platelet function in C3-deficient mice. *J Thromb Haemost*, 7, 865–70.

Gustavsson, S., Kinoshita, T. & Heyman, B. 1995. Antibodies to murine complement receptor 1 and 2 can inhibit the antibody response in vivo without inhibiting T helper cell induction. *J Immunol*, 154, 6524–8.

Haas, P. J. & Van Strijp, J. 2007. Anaphylatoxins: their role in bacterial infection and inflammation. *Immunol Res*, 37, 161–75.

Haeney, M. R. 1998. The role of the complement cascade in sepsis. *J Antimicrob Chemother*, 41 Suppl A, 41–6.

Haines, J. L., Hauser, M. A., Schmidt, S., Scott, W. K., Olson, L. M., Gallins, P., Spencer, K. L., Kwan, S. Y., Noureddine, M., Gilbert, J. R., Schnetz-Boutaud, N., Agarwal, A., Postel, E. A. & Pericak-Vance, M. A. 2005. Complement factor H variant increases the risk of age-related macular degeneration. *Science*, 308, 419–21.

Hajishengallis, G. 2009. Porphyromonas gingivalis-host interactions: open war or intelligent guerilla tactics? *Microbes Infect*, 11, 637–45.

Hajishengallis, G. 2010. Complement and periodontitis. *Biochem Pharmacol*, 80, 1992–2001.

Hajishengallis, G., Abe, T., Maekawa, T., Hajishengallis, E. & Lambris, J. D. 2013. Role of complement in host-microbe homeostasis of the periodontium. *Semin Immunol*, 25, 65–72.

Hajishengallis, G., Shakhatreh, M. A., Wang, M. & Liang, S. 2007. Complement receptor 3 blockade promotes IL-12-mediated clearance of Porphyromonas gingivalis and negates its virulence in vivo. *J Immunol*, 179, 2359–67.

Hajishengallis, G., Wang, M., Liang, S., Triantafilou, M. & Triantafilou, K. 2008. Pathogen induction of CXCR4/TLR2 cross-talk impairs host defense function. *Proc Natl Acad Sci U S A*, 105, 13532–7.

Hakobyan, S., Harris, C. L., Tortajada, A., Goicochea De Jorge, E., Garcia-Layana, A., Fernandez-Robredo, P., Rodriguez De Cordoba, S. & Morgan, B. P. 2008. Measurement of factor H variants in plasma using variant-specific monoclonal antibodies: application to assessing risk of age-related macular degeneration. *Invest Ophthalmol Vis Sci*, 49, 1983–90.

Hakobyan, S., Tortajada, A., Harris, C. L., De Cordoba, S. R. & Morgan, B. P. 2010. Variant-specific quantification of factor H in plasma identifies null alleles associated with atypical hemolytic uremic syndrome. *Kidney Int*, 78, 782–8.

Hamilton, K. K., Ji, Z., Rollins, S., Stewart, B. H. & Sims, P. J. 1990. Regulatory control of the terminal complement proteins at the surface of human endothelial cells: neutralization of a C5b-9 inhibitor by antibody to CD59. *Blood*, 76, 2572–7.
Harboe, M., Johnson, C., Nymo, S., Ekholt, K., Schjalm, C., Lindstad, J. K., Pharo, A., Hellerud, B. C., Nilsson Ekdahl, K., Mollnes, T. E. & Nilsson, P. H. 2017. Properdin binding to complement activating surfaces depends on initial C3b deposition. *Proc Natl Acad Sci U S A*, 114, E534–E539.
Harboe, M. & Mollnes, T. E. 2008. The alternative complement pathway revisited. *J Cell Mol Med*, 12, 1074–84.
Heeger, P. S., Lalli, P. N., Lin, F., Valujskikh, A., Liu, J., Muqim, N., Xu, Y. & Medof, M. E. 2005. Decay-accelerating factor modulates induction of T cell immunity. *J Exp Med*, 201, 1523–30.
Hill, J., Lindsay, T. F., Ortiz, F., Yeh, C. G., Hechtman, H. B. & Moore, F. D., Jr. 1992. Soluble complement receptor type 1 ameliorates the local and remote organ injury after intestinal ischemia-reperfusion in the rat. *J Immunol*, 149, 1723–8.
Hill, J. H. & Ward, P. A. 1971. The phlogistic role of C3 leukotactic fragments in myocardial infarcts of rats. *J Exp Med*, 133, 885–900.
Hofmeyer, T., Schmelz, S., Degiacomi, M. T., Dal Peraro, M., Daneschdar, M., Scrima, A., Van Den Heuvel, J., Heinz, D. W. & Kolmar, H. 2013. Arranged sevenfold: structural insights into the C-terminal oligomerization domain of human C4b-binding protein. *J Mol Biol*, 425, 1302–17.
Holers, V. M. 2014. Complement and its receptors: new insights into human disease. *Annu Rev Immunol*, 32, 433–59.
Holter, J. C., Pischke, S. E., De Boer, E., Lind, A., Jenum, S., Holten, A. R., Tonby, K., Barratt-Due, A., Sokolova, M., Schjalm, C., Chaban, V., Kolderup, A., Tran, T., Tollefsrud Gjolberg, T., Skeie, L. G., Hesstvedt, L., Ormasen, V., Fevang, B., Austad, C., Muller, K. E., Fladeby, C., Holberg-Petersen, M., Halvorsen, B., Muller, F., Aukrust, P., Dudman, S., Ueland, T., Andersen, J. T., Lund-Johansen, F., Heggelund, L., Dyrhol-Riise, A. M. & Mollnes, T. E. 2020. Systemic complement activation is associated with respiratory failure in COVID-19 hospitalized patients. *Proc Natl Acad Sci U S A*, 117, 25018–25.
Huber-Lang, M., Denk, S., Fulda, S., Erler, E., Kalbitz, M., Weckbach, S., Schneider, E. M., Weiss, M., Kanse, S. M. & Perl, M. 2012. Cathepsin D is released after severe tissue trauma in vivo and is capable of generating C5a in vitro. *Mol Immunol*, 50, 60–5.
Ikeda, K., Nagasawa, K., Horiuchi, T., Tsuru, T., Nishizaka, H. & Niho, Y. 1997. C5a induces tissue factor activity on endothelial cells. *Thromb Haemost*, 77, 394–8.
Jansen, P. M., Pixley, R. A., Brouwer, M., De Jong, I. W., Chang, A. C., Hack, C. E., Taylor, F. B., Jr. & Colman, R. W. 1996. Inhibition of factor XII in septic baboons attenuates the activation of complement and fibrinolytic systems and reduces the release of interleukin-6 and neutrophil elastase. *Blood*, 87, 2337–44.
Joseph, C. & Gattineni, J. 2013. Complement disorders and hemolytic uremic syndrome. *Curr Opin Pediatr*, 25, 209–15.
Junnikkala, S., Jokiranta, T. S., Friese, M. A., Jarva, H., Zipfel, P. F. & Meri, S. 2000. Exceptional resistance of human H2 glioblastoma cells to complement-mediated killing by expression and utilization of factor H and factor H-like protein 1. *J Immunol*, 164, 6075–81.
Kajikawa, T., Briones, R. A., Resuello, R. R. G., Tuplano, J. V., Reis, E. S., Hajishengallis, E., Garcia, C. A. G., Yancopoulou, D., Lambris, J. D. & Hajishengallis, G. 2017. Safety and efficacy of the complement inhibitor AMY-101 in a natural model of periodontitis in non-human primates. *Mol Ther Methods Clin Dev*, 6, 207–15.
Kanse, S. M., Gallenmueller, A., Zeerleder, S., Stephan, F., Rannou, O., Denk, S., Etscheid, M., Lochnit, G., Krueger, M. & Huber-Lang, M. 2012. Factor VII-activating protease is activated in multiple trauma patients and generates anaphylatoxin C5a. *J Immunol*, 188, 2858–65.
Karp, C. L., Wysocka, M., Wahl, L. M., Ahearn, J. M., Cuomo, P. J., Sherry, B., Trinchieri, G. & Griffin, D. E. 1996. Mechanism of suppression of cell-mediated immunity by measles virus. *Science*, 273, 228–31.
Kawasaki, N., Kawasaki, T. & Yamashina, I. 1983. Isolation and characterization of a mannan-binding protein from human serum. *J Biochem*, 94, 937–47.
Kemper, C., Pangburn, M. K. & Fishelson, Z. 2014. Complement nomenclature 2014. *Mol Immunol*, 61, 56–8.
Keshari, R. S., Silasi, R., Lupu, C., Taylor, F. B., Jr. & Lupu, F. 2017. In vivo-generated thrombin and plasmin do not activate the complement system in baboons. *Blood*, 130, 2678–81.
Kim, T. H., Bae, S. C., Lee, S. H., Kim, S. Y. & Baek, S. H. 2016. Association of complement receptor 2 gene polymorphisms with susceptibility to osteonecrosis of the femoral head in systemic lupus erythematosus. *Biomed Res Int*, 2016, 9208035.
Kim, A. H., Dimitriou, I. D., Holland, M. C., Mastellos, D., Mueller, Y. M., Altman, J. D., Lambris, J. D. & Katsikis, P. D. 2004. Complement C5a receptor is essential for the optimal generation of antiviral CD8+ T cell responses. *J Immunol*, 173, 2524–9.
Kim, D. D. & Song, W. C. 2006. Membrane complement regulatory proteins. *Clin Immunol*, 118, 127–36.
Kimura, Y., Zhou, L., Miwa, T. & Song, W. C. 2010. Genetic and therapeutic targeting of properdin in mice prevents complement-mediated tissue injury. *J Clin Invest*, 120, 3545–54.

Kinane, D. F., Shiba, H. & Hart, T. C. 2005. The genetic basis of periodontitis. *Periodontol 2000*, 39, 91–117.

Kinane, D. F., Stathopoulou, P. G. & Papapanou, P. N. 2017. *Periodontal diseases. Nat Rev Dis Primers*, 3, 17038.

Kinoshita, T., Thyphronitis, G., Tsokos, G. C., Finkelman, F. D., Hong, K., Sakai, H. & Inoue, K. 1990. Characterization of murine complement receptor type 2 and its immunological cross-reactivity with type 1 receptor. *Int Immunol*, 2, 651–9.

Klein, R. J., Zeiss, C., Chew, E. Y., Tsai, J. Y., Sackler, R. S., Haynes, C., Henning, A. K., Sangiovanni, J. P., Mane, S. M., Mayne, S. T., Bracken, M. B., Ferris, F. L., Ott, J., Barnstable, C. & Hoh, J. 2005. Complement factor H polymorphism in age-related macular degeneration. *Science*, 308, 385–9.

Kopf, M., Abel, B., Gallimore, A., Carroll, M. & Bachmann, M. F. 2002. Complement component C3 promotes T-cell priming and lung migration to control acute influenza virus infection. *Nat Med*, 8, 373–8.

Kopp, A., Strobel, S., Tortajada, A., Rodriguez De Cordoba, S., Sanchez-Corral, P., Prohaszka, Z., Lopez-Trascasa, M. & Jozsi, M. 2012. Atypical hemolytic uremic syndrome-associated variants and autoantibodies impair binding of factor h and factor h-related protein 1 to pentraxin 3. *J Immunol*, 189, 1858–67.

Koromantzos, P. A., Makrilakis, K., Dereka, X., Katsilambros, N., Vrotsos, I. A. & Madianos, P. N. 2011. A randomized, controlled trial on the effect of non-surgical periodontal therapy in patients with type 2 diabetes. Part I: effect on periodontal status and glycaemic control. *J Clin Periodontol*, 38, 142–7.

Krarup, A., Wallis, R., Presanis, J. S., Gal, P. & Sim, R. B. 2007. Simultaneous activation of complement and coagulation by MBL-associated serine protease 2. *PLoS One*, 2, e623.

Krych-Goldberg, M. & Atkinson, J. P. 2001. Structure-function relationships of complement receptor type 1. *Immunol Rev*, 180, 112–22.

Kuehn, M. H., Kim, C. Y., Jiang, B., Dumitrescu, A. V. & Kwon, Y. H. 2008. Disruption of the complement cascade delays retinal ganglion cell death following retinal ischemia-reperfusion. *Exp Eye Res*, 87, 89–95.

Kuno, N. & Fujii, S. 2011. Dry age-related macular degeneration: recent progress of therapeutic approaches. *Curr Mol Pharmacol*, 4, 196–232.

Kupp, L. I., Kosco, M. H., Schenkein, H. A. & Tew, J. G. 1991. Chemotaxis of germinal center B cells in response to C5a. *Eur J Immunol*, 21, 2697–701.

Kwan, W. H., Van Der Touw, W., Paz-Artal, E., Li, M. O. & Heeger, P. S 2013. Signaling through C5a receptor and C3a receptor diminishes function of murine natural regulatory T cells. *J Exp Med*, 210, 257–68.

Lachmann, P. J. 2009. The amplification loop of the complement pathways. *Adv Immunol*, 104, 115–49.

Lalli, P. N., Strainic, M. G., Yang, M., Lin, F., Medof, M. E. & Heeger, P. S. 2008. Locally produced C5a binds to T cell-expressed C5aR to enhance effector T-cell expansion by limiting antigen-induced apoptosis. *Blood*, 112, 1759–66.

Leffler, J., Bengtsson, A. A. & Blom, A. M. 2014. The complement system in systemic lupus erythematosus: an update. *Ann Rheum Dis*, 73, 1601–6.

Lesher, A. M., Zhou, L., Kimura, Y., Sato, S., Gullipalli, D., Herbert, A. P., Barlow, P. N., Eberhardt, H. U., Skerka, C., Zipfel, P. F., Hamano, T., Miwa, T., Tung, K. S. & Song, W. C. 2013. Combination of factor H mutation and properdin deficiency causes severe C3 glomerulonephritis. *J Am Soc Nephrol*, 24, 53–65.

Levi, M., Lowenberg, E. & Meijers, J. C. 2010. Recombinant anticoagulant factors for adjunctive treatment of sepsis. *Semin Thromb Hemost*, 36, 550–7.

Levi, M. & Van Der Poll, T. 2010. Inflammation and coagulation. *Crit Care Med*, 38, S26–34.

Li, Y. 2020. Human neutralizing antibodies to SARS-CoV-2: views and perspectives from Professor Linqi Zhang at Tsinghua University. *Antib Ther*, 3, 155–6.

Li, K., Anderson, K. J., Peng, Q., Noble, A., Lu, B., Kelly, A. P., Wang, N., Sacks, S. H. & Zhou, W. 2008. Cyclic AMP plays a critical role in C3a-receptor-mediated regulation of dendritic cells in antigen uptake and T-cell stimulation. *Blood*, 112, 5084–94.

Li, Q. & Chen, Z. 2021. An update: the emerging evidence of complement involvement in COVID-19. *Med Microbiol Immunol*, 210, 101–9.

Lidington, E. A., Haskard, D. O. & Mason, J. C. 2000. Induction of decay-accelerating factor by thrombin through a protease-activated receptor 1 and protein kinase C-dependent pathway protects vascular endothelial cells from complement-mediated injury. *Blood*, 96, 2784–92.

Lim, L. S., Mitchell, P., Seddon, J. M., Holz, F. G. & Wong, T. Y. 2012. Age-related macular degeneration. *Lancet*, 379, 1728–38.

Liszewski, M. K., Farries, T. C., Lublin, D. M., Rooney, I. A. & Atkinson, J. P. 1996. Control of the complement system. *Adv Immunol*, 61, 201–83.

Liu, J., Lin, F., Strainic, M. G., An, F., Miller, R. H., Altuntas, C. Z., Heeger, P. S., Tuohy, V. K. & Medof, M. E. 2008. IFN-gamma and IL-17 production in experimental autoimmune encephalomyelitis depends on local APC-T cell complement production. *J Immunol*, 180, 5882–9.

Liu, J., Miwa, T., Hilliard, B., Chen, Y., Lambris, J. D., Wells, A. D. & Song, W. C. 2005. The complement inhibitory protein DAF (CD55) suppresses T cell immunity in vivo. *J Exp Med*, 201, 567–77.

Liu, D. & Niu, Z. X. 2009. The structure, genetic polymorphisms, expression and biological functions of complement receptor type 1 (CR1/CD35). *Immunopharmacol Immunotoxicol*, 31, 524–35.

Loirat, C. & Fremeaux-Bacchi, V. 2011. Atypical hemolytic uremic syndrome. *Orphanet J Rare Dis*, 6, 60.

Lumry, W. R. 2018. Hereditary angioedema: the economics of treatment of an orphan disease. *Front Med (Lausanne)*, 5, 22.

Lunn, M., Santos, C. & Craig, T. 2010. Cinryze as the first approved C1 inhibitor in the USA for the treatment of hereditary angioedema: approval, efficacy and safety. *J Blood Med*, 1, 163–70.

Lupu, F., Keshari, R. S., Lambris, J. D. & Coggeshall, K. M. 2014. Crosstalk between the coagulation and complement systems in sepsis. *Thromb Res*, 133 Suppl 1, S28–31.

Lutz, H. U. & Jelezarova, E. 2006. Complement amplification revisited. *Mol Immunol*, 43, 2–12.

Maekawa, T., Abe, T., Hajishengallis, E., Hosur, K. B., Deangelis, R. A., Ricklin, D., Lambris, J. D. & Hajishengallis, G. 2014. Genetic and intervention studies implicating complement C3 as a major target for the treatment of periodontitis. *J Immunol*, 192, 6020–7.

Markiewski, M. M., Deangelis, R. A. & Lambris, J. D. 2008. Complexity of complement activation in sepsis. *J Cell Mol Med*, 12, 2245–54.

Markiewski, M. M., Deangelis, R. A., Strey, C. W., Foukas, P. G., Gerard, C., Gerard, N., Wetsel, R. A. & Lambris, J. D. 2009. The regulation of liver cell survival by complement. *J Immunol*, 182, 5412–8.

Markiewski, M. M. & Lambris, J. D. 2007. The role of complement in inflammatory diseases from behind the scenes into the spotlight. *Am J Pathol*, 171, 715–27.

Mastaglio, S., Ruggeri, A., Risitano, A. M., Angelillo, P., Yancopoulou, D., Mastellos, D. C., Huber-Lang, M., Piemontese, S., Assanelli, A., Garlanda, C., Lambris, J. D. & Ciceri, F. 2020. The first case of COVID-19 treated with the complement C3 inhibitor AMY-101. *Clin Immunol*, 215, 108450.

Mastellos, D. C., Ricklin, D., Yancopoulou, D., Risitano, A. & Lambris, J. D. 2014. Complement in paroxysmal nocturnal hemoglobinuria: exploiting our current knowledge to improve the treatment landscape. *Expert Rev Hematol*, 7, 583–98.

Master Sankar Raj, V., Gordillo, R. & Chand, D. H. 2016. Overview of C3 Glomerulopathy. *Front Pediatr, 4*, 45.

Matsushita, M. & Fujita, T. 1995. Cleavage of the third component of complement (C3) by mannose-binding protein-associated serine protease (MASP) with subsequent complement activation. *Immunobiology*, 194, 443–48.

Mcginley, E., Watkins, R., Mclay, A. & Boulton-Jones, J. M. 1985. Plasma exchange in the treatment of Mesangiocapillary glomerulonephritis. *Nephron*, 40, 385–90.

Mcharg, S., Clark, S. J., Day, A. J. & Bishop, P. N. 2015. Age-related macular degeneration and the role of the complement system. *Mol Immunol*, 67, 43–50.

Medof, M. E., Kinoshita, T. & Nussenzweig, V. 1984. Inhibition of complement activation on the surface of cells after incorporation of decay-accelerating factor (DAF) into their membranes. *J Exp Med*, 160, 1558–78.

Meyer, S., Leusen, J. H. & Boross, P. 2014. Regulation of complement and modulation of its activity in monoclonal antibody therapy of cancer. *MAbs*, 6, 1133–44.

Mikuls, T. R., Payne, J. B., Yu, F., Thiele, G. M., Reynolds, R. J., Cannon, G. W., Markt, J., Mcgowan, D., Kerr, G. S., Redman, R. S., Reimold, A., Griffiths, G., Beatty, M., Gonzalez, S. M., Bergman, D. A., Hamilton, B. C., 3rd, Erickson, A. R., Sokolove, J., Robinson, W. H., Walker, C., Chandad, F. & O'dell, J. R. 2014. Periodontitis and Porphyromonas gingivalis in patients with rheumatoid arthritis. *Arthritis Rheumatol*, 66, 1090–100.

Miller, D. J., Hemmrich, G., Ball, E. E., Hayward, D. C., Khalturin, K., Funayama, N., Agata, K. & Bosch, T. C. 2007. The innate immune repertoire in cnidaria–ancestral complexity and stochastic gene loss. *Genome Biol*, 8, R59.

Miller, D. P., Mcdowell, J. V., Bell, J. K., Goetting-Minesky, M. P., Fenno, J. C. & Marconi, R. T. 2014. Analysis of the complement sensitivity of oral treponemes and the potential influence of FH binding, FH cleavage and dentilisin activity on the pathogenesis of periodontal disease. *Mol Oral Microbiol*, 29, 194–207.

Miwa, T., Sato, S., Gullipalli, D., Nangaku, M. & Song, W. C. 2013. Blocking properdin, the alternative pathway, and anaphylatoxin receptors ameliorates renal ischemia-reperfusion injury in decay-accelerating factor and CD59 double-knockout mice. *J Immunol*, 190, 3552–9.

Morgan, E. L., Thoman, M. L., Weigle, W. O. & Hugli, T. E. 1983. Anaphylatoxin-mediated regulation of the immune response. II. C5a-mediated enhancement of human humoral and T cell-mediated immune responses. *J Immunol*, 130, 1257–61.

Morgan, E. L., Weigle, W. O. & Hugli, T. E. 1982. Anaphylatoxin-mediated regulation of the immune response. I. C3a-mediated suppression of human and murine humoral immune responses. *J Exp Med*, 155, 1412–26.

Muller, F. & Renne, T. 2008. Novel roles for factor XII-driven plasma contact activation system. *Curr Opin Hematol*, 15, 516–21.

Muller-Eberhard, H. J., Polley, M. J. & Calcott, M. A. 1967. Formation and functional significance of a molecular complex derived from the second and the fourth component of human complement. *J Exp Med*, 125, 359–80.

Nakae, H., Endo, S., Inada, K., Takakuwa, T., Kasai, T. & Yoshida, M. 1994. Serum complement levels and severity of sepsis. *Res Commun Chem Pathol Pharmacol*, 84, 189–95.

Nilsson, U. R., Funke, L., Nilsson, B. & Ekdahl, K. N. 2011a. Two conformational forms of target-bound iC3b that distinctively bind complement receptors 1 and 2 and two specific monoclonal antibodies. *Ups J Med Sci*, 116, 26–33.

Nilsson, S. C., Sim, R. B., Lea, S. M., Fremeaux-Bacchi, V. & Blom, A. M. 2011b. Complement factor I in health and disease. *Mol Immunol*, 48, 1611–20.

Nishimura, J., Yamamoto, M., Hayashi, S., Ohyashiki, K., Ando, K., Brodsky, A. L., Noji, H., Kitamura, K., Eto, T., Takahashi, T., Masuko, M., Matsumoto, T., Wano, Y., Shichishima, T., Shibayama, H., Hase, M., Li, L., Johnson, K., Lazarowski, A., Tamburini, P., Inazawa, J., Kinoshita, T. & Kanakura, Y. 2014. Genetic variants in C5 and poor response to eculizumab. *N Engl J Med*, 370, 632–9.

Nonaka, M. & Kimura, A. 2006. Genomic view of the evolution of the complement system. *Immunogenetics*, 58, 701–13.

Noris, M. & Remuzzi, G. 2013. Overview of complement activation and regulation. *Semin Nephrol*, 33, 479–92.

Nurnberger, J., Philipp, T., Witzke, O., Opazo Saez, A., Vester, U., Baba, H. A., Kribben, A., Zimmerhackl, L. B., Janecke, A. R., Nagel, M. & Kirschfink, M. 2009. Eculizumab for atypical hemolytic-uremic syndrome. *N Engl J Med*, 360, 542–4.

Nussenzweig, V. 1971. Complement-receptor lymphocytes. *Am J Pathol*, 65, 479–80.

O'neil, K. M., Ochs, H. D., Heller, S. R., Cork, L. C., Morris, J. M. & Winkelstein, J. A. 1988. Role of C3 in humoral immunity. Defective antibody production in C3-deficient dogs. *J Immunol*, 140, 1939–45.

Ochs, H. D., Wedgwood, R. J., Heller, S. R. & Beatty, P. G. 1986. Complement, membrane glycoproteins, and complement receptors: their role in regulation of the immune response. *Clin Immunol Immunopathol*, 40, 94–104.

Papadimitriou, J. C., Ramm, L. E., Drachenberg, C. B., Trump, B. F. & Shin, M. L. 1991. Quantitative analysis of adenine nucleotides during the prelytic phase of cell death mediated by C5b-9. *J Immunol*, 147, 212–7.

Papamichail, M., Gutierrez, C., Embling, P., Johnson, P., Holborow, E. J. & Pepys, M. B. 1975. Complement dependence of localisation of aggregated IgG in germinal centres. *Scand J Immunol*, 4, 343–47.

Parker, C., Omine, M., Richards, S., Nishimura, J., Bessler, M., Ware, R., Hillmen, P., Luzzatto, L., Young, N., Kinoshita, T., Rosse, W., Socie, G. & International, P. N. H. I. G. 2005. Diagnosis and management of paroxysmal nocturnal hemoglobinuria. *Blood*, 106, 3699–709.

Peerschke, E. I., Reid, K. B. & Ghebrehiwet, B. 1993. Platelet activation by C1q results in the induction of alpha IIb/beta 3 integrins (GPIIb-IIIa) and the expression of P-selectin and procoagulant activity. *J Exp Med*, 178, 579–87.

Peerschke, E. I., Yin, W., Grigg, S. E. & Ghebrehiwet, B. 2006. Blood platelets activate the classical pathway of human complement. *J Thromb Haemost*, 4, 2035–42.

Peng, Q., Li, K., Anderson, K., Farrar, C. A., Lu, B., Smith, R. A., Sacks, S. H. & Zhou, W. 2008. Local production and activation of complement up-regulates the allostimulatory function of dendritic cells through C3a-C3aR interaction. *Blood*, 111, 2452–61.

Pepys, M. B. 1972. Role of complement in induction of the allergic response. *Nat New Biol*, 237, 157–9.

Perl, M., Denk, S., Kalbitz, M. & Huber-Lang, M. 2012. Granzyme B: a new crossroad of complement and apoptosis. *Adv Exp Med Biol*, 946, 135–46.

Pickering, M. C. & Walport, M. J. 2000. Links between complement abnormalities and systemic lupus erythematosus. *Rheumatology (Oxford)*, 39, 133–41.

Pihlstrom, B. L., Michalowicz, B. S. & Johnson, N. W. 2005. Periodontal diseases. *Lancet*, 366, 1809–20.

Pinto, M. R., Chinnici, C. M., Kimura, Y., Melillo, D., Marino, R., Spruce, L. A., De Santis, R., Parrinello, N. & Lambris, J. D. 2003. CiC3-1a-mediated chemotaxis in the deuterostome invertebrate Ciona intestinalis (Urochordata). *J Immunol*, 171, 5521–8.

Pinto, M. R., Melillo, D., Giacomelli, S., Sfyroera, G. & Lambris, J. D. 2007. Ancient origin of the complement system: emerging invertebrate models. *Adv Exp Med Biol*, 598, 372–88.

Polley, M. J. & Nachman, R. 1978. The human complement system in thrombin-mediated platelet function. *J Exp Med*, 147, 1713–26.

Polley, M. J. & Nachman, R. L. 1983. Human platelet activation by C3a and C3a des-arg. *J Exp Med*, 158, 603–15.

Polycarpou, A., Howard, M., Farrar, C. A., Greenlaw, R., Fanelli, G., Wallis, R., Klavinskis, L. S. & Sacks, S. 2020. Rationale for targeting complement in COVID-19. *EMBO Mol Med*, 12, e12642.

Qin, X., Krumrei, N., Grubissich, L., Dobarro, M., Aktas, H., Perez, G. & Halperin, J. A. 2003. Deficiency of the mouse complement regulatory protein mCd59b results in spontaneous hemolytic anemia with platelet activation and progressive male infertility. *Immunity*, 18, 217–27.

Rathbone, J., Kaltenthaler, E., Richards, A., Tappenden, P., Bessey, A. & Cantrell, A. 2013. A systematic review of eculizumab for atypical haemolytic uraemic syndrome (aHUS). *BMJ Open*, 3, e003573.
Reid, R. R., Woodcock, S., Shimabukuro-Vornhagen, A., Austen, W. G., Jr., Kobzik, L., Zhang, M., Hechtman, H. B., Moore, F. D., Jr. & Carroll, M. C. 2002. Functional activity of natural antibody is altered in Cr2-deficient mice. *J Immunol*, 169, 5433–40.
Richards, A., Kemp, E. J., Liszewski, M. K., Goodship, J. A., Lampe, A. K., Decorte, R., Muslumanoglu, M. H., Kavukcu, S., Filler, G., Pirson, Y., Wen, L. S., Atkinson, J. P. & Goodship, T. H. 2003. Mutations in human complement regulator, membrane cofactor protein (CD46), predispose to development of familial hemolytic uremic syndrome. *Proc Natl Acad Sci U S A*, 100, 12966–71.
Ricklin, D., Barratt-Due, A. & Mollnes, T. E. 2017. Complement in clinical medicine: clinical trials, case reports and therapy monitoring. *Mol Immunol*, 89, 10–21.
Ricklin, D., Hajishengallis, G., Yang, K. & Lambris, J. D. 2010. Complement: a key system for immune surveillance and homeostasis. *Nat Immunol*, 11, 785–97.
Risitano, A. M. 2013. Paroxysmal nocturnal hemoglobinuria and the complement system: recent insights and novel anticomplement strategies. *Adv Exp Med Biol*, 735, 155–72.
Risitano, A. M. & Rotoli, B. 2008. Paroxysmal nocturnal hemoglobinuria: pathophysiology, natural history and treatment options in the era of biological agents. *Biologics,* 2, 205–22.
Ritis, K., Doumas, M., Mastellos, D., Micheli, A., Giaglis, S., Magotti, P., Rafail, S., Kartalis, G., Sideras, P. & Lambris, J. D. 2006. A novel C5a receptor-tissue factor cross-talk in neutrophils links innate immunity to coagulation pathways. *J Immunol*, 177, 4794–802.
Robak, E. & Robak, T. 2009. Monoclonal antibodies in the treatment of systemic lupus erythematosus. *Curr Drug Targets*, 10, 26–37.
Rollins, S. A. & Sims, P. J. 1990. The complement-inhibitory activity of CD59 resides in its capacity to block incorporation of C9 into membrane C5b-9. *J Immunol*, 144, 3478–83.
Rondelli, T., Risitano, A. M., Peffault De Latour, R., Sica, M., Peruzzi, B., Ricci, P., Barcellini, W., Iori, A. P., Boschetti, C., Valle, V., Fremeaux-Bacchi, V., De Angioletti, M., Socie, G., Luzzatto, L. & Notaro, R. 2014. Polymorphism of the complement receptor 1 gene correlates with the hematologic response to eculizumab in patients with paroxysmal nocturnal hemoglobinuria. *Haematologica*, 99, 262–6.
Ross, G. D., Cain, J. A. & Lachmann, P. J. 1985. Membrane complement receptor type three (CR3) has lectin-like properties analogous to bovine conglutinin as functions as a receptor for zymosan and rabbit erythrocytes as well as a receptor for iC3b. *J Immunol*, 134, 3307–15.
Rosse, W. F. 1997. Paroxysmal nocturnal hemoglobinuria as a molecular disease. *Medicine (Baltimore)*, 76, 63–93.
Saggu, G., Cortes, C., Emch, H. N., Ramirez, G., Worth, R. G. & Ferreira, V. P. 2013. Identification of a novel mode of complement activation on stimulated platelets mediated by properdin and C3(H2O). *J Immunol*, 190, 6457–67.
Schenkein, H. A. & Berry, C. R. 1991. Activation of complement by Treponema denticola. *J Dent Res*, 70, 107–10.
Seddon, J. M., Yu, Y., Miller, E. C., Reynolds, R., Tan, P. L., Gowrisankar, S., Goldstein, J. I., Triebwasser, M., Anderson, H. E., Zerbib, J., Kavanagh, D., Souied, E., Katsanis, N., Daly, M. J., Atkinson, J. P. & Raychaudhuri, S. 2013. Rare variants in CFI, C3 and C9 are associated with high risk of advanced age-related macular degeneration. *Nat Genet*, 45, 1366–70.
Selander, B., Martensson, U., Weintraub, A., Holmstrom, E., Matsushita, M., Thiel, S., Jensenius, J. C., Truedsson, L. & Sjoholm, A. G. 2006. Mannan-binding lectin activates C3 and the alternative complement pathway without involvement of C2. *J Clin Invest*, 116, 1425–34.
Semeraro, N., Ammollo, C. T., Semeraro, F. & Colucci, M. 2012. Sepsis, thrombosis and organ dysfunction. *Thromb Res*, 129, 290–5.
Semeraro, N., Ammollo, C. T., Semeraro, F. & Colucci, M. 2015. Coagulopathy of Acute Sepsis. *Semin Thromb Hemost*, 41, 650–8.
Shebuski, R. J. & Kilgore, K. S. 2002. Role of inflammatory mediators in thrombogenesis. *J Pharmacol Exp Ther*, 300, 729–35.
Silasi-Mansat, R., Zhu, H., Georgescu, C., Popescu, N., Keshari, R. S., Peer, G., Lupu, C., Taylor, F. B., Pereira, H. A., Kinasewitz, G., Lambris, J. D. & Lupu, F. 2015. Complement inhibition decreases early fibrogenic events in the lung of septic baboons. *J Cell Mol Med*, 19, 2549–63.
Silasi-Mansat, R., Zhu, H., Popescu, N. I., Peer, G., Sfyroera, G., Magotti, P., Ivanciu, L., Lupu, C., Mollnes, T. E., Taylor, F. B., Kinasewitz, G., Lambris, J. D. & Lupu, F. 2010. Complement inhibition decreases the procoagulant response and confers organ protection in a baboon model of Escherichia coli sepsis. *Blood*, 116, 1002–10.
Sims, P. J., Faioni, E. M., Wiedmer, T. & Shattil, S. J. 1988. Complement proteins C5b-9 cause release of membrane vesicles from the platelet surface that are enriched in the membrane receptor for coagulation factor Va and express prothrombinase activity. *J Biol Chem*, 263, 18205–12.

Sims, P. J. & Wiedmer, T. 1991. The response of human platelets to activated components of the complement system. *Immunol Today*, 12, 338–42.

Skoglund, C., Wettero, J., Tengvall, P. & Bengtsson, T. 2010. C1q induces a rapid up-regulation of P-selectin and modulates collagen- and collagen-related peptide-triggered activation in human platelets. *Immunobiology*, 215, 987–95.

Smith, K., Pace, A., Ortiz, S., Kazani, S. & Rottinghaus, S. 2020. A Phase 3 open-label, randomized, controlled study to evaluate the efficacy and safety of intravenously administered ravulizumab compared with best supportive care in patients with COVID-19 severe pneumonia, acute lung injury, or acute respiratory distress syndrome: a structured summary of a study protocol for a randomised controlled trial. *Trials*, 21, 639.

Smith, C. A., Pangburn, M. K., Vogel, C. W. & Muller-Eberhard, H. J. 1984. Molecular architecture of human properdin, a positive regulator of the alternative pathway of complement. *J Biol Chem*, 259, 4582–8.

Spitzer, D., Mitchell, L. M., Atkinson, J. P. & Hourcade, D. E. 2007. Properdin can initiate complement activation by binding specific target surfaces and providing a platform for de novo convertase assembly. *J Immunol*, 179, 2600–8.

Strainic, M. G., Liu, J., Huang, D., An, F., Lalli, P. N., Muqim, N., Shapiro, V. S., Dubyak, G. R., Heeger, P. S. & Medof, M. E. 2008. Locally produced complement fragments C5a and C3a provide both costimulatory and survival signals to naive CD4+ T cells. *Immunity*, 28, 425–35.

Strainic, M. G., Shevach, E. M., An, F., Lin, F. & Medof, M. E. 2013. Absence of signaling into CD4(+) cells via C3aR and C5aR enables autoinductive TGF-beta1 signaling and induction of Foxp3(+) regulatory T cells. *Nat Immunol*, 14, 162–71.

Sweigard, J. H., Yanai, R., Gaissert, P., Saint-Geniez, M., Kataoka, K., Thanos, A., Stahl, G. L., Lambris, J. D. & Connor, K. M. 2014. The alternative complement pathway regulates pathological angiogenesis in the retina. *FASEB J*, 28, 3171–82.

Takahashi, M., Ishida, Y., Iwaki, D., Kanno, K., Suzuki, T., Endo, Y., Homma, Y. & Fujita, T. 2010. Essential role of mannose-binding lectin-associated serine protease-1 in activation of the complement factor D. *J Exp Med*, 207, 29–37.

Tedesco, F., Pausa, M., Nardon, E., Introna, M., Mantovani, A. & Dobrina, A. 1997. The cytolytically inactive terminal complement complex activates endothelial cells to express adhesion molecules and tissue factor procoagulant activity. *J Exp Med*, 185, 1619–27.

Tolentino, M. J., Dennrick, A., John, E. & Tolentino, M. S. 2015. Drugs in Phase II clinical trials for the treatment of age-related macular degeneration. *Expert Opin Investig Drugs*, 24, 183–99.

Tuboly, E., Futakuchi, M., Varga, G., Erces, D., Tokes, T., Meszaros, A., Kaszaki, J., Suzui, M., Imai, M., Okada, A., Okada, N., Boros, M. & Okada, H. 2016. C5a inhibitor protects against ischemia/reperfusion injury in rat small intestine. *Microbiol Immunol*, 60, 35–46.

Ueda, Y., Miwa, T., Gullipalli, D., Sato, S., Ito, D., Kim, H., Palmer, M. & Song, W. C. 2018. Blocking Properdin Prevents Complement-Mediated Hemolytic Uremic Syndrome and Systemic Thrombophilia. *J Am Soc Nephrol*, 29, 1928–37.

Van Lookeren Campagne, M., Wiesmann, C. & Brown, E. J. 2007. Macrophage complement receptors and pathogen clearance. *Cell Microbiol*, 9, 2095–102.

Varela, J. C. & Brodsky, R. A. 2013. Paroxysmal nocturnal hemoglobinuria and the age of therapeutic complement inhibition. *Expert Rev Clin Immunol*, 9, 1113–24.

Verardi, S., Page, R. C., Ammons, W. F. & Bordin, S. 2007. Differential chemokine response of fibroblast subtypes to complement C1q. *J Periodontal Res*, 42, 62–8.

Veszeli, N., Kohalmi, K. V., Kajdacsi, E., Gulyas, D., Temesszentandrasi, G., Cervenak, L., Farkas, H. & Varga, L. 2018. Complete kinetic follow-up of symptoms and complement parameters during a hereditary angioedema attack. *Allergy*, 73, 516–520.

Volokhina, E. B., Bergseth, G., Van De Kar, N. C., Van Den Heuvel, L. P. & Mollnes, T. E. 2015. Eculizumab treatment efficiently prevents C5 cleavage without C5a generation in vivo. *Blood*, 126, 278–9.

Wada, K., Montalto, M. C. & Stahl, G. L. 2001. Inhibition of complement C5 reduces local and remote organ injury after intestinal ischemia/reperfusion in the rat. *Gastroenterology*, 120, 126–33.

Wagner, C., Ochmann, C., Schoels, M., Giese, T., Stegmaier, S., Richter, R., Hug, F. & Hansch, G. M. 2006. The complement receptor 1, CR1 (CD35), mediates inhibitory signals in human T-lymphocytes. *Mol Immunol*, 43, 643–51.

Walker, J. B. & Bajzar, L. 2004. The intrinsic threshold of the fibrinolytic system is modulated by basic carboxypeptidases, but the magnitude of the antifibrinolytic effect of activated thrombin-activable fibrinolysis inhibitor is masked by its instability. *J Biol Chem*, 279, 27896–904.

Walport, M. J. 2001. Complement. First of two parts. *N Engl J Med*, 344, 1058–66.

Walport, M. J., Davies, K. A. & Botto, M. 1998. C1q and systemic lupus erythematosus. *Immunobiology*, 199, 265–85.

Wang, H., Ricklin, D. & Lambris, J. D. 2017. Complement-activation fragment C4a mediates effector functions by binding as untethered agonist to protease-activated receptors 1 and 4. *Proc Natl Acad Sci U S A*, 114, 10948–53.

Ward, P. A. & Gao, H. 2009. Sepsis, complement and the dysregulated inflammatory response. *J Cell Mol Med*, 13, 4154–60.

Waters, A. M. & Licht, C. 2011. aHUS caused by complement dysregulation: new therapies on the horizon. *Pediatr Nephrol*, 26, 41–57.

Weisman, H. F., Bartow, T., Leppo, M. K., Marsh, H. C., Jr., Carson, G. R., Concino, M. F., Boyle, M. P., Roux, K. H., Weisfeldt, M. L. & Fearon, D. T. 1990. Soluble human complement receptor type 1: in vivo inhibitor of complement suppressing post-ischemic myocardial inflammation and necrosis. *Science*, 249, 146–51.

Welte, T., Arnold, F., Kappes, J., Seidl, M., Haffner, K., Bergmann, C., Walz, G. & Neumann-Haefelin, E. 2018. Treating C3 glomerulopathy with eculizumab. *BMC Nephrol*, 19, 7.

Wetsel, R. A. 1995. Structure, function and cellular expression of complement anaphylatoxin receptors. *Curr Opin Immunol*, 7, 48–53.

Wiedmer, T., Esmon, C. T. & Sims, P. J. 1986. Complement proteins C5b-9 stimulate procoagulant activity through platelet prothrombinase. *Blood*, 68, 875–80.

Wiedmer, T. & Sims, P. J. 1985. Effect of complement proteins C5b-9 on blood platelets. Evidence for reversible depolarization of membrane potential. *J Biol Chem*, 260, 8014–9.

Wingrove, J. A., Discipio, R. G., Chen, Z., Potempa, J., Travis, J. & Hugli, T. E. 1992. Activation of complement components C3 and C5 by a cysteine proteinase (gingipain-1) from Porphyromonas (Bacteroides) gingivalis. *J Biol Chem*, 267, 18902–7.

Wojta, J., Kaun, C., Zorn, G., Ghannadan, M., Hauswirth, A. W., Sperr, W. R., Fritsch, G., Printz, D., Binder, B. R., Schatzl, G., Zwirner, J., Maurer, G., Huber, K. & Valent, P. 2002. C5a stimulates production of plasminogen activator inhibitor-1 in human mast cells and basophils. *Blood*, 100, 517–23.

Yehoshua, Z., De Amorim Garcia Filho, C. A., Nunes, R. P., Gregori, G., Penha, F. M., Moshfeghi, A. A., Zhang, K., Sadda, S., Feuer, W. & Rosenfeld, P. J. 2014. Systemic complement inhibition with eculizumab for geographic atrophy in age-related macular degeneration: the COMPLETE study. *Ophthalmology*, 121, 693–701.

Zhang, X., Kimura, Y., Fang, C., Zhou, L., Sfyroera, G., Lambris, J. D., Wetsel, R. A., Miwa, T. & Song, W. C. 2007. Regulation of Toll-like receptor-mediated inflammatory response by complement in vivo. *Blood*, 110, 228–36.

Zhang, Y., Nester, C. M., Holanda, D. G., Marsh, H. C., Hammond, R. A., Thomas, L. J., Meyer, N. C., Hunsicker, L. G., Sethi, S. & Smith, R. J. 2013. Soluble CR1 therapy improves complement regulation in C3 glomerulopathy. *J Am Soc Nephrol*, 24, 1820–9.

Zipfel, P. F., Skerka, C., Chen, Q., Wiech, T., Goodship, T., Johnson, S., Fremeaux-Bacchi, V., Nester, C., De Cordoba, S. R., Noris, M., Pickering, M. & Smith, R. 2015. The role of complement in C3 glomerulopathy. *Mol Immunol*, 67, 21–30.

Zoja, C., Corna, D., Rottoli, D., Cattaneo, D., Zanchi, C., Tomasoni, S., Abbate, M. & Remuzzi, G. 2002. Effect of combining ACE inhibitor and statin in severe experimental nephropathy. *Kidney Int*, 61, 1635–45.

Zuber, J., Le Quintrec, M., Krid, S., Bertoye, C., Gueutin, V., Lahoche, A., Heyne, N., Ardissino, G., Chatelet, V., Noel, L. H., Hourmant, M., Niaudet, P., Fremeaux-Bacchi, V., Rondeau, E., Legendre, C., Loirat, C. & French Study Group For Atypical, H. U. S. 2012. Eculizumab for atypical hemolytic uremic syndrome recurrence in renal transplantation. *Am J Transplant*, 12, 3337–54.

Zuber, J., Le Quintrec, M., Sberro-Soussan, R., Loirat, C., Fremeaux-Bacchi, V. & Legendre, C. 2011. New insights into postrenal transplant hemolytic uremic syndrome. *Nat Rev Nephrol*, 7, 23–35.

23 Drug Discovery Approaches for Inflammatory Bowel Disease

Antibodies and Biosimilars

Leo R. Fitzpatrick and Ella Mokrushin

Corresponding author: Leo R. Fitzpatrick

Contents

DOI: 10.1201/9780429485626-23

23.1 INTRODUCTION

The discovery of anti-TNF antibodies has definitively altered the treatment algorithm for inflammatory bowel disease (IBD) over the last 15 years.[1] More recently, different classes of antibodies have been developed, and some (e.g., vedolizumab, ustekinumab) have already entered the market for the treatment of IBD. Vedolizumab is a specific $\alpha 4\beta 7$ integrin antibody, while ustekinumab is a non-selective antibody directed against the common p40 subunit of IL-12 and IL-23.[2–5] Over the last two years, biosimilars for two anti-TNF antibodies (infliximab and adalimumab) have been approved for marketing in the United States.[6,7] This book chapter will focus on IBD-related drug discovery approaches for new biologics (e.g., specific IL-23 and sphingosine-1-phosphate (S1P) receptor antibodies). The influence of T-cell plasticity on such antibody-related drug discovery approaches will also be discussed within this chapter.[8–9] Finally, some data will be presented related to new biosimilar drugs for the treatment of IBD.

23.2 DEVELOPMENT OF SPECIFIC ANTIBODIES DIRECTED AGAINST IL-23

Various murine colitis studies have suggested a prominent role for IL-23 in the pathogenesis of murine colitis and IBD.[10–13] IL-23, which is composed of p19 and p40 subunits, is a pro-inflammatory cytokine that contributes to the expansion of Th17 cells in inflammatory autoimmune diseases.[14] Moreover, the IL-23/IL-17 pathway has been the subject of review papers related to IBD.[15] Of note, IL-23 has been reported to play a role in both acute (Dextran Sulfate Sodium [DSS]) and chronic (T-cell transfer) models of IBD.[10–11] This cytokine is involved in the pathogenesis of murine colitis involving the adaptive (T-cell transfer) and innate (anti-CD40 antibody) immune systems.[11–13] Finally, IL-23 plays a role in murine models of ulcerative colitis (UC), such as DSS-induced colitis. IL-23 is also important in mouse models of Crohn's disease (CD), like the T-cell transfer model.[10,11]This preclinical profile makes IL-23 an attractive drug development target for the development of specific antibodies.[16–18] Furthermore, IL-23 responsive innate lymphoid cells are increased in the inflamed intestine of patients with CD.[19]

Such preclinical studies have led pharmaceutical companies to develop specific antibodies directed against the p19 subunit of IL-23. Some of these antibodies are currently advancing in defined clinical trials.[1] One of these drugs (risankizumab) potently binds to the p19 subunit of IL-23 and therefore prevents its binding to IL-23 receptor on pertinent cells like Th17 cells.[1] This monoclonal antibody (mAb) showed favorable results in a phase 2 trial conducted in patients with moderate to severe CD.[19]

23.3 DEVELOPMENT OF SPECIFIC ANTIBODIES DIRECTED AGAINST THE IL-23 RECEPTOR

An alternative approach to directly targeting IL-23 p19 could be the development of a human mAb directed against the IL-23 receptor (for example, on Th17 cells).[1,11,20,21] Immamura et al. found that a mAb directed against the IL-23 receptor normalized various parameters of colitis in mice using a T-cell transfer model of IBD. Moreover, these investigators suggested that an IL-23 receptor mAb may have higher efficacy than an anti-IL-23p19 mAB for the treatment of locally inflamed lesions in the intestine.[11] Of note, Astellas Pharma Inc. recently published a paper describing the development of a potent fully human mAb against the IL-23 receptor for the potential treatment of chronic inflammatory diseases like IBD.[20]

23.4 DEVELOPMENT OF SPHINGOSINE-1-PHOSPHATE RECEPTOR-SPECIFIC ANTIBODIES

Sphingosine-1-phosphate, a bioactive small lipid molecule, plays essential roles in cellular processes, including angiogenesis, migration, cytoskeleton rearrangement, proliferation, and cell survival through by utilizing five G protein-coupled receptors (S1P1 to S1P5).[22] Of relevance to IBD-related drug discovery, S1P-1 receptor expression was found to be increased on T cells from mice with chronic colitis or ileitis.[23] Fingolimod (FTY720) was the first S1P receptor agonist to be approved for clinical use. It is currently on the market for the treatment of multiple sclerosis. This drug acts by sequestering T-lymphocytes in secondary lymphoid tissues, which results in a long-lasting reduction of the lymphocyte count in peripheral blood and tissues.[22] To date, fingolimod has not been developed clinically for the treatment of IBD, although various preclinical studies demonstrated its efficacy in murine models of IBD.[23–25]

Ozanimod (Table 23.1) is an S1P1/S1P5 receptor agonist with a potency approximately 27 times higher for S1P1 receptors than for S1P5 receptors. It has shown efficacy in trinitrobenzene sulfonic acid (TNBS) and T-cell transfer models that mimic to some degree human CD.[21,26] Interestingly, ozanimod has also shown evidence of efficacy in UC patients.[27]

Recently, another selective S1P1 receptor modulator (erastimod) was reported to effectively attenuate parameters of T-cell transfer colitis in mice. Specifically, it reduced histological damage, as well as the levels of pro-inflammatory cytokines (TNF-α, IL-1β, and IL-17) in the colons of these mice.[28]

Based on the preclinical and clinical efficacy profiles of these small molecule drugs targeting the S1P receptor, it is not surprising that specific monoclonal antibodies have been developed that also target this receptor. Specifically, treatment of mice with a single dose of anti-S1P1 receptor antibody improved various parameters of DSS-induced colitis, including a reduction in colonic histological damage.[29] The authors emphasized the dosing regimen advantage of this antibody approach for the treatment of colitis.

It is evident from the relevant preclinical literature that specific targeting of a S1P receptor may represent a good pharmacological approach for the treatment of IBD. Evidence presented above suggests that targeting of the S1P1 or S1P5 receptor may be a logical approach.[20–25] However, genetic-based evidence (using S1P4 deficient mice) demonstrated that these animals showed decreased pathology in a DSS-induced colitis model.[30] Therefore, the most important drug development question is which SIP receptor represents the most optimal target as a therapeutic approach for IBD?

TABLE 23.1 Novel antibody approaches for the treatment of IBD: Effects in rodent models of colitis

ANTIBODY TARGET	*COLITIS MODEL(S)*	*REFERENCE NUMBER(S)*
IL-23 p19	T-cell transfer	16
IL-23 p19	Abcb 1a deficient mice with *Helicobacter bilis* infection	17
IL-23 p19	Winnie mice with spontaneous colitis	18
IL-23 receptor	T-cell transfer	11
S1P receptor-1 (FTY 720)	TNBS (mice)	23
	DSS, T-cell transfer	24
	T-cell transfer	25
S1P Receptor-1&5 (ozanimod)	TNBS (rats) T-cell transfer	26
S1P-1 receptor	DSS	29
TL-1A	DSS, T-cell transfer	36

23.5 DEVELOPMENT OF TNF-LIKE LIGAND 1A (TL1A) ANTIBODIES

Polymorphisms in the TNF family member, TL1A gene, is associated with the development of IBD and increased serum concentrations of TL1A have been demonstrated in patients with various chronic inflammatory disorders. Results from human disease, animal models (see below), and preclinical intervention studies delineate the development of anti-L1A therapies as a highly promising strategy for the treatment of chronic inflammatory disorders. Genome-Wide Association Studies (GWAS) analysis of human IBD has revealed that polymorphisms in the TL1A gene are associated with increased CD susceptibility in the Japanese, Korean, and Western European populations. An association with UC has also been described in some studies, but not all of them.[31] Results from chronic models of colitis have demonstrated a link between overexpression of TL1A and the development of colitis with fibrotic disease.[32–35] Of note, Shih and colleagues showed that treatment of mice with a TL1a antibody (using chronic DSS or T-cell transfer models), reversed various parameters of colonic fibrosis. Pertinent effects with TL1A antibody treatment included lowering the enhanced TGF-β and collagen contents associated with chronic colonic fibrosis.[36] Why is this IBD-related drug development approach interesting and important? Intestinal fibrosis is a common complication of IBD, and is usually defined as an excessive accumulation of scar tissue in the intestinal wall. Severe fibrosis is seen in up to 30% of patients with CD, a high proportion of who require surgery.[37] There is no current evidence that any medical intervention can modulate or reverse intestinal fibrosis.[38] Therefore, the TL1a antibody approach may potentially be a welcome addition, for the treatment of CD patients with the fibro-stenotic disease. A summary of the four antibody approaches described above is shown in Table 23.1; it also includes information related to efficacy in rodent models of IBD.

23.6 OTHER POTENTIAL ANTIBODY APPROACHES FOR IBD

Specific antibody approaches described here have already been summarized in recent review papers.[21,30,39] Most of these antibody approaches have already been tested in clinical trials. These approaches include antibodies directed against gut-specific adhesion molecules (e.g., Mad-CAM-1). PF-00547659 is an mAb directed against Mad-CAM-1, which has been tested in both CD and UC clinical trials.[21,40,41] Interestingly (like vedolizumab), the efficacy profile for PF-00547659 appeared to be more robust in UC patients, as compared to patients with CD.[40,41]

IL-6 is a pleiotropic pro-inflammatory cytokine that has been implicated in the pathophysiology of IBD.[21] PF-04236921 is a fully human anti-IL-6 antibody that has entered phase 2 clinical trials for patients with CD. This antibody (at a dose of 50 mg) showed a positive clinical response and remission rates compared to placebo treatment.[21,42]

Preclinical data was presented with an antibody directed against eotaxin-1 (CCL-11) that has reported the efficacy of this approach in a murine DSS-induced colitis model.[43] This data supported the subsequent clinical launch of bertlimumab, which is apparently in phase 2 clinical trials for both UC and CD.[43]

Fractalkine is expressed on vascular endothelial cells in patients with IBD. It is involved in inflammatory response when bound to fractalkine receptors (CX3CR1) expressed in immune cells. E6011 is an antibody that is thought to exhibit an anti-inflammatory effect by suppressing the migration of CX3CR1-positive immune cells. This antibody is currently in clinical trials targeting CD patients.[44]

IL-13 has been implicated in the pathogenesis of UC.[45,46] In this regard, treatment with an IL-13 receptor alpha 2 subunit fusion protein prevented the development of oxazolone-induced colitis (a model of UC) in mice.[46] Despite these positive preclinical results, a clinical trials with anrukinzumab (a humanized mAb directed against IL-13) was not effective in a phase 2a clinical trials with UC patients.[47] It has been suggested that dual inhibition of IL-13 and IL-4 may be a better therapeutic approach in UC, despite the fact that IL-4 is not consistently increased in UC patients.[48,49]

Recent evidence suggests that IL-9 produced by Th9 cells plays a prominent role in the etiology of UC.[50–52] Anti-IL-9 antibody treatment effectively reduced the severity of inflammation associated with DSS-induced colitis in mice.[51]Recently, IL-9 was shown to be a negative marker of colonic mucosal healing in patients with UC.[52] As a whole, these data suggest that clinical development of a specific IL-9 antibody is a rational future approach for the treatment of UC.

Granulocyte-macrophage colony-stimulating factor (GM-CSF) can play a role in inflammatory processes through activation and survival of macrophages and neutrophils. Mavrilimumab is a novel GM-CSF-receptor alpha mAb, which has shown efficacy in rheumatoid arthritis (RA) patients.[53] Of note, the production of GM-CSF by type 3 innate lymphoid cells is important in the pathogenesis of murine anti-CD40 colitis. Interestingly, administration of an anti-GM-CSF blocking antibody to mice significantly improved this colitis.[54]

23.7 POTENTIAL IMPACT OF T-CELL PLASTICITY ON NEW ANTIBODY APPROACHES FOR IBD

The cytokine responses characterizing IBD are the key pathophysiologic elements that govern the initiation and evolution of intestinal inflammation. Studies conducted during the last two decades now provide a detailed (but not yet complete) picture of the nature of these responses. These cytokine responses are controlled by the T-cell patterns dominating the specific type of disease. In CD, the Th1 and Th17 CD4+ T-cell differentiation processes result in T-lymphocytes producing the signature cytokines IFN-γ and IL-17, respectively. Recent studies have also identified T cells that produce both IFN-γ and IL-17.[55] Also, these dual Th17/Th1 cells can be converted into non-classic Th1 cells.[56] These T-lymphocytes may also be involved in the pathogenesis of CD. Th17 cells can shift into a Th1 phenotype, in the presence of cytokines like IL-12 and TNF-α, at sites of inflammation in diseases like CD. The Th17 derived Th1 cells have been named non-classic Th1 cells.[56] Of note, recent results from a murine T-cell transfer model of colitis suggest that IL-23 (acting through the transcription factor STAT4) could play a key role in Th17 plasticity and the formation of non-classic Th1 cells.[57] IL-1β also seems to play a role in the process.[55]

Other investigators referred to these non-classic Th1 cells as Th17.1 cells.[58] Interestingly, these T-lymphocytes can produce substantial amounts of IFN-γ, but minimal to no IL-17.[58,59] It has been suggested that these non-classic Th1 cells have a particularly aggressive pathogenic potential in chronic inflammatory diseases like CD.[60] Finally, non-classic Th1 cells are somewhat functionally distinct from both classic Th1 cells and their Th17 cell precursors.[61]

Some investigators showed that pro-inflammatory human Th17 cells were restricted to a subset of CD161+ cells that stably express P-glycoprotein (P-gp)/multi-drug resistance type 1 (MDR1).[62] These MDR1+Th17.1 cells are enriched and activated in the intestine of CD patients. Interestingly, these T-lymphocytes produced IFN-γ. Furthermore, this MDR1+Th17.1 cell subset was refractory to several glucocorticoids used to treat autoimmune diseases. Thus, these CD161+/MDR1+Th17.1 cells may be important mediators of chronic inflammation, particularly in clinical settings of steroid-resistant IBD.[62]

Novel data indicate that Th17 cells can transition into Th1 cells that produce IFN-γ and induce chronic colitis in mice.[63] Finally, new clinical data demonstrates that in patients with evidence of long-standing CD (i.e., undergoing bowel resection), there is evidence of T-cell plasticity in the inflamed ileum.[64]

Some of these T-cell plasticity concepts are illustrated in Figure 23.1. Figure 23.1 may also partially explain why drugs like infliximab and ustekinumab have shown efficacy in patients with CD.[21,30]

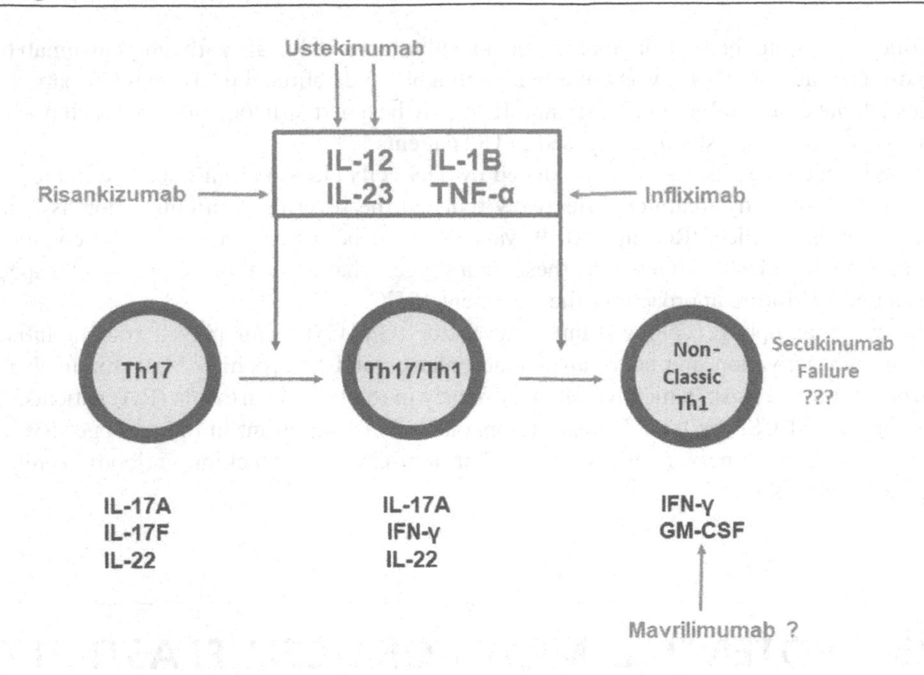

FIGURE 23.1 Th17-cell plasticity and effects of biologics targeting key cytokines.

Importantly, if the non-classic Th1-cell population is indeed prevalent during chronic CD, this may also help to partially explain the lack of efficacy for IL-17 directed antibody approaches like secukinumab and brodalumab in patients with well-established CD.[15,65,66] In this regard, a recent review paper suggests that Th17-cell plasticity to ex-Th17 cells are the main drivers of tissue pathology for autoimmune diseases (like CD).[8] From a drug development standpoint: Is it possible that a specific antibody directed against the IL-23p19 subunit or the IL-23 receptor (see above) may negate the Th17 cell plasticity process in CD patients with more chronic disease? Regarding IBD, GM-CSF is one of the signature cytokines produced by non-classic Th1 cells.[56] The question becomes whether mavrilimumab or an antibody directly targeting GM-CSF would be effective in certain patients with IBD (see Figure 23.1).

Interestingly, the paradigm became even more complicated, based on other pertinent publications.[67,68] The relevant literature suggests that CD patients have a higher prevalence of dual IL-17/IFN-γ secreting Th cells when exposed *in vitro* to a combination of IL-1β/TGF-β/IL-6. In contrast, UC patients have a higher proportion of these dual secreting cells when grown in the presence of IL-21/IL-23. These results suggest specific IBD-subtype plasticity of Th17 cells, which may impact antibody-directed approaches in these patients (see Figure 23.1).

23.8 BIOSIMILAR TESTING IN ANIMAL MODELS OF IBD

Biosimilars of infliximab being developed or approved for IBD include ABP710, PF-06438179, and infliximab BS.[69] Adalimumab biosimilars include MSB11022, GP2017, Imraldi, Exemptia, Adfrar, and GP2017.[70] The FDA Guidance for developing biosimilar drugs requires assessment of receptor binding and immunochemical properties.[70] However, strict and relevant preclinical efficacy testing is not typically required for biosimilars. Therefore, there is limited data from animal models related to IBD related to biosimilar drugs. One recent study showed that the biosimilar of infliximab (CT-p13, Remisma®) ameliorated cancer-related cachexia in mice. The investigators suggested that this therapeutic approach should be considered for the treatment of pediatric and adolescent CD patients with severe wasting disease.[71]

23.9 UPDATE ON RELEVANT CLINICAL TRIALS

23.9.1 Antibodies Directed Against IL-23 and the IL-23 Receptor

Recent clinical results showed that treatment with open-label intravenous risankizumab was effective in increasing clinical response and remission rates at week 26 in patients with CD.[72] Moreover, open-label subcutaneous risankizumab treatment maintained remission until week 52 in most patients who were in clinical remission at week 26. The authors concluded that selective blockade of IL-23 warrants further investigation as a treatment for CD.[72] In this regard, risankizumab is currently undergoing a phase 3 evaluation in CD patients. Moreover, this mAb is also currently undergoing clinical trials in patients with UC.[73]

Another phase 2a clinical trials in patients with moderate to severe CD indicated that MEDI2070 (brazikumab) treatment at 8 and 24 weeks resulted in clinical improvement in this patient population, who had previously failed anti-TNFαantibody treatments.[74] Recent results with mirikizumab, which is another IL-23p19 antibody, suggested that it resulted in endoscopic and clinical improvements in patients with active CD.[75] Another phase 2 clinical trials found that mirikizumab is also effective for induction of remission, as well as maintenance of remission, in patients with moderate-to-severe UC.[76,77] A new "expert opinion" paper suggested that IL-23 antagonists should be considered for first- or second-line therapy because of their effectiveness in both biologically naïve and experienced IBD patients.[78]

As alluded to previously, a fully human mAb against the IL-23 receptor (AS2762900-00) has recently been developed.[20] AS2762900-00 binds to both human and cynomolgus monkey IL-23 receptors. In a single-dose administration pharmacodynamics study in monkeys, 1 mg/kg of AS2762900-00 significantly inhibited (>85%) IL-23-induced STAT3 phosphorylation in blood for up to 84 days. Therefore, AS2762900-00 represents a potent novel IL-23-IL-17 pathway inhibitor with the potential to be developed into a new therapy for the treatment of autoimmune diseases like IBD.[20]

To our knowledge, IL-23 receptor (IL23R) antibodies have yet to be tested in clinical studies for IBD. Yet, the approach appears to be supported by previous scientific data, including recent results from patients who are resistant to anti-TNF therapy in CD.[79] These recent clinical results showed that responders to anti-TNF therapy displayed a significantly higher expression of TNF receptor 2 (TNFR2), but not IL23R, on T cells than non-responders prior to therapy. During anti-TNF therapy, there was a significant upregulation of mucosal IL-23p19, IL23R, and IL-17A in anti-TNF non-responders but not in responders. This upregulation resulted in the expansion of apoptosis-resistant intestinal TNFR2+IL23R+ T cells, which is associated with resistance to anti-TNF therapy in CD.[79] It can be concluded from this study that IL-23 (or perhaps the IL23R) is a reasonable target in patients with CD refractory to anti-TNF therapy.

23.9.2 Antibodies Directed Against the S1P Receptor

Recent clinical results suggest that etrasimod, a selective small-molecule S1P1 receptor modulator was effective in phase 2 clinical trials involving UC patients with moderate to severe disease. Specifically, etrasimod treatment resulted in significantly higher rates of endoscopic improvement, histological improvement, and mucosal healing compared to placebo.[80]

Despite promising preclinical results,[29] to our knowledge, no antibody approaches directed against an S1P receptor have currently progressed to clinical trials for IBD.

23.9.3 Antibodies Directed Against Mad-CAM-1

As mentioned above, a gut-specific adhesion molecule (e.g., Mad-CAM-1) represents a rationale therapeutic target for IBD.[40,41] During the past two years, follow-up IBD clinical trials have been conducted with PF-00547659, a fully human mAb directed against Mad-CAM-1.[81,82] Following the trend of earlier clinical trials, PF-00547659 was more effective in UC than CD patients.[81,82] Specifically, this antibody was better than placebo for inducing remission in patients with moderate to severe UC.[81] In contrast, clinical endpoint differences between PF-00547659 and placebo did not attain statistical significance in CD patients. Interestingly, this antibody was pharmacologically active. It decreased soluble Mad-CAM levels by >88% and also dose dependently increased circulating memory T cells.[82] These results suggest a pharmacodynamic disconnect between pharmacological target engagement and clinical efficacy, which needs to be better understood in future clinical trials of this ilk.

23.9.4 Antibodies Directed Against IL-6

Published results from 2019 have shown that an anti-IL-6antibody (PF-04236921) induced a significant clinical response and increased remission rate compared to placebo treatment in patients with anti-TNF refractory CD.[83] One potential safety concern from the study was the observation of apparent drug-related gastrointestinal perforation and abscess formation during the study.

23.9.5 Antibodies Directed Against IL-9

To our knowledge, antibodies directed against IL-9 (or the IL-9 receptor [IL-9R]) have not yet entered clinical trials for UC. Nevertheless, recent experimental clinical results reinforce the concept that IL-9 represents a logical therapeutic target for this form of IBD. Specifically, using both IHC and western blot analyses, investigators recently showed enhanced colonic expression of both IL-9 and the IL-9R in patients with UC.[84] Hopefully, these results will encourage future clinical trials with IL-9 antibodies in UC patients.

23.9.6 Antibodies Directed Against TLA-1

Duc to the potential for TL1A blockade in the treatment of inflammatory diseases, such as IBD, Teva Pharmaceuticals (Sydney, Australia) recently developed an anti-TL1A mAb termed C03V.[31–36,85] Based on its high affinity, selectivity, and potency with both *in vitro* and *in vivo* preclinical models, it was suggested that C03V is an attractive candidate for clinical development for IBD.[85] In this regard, future clinical trials with this antibody in IBD patients seem like a logical progression to the drug discovery/development process. In this regard, a recent clinical study with an anti-TL1A antibody in UC patients showed acceptable safety and evidence of histologic improvement.[86]

23.9.7 Antibodies Directed Against IL-12

Treatment with an mAb against IL-12 is associated with reduction in Th1-mediated inflammatory cytokines at the site of disease and was hypothesized to be a reasonable therapeutic approach in patients with active CD.[87] Prior studies suggested IL-12 as a key cytokine in the pathogenesis of IBD, but preclinical data indicated that antibodies against IL-12p40 provided their anti-inflammatory effect primarily by inhibition of IL-23. This is because the subunit p40 is common for both IL-12 and IL-23. Interestingly, selective IL-12 targeting with an mAb was not found to be an ideal treatment for CD patients.[87,88]

23.9.8 Antibodies Directed Against IL-13

Studies have shown that targeting of IL-13 in dermatitis and allergic diseases appears to be promising.[89,90] Monoclonal antibodies against IL-13 for the treatment of asthma have been studied since 2011. More recently, two phase 2 trials confirmed significant improvements with these antibodies in the treatment of atopic dermatitis.[90] Although the role of IL-13 in some intestinal Th2 disorders (i.e., parasitic infections) is well established, somewhat conflicting results have been noted for this cytokine in IBD.[48,91] Recent clinical trial data suggest that selective blockade of the IL-4Rα receptor (through which IL-13 signals) in UC patients is not effective in improving disease outcome, challenging the importance of IL-13 as a target for therapy. Nevertheless, the possible benefits of attenuating IL-13 production in UC remains an interesting approach for treating this disease. There is strong preclinical evidence demonstrating a reduction of inflammation, tumor size, and tumor number in association with IL-13 inhibition, which suggests its potential use in the prevention of colitis-associated colorectal cancer. Novel treatment approaches that selectively target IL-13 have the potential to be major in developing a successful treatment strategy in a subset of patients with increased mucosal IL-13.[89]

23.9.9 Antibodies Directed Against GM-CSF

Compelling effects of GM-CSF inhibition in some preclinical autoimmune/inflammation models have been reported.[54] Recent clinical rheumatoid arthritis trials, particularly phase 2 trials with namilumab (a human mAb against GM-CSF), demonstrate rapid and marked efficacy with no significant adverse effects.[92,93]

In contrast to these data, one study showed that CD patients with elevated neutralizing GM-CSF Ab exhibit an increase in bowel permeability relative to CD patients with lower levels of GM-CSF Ab, despite the absence of inflammatory differences.[94] Of note, clinical trials utilizing recombinant human GM-CSF showed improvements in patients with CD. An enhanced rate of remission as well as a decrease in mean CD activity index score was demonstrated in patients with moderate-to-severe CD who were treated with recombinant human GM-CSF. These results suggest that GM-CSF might be considered as an alternative to traditional immunosuppression for the treatment of CD.[95]

Despite these interesting results suggesting that GM-CSF might have a protective role in IBD, recent data cited above suggest that treatment with an anti-GM-CSF antibody improved innate immune colitis in mice.[54] Therefore, the exact role(s) of GM-CSF in patients with IBD needs to be carefully examined in future clinical trials.

23.9.10 Antibodies Directed Against IL-17

Studies show that IL-17A antagonists, including secukinumab and ixekizumab, are associated with disease exacerbations in IBD, and *de novo* cases of IBD have been reported. There is a possible correlation between therapy with IL-17 antagonist(s) and the risk of developing new-onset fulminant IBD.[96,97] One consideration in the interpretation of these studies is that strong/complete inhibition of IL-17 with antibody approaches may impair the normal "housekeeping" function of IL-17 (e.g., anti-fungal defense) and lead to unwanted clinical effects.[15,96–99] The results also point out a dichotomy between IL-23 and IL-17 inhibition in IBD patients[15].

23.9.11 Integrin Antagonists

As mentioned above, vedolizumab, which is a specific α4β7 integrin antibody, has already entered the market for the treatment of IBD. A recent phase 2 trial, which evaluated the safety and efficacy of etrolizumab (monoclonal Ab targeting the β7 subunits of the α4β7 and αEβ7 integrins), in patients with moderate-to-severe UC, demonstrated higher clinical remission rates in patients with higher baseline gene expression of integrin αE. An extensive phase 3 program is ongoing. In contrast, a phase 2b trial evaluating abrilumab (monoclonal Ab against the α4β7 integrin) in patients with moderate-to-severe CD failed to meet the primary endpoint (clinical remission, CDAI < 150, at week 8). In another phase 2b trial, however, abrilumab had a favorable safety and efficacy profile. Patients in the abrilumab treatment group had significantly enhanced remission rates in comparison with the placebo group. Of note, AJM-300 (a new, oral, small-molecule anti-α4 integrin) treatment contributed to enhanced clinical remission and mucosal healing compared to placebo in a phase 2 clinical trials. The rates of adverse events were similar between active-treatment and placebo groups. Further, the PROPEL study showed that PTG-100 (oral, gut-selective, small molecule that targets the α4β7 integrin) induced more clinical and endoscopic improvements than placebo.[100,101] As a whole, these data continue to support the development of integrin antagonist approaches for the treatment of IBD.[100,101]

23.10 SUMMARY AND CONCLUSIONS

An increased comprehension of IBD immunopathogenesis has contributed to the development of novel biologics. Some of these drugs (vedolizumab and ustekinumab) have been recently approved for the treatment of IBD. However, other agents (e.g., fingolimod) that have been approved for different indications are still being considered for testing in defined clinical trials. Recently, biosimilars for two anti-TNF antibodies (infliximab and adalimumab) have been approved. Novel drugs for IBD currently under investigation include specific antibodies directed against IL-23/IL-23R, S1P receptor-specific antibodies, TNF-like ligand 1A (TL1A) antibodies, and others. These agents could increase the treatment options available for the management of IBD patients in the future. Relevant clinical trials testing some of these novel mAb approaches (e.g., IL-23 antibodies) are ongoing.

REFERENCES

1. Verstockt, J., Ferrante, M., Vermeire, S. et al. 2018. New treatment options for inflammatory bowel diseases. *Journal of Gastroenterology* 53: 585–590.
2. Feagan, B.G., Rutgeerts, P., Sands, B.E. et al. 2013. Vedolizumab as induction and maintenance therapy for ulcerative colitis. *Lancet* 369: 699–710.
3. Sandborn, W.J., Feagan, B.G., Rutgeerts, P. et al. 2013. Vedolizumab as induction and maintenance therapy for Crohn's disease. *Lancet* 369: 711–721.
4. Sandborn, W.J., Gasink, C., Gao, L.L. et al. 2012. Ustekinumab induction and maintenance therapy in refractory Crohn's disease. *New England Journal of Medicine* 367: 1519–1528.
5. Feagan, B.G., Sandborn, W.J., Gasink, C. et al. Ustekinumab as induction and maintenance therapy for Crohn's disease. 2016. *New England Journal of Medicine* 375: 1519–1528.
6. Ha, C.Y., Kornbluth, A. 2016. A critical review of biosimilars in IBD: The confluence of biologic drug development, regulatory requirements, clinical outcomes and big business. *Inflammatory Bowel Diseases* 22: 2513–2526.
7. Delana, S. Gabbani, T., Annese, V. 2017. Biosimilars in inflammatory bowel disease. A review of post-marketing experience. *World Journal of Gastroenterology* 23: 197–203.
8. Stadhouders, R., Lubberts, E., Hendricks, R.W. 2018. A cellular and molecular view of T helper 17 cell plasticity in autoimmunity. *Journal of Autoimmunity* 87: 1–15.
9. Li, J., Iacucci, M., Fort Gassia, M. et al. 2017. Crossover subsets of CD4+ T lymphocytes in the intestinal lamina propria of patients with Crohn's disease and ulcerative colitis. *Digestive Diseases and Sciences* 62: 2357–2368.
10. Cox, J.H., Kljavin, N.M., Ota, N. et al. 2012. Opposing consequences of IL-23 signaling mediated by innate and adaptive cells in chemically induced colitis in mice. *Mucosal Immunology* 5: 99–109.
11. Imamura, E., Taguchi, K., Sasaki-Iwaoka, H. et al. 2018. Anti-IL-23 receptor monoclonal antibody prevents CD4+ T cell mediated colitis in association with decreased systemic Th1 and Th17 responses. *European Journal of Pharmacology* 824: 163–169.

12. Uhlig, H.H., McKenzie, B.S., Hue, S. et al. 2006. Differential activity of IL-12 and IL-23 in mucosal and systemic innate immune pathology. *Immunity* 25: 309–318.
13. Bunocore, S., Ahern, P.P., Uhlig, H.H. et al. 2010. Innate lymphoid cells drive IL-23 dependent innate intestinal pathology. *Nature* 46: 1371–1375.
14. Yago, T., Nanke, Y., Kawamoto, M. et al. 2017. IL-23 and Th17 disease in inflammatory arthritis. *Journal of Clinical Medicine* 6: 81.
15. Fitzpatrick, LR. 2012. Novel pharmacological approaches for inflammatory bowel disease: Targeting key intracellular pathways and the IL-23/IL-17 axis. *International Journal of Inflammation* 2012: 389404. doi: 10.1155/2012/389404.
16. Elson, C.O., Cong, Y., Weaver, C.T. et al. 2007. Monoclonal anti-interleukin 23 reverses active colitis in a T-cell mediated model in mice. *Gastroenterology* 132: 2359–2370.
17. Maxwell, J.R., Zhang, Y., Brown, W.A. et al. 2015. Differential roles for interleukin-23 and interleukin-17 in intestinal immunoregulation. *Immunity* 43: 739–750.
18. Wang, R., Hasnain, S.Z., Tong, H. et al. 2015. Neutralizing IL-23 is superior to blocking IL-17 in suppressing intestinal inflammation in a spontaneous murine colitis model. *Inflammatory Bowel Diseases* 21: 973–984.
19. Feagan, B.G., Sandborn, W.J., D'Haens, G. et al. 2017. Induction therapy with the selective interleukin-23 inhibitor risankizumab in patients with moderate-to-severe Crohn's disease. A randomized double-blind, placebo-controlled phase 2 study. *Lancet* 389: 1551–1560.
20. Sasaki-Iwaoka, H., Ohori, M., Imasato, A. et al. 2018. Generation and characterization of a fully human monoclonal antibody against the interleukin-23 receptor. *European Journal of Immunology* 828: 89–96.
21. Curro, D., Pugliese, D., Armuzzi, A 2017. Frontiers in drug research and development for inflammatory bowel disease. *Frontiers in Pharmacology* 8: 1–19.
22. Karuppuchamy, T., Behrens, E., Gonzalez-Cabrera, P. et al. 2017. Sphingosine-1-phosphate receptor-1 is expressed by lymphocytes, dendritic cells, and endothelium and modulated during inflammatory bowel disease. *Mucosal Immunology* 10: 162–171.
23. Daniel, C., Sartory, N.A., Zahn, N.A. et al. 2007. FTY720 ameliorates Th1-mediated colitis in mice by directly affecting the functional activity of CD4CCD25C regulatory T cells. *Journal of Immunology* 178: 2458–2468.
24. Deguchi, Y., Andoh, A., Yagi, Y. et al. 2006. The S1P receptor modulator FTY720 prevents the development of experimental colitis in mice. *Oncology Reports* 16: 699–703.
25. Fuji, T., Tomita, T., Kanai, T. et al. 2008. FTY720 suppresses the development of colitis in lymphoid-null mice by modulating the trafficking of colitogenic CD4C T cells in bone marrow. *European Journal of Immunology* 38: 3290–3303.
26. Scott, F.L., Clemons, B., Brooks, J. et al. 2016. Ozanimod (RPC1063) is a potent sphingosine-1-phosphate receptor-1 ($S1P_1$) and receptor-5 ($S1P_5$) agonist with autoimmune disease-modifying activity. *British Journal of Pharmacology* 173: 178–192.
27. Sandborn, W.J., Feagan, B., D'Haens, G. et al. 2016. Ozanimod induces histological response and remission: Results from the TOUCHSTONE study, a randomized, double-blind, placebo controlled trial of ozanimod, an oral S1P receptor modulator, in moderate to severe ulcerative colitis. *Journal of Crohn's and Colitis* 10: S59–S60.
28. Al-Shamma, H., Lehmann-Bruinsma, K., Carroll, C. et al. 2019. The selective sphingosine-1-phosphate receptor modulator etrasimod regulates lymphocyte trafficking and alleviates experimental colitis. *Journal of Pharmacology and Experimental Therapeutics* 369: 311–317.
29. Liao, J.J., Huang, M.C. Fast, K.K. et al. 2009. Immunosuppressive human anti-lymphocyte autoantibodies specific for the type 1 sphingosine 1-phosphate receptor. *FASEB Journal* 23: 1786–1796.
30. Schulze, T., Golfier, S., Tabeling, C. et al. 2011. Sphingosine-1-phosphate receptor 4 (S1P4) deficiency profoundly affects dendritic cell function and TH17-cell differentiation in a murine model. *FASEB Journal* 25: 4024–4036.
31. Tougaard, P., Zervides, K.A., Skov, S. et al. 2016. Biologics beyond TNF-α inhibitors and the effect of targeting the homologues TL1A-DR3 pathway in chronic inflammatory disorders. *Immunopharmacology and Immunotoxicology* 38: 29–38.
32. Bilsborough, J., Targan, S.R., Snapper, S.B. 2016. Therapeutic targets in inflammatory bowel disease: Current and future. *American Journal of Gastroenterology. Supplement* 3: 27.
33. Barrett, R., Zhang, X., Koon, H.W. et al. 2012. Constitutive TL1A expression under colitogenic conditions modulates the severity and location of gut mucosal inflammation and induces fibrostenosis. *American Journal of Pathology* 180: 636–649.
34. Shih, D.Q., Barrett, R., Zhang, X. et al. 2011. Constitutive TL1A (TNFSF15) expression on lymphoid or myeloid cells leads to mild intestinal inflammation and fibrosis. *PLoS One* 6: e16090.

35. Zheng, I., Zhang, X., Chen, J. et al. 2013. Sustained TL1a expression on both lymphoid and myeloid and cells leads to mild spontaneous intestinal inflammation and fibrosis. *European Journal of Microbiology and Immunology* 3: 11–20.
36. Shih, D.Q., Zheng, L., Zhang, H. et al. 2014. Inhibition of a novel fibrogenic factor TL1a reverses established colonic fibrosis. *Mucosal Immunology* 7: 1492–1503.
37. Rieder, E., Fiocchi, C. 2009. Intestinal fibrosis in IBD – a dynamic, multifactorial process. *Nature Reviews Gastroenterology & Hepatology* 6: 228–235.
38. Curciarello, R., Docena, G.H., MacDonald, T.T. 2017. The role of cytokines in the fibrotic responses in Crohn's disease. *Frontiers in Medicine* 4: 126. doi: 10.3389/fmed.2017.00126.
39. Lee, H.S., Park, S.K., Park, D.I. 2018. Novel treatments for inflammatory bowel disease. *Korean Journal of Internal Medicine* 33: 20–27.
40. Reich, W., Sandborn, W., Danes, S. et al. 2015. A randomized, multicenter double-blind, placebo-controlled study of the safety and efficacy of anti-MadCAM-1-antibody PF-00547659 (PF) severe ulcerative colitis: Results of the TURANDOT study. *Gastroenterology* 148(supplement 1): S-1193.
41. Sandborn, W., Lee, S.D., Tarabar, D. et al. 2015. Anti-MadCAM-1-antibody (PF-00547659) for active refractory Crohn's disease: Results of the OPERA study. *Gastroenterology* 148(supplement 1): S-162.
42. Danese, S., Vermeire, S., Hellstern, P. et al. 2016. Results of ADANTE, a randomized clinical study with an anti-IL6 antibody (PF-04236921) in subjects with Crohn's disease who are anti-tumor necrosis factor inadequate responders. *Journal of Crohn's and Colitis* 10: S12–S13.
43. Adar, T., Shteingart, S., Ben-Yaa'acov, A. et al., 2016. The importance of intestinal eotaxin-1 in inflammatory bowel disease: New insights and possible therapeutic applications. *Digestive Diseases and Sciences* 61: 1915–1924.
44. Matsuoka, K., Naganuma, M., Matsui, T. et al. 2016. Safety, tolerability and efficacy of E6011, anti-human fractaline monoclonal antibody, in the first patient study for Crohn's disease. *Gastroenterology* 150(supplement 1): S808.
45. Heller, E., Florian, P., Bojarski, C. et al. 2005. Interleukin-13 is the key effector Th2 cytokine in ulcerative colitis that affects epithelial tight junctions, apoptosis and cell restitution. *Gastroenterology* 129: 550–564.
46. Heller, F., Fuss, I.J., Nieuwenhuis, E.E. et al. 2002. Oxazalone colitis, a Th2 colitis model resembling ulcerative colitis, is mediated by IL-13 producing NK-T cells. *Immunity* 17: 629–638.
47. Renisch, W., Panes, J., Khurana, S. et al. 2015. Anrukinizumab, an anti-interleukin 13 monoclonal antibody, in active UC: Efficacy and safety from a phase IIa randomized multicenter study. *Gut* 64: 894–900.
48. Biancheri, P., Pender, S.L.F., Ammoscato, F. et al. 2013. Therapeutic activity of an interleukin-4/interleukin-13 dual antagonist on oxazolone-induced colitis in mice. *Immunology* 143: 116–127.
49. Fuss, I.J., Neurath, M., Boirvant, M. et al. 1996. Disparate CD4+ lamina propria (LP) lymphokine secretion profiles in inflammatory bowel disease. Crohn's disease LP cells manifest increased secretion of IFN-gamma whereas ulcerative colitis LP cells manifest increased secretion of IL-5. *Journal of Immunology* 157: 1261–1270.
50. Hufford, M., Kaplan, M.H. 2014. A gut reaction to IL-9. *Nature Immunology* 15: 599–600.
51. Yuan, A., Yang, H., Qi, H. et al. 2015. IL-9 antibody injection suppresses the inflammation in colitis mice. *Biochemical and Biophysical Research Communications* 468: 921–926.
52. Matusiewicz, M., Neubauer, K., Bednarz-Misa, I. et al. 2017. Systemic interleukin-9 in inflammatory bowel disease: Association with mucosal healing in ulcerative colitis. *World Journal of Gastroenterology* 23: 4039–4046.
53. Burmester, G.R., McInnes, I.B., Kremer, J. et al. 2017. A randomized phase IIB study of mavrilimumab, a novel GM-CSF receptor alpha monoclonal antibody, in the treatment of rheumatoid arthritis. *Annals of Rheumatic Diseases* 76: 1020–1030.
54. Pearson, C.P., Thornton, E.E., McKenzie, B. et al. 2015. ILC3 GM-CSF production and mobilization orchestrate acute intestinal inflammation. *eLIFE* e10066. doi: 10.7554.
55. Kleinscheck, M.A., Boniface, K., Sadevoka, S. et al. 2009. Circulating and gut resident human Th17 cells express CD161 and promote intestinal inflammation. *Journal of Experimental Medicine* 2006: 525–534.
56. Annunziato, F., Cosmi, L., Liotta, F. et al. 2015. Human T helper type 1 dichotomy: Origin, phenotype and biological activities. *Immunology* 144: 343–351.
57. Harbour, S.N., Maynard, C.L., Zindi, C.L. 2015. Th17 cells give rise to Th1 cells that are required for the pathogenesis of colitis. *Proceeding of National Academy of Science* 22: 7061–7066.
58. Paulissen, S.M., vanHamburg, J.P., Dankers, W. et al. 2015. The role and modulation of CCR6+ Th17 cell populations in rheumatoid arthritis. *Cytokine* 74: 43–53.
59. Annunziato, F., Santarlasci, V., Maggi, L. et al. 2013. Reasons for rarity of Th17 cells in inflammatory sites of human disorders. *Seminars in Immunology* 25: 299–304.

60. Cosmi, L., Liotta, F., Maggi, E. et al. 2014. Th17 and non-classic Th1 cells in chronic inflammatory disorders: Two sides of the same coin. *International Archives of Allergy and Immunology* 164: 171–177.
61. Basdeo, S.A, Cluxton, D. Sulaimani, J. et al. 2017. Ex-Th17 (nonclassical Th1) cells are functionally distinct from classical Th1 and Th17 cells and are not constrained by regulatory T cells. *Journal of Immunology* 198: 1–11.
62. Ramesh, R., Kozhaya, l., McKevitt, K. et al. 2014. Pro-inflammatory human Th17 cells selectively express P-glycoprotein and are refractory to glucocorticoids. *Journal of Experimental Medicine* 211: 89–104.
63. Morrison, P.J., Bending, D., Fouser, L.A. et al. 2013. Th17 cell plasticity in *Helicobacter hepaticus*-induced intestinal inflammation. *Mucosal Immunology* 6:1143–1156.
64. Li, J., Doty, A.L., Tang, Y. et al. 2017. Enrichment of IL-17A^{+} IFNγ^{+} and IL-22^{+} IFNγ^{+} T cell subsets is associated with reduction of NKp44+ ILC3s in the terminal ileum of Crohn's disease patients. *Clinical and Experimental Immunology* 190: 143–153.
65. Hueber, W., Sands, B.E., Lewitzky, S. et al. 2012. Secukinumab, a human anti-IL-17a monoclonal antibody for moderate to severe Crohn's disease: unexpected results of a randomized, double-blind placebo-controlled trial. *Gut* 61: 1693–1700.
66. Targan, S.R., Feagan, B., Vermeire, S. et al. 2016. A randomized, double-blind, placebo-controlled phase 2 study of brodalumab in patients with moderate-to-severe Crohn's disease. *American Journal of Gastroenterology* 111: 159–167.
67. Ueno, A., Ghosh, A., Hung, D. et al. 2015. Th17 plasticity and its changes associated with inflammatory bowel disease. *World Journal of Gastroenterology* 21: 12283–12295.
68. Ueno, A., Jijon, H., Chan, R. et al. 2013. Increased prevalence of circulating novel 1L-17 secreting Foxp3 expressing CD4+ T cells and defective suppressive function of circulating Foxp3+ regulatory cells support plasticity between Th17 and regulatory T cells in inflammatory bowel disease patients. *Inflammatory Bowel Diseases* 19: 2522–2534.
69. Rawla, P., Sunkara, T., Pradeep, J. 2018. Role of biologics and biosimilars in inflammatory bowel disease. Current trends and future perspectives. *Journal of Inflammation Research* 11: 215–226.
70. Ho, C.Y., Kornbluth, A. 2016. A critical review of biosimilars in IBD: The confluence of biologic drug development, regulatory requirements, clinical outcomes, and big business. *Inflammatory Bowel Diseases* 22: 2513–2526.
71. Hahm, K.B., Han, Y.M., Park, J.M. et al. 2018. PO89 adalimumab and infliximab biosimilar ameliorated cachexic syndrome of Crohn disease. *Journal of Crohn's and Colitis 12(supplement 1)*: S138.
72. Feagan, B., Panes, J., Ferrante, M. et al. 2018. Risankizumab in patients with moderate to severe Crohn's disease: An open-label extension study. *Lancet* 3: 671–680.
73. McKeage, K., Duggan, S. 2019. Risankizumab: First global approval. *Drugs* 79: 893–900.
74. Sands, B., Chen, J., Feagan, B.G. et al. 2017. Efficacy and safety of MEDI20170, an antibody against interleukin 23, in patients with moderate to severe Crohn's disease. *Gastroenterology* 153: 77–86.
75. Sands, BE, Sandborn, WJ, Biroulet, LP et al. 2019. Efficacy and safety of mirikizumab (LY3074828) in a phase 2 study of patients with Crohn's disease. *Gastroenterology* 156(supplement 1): S-216.
76. Sandborn, W.J., Ferrante, M., Bhandari, B.R. et al. 2019. Efficacy and safety of ant-interleukin 23 therapy with mirikizumab (LY3074828) in patients with moderate-to-severe ulcerative colitis in a phase 2 study. *Gastroenterology* 154(supplement 1): S-1360–1361.
77. D'Haens, G.R., Sandborn, W.J., Ferrante, M. et al. 2019. Maintenance treatment with mirikizumab, a p19 directed IL-23 antibody: Results in patients with moderately-to-severely active ulcerative colitis. *Gastroenterology* 156(supplement 1): S-216.
78. Wong, U., Cross, R.K. 2019. Expert opinion on interleukin-12/23 and interleukin-23 antagonists as potential therapeutic options for the treatment of inflammatory bowel disease. *Expert Opinion Investigational Drugs* 28: 473–479.
79. Schmitt, H., Billmeier, U., Dieterich, W. et al. 2019. Expansion of IL-23 receptor bearing TNFR2+T cells is associated with molecular resistance to anti-TNF therapy in Crohn's disease. *Gut* 68: 814–828.
80. Peyrin-Biroulte, L., Panes, J, Chiorean, M.V. et al. 2019. Histological remission and mucosal healing in a randomized placebo-controlled, phase 2 study of etrasimod in patients with moderately to severely active ulcerative colitis. *Gastroenterology* 156(supplement 1): S-217.
81. Vermiere, S., Sandborn, W.J., Danese, S. et al. 2017. Anti-MadCAM antibody (PF-00547659) for ulcerative colitis (TURANDOT): A phase 2, randomized, double-blind, placebo-controlled trial. *Lancet* 8: 135–144.
82. Sandborn, W.J., Lee, S.D., Tarabar, D. et al. 2018. Phase II evaluation of anti-MadCAM antibody PF-00547659 in the treatment of Crohn's disease: Report of the OPERA study. *Gut* 67: 1824–1835.
83. Danese, S., Vermiere, S, Hellstern, P. et al. 2019. Randomized trial and open label extension study of an anti-interleukin-6 antibody in Crohn's disease (ADANTE I and II). *Gut* 68: 40–48.

84. Tian, L., Li, Y., Zhang, J. et al. 2018. IL-9 promotes the pathogenesis of ulcerative colitis through STAT3/SOCS3 signaling. *Bioscience Reports* 38: BSR20181521.
85. Clarke, A.W., Poulton, L., Shim. D. et al. 2018. An anti-TL1A antibody for the treatment of asthma and inflammatory bowel disease. *MABS* 10: 664–677.
86. Danese, S., Klopocka, M., Scherl, E.J. et al. 2021. Anti-TL1A antibody PF-06480605 safety and efficacy for Ulcerative Colitis: A phase 2a single-arm study. *Clinical Gastroenterology & Hepatology* 19(11): 2324–2332.
87. Mannon, P.J., Fuss, I.J., Mayer, L. et al. 2004. Anti–interleukin-12 antibody for active Crohn's disease. *New England Journal of Medicine* 351(20): 2069–2079.
88. Kashani, A., Schwartz, D.A. 2019. The expanding role of anti–IL-12 and/or anti–IL-23 antibodies in the treatment of inflammatory bowel disease. *Gastroenterology & Hepatology* 15(5): 255–256.
89. Hoving, J.C. 2018. Targeting IL-13 as a host-directed therapy against ulcerative colitis. *Frontiers in Cellular and Infection Microbiology* 8: 395.
90. Hamann, C.R., Thyssen, J.P. 2018. Monoclonal antibodies against interleukin 13 and interleukin 31RA in development for atopic dermatitis. *Journal of the American Academy of Dermatology* 78(3): S37–S42.
91. Giuffrida, P., Caprioli, F., Facciotti, F. et al. 2019. The role of interleukin-13 in chronic inflammatory intestinal disorders. *Autoimmunity Reviews* 18(5): 549–555.
92. Taylor, P.C., Saurigny, D., Vencovsky, J. et al. 2019. Efficacy and safety of namilumab, a human monoclonal antibody against granulocyte-macrophage colony-stimulating factor (GM-CSF) ligand in patients with rheumatoid arthritis (RA) with either an inadequate response to background methotrexate therapy or an inadequate response or intolerance to an anti-TNF (tumour necrosis factor) biologic therapy: A randomized, controlled trial. *Arthritis Research & Therapy* 21(1): 101.
93. Hamilton, J.A. 2015. GM-CSF as a target in inflammatory/autoimmune disease: Current evidence and future therapeutic potential. *Expert Review of Clinical Immunology* 11(4): 457–465.
94. Nylund, C.M., D'Mello, S., Kim, M.O. et al. 2011. Granulocyte macrophage-colony-stimulating factor autoantibodies and increased intestinal permeability in Crohn disease. *Journal of Pediatric Gastroenterology and Nutrition* 52(5): 542–548.
95. Lotfi, N., Thome, R., Rezaei, N. et al. 2019. Roles of GM-CSF in the pathogenesis of autoimmune diseases: An update. *Frontiers in Immunology* 10: 1265.
96. Wang, J., Bhatia, A., Krugliak Cleveland, N. et al. 2018. Rapid onset of inflammatory bowel disease after receiving secukinumab infusion. *ACG Case Reports Journal* 5: e56.
97. Philipose, J., Ahmed, M., Idiculla, P.S. et al. 2018. Severe de novo ulcerative colitis following ixekizumab therapy. *Case Reports in Gastroenterology* 12(3): 617–621.
98. Yang, X.O., Chang, S.H., Park, H., Nurieva, R., Shah, B., Acero, L., Dong, C. 2008. Regulation of inflammatory responses by IL-17F. *The Journal of Experimental Medicine* 205(5): 1063–1075.
99. Zeng, H. 2018. A cytokine duet regulates inflammatory bowel disease. *Science Translational Medicine* 10(450): eaau4583.
100. Sabino, J., Verstockt, B., Vermeire, S. et al. 2019. New biologics and small molecules in inflammatory bowel disease: An update. *Therapeutic Advances in Gastroenterology* 12: 1–14.
101. Tamilarasan, A.G., Cunningham, G., Irving, P.M. et al. 2019. *Recent advances in monoclonal antibody therapy in IBD: Practical issues.* Frontline Gastroenterology 10: 409–416.

Transformative Stem Cell-Based Therapy in Neurological Diseases and Beyond

24

James Zhou and Yihui Shi

Corresponding author: Yihui Shi

Contents

DOI: 10.1201/9780429485626-24

24.1 INTRODUCTION

The term "stem cell" has been around since the 19th century, when German scientist Ernest Haeckel first coined the term. Although he initially thought these cells to be the origin of all multicellular organisms, he later described the fertilized egg as a stem cell, due to it later developing into all of the cells in an organism (Ramalho-Santos and Willenbring 2007). More commonly today, the term "stem cell" is used to denote cells that are capable of both self-renewal and asymmetric cell division, that is, they can divide into one copy of themselves and a progenitor cell (Figure 24.1a) (Weissman 2000). Depending on the cell environment and genetic programming factors, this progenitor gives rise to a host of differentiated cells which comprise the human body (Evans and Kaufman 1981; Morrison, Uchida, and Weissman 1995; Thomson et al. 1998; Takahashi and Yamanaka 2006; Ullah, Subbarao, and Rho 2015; Zhao and

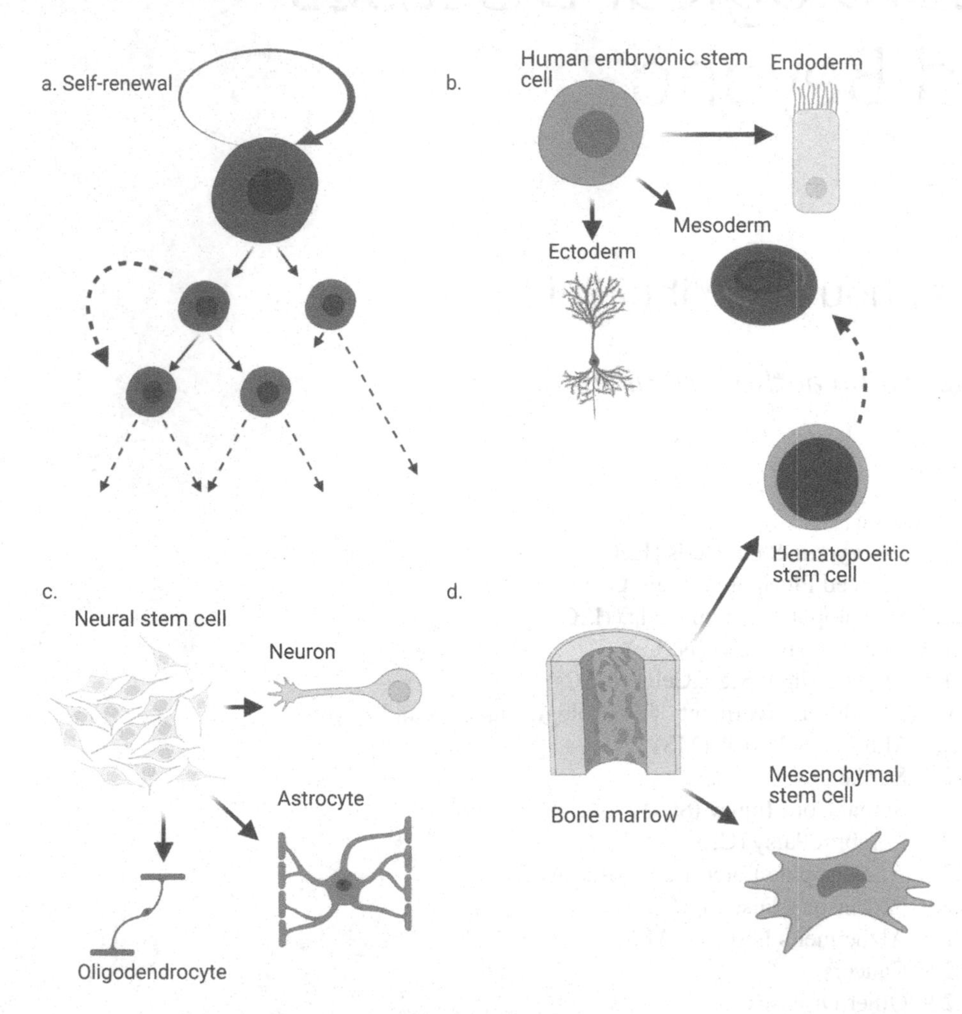

FIGURE 24.1 Stem cells and their regenerative properties and characteristics. (a) A stem cell's ability to regenerate target tissues is in part due to its ability to self-renew, generating a copy of itself as well as another copy toward a differentiated cell lineage. (b) Human embryonic stem cells are capable of regenerating all three germ layers: endoderm, mesoderm, and ectoderm. (c) Neural stem cells have great applications for their ability to regenerate different cells within the central and peripheral nervous systems. (d) Bone marrow-derived stem cells, such as hematopoietic stem cells and mesenchymal stem cells, are some of the cells with the highest current therapeutic use. Hematopoietic stem cells can replenish mesodermal blood cells.

Moore 2018). Cell surface markers, which help give identity to a cell, change over time as cells differentiate and can be used to determine which stage of potency a stem cell possesses. The ability to give rise to any of the cells types seen in human development, including the extra-embryonal tissues, is referred to as totipotency (Condic 2014). The most famous totipotent cell, which Haeckel correctly described in his later attempts, is the fertilized egg (Ramalho-Santos and Willenbring 2007). Due to their ability to regenerate not only themselves but also the cells of the surrounding tissue, stem cells, their use and application have been topics of intensive research since their discovery. This review will focus on the use of stem cells in treating neurological disease, reviewing the different types of stem cells currently used, the diseases they are currently being used to treat, and the progress being made.

24.1.1 Embryonic Stem Cells (ESCs)

Embryonic stem cells (ESCs) are cells derived from the inner cell mass of a developing blastocyst which possess pluripotency, i.e., the ability to divide asymmetrically and differentiate into multiple cell lineages (Figure 24.1b). First discovered in the early 80's (Evans and Kaufman 1981), these cells were shown to be able to differentiate into all of the cells of the developing embryo. Initially cultured from mice, human ESC's (hESCs) first became available in the late 1990s (Thomson et al. 1998). Since then, they have been used to study a variety of diseases. hESC lines with mutations for various diseases can be found in databases compiled by the NIH or the Human Pluripotent Stem Cell Registry (Ilic and Ogilvie 2017). However promising, hESC use in research has been limited by two separate factors: the ethical considerations of using human-derived embryonic cells and the development of induced-pluripotent stem cells by Shinya Yamanaka in 2006 (Annas, Caplan, and Elias 1999; Takahashi and Yamanaka 2006; Gottweis 2010; Ilic and Ogilvie 2017). Nonetheless, the discovery and description of hESCs offered an important perspective on cellular differentiation that has opened the door for scientific development afterward.

24.1.2 Induced Pluripotent Stem Cells (iPSCs)

Induced pluripotent stem cells (IPSCs) were a breakthrough discovery by Takahashi and Yamanaka in 2006, turning previously differentiated cells back into a precursor stem cell state by adding different transcription factors. The ability to transform somatic cells back into stem cells allowed for the therapeutic potential that hESCs did not possess, one of the most important being the ability to perform autologous cell transplants, avoiding problematic interactions with the host immune system (Morizane, Li, and Brundin 2008; Mandai et al. 2017). These transcription factors (Oct4, Sox2, c-Myc, and Klf4) would go onto be referred to as the "Yamanaka factors". The development of stem cells from embryo-free sources would be revolutionary for the field of stem cell biology as it directly addressed one of the main concerns with using hESCs, the use of human embryos (Annas, Caplan, and Elias 1999; Gottweis 2010). In addition, iPSCs proved to be just as powerful as a research tool, capable of differentiating into any of the three germ layers and their derivative cells (Brivanlou et al. 2003; Takahashi and Yamanaka 2006; Hentze et al. 2009). With the concurrent development of technologies that can more powerfully probe disease pathology and accurately remodel these phenotypes in iPSCs, they have become one of the best tools currently available to study many diseases (Shi et al. 2017). While these cells still possess concerns regarding their biological validity, the relative lack of ethical issues allows for greater opportunity in the study of regenerative medicine.

24.1.3 Hematopoietic Stem Cells (HSCs)

Hematopoietic stem cells (HSCs) are the stem cell which, through division and differentiation, form the hematopoietic system. First proposed as a means of protection against lethal doses of radiation and located in the bone marrow, the discovery of these cells provided a wealth of information about the process by which mature blood cells arise (Morrison, Uchida, and Weissman 1995; Pinho and Frenette 2019). Today, one of the primary purposes of HSCs from a therapeutic standpoint is for bone marrow transplantations – often used in the treatment of patients with leukemia (Wayne, Baird, and Egeler 2010). The use of HSCs to treat neurological disease is a more recent occurrence than their use in malignancy, yet HSCs have been used to treat various neurological illnesses (Greco et al. 2014; Mancardi et al. 2015; Burman et al. 2018). HSCs also represent incredibly powerful and interesting research tools and topics due to their critical importance in the body, particularly within the immune system.

24.1.4 Neural Stem Cells (NSCs)

Neural stem cells (NSCs) refer to the stem cell population which serve as precursors for the development of the specialized cells of the brain and spinal cord and are perhaps the cells with the highest therapeutic potential for neurological illnesses (Figure 24.1c). These precursors are most prominently found during normal brain development, whereby radial glia cells divide and differentiate into the three main cell types found in the adult brain: neurons, astrocytes and oligodendrocytes (Kriegstein and Alvarez-Buylla 2009; Zhao and Moore 2018). Self-renewing NSCs were first cultured in the late 1990s, when it was shown that NSCs from the sub-ventricular zone and sub-granular zone could differentiate into neurons, astrocytes and oligodendrocytes (Catavero, Bao, and Song 2018). Since then, the cell culture technique of NSCs has advanced, opening the door for these cells to be used in the treatment of diseases such as: ischemic stroke, spinal cord injury (SCI), Parkinson's, amyotrophic lateral sclerosis (ALS), age-related macular degeneration, and malignancies (Rossi and Cattaneo 2002; Ottoboni, von Wunster, and Martino 2020). Furthermore, developments in transforming both iPSCs and somatic cells into neuronal precursors offer the future potential to augment the therapeutic power of neuronal stem cell transplants (Marro et al. 2011; Pang et al. 2011; Sareen et al. 2014).

24.1.5 Mesenchymal Stem Cells (MSCs)

Mesenchymal stem cells (MSCs) are another category of stem cells that possess high therapeutic potential due to their ease of isolation and the ability to culture them easily *ex vivo* (Minguell, Erices, and Conget 2001). In normal development, these cells possess the ability to differentiate into many different mesodermal lineages such as osteocytes, adipocytes, chondrocytes, bone marrow stroma, and connective tissue (Caplan 1991; Ullah, Subbarao, and Rho 2015). MSCs can now be extracted from almost all tissues, including bone marrow and adipose tissue (Figure 24.1d) (Han et al. 2019). A critical component of the role MSCs play in treating disease is their interaction with the immune system. MSCs from allogeneic donors have been shown to evoke minimal immune responses in recipients and can be immuno-modulatory, stimulating release of cytokines and receptors (Le Blanc and Ringdén 2005; Fouillard et al. 2007). Currently, MSCs are also being used to treat a wide variety of cardiological, neurological, and oncological illnesses (Parekkadan and Milwid 2010). Importantly, in the treatment of neurological illnesses, MSCs have been utilized in clinical trials to treat diseases such as cerebral palsy, SCI, ALS, and traumatic brain injury (Wang et al. 2013; Petrou et al. 2016; Liu et al. 2017; Vaquero et al. 2017; Mukai, Tojo, and Nagamura-Inoue 2018).

24.2 STEM CELLS FOR THE TREATMENT OF NEUROLOGIC DISEASES AND BEYOND

Many of the issues regarding stem cells in the treatment of neurological illnesses overlap with the use of stem cells for the treatment of other illnesses. These include issues with both the ethicality and safety of the use of stem cells. In particular, the use of ESCs contracts a lot of political ire for the method by which they are produced (Annas, Caplan, and Elias 1999; King and Perrin 2014). Nevertheless, the scientific potential of stem cells ensures all types of stem cells can and will be utilized for further therapeutic development. From another perspective, keeping the safety of the patient, the ultimate recipient of any therapies, in mind remains of the highest importance in stem cell research. Due to their ability to differentiate into a multitude of cell types, coupled with the ability to divide almost indefinitely, stem cells possess the ability to develop into cancer-like growths very easily. It is telling that one of the gold standards for the identification of stem cells is the ability to form teratomas (Brivanlou et al. 2003; Nelakanti, Kooreman, and Wu 2015;). In addition, other issues remain, such as genetic stability of stem cells in culture or previous to implantation and biodistribution within the recipient (Goldring et al. 2011; Volarevic et al. 2018;). More research must be done in order to ensure the safety and future of these cell therapies for patients.

24.2.1 Multiple Sclerosis (MS)

Multiple sclerosis (MS) is a chronic inflammatory disease that affects the myelin sheaths of the central nervous system (Nourbakhsh 2013). It is most common in young adults and is the leading cause of neurodegenerative disorder in this population (Nourbakhsh 2013). There are four major sub-types of the disorder: relapsing-remitting (RRMS), in which episodes of neurological impairment, or relapses, occur intermittently, progressive-relapsing (PRMS), a progressive form of the illness with occasional relapses. There are also two additional forms of progressive MS, primary progressive (PPMS), in which the progressive nature of the illness is present from the onset of the symptoms, and secondary progressive, in which the progressive nature develops as a result of relapsing (Miljković and Spasojević 2013; Filippi et al. 2018). MS represents an interesting disease for the use of stem cells due to its complex pathophysiology. The exact causes of MS are unknown (Dobson and Giovannoni 2019), but it has been shown that the immune system, cells such as T-cells, B-cells, and microglia, plays a role in the development of lesions and correspondingly, neurological impairment (Figure 24.2a). In part due to the lack of understanding regarding an exact cause for MS, current treatment options are not typically seen as cures. However, there are still some disease-modifying therapies available. Ocrelizumab (Ocrevus), an anti-CD20 monoclonal antibody, is indicated for patients with PPMS. A wider variety of options are indicated for patients with the most common type of MS, RRMS, including immune-modulatory compounds, fumarate compounds, and monoclonal antibodies (Mayo 2020).

Stem cells represent a novel perspective on treatment for MS, in part due to their ability to regenerate the components of the myelin sheath that have been destroyed pathologically. Research into how these cells can treat more severe forms of MS already utilizes ESCs, iPSCs, HSCs, NSCs, and MSCs. Of these, both HSCs and MSCs have progressed to clinical trials in humans (Cuascut and Hutton 2019).

One therapeutic strategy involving HSCs is the use of immunoablative therapy immediately preceding HSC transplant (aHSCT). This strategy was first performed in the mid-1990s but was tested in a pilot study in 1997 (Fassas et al. 1997; Gavriilaki et al. 2019). The cellular mechanisms by which aHSCT is effective are not fully understood as of yet, but possible explanations include resetting the immune system or depleting its ability to mount a damaging immune response (Cuascut and Hutton 2019). aHSCT has been very successful clinically in small populations, with recommendations regarding its use and protocols being published in *JAMA Neurology* as of October 26, 2020 (Miller et al. 2021). A Phase III clinical trial, BEAT-MS, utilizing this therapy strategy was also initiated in August 2019 (Clinical trial identifier: NCT04047628).

In addition to HSCs, other types of stem cells are currently being studied in order to treat MS. MSCs have shown some promise in pre-clinical animal models, such as the experimental autoimmune encephalitis (EAE) mouse model (Bai et al. 2009; Kurte et al. 2015; Xin et al. 2020). Clinical trials involving MSCs have focused mostly on the safety concerning their transplantation (Karussis et al. 2010; Connick et al. 2011). NSCs are also being utilized in a similar vein to MSCs, albeit with the additional layer of safety in a reduced risk of mesodermal differentiation (Zhang et al. 2016, 2017). hESCs and iPSCs are also being considered for their ability to differentiate into cells from all germ layers (Izrael et al. 2007; Zhang et al. 2016).

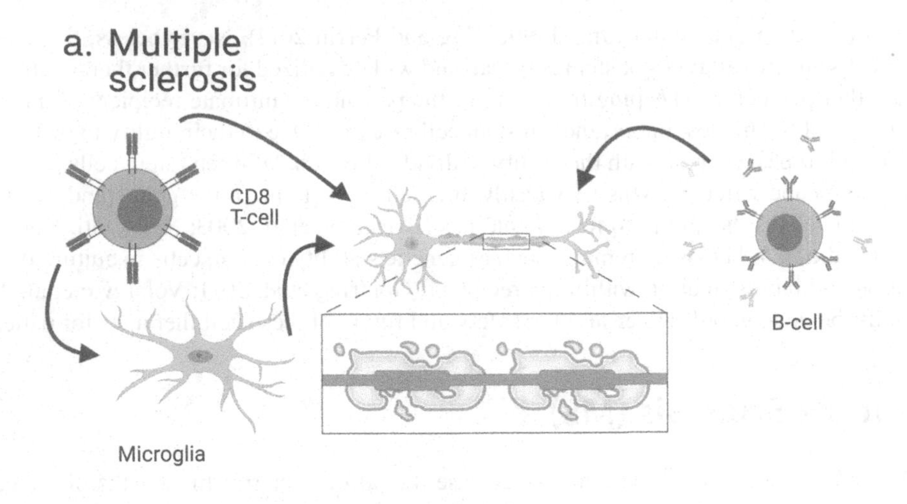

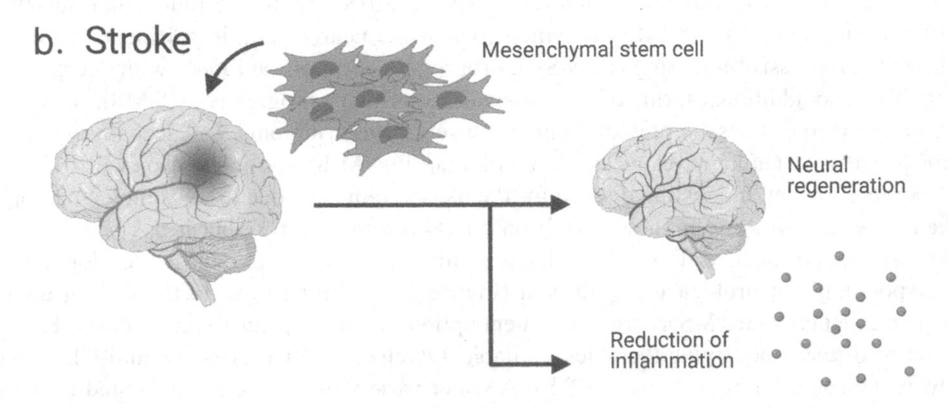

FIGURE 24.2 Disease pathogenesis and stem cell application. (a) A disease model of multiple sclerosis, showing the demyelinating effects of T-cells/microglia and B-cells on oligodendrocytes. Current stem cell applications involve the regeneration of myelin or inhibition of demyelination. (b) Ischemic or hemorrhagic stroke results in loss of perfusion to affected brain tissue. Mesenchymal stem cells represent one modality of treatment, whereby stem cells can inhibit inflammatory processes and potentially regenerate damaged neural tissue.

24.2.2 Stroke

Strokes are debilitating conditions brought upon by acute hypoxic injury to the brain. Globally, they are responsible for the most deaths behind ischemic heart disease (WHO 2021). The common image of a stroke is one of ischemic stroke, in which blood vessels of the brain are actively occluded (Campbell and Khatri 2020). Less common are hemorrhagic strokes, in which blood flow to the brain is disrupted by a rupture in efferent blood vessels. Hemorrhagic strokes constitute as few as 13% of total strokes (ASA 2020). From a high-level perspective, treating stroke can be as simple as restoring blood flow to a hypoxic area. And for most ischemic strokes, this principle holds true. First-line treatments for ischemic strokes are thrombolytic agents to clear any thromboembolic elements or endovascular thrombectomy using mechanical means to restore blood flow (Campbell et al. 2019). However, these treatments are time-dependent and may not result in the complete restoration of neurological activity in patients. Perhaps more relevant for the purposes of developing more complete and powerful treatments, and certainly for this review, is research into the development of stem cells as treatments post-stroke (Figure 24.2b).

The use of stem cells in therapy for stroke is fixated on their potential to restore the hypoxic tissue damage induced by the stroke, and not as strongly targeted toward the prevention and destruction of thromboembolic elements. To that end, NSCs and MSCs currently comprise the majority of clinical trials for stroke patients (www.Clinicaltrials.gov). We will go further into each cell type, its potential role in treatment and justification thereof.

In the treatment of ischemic stroke, NSCs are being used for their ability to regenerate damaged brain tissue. Transplantation of neural progenitor cultured cells derived from embryonal carcinoma cells into humans post-ischemic stroke first occurred in 2000, paving the way for stem cell-based therapies later (Kondziolka et al. 2000). In 2006, Pollock et al. succeeded in culturing NSCs from fetal tissue (Pollock et al. 2006). Testing this cell line in rodent models would later prove successful, opening the road to a Phase I clinical trial, which concluded in 2013 (Stroemer et al. 2009; Kalladka et al. 2016). The results of the clinical trial indicated no adverse immunological events induced by cell transplantation in addition to improved neurological function in 13 men. A follow-up Phase II trial has been started, which is ongoing at time of writing (Clinical trial identifier: NCT03629275). NR1, another neural stem cell line derived from embryonic tissue, has also entered Phase I and II clinical trials (Clinical trial identifier: NCT04631406).

In comparison to NSCs, MSCs have progressed even further in clinical trials and are potentially poised to be the first FDA-approved stem cell treatment for stroke. Typically derived from the bone marrow stroma, yet recently adipose tissue and umbilical cord blood have been found to be alternative sources of these cells (Minguell, Erices, and Conget 2001; Zuk et al. 2001). Pre-clinical rodent studies in the early 2000s established the efficacy of these cells in treating ischemic stroke (Chen et al. 2003; Shen et al. 2006). More recent studies hypothesize that the mechanism of action in MSC transplant is inhibition of the inflammatory environment post-ischemia (Dulamea 2015; Dabrowska et al. 2019). Currently, more than 30 clinical trials have been initiated with MSCs derived from different sources. Four clinical trials have progressed to Phase III as of writing (Clinicaltrials.gov).

Research into HSCs and endothelial precursor cells (EPCs) has also seen some success in treating stroke. HSCs are a natural target for their ability to re-vascularize an infarcted area. Pre-clinical trials in the 2000's show that injection of HSCs can reduce the severity of brain injury post-ischemic stroke (Willing et al. 2003; Schwarting et al. 2008). In addition to the efficacy, an additional advantage is the safety knowledge about HSCs previously derived from its applications in other diseases. Clinical trials in humans with HSCs to treat ischemic stroke have already begun (Clinical trial identifier: NCT01518231). EPCs are another possible treatment option, exerting their therapeutic properties by regulating the stroke inflammatory profile or vasculome (Ishikawa et al. 2013; Dabrowska et al. 2019). However, these cells have not progressed to clinical trials.

24.2.3 Spinal Cord Injury (SCI)

Traumatic spinal cord and spinal injury is a category of injury which describes mechanical damage to the spinal cord, resulting in deterioration of or loss of spinal cord function. This can present as sensory, motor, or autonomic dysfunction. It can be debilitating for patients, taking away from their personal autonomy and greatly increasing their risk for future medical complications (Eckert and Martin 2017). According to the 2020 SCI Data Sheet from the National SCI Statistical Center, approximately 17,810 new SCI cases are recorded each year (University of Alambama 2020). The vast majority (78%) of these are male (University of Alambama 2020). The difficulty in treating patients with SCI lies with the lack of effective treatment options. Methylprednisone, the only FDA-approved pharmaceutical therapy, has shown mixed results in previous clinical trials and there is an ongoing debate about its use in clinical settings (Fehlings, Wilson, and Cho 2014). Given the drastic reduction in quality of life as a result of this type of injury, the need for a cure is significant. The allure of stem cell therapy in SCI is the ability to regenerate the damaged spinal cord tissue and potentially restore the function it had before.

Perhaps the stem cell with the most progress in treating SCI is the MSC. Many clinical trials have already been initiated with MSCs as the treatment condition (Ra et al. 2011; Clinical trial identifiers:

NCT01624779; NCT01769872; NCT01274975). The rationale for the use of MSCs lies in their immune-modulatory properties, secreting cytokines that can delay or reduce local inflammation (Alessandro and Maria 2019; Lalu et al. 2012). This property is critical to delaying the development of further damage in the secondary phase of SCI, termed the secondary injury cascade. This cascade is initiated by primary trauma and is characterized by apoptosis, ischemic damage, reactive oxygen species formation, and local inflammation (Ahuja and Fehlings 2016; Gao et al. 2020). Reports of cancer-associated risk with the use of MSCs are relevant to this discussion as well, albeit lessened due to the pathogenesis and molecular mechanisms of each disease (Wong 2011). However, despite the safety concerns, three Phase III clinical trials have been initiated as of Dec 2020, 2 of which are still active, the other concluding in 2014 (www.Clinicaltrials.gov). There are still no FDA-approved MSC treatments for SCI as of writing.

An additional cell type to mention is the hESC, which saw use for its ability to differentiate into cells from all three germ layers (Liu et al. 2000; Sharp et al. 2010). These pre-clinical studies showed the efficacy of transforming hESCs into oligodendrocyte precursors. Implantation into mice rescued neurological deficit caused by SCI. These studies eventually progressed into clinical trials but ultimately, were stopped due to ethical and safety concerns (Goel 2016; Scott and Magnus 2014).

24.2.4 Cerebral Palsy (CP)

CP is a neurologic disorder characterized by non-progressive motor dysfunction whose pathogenesis starts in the developing infant (Cerebral Palsy Foundation [CPF] 2020a). It is a common pediatric disorder, affecting approximately 2–3 per 1000 live births (Jan 2006; Patel et al. 2020). There are varying types of CP, typically presenting differently depending on the brain areas affected: spastic, dyskinetic, ataxic and mixed types (CPF 2020b). While not typically life-threatening, CP is associated with secondary disturbances in other neurological functions, such as sensation, seizures, or communication and behavior (Graham et al. 2016). These impairments can make like much more difficult for the patient despite the lack of more adverse outcomes present in other neurological diseases. One of the major symptoms of CP that is the target of many pharmaceutical therapeutic strategies is spasticity, defined as muscle contraction, or an increase in muscle tone, that limits movement or elicits discomfort or pain (NINDS 2019). Botulinum toxin and systemic medications such as diazepam or dantrolene are indicated for the treatment of spasticity in CP patients (Delgado et al. 2010; Shamsoddini et al. 2014). However, these treatment options only treat the symptoms associated with CP and do not modify the disease pathology itself. Although CP has been associated with neuropathological findings in the developing brain, there are no FDA-approved pharmaceutical agents which target the disease pathology (Abi Chahine et al. 2016; Patel et al. 2020). Stem cells could play a large role in the treatment of this disease through their regenerative properties, restoring sensation and motor function to patients, even after their illness has been diagnosed.

Currently, over 30 clinical trials have been registered through the FDA using stem cells as a treatment option for CP. The vast majority of these trials utilize MSCs. MSCs are utilized in CP for their ability to regulate the microenvironment surrounding the tissue damage, increasing cellular processes such as neurogenesis, neuron regeneration, and also immunomodulation (Jantzie, Scafidi, and Robinson 2018). One complication of using MSCs is sourcing the cells. CP is often associated with pre-term births (Hafström et al. 2018), which can be associated with a reduced yield of umbilical cord MSCs (UC-MSCs) (Kern et al. 2006). The preferential nature and increased availability during the childhood of UC-MSCs hinder the therapeutic potential of adult-derived MSCs. As of writing, two studies have progressed to Phase III clinical trials (Clinical trial identifiers: NCT01929434; NCT01832454).

Other than MSCs, bone-marrow mononuclear cells are another treatment option (Purandare et al. 2012; Sharma et al. 2015; Nguyen, Nguyen, and Nguyen 2018). Consisting of both the hematopoietic and mesenchymal elements in bone marrow stroma, these stem cells have also advanced to clinical trials (Clinical trial identifier: NCT02574923). Additional research into the efficacy of NPCs derived from fetal tissue has progressed to small-scale clinical studies (Luan et al. 2012).

24.2.5 Amyotrophic Lateral Sclerosis (ALS)

ALS is a devastating neurodegenerative disorder that is characterized by the progressive degeneration of motor neurons throughout the central nervous system (Association). There are two forms of the disease: sporadic and familial, with the former making up around 90%–95% of all cases (Zarei et al. 2015; Association). Both forms ultimately terminate in fatal illness by respiratory failure (Zarei et al. 2015; Association). In the early 1990s, familial ALS was found to be associated with mutations in the gene for superoxide dismutase 1 (SOD1) gene but the genetic basis for sporadic ALS has not yet been elucidated (Boylan 2015). There is also growing evidence that inflammatory processes initiated by immune cells play a role in the pathology of ALS (McCombe and Henderson 2011). Likely owing to the complexity of the disease and its underlying physiology, a true cure does not yet exist, but progress has been made in drug discovery to treat the symptoms of ALS. Four drugs are currently approved in the United States to treat ALS: riluzole, Nuedexta, Radicava, and Tiglutik (NINDS 2020). At the time of writing, no cellular therapies have been approved. However, over 100 clinical trials have been initiated (www.Clinicaltrials.gov). Like with many of the diseases we have reviewed, the promise of stem cells in ALS treatment stems from either the regeneration of motor neurons or the secretion of neuroprotective factors (Lunn, Sakowski, and Feldman 2014).

One of the cell types showing the most promise in clinical trials and small patient cohorts are NSCs (Mazzini et al. 2015; Mazzini et al. 2019; Clinical trial identifier: NCT01640067). These studies tested the feasibility and safety of transplanting NSCs derived from fetal tissue into patients with ALS. Injections were made into the lumbar or cervical spinal cord. Patient outcomes were modest but were generally well tolerated and show that these cells can help reverse some of the damage caused by ALS. In order to better understand the cellular mechanisms that were responsible for these effects, researchers implanted the same cells into SOD1 deficient mice, observing that NSC delayed disease progression and decreased neuroinflammation (Zalfa et al. 2019). Recent advances in the ability to culture NSCs from iPSCs, rather than hESCs, only serve to improve the viability of this treatment modality (Rosati et al. 2018). Another clinical trial has also been initiated studying the safety of astrocytes derived from hESCs (Clinical trial identifier: NCT03482050).

Another cell type that has progressed to clinical trials is MSCs. One of the main advantages of MSCs is their ability to be directly differentiated into motor neuron-like cells (Faghihi et al. 2016). However, pre-clinical data using SOD1 mutant mice has shown that the beneficial effects produced by MSC implantation are due to a decrease in neuroinflammation as opposed to direct regeneration of motor neurons (Uccelli et al. 2012). More recent research in humans supports this claim (Oh et al. 2018). This is consistent with previously held beliefs about the role and function MSCs play in regenerative medicine (Le Blanc and Ringdén 2005; Fouillard et al. 2007). In addition to standardized MSC injection, other groups have shown that MSCs cultured to secrete neurotrophic factors, such as GDNF, NTF, and VEGF, could be feasible in humans (Petrou et al. 2016). Both types of MSCs are currently in clinical trials (Cudkowicz et al. 2022; Nabavi et al. 2019; Clinial trial identifiers: NCT01771640; NCT03280056).

Studies in Turkey have also shown the feasibility of treating patients with HSCs derived from patient bone marrow (Deda et al. 2009). Pre-clinical data in SOD1 mutant mice suggest a similar mechanism of action as compared to MSCs, primarily affecting the inflammatory response and possibly permeability of the blood–brain barrier (Epperly et al. 2019). Hematopoietic stem cells have also progressed to clinical trials (Clinical trial identifier: NCT01933321).

24.2.6 Parkinson's Disease (PD)

PD is a common neurodegenerative disorder that affects roughly 930,000 people in the United States currently (Elkouzi 2020). PD is characterized by loss of neurons in the substantia nigra, consequently causing dopamine deficiency, in part due to the deposition of a-synuclein aggregates intracellularly (Poewe et al. 2017). Patients with PD typically present with tremors, bradykinesia or slowness of movement, limb rigidity, or gait problems (Elkouzi 2020). Not only is PD insidious and fatal for the patient, but it also presents an economic problem due to the large prevalence rate. Recent literature estimates that PD represents $51.9 billion yearly economic burden to the United States alone (Yang et al. 2020). Together with the disease presentation, it becomes clear that a cure is necessary for those suffering. Current therapies to treat symptoms of PD include pharmaceutical agents such as levodopa, a dopamine precursor, dopaminergic agonists, monoamine oxidase-B inhibitors or physical therapies such as deep brain stimulation (Armstrong and Okun 2020). Pharmaceutical agents used in treatment typically increase the availability of dopamine in the brain while deep brain stimulation uses electrodes to electrically stimulate certain areas of the brain associated with PD, such as the subthalamic nucleus or the globus pallidus (Okun 2012). However, no disease-modifying therapies have yet been developed and the need for a cure still remains (Armstrong and Okun 2020). Stem cells can play a critical role in PD treatment due to their ability to potentially regenerate the neurons which are being damaged and restore dopaminergic function in the brain.

As of writing, there are 27 studies testing the safety and efficacy of stem cells in Parkinson's disease (www.Clinicaltrials.gov). Cell types currently undergoing clinical testing are: amniotic epithelial stem cells, iPSCs, NSCs, and MSCs. We will review each of these cell types and the mechanism by which they can potentially treat PD.

First characterized in the 1980s by Akle and group, human amniotic epithelial cells are derived from the human chorionic syncytiotrophoblast (Akle et al. 1981; Qiu et al. 2020). In 2000, Kakishita and group showed that these cells, when cultured, can produce dopamine and are implantable in rats, opening up a new avenue for the treatment of PD (Kakishita et al. 2000). Another cell type currently seeing use are NSCs, which were first tested in rats in 2006 by Yasuhara et al. (2006). These cells contain tyrosine hydroxylase (TH), one of the most important enzymes for dopamine synthesis, restoring function to treated rats. This study opened the door for NSC clinical trials.

MSCs have also been utilized in treating PD, for both their growth-factor secreting properties, as well as their ability to differentiate into dopaminergic neurons (Kitada and Dezawa 2012). In 1999, Schwarz and group showed that human bone marrow stromal cells containing MSCs could be transduced to produce TH. These cells, when grafted, were capable of partially extending survival for short periods of time in rats (Schwarz et al. 1999). In the same year, Dezawa and group discovered a method of inducing dopaminergic differentiation from naïve MSCs using transfected Notch intracellular domain and treatment with growth factors such as bFGF, ciliary neurotrophic factor, and forskolin (Dezawa et al. 2004). These cells, when implanted into 6-hydroxy dopamine lesioned mice, improved rotational behavior associated with loss of dopaminergic pathway integrity. Pre-clinical studies such as these paved the way for clinical trials in humans using MSCs.

Both ESCs and ISCs can also be used for cellular replacement therapy in Parkinson's disease. Both capable of differentiating into dopaminergic neurons, the recent research into these cell types has centered around the ability to differentiate them safely and efficiently. In 2002, Kim and group showed that it was possible to efficiently culture dopamine neurons from ESCs and that the physiologic properties of these cells, coupled with their increased availability, when compared to fetal midbrain neurons, made them an attractive choice for clinical trials in humans (Kim et al. 2002). iPSC differentiation has also been shown through SMAD transcription factor inhibition, allowing iPSC to enter pre-clinical and now, nascent human clinical trials (Chambers et al. 2009; Kikuchi et al. 2017; NCT00874783). Although neither cell type is without safety risks, their use as extremely powerful tools for regenerative medicine cannot be denied and the potential for a cure for Parkinson's does not seem out of reach.

24.2.7 Alzheimer's Disease (AD)

The most prevalent neurodegenerative disorder in the United States and globally, AD represents both a personal health and public health issue for those it affects and their caretakers (Salawu, Umar, and Olokoba 2011). First described over 100 years ago by Dr. Alois Alzheimer, who submitted a case report of a woman with peculiar disease pathology in the cerebral cortex (Hippius and Neundörfer 2003). Since then, diagnosis and healthcare costs associated with diagnosis have ballooned, with Alzheimer's representing a $300 billion burden in direct care costs in 2020 (AIM 2020). The combination of disease presentation, which typically involves gradual erosion of memory (Galton et al. 2000) and the mounting cost of caring for patients diagnosed with Alzheimer's mean the importance of finding a cure is ever-growing. Although the disease was identified over 100 years ago, characterization of disease presentation for diagnosis and pathology for elucidation of cellular mechanisms of disease damage has only started recently (Bondi, Edmonds, and Salmon 2017). Amyloid protein has been the hallmark for Alzheimer's disease histopathologically, but clinical trials with treatments targeting this protein have all historically failed (Hardy and Selkoe 2002; Selkoe and Hardy 2016; Kametani and Hasegawa 2018). Tau protein, a microtubule-associated protein, and the neuropathological cascade misfolded tau initiates have been the subject of much study in recent years (Maccioni et al. 2010; Kametani and Hasegawa 2018). Due to the incredibly complex pathology in AD, a cure has not yet been found. Donepezil, considered the first-line pharmaceutical therapy to treat AD, does not modify the disease course and only slightly improves symptoms (Burns et al. 1999).

Stem cells have been looked at as an alternative treatment for the neurodegeneration associated with Alzheimer's disease (Liu, Yang, and Zhao 2020). Most clinical trials involving humans utilize MSCs (www.Clinicaltrials.gov). The reasoning for their popularity relates to pre-clinical work by Lee and group, who showed that co-culture of MSCs with neurons treated with amyloid protein reduced apoptosis and additional *in vivo* studies showed decreased inflammation as well as apoptosis (Lee et al. 2010). More recent literature sheds light on a possible mechanism for this rescue, the increased autophagy of amyloid protein in the presence of MSCs (Shin et al. 2014). This is consistent with previous studies in other diseases which have utilized MSCs for their immunomodulatory profiles.

Other cell types such as NSCs, ESCs, and iPSCs have also seen pre-clinical success in treating Alzheimer's disease, although their utilization in human clinical trials has not followed. In 2009, Blurton-Jones and group showed that NSC transplantation in the 3×Tg-AD mouse model of Alzheimer's could alleviate spatial learning and memory deficits through brain-derived neurotrophic factor secretion without targeting amyloid or tau proteins directly (Blurton-Jones et al. 2009). A more recent study showed NSCs have the capability to modify amyloid phagocytosis via immunomodulation of surrounding microglia (McGinley et al. 2018).

IPSCs and ESCs are also the subjects of pre-clinical research. Likely utilized for their ability to differentiate into any of the three germ layers, iPSCs represent an important way to examine the pathology of AD by generating more realistic cellular models of disease (Yagi et al. 2011; Israel et al. 2012). This is significant due to the difficulty of obtaining disease-state neurons from patients (Mungenast, Siegert, and Tsai 2016). ESCs can function similarly to iPSCs but pre-clinical research utilizing ESCs has looked into their transplantation into rats (Moghadam et al. 2009).

24.2.8 Cancers

Cancer, caused by uncontrolled cellular proliferation and eventual metastasis, is one of the leading causes of mortality worldwide (NCI 2020). It is estimated that in 2020, around 1.8 million people were diagnosed with cancer and roughly 600,000 patients with cancer died (ACS 2020). While there are many treatment modalities currently available to treat a wide variety of cancers, this review will focus on stem cells and their utility in treating cancer. With both fields of oncology and stem cell regenerative medicine boasting an incredibly rich inventory of literature to review, entire papers can and have been written about both topics (Sylvester and Longaker 2004; Orkin and Zon 2008; Omuro and DeAngelis 2013; Litwin and Tan 2017; Zakrzewski et al. 2019). We will briefly review some of the cell types most commonly utilized in cancer therapy and innovations in how stem cells can be mobilized to fight cancers.

One of the earliest forms of stem cell therapy for treating cancers was hematopoietic stem cell transfers (HSCT), first performed by Dr. E. Donnall Thomas in 1957 (Henig and Zuckerman 2014). An allogeneic transfer occurred, meaning bone marrow was transferred between individuals. Current medicine has improved upon the protocol where now two options are readily available for HSCT: autologous, or from the same individual, or allogeneic, whereby a patient receives a graft from a donor. Transplantation success depends on a variety of factors, including age, health, HLA-matching, disease type, and progression (Majhail et al. 2015) and both types of transplants have advantages and disadvantages (van Besien et al. 2003). However, the number of procedures occurring every year is rising. Organizations such as Be The Match have created libraries of bone marrow donors that allow for increased access to HSCT therapy for those who need allogeneic transfers (BeTheMatch).

MSCs have also emerged as a possible therapeutic for treating cancers due to their tumor tropism properties, i.e., the ability to home into tumor cells via the chemical process of chemotaxis. Pre-clinical studies in rats showed the ability of MSCs to migrate toward glioma cells when injected both intracranially as well as in the carotid artery (Nakamura et al. 2004; Nakamizo et al. 2005). NSCs also display tropism for glioma cells (Aboody et al. 2000; Schmidt et al. 2005). These studies reveal the possibility of using stem cells as tracers for tumor growth and metastasis, even for difficult to resect tumors, such as gliomas.

Augmenting the power of stem cells to track tumors through chemoattractive means, gene editing allows successful tropism to deliver various pharmaceutical agents to cancer cells directly. Aboody and group first showed this by delivering stem cells genetically modified to express cytosine deaminase enzyme, enabling conversion of inactive pro-drug 5-FC to the active, cytotoxic 5-FU directly at the site of the tumor (Aboody et al. 2000). In addition to drug-converting enzymes, apoptosis inducers, oncolytic viruses, and checkpoint inhibitors are also being carried by stem cells directly to tumors (Kimbrel and Lanza 2020). Another modality of cancer treatment currently undergoing pre-clinical testing utilizes genetically edited cancer cell stems that combine chimeric antigen receptor T-cell therapy with the unique regenerative potential of stem cells to produce long-lasting cell populations that enact antigen-specific antitumor immunity (Kimbrel and Lanza 2020). These cells, HSCs, MSCs, and iPSC-derived immune effector cells, all are capable of producing cells with chimeric antigen receptors and improving a previously successful treatment strategy (Li et al. 2018; Golinelli et al. 2020).

In order to create new therapies and pharmaceutical treatments for cancer, it is critical to understand the molecular biology underlying the individual diseases. To that end, cellular modeling of disease processes using iPSCs and ESCs can provide important information to aid in drug discovery (Funato et al. 2014). While there have been concerns iPSCs are genetically unstable, more recent literature seems to refute that idea (Young et al. 2012; De Los Angeles et al. 2015), establishing a model system that allows for novel insights into cancer research.

24.2.9 Other Diseases

Stem cells are both powerful therapeutic options and research tools for fighting many different illnesses. In this review, we have looked at how stem cells are being utilized to treat neurological disorders, such as MS, PD, AD, and malignancy. However, these applications only scratch the surface of the entire range of applications for stem cells. Other diseases such as heart disease, diabetes, and wound healing are all major illnesses that stand to benefit from stem cell research. For example, in 2012, the CADUCEUS clinical trial examined the effects of cardiosphere-derived cardiac stem cells in treating patients with previous myocardial infarction (Makkar et al. 2012; Faiella and Atoui 2016). ESCs and iPSCs have also been used to treat and model type I diabetes, and clinical trials with pancreatic precursor cells derived from ESCs have been initiated (Soria et al. 2000; Maehr et al. 2009; ViaCyte 2020). With regards to wound healing, diseases like chronic lung fibrosis and liver cirrhosis represent over-healing responses that can be treated with stem cell therapies (Du et al. 2014; Dinh et al. 2017; Jones et al. 2019).

24.3 SUMMARY AND FUTURE PERSPECTIVES

Since their discovery, stem cells have represented a source of incredibly untapped potential for therapies for untreatable diseases. Research in recent decades has made use of more advanced techniques and technology in order to unlock this potential. The same innovations drive further research into the basis of neurologic, cardiologic, endocrine, cancer, and fibrotic disease. As more and more discoveries are made about both the biology of these cells and the pathologies of the diseases they are aimed at treating, improved treatment modalities spring forth. Pre-clinical trials, and consequently, clinical trials with humans, reflect the culmination of these efforts. Stem cell banking, including bone marrow, umbilical cord blood and tissue, and adipose tissue are increasing and becoming popular. Those cryogenically preserved stem cells provide an opportunity for later use in stem cell therapy applications. The capabilities and development of stem cell therapies are growing rapidly, although obstacles are present which require continued effort and breakthroughs in research to overcome. The future of regenerative medicine and personalized medicine, particularly in the treatment of incurable neurological disease, is extremely bright.

ACKNOWLEDGMENT

Figures 24. 1 and 24. 2 were created with Biorender.com

REFERENCES

Abi Chahine, N. H., T. W. Wehbe, R. A. Hilal, V. V. Zoghbi, A. E. Melki, and E. B. Habib. 2016. Treatment of cerebral palsy with stem cells: A report of 17 cases. *Int J Stem Cells* 9 (1):90–5.

Aboody, Karen S., Alice Brown, Nikolai G. Rainov, et al. 2000. Neural stem cells display extensive tropism for pathology in adult brain: Evidence from intracranial gliomas. *Proc Natl Acad Sci* 97 (23):12846–51.

ACS. 2020. Cancer Facts & Figures. American Cancer Society. Available from https://www.cancer.org/content/dam/cancer-org/research/cancer-facts-and-statistics/annual-cancer-facts-and-figures/2020/cancer-facts-and-figures-2020.pdf.

Ahuja, C. S., and M. Fehlings. 2016. Concise review: Bridging the gap: Novel neuroregenerative and neuroprotective strategies in spinal cord injury. *Stem Cells Transl Med* 5 (7):914–24.

AIM. 2020. *2020 Alzheimer's Disease Facts and Figures*. Alzheimer's Impact Movement. Available from https://www.alz.org/aaic/downloads2020/2020_Facts_and_Figures_Fact_Sheet.pdf.

Akle, C. A., M. Adinolfi, K. I. Welsh, S. Leibowitz, and I. McColl. 1981. Immunogenicity of human amniotic epithelial cells after transplantation into volunteers. *Lancet* 2 (8254):1003–5.

Alessandro, Poggi, and R. Zocchi Maria. 2019. Immunomodulatory properties of mesenchymal stromal cells: Still unresolved "Yin and Yang". *Curr Stem Cell Res Ther* 14 (4):344–50.

ALS Association. 2020. Understanding ALS. Available from https://www.als.org/understanding-als.

Annas, G. J., A. Caplan, and S. Elias. 1999. Stem cell politics, ethics and medical progress. *Nat Med* 5 (12):1339–41.

Armstrong, Melissa J., and Michael S. Okun. 2020. Diagnosis and treatment of Parkinson disease: A review. *JAMA* 323 (6):548–60.

ASA. 2020. Hemorrhagic Stroke (Bleeds). Available from https://www.stroke.org/en/about-stroke/types-of-stroke/hemorrhagic-strokes-bleeds.

Bai, L., D. P. Lennon, V. Eaton, et al. 2009. Human bone marrow-derived mesenchymal stem cells induce Th2-polarized immune response and promote endogenous repair in animal models of multiple sclerosis. *Glia* 57 (11):1192–203.

BeTheMatch. 2020. Available from https://my.bethematch.org/s/?language=en_US.

Blurton-Jones, Mathew, Masashi Kitazawa, Hilda Martinez-Coria, et al. 2009. Neural stem cells improve cognition via BDNF in a transgenic model of Alzheimer disease. *Proc Natl Acad Sci* 106 (32):13594–9.

Bondi, M. W., E. C. Edmonds, and D. P. Salmon. 2017. Alzheimer's disease: Past, present, and future. *J Int Neuropsychol Soc* 23 (9–10):818–31.

Boylan, K. 2015. Familial amyotrophic lateral sclerosis. *Neurol Clin* 33 (4):807–30.

Brivanlou, A. H., F. H. Gage, R. Jaenisch, T. Jessell, D. Melton, and J. Rossant. 2003. Stem cells. Setting standards for human embryonic stem cells. *Science* 300 (5621):913–6.

Burman, Joachim, Andreas Tolf, Hans Hägglund, and Håkan Askmark. 2018. Autologous haematopoietic stem cell transplantation for neurological diseases. *J Neurol Neurosurg Psychiatry* 89 (2):147–55.

Burns, A., M. Rossor, J. Hecker, et al. 1999. The effects of donepezil in Alzheimer's disease – results from a multinational trial. *Dement Geriatr Cogn Disord* 10 (3):237–44.

Campbell, B. C. V., D. A. De Silva, M. R. Macleod, et al. 2019. Ischaemic stroke. *Nat Rev Dis Primers* 5 (1):70.

Campbell, B. C. V., and P. Khatri. 2020. *Stroke. Lancet* 396 (10244):129–42.

Caplan, A. I. 1991. Mesenchymal stem cells. *J Orthop Res* 9 (5):641–50.

Catavero, C., H. Bao, and J. Song. 2018. Neural mechanisms underlying GABAergic regulation of adult hippocampal neurogenesis. *Cell Tissue Res* 371 (1):33–46.

Chambers, S. M., C. A. Fasano, E. P. Papapetrou, M. Tomishima, M. Sadelain, and L. Studer. 2009. Highly efficient neural conversion of human ES and iPS cells by dual inhibition of SMAD signaling. *Nat Biotechnol* 27 (3):275–80.

Chen, J., Y. Li, M. Katakowski, et al. 2003. Intravenous bone marrow stromal cell therapy reduces apoptosis and promotes endogenous cell proliferation after stroke in female rat. *J Neurosci Res* 73 (6):778–86.

Clinicaltrials.gov. 2020. Available from https://clinicaltrials.gov/ct2/results?cond=Parkinson&term=stem+cell&cntry=&state=&city=&dist=&Search=Search.

Condic, M. L. 2014. Totipotency: What it is and what it is not. *Stem Cells Dev* 23 (8):796–812.

Connick, P., M. Kolappan, R. Patani, et al. 2011. The mesenchymal stem cells in multiple sclerosis (MSCIMS) trial protocol and baseline cohort characteristics: An open-label pre-test: post-test study with blinded outcome assessments. *Trials* 12:62.

CPF. 2020a. *Cause and Timing*. Cerebral Palsy Foundation. Available from https://www.yourcpf.org/cause-and-timing/.

———. 2020b. Types of CP. Cerebral Palsy Foundation. Available from https://www.yourcpf.org/types-of-cp/.

Cuascut, F. X., and G. J. Hutton. 2019. Stem cell-based therapies for multiple sclerosis: Current perspectives. *Biomedicines* 7 (2):26.

Cudkowicz, Merit E., Stacy R. Lindborg, Namita A. Goyal, et al. 2022. A randomized placebo-controlled phase 3 study of mesenchymal stem cells induced to secrete high levels of neurotrophic factors in amyotrophic lateral sclerosis. *Muscle & Nerve* 65 (3):291–302.

Dabrowska, Sylwia, Anna Andrzejewska, Barbara Lukomska, and Miroslaw Janowski. 2019. Neuroinflammation as a target for treatment of stroke using mesenchymal stem cells and extracellular vesicles. *J Neuroinflammation* 16 (1):178.

De Los Angeles, A., F. Ferrari, R. Xi, et al. 2015. Hallmarks of pluripotency. *Nature* 525 (7570):469–78.

Deda, H., M. C. Inci, A. E. Kürekçi, et al. 2009. Treatment of amyotrophic lateral sclerosis patients by autologous bone marrow-derived hematopoietic stem cell transplantation: A 1-year follow-up. *Cytotherapy* 11 (1):18–25.

Delgado, M. R., D. Hirtz, M. Aisen, et al. 2010. Practice parameter: Pharmacologic treatment of spasticity in children and adolescents with cerebral palsy (an evidence-based review): Report of the Quality Standards Subcommittee of the American Academy of Neurology and the Practice Committee of the Child Neurology Society. *Neurology* 74 (4):336–43.

Dezawa, M., H. Kanno, M. Hoshino, et al. 2004. Specific induction of neuronal cells from bone marrow stromal cells and application for autologous transplantation. *J Clin Invest* 113 (12):1701–10.

Dinh, P. C., J. Cores, M. T. Hensley, et al. 2017. Derivation of therapeutic lung spheroid cells from minimally invasive transbronchial pulmonary biopsies. *Respir Res* 18 (1):132.

Dobson, R., and G. Giovannoni. 2019. Multiple sclerosis – a review. *Eur J Neurol* 26 (1):27–40.

Du, Y., J. Wang, J. Jia, et al. 2014. Human hepatocytes with drug metabolic function induced from fibroblasts by lineage reprogramming. *Cell Stem Cell* 14 (3):394–403.

Dulamea, Adriana Octaviana. 2015. The potential use of mesenchymal stem cells in stroke therapy – from bench to bedside. *J Neurol Sci* 352 (1):1–11.

Eckert, M. J., and M. J. Martin. 2017. Trauma: Spinal cord injury. *Surg Clin North Am* 97 (5):1031–45.

Elkouzi, Dr. Ahmad. 2020. *What Is Parkinson's?* Parkinson's Foundation. Available from https://www.parkinson.org/understanding-parkinsons/what-is-parkinsons.

Epperly, M. W., R. Fisher, L. Rigatti, et al. 2019. Amelioration of amyotrophic lateral sclerosis in SOD1(G93A) mice by M(2) microglia from transplanted marrow. *In Vivo* 33 (3):675–88.

Evans, M. J., and M. H. Kaufman. 1981. Establishment in culture of pluripotential cells from mouse embryos. *Nature* 292 (5819):154–6.

Faghihi, G., S. Keyvan, A. Asilian, S. Nouraei, S. Behfar, and M. A. Nilforoushzadeh. 2016. Efficacy of autologous platelet-rich plasma combined with fractional ablative carbon dioxide resurfacing laser in treatment of facial atrophic acne scars: A split-face randomized clinical trial. *Indian J Dermatol Venereol Leprol* 82 (2):162–8.

Faiella, W., and R. Atoui. 2016. Therapeutic use of stem cells for cardiovascular disease. *Clin Transl Med* 5 (1):34.

Fassas, A., A. Anagnostopoulos, A. Kazis, et al. 1997. Peripheral blood stem cell transplantation in the treatment of progressive multiple sclerosis: First results of a pilot study. *Bone Marrow Transplant* 20 (8):631–8.

Fehlings, Michael G., Jefferson R. Wilson, and Newton Cho. 2014. Methylprednisolone for the treatment of acute spinal cord injury: Counterpoint. *Neurosurgery* 61 (CN_suppl_1):36–42.

Filippi, Massimo, Amit Bar-Or, Fredrik Piehl, et al. 2018. Multiple sclerosis. *Nat Rev Dis Primers* 4 (1):43.

Fouillard, L., A. Chapel, D. Bories, et al. 2007. Infusion of allogeneic-related HLA mismatched mesenchymal stem cells for the treatment of incomplete engraftment following autologous haematopoietic stem cell transplantation. *Leukemia* 21 (3):568–70.

Funato, K., T. Major, P. W. Lewis, C. D. Allis, and V. Tabar. 2014. Use of human embryonic stem cells to model pediatric gliomas with H3.3K27M histone mutation. *Science* 346 (6216):1529–33.

Galton, C. J., K. Patterson, J. H. Xuereb, and J. R. Hodges. 2000. Atypical and typical presentations of Alzheimer's disease: A clinical, neuropsychological, neuroimaging and pathological study of 13 cases. *Brain* 123 Pt 3:484–98.

Gao, L., Y. Peng, W. Xu, et al. 2020. Progress in stem cell therapy for spinal cord injury. *Stem Cells Int* 2020:2853650.

Gavriilaki, M., I. Sakellari, E. Gavriilaki, V. K. Kimiskidis, and A. Anagnostopoulos. 2019. Autologous hematopoietic cell transplantation in multiple sclerosis: Changing paradigms in the era of novel agents. *Stem Cells Int* 2019:5840286.

Goel, A. 2016. Stem cell therapy in spinal cord injury: Hollow promise or promising science? *J Craniovertebr Junction Spine* 7 (2):121–6.

Goldring, C. E., P. A. Duffy, N. Benvenisty, et al. 2011. Assessing the safety of stem cell therapeutics. *Cell Stem Cell* 8 (6):618–28.

Golinelli, G., G. Grisendi, M. Prapa, et al. 2020. Targeting GD2-positive glioblastoma by chimeric antigen receptor empowered mesenchymal progenitors. *Cancer Gene Ther* 27 (7–8):558–70.

Gottweis, H. 2010. The endless hESC controversy in the United States: History, context, and prospects. *Cell Stem Cell* 7 (5):555–8.

Graham, H. K., P. Rosenbaum, N. Paneth, et al. 2016. Cerebral palsy. *Nat Rev Dis Primers* 2:15082.

Greco, Raffaella, Attilio Bondanza, Luca Vago, et al. 2014. Allogeneic hematopoietic stem cell transplantation for neuromyelitis optica. *Annals of Neurology* 75 (3):447–53.

Hafström, Maria, Karin Källén, Fredrik Serenius, et al. 2018. Cerebral palsy in extremely preterm infants. *Pediatrics* 141 (1):e20171433.

Han, Y., X. Li, Y. Zhang, Y. Han, F. Chang, and J. Ding. 2019. Mesenchymal stem cells for regenerative medicine. *Cells* 8 (8):886.

Hardy, J., and D. J. Selkoe. 2002. The amyloid hypothesis of Alzheimer's disease: progress and problems on the road to therapeutics. *Science* 297 (5580):353–6.

Henig, I., and T. Zuckerman. 2014. Hematopoietic stem cell transplantation-50 years of evolution and future perspectives. *Rambam Maimonides Med J* 5 (4):e0028.

Hentze, H., P. L. Soong, S. T. Wang, B. W. Phillips, T. C. Putti, and N. R. Dunn. 2009. Teratoma formation by human embryonic stem cells: Evaluation of essential parameters for future safety studies. *Stem Cell Res* 2 (3):198–210.

Hippius, H., and G. Neundörfer. 2003. The discovery of Alzheimer's disease. *Dialogues Clin Neurosci* 5 (1):101–8.

Ilic, D., and C. Ogilvie. 2017. Concise review: Human embryonic stem cells-what have we done? What are we doing? Where are we going? *Stem Cells* 35 (1):17–25.

Ishikawa, H., N. Tajiri, K. Shinozuka, et al. 2013. Vasculogenesis in experimental stroke after human cerebral endothelial cell transplantation. *Stroke* 44 (12):3473–81.

Israel, M. A., S. H. Yuan, C. Bardy, et al. 2012. Probing sporadic and familial Alzheimer's disease using induced pluripotent stem cells. *Nature* 482 (7384):216–20.

Izrael, M., P. Zhang, R. Kaufman, et al. 2007. Human oligodendrocytes derived from embryonic stem cells: Effect of noggin on phenotypic differentiation in vitro and on myelination in vivo. *Mol Cell Neurosci* 34 (3):310–23.

Jan, M. M. 2006. Cerebral palsy: Comprehensive review and update. *Ann Saudi Med* 26 (2):123–32.

Jantzie, Lauren L., Joseph Scafidi, and Shenandoah Robinson. 2018. Stem cells and cell-based therapies for cerebral palsy: A call for rigor. *Pediatr Res* 83 (1):345–55.

Jones, R. E., D. S. Foster, M. S. Hu, and M. T. Longaker. 2019. Wound healing and fibrosis: Current stem cell therapies. *Transfusion* 59 (S1):884–92.

Kakishita, Koji, Mohamed Elwan, Naoyuki Nakao, Toru Itakura, and Norio Sakuragawa. 2000. Human amniotic epithelial cells produce dopamine and survive after implantation into the striatum of a rat model of Parkinson's disease: A potential source of donor for transplantation therapy. *Exp Neurol* 165:27–34.

Kalladka, D., J. Sinden, K. Pollock, et al. 2016. Human neural stem cells in patients with chronic ischaemic stroke (PISCES): A phase 1, first-in-man study. *Lancet* 388 (10046):787–96.

Kametani, F., and M. Hasegawa. 2018. Reconsideration of amyloid hypothesis and tau hypothesis in Alzheimer's *disease. Front Neurosci* 12:25.

Karussis, D., C. Karageorgiou, A. Vaknin-Dembinsky, et al. 2010. Safety and immunological effects of mesenchymal stem cell transplantation in patients with multiple sclerosis and amyotrophic lateral sclerosis. *Arch Neurol* 67 (10):1187–94.

Kern, Susanne, Hermann Eichler, Johannes Stoeve, Harald Klüter, and Karen Bieback. 2006. Comparative analysis of mesenchymal stem cells from bone marrow, umbilical cord blood, or adipose tissue. *Stem Cells* 24 (5):1294–1301.

Kikuchi, Tetsuhiro, Asuka Morizane, Daisuke Doi, et al. 2017. Human iPS cell-derived dopaminergic neurons function in a primate Parkinson's disease model. *Nature* 548 (7669):592–6.

Kim, J. H., J. M. Auerbach, J. A. Rodríguez-Gómez, et al. 2002. Dopamine neurons derived from embryonic stem cells function in an animal model of Parkinson's disease. *Nature* 418 (6893):50–6.

Kimbrel, E. A., and R. Lanza. 2020. Next-generation stem cells – ushering in a new era of cell-based therapies. *Nat Rev Drug Discov* 19 (7):463–79.

King, N. M., and J. Perrin. 2014. Ethical issues in stem cell research and therapy. *Stem Cell Res Ther* 5 (4):85.

Kitada, M., and M. Dezawa. 2012. Parkinson's disease and mesenchymal stem cells: Potential for cell-based therapy. *Parkinsons Dis* 2012:873706.

Kondziolka, D., L. Wechsler, S. Goldstein, et al. 2000. Transplantation of cultured human neuronal cells for patients with stroke. *Neurology* 55 (4):565–9.

Kriegstein, A., and A. Alvarez-Buylla. 2009. The glial nature of embryonic and adult neural stem cells. *Annu Rev Neurosci* 32:149–84.

Kurte, M., J. Bravo-Alegría, A. Torres, et al. 2015. Intravenous administration of bone marrow-derived mesenchymal stem cells induces a switch from classical to atypical symptoms in experimental autoimmune encephalomyelitis. *Stem Cells Int* 2015:140170.

Lalu, M. M., L. McIntyre, C. Pugliese, et al. 2012. Safety of cell therapy with mesenchymal stromal cells (SafeCell): A systematic review and meta-analysis of clinical trials. *PLoS One* 7 (10):e47559.

Le Blanc, K., and O. Ringdén. 2005. Immunobiology of human mesenchymal stem cells and future use in hematopoietic stem cell transplantation. *Biol Blood Marrow Transplant* 11 (5):321–34.

Lee, J. K., H. K. Jin, S. Endo, E. H. Schuchman, J. E. Carter, and J. S. Bae. 2010. Intracerebral transplantation of bone marrow-derived mesenchymal stem cells reduces amyloid-beta deposition and rescues memory deficits in Alzheimer's disease mice by modulation of immune responses. *Stem Cells* 28 (2):329–43.

Li, Ye, David L. Hermanson, Branden S. Moriarity, and Dan S. Kaufman. 2018. Human iPSC-derived natural killer cells engineered with chimeric antigen receptors enhance anti-tumor activity. *Cell Stem Cell* 23 (2):181–192.e5.

Litwin, M. S., and H. J. Tan. 2017. The diagnosis and treatment of prostate cancer: A review. *Jama* 317 (24):2532–42.

Liu, S., Y. Qu, T. J. Stewart, et al. 2000. Embryonic stem cells differentiate into oligodendrocytes and myelinate in culture and after spinal cord transplantation. *Proc Natl Acad Sci U S A* 97 (11):6126–31.

Liu, Xuebin, Xiaojun Fu, Guanghui Dai, et al. 2017. Comparative analysis of curative effect of bone marrow mesenchymal stem cell and bone marrow mononuclear cell transplantation for spastic cerebral palsy. *J Transl Med* 15 (1):48.

Liu, X. Y., L. P. Yang, and L. Zhao. 2020. Stem cell therapy for Alzheimer's disease. *World J Stem Cells* 12 (8):787–802.

Luan, Z., W. Liu, S. Qu, et al. 2012. Effects of neural progenitor cell transplantation in children with severe cerebral palsy. *Cell Transplant* 21 Suppl 1:S91–8.

Lunn, J. S., S. A. Sakowski, and E. L. Feldman. 2014. Concise review: Stem cell therapies for amyotrophic lateral sclerosis: Recent advances and prospects for the future. *Stem Cells* 32 (5):1099–109.

Maccioni, R. B., G. Farías, I. Morales, and L. Navarrete. 2010. The revitalized tau hypothesis on Alzheimer's disease. *Arch Med Res* 41 (3):226–31.

Maehr, René, Shuibing Chen, Melinda Snitow, et al. 2009. Generation of pluripotent stem cells from patients with type 1 diabetes. *Proc Natl Acad Sci* 106 (37):15768–73.

Majhail, N. S., S. H. Farnia, P. A. Carpenter, et al. 2015. Indications for autologous and allogeneic hematopoietic cell transplantation: Guidelines from the American Society for Blood and Marrow Transplantation. *Biol Blood Marrow Transplant* 21 (11):1863–9.

Makkar, R. R., R. R. Smith, K. Cheng, et al. 2012. Intracoronary cardiosphere-derived cells for heart regeneration after myocardial infarction (CADUCEUS): A prospective, randomised phase 1 trial. *Lancet* 379 (9819):895–904.
Mancardi, G. L., M. P. Sormani, F. Gualandi, et al. 2015. Autologous hematopoietic stem cell transplantation in multiple sclerosis: A phase II trial. *Neurology* 84 (10):981–8.
Mandai, M., A. Watanabe, Y. Kurimoto, et al. 2017. Autologous induced stem-cell-derived retinal cells for macular degeneration. *N Engl J Med* 376 (11):1038–46.
Marro, S., Z. P. Pang, N. Yang, et al. 2011. Direct lineage conversion of terminally differentiated hepatocytes to functional neurons. *Cell Stem Cell* 9 (4):374–82.
Mayo. 2021. Multiple Sclerosis. Available from https://www.mayoclinic.org/diseases-conditions/multiple-sclerosis/diagnosis-treatment/drc-20350274.
Mazzini, L., M. Gelati, D. C. Profico, et al. 2015. Human neural stem cell transplantation in ALS: Initial results from a phase I trial. *J Transl Med* 13:17.
Mazzini, L., M. Gelati, D. C. Profico, et al. 2019. Results from phase I clinical trial with intraspinal injection of neural stem cells in amyotrophic lateral sclerosis: A long-term outcome. *Stem Cells Transl Med* 8 (9):887–97.
McCombe, P. A., and R. D. Henderson. 2011. The Role of immune and inflammatory mechanisms in ALS. *Curr Mol Med* 11 (3):246–54.
McGinley, Lisa M., Osama N. Kashlan, Elizabeth S. Bruno, et al. 2018. Human neural stem cell transplantation improves cognition in a murine model of Alzheimer's disease. *Sci Rep* 8 (1):14776.
Miljković, D., and I. Spasojević. 2013. Multiple sclerosis: Molecular mechanisms and therapeutic opportunities. *Antioxid Redox Signal* 19 (18):2286–334.
Miller, Aaron E., Tanuja Chitnis, Bruce A. Cohen, et al. 2021. Autologous hematopoietic stem cell transplant in multiple sclerosis: Recommendations of the National Multiple Sclerosis Society. *JAMA Neurol* 78 (2):241–6.
Minguell, J. J., A. Erices, and P. Conget. 2001. Mesenchymal stem cells. *Exp Biol Med (Maywood)* 226 (6):507–20.
Moghadam, F. H., H. Alaie, K. Karbalaie, S. Tanhaei, M. H. Nasr Esfahani, and H. Baharvand. 2009. Transplantation of primed or unprimed mouse embryonic stem cell-derived neural precursor cells improves cognitive function in Alzheimerian rats. *Differentiation* 78 (2–3):59–68.
Morizane, A., J. Y. Li, and P. Brundin. 2008. From bench to bed: The potential of stem cells for the treatment of Parkinson's disease. *Cell Tissue Res* 331 (1):323–36.
Morrison, S. J., N. Uchida, and I. L. Weissman. 1995. The biology of hematopoietic stem cells. *Annu Rev Cell Dev Biol* 11:35–71.
Mukai, T., A. Tojo, and T. Nagamura-Inoue. 2018. Mesenchymal stromal cells as a potential therapeutic for neurological disorders. *Regen Ther* 9:32–37.
Mungenast, A. E., S. Siegert, and L. H. Tsai. 2016. Modeling Alzheimer's disease with human induced pluripotent stem (iPS) cells. *Mol Cell Neurosci* 73:13–31.
Nabavi, S. M., L. Arab, N. Jarooghi, et al. 2019. Safety, feasibility of intravenous and intrathecal injection of autologous bone marrow derived mesenchymal stromal cells in patients with amyotrophic lateral sclerosis: An open label phase I clinical trial. *Cell J* 20 (4):592–98.
Nakamizo, A., F. Marini, T. Amano, et al. 2005. Human bone marrow-derived mesenchymal stem cells in the treatment of gliomas. *Cancer Res* 65 (8):3307–18.
Nakamura, K., Y. Ito, Y. Kawano, et al. 2004. Antitumor effect of genetically engineered mesenchymal stem cells in a rat glioma model. *Gene Ther* 11 (14):1155–64.
National Spinal Cord Injury Statistical Center. 2020. *Spinal Cord Injury – Facts and Figures at a Glance*. University of Alabama, Birmingham. Available from https://www.nscisc.uab.edu/Public/Facts%20and%20Figures%202020.pdf.
NCI. 2020. *Cancer Statistics*. Available from https://www.cancer.gov/about-cancer/understanding/statistics.
NCT00874783. 2009. Development of iPS From Donated Somatic Cells of Patients With Neurological Diseases. Available from https://ClinicalTrials.gov/show/NCT00874783.
NCT01274975. 2011. Autologous Adipose Derived MSCs Transplantation in Patient With Spinal Cord Injury. Available from https://ClinicalTrials.gov/show/NCT01274975.
NCT01518231. 2012. Autologous Hematopoietic Stem Cell Transplantation in Ischemic Stroke. Available from https://ClinicalTrials.gov/show/NCT01518231.
NCT01624779. 2012. Intrathecal Transplantation of Autologous Adipose Tissue Derived MSC in the Patients with Spinal Cord Injury. Available from https://ClinicalTrials.gov/show/NCT01624779.
NCT01640067. 2012. Human Neural Stem Cell Transplantation in Amyotrophic Lateral Sclerosis (ALS). Available from https://ClinicalTrials.gov/show/NCT01640067.
NCT01769872. 2013. Safety and Effect of Adipose Tissue Derived Mesenchymal Stem Cell Implantation in Patients with Spinal Cord Injury. Available from https://ClinicalTrials.gov/show/NCT01769872.
NCT01771640. 2013. Intrathecal Transplantation of Mesenchymal Stem Cell in Patients With ALS. Available from https://ClinicalTrials.gov/show/NCT01771640.

NCT01832454. 2013. Safety and Efficacy of Bone Marrow MNC for the Treatment of Cerebral Palsy in Subjects Below 15 Years. Available from https://ClinicalTrials.gov/show/NCT01832454.

NCT01929434. 2013. Efficacy of Stem Cell Transplantation Compared to Rehabilitation Treatment of Patients With Cerebral Paralysis. Available from https://ClinicalTrials.gov/show/NCT01929434.

NCT01933321. 2013. Effect of Intrathecal Administration of Hematopoietic Stem Cells in Patients With Amyotrophic Lateral Sclerosis (ALS). Available from https://ClinicalTrials.gov/show/NCT01933321.

NCT02574923. 2015. Outcomes of Stem Cells for Cerebral Palsy. Available from https://ClinicalTrials.gov/show/NCT02574923.

NCT03280056. 2017. Safety and Efficacy of Repeated Administrations of NurOwn® in ALS Patients. Available from https://ClinicalTrials.gov/show/NCT03280056.

NCT03482050. 2018. A Study to Evaluate Transplantation of Astrocytes Derived From Human Embryonic Stem Cells, in Patients With Amyotrophic Lateral Sclerosis (ALS). Available from https://ClinicalTrials.gov/show/NCT03482050.

NCT03629275. 2018. Investigation of Neural Stem Cells in Ischemic Stroke. Available from https://ClinicalTrials.gov/show/NCT03629275.

NCT04047628. 2019. Best Available Therapy versus Autologous Hematopoetic Stem Cell Transplant for Multiple Sclerosis (BEAT-MS). Available from https://ClinicalTrials.gov/show/NCT04047628.

NCT04631406. 2020. A Safety and Tolerability Study of Neural Stem Cells (NR1) in Subjects with Chronic Ischemic Subcortical Stroke (ISS). Available from https://ClinicalTrials.gov/show/NCT04631406.

Nelakanti, R. V., N. G. Kooreman, and J. C. Wu. 2015. Teratoma formation: A tool for monitoring pluripotency in stem cell research. *Curr Protoc Stem Cell Biol* 32:4A 8 1-4A 8 17.

Nguyen, Thanh Liem, Hoang Phuong Nguyen, and Trung Kien Nguyen. 2018. The effects of bone marrow mononuclear cell transplantation on the quality of life of children with cerebral palsy. *Health Qual Life Outcomes* 16 (1):164.

NIH. 2020. *NIH Human Embryonic Stem Cell Registry.* Available from https://grants.nih.gov/stem_cells/registry/current.htm.

NINDS. 2019. *Spasticity Information Page.* Available from https://www.ninds.nih.gov/Disorders/All-Disorders/Spasticity-Information-Page#:~:text=Spasticity%20is%20a%20condition%20in, cord%20that%20control%20muscle%20movement.

———. 2020. *Amyotrophic Lateral Sclerosis (ALS) Information Page.* Available from https://www.ninds.nih.gov/Disorders/All-Disorders/Amyotrophic-Lateral-Sclerosis-ALS-Information-Page.

Nourbakhsh, Bardia. 2013. Review of the biology of multiple sclerosis. *JAMA Neurol* 70 (11):1461.

Oh, K. W., M. Y. Noh, M. S. Kwon, et al. 2018. Repeated intrathecal mesenchymal stem cells for amyotrophic lateral sclerosis. *Ann Neurol* 84 (3):361–73.

Okun, Michael S. 2012. Deep-brain stimulation for Parkinson's disease. *N Eng J Med* 367 (16):1529–38.

Omuro, A., and L. M. DeAngelis. 2013. Glioblastoma and other malignant gliomas: A clinical review. *JAMA* 310 (17):1842–50.

Orkin, S. H., and L. I. Zon. 2008. Hematopoiesis: An evolving paradigm for stem cell biology. *Cell* 132 (4):631–44.

Ottoboni, L., B. von Wunster, and G. Martino. 2020. Therapeutic plasticity of neural stem cells. *Front Neurol* 11:148.

Pang, Z. P., N. Yang, T. Vierbuchen, et al. 2011. Induction of human neuronal cells by defined transcription factors. *Nature* 476 (7359):220–3.

Parekkadan, B., and J. M. Milwid. 2010. Mesenchymal stem cells as therapeutics. *Annu Rev Biomed Eng* 12:87–117.

Patel, D. R., M. Neelakantan, K. Pandher, and J. Merrick. 2020. Cerebral palsy in children: A clinical overview. *Transl Pediatr* 9 (Suppl 1):S125–35.

Petrou, P., Y. Gothelf, Z. Argov, et al. 2016. Safety and clinical effects of mesenchymal stem cells secreting neurotrophic factor transplantation in patients with amyotrophic lateral sclerosis: Results of phase 1/2 and 2a clinical trials. *JAMA Neurol* 73 (3):337–44.

Pinho, S., and P. S. Frenette. 2019. Haematopoietic stem cell activity and interactions with the niche. *Nat Rev Mol Cell Biol* 20 (5):303–20.

Poewe, Werner, Klaus Seppi, Caroline M. Tanner, et al. 2017. Parkinson disease. *Nat Rev Dis Primers* 3 (1):17013.

Pollock, K., P. Stroemer, S. Patel, et al. 2006. A conditionally immortal clonal stem cell line from human cortical neuroepithelium for the treatment of ischemic stroke. *Exp Neurol* 199 (1):143–55.

Purandare, C., D. G. Shitole, V. Belle, A. Kedari, N. Bora, and M. Joshi. 2012. Therapeutic potential of autologous stem cell transplantation for cerebral palsy. *Case Rep Transplant* 2012:825289.

Qiu, C., Z. Ge, W. Cui, L. Yu, and J. Li. 2020. Human amniotic epithelial stem cells: A promising seed cell for clinical applications. *Int J Mol Sci* 21 (20):7730.

Ra, J. C., I. S. Shin, S. H. Kim, et al. 2011. Safety of intravenous infusion of human adipose tissue-derived mesenchymal stem cells in animals and humans. *Stem Cells Dev* 20 (8):1297–308.

Ramalho-Santos, M., and H. Willenbring. 2007. On the origin of the term "stem cell". *Cell Stem Cell* 1 (1):35–8.

Rosati, Jessica, Daniela Ferrari, Filomena Altieri, et al. 2018. Establishment of stable iPS-derived human neural stem cell lines suitable for cell therapies. *Cell Death Dis* 9 (10):937.

Rossi, F., and E. Cattaneo. 2002. Opinion: Neural stem cell therapy for neurological diseases: Dreams and reality. *Nat Rev Neurosci* 3 (5):401–9.

Salawu, Fatai, Joel Umar, and Abdulfatai Olokoba. 2011. Alzheimer's disease: A review of recent developments. *Annals of African Medicine* 10 (2):73–9.

Sareen, D., G. Gowing, A. Sahabian, et al. 2014. Human induced pluripotent stem cells are a novel source of neural progenitor cells (iNPCs) that migrate and integrate in the rodent spinal cord. *J Comp Neurol* 522 (12):2707–28.

Schmidt, N. O., W. Przylecki, W. Yang, et al. 2005. Brain tumor tropism of transplanted human neural stem cells is induced by vascular endothelial growth factor. *Neoplasia* 7 (6):623–9.

Schwarting, S., S. Litwak, W. Hao, M. Bähr, J. Weise, and H. Neumann. 2008. Hematopoietic stem cells reduce postischemic inflammation and ameliorate ischemic brain injury. *Stroke* 39 (10):2867–75.

Schwarz, Emily J., Guillermo M. Alexander, Darwin J. Prockop, and S. Ausim Azizi. 1999. Multipotential marrow stromal cells transduced to produce L-DOPA: Engraftment in a rat model of Parkinson disease. *Hum Gene Ther* 10 (15):2539–49.

Scott, C. T., and D. Magnus. 2014. Wrongful termination: lessons from the Geron clinical trial. *Stem Cells Transl Med* 3 (12):1398–401.

Selkoe, D. J., and J. Hardy. 2016. The amyloid hypothesis of Alzheimer's disease at 25 years. *EMBO Mol Med* 8 (6):595–608.

Shamsoddini, A., S. Amirsalari, M. T. Hollisaz, A. Rahimnia, and A. Khatibi-Aghda. 2014. Management of spasticity in children with cerebral palsy. *Iran J Pediatr* 24 (4):345–51.

Sharma, Alok, Hemangi Sane, Nandini Gokulchandran, et al. 2015. A clinical study of autologous bone marrow mononuclear cells for cerebral palsy patients: A New Frontier. *Stem Cells Int* 2015:905874.

Sharp, J., J. Frame, M. Siegenthaler, G. Nistor, and H. S. Keirstead. 2010. Human embryonic stem cell-derived oligodendrocyte progenitor cell transplants improve recovery after cervical spinal cord injury. *Stem Cells* 28 (1):152–63.

Shen, L. H., Y. Li, J. Chen, et al. 2006. Intracarotid transplantation of bone marrow stromal cells increases axon-myelin remodeling after stroke. *Neuroscience* 137 (2):393–9.

Shi, Y., H. Inoue, J. C. Wu, and S. Yamanaka. 2017. Induced pluripotent stem cell technology: A decade of progress. *Nat Rev Drug Discov* 16 (2):115–30.

Shin, J. Y., H. J. Park, H. N. Kim, et al. 2014. Mesenchymal stem cells enhance autophagy and increase β-amyloid clearance in Alzheimer disease models. *Autophagy* 10 (1):32–44.

Soria, B, E Roche, G Berná, T León-Quinto, J A Reig, and F Martín. 2000. Insulin-secreting cells derived from embryonic stem cells normalize glycemia in streptozotocin-induced diabetic mice. *Diabetes* 49 (2):157–62.

Stroemer, P., S. Patel, A. Hope, C. Oliveira, K. Pollock, and J. Sinden. 2009. The neural stem cell line CTX0E03 promotes behavioral recovery and endogenous neurogenesis after experimental stroke in a dose-dependent fashion. *Neurorehabil Neural Repair* 23 (9):895–909.

Sylvester, K. G., and M. T. Longaker. 2004. Stem cells: Review and update. *Arch Surg* 139 (1):93–9.

Takahashi, K., and S. Yamanaka. 2006. Induction of pluripotent stem cells from mouse embryonic and adult fibroblast cultures by defined factors. *Cell* 126 (4):663–76.

Thomson, J. A., J. Itskovitz-Eldor, S. S. Shapiro, et al. 1998. Embryonic stem cell lines derived from human blastocysts. *Science* 282 (5391):1145–7.

Uccelli, Antonio, Marco Milanese, Maria Cristina Principato, et al. 2012. Intravenous mesenchymal stem cells improve survival and motor function in experimental amyotrophic lateral sclerosis. *Mol Med* 18 (5):794–804.

Ullah, I., R. B. Subbarao, and G. J. Rho. 2015. Human mesenchymal stem cells – current trends and future prospective. *Biosci Rep* 35 (2):e00191.

van Besien, K., F. R. Loberiza, Jr., R. Bajorunaite, et al. 2003. Comparison of autologous and allogeneic hematopoietic stem cell transplantation for follicular lymphoma. *Blood* 102 (10):3521–9.

Vaquero, Jesús, Mercedes Zurita, Miguel A. Rico, et al. 2017. Repeated subarachnoid administrations of autologous mesenchymal stromal cells supported in autologous plasma improve quality of life in patients suffering incomplete spinal cord injury. *Cytotherapy* 19 (3):349–59.

ViaCyte. 2020. Our Clinical Trials. ViaCyte. Available from https://viacyte.com/clinical-trials/.

Volarevic, V., B. S. Markovic, M. Gazdic, et al. 2018. Ethical and safety issues of stem cell-based therapy. *Int J Med Sci* 15 (1):36–45.

Wang, S., H. Cheng, G. Dai, et al. 2013. Umbilical cord mesenchymal stem cell transplantation significantly improves neurological function in patients with sequelae of traumatic brain injury. *Brain Res* 1532:76–84.

Wayne, A. S., K. Baird, and R. M. Egeler. 2010. Hematopoietic stem cell transplantation for leukemia. *Pediatr Clin North Am* 57 (1):1–25.

Weissman, I. L. 2000. Stem cells: units of development, units of regeneration, and units in evolution. *Cell* 100 (1):157–68.

WHO. 2021. The top 10 causes of death. Available from https://www.who.int/news-room/fact-sheets/detail/the-top-10-causes-of-death.

Willing, A.E., J. Lixian, M. Milliken, et al. 2003. Intravenous versus intrastriatal cord blood administration in a rodent model of stroke. *J Neurosci Res* 73 (3):296–307.

Wong, Rebecca S. Y. 2011. Mesenchymal stem cells: Angels or demons? *J Biomed Biotechnol* 2011:459510.

Xin, Ying, Jie Gao, Rong Hu, et al. 2020. Changes of immune parameters of T lymphocytes and macrophages in EAE mice after BM-MSCs transplantation. *Immunol Lett* 225:66–73.

Yagi, T., D. Ito, Y. Okada, et al. 2011. Modeling familial Alzheimer's disease with induced pluripotent stem cells. *Hum Mol Genet* 20 (23):4530–9.

Yang, Wenya, Jamie L. Hamilton, Catherine Kopil, et al. 2020. Current and projected future economic burden of Parkinson's disease in the U.S. *NPJ Parkinson's Dis* 6 (1):15.

Yasuhara, T., N. Matsukawa, K. Hara, et al. 2006. Transplantation of human neural stem cells exerts neuroprotection in a rat model of Parkinson's disease. *J Neurosci* 26 (48):12497–511.

Young, M. A., D. E. Larson, C. W. Sun, et al. 2012. Background mutations in parental cells account for most of the genetic heterogeneity of induced pluripotent stem cells. *Cell Stem Cell* 10 (5):570–82.

Zakrzewski, Wojciech, Maciej Dobrzyński, Maria Szymonowicz, and Zbigniew Rybak. 2019. Stem cells: past, present, and future. *Stem Cell Res Ther* 10 (1):68.

Zalfa, Cristina, Laura Rota Nodari, Elena Vacchi, et al. 2019. Transplantation of clinical-grade human neural stem cells reduces neuroinflammation, prolongs survival and delays disease progression in the SOD1 rats. *Cell Death Dis* 10 (5):345.

Zarei, S., K. Carr, L. Reiley, et al. 2015. A comprehensive review of amyotrophic lateral sclerosis. *Surg Neurol Int* 6:171.

Zhang, Y., X. Li, B. Ciric, et al. 2017. Effect of fingolimod on neural stem cells: A novel mechanism and broadened application for neural repair. *Mol Ther* 25 (2):401–15.

Zhang, Mingliang, Yuan-Hung Lin, Yujiao Jennifer Sun, et al. 2016. Pharmacological reprogramming of fibroblasts into neural stem cells by signaling-directed transcriptional activation. *Cell Stem Cell* 18 (5):653–67.

Zhao, X., and D. L. Moore. 2018. Neural stem cells: Developmental mechanisms and disease modeling. *Cell Tissue Res* 371 (1):1–6.

Zuk, P. A., M. Zhu, H. Mizuno, et al. 2001. Multilineage cells from human adipose tissue: Implications for cell-based therapies. *Tissue Eng* 7 (2):211–28.

Biologics and Biosimilars

25

Clinical Applications and Biomarker Testing

Jianbo Song

Corresponding author: Jianbo Song

Contents

DOI: 10.1201/9780429485626-25

25.1 INTRODUCTION

Biologics (also known as biological drugs, biopharmaceuticals, or biological therapies) are a class of complex large-molecule drugs manufactured or derived from the living system (such as human cells, animal cells, plants, or microorganisms) or that contain components of living organisms, used for the treatment, prevention or cure of human diseases (www.fda.gov). There are a broad range of biological products including blood-derived products, *in vivo* diagnostic allergenic products, immunoglobulin products, products containing cells or microorganisms, proteins such as antibodies and cell signaling proteins, and vaccines. Biological products are typically composed of sugars, proteins, nucleic acids, or complex combinations of these substances. Compared with the small-molecule drugs produced by the process of chemical synthesis with low molecular weight, well-defined structure, and non-immunogenic, biological products are produced by process-dependent methods using living organisms, with high molecular weight, complex heterogeneous structures, immunogenic, and usually unstable due to sensitivity to external condition such as high temperature and microbial contamination. Biologics have revolutionized the treatment of many serious and chronic diseases, including autoimmune diseases (such as rheumatoid arthritis [RA]), diabetes, certain cancers, and some rare genetic diseases. Most biologics are administered via injection or infusion usually and require close supervision of a medical professional.

There are many types of biologics or biological therapies, including blood or products derived from blood, steroid hormones (such as estrogen and testosterone), antitoxins (such as antivenom for venomous bites and stings), recombinant proteins (such as insulin), cytokines (such as interleukins), monoclonal antibodies (mAbs, such as those used for the treatment of autoimmune diseases and cancers), stem cell therapy, gene therapy, and vaccine. Vaccines are the oldest form of biologics that have been around since they were developed in the 19th century [1]. The first licensed biologic manufactured using recombinant DNA technology was human insulin developed by Genentech and licensed to Eli Lilly to market in 1982 [2]. The number of biologics available on the market has significantly increased since 1990s. Currently, there are many types of biologics that revolutionized the treatments of a variety of different diseases. Biologics have become an increasingly important part of health care. With the increasing attention to the cost of drugs, it became apparent that there is the need to create competition and option that would lead to better access in the biologic marketplace.

Biosimilar is a special class of biologics developed as a version of previously FDA-approved biologic known as the reference product with patent expired. In the United States, biologics have a 12-year exclusivity (Patient Protection and Affordable Care Act, 42 U.S.C. § 262(k) (2010)). A biosimilar is very similar to, but not an exact copy of, a biologic drug. There is much controversy regarding the non-proprietary naming of biosimilars. The utilization, marketing m market penetration, and substitution would be significantly facilitated if biosimilars can share the names with the originator biologics. On the other hand, pharmacovigilance, tracking, and preventing unintended substitution can be achieved easier with unique naming. The FDA has taken the middle in the naming of biosimilars. A biosimilar is named after the generic name, followed by four letters [3]. A biosimilar is based on a biologic drug that's already been researched, developed, evaluated for safety and effectiveness through clinical trials, and approved by the FDA. The biologic drug is often called a "reference drug" because the biosimilar is based on the approved drug. The biosimilar works the same way as its reference drug and also has to be approved by the FDA.

A biosimilar is somewhat like a generic version of a biologic drug, although the two are not entirely the same. A generic drug has the same active ingredients as its brand-name drug. Biosimilars are not considered generic because there may be small differences between the biosimilar and its reference drug. However, these variations do not cause clinical differences in the biosimilar's safety or how well it works. Table 25.1 shows the major differences among biologics, biosimilars, and generic drugs.

TABLE 25.1 Major differences between biologics, biosimilars, and generic drugs

PROCESS	*BIOLOGICS*	*BIOSIMILARS*	*GENERIC DRUGS*
Source	Living organisms	Living organisms	Chemical synthesis
Size	Large molecule	Large molecule	Small molecule
Chemical structure	Complex, heterogeneous	Highly similar to the reference product	Chemically identical to the reference product
Analytical characterization	Molecular characterization	Physiochemical and biological characterization	Currently available techniques can establish that the generic and reference product are identical
Manufacturing	Difficult; sensitive to production process; expensive and specialized facilities; reproducibility difficult to establish	Difficult; sensitive to production process; expensive and specialized facilities; reproducibility difficult to establish	Relatively simple; less sensitive to production process; reproducibility easy to establish
Stability	Unstable, sensitive to external conditions	Unstable, sensitive to external conditions	Stable
Immunogenicity	Immunogenic	Immunogenic	Mostly non-immunogenic
Bio-equivalence with reference product	N/A	No	Yes
Interchangeable with reference product	N/A	None approved yet	Yes
Regulation	PHS ACT, BLA under Section 351(a)	PHS ACT, Abbreviated BLA Section 351(k)	FDCA, Abbreviated NDA under Section 505(b)(j)
Clinical studies	Phase I–III studies Efficacy and safety	PK/PD studies in Phase I in health volunteers; comparison studies in Phase III	Bioequivalence studies in healthy volunteers
Patients required	800–1000	~100–500	20–50
Post-authorization activities	Phase IV, risk management plan including pharmacovigilance	Phase IV, risk management plan including pharmacovigilance	Pharmacovigilance
Development cost ($USD)	~900 million–1 billion	100–300 million	2–3 million
Time to market (years)	8–10	7–8	2–3
Patent licensing, exclusivity period	Yes, 12 years	No	No

Interchangeable biosimilar is a specific subcategory of biosimilars that FDA has designated as interchangeable with the original product, meaning these products are safe to swap. The purpose of interchangeable designation is to allow pharmacy-level substitution without prescriber intervention. For most generic drugs, pharmacist can switch branded drugs for generics at the point of purchase, subject to state law, once the generic drugs were approved based on demonstrating enough similarity to the originator branded drugs using bioequivalence under the abbreviated New Drug Application (NDA). However, there are currently no interchangeable biological products approved by FDA because FDA is finalizing the guidelines for drugmakers.

In the United States, biologic is under the regulation of the Food and FDA. The regulation of biologics and biosimilars is an evolving process in the United States. Both the FDA's Center for Drug Evaluation and Research (CDER) and Center for Biologics Evaluation and Research (CBER) have regulatory responsibility for therapeutic biologics, including premarket review and oversight. Figure 25.1 showed the four FDA approval pathways for chemical drugs and generics as well as biologics and biosimilar, respectively. Chemical drugs are approved via New Drug Application (NDA) under Section 505(b)(1) and 505(b)(2) of the Federal Food, Drug, and Cosmetic (FDC) Act of 1938, which was updated by the Hatch-Waxman Amendments in 1984 (Figure 25.1A). Generic drugs are approved via abbreviated NDA under Section 505(b)(j). Approval pathways for biologics and biosimilars were via NDA under the FDC ACT before the Biologics Price Competition and Innovation (BPCI) Act of 2009 (Figure 25.1B, left), then via the Biologic License Application (BLA) under Section 351(a) for biologics and 351(k) for biosimilars (Figure 25.1B, right) after the BPCI was signed into US law in 2010 as part of the Patient Protection Affordable Care Act. BLA via 351(k) is an abbreviated pathway designed to reduce the amount of testing required in animals and humans compared with innovator biologics via 351(a). Currently, most biologics are regulated under Section 351 of the Public Health Service (PHS) Act. It is worth noting that some hormones such as insulin, glucagon, and human growth hormone, are regulated as drugs under the Federal Food, Drug and Cosmetic (FDC) Act, not the biological products under the PHS Act. An NDA is used for drugs subject to the drug approval provision of the FDC Act, whereas a BLA is required for biological products subject

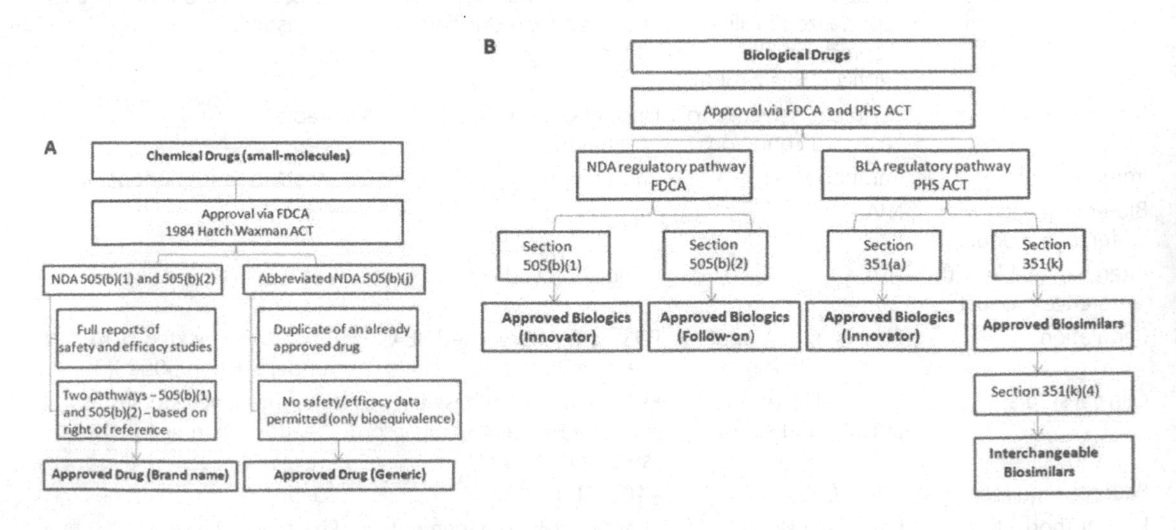

FIGURE 25.1 FDA Approval pathways for chemical drugs (small molecules) and biological drugs (biologics and biosimilars). (A) Approval pathways for chemical drugs (left) and generic drugs (right). Chemical drugs are approved via New Drug Application (NDA) under Section 505(b)(1) and 505(b)(2) of the Federal Food, Drug, and Cosmetic (FDC) Act of 1938, which was updated by the Hatch-Waxman Amendments in 1984. (B) Approval pathways for biologics and biosimilars were via NDA under the FDC ACT before the Biologics Price Competition and Innovation (BPCI) Act of 2009, then via the Biologic License Application (BLA) under Section 351(a) for biologics and 351(k) for biosimilars after the BPCI was signed into US law in 2010 as part of the Patient Protection Affordable Care Act.

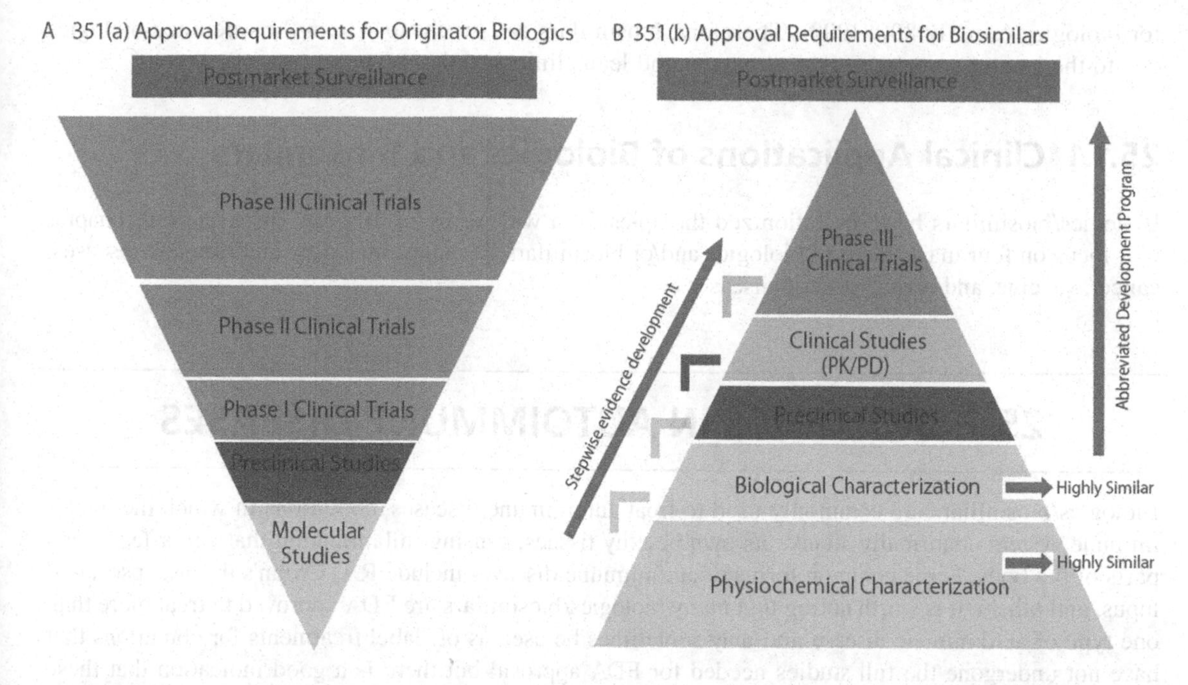

FIGURE 25.2 Evidence required for FDA approval of biologics and biosimilars. (A) The FDA approval requirements for innovator biologics. The BLA for the innovator biologic consists of reports of all studies including analytical, preclinical, and clinical studies along with other pertinent information for evaluation of the product's purity, safety, potency, and effectiveness. (B). Totality of evidence approach for biosimilar approval by stepwise evidence development. At each step, the FDA assesses totality of evidence to decide if additional studies are needed to address residual uncertainty about biosimilarity of the candidate product.

to licensure under the PHS Act. FDA approval to market a biologic is granted by issuance of a biologics license based on whether the product, the manufacturing process, and the manufacturing facilities meet the applicable requirements to ensure the continued safety, purity, and potency of the biologics.

The requirements for innovator biologics are similar to the traditional drug application (Figure 25.2A). This whole process can take up to 10 years and cost up to $800 million (Table 25.1). Thus, biologics are very expensive, often in the range of tens of thousands of dollars annually, because of the complicated development and manufacturing process. Typically, patients need to access these drugs through a specialty pharmacy.

Biosimilar can be approved through this abbreviated pathway based on a totality-of-evidence approach to demonstrate there are no clinically significant differences between a biosimilar and the reference product in terms of safety, purity, and potency for treating the approved clinical conditions (Figure 25.2B).

The bulk of a biosimilar development program is composed of stepwise comparative studies, including physiochemical characterization, biological characterization, preclinical studies, clinical studies, and clinical trials. The comprehensive analytical (biochemical and biophysical) comparative studies demonstrate equivalence or similarity including evidence such as protein structure (primary, secondary, tertiary, and quaternary), enzymatic posttranslational modification (glycosylation, phosphorylation). The next step is biological characterization to demonstrate similarity by functional studies such as bioassays, biological assays, binding assays, and enzyme kinetics. The preclinical studies characterize PK and PD as well as toxicity and immunogenicity using animals. Similarity in PK and PD then is further demonstrated by Phase I bridging study in healthy volunteers. The final demonstration is from one Phase III confirmatory efficacy and efficacy study in patients with the most sensitive indication. The development cost is cheaper for biosimilars ($100–$300 million) than biologics ($900 million–$1 billion). The clinical trials for biosimilars require less patients (~100–500) than that for biologics (usually 800–1000). Therefore, biosimilars can be cheaper than the reference biologics due to the lack of complicated development and lean clinical trials.

25.1.1 Clinical Applications of Biologics and Biosimilars

Biologics/biosimilars have revolutionized therapies for a wide array of different diseases. This chapter will focus on four major fields of biologics and/or biosimilars therapies, including autoimmune diseases, cancer, vaccine, and certain genetic diseases.

25.2 BIOLOGICS IN AUTOIMMUNE DISEASES

Biologics/biosimilars are commonly used to treat autoimmune diseases, conditions in which the body's immune system abnormally attacks its own healthy tissues, causing inflammation that can affect many parts of the body. Some common forms of autoimmune diseases include RA, Crohn's disease, psoriasis, lupus, and others. It is worth noting that many biologics/biosimilars are FDA approved to treat more than one type of autoimmune disease and may sometimes be used as off-label treatments for conditions that have not undergone the full studies needed for FDA approval but there is a good indication that these drugs may be effective [4].

Currently, the exact causes of autoimmune diseases such as RA are not clear. However, both genetic factors such as genes in the human leukocyte antigen (HLA) system and environmental factors including obesity, coffee consumption, cigarette use, and occupational hazards, in combination with a triggering event, contribute to RA [5]. Dysregulation of the inflammatory pathway plays an important role in RA. Multiple factors in the inflammatory pathway including macrophage, T cell, and B cell, contribute to the development of RA [5]. Macrophages triggered by the immune system can increase inflammation by stimulating T cells and osteoclasts. It is known that RA patients have T cells with a high allele expression of HLA and CD-27, resulting in increased affinity for lymphokines. The activated T cells can stimulate B cells to produce autoantibodies and osteoclasts that destroy bone. Both activated T cells and macrophages can release factors that lead to tissue destruction, increased blood flow, invasion of cells into the synovial tissue and joint fluid, and inflammation [6].

Biologics/biosimilar drugs to treat autoimmune diseases are designed to target different components of the immune system including both innate and adaptive immune systems. These targets might be cytokines released by the immune cells, such as tumor necrosis factor (TNF), interleukin-1 (IL-1), and interleukin-6 (IL-6), T cell or B cell. Unlike the traditional disease-modifying antirheumatic drugs (DMARDs) such as hydroxychloroquine (HCQ), methotrexate (MTX), and sulfasalazine (SFZ) that are chemically produced, biologics/biosimilars or biologic DMARDs (bDMARDs) that target different components of the immune systems are manufactured using recombinant DNA technology.

25.2.1 American College of Rheumatology (ACR) Guidelines

Treatment for RA is based on the disease activity and prior therapies. The most currently available American College of Rheumatology (ACR) guidelines recommend that RA therapy typically starts with pharmacological treatment with DMARD monotherapy, using agents such as HCQ, MTX, and SFZ. Then the treatment can be progressed to DMARD combinations, TNF inhibitors, or non-TNF biologics, with or without MTX, depending on disease activity and lack of response to initial therapy [7].

There are six major types of biologics/biosimilars for the treatment of RA based on their targets of the component of the immune system (see Table 25.2).

TABLE 25.2 Biologics and biosimilars in autoimmune diseases

CLASS	*BIOLOGICS*	*BIOSIMILARS*	*SUBSTANCE TARGETED*	*INDICATIONS*
B-cell inhibitors	Rituximab (Rituxan)	Rituximab-abbs (Truxima) Rituximab-pvvr (Ruxience) Rituximab-arrx (Riabni)	CD20	RA, hematological malignancy (NHL, CLL), ANCA-associated vasculitis
	Ocrelizumab (Ocrevus)		CD20	MS
	Ofatumumab (Arzerra)		CD20	CLL, MS
	Obinutuzumab (Gazyva)		CD20	CLL, FL
B-cell growth factor antibody	Belimumab (Benlysta)		BAFF	SLE
IL-1 blocker	Anakinra (Kineret)		IL-1R	RA, CAPS
	Canakinumab (Ilaris)		IL-1 beta	RA, CAPS, etc.
IL-6 blockers	Tocilizumab (Actemra)		IL-6R	RA, SJIA
	Sarilumab (Kevzara)		IL-6R	RA
IL-17 blockers	Secukinumab (Cosentyx)		IL-17A	Psoriasis, PsA, AS
	Ixekizumab (Taltz)		IL-17A	Plaque psoriasis
	Brodalumab (Siliq)		IL-17R	Plaque psoriasis
T-cell inhibitor	Abatacept (Orencia)		B7-1 (CD80) and B7-2 (CD86)	RA, JIA

RA: rheumatoid arthritis; CAPS: cryopyrin-associated periodic syndromes; SJIA: systemic juvenile idiopathic arthritis; AS: ankylosing spondylitis; PsA: psoriatic arthritis; JIA: juvenile idiopathic arthritis; NHL: non-Hodgkin's lymphoma; CLL: chronic lymphocytic leukemia; MS: multiple sclerosis.

25.2.1.1 The B-cell inhibitor

Targets for B-cell inhibitor include antibodies directly targeting B cell, such as Rituximab (Rituxan) and antibodies targeting B-cell growth factor such as Belimumab (Benlysta) that targets cytokine BLyS, which is an important protein for B cell survival and growth.

Rituximab (sold under the brand name Rituxan) is a chimeric mAb targeting the cell surface protein CD20 antigen on normal and malignant B-cells to treat certain autoimmune diseases and some types of cancer. Rituximab was developed by IDEC Pharmaceuticals under the name IDEC-C2B8 and approved for medical use in 1997 [8]. Rituximab is indicated to treat RA and other diseases, including hematological cancers such as non-Hodgkin's lymphoma (NHL) and chronic lymphocytic leukemia (CLL), as well as vasculitis, including granulomatosis with polyangiitis and microscopic polyangiitis and moderate to severe pemphigus vulgaris. There are three independent mechanisms of action for rituximab: antibody-dependent cellular cytotoxicity (ADCC), complement-mediated cytotoxicity (CMC), and programmed cell death (or Apoptosis) [9]. In the United States, rituximab has been used in combination with MTX to treat adult patients with moderately to severely active RA with inadequate response to one or more anti-TNF-alpha therapy [10]. The common side-effects of rituximab include rash, hypotension, itchiness, and shortness of breath. There are also adverse side-effects, including severe infusion reaction, cardia arrest, cytokine release syndrome, tumor lysis syndrome that results in acute kidney injury, and infections, including hepatitis B and other viral infections, immune toxicity with depletion of B cells in 70%–80% of lymphoma patients, pulmonary toxicity [11].

There are other anti-CD20 mAbs developed to treat similar indications. These biologics include (1) ocrelizumab (sold under brand name Ocrevus), humanized (90%–95% human) B-cell-depleting agent developed by Genentech to treat multiple sclerosis (MS) [12]; (2) ofatumumab (sold under the brand name Arzerra), a fully human mAb targeting CD20 developed by Novartis, to treat CLL and MS [13]; (3) third-generation anti-CD20s such as obinutuzumab, a humanized anti-CD20 mAb originated by GlyArt

Biotechnology AG and developed by Roche to treat hematological cancers including CLL and follicular lymphoma [14, 15].

The patent of rituximab is expired in February 2013 in Europe and in September 2016 in the United States. There are several biosimilars approved in the United States, India, the European Union, Switzerland, Japan, and Australia after the patent expired. Rituximab-abbs (Truxima) is the first biosimilar to rituximab approved by the FDA in 2018 to treat NHL and CLL [16]. It is also the only FDA-approved biosimilar rituximab indicated for RA. Rituximab-pvvr (Ruxience) is another biosimilar to rituximab developed by Pfizer and approved by FDA in 2019 to treat NHL and CLL as well as autoimmune conditions, such as granulomatosis with polyangiitis (GPA) and microscopic polyangiitis (MPA) [17]. The most recent biosimilar to rituximab is the Amgen's rituximab-arrx (Riabni) which is FDA approved in December 2020 for the treatment of adult patients with NHL, CLL, GPA (Wegener's granulomatosis), and MPA [18].

25.2.1.1.1 B-cell growth factor-targeted biologics

This class of biologics targets cytokines that are important for B-cell survival and growth. Belimumab (Benlysta) is a human mAB targeting B-cell activating factor (BAFF) [19], which is also known as B-lymphocyte stimulator (BLys) [20]. BAFF plays a critical role in B-cell differentiation, survival, and activation. In certain disease conditions such as systemic lupus erythematosus (SLE), BAFF is overexpressed and leads to autoimmune B cell proliferation and survival. Belimumab acts by preventing BAFF binding to B cells and results in B cells suicide to reduce autoimmune damage of SLE. In 2011, FDA approved Belimumab (Benlysta) for the treatment of lupus [21] and it was the first new drug approved for treating lupus in 56 years [22].

25.2.1.2 The interleukin (IL) blockers

25.2.1.2.1 The IL-1 blocker

Anakinra (Kineret) is a recombinant and slightly modified version of the human IL-1 receptor antagonist protein marketed by Swedish Orphan Biovitrum [23]. Anakinra was approved in the United States in 2001 [24] and Europe in 2002 [23] to treat RA. It can be used as a second-line treatment to reduce symptoms of RA after treatment with a DMARD failed. Anakinra can be used in combination with some other DMARDs [23–25]. In addition to treating RA, it can also be used to treat other conditions such as a cryopyrin-associated periodic syndrome including neonatal-onset multisystem inflammatory disease, and cytokine storm, such as macrophage activation syndrome (MAS) [26]. Anakinra is also used for the off-label treatment of Schnitzler's syndrome, a rare disease with autoinflammatory disorder [27].

Canakinumab (Ilaris) is another IL-1 blocker that specifically blocks IL-1 beta with no cross-reactivity with other members of the IL-1 family, including IL-1 alpha. Canakinumab is a human mAb developed by Novartis for the treatment of RA. In 2009, FDA approved Canakinumab for the treatment of cryopyrin-associated periodic syndromes (CAPS), a spectrum of autoinflammatory syndromes including familial cold autoinflammatory syndrome (FCAS), Muckle-Wells syndrome (MWS), and neonatal-onset multisystem inflammatory disease (NOMID) [28]. Later, canakinumab was also approved for three additional rare and serious autoinflammatory diseases: TNF receptor-associated periodic syndrome (TRAPS), hyperimmunoglobulin D syndrome (HIDS)/mevalonate kinase deficiency (MKD), and familial Mediterranean fever (FMF) [29], as well as Still's disease, including adult-onset Still's disease (AOSD). The common side effect of canakinumab includes increased risk of serious infections due to IL-1 blockade.

25.2.1.2.2 IL-6 blockers

Tocilizumab (Actemra), also known as atlizumab, is a humanized mAb targeting the IL-6 receptor (IL-6R), mainly used for the treatment of RA and systemic juvenile idiopathic arthritis, a severe form of arthritis in children. Tocilizumab could be used to treat moderate to severe RA in combination with MTX if other DMARDs and TNF-alpha inhibitors failed. However, it can also be used as monotherapy for RA patients who cannot tolerate MTX. On 30 August 2017, the FDA approved tocilizumab for cytokine release syndrome, a side effect of CAR-T cell therapy [30]. Recently, tocilizumab obtained an emergency use authorization (EUA) from FDA to treat COVID-19 in the United States in June 2021 [31].

Sarilumab (Kevzara) is another human mAb targeting IL-6R developed by Regeneron Pharmaceuticals and Sanofi to treat RA. This drug was approved by FDA in May 2017.

25.2.1.2.3 Anti-IL-17 biologics

Several anti-IL-17 biologics including secukinumab (Cosentyx), ixekizumab (Taltz), and brodalumab (Siliq) were developed. Secukinumab (Cosentyx) is a human IgG1k mAb targeting IL-17A, marketed by Novartis to treat autoimmune diseases, including psoriasis, ankylosing spondylitis, and psoriatic arthritis [32]. The very common (greater than 10% of people receiving it) adverse effects of secukinumab include upper respiratory tract infections and the common (between 1% and 10% of receiving it) adverse effects include oral herpes, runny nose, and diarrhea. Ixekizumab (Taltz) is another humanized mAb targeting IL-17A, developed by Eli Lilly for the treatment of plaque psoriasis [33]. Brodalumab (Siliq) is a human mAb targeting the IL-17 receptor, developed by Amgen under the name AMG 827. It was approved by FDA in 2017 to treat moderate to severe plague psoriasis in patients with no improvement by other therapies [34]. The side effects of IL-17 blocking treatments are increased risk of infections, mostly including upper respiratory tract infections. Therefore, live vaccines should be avoided while using an IL-17 blocking therapy.

25.2.1.3 The T-cell inhibitor

RA has characteristic chronic joint inflammation and extra-articular involvement. The immunopathology of RA is polygenic and involves different cell populations including macrophages, B-cells, and T-cells, as well as osteoclasts and endothelial cells. Biologics aimed at blocking T-cell activation have been developed to treat RA. T-cell activation requires two signals: signal one and signal two or co-stimulatory signal. Signal one is provided by the antigen-presenting cells (APC) with antigen (foreign particle) held in a structure called the major histocompatibility complex (MHC) on the surface of APC [35]. The T cell receptor (TCR) on both CD4+ helper T cells and CD8+ cytotoxic T cells binds to the antigen on the surface of the APC as well as to the MHC molecule. This initial binding between a T cell specific for the antigen and the antigen-MHC is matched sets in the whole immune response in motion. The process usually happens in the secondary lymphoid organs. Signal two or co-stimulating signal is the secondary signal required for both helper T cells and cytotoxic T cells to become activated [36, 37]. For helper T cells, CD28 on the T cells binds to one of the two molecules on the surface of APC: B7.1 (CD80) or B7.2 (CD86) and initiates T-cell proliferation, leading to the production of many millions of T cells recognizing the antigen. Stimulation of CD28 by B7 induces the production of CTLA-4 (CD152), a protein receptor constitutively expressed in Treg cells but only upregulated in conventional T cells after activation. CTLA-4 competes with CD28 for B7 and so reduces activation signals to the T cells and winds down the immune response [38]. Abatacept (Orencia) is developed by Bristol-Myers Squibb as a fusion protein composed of the Fc region of the IgG1 fused to the extracellular domain of CTLA-4. Abatacept (Orencia) can bind B7-1 and B7-2 on APC and prevent the co-stimulating signal for T cell activation. Abatacept (Orencia) received FDA approval in 2005 for the treatment of RA with inadequate response to anti-TNFα therapy [39]. Adverse effects of abatacept (Orencia) include increased risk for serious infections due to suppression of the immune system, some of which can be fatal. Before starting treatment by abatacept (Orencia), patient should be tested for tuberculosis and hepatitis B and C because they can become active when the activity of T cells is suppressed. Live vaccines should not be given concurrently or within three months of stopping this medication. Furthermore, abatacept treatment may cause otherwise slow-growing cancers to proliferate and spread due to suppression of the immune system.

25.2.1.4 The TNF inhibitor

TNF is a cytokine released by macrophage when an infection is detected to alert the other immune system cells as part of an inflammatory response. TNFα and other cytokines such as IL-1, IL-6, and potentially IL-17 amplify osteoclast differentiation and activation that contribute to the cartilage damage in RA [40].

TABLE 25.3 TNF inhibitors in autoimmune diseases

BIOLOGICS (BRAND NAMES)	*TYPE OF MOLECULE*	*BIOSIMILARS (BRAND NAME)*	*INDICATION*
Adalimumab (Humira)	Human mAb targeting TNFα	Adalimumab-atto (Amjevita)	RA, AS, PsA, CD, psoriasis
		Adalimumab-adbm (Cyltezo)	
		Adalimumab-bwwd (Hadlima)	
		Adalimumab-afzb (Abrilada)	
		Adalimumab-fkjp (Hulio)	
Certolizumab (Cimzia)	MAb targeting TNFα		CD, RA, PsA, AS
Etanercept (Enbrel)	Fusion protein of TNF receptor to Fc of IgG1	Etanercept-szzs (Erelzi) Etanercept-ykro (Eticovo)	RA, PsA, JIA, AS, plaque psoriasis
Golimumab (Simponi)	Human mAb targeting TNFα		UC
Infliximab (Remicade)	Chimeric human-mouse mAbtargeting TNFα	Infliximab-dyyb (Inflectra)	CD, UC
		Remsima	
		Infliximab-abda (Renflexis)	
		Infliximab-axxq (Avsola)	
		Infliximab-qbtx (Ixifi)	

AS: ankylosing spondylitis; CD: Crohn's disease; JCA: juvenile chronic arthritis; PsA: psoriatic arthritis; RA: rheumatoid arthritis; UC: ulcerative colitis; JIA: juvenile idiopathic arthritis; UC: ulcerative colitis.

Many TNFα inhibitors have been developed to treat RA because of its important role in the development of this disease (Table 25.3). Adalimumab (Humira) was the first fully human mAb approved by the US FDA [41, 42]. The brand name Humira stands for "**h**uman **mAbi**n **RA**". Adalimumab (Humira) derived from phage display was discovered by collaboration between BASF Bioresearch Corporation and Cambridge Antibody Technology, UK with an initial name D2E7 [43]. It was ultimately manufactured and marketed by Abbott Laboratories after Abbott's acquisition of BASF Pharma. In 2013, AbbVie spined off from Abbott and took over the development and marketing of adalimumab (Humira). Adalimumab (Humira) has been used alone or in combination with MTX or similar medicines to treat RA in the United States since 2002. It is also indicated for the treatment of many other autoimmune diseases, including juvenile idiopathic arthritis, psoriatic arthritis, ankylosing spondylitis, adult Crohn's disease, pediatric Crohn's disease, ulcerative colitis, plaque psoriasis, hidradenitis suppurativa, and uveitis. The adverse effects of adalimumab include increased risk of serious infections such as tuberculosis as well as increased risk of developing various cancers.

Many biosimilars of adalimumab (Humira) have been developed (Table 25.3). The first adalimumab biosimilar was developed by Indian drugmaker in 2014 and the second one was also declared by Indian drugmaker in 2016. Amgen's biosimilar adalimumab-atto (Amjevita) received FDA approval in 2016 [44] but it will not be available in the United States until at least February 2023. In 2017, the FDA approved German pharmaceutical company Boehringer Ingelheim's biosimilar, adalimumab-adbm (Cyltezo) [45]. In 2018, adalimumab-adaz (Hyrimoz) was approved for use in the United States [46]. In July 2019, adalimumab-bwwd (Hadlima), produced by Samsung Bioepsis, was approved for use in the United States [47]. However, it will not be available until at least June 2023, after the availability of Amgen's offering as a result of a negotiated intellectual property settlement with AbbVie. In November 2019, adalimumab-afzb (Abrilada) was approved for use in the United States [48]. It was the 25th biosimilar to be approved by the FDA. In February 2020, the biosimilar Amsparity was approved for use in the European Union. In June 2020, the biosimilar Idacio was approved for use in Australia. In July 2020, adalimumab-fkjp (Hulio) was approved for use in the United States [49].

Certolizumab (Cimzia) is a fragment of a mAb targeting TNFα manufactured by UCB. It is indicated to treat autoimmune diseases including Crohn's disease, RA, psoriatic arthritis, and ankylosing spondylitis [50].

In addition to antibodies targeting TNFα, fusion proteins produced by recombinant DNA technology have been developed to treat autoimmune diseases as TNF inhibitors. Etanercept (Enbrel) is a fusion protein of TNF receptor to the Fc portion of the IgG1 antibody and it can bind TNFα and decrease its role in diseases involving excessive inflammation [51]. This fusion protein was developed by Bruce A. Beutler and colleagues at University of Texas Southwestern Medical Center at Dallas in 1990s and the patent was sold to Immunex, which was acquired by Amgen in 2002. Etanercept received FDA approval in 1998 and European Union approval in 2000. Several biosimilars to etanercept were developed and approved in the European Union. In January 2016, Benepali, a biosimilar of etanercept developed by Samsung Bioepis was approved for use in the European Union [52]. In June 2017, etanercept-szzs (Erelzi) developed by Sandoz, a Novartis division, was approved by FDA for indications including moderate to severe RA either as monotherapy or in combination with MTX, moderate to severe polyarticular juvenile idiopathic arthritis in patients aged 2 years and older, active psoriatic arthritis, including use in combination with MTX in patients who have not adequately responded to MTX monotherapy, active ankylosing spondylitis, and chronic, moderate to severe plaque psoriasis in adult patients aged 18 years and older who are candidates for systemic therapy or phototherapy [53]. In April 2019, etanercept-ykro (Eticovo) developed by Samsung Bioepis, received FDA approval for the treatment of RA, ankylosing spondylitis, plaque psoriasis, psoriatic arthritis, and polyarticular juvenile idiopathic arthritis [54]. Etanercept-ykro (Eticovo) is already approved in European Union and marketed under the name of Benepali.

Golimumab (Simponi, Simponi Aria) is a human mAb targeting TNFα developed by Janssen Biotech, Inc. (formerly Centocor Biotech, Inc.). Golimumab was approved for the treatment of ulcerative colitis by the US Food and FDA as well as the European Medicines Agency (EMA) in 2013 [55]. Currently, clinical trials for the treatment of RA by golimumab are actively recruiting in Canada (NCT03729349) and in China (NCT04188249).

Infliximab (Remicade) is a purified recombinant DNA-derived chimeric human-mouse IgG mAb composed of mouse heavy and light chain variable regions and human heavy and light chain constant regions, binding and neutralizing TNFα, originally developed by Junming Le (b.1940) and Jan Vilček (b.1933) at New York University School of Medicine and in collaboration with Centocor (now Janssen Biotech, Inc.) [56]. Infliximab received FDA approval in 1998 [57] and the European Union approval in 1999. Infliximab is used to treat autoimmune diseases including Crohn's disease, Fistulizing disease, inflammatory disease, and ulcerative colitis. Infliximab has adverse effects common to the TNF inhibitor drugs, such as serious infections, reactivation of hepatitis B and tuberculosis, lethal hepatosplenic T-cell lymphoma, drug-induced lupus, demyelinating central nervous system disorders, and more. Several biosimilars of infliximab have been developed and approved for medical uses (Table 25.3). The Celltrion/Hospira/Pfizer's infliximab-dyyb (Inflectra, in the United States) and Remsima (brand name marketed in the European Union), a biosimilar version of Infliximab (Remicade), developed by Celltrion Inc., a South Korea biopharmaceutical company, received FDA approval in 2016 [58], for the treatment of patients with RA, ankylosing spondylitis, ulcerative colitis, pediatric and adult Crohn's disease, psoriasis, and psoriatic arthritis.

25.3 BIOLOGICS AND BIOSIMILARS IN CANCERS

Traditional cancer treatments include surgery, chemotherapy, and radiation therapy. Biological therapies have revolutionized the field of cancer treatments. Targeted therapy started with the seminal invention of Imatinib (Gleevec) in the late 1990s as inhibitor of the Bcr-Abl tyrosine kinase to treat chronic myelogenous leukemia (CML) and later for other malignancies. The first biologic for targeted therapy is trastuzumab (Herceptin), a mAb targeting HER2 receptor to treat breast cancer and stomach cancer. Over the past two decades, targeted therapies have emerged as standard treatments for many cancers. Further, immunotherapy has emerged as the "fifth pillar" of cancer treatment over the past few years. Biologics can sometimes be used as a first-line treatment or after other treatments have failed, or in advanced cancers. They are often used in combination with other treatments. There are three major types of biologics/biosimilars treatments of cancers, including immunotherapy, targeted therapy, and treatment of side effects of cancer therapies.

25.3.1 Immunotherapies

Immunotherapy refers to the utilization of immune system to treat cancer. There are three major types of immunotherapies, including immune checkpoint inhibitors, adoptive cell transfer (ACT), and combination immunotherapy.

25.3.1.1 Immune checkpoint inhibitors

The human immune system has mechanisms to keep the immune response in check to avoid overreaction that may result in diseases, such as autoimmune diseases. The discovery of programmed cell death protein 1 (PD-1) by Tsuku Honjo and colleagues at the Kyoto University in 1992 [59] and its ligand PD-L1 that was originally characterized as an immune regulatory molecule B7-H1 by Lieping Chen and colleagues at Mayo Clinic in 1999 [60] laid the foundation of immune checkpoint inhibitor therapy for cancer. PD-1 is a type I membrane protein of 288 amino acid residues belonging to the CD28/CTLA-4 family of T cell regulators. The protein has an extracellular IgV domain, a transmembrane region, and an intracellular domain. The intracellular domain of PD-1 contains two phosphorylation sites located at two motifs: the immunoreceptor tyrosine-based inhibitory motif (ITIM) and the immunoreceptor tyrosine-based switch motif (ITSM), suggesting that PD-1 negatively regulates T-cell receptor (TCR) signals [59, 61, 62]. Unlike CTLA-4 which is constitutively expressed in Treg cells but only upregulated in conventional T cells after activation, PD-1 is expressed on the surface of activated T cells, B cells, and macrophages [63], suggesting that PD-1 more broadly negatively regulates immune responses. PD-L1 (also known as CD274 or B7 homolog 1) is a 40kDa type-I transmembrane protein with a major role in suppressing the adaptive immune system in conditions such as autoimmune disease and hepatitis. Upon binding to PD-L1, the ITIM and ITSM of the PD-1 protein are phosphorylated, leading to recruitment of the SRC homology phosphatases SHP1 and SHP2, which transmit the signal that inhibits TCR-mediated activation of IL-2 production and drives T cell in exhaustion state in which T cells are unable to proliferate or perform their effector functions [64]. The PD-1/PD-L1 interaction minimizes the potential of chronic autoimmune inflammation by safeguarding the immune system only when activated at the appropriate time.

Under normal circumstances, the immune system has an anticancer immune response that can lead to cancer cell death through the cancer immunity cycle [65]. The dendritic cells can capture mutated antigens produced by the tumor cells and prime T cell with tumor antigen to stimulate the activation of cytotoxic T cells. The activated T cells then travel to the tumor site and infiltrate the tumor environment to recognize and bind to the cancer cells. The bound effector T cells then release cytotoxins to induce apoptosis in their target cancer cells.

However, the PD-1/PD-L1 pathway is also the adaptive immune resistance mechanism utilized by tumor cells to inhibit the endogenous immune anti-tumor activity [66]. PD-L1 is overexpressed on tumor cells of certain types of cancers such as melanoma, non-small cell lung cancer (NSCLC), renal cell carcinoma, hand and neck cancer, breast cancer, pancreatic cancer, and ovarian cancer [67–72] or on non-transformed cells in the tumor microenvironment (TME) and binds to PD-1 receptors on the activated T cells, leading to the inhibition of the cytotoxic T cells [73]. It is known that these deactivated T cells remain inactive in the TME.

Immune checkpoint inhibitors act by blocking immune checkpoint proteins including PD-1, PD-L1, and CTLA-4 with mAbs to overcome cancer's ability to resist the immune responses and to stimulate the immune system to fight against cancer. The FDA has approved 3 PD-1 immune checkpoint inhibitors and 3 PD-L1 immune checkpoint inhibitors, covering a total of 16 cancer indications to use as monotherapy or in combination with other drugs (Table 25.4).

TABLE 25.4 Biologics in PD-1/PD-L1 pathway

TARGET	*BIOLOGIC (BRAND NAMES)*	*DRUGMAKER*	*APPROVED INDICATIONS*
PD-1	Pembrolizumab (Keytruda)	Merck	Cervical cancer, CRC, endometrial cancer, esophageal cancer, gastric cancer, HNSCC, HCC, Hodgkin's lymphoma, melanoma, Merkel cell carcinoma, MSI high cancer, NSCLC, PMBCL, RCC, SCLC, UC
	Nivolumab (Opdivo)	Bristol-Meyers Squibb/Ono	CRC (MSI high), HNSCC, HCC, Hodgkin's lymphoma, melanoma, NSCLC, RCC, SCLC, UC
	Cemiplimab (Libtayo)	Sanofi	Cutaneous squamous cell carcinoma
PD-L1	Atezolizumab (Tecentriq)	Roche/Genentech	Breast cancer, NSCLC, SCLC, UC
	Avelumab (Bavencio)	Merck Serono/Pfizer	Merkel cell carcinoma, RCC, UC
	Durvalumab (Imfinzi)	MedImmune/ AstraZeneca/Celgene	NSCLC, UC

CRC: colorectal cancer; HCC: hepatocellular carcinoma; HNSCC: head and neck squamous cell carcinoma; MSI: microsatellite instability; NSCLC: non-small cell lung cancer; PMBCL: primary mediastinal large B cell lymphoma; RCC: renal cell carcinoma; SCLC: small cell lung cancer; UC: urothelial carcinoma.

Pembrolizumab (Keytruda) is a humanized mAb against the human PD-1 receptor developed by Merck. It was approved by FDA in September 2014 for use following treatment with ipilimumab or after treatment with ipilimumab and a BRAF inhibitor in advanced melanoma patients harboring a BRAF mutation [74]. In 2015, FDA then approved pembrolizumab for the treatment of metastatic non-small cell lung cancer (NSCLC) in patients with expression of PD-L1 in the tumor cells after failed treatment with other chemotherapeutic agents [75]. In 2017, pembrolizumab was approved for any unresectable or metastatic solid tumors with certain genetic abnormalities, such as mismatch repair deficiency or microsatellite instability (MSI) [76]. Therefore, pembrolizumab is a so-called tissue-agnostic drug since this is the first time a cancer drug approved based on tumor genetics rather than tissue type of tumor site. Pembrolizumab can be used as monotherapy or in combination with other drugs. Pembrolizumab is a first-line treatment for NSCLC if the cancer overexpresses PD-L1 and no mutations in EGFR or in ALK detected [77]. Evaluation of PD-L1 expression status must be conducted with a validated and approved companion diagnostic. Pembrolizumab can be used as a second-line treatment if chemotherapy has already been administered. However, targeted therapy for EGFR or ALK should be used first if mutations of EGFR or ALK detected.

Nivolumab (Opdivo) is a humanized mAb targeting PD-1, originally developed as MDX-1106/ONO-4538 by Changyu Wang and colleagues at Medarex in collaboration with Ono in 2005 that published in 2014 [78]. Bristol-Myers Squibb acquired Medarex in 2009 and Nivolumab received FDA approval for the treatment of patients with unresectable or metastatic melanoma who no longer respond to other treatments in December 2014 [79], just a few months after the first FDA approval for Merck's PD-1 inhibitor, pembrolizumab (Keytruda), in September 2014. Nivolumab is the first FDA-approved immunotherapy for the first-line treatment of gastric cancer [80] and the second FDA-approved systemic therapy for mesothelioma [81]. Currently, nivolumab is indicated for the treatment of a wide array of cancers, including melanoma, lung cancer, malignant pleural mesothelioma, renal cell carcinoma, Hodgkin lymphoma, head and neck cancer, urothelial carcinoma, colon cancer, esophageal squamous cell carcinoma, liver cancer, and gastric cancer.

Cemiplimab (Libtayo) is a mAb targeting PD-1 developed by Sanofi for the treatment of squamous cell skin cancer [82]. In September 2018, cemiplimab received FDA approval for treatment of metastatic cutaneous squamous cell carcinoma (CSCC) or locally advanced CSCC not suitable for surgery or radiation therapy. Unlike the broad indications in a variety of different cancers for nivolumab and pembrolizumab, cemiplimab is the first FDA-approved medication specifically for advanced CSCC. However, clinical trials of cemiplimab in the treatment of many cancers including myeloma, advanced squamous skin cancer, metastatic pancreatic cancer, lung cancer, and other advanced solid tumors, either as monotherapy or in combination with other treatments, are currently actively recruiting (www.clinicaltrials.org, accessed June 2021).

25.3.1.1.1 PD-L1 inhibitors

FDA-approved PD-L1 inhibitors include atezolizumab, durvalumab, and velumab. Atezolizumab (Tecentriq) is a fully humanized mAb targeting PD-L1 developed by Roche. It was initially approved by FDA in October 2016 for the treatment of metastatic NSCLC progressed during or following platinum-based chemotherapy. Currently, it is indicated for the treatment of several cancers including urothelial carcinoma, NSCLC, triple-negative breast cancer (TNBC), small cell lung cancer (SCLC), and hepatocellular carcinoma (HCC) [83]. Durvalumab (Imfinzi) is another humanized mAb targeting PD-L1 developed by MedImmune/AstraZeneca/Celgene. It received FDA approval for the treatment of certain types of bladder and lung cancer [84]. Specifically, durvalumab is used in combination with etoposide and either carboplatin or cisplatin as first-line treatment for adults with extensive-stage (meaning widely spread) SCLC. It is also used to treat locally advanced or metastatic urothelial carcinoma with disease progression during or following platinum-based chemotherapy or disease progression within 12 months of neoadjuvant or adjuvant treatments with platinum-based chemotherapy. Durvalumab can also be used to treat unresectable, stage-III NSCLC with no disease progression following concurrent platinum-based chemotherapy or radiation therapy.

25.3.1.2 Combination immunotherapy

Although remarkable efficiency of the immune checkpoint inhibition therapy with mAbs is achieved in cancer treatment when used as monotherapy or in combination with chemotherapy, not all patients respond to a single immune checkpoint inhibitor therapy. Combination immunotherapy using synergistic mechanisms of action is currently being investigated to enhance and broaden the anti-tumor activity of immune checkpoint inhibition. This can be achieved by combining two different PD-1 inhibitors. For example, FDA-approved nivolumab and pembrolizumab, two PD-1 inhibitors, to treat advanced or metastatic melanoma [85]. Combination immunotherapy can also be achieved by combining PD-1/PD-L1 inhibitors with other biologics that block immune checkpoint by different mechanisms, such as cytotoxic T-lymphocyte–associated antigen 4 (CTLA-4) inhibitor. Both CTLA-4 and PD-1 are negative regulators of T-cell activation but they play distinct roles in inhibiting immune responses. CTLA-4 regulates T cell proliferation early in an immune response primarily in lymph nodes [86]. However, PD-1/PD-L1 pathway inhibits T cell activation at a later stage in an immune response and acts primarily in peripheral tissues [86]. Ipilimumab (Yervoy) is a mAb targeting CTLA-4 initially developed by James Allison, when worked at University of California, Berkeley, and clinically developed by Medarex, which was later acquired by Bristol-Myers Squibb. In March 2011, ipilimumab received FDA approval for the treatment of unresectable melanoma or advanced (metastatic) melanoma [87]. More importantly, ipilimumab can be used as a complementary checkpoint inhibitor with PD-1/PD-L1 inhibitors to treat different cancers. For example, the combination of ipilimumab with nivolumab is indicated for the first-line treatment of metastatic NSCLC with expression of PD-L1 (≥ 1%) as determined by an FDA-approved companion test [88]. The combination of ipilimumab and nivolumab is also approved to treat BRAF wild-type (WT) metastatic or unresectable melanoma. The combination of nivolumab and ipilimumab is also indicated in the treatment of other cancers including intermediate or poor risk previously untreated advanced renal cell carcinoma, adults and adolescents aged 12 and older with MSI-high or mismatch repair deficient metastatic colorectal cancer (CRC) that has progressed following treatment with a fluoropyrimidine, oxaliplatin, and irinotecan, and HCC previously treated with sorafenib [89]. More recently, FDA approved the combination of nivolumab and ipilimumab as the first-line treatment of unresectable malignant pleural mesothelioma in October 2020 [90]. This is the first drug regimen approved for mesothelioma in sixteen years and the second FDA-approved systemic therapy for mesothelioma.

In summary, combination of immune checkpoint inhibitors and different mechanisms of action significantly increased the immune system activation leading to new immunotherapies for different cancers including melanoma, NSCLC, and other cancers.

25.3.1.1.2 PD1/PD-L1 antibody development

The clinical success of immune checkpoint inhibitors (ICIs) therapy through mAbs targeting PD-1 and PD-L1 revolutionized the field of cancer treatment and attracted tremendous efforts from numerous drug companies all over the world to develop their own mAbs. There are currently about 3000 active clinical trials testing numerous PD-1/PD-L1 inhibitors on participants across a broad array of cancers as of June 2021 (www.clinicaltrials.gov). There are also tremendous efforts focusing on improving efficacy by combination therapy, rather than monotherapy, immediately from the start. Some popular additions to PD-1/PD-L1 inhibitors including VEGF inhibitors (will be discussed under targeted therapy), other ICIs, and costimulatory receptor agonists, other than chemotherapy and radiation therapy.

25.3.1.3 Adoptive T cell therapy

Adoptive T cell therapy, also known as adoptive T cell transfer (ACT) therapy, adoptive immunotherapy, or immune cell therapy, is a rapidly emerging immunotherapy involving collecting and using patient's own immune cells to treat cancer [91]. The process involves first isolating tumor-specific T cells from patients followed by expanding *ex vivo*. These tumor-specific T cells can then be infused back into the patients to enable the immune system to overwhelm the remaining tumor cells. There are several types of ACT including TILs, T-cell receptors (TCRs), and chimeric antigen receptor T cells (CAR-T) [91]. The therapeutic T cells can be harvested either from tumor-infiltrating lymphocytes (TILs) or from peripheral blood lymphocytes (PBLs). However, tumor specificity must be induced in PBLs either through antigen-specific expansion or genetic engineering. In CAR T cell therapy, T cells are engineered to express chimeric antigen receptors (CARs) recognizing cancer-specific antigens, such as CD19 to prime these cells to recognize and kill tumor cells that would otherwise escape immune detection [92]. CAR-T cells and T cells with engineered tumor-specific TCRs show anti-tumor activity in some solid tumors and hematological malignances. Thus far, CAR T-cell therapy is the most advanced ACT in clinical development.

CAR T-cell therapy modifies T cells with CARs (also known as chimeric immunoreceptors, chimeric T cell receptors, or artificial T cell receptors) that combine both antigen-binding and T-cell activating functions into a single receptor. Typically, CARs are composed of four functional domains, including an antigen recognition domain, an extracellular hinge region (spacer), a transmembrane domain, and an intracellular T-cell signaling domain [93, 94]. The antigen recognition domain is responsible for targeting the CAR T cell to tumor cell expressing a matching antigen. It is actually a chimeric protein composed of the light (VL) and heavy (VH) chains derived from the variable regions of a mAb linked together by a short linker peptide to form a single-chain variable fragment (scFv). VL and VH regions provide specific binding ability to the target antigen by selection in advance. The linker region gives flexibility and solubility to the CARs. Non-antibody-based methods have also been taken to direct CAR specificity such as using ligand/receptor pairs. Successful CAR antigen recognition domains include cytokines, innate immune receptors, TNF receptors, growth factors, and structural proteins. The hinge region (spacer), usually based on membrane-proximal regions from other immune molecules such as IgG, CD8, and CD28, can promote antigen binding and synapse formation between the CAR T-cells and target cells by enhancing the flexibility of the scFv receptor head and reducing the spatial constraints between the CAR and targeting antigen. The transmembrane domain plays an essential role in the stability of CAR by anchoring it to the plasma membrane and bridging the extracellular hinge and antigen recognition domains with the intracellular signaling region.

The intracellular T-cell signaling domain is critical for transmitting an activation signal upon binding to the antigen of the antigen recognition domain of CAR. CD3-zeta's cytoplasmic domain is commonly used as the main CAR intracellular T-cell signaling component to mimic the normal T cell activation signaling through the phosphorylation of immunoreceptor tyrosine-based activation motifs (ITAMs) of the cytoplasmic domain of CD3-zeta since other ITAM-containing domains are not as effective [94]. Currently, there are three generations of CARs based on the components of the intracellular domains [95]. First-generation CARs use only a CD3-zeta cytoplasmic domain. However, to persist after activation, T-cells also need co-stimulatory signals from a wide variety of co-stimulatory molecules, such as CD27, CD28, CD134 (OX40), and CD137 (4-1BB). Thus, second-generation CARs add a co-stimulatory domain to the cytoplasmic component, such as CD28 or CD137 (4-1BB), to improve T cell proliferation, cytokine secretion, resistance to apoptosis, and *in vivo* persistence. To further enhance T cell activity, third-generation CARs combine multiple co-stimulatory domains, such as CD28-41BB or CD28-OX40. Compared to second-generation CARs, the third-generation CARs showed significantly improved effector functions and better persistence *in vivo* [95].

The use of CAR T-cell therapy has been limited to small clinical trials on patients with advanced hematological cancers until several CAR T-cell therapies received FDA approval in the recent few years (Table 25.5). In 2017, two CAR T-cell therapies, Novartis' tisagenlecleucel (Kymriah) and Kite Pharma/Gilead's axicabtagene ciloleucel (Yescarta), were approved by the FDA. Tisagenlecleucel (Kymriah) is a CAR T-cell therapy using CAR targeting CD19 and second-generation intracellular T-cell signaling component 41BB – CD3ζ to treat pediatric patients with B-cell acute lymphoblastic leukemia (ALL), the most common cancer in children [96]. Relapsed ALL is a leading cause of death from pediatric cancer [97, 98]. Tisagenlecleucel was originally developed by Dario Campana at St. Jude's Children's Research Hospital and eventually licensed to Novartis. In August 2017, it became the first FDA-approved CAR T-cell therapy in the United States for the treatment of pediatric patients (under 25 years of age) with relapsed or refractory B-cell ALL [99]. It is now also indicated for the treatment of adult patients with relapsed or refractory large B-cell lymphoma after two or more lines of systemic therapy, including diffuse large B-cell lymphoma (DLBCL) not otherwise specified, high-grade B-cell lymphoma and DLBCL arising from follicular lymphoma.

Axicabtagene ciloleucel (Yescarta) is another CAR T-cell therapy using CAR targeting CD19 and second-generation intracellular T-cell signaling component CD28 – CD3ζ, developed by Los Angeles, California-based Kite Pharma. It received FDA approval in October 2017 as the second-line treatment for DLBCL [100], just two months after the approval of Novartis' tisagenlecleucel (Kymriah). Kite Pharma's second CART-cell therapy, Brexucabtagene autoleucel (Tecarta) received FDA approval in July 2020 for the treatment of adults with relapsed or refractory mantle cell lymphoma [101].

More recently, FDA approved two new CAR T-cell therapies in 2021 (Table 25.5). Lisocabtagene maraleucel (Breyanzi) developed by Juno, a Bristol-Myers Squibb Company, using CAR targeting CD19

TABLE 25.5 CAR T cell biologics

CAR T CELL (BRAND)	*COMPANY*	*TARGET*	*INTRACELLULAR SIGNALING DOMAIN*	*APPROVAL DATE*	*INDICATION*
Tisagenlecleucel (Kymriah)	Novartis	CD19	41BB – CD3ζ	08/30/2017	B-cell acute lymphoblastic leukemia (ALL) Diffuse large B-cell lymphoma (DLBCL)
Axicabtagene ciloleucel (Yescarta)	Kite Pharma/ Gilead	CD19	CD28 – CD3ζ	10/18/2017	Diffuse large B-cell lymphoma (DLBCL) Follicular lymphoma
Brexucabtagene autoleucel (Tecartus)	Kite Pharma/ Gilead	CD19	CD28 – CD3ζ	07/24/2020	Mantle cell lymphoma (MCL)
Lisocabtagene maraleucel (Breyanzi)	Juno Therapeutics/ BMS	CD19	41BB – CD3ζ	02/05/2021	Diffuse large B-cell lymphoma (DLBCL)
Idecabtagene vicleucel (Abecma)	Celgene/BMS	BCMA	41BB – CD3ζ	03/26/2021	Multiple myeloma

and second-generation intracellular T-cell signaling component 41BB-CD3ζ, received FDA approval on February 5, 2021, to treat adult patients with relapsed or refractory large B-cell lymphoma [102]. Unlike the other four CAR T-cell therapies with CAR targeting CD19, Idecabtagene vicleucel (Abecma) is a B-cell maturation antigen (BCMA)-directed CAR T-cell therapy developed by Celgene, a Bristol-Myers Squibb Company. On March 26, 2021, idecabtagene vicleucel (Abecma) received FDA approval for the treatment of adult patients with relapsed or refractory multiple myeloma (MM) after four or more prior lines of therapy, including an immunomodulatory agent, a proteasome inhibitor, and an anti-CD38 mAb. This is the first FDA-approved CAR T-cell therapy for MM.

CAR T-cell therapy may cause some major adverse side effects, such as cytokine release syndrome (CRS), B-cell aplasia, cerebral edema, and neurotoxicities. Cerebral edema seems to be a limited side effect reported only in some of the larger CAR T-cell therapy clinical trials. Neurotoxicities are more often, in most CAR T-cell therapy trials, but this problem is transit and reversible [103]. The major side effects that need treatment are cytokine release syndrome (CRS) and B-cell aplasia. CRS, also called cytokine syndrome, is the most frequent side effect associated with CAR T-cell therapy [103]. It is a potentially life-threatening condition resulted from hypersecretion of cytokines by T cells leading to an acute systemic inflammatory syndrome characterized by fever and multiple organ dysfunction. CRS can be managed by monitoring and supportive care to control symptoms. IL-6 is a common target for cytokines. Drugs that interfering with IL-6, such as tocilizumab and siltuximab, are often used to manage severe CRS [104]. Immunosuppressive drugs, such as corticosteroids including methylprednisolone or dexamethasone, may be used to reduce inflammatory and immune response, but these drugs are not used in all cases due to the potential to interfere with the anti-cancer effect of immunotherapy [105]. B-cell aplasia is a well-known "off tumor and on target" phenomenon associated with anti-CD19 CAR T cell therapy due to inadvertently damage of normal B-cells that express CD19. Patients are typically at high risk of developing infections due to hypogammaglobulinemia. However, this can be treated with intravenous immunoglobulin (IVIG) replacement therapy [106].

The impressive success of CAR T cell therapy for hematological malignances is very encouraging. There are some major obstacles in expanding CAR T cell therapy to solid tumors. The first challenge is the difficulty to find suitable antigens to target solid tumors. There are some potential antigens, including mesothelin (MSLN) and EGFRvIII are currently under investigation for CAR T cell therapy. MSLN is a tumor differentiation antigen normally only expressed in mesothelial surfaces, but significantly overexpressed in some of the cancers such as pancreatic and lung cancer. MSLN has emerged as a promising target for anti-MSLN CAR-T cell therapy [107]. Currently, clinical trials are conducted for a broad range of solid tumors, including malignant pleural mesothelioma (MPM), lung cancer, ovarian cancer, pancreatic ductal adenocarcinoma (PDAC), triple-negative breast cancer (TNBC), endometrial cancer, gastric cancer, CRC, hepatoma (HCC), glioma, esophagus cancer, neuroendocrine tumors/Merkel cell carcinoma, and squamous cell cancer. Currently, several antigens including IL13R2 [108, 109], HER2 [109], and EGFRvIII [110] are the most commonly used targets for CAR T cell therapy for the treatment of the aggressive brain cancer glioblastoma (GBM). EGFRvIII is the most common EGFR variant in GBM (present in ~31%–64% of patients) [111], characterized by an in-frame deletion of 267 amino acids in the extracellular domain leading to a constitutively active receptor.

The second obstacle of CAR T therapy for solid tumor is that the overwhelming majority of tumor antigens reside inside the tumor cells. CAR T cells can only bind antigens on the cell surface. Therefore, other forms of T cell therapy, such as engineered T cell receptor (TCR) therapy, have been developed. Unlike CARs that only recognize cell surface antigens, engineered TCRs can recognize intracellular tumor-related antigens that are shuttled to the cell surface in the context of a specific peptide-major histocompatibility complex (pMHC) [112]. Thus, TCRs have shown greater promises against solid tumor than CAR-T cells.

Another major obstacle is the immunosuppressive TME composed of desmoplastic stroma (cellular part) and extracellular matrix (CEM) components (non-cellular) [113]. This hostile microenvironment makes it difficult for the CAR T cells to infiltrate the tumor, recognize their cognate antigen and perform their effector function. One approach to solve this problem is to use TIL (tumor-infiltrating lymphocyte), immune cells already penetrated the environment and surrounding the tumor [114]. In TIL therapy, TILs are first collected from the tumor during a biopsy or surgical resection. Then TILs are grown to very large numbers (billions) in a laboratory with IL-2, a protein that promotes rapid TIL growth. Finally, TILs are infused back into the patient to actively attack cancer cells but not normal cells. TIL therapy has been used to treat several solid tumors including melanoma, lung cancer, head and neck squamous cell carcinoma, and other malignancies [115–118].

25.3.2 Targeted Therapy for Cancers

Targeted cancer therapy, also known as molecular targeted drug, molecularly targeted therapy, precision medicine, or personalized medicine, refers to the use of drugs or other molecules to treat cancer by blocking the growth and spread of cancer through interfering with specific targets involved in controlling or regulating cancer growth, progression, and spread. The commonly used targets include growth factors or growth factor receptors, signaling molecules, cell-cycle proteins, modulators of programmed cell death (apoptosis), molecules promoting angiogenesis, and many others. Targeted cancer therapy can directly target cancer cells or TME [119].

Targeted cancer therapy can be considered as one type of chemotherapy. However, it is very different from standard chemotherapy in several aspects. First, it is more specific by acting on cancer-specific molecular targets, whereas most standard chemotherapies act on both cancer cells and rapidly dividing normal cells. Second, targeted therapies are usually cytostatic, meaning they just block cancer cell proliferation, but standard chemotherapies are typically cytotoxic, meaning they kill tumor cells.

Based on the molecular nature, targeted cancer therapy can be categorized as conventional small molecule drugs and novel biologics with large complex structure. The mechanisms of action of targeted cancer therapy include inhibiting angiogenesis, such as mAb targeting vascular endothelial growth factor (VEGF) and VGEFR2, blocking signals that promote cell growth and spread, interfering with cell-cycle regulation, inducing cell death, and carrying toxins specifically to the targeted cancer cells. Biologics can also be used to trigger the immune system to kill cancer cells, which is discussed in the previous section.

25.3.2.1 Angiogenesis inhibitors

Biologics inhibit cancer angiogenesis by blocking the formation of new blood vessels that feed and nourish the cancer cells. Angiogenesis plays a critical role in cancer growth because solid tumors need a blood supply to provide oxygen and nutrients to grow beyond a few millimeters in size [120]. Cancer cells can release chemical signals to simulate angiogenesis or stimulate nearby normal cells to produce angiogenesis signaling molecules [121]. Chemical signals that control the process of angiogenesis include VEGF, which can bind to its receptors on the surface of normal endothelial cells and initiate signals within these cells to promote the growth and survival of new blood vessels [122]. Angiogenesis inhibitors can interfere with the process of angiogenesis by blocking VEGF or its receptors as well as other receptors or downstream signaling pathway proteins. FDA approved several mAbs targeting VEGF or its receptors for cancer therapy (Table 25.6).

Bevacizumab (Avastin) is a recombinant humanized mAb targeting VEGF-A developed by Genentech based on the discovery of human VEGF by Napoleone Ferrara [123, 124]. It is the first clinically used angiogenesis inhibitor approved by FDA in 2004 to treat metastatic CRC when used with standard chemotherapy as first-line treatment [125]. It then received FDA approval for treatment of several different cancers including advanced non-squamous NSCLC used as first-line treatment in combination with carboplatin/paclitaxel chemotherapy in 2006, metastatic renal cell cancer in 2009, recurrent glioblastoma multiforme in 2009, and stage III or IV ovarian cancer in combination with chemotherapy. The treatment for breast cancer by bevacizumab is complicated. FDA approved bevacizumab to treat breast cancer in 2008 but revoked the approval in 2011 due to lack of significant benefit (not been shown to be safe and effective), although there was evidence demonstrating that it slowed progression of metastatic breast

TABLE 25.6 Targeted therapy in cancer treatment

AGENT CATEGORY	*APPROVE AGENTS*	*DRUGMAKER*	*INITIAL FDA APPROVAL TIME*	*MOLECULAR TARGETS*	*MECHANISM OF ACTION*	*FDA APPROVED INDICATION*
Monoclonal antibody	Bevacizumab (Avastin)	Genentech/Roche	2004	VEGF	VEGFR PI3k, PLC-γ Prevents neovascularization	Colorectal cancer, lung cancer, breast cancer, renal cancer, brain cancer, ovarian cancer
	Ranibizumab (Lucentis)	Genentech/Roche	2006			wet macular degeneration and metastatic colorectal cancer
	Cetuximab (Erbitux)	Eli Lilly	2009	EGFR	OI3K, RAS, STAT Signaling inhibition	colon cancer, head and neck cancer
	Panitumumab (Vectibix)	Amgen	2006			metastatic colorectal cancer
	Trastuzumab (Herceptin)	Genentech/Roche	1998	HER2, HER3	RAS/Raf/MAPK inhibition	Breast cancer, gastric cancer
	Pertuzumab (Perjeta)	Genentech/Roche	2012	HER2, HER3	RAS/Raf/MAPK inhibition	Metastatic breast cancer
	Ramucirumab (Cryramza)	ImClone/Eli Lilly	2014	VEGFR2		NSCLC, colorectal cancer, gastric cancer
	Obinutuzumab	Roche	2013	CD20 (encoded by MS4A1 gene)		Chronic lymphocytic leukemia
	Ofatumumab (Arzerra)	Genmab	2009	CD20		Refractory CLL
	Rituximab (Rituxan)		1997	CD20		NHL, CLL
Biosimilars to Trastuzumab	Trastuzumab-dkst (Ogivri)	Mylan GmbH	2017	HER2, HER3	RAS/Raf/MAPK inhibition	Breast cancer, gastric cancer
	Trastuzumab-pkrb (Herzuma)	Celltrion Inc.	2018			
	Trastuzumab-qyyp (Trazimera)	Pfizer	2019			
	Trastuzumab-dttb (Ontruzant)	Merck	2019			
	Trastuzumab-anns (Kanjinti)	Amgen	2019			

(*Continued*)

TABLE 25.6 Targeted therapy in cancer treatment (*Continued*)

AGENT CATEGORY	*APPROVE AGENTS*	*DRUGMAKER*	*INITIAL FDA APPROVAL TIME*	*MOLECULAR TARGETS*	*MECHANISM OF ACTION*	*FDA APPROVED INDICATION*
Antibody-drug conjugate	Trastuzumab emtansine (Kadcyla), also known as ado-trastuzumab emtansine and abbreviated T-DM1	Genentech/Roche	2019	HER2	Cytotoxic agent vinca alkaloid	HER2-positive metastatic breast cancer
	Ibritumomab tiuxetan (Zevalin)	Biogen Idec	2002	CD20	Radioactive isotope (either yttrium-90 or indium-111)	B-cell non-Hodgkin's lymphoma (NHL), follicular NHL
	Tositumomab (Bexxar)	GlaxoSmithKline	2003	CD20	Radioactive iodine 131 tositumomab	NHL
	Brentuximab vedotin (Adcetris)	Millennium Pharmaceuticals/ Seattle Genetics	2011	CD30	Antimitotic agent MMAE	Hodgkin's lymphoma and ALCL

cancer [126]. However, bevacizumab remains approved for the treatment of breast cancer in other countries including Australia. Biosimilars to bevacizumab have been developed and approved in the United States and European. Amgen's biosimilar bevacizumab-awwb (Mvasi) received FDA approval as the first biosimilar [127] to treat multiple types of cancer including metastatic CRC, non-squamous NSCLC, glioblastoma multiforme, metastatic renal cell carcinoma, and recurrent or metastatic cervical cancer.

Ranibizumab (Lucentis) is another mAb developed by Genentech to target VEGF from the same parent mouse antibody as bevacizumab. Unlike bevacizumab, ranibizumab is approved by FDA to treat neovascular (wet) age-related macular degeneration, macular edema following retinal vein occlusion, diabetic macular edema, diabetic retinopathy, and myopic choroidal neovascularization. Ranibizumab and bevacizumab are structurally similar but bevacizumab has a lower cost. However, only ranibizumab has received FDA approval for the treatment of macular degeneration, although many studies have indicated that ranibizumab and bevacizumab are of roughly equal short-term efficacy and safety [128].

Biologics targeting VEGF receptors have also been developed. Ramucirumab (Cryramza) is a fully human mAb (IgG1) targeting VEGFR2 developed by ImClone System Inc., which was later acquired by Eli Lilly. Ramucirumab can bind to the extracellular domain of VEGFR2 with high affinity to block the binding of its natural ligands including VEGF-A, VEGF-C, and VEGF-D, to inhibit VEGF-mediated tumor angiogenesis. It received FDA approval in 2014 to treat advanced gastric cancer of gastro-esophageal junction (GEJ) adenocarcinoma after prior treatment with fluoropyrimidine- or platinum-containing chemotherapy [129]. A few months later, FDA approved ramucirumab in combination with docetaxel for treatment of metastatic NSCLC with disease progression during or after first-line platinum-containing chemotherapy. It is also approved by FDA to treat metastatic CRC (in 2015) and HCC (in 2019). On May 29, 2020, FDA approved ramucirumab in combination with erlotinib for first-line treatment of metastatic NSCLC with EGFR exon 19 deletions or exon 21 (L858R) mutations [130].

25.3.2.2 HER2 inhibitors

The mAbs targeting growth factor receptors on the cell surface inhibiting cancer cell proliferation have been developed to treat cancer. HER2 is the most prominent example of such a target for cancer treatment. HER2, also known as CD340, proto-oncogene Neu, Erbb2 (rodent), or ERBB2 (human), is a member of the human epidermal growth factor receptor (HER/EGFR/ERBB) family containing an extracellular domain for ligand binding, a transmembrane domain, and an intracellular signaling transduction domain [131]. However, no ligands for HER2 have been identified yet. HER2 can heterodimerize with any of the other three receptors to activate signaling pathways, including mitogen-activated protein kinase (MAPK), phosphoinositide 3-kinase (PI3K/Akt), phospholipase C γ, protein kinase C (PKC), and signal transducer and activator of transcription (STAT) to promote cell growth and proliferation [132]. The HER2 gene is amplified in about 10%–30% of invasive breast cancers [133, 134] and also in other types of cancer such as colon, lung, and gastric carcinomas. Trastuzumab (Herceptin) is a mAb that specifically targets HER2 protein by directly binding the extracellular domain of the receptor. Trastuzumab enhances survival rates in both primary and metastatic HER2-positive breast cancer patients [135].

Trastuzumab (Herceptin) is one of the earliest biologics for cancer treatment. It was first discovered by Genentech scientists, including Axel Ullrich, H. Michael Shepard, and colleagues. Other earlier discoveries including neu oncogene discovered by Robert Weinberg's laboratory [136] and the mAb recognizing the oncogenic receptor by Mark Greene's laboratory also contributed to the establishment of HER2 targeted therapies. Dr. Dennis Slamon subsequently worked on trastuzumab's development. Genentech developed trastuzumab jointly with UCLA, beginning the first clinical trial with 15 women in 1992. By 1996, clinical trials had expanded to over 900 women, but due to pressure from advocates based on early success, Genentech worked with the FDA to begin a lottery system allowing 100 women each quarter access to the medication outside the trials. Herceptin was fast-tracked by the FDA and gained approval in September 1998. It is indicated for the treatment of breast cancer and gastric cancer that is HER2-positive either alone or in combination with chemotherapy medication. However, trastuzumab does not have a beneficial effect if the breast cancer does not overexpress HER2 and it may even cause harm. Therefore, laboratory test for HER2 status is required. The test will be discussed in more detail in the section of accompanied test for cancer treatment.

There are several biosimilars of trastuzumab that received FDA approval during the past few years to treat breast cancer and gastric cancer. Trastuzumab-dkst (Ogivri, Mylan GmbH) was approved in 2017 for the treatment of breast cancer or gastric or gastroesophageal junction adenocarcinoma, that positive for HER2 overexpression. It was also approved in the European Union in 2018. Trastuzumab-pkrb (Herzuma) from Celltrion Inc. was approved by FDA in 2018 after demonstrating that it is biosimilar to trastuzumab (Herceptin) by comparisons of extensive structural and functional product characterization, animal data, human PK, clinical immunogenicity, and other clinical data. Pfizer's biosimilar to trastuzumab (Herceptin), Trastuzumab-qyyp (Trazimera) received FDA approval in 2019. Trastuzumab-dttb (Ontruzant) is Merck's biosimilar to trastuzumab that received FDA approval in 2019. Amgen's biosimilar to trastuzumab, Trastuzumab-anns (Kanjinti) also received FDA approval in 2019 after it was approved in the European Union in 2018.

The mAbs targeting epidermal growth factor receptor (EGFR) have been developed to treat cancer. Cetuximab (Erbitux) is a chimeric (mouse/human) mAb targeting EGFR. It received FDA approval in July 2009 to treat colon cancer with wild-type KRAS. KRAS mutation status needs to be tested since cetuximab had little or no effect in CRC harboring KRAS mutations. This is the first genetic test guided treatment of cancer. The companion diagnostic will be discussed in detail in the section of Biomarker Testing. Originally developed by Abgenix Inc, Amgen's panitumumab (Vectibix) is a fully human mAb-specific to EGFR. Panitumumab (Vectibix) received FDA approval in September 2006 to treat EGFR-expressing metastatic CRC with disease progression despite prior chemotherapies. Panitumumab was the first mAb to demonstrate the use of KRAS as a predictive biomarker. In 2014, Amgen and Illumina entered into an agreement to develop a companion diagnostic to accompany panitumumab.

25.3.2.3 Antibody-drug conjugates

Antibody-drug conjugates (ADCs) are a class of biopharmaceutical drugs designed as a targeted therapy for treating cancer. ADCs are complex molecules composed of an antibody linked to a biologically active cytotoxic (anticancer) agent to specifically deliver to the targeted cancer cells [137]. Thus, ADCs can discriminate between cancer cells and normal tissue by combining the targeting capabilities of mAbs with the cancer-killing ability of cytotoxic drugs. The first ADC approved by FDA is Gemtuzumab ozogamicin (Mylotag), a mAb to CD33 that is expressed in most leukemic blast cells and normal hematopoietic cells, linked to a cytotoxic agent from the class of calicheamicins [138]. This drug was initially created in a collaboration between Celltech and Wyeth in 1991 and approved by FDA in 2000 for the treatment of newly diagnosed CD33-positive acute myeloid leukemia (AML). There are a couple of ADCs developed based on the mAb targeting HER2. Trastuzumab emtansine (Kadcyla) developed by Genentech is an antibody-drug conjugate composed of trastuzumab (Herceptin) covalently linked to the cytotoxic agent DM1, a tubulin inhibitor [139]. In 2013, Trastuzumab emtansine (Kadcyla) received FDA approval for treatment of HER2-positive metastatic breast cancer patients previously treated with trastuzumab and a taxane (paclitaxel or docetaxel) and relapsed within six months of adjuvant therapy. Another trastuzumab-based ADC is trastuzumab deruxtecan (Enhertu) that linked trastuzumab to deruxtecan (a derivative of exatecan), a topoisomerase I inhibitor. This drug was developed by AstraZeneca/Daiichi Sankyo and received FDA approval in 2019 for treatment of unresectable or metastatic HER2-positive breast cancer patients who have received two or more prior anti-HER2-based therapies in the metastatic setting [140]. It is also indicated for the treatment of locally advanced or metastatic HER2-positive gastric or gastroesophageal junction adenocarcinoma patients who have received a prior trastuzumab-based therapy. Other antitumor agents such as antimitotic agent monomethyl auristatin E (MMAE) is also used for ADC. Brentuximab vedotin (adcetris) developed by Millennium Pharmaceuticals and Seattle Genetics is an example of such an ADC that is composed of brentuximab (a chimeric mAb targeting CD30) linked with maleimide attachment groups, cathepsin-cleavable linkers (valine-citrulline), and para-aminobenzylcarbamate spacers to three to five units of the antimitotic agent MMAE. The ADC is designed to be stable so the drug is not easily released from the antibody under physiologic conditions to prevent toxicity to normal cells and ensure dosage efficiency. However, the drug can be released inside the tumor cells after the antibody part binds to CD30 and is internalized by endocytosis (selectively taken up by the target cancer cells not the normal cells). Brentuximab vedotin (adcetris) received FDA approval in 2011 to treat relapsed Hodgkin's lymphoma (HL) and anaplastic large cell lymphoma (ALCL) in 2017 [141].

ADC can also be designed to deliver radioactive isotopes specifically to the targeted cancer cells. Ibritumomab tiuxetan (Zevalin) is a mAb (mouse IgGl) targeting CD20 covalently linked to chelator tiuxetan attached with a radioactive isotope such as yttrium-90 or indium-111 [142]. This ADC can specifically deliver the radioactive isotope to malignant B cells through the binding of ibritumomab to the CD20 antigen. Ibritumomab tiuxetan was developed by IDEC Pharmaceuticals (now part of Biogen Idec) and it was the first radioimmunotherapy drug approved by the FDA in 2002 to treat relapsed or refractory low-grade or follicular B-cell non-Hodgkin's lymphoma (NHL), including rituximab refractory follicular NHL. In September 2009, ibritumomab received approval from the FDA for an expanded label to include previously untreated patients with a chemotherapy response.

25.4 TREATMENTS OF SIDE EFFECTS OF CANCER THERAPIES

In March 2015, the FDA approved the first biosimilar, called filgrastim-sndz (Zarxio) [143], which is the first biosimilar approved by FDA in the United States. It is a biosimilar that helps the body fight infections. Filgrastim-sndz stimulates the body to make white blood cells. People with cancer who receive chemotherapy, bone marrow transplants, and other treatments can often have low levels of white blood cells. Filgrastim-sndz's reference drug is called filgrastim (Neupogen). Filgrastim-aafi (Nivestym) is another FDA-approved biosimilar to filgrastim. From 2018 to 2019, the FDA approved pegfilgrastim-jmdb (Fulphila), pegfilgrastim-cbqv (Udenyca), and pegfilgrastim-bmez (Ziextenzo), which are biosimilars that help fight infections, specifically in people with non-myeloid cancer treated with chemotherapy. Their reference drug is pegfilgrastim (Neulasta).

25.5 BIOLOGICS IN RARE GENETIC DISEASES

A rare disease refers to a condition affecting less than 200,000 people in the United States (www.fda.gov) and is defined by an incidence rate of less than 1 in 2,000 people in the European Union [144]. There are over 7,000 rare diseases identified and about 80% of which are genetic diseases [145]. Collectively, rare diseases affect about 25–30 million people in the United States, some of which can be life-threatening. Rare diseases are often not considered as a group by the medical profession because they can affect people at different times of their lives, affect a single system or multiple organs. Diagnosis of rare diseases can be achieved by biochemical and/or molecular testing. Genome or whole exome sequencing can be used to identify the etiologies of rare diseases. However, there is a significant gap between basic research and clinical treatments. State-of-the-art drug discovery strategies for the treatment of rare diseases using small molecules and biologics have been developed in the past decade.

Biologics are very important as effective therapeutics for rare genetic diseases. In the future, biologics will become more important as more and more genetic therapies become available. Biologics for rare disease treatment can be categorized as recombinant proteins (including enzymes, peptides, and antibodies), stem cell transplant, and gene therapy. Natural protein isolated from an animal or recombinant protein produced in cells can be used to treat a rare disease caused by the deficiency of that protein.

Enzyme replacement therapy is a good example of recombinant protein biologics [146]. There are two major challenges of recombinant protein therapy. First, the production of recombinant proteins can be very difficult due to unique post-translational modification such as glycosylation typically employed in human cells but not easy to recapitulate in non-human cell culture [147]. Second, biologics can be less effective or ineffective for the treatment of rare diseases affecting the brain and central nervous system (CNS) because they cannot usually cross the blood-brain barrier (BBB). In a case like this, biologics may need to be administrated by intracerebroventricular or intrathecal injection despite its risks. Aseptic conditions must be maintained during the manufacturing of biologics because biologics may be susceptible to microbial contamination.

25.5.1 Enzyme Replacement Therapy (ERT)

The foundation for ERT can be traced back to a few decades ago by the work of Christian de Duve on the discoveries of lysosomes and peroxisomes [148] and Roscoe Brady and his colleagues on the identification of the enzymatic defects in Gaucher's disease [149, 150] and other rare diseases. But the clinical application was not in practice until the first FDA approval of alglucerase (Ceredase) in 1991 as an orphan drug to treat Gaucher disease, which was the first drug approved as an enzyme replacement therapy [151]. ERTs were initially isolated from the human placenta but FDA approved ERTs derived from other human cells, animal cells such as Chinese hamster ovary cells (CHO), and plant cells. Currently, ERT has been approved for eight lysosomal storage diseases including Gaucher disease, Fabry disease, mucopolysaccharidosis types I, II, IVA, and VI, Pompe disease, and lysosomal acid lipase deficiency (Table 25.7). ERT has also been approved for the treatment of patients with severe combined immunodeficiency (SCID) resulting from an adenosine deaminase deficiency (ADA-SCID), and infantile or juvenile-onset hypophosphatasia.

Gaucher disease is a genetic disorder caused by the accumulation of glucocerebroside (a sphingolipid, also known as glucosylceramide) in cells(leukocyte particularly macrophages) and certain organs (spleen, liver, kidney, lungs, brain, and bone marrow), due to the deficiency of the enzyme glucocerebrosidase (also known as glucosylceramidase) caused by autosomal recessive mutation with high frequency among Ashkenazi Jewish population [152]. There are three types of Gaucher disease. Type 1 (N370S homozygote) Gaucher disease is the most common form and the symptoms may include abdominal pain due to liver and spleen enlargement, skeletal abnormalities, and blood disorders [153]. Type 2 (one or two alleles L444P) Gaucher disease is rare but it begins in infancy and typically results in death by two years of age. Type 3 (also one or two copies of L444P, possibly delayed by protective polymorphisms) occurs in Swedish patients from the Norrbotten region with somewhat later onset but most patients die before 30 years old [154]. ERT has been developed for the treatment of Gaucher disease, Type 1 and Type 3. The first ERT is alglucerase (Ceredase) isolated from human placental tissue. But it has been withdrawn from the market because the availability of similar drugs produced by using recombinant DNA technology instead of being isolated from human tissue. The FDA has approved ERT treatments for Gaucher disease including the following ERT drugs: imiglucerase (Cerezyme) manufactured by Genzyme Corporation, velaglucerase alfa (Vpriv) manufactured by Shire plc, taliglucerase alfa (Elelyso) manufactured by Protalix (An Israeli biotherapeutics company) but marketed by Pfizer.

Pompe disease (glycogen storage disease type II) is an autosomal recessive metabolic disorder caused by the accumulation of glycogen in the lysosome due to deficiency of the lysosomal acid alpha-glucosidase enzyme [155]. Pompe disease is the first glycogen storage disease. It was named after the Dutch pathologist J.C. Pompe who identified this disease in 1932. The symptoms include progressive myopathy throughout the body and the build-up of glycogen also affects various organs include heart, liver, and the nervous system. The first treatment for Pompe disease was Myozyme (alglucosidase alfa, rhGAA), a recombinant protein produced in Chinese Hamster Ovary Cells, which received FDA approval in 2006 [156]. Later, the manufacturer Genzyme Corporation made a similar version of Myozyme, Lumizyme to treat late-onset Pompe disease.

TABLE 25.7 Enzyme replacement therapy for rare genetic diseases

LYSOSOMAL STORAGE DISORDER	*DEFICIENT ENZYME*	*ENZYME REPLACEMENT THERAPIES (BRAND NAMES)*	*MANUFACTURER*	*YEAR OF INITIAL MARKETING APPROVAL*
Type 1 Gaucher disease	β-glucocerebrosidase (β-GCase)	Alglucerase (Ceredase)	Genzyme-Sanofi	1991
		Imiglucerase (Cerezyme)	Genzyme-Sanofi	1995
		Velaglucerase alfa (VPRIV)	Shire	2010
		Taliglucerase alfa (Elelyso)	Protalix-Pfizer	2013
Fabry disease	α-galactosidase A	Agalsidase beta (Fabrazyme)	Genzyme-Sanofi	2001
MPS I (Hurler-Scheie and Scheie syndrome)	α-iduronidase	Laronidase (Aldurazyme)	Biomarin (manufacturer) Genzyme-Sanofi (marketing)	2003
MPS VI (Maroteaux-Lamy syndrome)	N-acetylgalactosamine-4-sulfatase (arylsulfatase B)	Galsulfase (Naglazyme)	Biomarin	2005
Pompe disease (glycogen storage disease type II)	α-glucosidase	Algulucosidase alfa (Myozyme)	Genzyme-Sanofi	2006
MPS II (Hunter syndrome)	Iduronidase-2-sulfatase	Idursulfase (Elaprase)	Shire	2007
MPS IV (Morquio syndrome A)	N-acetylgalactosamine-6-sulfate sulfatase	Elosulfase alfa (Vimizim)	Biomarin	2014
Wolman disease	Lysosomal acid lipase deficiency	Sebelipase alfa (Kanuma)	Synageva/Alexion	2015
MPS VII (Sly syndrome)	β-glucoronidase	Vestronidase alfa (Mepsevii)	Ultragenyx	2017
Late infantile neuronal ceroid lipofuscinosis type 2 (CLN disease)	Tripeptidyl-peptidase 1	Cerliponase alfa (Brineura)	Biomarin	2017

Other forms of recombinant protein as biologic for disease treatment include recombinant blood clotting factors for hemophilia and immunoglobulins for patients with certain genetic immune disorders. Hemophilia is a mostly inherited genetic disorder with impaired blood clotting to stop bleeding. There are two main types of hemophilia: hemophilia A caused by deficiency of clotting factor VIII, and hemophilia B caused by low levels of clotting factor IX; both are X-linked recessive disorders [157]. Patients with hemophilia can suffer bleeding for a longer time after an injury, easy bruising, and increased risk of bleeding inside joints or brain that result in long-term headaches, seizures, or decreased level of consciousness. Factor VIII (used in hemophilia A) and factor IX (used in hemophilia B) are both used for replacement therapy. These factors can be either isolated from human blood serum or recombinant product typically manufactured from CHO tissue culture cells or a combination of the two [158]. The main limitation with these recombinant proteins is their short half-lives (8–12 hours for factor VIII), making repeated administrations necessary. A serious drawback of this and other replacement therapies is the development of antibodies directed against infused proteins, which reduces the efficacy of future treatment. One strategy for preventing antibody formation is to design genetically engineered proteins to better match the native proteins and to perform the intravenous infusion very slowly to minimize the immune reactions.

25.5.2 Stem Cell-Based Therapy

Stem cell therapy, also known as stem cell transplantation or regenerative medicine, is a procedure of using stem cells or their derivatives to repair or replace diseased, dysfunctional, or injured tissues. Stem cells are special human cells with the ability to differentiate into many different cell types. There are two main forms of stem cells: embryonic stem cells (ESCs) and adult stem cells [159]. ESCs derived from unused embryos are pluripotent, meaning they can differentiate into more than one type of cell [159]. Adult stem cells include those from fully developed tissues such as the brain, skin, and bone marrow, with limited potent to differentiate into only certain types of cells, and induced pluripotent stem cells (iPSCs) that are reprogramed to an embryonic stem cell-like state by introducing genes critical for maintaining the essential properties of ESCs [159]. The discovery of iPSCs by Shinya Yamanaka and colleagues in 2006 opened a new avenue for the resource of stem cell-based therapy [160]. However, as the FDA stated, "Currently, the only stem cell products that are FDA approved for use in the United States consist of blood-forming stem cells (also known as hematopoietic progenitor cells) that are derived from umbilical cord blood. These products are approved for use in patients with disorders that affect the production of blood (i.e., the "hematopoietic" system) but they are not approved for other uses" [161]. For patient's safety, stem cell treatment should be either FDA approved or being studied under an Investigational New Drug Application (IND), which is a clinical investigation plan submitted and allowed to proceed by the FDA.

For clinical applications, stem cell products require FDA approval. Currently, the only stem cell products with FDA approval for clinical use in the United States are hematopoietic stem cells derived from umbilical cord blood to treat diseases affecting the hematopoietic system (Table 25.8). The first such approval of a stem cell product in the world is the New York Blood Center's HEMACORD™, which received FDA approval in November 2001 for treatments of hematopoietic system diseases "either inherited, acquired, or result from myeloablative treatment", such as hematological cancer, primary immunodeficiency diseases, bone marrow failure, and beta-thalassemia. Currently, there are a total of 8 FDA approved umbilical cord blood derivatives including HEMACORD™ (New York Blood Center), ALLOCORD™ (SSM Cardinal Glennon Children's Medical Center), CLEVECORD™ (Cleveland Cord Blood Center), and Ducord™ (Duke University School of Medicine), as well as four hematopoietic progenitor cell (HPC) products developed by Clinimmune Labs, MD Anderson Cord Blood Bank, LifeSouth Community Blood Centers, and Bloodworks, respectively. In addition to these FDA-approved therapies,

TABLE 25.8 FDA approved umbilical cord blood products

BIOLOGIC	*CELL TYPE*	*COMPANY*	*INDICATION*	*FDA APPROVAL YEAR*
HEMACORD (HPC, cord blood)	Allogeneic hematopoietic progenitor cell from cord blood	New York Blood Center	Disorders affecting the hematopoietic system that are inherited, acquired, or result from myeloablative treatment	2011
HPC, cord blood		Clinimmune Labs, University of Colorado Cord Blood Bank		2012
DUCORD (HPC, cord blood)		Duke University School of Medicine		2012
HPC, cord blood		LifeSouth Community Blood Centers, Inc.		2013
ALLOCORD (HOC, cord blood)		SSM Cardinal Glennon Children's Medical Center		2013
CLEVECORD (HPC, cord blood)		Cleveland Cord Blood Center		2016
HPC, cord blood		Bloodworks		2016
HPC, cord blood		MD Anderson Cord Blood Bank		2017

a diverse range of stem cell therapies are being offered directly to patients. Today, there are more than 400 medical clinics that provide stem cell-based therapy directly to patients (https://bioinformant.com/product/stem-cell-treatment-clinics/).

There are two main types of bone marrow/stem cell transplants including autologous transplant and allogenic transplant. Autologous transplant uses stem cells from patient's own body, whereas allogenic transplant uses stem cells from another person known as a donor. Although cell-based therapy offers the opportunity for long-term correction, there are some problems associated with this procedure. Allogenic bone marrow transplantation can cause severe issues including failure of reconstitution, graft versus host disease (GVHD), severe infections due to immune suppression, and even death [162]. Therefore, strategies to reduce or eliminate these side effects are urgently needed.

Recent advances in iPSC technology have enabled the conversion of patient cells such as skin fibroblasts and peripheral blood monocytes to iPSCs [159]. Once expanded, the iPSC can then be further differentiated into mature cells, thereby reducing the adverse effects in stem cell-based therapy by using an autologous graft. However, the development of each cell product is different for each disease and the procedures are not standardized yet. Additionally, the process of iPSC generation, scaling-up, and differentiation can take up to several months, and it is presently associated with low yield, high variability, and very high costs. There are several advantages of using iPSCs in cell therapy. Firstly, iPSCs can be produced in virtually any amount and subsequently differentiated to any cell type *in vitro* compared to the limited availability of the other cell types used in cell therapy. Secondly, iPSCs provide autologous patient-derived cells, which negates the need to find an HLA-compatible cell donor and the need for immunosuppression. However, many obstacles have to be overcome before iPSC-based cell therapy can be used in humans, including the challenges of differentiation to many mature tissue types, short *in vivo* surviving time after injection of the cells, low integration into host tissue *in vivo*, and high costs. In addition, the genetic integrity and stability of iPSC have to be better controlled, a problem that contributed to the suspension of the first iPSC clinical trial based on cells differentiated from iPSCs. Currently, there are no FDA-approved iPSCs-based therapy in the United States.

25.5.3 Gene Therapy

Gene therapy is an experimental technique that modifies or manipulates genes to treat, cure or prevent disease. There are several different mechanisms of gene therapy including replacing a mutated gene that causes disease with a normal (wild-type) copy of the gene, inactivating or "knocking out" a mutated gene that is functioning improperly, and introducing a new gene into the body to help fight a disease. Thus, gene therapy is ideal for the treatment of genetic diseases, most readily those caused by loss-of-function mutations in a single gene. Although gene therapy is a promising treatment option for a number of diseases including cancer, genetic diseases, and infectious diseases, the technique remains risky and is still under study to make sure that it will be safe and effective. Currently, the lack of safe and effective methods to permanently deliver a gene to patients prevents the widespread application of gene therapy for the treatment of genetic diseases. Gene therapy is currently being tested only for diseases that have no other cures.

There are a variety of types of gene therapy products including plasmid DNA, viral vector, bacterial vector, patient-derived cellular gene therapy products, and the most recent human gene-editing technology [163]. Circular plasmid DNA molecules can be genetically engineered to carry therapeutic genes into human cells. However, special carriers are needed because naked plasmid DNA does not offer good therapeutic efficacy due to premature degradation, poor cellular uptake, and low protein expression. There are generally two types of carriers (vectors) developed for gene therapy: non-viral vectors and viral vectors. Non-viral gene delivery carriers include lipoplex-mediated (liposome or phospholipids vesicle), polyplex-mediated (polymer), dendrimer-mediated (repetitively branched molecules), and graphene-mediated (a thin layer of pure carbon) gene delivery systems. The advantages of non-viral carriers include low immunogenicity, low cost, ability to deliver large-sized DNA, lack of incorporation into host chromosomes, and lower mutation risk compared with viral carriers [164]. Hydrophilic polymers like polyethylene glycol (PEG) have been conjugated to non-viral carriers to combat RES uptake and enhance the circulation time in blood. However, the gene delivery efficiency of the non-viral delivery method needs to be further improved.

The second method employs viral gene delivery carriers [165, 166]. Viruses offer a promising approach to deliver genes. The natural mechanisms of infection and transduction in viruses are very efficient, so two to three orders of magnitude less DNA is needed compared to non-viral carriers of similar efficacy. Adenoviruses or adeno-associated virus (AAV) carriers are currently better choices for gene delivery carriers because the genes delivered are not integrated into the host chromosomes. Despite improvements, adenovirus carriers suffer from short duration of expression and immunogenicity. They evoke a mild immune response (to the viral carries), transduce a broad range of cell types, are non-pathogenic in humans, and provide significantly longer transgene expression. AAVs have different serotypes based on amino acid sequence of the capsid proteins, which confer different tropisms for different organs, a property that can help to reduce off-target effects. AAVs can typically only carry about 4.7 kb of DNA, making it very difficult to deliver a gene that encodes larger proteins. To overcome this limitation, it has been shown that a large gene can be fragmented into smaller pieces, each transported by their own AAV vector and co-administered, although such modifications increase the complexity in the resulting system. The disadvantages of viral carriers include the high cost and an immune response to the viral capsid proteins. The reported adverse reactions in clinical trials using viral carriers include a massive and uncontrolled immune response and the induction of new T-cell lymphomas. As some patients may have pre-existing neutralizing antibodies against the specific viral capsid protein due to prior exposure to the virus in the community, individuals who are considering viral vector-mediated gene therapy will need to be screened to determine the baseline neutralizing antibody status. Additionally, once a patient has received a gene therapy product through a given viral vector, he/she will no longer be able to receive further doses of the product or other therapies using the same vector due to the inevitable induction of neutralizing antibodies, which is a long-term limitation of gene therapy.

Human gene editing technology: The goals of gene editing are to disrupt harmful genes or to repair mutated genes. The use of zinc finger nucleases (ZFNs), transcription activator-like effector nucleases (TALEN), and most recently, clustered regularly interspaced short palindromic repeats (CRISPR)/Cas9, is a rapidly evolving field.

Gene transfer efficacy is typically limited by insufficient delivery to the target tissue, negative immune response (autoantibody) to the treatment, and loss of therapeutic effect over time. The AAV is the most commonly used vector for the delivery of genes in gene therapy [167]. While AAVs are currently used as carriers in gene therapy, a more efficient and safer gene delivery vector remains to be discovered and developed. To deliver nucleic acids into nuclei of cells, a number of barriers must be overcome. Extracellular barriers include inactivation by enzymatic degradation and recognition by the RES. After the genes penetrate the cell membrane, they encounter many intracellular barriers. Most exogenous genetic material is internalized through endocytosis pathways such as clathrin-mediated, caveolae-mediated, macropinocytosis, and phagocytosis. The associated vesicles are then trafficked from early endosomes to late endosomes to lysosomes, where the nucleic acid cargo is degraded. The efficiency of the gene therapy platform is significantly reduced by vesicle trafficking. One of the strategies used to evade this endosomal entrapment is the design of carriers that release the nucleic acid cargo into the cytoplasm. Finally, the genes delivered to the cytoplasm need to traverse the nuclear membrane to enter the nucleus and integrate with a chromosome. The nuclear transport is aided by nuclear pore complexes and is often a barrier for larger nucleic acids such as plasmid DNA. A nuclear localization signal peptide is used for the active transport of DNA to the nucleus and as a method of restricting transgene expression to the desired tissues.

Another issue in gene therapy is the delivery of genes across the blood-brain barrier (BBB) to the CNS [168]. In this context, intracranial injections of viral gene carriers have been used to treat neuronal diseases such as aromatic L-amino acid decarboxylase (AADC) deficiency. Although it has been reported that some AAV serotypes (e.g., AAV9) can cross the BBB to deliver genes to the CNS, new methods are still needed to increase the delivery efficiency of genes to the brain and to simplify the overall gene delivery procedures.

There is a continual effort to improve the overall gene therapy platform. In particular, the use of gene-editing technologies, including ZFNs, TALEN, and most recently, clustered regularly interspaced short palindromic repeats (CRISPR)/Cas9, is a rapidly evolving field. Tebas *et al.* used ZFNs to edit the CCR5 gene (encoding an HIV co-receptor) in autologous T-cells, which were then infused into patients with HIV [169]. This procedure was shown to be safe within the limited number of subjects in the study. More recently, a few rare disorders have been corrected by the editing of the mutated genes in the patient's hematopoietic stem cells and returning these edited cells to the patient. This approach was successful in treating X-linked severe combined immunodeficiency, but the trial was halted due to a serious adverse event in which malignant transformation of lymphocytes was found in several of the treated patients.

There are over 2900 clinical trials conducted for gene therapy since 1989, of which more than half are in Phase I (www.clinicaltrials.gov). However, only a limited number of gene therapy products received FDA approval (Table 25.7). The first *in vivo* gene therapy approved by FDA is Voretigene neparvovec (Luxturna) developed by Spark Therapeutics and Children's Hospital of Philadelphia for the treatment of Leber congenital amaurosis (LCA) [170]. LCA is a rare inherited eye disease characterized by nystagmus, sluggish or absent pupillary responses, and severe vision loss of blindness, onset from birth or within the first few months of life, with a prevalence of about 1:80,000 [171]. LCA is a heterogeneous disease with online Mendelian Inheritance in Man (OMIM) recognized 18 types due to different genes affected (www.omim.org/entry/204000). One form of LCA, LCA2 is an autosomal recessive disease due to homozygous mutations in the *RPE65* gene (www.omim.org/entry/204100). Voretigene neparvovec (Luxturna) is a gene therapy using an AAV2 vector containing human RPE65 cDNA to treat LCA and retinitis pigmentosa. It was granted orphan drug and a biological license application (BLA) was submitted to the FDA in July 2017. The product was approved in December 2017. It is the first FDA-approved gene therapy product in the United States. The product is expensive at $425,000 per eye.

Patisiran (Onpattro) is the first small interfering (si) RNA-based drug approved by the FDA [172] and the first drug to treat polyneuropathy in people with hereditary transthyretin-mediated amyloidosis (also known as familial amyloid polyneuropathy or FAP), a fatal rare disease affecting about 50,000 people worldwide [173]. FAP is characterized by pain, paresthesia, muscular weakness, and autonomic dysfunction due to systemic deposition of amyloidogenic variants of the transthyretin protein. FAP is caused by mutations in the TTR gene, with TTR V30M as the most common mutation [173]. Patisiran is an siRNA-based treatment developed by Alnylam Pharmaceuticals Inc., a private company headquartered in Cambridge, MA, USA. It is a gene-silencing drug that interferes with the production of the abnormal form of TTR gene. It received FDA approval in August 2018, at an expected cost of up to $450,000 per year.

Onasemnogene abeparvovec (Zolgensma) is a gene therapy developed by a biotechnology startup company AveXis acquired by Novartis in 2018. It received FDA approval in May 2019 to treat spinal muscular atrophy (SMA) [174]. SMA is a rare neuromuscular disorder with loss of motor neurons and progressive muscle wasting, due to mutation in SMN1 gene encoding the SMN protein [175]. SMA is an autosomal recessive disease with an incidence of about 1 in 10,000 in the United States. Onasemnogene abeparvovec (Zolgensma) consists of AAV9 virus capsids containing a SMN1 transgene and synthetic promoters. It was approved to treat children with SMA less than two years old, including at the presymptomatic stage of disease, by a one-time injection into a vein with at least two months of corticosteroids. The AAV9 viral vector delivers the SMN1 transgene to the affected motor neurons to increase the level SMN protein. This drug is very expensive, with a list price of $2.125 million per treatment. It was in fact the most expensive drug in the world as of 2019.

25.6 VACCINES

Vaccine is a biological product administered by different methods including needle injection or oral or nasal spray to stimulate the body's immune system to obtain acquired immunity to a specific infectious disease. There are several different types of biological products commonly used as vaccines including an antigenic preparation of live attenuated or deactivated pathogenic agent such as a virus or bacterium, or one of its components or products such as a protein or toxin, or a preparation of genetic material such as DNA or RNA that can be used by the cells to generate an antigenic substance such as a virus spike protein. Depending on the purpose, vaccines can be prophylactic such as preventing future infection or therapeutic such as cancer treatment vaccine. This section will discuss cancer vaccines and SARS-CoV-2 vaccines.

25.6.1 Cancer Vaccines

Vaccine for cancer treatment or prevention is a very active field. Vaccines can be used to prevent healthy people from getting certain types of cancer caused by viruses. Similar to other infectious diseases, this type of vaccine will only work when administered before being infected with virus. Currently, there are two types of vaccines approved by the FDA to prevent cancer, HPV vaccine and hepatitis B vaccine.

25.6.1.1 HPV vaccine

Human papillomavirus (HPV) is a DNA virus that belongs to the *Papillomaviridae* family. There are over 170 types of HPV that have been described [176] and more than 40 types may be sexually transmitted. It is estimated that almost every individual is infected by HPV at some point in their lives, many of which cause no symptoms and 90% resolved spontaneously within two years. However, some HPV infections can persist and result in either warts or precancerous lesions that increase the risk of cancer of the cervix, vulva, vagina, or throat and other sites. About a dozen of HPV types including types 16, 18, 31, 45, 33, 35, 39, 51, 52, 56, 58, and 59 are high-risk types because persistent infection of these types has been associated with genital cancers [177]. Nearly all cervical cancer is caused by HPV and the two high risk types HPV 16 and HPV 18 account for >70% of all cervical cancer diagnosed worldwide [178]. HPV16 is responsible for almost 90% of HPV-positive oropharyngeal cancers. HPV6 and HPV11 are common causes of genital warts and laryngeal papillomatosis.

HPV genome contains early genes (E1–E7) and late genes (L1–L2) that encode capsid proteins. HPV is believed to cause cancer by integrating its genome into nuclear DNA. Some of the early genes expressed by HPV, such as E6 and E7 can act as oncogenes to promote tumor growth and malignant transformation. E6 protein can bind to and inactivate the tumor suppressor gene TP53 protein and E7 protein inactivates another tumor suppressor gene pRB [179].

There are currently three vaccines available to prevent infection by some HPV types including Gardasil, Gardasil 9, and Cervarix. Gardasil is a quadrivalent HPV vaccine manufactured by Merck Sharp & Dohme. The biological vaccine product of Gardasil is the major capsid protein L1 epitope of HPV types 6, 11, 16, and 18. It initially received FDA approval in 2006 [180] for use in females aged 9–26 years and expanded for routine vaccination of girls aged 11 and 12 years in 2007. Gardasil then was recommended for both males and females before adolescence and the beginning of potential sexual activity since 2009. Currently, the vaccine has been approved in 120 other countries. Gardasil 9 is a nine-valent version of the original version of Gardasil with the addition of five other HPV strains, including types 31, 33, 45, 52, and 58, that collectively responsible for 20% of cervical cancers. Gardasil 9 received FDA approval in 2014 [181] and in 2018 FDA approved Gardasil 9 for women and men aged 27–45 years based on the vaccine being 88% effective against persistent HPV infections that cause certain types of genital warts and cancers in females.

Cervarix is a bivalent HPV vaccine designed to prevent infection from HPV types 16 and 18, manufactured by GlaxoSmithKline. Cervarix was approved by FDA in 2009. Currently, the only HPV vaccine available in the United States is Gardasil 9 because it was shown to be more cost-effective owing to its additional benefit in protecting against genital warts.

25.6.1.2 Hepatitis B vaccine

Hepatitis B is a serious disease caused by infection of hepatitis B virus (HBV) [182]. Infection of HBV can cause lifelong infection, cirrhosis (scarring) of the liver, liver cancer, liver failure, and death.

Hepatitis B can be acute and later become chronic, leading to other diseases and health conditions. In addition to causing hepatitis, infection with HBV can lead to cirrhosis, hepatocellular carcinoma, and maybe an increased risk of pancreatic cancer. Despite there being a vaccine to prevent hepatitis B, HBV remains a global health problem.

HBV is a partially double-stranded DNA virus belongs to the *Hepadnaviridae* family of viruses. The virus particle, called Dane particle (virion), consists of an outer lipid envelope and an icosahedral nucleocapsid core composed of protein. The nucleocapsid encloses the viral DNA and a DNA polymerase that has reverse transcriptase activity similar to retroviruses. The outer envelope contains embedded proteins that are involved in viral binding of, and entry into, susceptible cells. The protein components of the virus include hepatitis B surface antigen (HBsAg), hepatitis B core antigen (HBcAg), HBV DNA polymerase, hepatitis B envelope antigen (HBeAg), and HBx. HBsAg was the first hepatitis B virus protein to be discovered. HBeAg and HBcAg are made from the same reading frame [183].

The first generation of hepatitis B vaccine was blood-derive surface proteins approved in 1981 [184] based on the pioneering discovery of the "Australia Antigen" (now called HBsAg) in the serum of an Australian Aboriginal person by the American physician/geneticist Baruch Blumberg in 1968. RECOMBIVAX HB, a recombinant hepatitis B vaccine based on HBsAg gene, developed by Maurice Hilleman and colleagues at Merck in collaboration with other scientists, received FDA approval in 1986 [185]. It was in fact the first human vaccine produced by recombinant DNA technology. Since then, the blood-derived hepatitis B vaccine was withdrawn from the marketplace and replaced by the recombinant hepatitis B vaccine. A second recombinant hepatitis B vaccine, Engerix-B manufactured by GlaxoSmithKline Biologicals, received the FDA approval in 1989 [186]. Both RECOMBIVAX HB and Engerix-B are indicated for immunization against infection caused by all known subtypes of HBV for all age groups. HEPLISAV-B is indicated for prevention of infection caused by all known subtypes of HBVs [187]. HEPLISAV-B is approved for use in adults 18 years of age and older. Engerix-B vaccine, which consists of a three-dose regimen over a six-month time period. A new hepatitis B vaccine, Heplisav-B, has been approved for adults in the United States and requires only two doses over one month. The unique dosing schedule of Heplisav-B provides the potential for increasing patient compliance and therefore can aid in the effort toward protecting individuals from developing an HBV infection.

25.6.1.3 Cancer treatment vaccine

Cancer treatment vaccines or therapeutic vaccines are different from the previously discussed cancer prevent vaccines working against viruses. Instead of preventing cancer, these vaccines work to boost the immune system to attack against cancer cells already present in the body. Cancer treatment vaccines can be made up of cancer cells, parts of cells, or pure antigens, often combined with other substances or cells called adjuvants that help boost the immune response even further. Alternatively, a patient's own immune cells can be used to create vaccine by a process including isolation, exposure to cancer cells or antigens in the laboratory, multiplication, and injection back to the body to fight cancer. Cancer treatment vaccines can work in different ways including keeping cancer from relapse (growing back), killing cancer cells, or stopping tumor growing or spreading.

Although holding a promising option to treat cancer, this field is still at an early stage and most cancer vaccines are only offered through clinical trials. Currently, there are only three FDA approved cancer treatment vaccines including sipuleucel-T (Provenge) for treatment of metastatic prostate cancer, Bacillus Calmette-Guérin (BCG) for treatment of carcinoma *in situ* (CIS) of the urinary bladder, and talimogene laherparepvec (T-VEC) for treatment of advanced melanoma skin cancer.

In 2010, the FDA approved Sipuleucel-T (Provenge) (Sip-T), an autologous cellular immunological agent developed by Dendreon pharmaceuticals, to treat men with asymptomatic or minimally symptomatic castrate-resistant metastatic prostate cancer [188]. This is the first-in-class therapeutic autologous vaccine approved by FDA. Sipuleucel-T works through antigen-presenting cells (APCs) to stimulate T-cell immune response specifically targeting prostatic acid phosphatase (PAP), which is highly expressed in most prostate cancer cells [189]. Sipuleucel-T is tailored to each patient through a series of steps: first, collecting patient's blood containing the APCs by leukapheresis at physician's office, or blood collection center, or the laboratory; second, sending blood to Dendreon's manufacturing facility in New Jersey and incubating the harvested APCs with recombinant fusion protein antigen containing both PAP and GM-CSF; transporting the activated antigen-loaded APCs back to the infusion center and infusing the patient.

The complete process takes up to four days. Because Dendreon is the only manufacturer for this vaccine, transportation, sterility, and stability of the cells must be ensured before the process begins. Only a few doctors and 50 centers are using this therapy. Sipuleucel-T is intended for patients with metastatic, symptomatic, or minimally symptomatic, castrate-resistant (hormone-refractory) prostate cancer. This drug is used to treat advanced prostate cancer that is no longer being helped by hormone therapy. Side effects are usually mild and can include fever, chills, fatigue, back and joint pain, nausea, and headache. A few men may have more severe symptoms, including problems breathing and high blood pressure.

Another cancer treatment vaccine uses a weakened bacteria called Bacillus Calmette-Guérin (BCG) to activate the immune system to treat early-stage bladder cancer [190]. *Mycobacterium bovis* BCG is a species that originated after 230 recultures of the pathogen *M. bovis*. Over a period of 13 years, Albert Calmette and Camille Guérin recultured isolated colonies from the originally pathogenic *M. bovis*. In 1921, they demonstrated that the obtained bacillus was not only non-pathogenic in animal models but also protected against tuberculosis challenge in vaccinated animals. Afterward, the massive production of BCG was initiated for use in tuberculosis prevention in humans, and it is still the only commercially available vaccine against tuberculosis. At that time, the use of a mixture of two bacteria, *Serratia marcescens* and *Streptococcus pyogenes*, was investigated for cancer treatment, and the possibility to use the newly developed and safe BCG offered a novel therapeutic option for some cancer patients. Although some studies demonstrated the potential efficacy of the new BCG as a treatment for diverse types of cancer, it was not until the 1970s that BCG was approved as an immunotherapeutic treatment for bladder cancer (BC) patients.

Since then, BCG has been the standard therapy for treating high-risk non-muscle-invasive bladder cancer (NMIBC) patients to avoid the recurrence and progression of the disease. Intravesical instillations of hundreds of millions of bacilli are applied weekly in those patients over the course of six weeks ("induction treatment") after the transurethral resection of tumors (TURBT) visible at the lumen surface of the bladder. If the patient responds appropriately to the therapy, a "maintenance treatment" consisting of six-week periods of instillation every three months for one to three years is then undertaken to reach the optimum effect for avoiding recurrence and progression episodes.

Talimogene laherparepvec (T-VEC) is a vaccine approved to treat advanced melanoma skin cancer [191]. It is made from a herpes virus that has been altered in the lab to produce a substance that the body normally produces, called a cytokine. This cytokine boosts the immune system and can cause flu-like symptoms for a short time.

Talimogene laherparepvec (T-VEC; Imlygic™) is a genetically modified herpes simplex virus, type 1, and is the first oncolytic virus therapy to be approved for the treatment of advanced melanoma by the US FDA. T-VEC is attenuated by the deletion of the herpes neurovirulence viral genes and enhanced for immunogenicity by the deletion of the viral ICP47 gene. Immunogenicity is further supported by the expression of the human granulocyte-macrophage colony-stimulating factor (GM-CSF) gene, which helps promote the priming of T cell responses. T-VEC demonstrated significant improvement in durable response rate, objective response rate, and progression-free survival in a randomized Phase III clinical trial for patients with advanced melanoma.

25.6.2 COVID-19 Vaccine

Emerged in December of 2019, the severe acute respiratory syndrome coronavirus 2 (SARS-CoV-2) has caused a once in a century pandemic of acute respiratory disease, named 'coronavirus disease 2019' (COVID-19) by the World Health Organization (WHO). To date, there are over 240 million people infected and over 4.9 million death worldwide as of October 2021 (https://covid19.who.int/). The speed of the development of COVID-19 vaccine is unprecedented with the great support from the governments and collaboration from the industry. This section will discuss the main types of COVID-19 vaccines and testing of SARS-CoV-2 since it is important to detect and monitor the mutant strains that may render the efficacy of the vaccine protection.

There are four main types of COVID-19 vaccines classified by the methods of exposure used, including whole virus vaccine, non-replicating viral vector, protein subunit, and mRNA vaccine.

25.6.2.1 Whole virus vaccine

Vaccine using the whole virus either attenuated live virus or inactivated virus is a traditional method with well-established technology. The inactivated virus vaccine is more popular because it is suitable for people with compromised immune systems and it is relatively simple to manufacture. Historically, many licensed vaccines using inactivated type, such as hepatitis A, polio, and rabies. There are two COVID-19 vaccines using inactivated virus, Sinopharm and Sinovac, both developed in China (Table 25.9). Sinopharm BBIBP-CorV uses traditional technology similar to the inactivated polio vaccine [192]. The initial step is to grow large quantities of a sample of SARS-CoV-2 from China with the capability of rapid multiplication using vero cells, a cell line isolated from the kidney epithelial cells extracted from an African green monkey. The viruses are then inactivated by soaking in beta-propiolactone that deactivates the virus by binding to their genes but leaving other viral particles intact. Finally, the inactivated viruses are then mixed with an aluminum-based adjuvant for injection. Different levels of efficacy against COVID-19 infection by Sinopharm BBIBP-CorV have been reported. The Phase III clinical trial in the United Arab Emirates (UAE) reported an 86% efficacy against COVID-19 infection [193] and Sinopharm claimed

TABLE 25.9 COVID-19 vaccines

		VACCINE EFFICACY				
TYPE	*COMPANY*	*SYMPTOMATIC COVID-19*	*SEVERE COVID-19*	*DOSES*	*STORAGE*	*OTHER VACCINES USING THIS TECHNOLOGY*
mRNA in lipid nanoparticles	Pfizer/BioNTech	95%	100%	2 doses 21 days apart	−70°C ± 10°C	None
	Moderna	94%	100%	2 doses 28 days apart	−20°C	
Non-replicating Viral vector	Oxford-AstraZeneca	70%	100%	2 doses 28 days apart	2°C–8°C	Ebola
	Johnson & Johnson	66%	85%	One dose	2°C–8°C	
	Sputnik (Gamaleya Research Institute)	92%	100%	2 doses 21 days apart	2°C–8°C	
	CanSinoBIO	66%	Unknown	2 doses	2°C–8°C	
Whole virus	SinoVac (inactivated)	50%–83%	100%	2 doses	2°C–8°C	Whooping cough (inactivated)
	Sinopharm (inactivated)	79%–86%	100%	2 doses	2°C–8°C	Rabbies (inactivated) Hepatitis A (inactivated) HPV/cervical particle (virus-like particle)
Protein Subunit	Novavax[a]	89%	100%	2 doses	2°C–8°C	Hepatitis B

[a] Not approved for emergency use yet.

a 79% efficacy in this category by its internal analysis [194]. Nevertheless, the vaccine showed 100% effectiveness in preventing moderate to severe cases of the disease. Sinopharm's BBIBP-CorV is currently being used in vaccination campaigns by several countries in Asia, Africa, South America, and Europe. The WHO added the Sinopharm COVID-19 vaccine to the list of vaccines authorized for emergency use for the COVID-19 Vaccines Global Access (COVAX) on May 7, 2021 (who.int/news).

Sinovac COVID-19 vaccine is CoronaVac developed by the Chinse company Sinovac Biotech. It is manufactured using the same technology as Sinopharm's COVID-19 vaccine. There are some contradictory reports regarding the efficacy of Sinovac's COVID-19 vaccine. Based on the Phase III clinical trial from Brazil, Sinovac claimed an efficacy of 50.65% (95% CI, 35.66%–62.15%) against all symptomatic cases, 83.70% (57.99–93.67%) against cases that require medical treatment, and 100.00% (56.37–100.00%) against severe, hospitalized and fatal cases. Final Phase III trial from Turkey showed an efficacy of 83.5%, while they previously announced an efficacy at 91.25% from an interim result with 29 infections [195].

25.6.2.2 Non-replicating viral vector

A viral vector vaccine is a type of vaccine using a viral vector, such as adenovirus, to deliver genetic material coding for a desired antigen into the recipient's host cells. The COVID-19 vaccine in this category uses the gene coding for the spike proteins found on the surface of the coronavirus as the antigen. There are four COVID-19 based on a viral vector including the Oxford-AstraZeneca vaccine, the Sputnik V from Russia, the Johnson & Johnson vaccine, and the Convidecia (CanSinoBIO) from China (Table 25.9). Adenovirus vectors have the advantages of high transduction efficiency, transgene expression, and broad viral tropism, and can infect both dividing and non-dividing cells [196]. However, a disadvantage is that many people have pre-existing immunity to adenovirus due to previous exposure [197]. The Oxford–AstraZeneca vaccine uses the modified chimpanzee adenovirus ChAdOx1, The Johnson & Johnson vaccine uses human adenovirus serotype 26, Convidecia uses serotype 5, Sputnik V uses human adenovirus serotype 26 for the first shot and serotype 5 for the second.

The efficacy of the Oxford-AstraZeneca COVID-19 vaccine (codenamed AZD1222 and sold under the brand names Covishield and Vaxzevria) developed by Oxford University and AstraZeneca. The efficacy of this vaccine is 76.0% at preventing symptomatic COVID-19 infection beginning at 22 days after the first dose and 81.3% after the second dose [198]. This vaccine was first approved for use in the United Kingdom on December 30, 2020 and the first vaccination was administered on January 4, 2021. The vaccine has since been approved by several medicine agencies including the European Medicines Agency (EMA) and the Australian Therapeutic Goods Administration, as well as approved for the Emergency Use Listing by WHO. Generally, the side effects including injection-site pain, headache, and nausea can resolve within a few days. Serious allergic reaction such as anaphylaxis may occur but it is rare. More importantly, this vaccine has been linked with an increased risk of blood clots in combination with low levels of blood platelets in very rare cases (around 1 in 100,000 people). However, specific risk factors for this side effect have not been confirmed based on the currently available evidence although most cases occurred in women under 60 years of age within 2 weeks of vaccination.

The Johnson & Johnson COVID-19 vaccine is developed by Janssen Vaccines (Netherlands) and Jassen Pharmaceuticals (Belgian company), a subsidiary of American company Johnson & Johnson, received FDA approval on February 27, 2021. Unlike the Oxford-AstraZeneca COVID-19 vaccine that uses the chimpanzee adenovirus ChAdOx1, the Johnson & Johnson COVID-19 vaccine uses a human adenovirus serotype 26 as the viral vector to carry the gene for the spike protein of the SARS-CoV-2. The advantage of this vaccine is that it only requires one dose and does not need to be stored frozen. However, the efficacy is relatively lower compared to the Oxford-AstraZeneca vaccine, with 66% efficacy in preventing symptomatic infection and 85% efficacy in preventing severe COVID-19, but it is also 100% effective in preventing hospitalization or death caused by the disease [199]. Common side effects include coughing, joint pain, fever, chills, redness, and swelling at the injection site usually occur in less than 1 in 10 people and most of these can get better within one or two days. Similar to the Oxford-AstraZeneca vaccine, it was also associated with a few cases of blood clots and the CDC and FDA issued a joint statement recommending suspension of the use of this vaccine on April 13, 2021. However, on April 23, FDA and CDC lifted the recommended pause after examination of the available data and concluded that this vaccine is safe and effective in preventing COVID-19 and the blood clot occurring is very low. FDA and CDC will continue to investigate the risk of blood clots.

Russia's COVID-19 vaccine Sputnik V (Gam-COVID-Vac) developed by Gamaleya Research Institute of Epidemiology and Microbiology is different from Oxford-AstraZeneca and Johnson & Johnson's viral vector vaccine because it used a viral two-vector based on two human adenoviruses: serotype 26 for the first shot and serotype 5 for the second with 21 days in between. Both recombinant adenoviruses were biotechnology-derived and contain the SRAS-CoV-2 S protein cDNA. Sputnik V was approved in early August 2020 in Russia based on the preliminary results of Phase I–II studies, which caused media criticism in mass media and discussions in the scientific community as to whether approval was justified in the absence of robust scientific research confirming safety and efficacy. The interim analysis from the trial was finally published on February 2021 in The Lancet, claiming a 91.6% efficacy without unusual side effects [200]. Emergency mass distribution of the vaccine began in December 2020 in countries including Russia, Argentina, Belarus, Hungary, Serbia, and the United Arab Emirates. By February 2021, over a billion doses of the vaccine had been ordered for immediate distribution globally.

Convidecia (AD5-nCOV) is another single-dose viral vector vaccine for COVID-19 developed by CanSino Biologics in China using the recombinant human adenovirus serotype 5 as the viral vector. In February 2021, global data from Phase III trials with 30,000 participants 101 COVID cases showed that the vaccine had an efficacy of 65.7% at preventing moderate cases of COVID-19 and 90.98% efficacy in preventing severe cases [201]. But the data from Pakistan trial subset showed a little better efficacy, with 74.8% in preventing symptomatic cases and 100% in preventing severe disease. While the efficacy rates were lower than the Pfizer-BioNTech and Moderna vaccines, its single-dose regimen and normal refrigerator storage requirement (2 to 8°C) could make it a favorable option for many countries. It has similar efficacy to Johnson & Johnson's Ad26.COV2.S, another one-shot adenovirus vaccine found to be 66% effective in a global trial. Convidecia is authorized for use in China, Mexico, Pakistan, Hungary, and Chile.

25.6.2.3 Protein subunit

The Novavax COVID-19 vaccine (codenamed NVX-CoV2373), also known as SARS-CoV-2 rS, is recombinant spike protein nanoparticle with Matrix-M1 adjuvant, developed by Novavax and the Coalition for Epidemic Preparedness Innovations (CEPI). The company also announced that it might receive $1.6 billion from Operation Warp Speed to expedite the development of its COVID-19 vaccine candidate by 2021 if clinical trials show effectiveness. This vaccine is currently undergoing clinical trials in India under the brand name Covovax. It requires two doses and is stable at 2 to 8°C (36 to 46°F). The company announced in January 2021 that the Phase III clinical trial in the United Kingdom showed an efficacy of 89.3% (95% CI: 75.2–95.4) during a period of high transmission and with the United Kingdom variant strain emerging and circulating widely. The company announced on June 14, 2021 that the vaccine candidate demonstrated "100% protection against moderate and severe disease, 90.4% efficacy overall, and met the primary endpoint in its PREVENT-19 pivotal Phase 3 trial" [202].

Vaccines using a protein subunit is a well-established technology. There are other licensed vaccines using this technology, including Hepatitis B, meningococcal disease, pneumococcal disease, and shingles. The advantage of this technology is to minimize the risk of side effects by using a purified piece of a pathogen rather than the whole pathogen to trigger an immune response. But it is relatively complex to manufacture this type of vaccine. Adjuvants and booster shots may be required, particularly for the fast-evolving SARS-CoV-2.

25.6.2.4 mRNA vaccine

mRNA vaccines are a new type of vaccine that some people call the third-generation vaccine, in comparison to the traditional whole virus vaccine and the recombinant viral vector vaccine. Unlike the traditional methods including whole virus, recombinant antigen encoding viral vector, or protein antigen, that prepare and grow outside of the body, mRNA vaccines introduce a short-lived synthetically created RNA sequence of a virus gene into the vaccinated individual to use the host cell, primarily dendritic cells, to produce the viral antigen using their own internal machinery (ribosomes). There are mRNA vaccines developed and tested for human use for diseases such as Zika, rabies, cytomegalovirus, and influenza, but none of which has been approved yet. However, mRNA vaccine technology has been developed for the past three decades. Early studies showed that *in vitro* transcribed (IVT) mRNA could successfully deliver genetic material into living cell tissue and produce proteins using their own machinery. The use of mRNA vaccines was proposed and developed in 1990s and studies demonstrated that mRNA vaccine could stimulate both humoral and cellular immune response against a pathogen.

Before the success of COVID-19 vaccine, there were no vaccines developed and licensed in less than a few years. However, the development and approval of COVID-19 vaccines are expedited unprecedently with the support from governments and the unprecedented flow of funding from the industry. BioNTech received a $135 million investment from Fosun Pharma of China in March 2020, for development, marketing, and distribution rights of the vaccine in China, followed by $185 from a partnership with Pfizer, including an equity investment of about $113 million, for support with clinical trials, logistics, and manufacturing of the vaccine. BioNTech also received funding from other sources such as the European Commission and European Investment Bank, as well as the German government. Pfizer did not receive funding directly from the US government's Operation Warp Speed but did enter into an agreement with the United States for the distribution of the vaccine. Moderna received US$955 million from the Biomedical Advanced Research and Development Authority (BARDA), an office of the US Department of Health and Human Services. BARDA funded 100% of the cost of bringing the Moderna vaccine to FDA licensure. In total, the US government provided US$2.5 billion funding for the Moderna COVID-19 vaccine.

BioNTech began the COVID-19 development on January 10, 2020 right after the SARS-CoV-2 genetic sequences were released by the Chinese Center for Disease Control and Prevention via GISAID (Global Initiative on Sharing All Influenza Data), a global science initiative and primary source established in 2008 that provides open-access to genomic data of influenza viruses and the coronavirus responsible for the COVID-19 pandemic. The vaccine contains a 4,284 nucleotide modRNA sequence and several inactive ingredients (excipients). The modRNA sequence consists of a five-prime cap, a five prime untranslated region derived from the human alpha globin, a signal peptide (bases 55–102) and two proline substitutions (K986P and V987P) that cause the spike to adopt a prefusion-stabilized conformation to reduce the membrane fusion ability, increase expression and stimulate neutralizing antibodies, a codon-optimized gene of the full-length spike protein of SRAS-CoV-2 (bases 103–3879), followed by a three prime untranslated region (bases 3880–4174) combined from AES and mtRNR1 selected for increased protein expression and mRNA stability and a poly(A) tail comprising 30 adenosine residues, a 10-nucleotide linker sequence, and 70 other adenosine residues (bases 4175–4284). The sequence contains no uridine residues; they are replaced by 1-methyl-3′-pseudouridylyl. The 2P proline substitutions in the spike proteins were originally developed for a MERS vaccine by researchers at the National Institute of Allergy and Infectious Diseases' Vaccine Research Center, Scripps Research, and Jason McLellan's team (at the University of Texas at Austin, previously at Dartmouth College).

There are two obstacles to mRNA vaccine [203]. First, unprotected mRNA can be degraded by the nearly-ubiquitous ribonuclease (RNase). This problem can be solved by packing the mRNA molecule into a lipid nanoparticle, which also increases the delivery efficiency of penetrance into the cells. Second, the heterologous mRNA can be recognized by the immune system as foreign mRNA through innate immune system receptors including toll-like receptor (TLR) 7 and TLR8 located in endosomal membranes, which can dramatically reduce protein production and trigger the release of cytokines such as interferon and TNFα, which may lead to programmed cell death. This major obstacle of mRNA vaccine was solved by the groundbreaking work of Hungarian biochemist Katalin Karikó and American immunologist Drew Weissman, published in 2005, that showed RNA recognition by toll-like receptors can be suppressed or avoided by incorporation of modified nucleosides such as m5C, m6A, m5U, s2U, or pseudouridine [204]. Both Moderna and BioNTech licensed Karikó and Weissman's work to develop their COVID-19 vaccines. Karikó is currently a senior vice president at BioNTech, overseeing its mRNA work.

Pfizer-BioNTech COVID-19 vaccine started clinical trials in April 2020 and entered Phase III clinical trials with over 40,000 participants in November 2020. The Phase III trial assesses the safety, efficacy, tolerability, and immunogenicity of BNT162b2 at a mid-dose level (two doses injected with 21 days apart) in three age groups: 12–15 years, 16–55 years, or above 55 years. The overall efficacy of 95% in preventing symptomatic infection and 100% in preventing severe disease. The most common side effects include mild to moderate pain at the injection site, fatigue, and headache, but no long-term complications were reported. Some serious side effects such as allergic reactions have been reported but are very rare. The Pfizer-BioNTech COVID-19 vaccine (Comirnaty) received an expedited review by regulatory agencies. On December 2, 2020, the United Kingdom's Medicines and Healthcare products Regulatory Agency (MHRA) gave the vaccine "rapid temporary regulatory approval to address significant public health issues such as a pandemic". Thus, the Pfizer-BioNTech COVID-19 vaccine (Comirnaty) became the first COVID-19 vaccine approved for national use after undergoing large-scale trials. More importantly, it is the first mRNA vaccine to be approved for use in humans. The US FDA approved the Pfizer-BioNTech COVID-19 vaccine for emergency use on December 11, 2020. Many other countries also expedited the approval process for this vaccine. Currently, it is authorized for use at some level in 84 countries and the World Health Organization (WHO) authorized it for emergency use.

Similar to the Pfizer-BioNTech COVID-19 vaccine, the Moderna COVID-19 vaccine (codenamed mRNA-1273) also composed of nucleoside-modified mRNA (modRNA) encoding the spike protein of SARS-CoV-2 and the encapsulating lipid nanoparticles. There are some differences in the non-modRNA ingredients between these two vaccines. Large scale Phase III clinical trials showed 94.1% efficacy of the Moderna COVID-19 vaccine in preventing symptomatic infection and 100% in preventing severe disease [205]. The vaccine received the FDA approval for emergency use on December 18. On May 10, 2021, the US FDA approved the Pfizer-BioNTech COVID-19 vaccine for emergency use for children 12 to 15 years old after a rigorous and thorough review of all available data.

The advantages of mRNA vaccine include easier for vaccine creators to produce than attenuated virus or proteins, speed of design and production, and stimulation of both cellular immunity and humoral immunity [203]. For example, Moderna designed its mRNA-1273 vaccine for COVID-19 just in two days. The major disadvantage for the Pfizer-BioNTech COVID-19 vaccine is the ultracold storage temperatures required between −80°C and −60°C (−112°F and −76°F), which is a big logistics challenge, particularly for low-income countries which usually only have cold chain transport for standard refrigerator storage. In February 2021, the US FDA updated the emergency use authorization (EUA) to permit undiluted frozen vials of the vaccine to be transported and stored at between −25°C and −15°C (−13°F and 5°F) for up to two weeks before use. The Moderna vaccine only requires storage temperature of a standard medical refrigerator of 2°C–8°C (36°F–46°F).

As of May 2021, there are 3 COVID-19 vaccines approved by the US FDA for emergency use. Many other countries also approved a few vaccines for some level of emergency use. However, there are currently multiple SARS-CoV-2 variants circulating globally and most likely more mutant strains are emerging or will emerge, particularly in high-risk countries such as India. The most notable mutant strains include the United Kingdom strain B.1.1.7 (also known as 201/501Y.V1, VOC 202012/01), the South Africa strain B.1.351, and the Brazil strain P.1 [206]. The UK B.1.1.7 lineage has the following key mutations: N501Y, P618H, and H69V70 deletion, and Y144/145 deletion. The 69/70 deletion occurred spontaneously many times and likely led to a conformational change in the spike protein. P681H located near the S1/S2 furin cleavage site, a site with high variability in coronaviruses, has also emerged spontaneously multiple times. This variant is estimated to emerge in the United Kingdom during September 2020 and spread to several countries including the United States since December 2020. It is associated with increased transmissibility and may be associated with an increased risk of death compared with other variants. The South Africa strain B.1.351 also has multiple mutations in the spike protein including K417N, E484K, N501Y.

Unlike the B.1.1.7 lineage detected in the UK, this variant does not contain the deletion at 69/70. This variant was first identified in Nelson Mandela Bay, South Africa, in samples dating back to the beginning of October 2020, and cases have since been detected outside of South Africa, including the United States. There is some evidence to indicate that one of the spike protein mutations, E484K, may affect neutralization by some polyclonal and monoclonal antibodies. The Brazil strain P.1 lineage is a branch off the B.1.1.28 lineage that was first reported by the National Institute of Infectious Diseases (NIID) in Japan in four travelers from Brazil, sampled during routine screening at Haneda airport outside Tokyo. The P.1 lineage contains three mutations in the spike protein receptor-binding domain: K417T, E484K, and N501Y. There is evidence to suggest that some of the mutations in the P.1 variant may affect its transmissibility and antigenic profile, which may affect the ability of antibodies generated through a previous natural infection or through vaccination to recognize and neutralize the virus.

A recent study reported on a cluster of cases in Manaus, the largest city in the Amazon region, in which the P.1 variant was identified in 42% of the specimens sequenced from late December [5]. In this region, it is estimated that approximately 75% of the population had been infected with SARS-CoV2 as of October 2020. However, since mid-December, the region has observed a surge in cases. The emergence of this variant raises concerns of a potential increase in transmissibility or propensity for SARS-CoV-2 re-infection of individuals. This variant was identified in the United States at the end of January 2021.

Some of the potential consequences of emerging variants are the following:

Ability to spread more quickly in people. There is already evidence that one mutation, D614G, confers increased ability to spread more quickly than the wild-type SARS-CoV-2 [207]. In the laboratory, 614G variants propagate more quickly in human respiratory epithelial cells, outcompeting 614D viruses. There also is epidemiologic evidence that the 614G variant spreads more quickly than viruses without the mutation.

Ability to cause either milder or more severe disease in people. In January 2021, experts in the United Kingdom reported that B.1.1.7 variant might be associated with an increased risk of death compared to other variants [208]. More studies are needed to confirm this finding.

Ability to evade detection by specific viral diagnostic tests. Most commercial reverse-transcription polymerase chain reaction (RT-PCR)-based tests have multiple targets to detect the virus, such that even if a mutation impacts one of the targets, the other RT-PCR targets will still work.

Decreased susceptibility to therapeutic agents such as monoclonal antibodies.

Ability to evade natural or vaccine-induced immunity.

There is some evidence that the mutant strains may reduce the efficacy of COVID-19 vaccines. A study from Qatar with the vaccination of the Pfizer-BioNTech COVID-19 vaccine showed the estimated efficacy against B.1.1.7 was 89.5% (95% confidence interval [CI], 85.9 to 92.3) and efficacy against B.1.351 variant was further reduced to 75% (95% CI, 70.5 to 78.9) [209]. Therefore, it is very important to detect SARS-CoV-2, particularly to monitor the mutant strains. Based on the principal of tests, there are different types of detection methods including nucleic acid amplification tests (NAAT)s such as real-time reverse-transcription polymerase chain reaction (rRT-PCR), host antibody detection, whole virus or viral protein detection (antigen detection), and genomic sequencing.

Most of the viral RNA detection methods are based on nucleic acid amplification [210]. Since the SARS-CoV-2 is an RNA virus, the first step of NAAT is to reverse transcribe the extracted target RNA to form the cDNA. The target cDNA is then amplified through PCR. The amplified cDNA can be detected after the completion of the reaction (RT-PCR) or by combining the amplification with detection (real-time RT-PCR or rRT-PCR). The World Health Organization recommends rRT-PCR as a standard method to detect SARS-CoV-2 and developed a technical guidance including protocols from different countries to help COVID-19 diagnosis [211]. The optimal diagnostics of rRT-PCR should consist of two independent targets including the commonly used E, N, and RNA-dependent RNA polymerase (RdRP) on the SARS-CoV-2 genome although a simple algorithm might be adopted with a single discriminatory target in areas with widespread transmission [212]. rRT-PCR is relatively expensive, time-consuming, and requires expert technician and specialized equipment. There are two types of detection methods in rRT-PCR, non-specific fluorescent DNA label such as SYBR Green dye that binds to any double-strands or specific fluorescence probe such as the TaqMan probe. The latter is recommended by WHO since the non-specific one can give rise to false-positive results due to the binding of SYBR Green dye to primer-dimers or other DNA double strands in the mixture. Interpretation of weak positive NAAT result is challenging because it may produce false signals at high Ct values. The patient should be resampled and retested in case of invalid or questionable results. RNA can be re-extracted from the original samples and retested by a highly experienced technician if additional samples from the patient are not available. It is important to understand that negative results do not necessarily rule out the SARS-CoV-2 infection due to various factors that could lead to false-negative result in an infected individual such as poor quality of the specimen that contains too little patient material, specimen collected late in the course of the disease, or specimen taken from the body compartment that did not contain the virus at that given time, specimen not handled or shipped appropriately, or technical reasons inherent in the test such as PCR inhibition or virus mutation.

In addition to PCR, there are other different nucleic acid amplification methods including nucleic acid sequence-based amplification (NASBA), loop-mediated isothermal amplification (LAMP), isothermal multiple displacement amplification (IMDA), transcription-mediated amplification (TMA), signal-mediated amplification of RNA technology (SMART), etc. LAMP is the second most common method after PCR, particularly for the detection purposes [213]. In contrast to two primers (forward and reverse) for PCR, the design of primers is more challenging for LAMP because it requires four (or six) primers to divide the target regions into six regions (three in the forward section and three in the backward section). But LAMP has its advantages, such as being cheaper and faster compared to PCR because it is carried out in an isothermal condition [213].

Antibody testing is a serological assay to detect antibodies produced by the human body in response to infection with the SARS-CoV-2. It can be used for serosurveillance studies to support the investigation of an ongoing outbreak and retrospective assessment of the attack rate or the size of an outbreak. However, antibody detection tests should be used with caution and not used to determine acute infections since the understanding of the antibody response is still emerging since SARS-CoV-2 is a novel antigen. Commercial and non-commercial antibody testing are developed to measure binding antibodies including total immunoglobulins (Ig), IgG, IgM, and/or IgA using a variety of detecting techniques such as Lateral Flow immunoassay (LFI), enzyme-linked immunoassay (ELISA), and chemiluminescence immunoassay (CLIA) [214]. It is important to understand that the performance of serologic assays can vary widely in different testing groups, such as in patients with mild versus moderate-to-severe symptoms or young versus old patients. The timing of testing and the target viral proteins can also affect the performance. Cross-reaction to other pathogens such as other human coronaviruses or with pre-existing conditions such as pregnancy and autoimmune diseases may yield false-positive results.

Antigen detection tests have been developed to rapidly detect the presence of SARS-CoV-2 viral proteins or RNA. Biosensors for RNA detection provide simple, cost-effective, rapid, and highly sensitive and specific tests since they can eliminate the amplification process. The basic idea is to use binding of synthetic single-stranded DNA probes tagged with detection sensor to the target viral RNA through hybridization reaction to allow detection of the presence of viral RNA by measuring the biosensor signals. Both electrochemical biosensors such as cyclic voltammetry (CV), square-wave voltammetry (SWV), chronoamperometry (CA), electrochemical impedance spectroscopy (EIS), and differential pulse voltammetry (DPV), and optical biosensors such as SPR, fluorescence, SMR, and colorimetry have been developed for viral RNA detection [215–217]. Similarly, whole virus or viral protein can also be detected using biosensor-based methods.

Genomic sequencing for SARS-CoV-2 is not used as a routine detection method due to its high cost. But it can be used to investigate the dynamics of the outbreak, including changes in the size of an epidemic over time, its spatiotemporal spread, and testing hypotheses about transmission routes. It can also be used to decide which diagnostic assays, drugs, and vaccines may be suitable candidates for further exploration.

25.7 BIOMARKER TESTING

Biological therapies such as targeted therapy and immunotherapy have significantly transformed the treatment and management of many cancers. However, the majority of patients have primary or acquired resistance to immunotherapies [218]. Predictive biomarkers that can reliably identify patients with clinically meaningful response to PD-1/PD-L1 immune checkpoint inhibition therapy is in great desire. Currently, the predictive biomarkers for immunotherapy include PD-L1 expression, high tumor mutation burden (TMB-H), and microsatellite instability-high (MSI-H). These predictive biomarkers can be detected to select patients for the appropriate treatment options using different methods including immunohistochemistry (IHC), fluorescence *in situ* hybridization (FISH), polymerase chain reaction (PCR), and next-generation sequencing (NGS). These tests also need to be performed to identify patients suitable for the targeted therapies such as HER2+ breast cancer patients for trastuzumab (Herceptin) treatment. This section will focus on the discussion of testing for the selected biomarkers. The regulation of clinical testing will be discussed first.

25.7.1 Clinical Testing Regulations

In the United States, all tests using materials derived from the human body for the purpose of assessing health condition, diagnosing, preventing, or directing treatment, are regulated by the Clinical Laboratory Improvement Amendments of 1988 (CLIA), which were published in 1992, phased in through 1994, and amended in 1993, 1995 and 2003. Currently, CLIA covers about 260,000 laboratory entities in the United States. CLIA regulations establish federal quality standards for clinical laboratory testing. The CLIA regulations are overseen by three federal agencies: the Center for Medicare and Medicaid Services (CMS), the US Food and Drug Administration (FDA), and the Center for Disease Control and Prevention (CDC).

CLIA regulations require all clinical laboratories to be certified by the Center for Medicare and Medicaid Services (CMS) before they can accept human samples for diagnostic testing. CMS is in charge of enforcing CLIA compliance by ensuring basic CLIA requirements within a laboratory, such as having an appropriate facility site, qualified laboratory director, and adequate description of assays being performed. CMS will issue a CLIA certificate in line with the complexity level of tests once the compliance is confirmed by inspection of CMA-approved accreditation organizations. There are seven CMS-approved accreditation organizations including American Association of Blood Banks (AABB), American Association for Laboratory Accreditation (A2LA), Accreditation Association for Hospital and Health Systems/Healthcare Facilities Accreditation Program (AAHHS/HFAP), American Society for Histocompatibility and Immunogenetics (ASHI), COLA Inc., College of American Pathologists (CAP), and The Joint Commission.

The CDC manages the Clinical Laboratory Improvement Advisory Committee (CLIAC), which includes diverse membership across laboratory specialties, professional roles (laboratory management, technical, physicians, nurses), and practice settings (academic, clinical, public health) and includes a consumer representative. CLIAC provides scientific and technical advice and guidance to the Department of Health and Human Services (HHS). The advice and guidance CLIAC provides to HHS pertains to general issues related to improvement in clinical laboratory quality and laboratory medicine practice. In addition, the Committee provides advice and guidance on specific questions related to the possible revision of the CLIA standards.

The FDA categorizes clinical tests as waived, moderate, or high complexity based on seven categorization criteria including (1) knowledge; (2) training and experience; (3) reagents and materials preparation; (4) characteristics of operational steps; (5) calibration, quality control, and proficiency testing materials; (6) test system troubleshooting and equipment maintenance; and (7) interpretation and judgment. Generally, the more complex the tests will need the more stringent requirements. The high complex testing requires advanced education, training, and experience for both testing personnel and management. There are two types of clinical tests including FDA cleared or approved commercially distributed tests and laboratory developed tests (LDTs). Commercially distributed tests are regulated by FDA under the Medical Device Amendments (MDA) (Pub. L. 94–295) to the Federal Food, Drug, and Cosmetic (FD&C) Act enacted on May 28, 1976. The MDA directed FDA to issue regulations that classify all devices into three categories: Class I, II, or III depending on the degree of regulation to ensure their safety and effectiveness [219].

Class I: Devices are subject to a comprehensive set of regulatory authorities called general controls that are applicable to all classes of devices.

Class II: Devices for which general controls, by themselves, are insufficient to provide reasonable assurance of the safety and effectiveness of the device, and for which there is sufficient information to establish special controls to provide such assurance.

Class III: Devices for which general controls, by themselves, are insufficient and for which there is insufficient information to establish special controls to provide reasonable assurance of the safety and effectiveness of the device. Class III devices typically require premarket approval.

The companion diagnostic tests discussed in this chapter belong to the class III devices that require premarket approval. The manufacturer can apply for Premarket Notification or 510(K) to FDA to demonstrate the intended marketing device is as safe and effective, or substantially equivalent, to a legally marketed device (Section 513(i)(1)(A) FD&C Act). There are three types of Premarket Notification 510(K)s: Traditional, Special, and Abbreviated. The Special and Abbreviated 510(k) Programs were developed by FDA in 1998 to facilitate the review of certain types of submissions. These programs were previously described in The New 510(k) Paradigm guidance, which was split into two distinct guidance documents in 2019: The Special 510(k) Program and The Abbreviated 510(k) Program that can be used when a 510(k) submission meets certain requirements. The Traditional 510(k) Program can be used under any circumstance for a manufacturer to seek FDA clearance or approval. There are some FDA cleared or approved companion diagnostic devices including FISH probes for HER2 testing, antibodies for PD-L1 testing, NGS panels for targeted therapy testing, and others. Some of these tests will be discussed in the following sections.

The vast majority of clinical tests are laboratory developed tests (LDTs). LDTs are a type of *in vitro* diagnostic tests that are designed, developed, manufactured, and used within a laboratory in a healthcare setting. Most LDTs are either not commercially available or developed in response to unmet clinical needs. However, some LDTs use modified FDA-cleared or approved tests. All LDTs are in the category of high complex tests that are currently regulated by CMS under CLIA. LDTs can provide safe, accurate, and timely patient testing for many conditions with no commercially available tests or the existing tests not meeting the current clinical needs. The US FDA asserted that LDTs were subject to the same regulatory oversight as other *in vitro* diagnostic devices (IVDs) [220]. In the past, FDA has generally not enforced premarket review and other applicable FDA requirements for LDTs because these tests were relatively simple laboratory tests and usually used within a single laboratory. However, LDTs are a fast-evolving field due to advances in technology and business models, particularly in the areas of genetic testing that are more complex with higher risks. The FDA announced that all LDTs should be under the agency oversight in 2010 and held a workshop to gather input from stakeholders on this issue. In 2014, FDA published an initial draft approach for LDT oversight. After obtaining feedback on the LDT draft guidances issued in 2014, the FDA issued a discussion paper (https://www.fda.gov/media/102367/download) on LDTs but this is neither a formal position nor is enforceable currently. In September 2020, the US Department of Health and Human Services (HHS) stated that future FDA policy on LDT will need to go through the formal rulemaking process. Currently, LDTs are regulated by CMS under CLIA with requirements for qualification of testing personnel, quality control, and proficiency testing standards. The CLIA laboratories must document the analytic validation of the LDTs and make the information available for inspection. In addition, many clinical laboratories performing LDTs are under the oversight of state and private organization such as College of American Pathologists (CAP). New York State requires laboratories to document both analytic and clinical validity of LDTs before introducing the tests. CAP also requires the laboratories to demonstrate analytic validity and document clinical validation. Clinical validity can be established by a variety of approaches including the use of existing literature review, documenting that the test is "standard of care", or providing evaluation of patient benefit. All LDTs are classified as high complexity tests by FDA under CLIA, meaning all new tests need approval by an MD or PhD with specific board certification, who is also responsible for ensuring appropriate test validation, overseeing test performance, and providing interpretative assistance to the ordering clinician when needed. The development of LDTs is essential for providing new innovative technologies to assist patients and to respond in a timely fashion to clinical needs. For example, clinical laboratories rapidly developed LDTs to address world health crises such as COVID-19.

25.7.1.1 HER2 testing

HER2, also known as ERBB2, a member of the human epidermal growth factor receptor (HER/EGFR/ERBB) family, is an important prognostic and predictive biomarker for breast cancer. HER2 overexpression/amplification is present in about 15%–20% of all breast cancers [221] and approximately 7%–34% of gastric cancer [222, 223]. HER2 overexpression is also detected in other cancers such as ovarian, adenocarcinoma of lung, uterine serous endometrial carcinoma, and salivary duct carcinomas [224]. HER2 overexpression/amplification is strongly associated with disease recurrence and a poor prognosis before the discovery of targeted therapies. The discovery of anti-HER2 drugs including mAbs such as trastuzumab (Herceptin), tyrosine kinase inhibitors such as lapatinib (L), and antibody drug conjugate (ADC) impressively improved the clinical outcome of breast cancer at all disease stages. Therefore, it is very important to test HER2 amplification status to identify patients for the targeted therapy. HER2 amplification can be detected by two methods: immunohistochemistry (IHC) and fluorescence *in situ* hybridization (FISH). College of American Pathology (CAP) published a consensus statement regarding HER2 testing in breast cancer patients in 2000 [225] and the American Society of Clinical Oncology (ASCO) subsequently recommended HER2 testing for all invasive breast cancer patients in 2001. These two professional societies then jointly published guidelines on HER2 testing in breast cancer in 2007 [226]. The guidelines were updated in 2013 [227] and in 2018 [228] to ensure reliable identification of patients who will benefit from HER2-targeted therapies. The current guidelines on both IHC and FISH are summarized as the following:

HER2 PATHWAY TEST INTERPRETATION (2018 ASCO/CAP CRITERIA)		
TEST RESULTS		
STAINING PATTERN	*SCORE*	*HER2 OVEREXPRESSION*
No staining observed or membrane staining that is incomplete and is faint/barely perceptible in ≤10% of tumor cells	0	Negative
Incomplete membrane staining that is faint/barely perceptible in >10% of tumor cells	1+	Negative
Weak to moderate complete membrane staining in >10% of tumor cells or intense circumferential membrane staining in ≤10% of tumor cells	2+	Equivocal
Complete, intense, circumferential membrane staining in >10% of tumor cells	3+	Positive

All fluorescence *in situ* hybridization (FISH) tests:

- are analyzed with Abbott probes specific for the centromere of chromosome 17 (control probe) and HER2 (17q12). It was performed and scored by a technologist, and results were interpreted by a pathologist.

- use Abbott PathVysion HER-2 DNA Probe kit for amplification.
- are analyzed with adequate number of tumor cells.
- the nuclei counted were within the region selected by a pathologist.
- are run with positive controls that gave expected results.

HER2 FISH interpretation:

HER2 FISH is considered:

a. Positive if:
   - HER2/CEP17 ratio ≥ 2.0 with an average HER2 copy number ≥ 4.0 signals/cell
   - HER2/CEP17 ratio < 2.0 with an average HER2 copy number ≥ 6.0 signals/cell

b. Negative if:
   - HER2/CEP17 ratio < 2.0 with an average HER2 copy number < 4.0 signals
   - HER2/CEP17 ratio ≥ 2.0 with an average HER2 copy number < 4.0 signals/cell (SEE *COMMENT*)
   - *COMMENT: Evidence is limited on the efficacy of HER2-targeted therapy in the small subset of cases with HER2/CEP17 ratio ≥ 2.0 and an average HER2 copy number < 4.0/cell. In the first generation of adjuvant trastuzumab trials, patients in this subgroup were randomized to the trastuzumab arm did not appear to derive an improvement in disease free or overall survival, but there were too few such cases to draw definitive conclusions. IHC expression for HER2 should be used to complement ISH and define HER status. IF IHC is not 3+ positive, it is recommended that the specimen be considered HER2 negative because of the low HER2 copy number by ISH and lack of protein overexpression.*
   - HER2/CEP17 ratio < 2.0 with an average HER2 copy number ≥ 4.0 and < 6.0 signals/cell (SEE *COMMENT*)
   - *COMMENT: It is uncertain whether patients with ≥ 4.0 and < 6.0 average HER2 signals/cell and HER2/CEP17 ratio < 2.0 benefit from HER2 targeted therapy in the absence of protein overexpression (IHC 3+). If the specimen test result is close to the ISH ratio threshold for positive, there is a high likelihood that repeat testing will result in different results by chance alone. Therefore, when IHC results are not 3+ positive, it is recommended that the sample be considered HER2 negative without additional testing on the same specimen.*

There are ten FDA-cleared or approved tests for HER2 including FISH, ISH, CISH, and IHC (Table 25.10).

Figure 25.3C showed IHC staining HER2 using the VENTANA pathway anti-HER2 (4B5) rabbit monoclonal primary antibody; this case is positive for HER2 overexpression with IHC score 3+. This is an 81 years old patient with metastatic carcinoma found in the brain. The tumor is also ER+ but PR-. In contrast, Figure 25.3A is IHC staining for a 70 years old female with newly diagnosed left breast cancer (left breast cT1cN0 invasive ductal carcinoma) biopsy. IHC Score 2+ is equivocal. This tumor is also ER+, PR+. For equivocal IHC testing result, FISH is performed. The FISH result showed positive for HER2 amplification, with HER2/Chr17 ratio of 2.7 and average HER2 copies per tumor cell 5.9 and average chr17 copies per tumor cell 2.2. This patient was put on trastuzumab-pertuzumab adjuvant setting.

In addition to HER2 amplification, HER2 mutations have been detected in NSCLC, which can direct treatment. An insertion in ERBB2, A771_M774dup, is identified in this patient. ERBB2 A771_M774dup has been reported in multiple neoplasms including lung adenocarcinoma (COSMIC, March 2019). In NSCLC, emerging biomarkers include HER2 mutations (NCCN, NSCLC v3.2018). HER2 mutations, largely exon 20 in-frame insertions, have been described as an oncogenic driver alteration in 1% to 4% of NSCLC, exclusively in adenocarcinoma histology [229]. HER2-mutant patients with lung adenocarcinoma

TABLE 25.10 FDA-cleared or approved HER2 tests

DIAGNOSTIC NAME	*DIAGNOSTIC MANUFACTURER*	*PMA/510(K) HDE*	*INDICATIONS TRADE NAME (GENERIC)-NDA/BLA*
INFORM HER-2/neu	Ventana Medical Systems, Inc.	P94004	Breast cancer Herceptin (trastuzumab)-BLA 103792
PathVysion HER-2 DNA Probe Kit	Abbott Molecular Inc.	P980024	Breast cancer Herceptin (trastuzumab)-BLA 103792
PATHWAY anti-Her2/neu (4B5) Rabbit Monoclonal Primary Antibody	Ventana Medical Systems, Inc.	P990081/S001-S028 P990081/S039	Breast cancer Herceptin (trastuzumab)-BLA 103792 Kadcyla (ado-trastuzumab emtansine)-BLA 125427
InSite-LIGHT HER2 IHC System	Biogenex Laboratories, Inc.	P040030	Breast cancer Herceptin (trastuzumab)-BLA 103792
SPOT-LIGHT HER2 CISH Kit	Life Technologies Corporation	P050040/S001-S003	Breast cancer Herceptin (trastuzumab)-BLA 103792
Bond Oracle HER2 IHC System	Leica Biosystems	P090015	Breast cancer Herceptin (trastuzumab)-BLA 103792
HER2 CISH pharmDx Kit	Dako Denmark A/S	P100024	Breast cancer Herceptin (trastuzumab)-BLA 103792
INFORM HER2 Dual ISH DNA Probe Cocktail	Ventana Medical Systems, Inc.	P100027 P100027/S030	Breast cancer Herceptin (trastuzumab)-BLA 103792
HercepTest	Dako Denmark A/S	P980018/S018	Breast cancer Herceptin (trastuzumab)-BLA 103792 Perjeta (pertuzumab)-BLA 125409 Kadcyla (ado-trastuzumab emtansine)-BLA 125427 Gastric and gastroesophageal cancer Herceptin (trastuzumab)-BLA 103792
HER2 FISH pharmDx Kit	Dako Denmark A/S	P040005 P040005/S005 P040005/S006 P040005/S009	Breast cancer Herceptin (trastuzumab)-BLA 103792 Perjeta (pertuzumab)-BLA 125409 Kadcyla (ado-trastuzumab emtansine)-BLA 125427 Gastric and gastroesophageal cancer Herceptin (trastuzumab)-BLA 103792

HDE: Humanitarian Device Exemption. NDA: New Drug Application. BLA: Biologics License Application.

stage-IIIB/IV disease showed a better overall survival (OS) compared to patients lacking driver mutations including HER2 [230].

The NCCN Panel recommends ado-trastuzumab emtansine (category 2A) for patients with HER2 mutations based on preliminary results from a recent Phase II basket trial (NCCN, NSCLC v3.2019). The NCCN Panel does not recommend single-agent therapy with trastuzumab or afatinib (both for HER2 mutations), because response rates are lower and treatment is less effective when these agents are used for patients with HER2 mutations (NCCN, NSCLC v3.2019). In addition, small molecule drug such as poziotinib, a pan-epidermal growth factor receptor (EGFR or HER) inhibitor is currently under investigation in clinical trials in NSCLC patients harboring HER2 exon 20 insertion mutations (NCT03318939, Phase II; NCT03066206, Phase II). ERBB2 inhibitors such as TAS0728, neratinib (alone or in combinations) are presently under investigation in clinical trials in solid tumor patients harboring HER2 alterations/mutations (NCT03410927, Phase I/II; NCT01953926, Phase II; NCT03065387, Phase I).

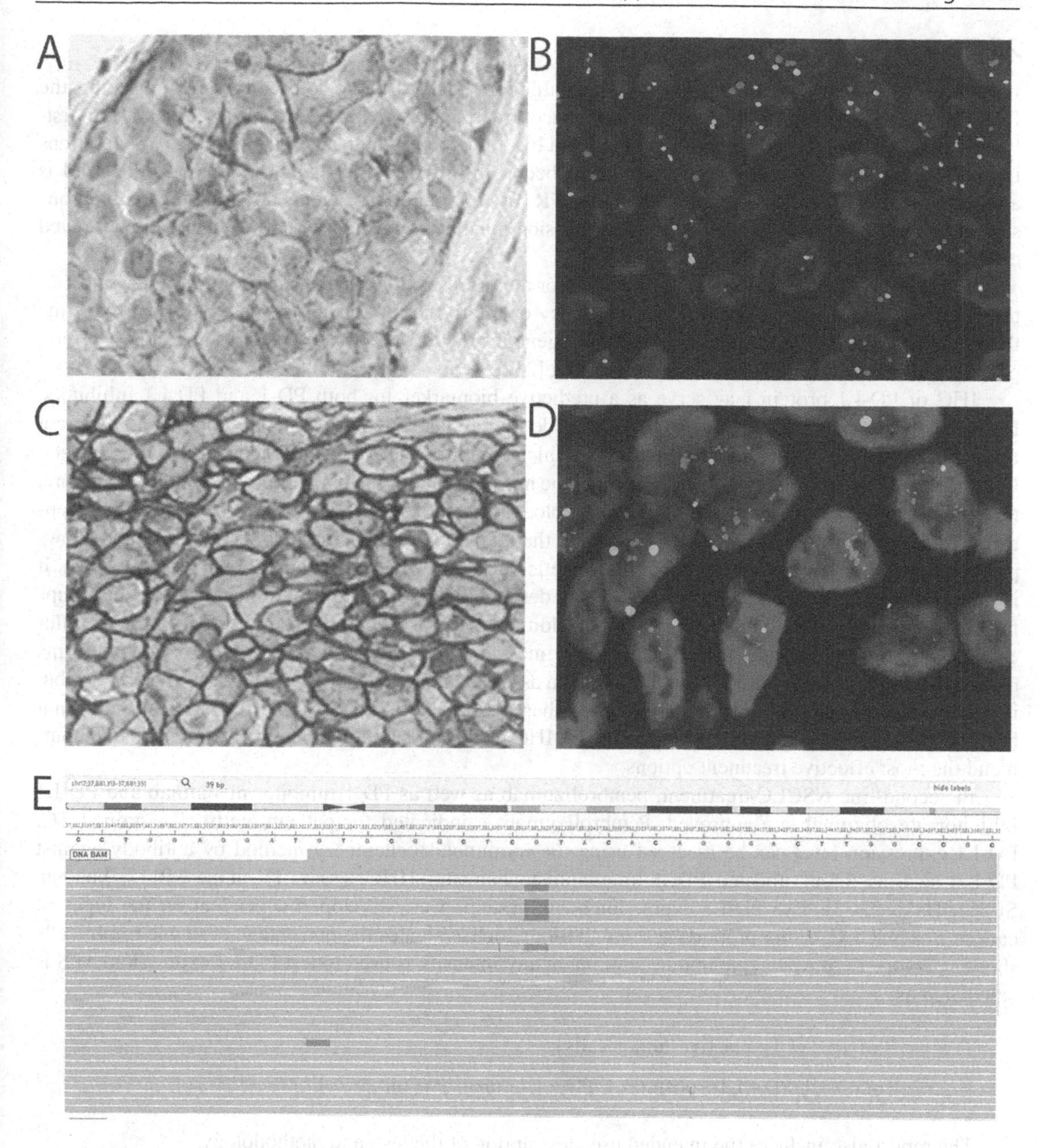

FIGURE 25.3 HER2 (ERBB2) testing. (A). Immunohistochemistry staining of HER2 in an equivocal case (2+). (B). Fluorescence *in situ* hybridization (FISH) using PathVysion HER-2 DNA Probe Kit consists of LSI HER-2 probe (gray white) and CEP 17 probe (bright white) in a HER2-negative case. (C). Immunohistochemistry staining of HER2 in a HER2 positive case (3+). (D). FISH for HER2 of the IHC equivocal case in (E) showed HER2 gene amplification.

HER2 mutations have also been found in other cancers such as in uterine endometrial carcinoma [231]. Figure 25.3E showed a HER2 mutation, a missense alteration in ERBB2, c. 2524G>A (V842I), identified in uterine endometrial carcinoma. Trastuzumab (FDA approved, HER2/neu receptor antagonist for HER2-positive stomach and breast cancers) and pertuzumab (FDA approved mAb against HER2 for HER2-positive breast cancer) are in Phase II trial for advanced solid tumors with specific ERBB2 mutations.

25.7.1.2 PD-L1 expression testing

Immunotherapy using the immune checkpoint inhibitors such as monoclonal antibodies targeting the PD1/PD-L1 pathway significantly transformed cancer treatment. For NSCLC, pembrolizumab is the first-line treatment if the cancer overexpresses PD-L1, a PD-1 receptor ligand, and the cancer has no mutations in EGFR or in ALK; if chemotherapy has already been administered, then pembrolizumab can be used as a second-line treatment, but if the cancer has EGFR or ALK mutations, agents targeting those mutations should be used first. Assessment of PD-L1 expression must be conducted with a validated and approved companion diagnostic.

In 2017, the FDA approved pembrolizumab for any unresectable or metastatic solid tumor with certain genetic anomalies (mismatch repair deficiency or microsatellite instability). This was the first time the FDA approved a cancer drug based on tumor genetics rather than tissue type or tumor site; therefore, pembrolizumab is a so-called tissue-agnostic drug [232].

IHC of PD-L1 protein may serve as a predictive biomarker for both PD-1 and PD-L1 inhibitors. However, the mechanisms of PD-1/PD-L1 blockade therapy and PD-L1 testing are completely different from those of EGFR, ALK, and ROS1 testing, which inhibit addictive driver oncogenes in lung cancer. ICIs block only the interaction, which is a part of the normal functioning of the immune system. Therefore, the clinical effect or duration of the PD-1/PD-L1 blockade response will be different from those of receptor tyrosine kinase inhibitors. PD-L1 is a protein that is expressed with biological continuity and shows profound intra-tumoral heterogeneity, unlike genetic variation, which is separated by a binary system. It is important to choose the correct cutoff levels to define biomarker-positive and -negative patient groups for PD-L1 testing to have a predictive value. In addition, IHC for detecting protein activity may be influenced by the choice of various factors including primary antibody clones, detection system, and platforms related to complex biochemistry. PD-L1 expression assessment is now established as a routine practice but is not without challenges. Understanding these inherent characteristics of PD-L1 testing is an important basis for pathologists to correctly interpret PD-L1 IHC results and communicate with clinicians to recommend the most effective treatment options.

In second-line NSCLC-treatment, pembrolizumab as well as PD-1 inhibitor nivolumab and PD-L1 inhibitor atezolizumab are approved. Pembrolizumab is indicated for patients with ≥1% stained TC. PD-L1 expression level can be assessed using the immunohistochemistry method by antibody against PD-L1, such as three validated PDL1 immunohistochemistry (IHC) assays (Ventana SP142, Ventana SP263, Dako 22C3). Two PD-L1 expression score systems were developed dependent on the types of cancer. For NSCLC, tumor proportion score (TPS), which evaluates the percentage of viable tumor cells showing complete or partial membrane staining at any intensity is used to test PD-L1 expression. TPS is calculated as:

$$\begin{aligned}\text{TPS} = &\left(\#\text{ of PD}-\text{L1}-\text{positive tumor cells}\right)/\\&\left(\text{Total \# of PD}-\text{L1}-\text{positive} + \text{PD}-\text{L1}-\text{negative tumor cells}\right) \times 100\%\end{aligned}$$

The report also includes the intended use, description of the test, and methodology.

Intended use

Stains were scored by a pathologist using manual microscopy.

All controls were reviewed and showed appropriate positive and negative reactivity.

PD-L1 22C3 FDA (Keytruda®) for NSCLC

PD-L1 IHC 22C3 pharmDx is a qualitative immunohistochemical assay using Monoclonal Mouse Anti-PD-L1, clone 22C3 intended for use in the detection of PD-L1 protein in formalin-fixed paraffin-embedded (FFPE) non-small cell lung cancer (NSCLC) tissue. PD-L1 IHC 22C3 pharmDx is indicated as an aid in identifying NSCLC patients for treatment with KEYTRUDA® (pembrolizumab) with no EGFR or ALK genomic aberrations. The performance characteristics of this assay have not been validated for decalcified specimens. Results should be interpreted with caution given the likelihood of false negativity on decalcified specimens. A minimum of 100 viable tumors cells is needed to determine PD-L1 expression.

Methodology

PD-L1 22C3 FDA (Keytruda®) PD-L1 staining was performed utilizing the DAKO FDA-approved PD-L1, 22C3 pharm Dx™ protocol using EnVision FLEX visualization system on the Dako Automated Link 48 platform. PD-L1 protein expression is determined by using Tumor Proportion Score (TPS), which is the percentage of viable tumor cells showing partial or complete membrane staining at any intensity. The specimen should be considered to have PD-L1 expression if TPS >=1% and high PD-L1 expression if TPS >=50%.

Figure 25.4A showed PD-L1 expression profile for a 51-year-old male with metastatic lung adenocarcinoma (cT3N3M1c adenocarcinoma of the lung with right upper lobe primary and metastatic disease to the brain, bone, pleura, lymph nodes). The result showed a TPS score of PDL1 >90%, which is a high expression. Therefore, this patient is indicated for immunotherapy. Currently, this patient is in cycle 3 of carboplatin, pemetrexed, and pembrolizumab (pembrolizumab started with cycle 2).

For other types of cancers including head and neck squamous cell carcinoma (HNSCC), urothelial carcinoma, gastric or GEJ cancer, esophageal carcinoma, and cervical cancer, a combined positive score (CPS) is used, which is to evaluate the number of PD-L1 staining cells including tumor cells, lymphocytes, and macrophages, relative to all viable tumor cells. The score is generated by the number of PD-L1 staining cells divided by the total number of viable tumor cells then multiplied by 100. Figure 25.4B showed a CPS score of PD-L1 expression for a 76-year-old male with invasive poorly differentiated gastric adenocarcinoma. The result is CPS<1, which is interpreted as no PD-L1 expression. This patient is unlikely to respond to immunotherapy.

25.7.1.3 Tumor mutational burden (TMB)

Tumor mutational burden (TMB), also known as tumor mutational load (TML), refers to the number of mutations per megabase of the genome (muts/Mb) harbored by the tumor cells. TMB is a predictive biomarker for immunotherapy response in multiple types of cancers [218]. Tumors with high TMB (TMB-H) usually have more immunogenic neoantigens compared with tumors with low TMB. It has been shown

A

Study / Test (Clone)	Block	Tumor Proportion Score (TPS)	Reference Ranges
PD-L1 (22C3) * FDA (Keytruda®) for NSCLC	A2	TPS: >90%	High expression: >= 50% TPS Expressed: 1-49% TPS No expression: <1% TPS

B

Study / Test (Clone)	Block	Combined Positive Score (CPS)	Reference Ranges (Positive)
PD-L1 (22C3) * FDA (Keytruda®)	A1	CPS: <1	**CPS >=1**: Gastric/Gastroesophageal junction adenocarcinoma, Uterine cervical cancer, Head and Neck squamous cell carcinoma **CPS >=10**: Urothelial carcinoma, Esophageal squamous cell carcinoma

FIGURE 25.4 Examples for ICH staining to detect PD-L1 expression. (A) A non-small cell lung carcinoma (NSCLC) case with high expression of PD-L1 by immunohistochemistry staining using PD-L1 IHC 22C3 pharmDx. Tumor Proportion Score (TPS) was used to evaluate the percentage of viable tumor cells showing complete or partial membrane staining. TPS = (# of PD-L1-positive tumor cells)/(Total # of PD-L1-positive + PD-L1-negative tumor cells) × 100%. (B) A gastric adenocarcinoma case negative for PD-L1 expressing using PD-L1 IHC 22C3 pharmDx staining and combined positive score (CPS). CPS = (# of PD-L1-positive cells)/(Total # of PD-L1-positive + PD-L1-negative tumor cells) × 100.

that tumor neoantigens can be recognized by the host T cells, thus predicting immunotherapy response. Generally, TMB correlates with anti-PD1 immunotherapy response in many tumor types.

TMB can be determined by different methods. Initially, TMB is typically calculated from whole genome sequencing or whole exome sequencing (WES) data using the next-generation sequencing technology. The human exome contains about 30–50 megabases of coding sequences (<2% of the genome) [233]. Whole exome sequencing (WES) can provide a clear landscape of most mutations in the coding region that can contribute to tumor progression. Compared to the whole genome sequencing, WES is cost-effective. WES requires sequencing of both tumor sample and matched normal samples to identify tumor-specific variants. It usually requires a minimum of 150–200 ng of genomic DNA, not suitable for a small biopsy. The ability of detecting somatic variants using NGS technology is essentially dependent on the variant frequency and the sequencing depth. WES can usually achieve 50X coverage, which can consistently detect 95% of somatic variants including single nucleotide variants (SNVs) and short insertions or deletions (INDELs) with a variant frequency of 15% or higher [234]. Factors that can affect the calculation include contamination of non-tumor cells, tumor heterogeneity, and aneuploidy. The relatively higher cost and the relatively low sequencing depth as well as longer turnaround time (TAT) of WES make it difficult to use for TMB calculation in the clinical setting. Thus, TMB estimation using NGS panels of selected cancer genes is more feasible.

Recently, several gene panels showed comparable results for TMB estimation. Compared to WES, gene panels are preferable due to lower sequencing cost, lower DNA input requirement, deeper sequencing depth, and shorter TAT. FoundationOne CDx is a cancer sequencing panel of 315 genes targeting about 1.1 Mb of coding genome that can generate TMB estimation by counting synonymous and non-synonymous variants present at 5% allele frequency or higher [235]. The Memorial Sloan Kettering Cancer Center (MSKCC) MSK-IMPACT targeted NGS assay of over 400 genes is another NGS panel that can provide TMB estimation. A third NGS panel is a commercially available gene panel, the Oncomine Tumor Mutation Load Assay from ThermoFisher Scientific, targeting 1.7 Mb of the genome covering 409 key cancer genes. TMB is determined from 20 ng of DNA by counting the somatic SNVs per Mb using custom filtering criteria including exonic variants with an allele frequency of 10% or higher to avoid counting the variants caused by FFPE artifact, not found in the dbSNP to exclude germline mutations, not present in the Catalogue of Somatic Mutations in Cancer (COSMIC) database to eliminate the potential bias of studying cancer-related genes in the panel. The major problem of this Oncomine Tumor Mutation Load Assay is the difficulties in the analysis of some FFPE samples with many artifacts. This can be solved by using duplicates and improving the filtering pipeline.

Despite the methods to calculate TMB, it is emerging as an important biomarker to identify patients who may benefit from immunotherapy. However, many factors including the site of tumor biopsy, specimen tumor content, amount of genome interrogated, filtering of variants, tumor heterogeneity, and sequencing read depth, can impact the TMB score. In June 2020, the FDA approved the anti-PD-1 therapy pembrolizumab for treating patients with advanced and refractory cancers with a high TMB, as indicated by a defined threshold level of mutations [236].

25.7.1.4 Microsatellite instability (MSI) testing

The FDA approved pembrolizumab (Keytruda) as the first-line treatment for patients with unresectable or metastatic colorectal cancer with microsatellite instability-high (MSI-H) or mismatch repair deficiency (dMMR) on June 29, 2020 [237]. This is the first immunotherapy approved for the treatment of this patient population as the first-line treatment.

Microsatellites, also known as short tandem repeats, are repetitive DNA sequences ranging in length from one to six bases distributed along the genome. Microsatellites are highly polymorphic, meaning the number of repeats of sequences varies between individuals, but they are typically the same length in the germline DNA and the somatic DNA of the tumor in the same patient. Microsatellites are very sensitive to DNA mismatching repair (MMR) errors during DNA replication or iatrogenic damage due to their repetitive nature [238]. DNA MMR is a highly conserved mechanism used by the cell to maintain or repair

DNA integrity after mismatching errors such as single-base mismatches or short insertion and deletions (INDELs). There are four genes that play critical roles in DNA mismatch repair, including MLH1 (mutL homologue 1), MSH2 (mutS homologue 2), MSH6 (mutS homologue 6) and PMS2 (postmeiotic segregation increased 2) [239]. MLH1 and MSH2 are obligator partners for heterodimers. PMS2 can only form a heterodimer with MLH1 and MSH6 can only form a heterodimer with MSH2. Mutations in MLH1 and MSH2 generally result in subsequent proteolytic degradation of the mutated protein and its respective secondary partner, PMS2 and MSH6. However, mutations in PMS2 or MSH6 may not result in proteolytic degradation of its primary partners since MSH6 can be substituted in the heterodimer by MSH3 and PMS2 can be substituted by PMS1 or MLH3 in the heterodimer [240]. Mutations in the MMR gene can result in defective MMR (dMMR) tumor that accumulates thousands of mutations, particularly clustered in microsatellites with altered numbers of repeats. Microsatellite instability (MSI) refers to the condition of genetic hypermutability resulting from defective DNA MMR. Therefore, MSI is a marker of dMMR.

DMMR can be identified by immunohistochemistry to detect loss of MMR proteins or sequencing to identify mutations in the MMR genes. MSI can be detected by PCR or NGS. dMMR/MSI testing is recommended for any sporadic cancer type belonging to the spectrum of cancers found in Lynch syndrome including colorectal, endometrial, small intestine, urothelial, central nervous system (gliomas/glioblastomas), and sebaceous gland. The IHC method is to detect the four MMR proteins using antibodies. The MMR proteins are ubiquitously expressed in cell nuclei and most mutations of the MMR genes interfere with the dimerization to result in the proteolytic degradation of the heterodimers and subsequent loss of both obligatory and secondary partner proteins. To be specific, mutations in MLH1 can cause loss of both MLH1 and PMS2, while mutations in MSH2 can cause loss of both MSH2 and MSH6. Both technical and biological reasons may lead to false-negative immunostaining of the MMR proteins. Technical problems are mainly due to pre-analytic issues such as tissue fixation. Biological problems for false-positive immunostaining include missense mutations in any MMR gene resulting in mutant primary proteins antigenically intact but catalytically inactive, or PMS2 or MSH6 substituted by another secondary MMR partner such as MSH3 replacing MSH6 and MLH3 or PMS1 replacing PMS2. Thus, it is recommended to use all four antibodies and to include an internal positive control such as normal mucosa, lymphocytes, or stromal cells. For any doubt in IHC interpretation, MSI-PCR should be performed.

MSI-PCR molecular testing is based on PCR amplification of microsatellite markers. There are two possible panels of microsatellite markers. The "Bethesda guidelines" for colorectal cancer recommended a panel including two mononucleotide repeats (BAT-26 and BAT-26) and three dinucleotide markers (D5S346, D2S123, and D17S250). However, the alternative panel with five poly-A mononucleotide repeats (BAT-25, BAT-26, NR-21, NR-24, NR-27, or Mono-27) is currently considered standard due to its higher sensitivity and specificity. For the MSI-PCR test, MSI is defined as loss of stability in 2 or more of the five microsatellite markers. A commercially available MSI-PCR test kit, Promega MSI Analysis System, uses the five poly-A mononucleotide markers (BAT-25, BAT-26, NR-21, NR-24, and MONO-27) and two pentanucleotide repeat markers (Penta C and Penta D) with fluorescently labeled primers for tumor and matched normal samples. A PCR positive control (K562 cell line) and a negative control (no template control) are included in each run to confirm the validity of the run. Figure 25.5 showed the MSI-PCR result for a 68 years old male diagnosed with pT4bN1b colonic adenocarcinoma with liver metastatic using the Promega™ MSI PCR Test kit. All five poly-A mononculeotide repeat markers showed a shift of the length due to dMMR. The dMMR status was evaluated by IHC using all four antibodies, which showed loss of immunostaining of both MLH1 and PMS2 but MSH2 and MSH6 immunostaining remain intact in the nuclei (Table 25.11).

This result is consistent with a missense mutation in the MLH1 (c.440G>A, p.G147E). This patient was treated with pembrolizumab (Keytruda) and the carcinoembryonic antigen (CEA) has decreased.

To improve the finding of treatment option for cancer patient, multiple tests need to be performed and the results should be considered accordingly. MSI testing can also be performed by NGS. For example, the MSK-IMPACT NGS cancer panel also integrates MSI testing by including 25 microsatellite markers. It becomes standard to include PD-L1, NGS panel, MSI (either alone or in the NGS panel), and TMB (either alone or in the NGS panel) for most cancer patients testing. For example, IHC for MMR proteins, MSI, TMB, and NGS panel testing were performed for an 82-year-old male patient with metastatic colorectal

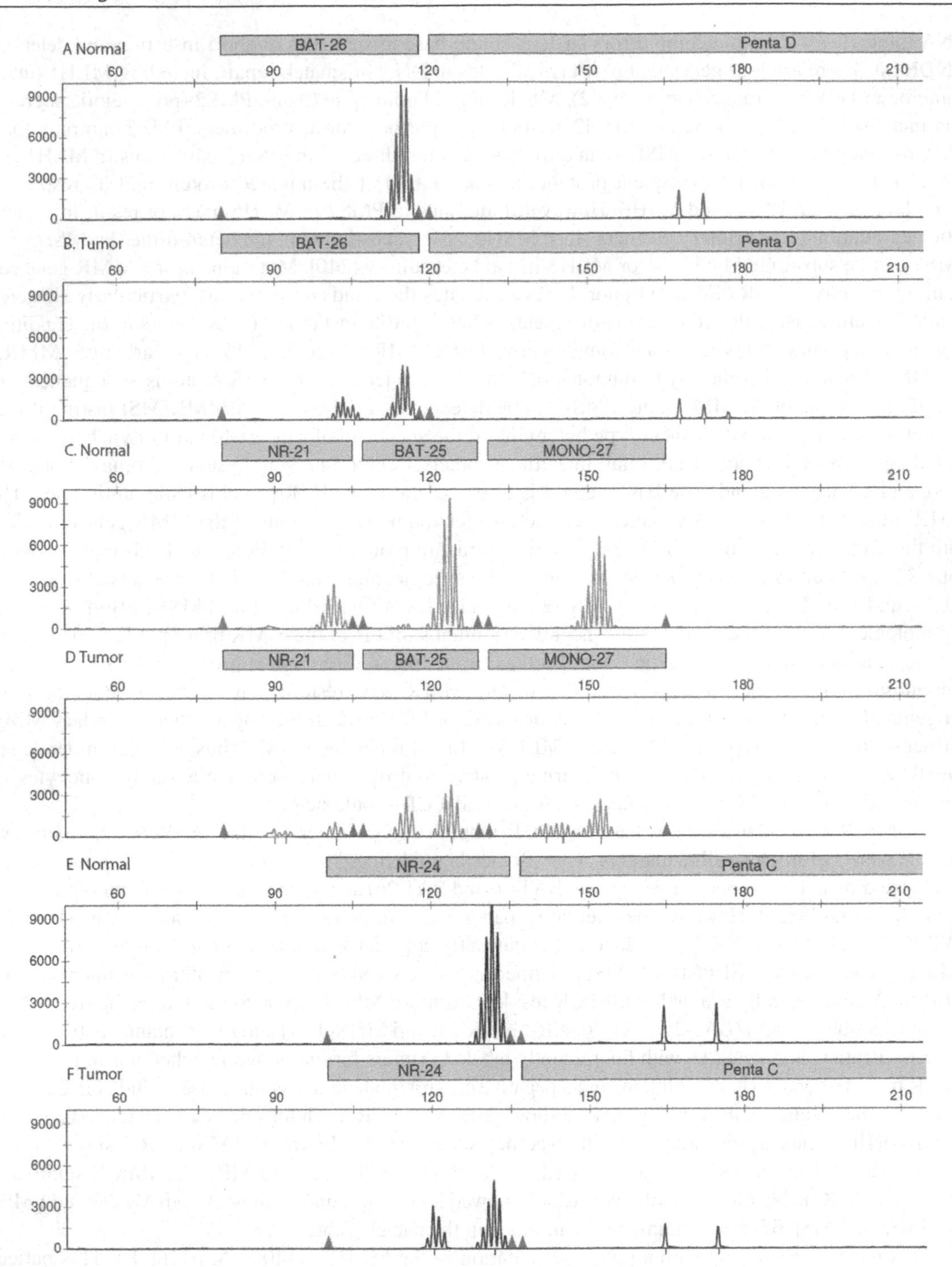

FIGURE 25.5 An example of MSI-H results by Promega™ MSI PCR Test Kit. (a) BAT-26 (114 bp) and Penta D (167 bp and 172 bp) in the normal control tissue. (b) BAT-26 exhibits a shift in the tumor tissues with a 103 bp peak in addition to the normal 114 bp peak. A tiny extra peak of 177 bp is shown in the Penta D marker. (c) NR-21 (101 bp), BAT-25 (124 bp), and MONO-27 (152 bp) in normal tissue. (d) NR-21 (89bp and 101 bp), BAT-25 (115 bp and 124 bp), and MONO-27 (141 bp and 152 bp) shifted in the tumor tissue. (e) NR-24 (132 bp) and Penta C (165 bp and 175 bp) in the normal tissue. (f) NR-24 (120 bp and 132 bp) and Penta C (165 bp and 175 bp) in the tumor tissue.

TABLE 25.11 MMR proteins immunohistochemistry staining profile

PROTEIN	*BLOCK*	*RESULT*
MLH1	A10	Loss of nuclear protein expression
PMS2	A10	Loss of nuclear protein expression
MSH2	A10	Intact nuclear protein expression
MSH6	A10	Intact nuclear protein expression

adenocarcinoma (Figure 25.6). The result of IHC for MMR protein staining showed loss of nuclear protein expression for MLH1 and PMS2. TMB result showed this cancer is TMB high with 28 muts/Mb, therefore this patient is indicated for Pembrolizumab treatment. MSI testing by PCR method showed 5 of 5 markers with shifted peaks, indicating MSI high, which is consistent with the loss of MLH1 and PMS2 nuclear expression. NGS panel result detected multiple genomic alterations including a BRAF V600E,

A

Study / Test (Clone)	Block	Result
MLH-1 QL (G168-15)	A1	Loss of nuclear protein expression
PMS2 QL (A16-4)	A1	Loss of nuclear protein expression
MSH-2 QL (G219-1129)	A1	Intact nuclear protein expression
MSH-6 QL (SP93)	A1	Intact nuclear protein expression

These IHC studies were interpreted in conjunction with appropriate positive and negative controls which demonstrated the expected positive and negative reactivity.

B

IMMUNOTHERAPY

TMB High: 28 muts/Mb

Tumor mutational burden (TMB) is a measurement of the amount of nonsynonymous somatic mutations present within a tumor sample. High TMB is associated with elevated neoantigen levels, which leads to an increased probability of T lymphocyte response (PMID 26258412). Pembrolizumab has recently been approved for tumor with high TMB (>10 muts/Mb). Additionally, the KEYNOTE-158 study is recruiting patients with high TMB for treatment with pembrolizumab (NCT02628067).

MSI-High: 5 of 5 markers show instability

C

NGS RESULT SUMMARY

Variant Detected	FDA Approved Therapy Within Indication	FDA Approved Therapy Outside Indication	Resistance to Therapies	Guidelines	Clinical Trial Opportunity
BRAF p.V600E 17.8%	See MPI	See Variant Details	Panitumumab, Cetuximab without BRAF inhibitor	See Variant Details	Yes - see below
PTEN p.K267Rfs*9 34%	No	No	No	No	Yes - see below
PTEN p.M134T 16.8%	No	No	No	No	Yes -see below
ARID1A p.M890Hfs*46 12.2%	No	No	No	No	Yes - see below

FIGURE 25.6 Integrated reports for a metastatic colorectal carcinoma. (A) IHC staining profile for MMR protein. (B) TMB and MSI testing results. No PD-L1 expression was performed for this case. (C). Four genomic alterations were identified by NGS analysis.

two inactivating alterations of PTEN, an inactivating alteration of ARID1A, and a high tumor mutational burden in an MSI-High tumor. Per the surgical pathology report, methylation studies were performed on a prior biopsy which showed hypermethylation of MLH1, therefore making this most likely a sporadic microsatellite unstable tumor. Pembrolizumab is FDA approved for the treatment of MSI-High colorectal cancer that has progressed following treatment with fluoropyrimidine, oxaliplatin, and irinotecan. Pembrolizumab was also FDA approved recently as first-line monotherapy for MSI-High colorectal cancer based on the results of the KEYNOTE-177 trial, which showed an increase of progression-free survival in metastatic colorectal cancer patients treated with pembrolizumab of 16.5 months vs. 8.2 months with chemotherapy. Additionally, based on the results of the BEACON trial, encorafenib with cetuximab was recently FDA approved for the treatment of metastatic colorectal cancer patients with a BRAF V600E mutation. The BEACON trial showed an increase in median overall survival of 8.4 months for patients treated with encorafenib and cetuximab vs. 5.4 months for the control group. Dabrafenib and trametinib in combination PDR001 (anti-PD-1 monoclonal antibody) are currently being tested in a phase II clinical trial for advanced or metastatic colorectal cancer patients with V600E mutation (NCT03668431). Other clinical trials are also available (see below). These findings require integration with all other relevant clinical and laboratory information.

REFERENCES

1. Plotkin, S., *History of vaccination*. Proc Natl Acad Sci U S A, 2014. **111**(34): p. 12283–7.
2. George, K. and G. Woollett, *Insulins as Drugs or biologics in the USA: what difference does it make and why does it matter?* BioDrugs, 2019. **33**(5): p. 447–51.
3. FDA.gov/files/drugs/published/Nonproprietary-Naming-of-Biological-Products-Guidance-for-industry.pdf. 2017.
4. Rosman, Z., Y. Shoenfeld, and G. Zandman-Goddard, *Biologic therapy for autoimmune diseases: an update.* BMC Med, 2013. **11**: p. 88.
5. Wang, L., F.S. Wang, and M.E. Gershwin, *Human autoimmune diseases: a comprehensive update.* J Intern Med, 2015. **278**(4): p. 369–95.
6. Duan, L., X. Rao, and K.R. Sigdel, *Regulation of inflammation in autoimmune disease.* J Immunol Res, 2019. **2019**: p. 7403796.
7. Singh, J.A., et al., *2015 American College of Rheumatology Guideline for the treatment of RA*. Arthritis Rheumatol, 2016. **68**(1): p. 1–26.
8. Bosch, X., R.-C. Manuel, and M.A. Khamashta, *Drugs Targeting B-Cells in Autoimmune Diseases.* 2013, Springer Science & Business Media: p. 1–4.
9. Shaw, T., J. Quan, and M.C. Totoritis, *B cell therapy for RA: the rituximab (anti-CD20) experience.* Ann Rheum Dis, 2003. **62**(Suppl 2): p. ii55–9.
10. Edwards, J.C., et al., *Efficacy of B-cell-targeted therapy with rituximab in patients with RA*. N Engl J Med, 2004. **350**(25): p. 2572–81.
11. Burton, C., R. Kaczmarski, and R. Jan-Mohamed, *Interstitial pneumonitis related to rituximab therapy.* N Engl J Med, 2003. **348**(26): p. 2690–1; discussion 2690–1.
12. McGinley, M.P., B.P. Moss, and J.A. Cohen, *Safety of monoclonal antibodies for the treatment of multiple sclerosis.* Expert Opin Drug Saf, 2017. **16**(1): p. 89–100.
13. Zhang, B., *Ofatumumab.* MAbs, 2009. **1**(4): p. 326–31.
14. Evans, S.S. and A.B. Clemmons, *Obinutuzumab: a novel anti-CD20 MAbfor chronic lymphocytic leukemia.* J Adv Pract Oncol, 2015. **6**(4): p. 370–4.
15. Shah, A., *Obinutuzumab: a novel anti-CD20 mAbfor previously untreated chronic lymphocytic leukemia.* Ann Pharmacother, 2014. **48**(10): p. 1356–61.
16. FDA approves first biosimilar for treatment of adult patients with non-Hodgkin's lymphoma. U.S. Food and Drug Administration (FDA) (Press release). November 28, 2018. Retrieved June 25, 2021. https://www.fda.gov/news-events/press-announcements/fda-approves-first-biosimilar-treatment-adult-patients-non-hodgkins-lymphoma#:~:text=The%20U.S.%20Food%20and%20Drug,or%20in%20combination%20with%20chemotherapy.

17. https://www.centerforbiosimilars.com/view/fda-approves-pfizers-rituximab-biosimilar-ruxience. FDA Approves Pfizer's Rituximab Biosimilar, Ruxience. 2019, The Center for Biosimilars.
18. Amgen Rituximab Biosimilar Gains FDA Approval. The Center for Biosimilars, 2020. https://www.centerforbiosimilars.com/view/fda-approves-rituximab-biosimilar-from-amgen.
19. Bossen, C. and P. Schneider, *BAFF, APRIL and their receptors: structure, function and signaling.* Semin Immunol, 2006. **18**(5): p. 263–75.
20. Kaveri, S.V., L. Mouthon, and J. Bayry, *Basophils and nephritis in lupus.* N Engl J Med, 2010. **363**(11): p. 1080–2.
21. Horowitz, D.L. and R. Furie, *Belimumab is approved by the FDA: what more do we need to know to optimize decision making?* Curr Rheumatol Rep, 2012. **14**(4): p. 318–23.
22. *Belimumab: The First Drug to be FDA Approved for the Treatment of Lupus since 1955.* 2011, by Rebecca Manno, Johns Hopkins Arthritis Center. https://www.fda.gov/media/140599/download#:~:text=Belimumab%20is%20the%20first%20drug,seropositive%20SLE%20receiving%20standard%20care.
23. UK Anakinra label. UK Electronic Medicines Compendium. October 5, 2017. Retrieved June 25, 2021. https://www.ema.europa.eu/en/medicines/human/EPAR/kineret
24. US Anakinra label. FDA. May 2016. Retrieved June 25, 2021. For label updates see FDA index page for BLA 103950, 2016.
25. Singh, J.A., et al., *Biologics or tofacitinib for RA in incomplete responders to methotrexate or other traditional disease-modifying anti-rheumatic drugs: a systematic review and network meta-analysis.* Cochrane Database Syst Rev, 2016. **2016**(5): p. Cd012183.
26. Canna, S.W. and E.M. Behrens, *Making sense of the cytokine storm: a conceptual framework for understanding, diagnosing, and treating hemophagocytic syndromes.* Pediatr Clin North Am, 2012. **59**(2): p. 329–44.
27. Gusdorf, L. and D. Lipsker, *Schnitzler syndrome: a review.* Curr Rheumatol Rep, 2017. **19**(8): p. 46.
28. Dhimolea, E., *Canakinumab.* MAbs, 2010. **2**(1): p. 3–13.
29. FDA approves expanded indications for Ilaris for three rare diseases. FDA. September 23, 2016. Retrieved June 25, 2021
30. Le, R.Q., et al., *FDA approval summary: tocilizumab for treatment of chimeric antigen receptor t cell-induced severe or life-threatening cytokine release syndrome.* Oncologist, 2018. **23**(8): p. 943–7.
31. *Coronavirus (COVID-19) Update: FDA Authorizes Drug for Treatment of COVID-19.* U.S. Food and Drug Administration (FDA) (Press release). June 24, 2021. Retrieved June 25, 2021.
32. *Cosentyx (secukinumab) injection.* FDA. September 2017. For label updates see FDA index page for BLA 125504, 2017.
33. Taltz (ixekizumab) Injection. U.S. Food and Drug Administration (FDA). May 3, 2016. Retrieved June 2021. https://www.accessdata.fda.gov/drugsatfda_docs/appletter/2016/125521Orig1s000ltr.pdf.
34. Office of the Commissioner (February 15, 2017). *Press Announcements – FDA approves new psoriasis drug.* www.fda.gov. Retrieved June 2021.
35. Weyand, C.M. and J.J. Goronzy, *T-cell-targeted therapies in RA.* Nat Clin Pract Rheumatol, 2006. **2**(4): p. 201–10.
36. Acuto, O. and F. Michel, *CD28-mediated co-stimulation: a quantitative support for TCR signalling.* Nat Rev Immunol, 2003. **3**(12): p. 939–51.
37. Snyder, M.R., C.M. Weyand, and J.J. Goronzy, *The double life of NK receptors: stimulation or co-stimulation?* Trends Immunol, 2004. **25**(1): p. 25–32.
38. Greenwald, R.J., G.J. Freeman, and A.H. Sharpe, *The B7 family revisited.* Annu Rev Immunol, 2005. **23**: p. 515–48.
39. Drug Approval Package: *Orencia (Abatacept) NDA #125118.* 2020. www.accessdata.fda.gov. Retrieved June 25, 2021.
40. Lubberts, E. and W.B. van den Berg, *Cytokines in the pathogenesis of RA and collagen-induced arthritis.* Adv Exp Med Biol, 2003. **520**: p. 194–202.
41. Drug Approval Package: Humira (adalimumab). U.S. Food and Drug Administration (FDA). April 7, 2017. Retrieved June 2021. https://www.accessdata.fda.gov/drugsatfda_docs/label/2017/125057s399lbl.pdf.
42. Frenzel, A., T. Schirrmann, and M. Hust, *Phage display-derived human antibodies in clinical development and therapy.* MAbs, 2016. **8**(7): p. 1177–94.
43. Kempeni, J., *Preliminary results of early clinical trials with the fully human anti-TNFalpha mAbD2E7.* Ann Rheum Dis, 1999. **58**(Suppl 1): p. I70–2.
44. FDA approves Amjevita, a biosimilar to Humira. 2016. www.accessdata.fda.gov. Retrieved June 25, 2021.
45. Cyltezo: FDA-Approved Drugs. U.S. Food and Drug Administration (FDA). Retrieved June 21, 2021. https://www.fda.gov/news-events/press-announcements/fda-approves-cyltezo-first-interchangeable-biosimilar-humira#:~:text=The%20FDA%20granted%20approval%20of,a%20variety%20of%20health%20conditions.

46. "Drug Approval Package: Hyrimoz". U.S. Food and Drug Administration (FDA). March 21, 2019. Retrieved June 21, 2021. https://www.accessdata.fda.gov/drugsatfda_docs/label/2018/761071lbl.pdf.
47. Hadlima: FDA-Approved Drugs. U.S. Food and Drug Administration (FDA). Retrieved June 21, 2021. https://www.accessdata.fda.gov/drugsatfda_docs/nda/2019/761059Orig1s000TOC.cfm.
48. Abrilada: FDA-Approved Drugs. U.S. Food and Drug Administration (FDA). November 15, 2019. Retrieved June 22, 2021. https://www.accessdata.fda.gov/drugsatfda_docs/nda/2019/761118Orig1s000TOC.cfm.
49. Hulio: FDA-Approved Drugs. U.S. Food and Drug Administration (FDA). Retrieved 7 July 2020. Retrieved June 25, 2021. https://www.accessdata.fda.gov/drugsatfda_docs/nda/2020/761154Orig1s000TOC.cfm.
50. Sandborn, W.J., et al., *Certolizumab pegol for the treatment of Crohn's disease.* N Engl J Med, 2007. **357**(3): p. 228–38.
51. Feldmann, M. and R.N. Maini, *Lasker Clinical Medical Research Award. TNF defined as a therapeutic target for RA and other autoimmune diseases.* Nat Med, 2003. **9**(10): p. 1245–50.
52. Benpali EPAR. European Medicines Agency (EMA). Retrieved June 20, 2021. https://www.ema.europa.eu/en/medicines/human/EPAR/benepali.
53. FDA approves Erelzi, a biosimilar to Enbrel. The US FDA. Retrieved June 22, 2021. https://www.fda.gov/news-events/press-announcements/fda-approves-erelzi-biosimilar-enbrel
54. Drug Approval Package: Eticovo. U.S. Food and Drug Administration (FDA). April 25, 2019. Retrieved June 2021. https://www.accessdata.fda.gov/drugsatfda_docs/nda/2019/761066Orig1s000TOC.cfm.
55. Löwenberg, M., N. de Boer, and F. Hoentjen, *Golimumab for the treatment of ulcerative colitis.* Clin Exp Gastroenterol, 2014. **7**: p. 53–9.
56. Knight, D.M., et al., *Construction and initial characterization of a mouse-human chimeric anti-TNF antibody.* Mol Immunol, 1993. **30**(16): p. 1443–53.
57. *Infliximab, Infliximab-dyyb Monograph for Professionals.* 1998. Drugs.com. American Society of Health-System Pharmacists. Retrieved May 25, 2021.
58. "Inflectra: FDA-Approved Drugs". U.S. Food and Drug Administration (FDA). December 6, 2019. Retrieved May 27, 2021.
59. Ishida, Y., et al., *Induced expression of PD-1, a novel member of the immunoglobulin gene superfamily, upon programmed cell death.* Embo J, 1992. **11**(11): p. 3887–95.
60. Dong, H., et al., *B7-H1, a third member of the B7 family, co-stimulates T-cell proliferation and interleukin-10 secretion.* Nat Med, 1999. **5**(12): p. 1365–9.
61. Sharpe, A.H. and K.E. Pauken, *The diverse functions of the PD1 inhibitory pathway.* Nat Rev Immunol, 2018. **18**(3): p. 153–167.
62. Blank, C. and A. Mackensen, *Contribution of the PD-L1/PD-1 pathway to T-cell exhaustion: an update on implications for chronic infections and tumor evasion.* Cancer Immunol Immunother, 2007. **56**(5): p. 739–45.
63. Agata, Y., et al., *Expression of the PD-1 antigen on the surface of stimulated mouse T and B lymphocytes.* Int Immunol, 1996. **8**(5): p. 765–72.
64. Chemnitz, J.M., et al., *SHP-1 and SHP-2 associate with immunoreceptor tyrosine-based switch motif of programmed death 1 upon primary human T cell stimulation, but only receptor ligation prevents T cell activation.* J Immunol, 2004. **173**(2): p. 945–54.
65. Chen, D.S. and I. Mellman, *Oncology meets immunology: the cancer-immunity cycle.* Immunity, 2013. **39**(1): p. 1–10.
66. Dosset, M., et al., *PD-1/PD-L1 pathway: an adaptive immune resistance mechanism to immunogenic chemotherapy in colorectal cancer.* Oncoimmunology, 2018. **7**(6): p. e1433981.
67. Gupta, H.B., et al., *Tumor cell-intrinsic PD-L1 promotes tumor-initiating cell generation and functions in melanoma and ovarian cancer.* Signal Transduct Target Ther, 2016. **1**: p. 16030.
68. Chen, L., et al., *PD-L1 expression promotes epithelial to mesenchymal transition in human esophageal cancer.* Cell Physiol Biochem, 2017. **42**(6): p. 2267–80.
69. Li, Y., et al., *B7-H3 promotes gastric cancer cell migration and invasion.* Oncotarget, 2017. **8**(42): p. 71725–35.
70. Dai, W., et al., *Aberrant expression of B7-H3 in gastric adenocarcinoma promotes cancer cell metastasis.* Oncol Rep, 2014. **32**(5): p. 2086–92.
71. Kraft, S., et al., *PDL1 expression in desmoplastic melanoma is associated with tumor aggressiveness and progression.* J Am Acad Dermatol, 2017. **77**(3): p. 534–42.
72. Lyford-Pike, S., et al., *Evidence for a role of the PD-1:PD-L1 pathway in immune resistance of HPV-associated head and neck squamous cell carcinoma.* Cancer Res, 2013. **73**(6): p. 1733–41.
73. Jiang, X., et al., *Role of the tumor microenvironment in PD-L1/PD-1-mediated tumor immune escape.* Mol Cancer, 2019. **18**(1): p. 10.

74. Drug Approval Package: Pembrolizumab. U.S. Food and Drug Administration (FDA) (Press release). September 4, 2014. Retrieved June 25, 2021. https://www.fda.gov/drugs/resources-information-approved-drugs/fda-approves-pembrolizumab-combination-first-line-treatment-cervical-cancer#:~:text=On%20October%2013%2C2021%2C%20the,by%20an%20FDA%2Dapproved%20test.
75. Sul, J., et al., *FDA approval summary: pembrolizumab for the treatment of patients with metastatic non-small cell lung cancer whose tumors express programmed death-ligand 1*. Oncologist, 2016. **21**(5): p. 643–50.
76. FDA grants accelerated approval to pembrolizumab for first tissue/site agnostic indication. U.S. Food and Drug Administration (FDA) (Press release). March 23, 2017. Retrieved June 25, 2021. https://www.fda.gov/drugs/resources-information-approved-drugs/fda-grants-accelerated-approval-pembrolizumab-first-tissuesite-agnostic-indication.
77. FDA expands pembrolizumab indication for first-line treatment of NSCLC (TPS ≥ 1%). U.S. Food and Drug Administration (FDA) (Press release). April 11, 2019. Retrieved June 25, 2021. https://www.fda.gov/drugs/fda-expands-pembrolizumab-indication-first-line-treatment-nsclc-tps-1.
78. Wang, C., et al., *In vitro characterization of the anti-PD-1 antibody nivolumab, BMS-936558, and in vivo toxicology in non-human primates*. Cancer Immunol Res, 2014. **2**(9): p. 846–56.
79. Drug Approval Package: Nivolumab. U.S. Food and Drug Administration (FDA) (Press release). December 22, 2014. Retrieved June 25, 2021. https://www.accessdata.fda.gov/drugsatfda_docs/nda/2014/125554Orig1s000TOC.cfm.
80. FDA approves nivolumab for resected esophageal or GEJ cancer. U.S. Food and Drug Administration (FDA) (Press release). May 20, 2020. Retrieved June 25, 2021. https://www.fda.gov/drugs/resources-information-approved-drugs/fda-approves-nivolumab-resected-esophageal-or-gej-cancer.
81. FDA approves nivolumab and ipilimumab for unresectable malignant pleural mesothelioma. U.S. Food and Drug Administration (FDA) (Press release). October 2, 2020. Retrieved June 25, 2021. https://www.fda.gov/drugs/resources-information-approved-drugs/fda-approves-nivolumab-and-ipilimumab-unresectable-malignant-pleural-mesothelioma.
82. FDA approves first treatment for advanced form of the second most common skin cancer U.S. Food and Drug Administration. September 28, 2018. Retrieved June 28, 2021. https://www.fda.gov/news-events/press-announcements/fda-approves-first-treatment-advanced-form-second-most-common-skin-cancer.
83. Tecentriq-atezolizumab injection, solution. DailyMed. June 3, 2020. Retrieved June 28, 2021. https://dailymed.nlm.nih.gov/dailymed/drugInfo.cfm?setid=6fa682c9-a312-4932-9831-f286908660ee.
84. Approved Drugs – Durvalumab (Imfinzi). www.fda.gov. Retrieved June 2021.
85. Eggermont, A.M.M., M. Crittenden, and J. Wargo, *Combination immunotherapy development in melanoma*. Am Soc Clin Oncol Educ Book, 2018. **38**: p. 197–207.
86. Buchbinder, E.I. and A. Desai, *CTLA-4 and PD-1 pathways: similarities, differences, and implications of their inhibition*. Am J Clin Oncol, 2016. **39**(1): p. 98–106.
87. FDA approves new treatment for a type of late-stage skin cancer. U.S. Food and Drug Administration (FDA). March 25, 2011. Retrieved June 28, 2021. https://www.formularywatch.com/view/fda-approves-new-treatment-type-late-stage-skin-cancer.
88. FDA approves nivolumab plus ipilimumab for first-line mNSCLC (PD-L1 tumor expression ≥1%). U.S. Food and Drug Administration (FDA). May 15, 2020. Retrieved June 28, 2021. https://www.fda.gov/drugs/resources-information-approved-drugs/fda-approves-nivolumab-plus-ipilimumab-first-line-mnsclc-pd-l1-tumor-expression-1.
89. FDA grants accelerated approval to nivolumab and ipilimumab combination for hepatocellular carcinoma. U.S. Food and Drug Administration (FDA). March 10, 2020. Retrieved May 26, 2021. https://www.fda.gov/drugs/resources-information-approved-drugs/fda-grants-accelerated-approval-nivolumab-hcc-previously-treated-sorafenib.
90. FDA Approves Drug Combination for Treating Mesothelioma. U.S. Food and Drug Administration (FDA) (Press release). October 2, 2020. Retrieved June 25, 2021. https://www.fda.gov/drugs/resources-information-approved-drugs/fda-approves-nivolumab-and-ipilimumab-unresectable-malignant-pleural-mesothelioma.
91. Rohaan, M.W., S. Wilgenhof, and J. Haanen, *Adoptive cellular therapies: the current landscape*. Virchows Arch, 2019. **474**(4): p. 449–61.
92. Sermer, D. and R. Brentjens, *CAR T-cell therapy: full speed ahead*. Hematol Oncol, 2019. **37**(Suppl 1): p. 95–100.
93. Chandran, S.S. and C.A. Klebanoff, *T cell receptor-based cancer immunotherapy: emerging efficacy and pathways of resistance*. Immunol Rev, 2019. **290**(1): p. 127–47.
94. Dotti, G., et al., *Design and development of therapies using chimeric antigen receptor-expressing T cells*. Immunol Rev, 2014. **257**(1): p. 107–26.

95. Hartmann, J., et al., *Clinical development of CAR T cells-challenges and opportunities in translating innovative treatment concepts.* EMBO Mol Med, 2017. **9**(9): p. 1183–97.
96. FDA approval brings first gene therapy to the United States. U.S. Food & Drug Administration (FDA) (Press release). Retrieved May 28, 2021. https://www.fda.gov/news-events/press-announcements/fda-approval-brings-first-gene-therapy-united-states.
97. Ghaderi, S., et al., *Cancer in childhood, adolescence, and young adults: a population-based study of changes in risk of cancer death during four decades in Norway.* Cancer Causes Control, 2012. **23**(8): p. 1297–305.
98. Ceppi, F., et al., *Improvement of the outcome of relapsed or refractory acute lymphoblastic leukemia in children using a risk-based treatment strategy.* PLoS One, 2016. **11**(9): p. e0160310.
99. Novartis gets second CAR-T candidate FDA 'breakthrough' tag. www.fiercebiotech.com. Fierce Biotech. Retrieved June 25, 2021.
100. *FDA approves CAR-T cell therapy to treat adults with certain types of large B-cell lymphoma.* U.S. Food and Drug Administration (Press release). Retrieved June 25, 2021. https://www.fda.gov/news-events/press-announcements/fda-approves-car-t-cell-therapy-treat-adults-certain-types-large-b-cell-lymphoma.
101. *FDA Approves First Cell-Based Gene Therapy For Adult Patients with Relapsed or Refractory MCL.* U.S. Food & Drug Administration (FDA) (Press release). Retrieved May 28, 2021. https://www.fda.gov/news-events/press-announcements/fda-approves-first-cell-based-gene-therapy-adult-patients-multiple-myeloma.
102. "FDA Approves New Treatment for Adults with Relapsed or Refractory Large-B-Cell Lymphoma" U.S. Food & Drug Administration (FDA) (Press release). Retrieved May 28, 2021. https://www.fda.gov/news-events/press-announcements/fda-approves-new-treatment-adults-relapsed-or-refractory-large-b-cell-lymphoma.
103. Acharya, U.H., et al., *Management of cytokine release syndrome and neurotoxicity in chimeric antigen receptor (CAR) T cell therapy.* Expert Rev Hematol, 2019. **12**(3): p. 195–205.
104. Si, S. and D.T. Teachey, *Spotlight on tocilizumab in the treatment of car-t-cell-induced cytokine release syndrome: clinical evidence to date.* Ther Clin Risk Manag, 2020. **16**: p. 705–14.
105. Hay, K.A., *Cytokine release syndrome and neurotoxicity after CD19 chimeric antigen receptor-modified (CAR-) T cell therapy.* Br J Haematol, 2018. **183**(3): p. 364–74.
106. Wudhikarn, K., et al., *Infection during the first year in patients treated with CD19 CAR T cells for diffuse large B cell lymphoma.* Blood Cancer J, 2020. **10**(8): p. 79.
107. Klampatsa, A., V. Dimou, and S.M. Albelda, *Mesothelin-targeted CAR-T cell therapy for solid tumors.* Expert Opin Biol Ther, 2021. **21**(4): p. 473–86.
108. Brown, C.E., et al., *Stem-like tumor-initiating cells isolated from IL13Rα2 expressing gliomas are targeted and killed by IL13-zetakine-redirected T Cells.* Clin Cancer Res, 2012. **18**(8): p. 2199–209.
109. Bielamowicz, K., et al., *Trivalent CAR T cells overcome interpatient antigenic variability in glioblastoma.* Neuro Oncol, 2018. **20**(4): p. 506–18.
110. Yang, J., J. Yan, and B. Liu, *Targeting EGFRvIII for glioblastoma multiforme.* Cancer Lett, 2017. **403**: p. 224–30.
111. Ekstrand, A.J., et al., *Amplified and rearranged epidermal growth factor receptor genes in human glioblastomas reveal deletions of sequences encoding portions of the N- and/or C-terminal tails.* Proc Natl Acad Sci U S A, 1992. **89**(10): p. 4309–13.
112. Fesnak, A.D., C.H. June, and B.L. Levine, *Engineered T cells: the promise and challenges of cancer immunotherapy.* Nat Rev Cancer, 2016. **16**(9): p. 566–81.
113. Anderson, K.G., I.M. Stromnes, and P.D. Greenberg, *Obstacles posed by the tumor microenvironment to t cell activity: a case for synergistic therapies.* Cancer Cell, 2017. **31**(3): p. 311–25.
114. Met, Ö., et al., *Principles of adoptive T cell therapy in cancer.* Semin Immunopathol, 2019. **41**(1): p. 49–58.
115. Geukes Foppen, M.H., et al., *Tumor-infiltrating lymphocytes for the treatment of metastatic cancer.* Mol Oncol, 2015. **9**(10): p. 1918–35.
116. Lee, N., et al., *Tumour-infiltrating lymphocytes in melanoma prognosis and cancer immunotherapy.* Pathology, 2016. **48**(2): p. 177–87.
117. Bruno, T.C., et al., *Antigen-presenting intratumoral B cells affect CD4(+) TIL phenotypes in non-small cell lung cancer patients.* Cancer Immunol Res, 2017. **5**(10): p. 898–907.
118. Hendry, S., et al., *Assessing tumor-infiltrating lymphocytes in solid tumors: a practical review for pathologists and proposal for a standardized method from the international immuno-oncology biomarkers working group: Part 2: TILs in melanoma, gastrointestinal tract carcinomas, non-small cell lung carcinoma and mesothelioma, endometrial and ovarian carcinomas, squamous cell carcinoma of the head and neck, genitourinary carcinomas, and primary brain tumors.* Adv Anat Pathol, 2017. **24**(6): p. 311–35.
119. Lee, Y.T., Y.J. Tan, and C.E. Oon, *Molecular targeted therapy: treating cancer with specificity.* Eur J Pharmacol, 2018. **834**: p. 188–96.

120. Nishida, N., et al., *Angiogenesis in cancer.* Vasc Health Risk Manag, 2006. **2**(3): p. 213–9.
121. Gupta, M.K. and R.Y. Qin, *Mechanism and its regulation of tumor-induced angiogenesis.* World J Gastroenterol, 2003. **9**(6): p. 1144–55.
122. Carmeliet, P., *VEGF as a key mediator of angiogenesis in cancer.* Oncology, 2005. **69**(Suppl 3): p. 4–10.
123. Palmer, A.M., F.A. Stephenson, and R.J. Williams, *Society for medicines research: 40th anniversary symposium.* Drug News Perspect, 2007. **20**(3): p. 191–6.
124. Ferrara, N., *Anti-angiogenic drugs to treat human disease: an interview with Napoleone Ferrara by Kristin H. Kain.* Dis Model Mech, 2009. **2**(7–8): p. 324–5.
125. Drug Approval Package: Avastin. U.S. Food & Drug Administration (FDA) (Press release). February 26, 2004. Retrieved May 28, 2021. https://www.accessdata.fda.gov/drugsatfda_docs/nda/2004/STN-125085_Avastin.cfm.
126. Montero, A.J., et al., *Bevacizumab in the treatment of metastatic breast cancer: friend or foe?* Curr Oncol Rep, 2012. **14**(1): p. 1–11.
127. "FDA approves first biosimilar for the treatment of cancer". U.S. Food & Drug Administration (FDA) (Press release). September 14, 2017. Retrieved May 28, 2021. https://www.fda.gov/drugs/resources-information-approved-drugs/fda-approves-first-biosimilar-cancer-treatment#:~:text=On%20Sept.,for%20the%20treatment%20of%20cancer.
128. Drug Approval Package: Lucentis. U.S. Food & Drug Administration (FDA). June 30, 2006. Retrieved May 28, 2021. https://www.accessdata.fda.gov/drugsatfda_docs/nda/2006/125156s0000_LucentisTOC.cfm.
129. Casak, S.J., et al., *FDA approval summary: ramucirumab for gastric cancer.* Clin Cancer Res, 2015. **21**(15): p. 3372–6.
130. FDA approves ramucirumab plus erlotinib for first-line metastatic NSCLC. U.S. Food & Drug Administration (FDA) (Press release). May 29, 2020. Retrieved May 28, 2021. https://www.fda.gov/drugs/resources-information-approved-drugs/fda-approves-ramucirumab-plus-erlotinib-first-line-metastatic-nsclc#:~:text=On%20May%2029%2C%202020%2C%20the,exon%2021%20(L858R)%20mutations.
131. Sheng, Q. and J. Liu, *The therapeutic potential of targeting the EGFR family in epithelial ovarian cancer.* Br J Cancer, 2011. **104**(8): p. 1241–5.
132. Keshamouni, V.G., R.R. Mattingly, and K.B. Reddy, *Mechanism of 17-beta-estradiol-induced Erk1/2 activation in breast cancer cells. A role for HER2 AND PKC-delta.* J Biol Chem, 2002. **277**(25): p. 22558–65.
133. Slamon, D.J., et al., *Human breast cancer: correlation of relapse and survival with amplification of the HER-2/neu oncogene.* Science, 1987. **235**(4785): p. 177–82.
134. Sjögren, S., et al., *Prognostic and predictive value of c-erbB-2 overexpression in primary breast cancer, alone and in combination with other prognostic markers.* J Clin Oncol, 1998. **16**(2): p. 462–9.
135. Cameron, D., et al., *11 years' follow-up of trastuzumab after adjuvant chemotherapy in HER2-positive early breast cancer: final analysis of the HERceptin Adjuvant (HERA) trial.* Lancet, 2017. **389**(10075): p. 1195–205.
136. Drebin, J.A., et al., *Inhibition of tumor growth by a mAbreactive with an oncogene-encoded tumor antigen.* Proc Natl Acad Sci U S A, 1986. **83**(23): p. 9129–33.
137. Drago, J.Z., S. Modi, and S. Chandarlapaty, *Unlocking the potential of antibody-drug conjugates for cancer therapy.* Nat Rev Clin Oncol, 2021. **18**(6): p. 327–44.
138. FDA Approves Gemtuzumab Ozogamicin for CD33-positive AML. U.S. Food & Drug Administration (FDA) (Press release). September 1, 2017. Retrieved May 28, 2021. https://www.fda.gov/drugs/drug-approvals-and-databases/fda-approves-gemtuzumab-ozogamicin-cd33-positive-aml-pediatric-patients.
139. Teicher, B.A. and J.H. Doroshow, *The promise of antibody-drug conjugates.* N Engl J Med, 2012. **367**(19): p. 1847–8.
140. "FDA approves fam-trastuzumab deruxtecan-nxki for HER2-positive gastric adenocarcinomas". U.S. Food & Drug Administration (FDA) (Press release). January 15, 2021. Retrieved May 28, 2021. https://www.fda.gov/drugs/resources-information-approved-drugs/fda-approves-fam-trastuzumab-deruxtecan-nxki-her2-positive-gastric-adenocarcinomas.
141. FDA approves Brentuximab vedotin for the treatment of adult patients with primary cutaneous anaplastic large cell lymphoma. U.S. Food & Drug Administration (FDA) (Press release). September 9, 2017. Retrieved May 28, 2021. https://www.fda.gov/drugs/resources-information-approved-drugs/fda-approves-brentuximab-vedotin-treatment-adult-patients-primary-cutaneous-anaplastic-large-cell#:~:text=On%20November%209%2C%202017%2C%20the,have%20received%20prior%20systemic%20therapy.
142. Milenic, D.E., E.D. Brady, and M.W. Brechbiel, *Antibody-targeted radiation cancer therapy.* Nat Rev Drug Discov, 2004. **3**(6): p. 488–99.
143. Raedler, L.A., *Zarxio (filgrastim-sndz): first biosimilar approved in the United States.* Am Health Drug Benefits, 2016. 9(Spec Feature): p. 150–4.

144. Moliner, A.M. and J. Waligora, *The European Union Policy in the field of rare diseases*. Adv Exp Med Biol, 2017. **1031**: p. 561–87.
145. Puiu, M. and D. Dan, *Rare diseases, from European resolutions and recommendations to actual measures and strategies*. Maedica (Bucur), 2010. **5**(2): p. 128–31.
146. Moore, D.F., et al., *Enzyme replacement therapy in orphan and ultra-orphan diseases: the limitations of standard economic metrics as exemplified by Fabry-Anderson disease*. Pharmacoeconomics, 2007. **25**(3): p. 201–8.
147. Jenkins, N., *Modifications of therapeutic proteins: challenges and prospects*. Cytotechnology, 2007. **53**(1–3): p. 121–5.
148. Bowers, W.E., *Christian de Duve and the discovery of lysosomes and peroxisomes*. Trends Cell Biol, 1998. **8**(8): p. 330–3.
149. Brady, R.O., J.N. Kanfer, and D. Shapiro, *Metabolism of glucocerebrosides. ii. Evidence of an enzymatic deficiency in Gaucher's disease*. Biochem Biophys Res Commun, 1965. **18**: p. 221–5.
150. Brady, R.O., et al., *Demonstration of a deficiency of glucocerebroside-cleaving enzyme in Gaucher's disease*. J Clin Invest, 1966. **45**(7): p. 1112–5.
151. Deegan, P.B. and T.M. Cox, *Imiglucerase in the treatment of Gaucher disease: a history and perspective*. Drug Des Devel Ther, 2012. **6**: p. 81–106.
152. Zimran, A., et al., *High frequency of the Gaucher disease mutation at nucleotide 1226 among Ashkenazi Jews*. Am J Hum Genet, 1991. **49**(4): p. 855–9.
153. Weinreb, N.J., et al., *Life expectancy in Gaucher disease type 1*. Am J Hematol, 2008. **83**(12): p. 896–900.
154. Dahl, N., et al., *Gaucher disease type III (Norrbottnian type) is caused by a single mutation in exon 10 of the glucocerebrosidase gene*. Am J Hum Genet, 1990. **47**(2): p. 275–8.
155. Bembi, B., et al., *Diagnosis of glycogenosis type II*. Neurology, 2008. **71**(23 Suppl 2): p. S4–11.
156. Drug Approval Package: Myozyme. U.S. Food & Drug Administration (FDA). April 28, 2006. Retrieved June 28, 2021. https://www.accessdata.fda.gov/drugsatfda_docs/nda/2006/125141s000_MyozymeTOC.cfm.
157. Bolton-Maggs, P.H. and K.J. Pasi, *Haemophilias A and B*. Lancet, 2003. **361**(9371): p. 1801–9.
158. Mannucci, P.M., *Hemophilia therapy: the future has begun*. Haematologica, 2020. **105**(3): p. 545–53.
159. Zakrzewski, W., et al., *Stem cells: past, present, and future*. Stem Cell Res Ther, 2019. **10**(1): p. 68.
160. Takahashi, K. and S. Yamanaka, *Induction of pluripotent stem cells from mouse embryonic and adult fibroblast cultures by defined factors*. Cell, 2006. **126**(4): p. 663–76.
161. "Consumer Alert on Regenerative Medicine Products Including Stem Cells and Exosomes". U.S. Food & Drug Administration (FDA) (Press release). July 22, 2020. Retrieved May 28, 2021. https://www.fda.gov/vaccines-blood-biologics/consumers-biologics/consumer-alert-regenerative-medicine-products-including-stem-cells-and-exosomes.
162. Tabbara, I.A., *Allogeneic bone marrow transplantation: acute and late complications*. Anticancer Res, 1996. **16**(2): p. 1019–26.
163. Athanasopoulos, T., M.M. Munye, and R.J. Yáñez-Muñoz, *Nonintegrating gene therapy vectors*. Hematol Oncol Clin North Am, 2017. **31**(5): p. 753–70.
164. Yin, H., et al., *Non-viral vectors for gene-based therapy*. Nat Rev Genet, 2014. **15**(8): p. 541–55.
165. Benskey, M.J., et al., *Basic concepts in viral vector-mediated gene therapy*. Methods Mol Biol, 2019. **1937**: p. 3–26.
166. Lukashev, A.N. and A.A. Zamyatnin, Jr., *Viral vectors for gene therapy: current state and clinical perspectives*. Biochemistry (Mosc), 2016. **81**(7): p. 700–8.
167. Naso, M.F., et al., *Adeno-associated virus (AAV) as a vector for gene therapy*. BioDrugs, 2017. **31**(4): p. 317–34.
168. Stanimirovic, D.B., J.K. Sandhu, and W.J. Costain, *Emerging technologies for delivery of biotherapeutics and gene therapy across the blood-brain barrier*. BioDrugs, 2018. **32**(6): p. 547–59.
169. Tebas, P., et al., *Gene editing of CCR5 in autologous CD4 T cells of persons infected with HIV*. N Engl J Med, 2014. **370**(10): p. 901–10.
170. FDA approves novel gene therapy to treat patients with a rare form of inherited vision loss. U.S. Food & Drug Administration (FDA) (Press release). December 18, 2017. Retrieved May 28, 2021. https://www.fda.gov/news-events/press-announcements/fda-approves-novel-gene-therapy-treat-patients-rare-form-inherited-vision-loss.
171. Tsang, S.H. and T. Sharma, *Leber congenital amaurosis*. Adv Exp Med Biol, 2018. **1085**: p. 131–7.
172. FDA approves first-of-its kind targeted RNA-based therapy to treat a rare disease. U.S. Food & Drug Administration (FDA) (Press release). August 10, 2018. Retrieved May 28, 2021. https://www.fda.gov/news-events/press-announcements/fda-approves-first-its-kind-targeted-rna-based-therapy-treat-rare-disease.
173. Planté-Bordeneuve, V. and G. Said, *Familial amyloid polyneuropathy*. Lancet Neurol, 2011. **10**(12): p. 1086–97.

174. FDA approves innovative gene therapy to treat pediatric patients with spinal muscular atrophy, a rare disease and leading genetic cause of infant mortality. U.S. Food & Drug Administration (FDA) (Press release). May 24, 2019. Retrieved May 28, 2021. https://www.fda.gov/news-events/press-announcements/fda-approves-innovative-gene-therapy-treat-pediatric-patients-spinal-muscular-atrophy-rare-disease.
175. Kolb, S.J. and J.T. Kissel, *Spinal muscular atrophy.* Neurol Clin, 2015. **33**(4): p. 831–46.
176. de Villiers, E.M., *Cross-roads in the classification of papillomaviruses.* Virology, 2013. **445**(1–2): p. 2–10.
177. Bouvard, V., et al., *A review of human carcinogens–Part B: biological agents.* Lancet Oncol, 2009. **10**(4): p. 321–2.
178. Smith, J.S., et al., *Human papillomavirus type distribution in invasive cervical cancer and high-grade cervical lesions: a meta-analysis update.* Int J Cancer, 2007. **121**(3): p. 621–32.
179. Graham, S.V., *Human papillomavirus: gene expression, regulation and prospects for novel diagnostic methods and antiviral therapies.* Future Microbiol, 2010. **5**(10): p. 1493–506.
180. "FDA Licenses New Vaccine for Prevention of Cervical Cancer and Other Diseases in Females Caused by Human Papillomavirus". U.S. Food & Drug Administration (FDA) (Press release). June 8, 2006. Retrieved June 28, 2021. https://www.fda.gov/vaccines-blood-biologics/safety-availability-biologics/gardasil-vaccine-safety#:~:text=FDA%20approved%20Gardasil%20on%20June,HPV%20types%206%20and%2011.
181. FDA Approves Gardasil 9 for Prevention of Certain Cancers Caused by Five Additional Types of HPV. U.S. Food & Drug Administration (FDA) (Press release). December 10, 2014. Retrieved May 28, 2021. https://www.fda.gov/vaccines-blood-biologics/vaccines/gardasil-9.
182. Tripathi, N. and O.Y. Mousa, Hepatitis B, in StatPearls. 2021, StatPearls Publishing Copyright © 2021, StatPearls Publishing LLC.: Treasure Island (FL).
183. Karayiannis, P., *Hepatitis B virus: virology, molecular biology, life cycle and intrahepatic spread.* Hepatol Int, 2017. **11**(6): p. 500–8.
184. Edey, M., K. Barraclough, and D.W. Johnson, *Review article: hepatitis B and dialysis.* Nephrology (Carlton), 2010. **15**(2): p. 137–45.
185. Venters, C., W. Graham, and W. Cassidy, *Recombivax-HB: perspectives past, present and future.* Expert Rev Vaccines, 2004. **3**(2): p. 119–29.
186. Keating, G.M. and S. Noble, *Recombinant hepatitis B vaccine (Engerix-B): a review of its immunogenicity and protective efficacy against hepatitis B.* Drugs, 2003. **63**(10): p. 1021–51.
187. Splawn, L.M., et al., *Heplisav-B vaccination for the prevention of hepatitis B virus infection in adults in the United States.* Drugs Today (Barc), 2018. **54**(7): p. 399–405.
188. Thomas, S. and G.C. Prendergast, *Cancer vaccines: a brief overview.* Methods Mol Biol, 2016. **1403**: p. 755–61.
189. Kantoff, P.W., et al., *Sipuleucel-T immunotherapy for castration-resistant prostate cancer.* N Engl J Med, 2010. **363**(5): p. 411–22.
190. Larsen, E.S., et al., *Bacillus Calmette-Guérin immunotherapy for bladder cancer: a review of immunological aspects, clinical effects and BCG infections.* Apmis, 2020. **128**(2): p. 92–103.
191. Johnson, D.B., I. Puzanov, and M.C. Kelley, *Talimogene laherparepvec (T-VEC) for the treatment of advanced melanoma.* Immunotherapy, 2015. **7**(6): p. 611–9.
192. Xia, S., et al., *Safety and immunogenicity of an inactivated SARS-CoV-2 vaccine, BBIBP-CorV: a randomised, double-blind, placebo-controlled, phase 1/2 trial.* Lancet Infect Dis, 2021. **21**(1): p. 39–51.
193. *UAE says Sinopharm vaccine has 86% efficacy against COVID-19.* Reuters. December 8, 2020. Retrieved June 22, 2021. https://www.reuters.com/article/health-coronavirus-emirates/uae-says-sinopharm-vaccine-has-86-efficacy-against-covid-19-idUSKBN28J0G4.
194. *China's Sinopharm Claims COVID-19 Vaccine has 79% Efficacy.* BioSpace.com. December 30, 2020. Retrieved June 22, 2021.
195. *Sinovac's COVID-19 shot is 83% effective, not 91%, Turkey says.* asia.nikkei.com. March 4, 2021. Retrieved June 29, 2021.
196. de Vries, R.D. and G.F. Rimmelzwaan, *Viral vector-based influenza vaccines.* Hum Vaccin Immunother, 2016. **12**(11): p. 2881–901.
197. Mennechet, F.J.D., et al., *A review of 65 years of human adenovirus seroprevalence.* Expert Rev Vaccines, 2019. **18**(6): p. 597–613.
198. Knoll, M.D. and C. Wonodi, *Oxford-AstraZeneca COVID-19 vaccine efficacy.* Lancet, 2021. **397**(10269): p. 72–4.
199. *FDA authorizes Johnson & Johnson COVID-19 vaccine.* Med Lett Drugs Ther, 2021. **63**(1620): p. 41–2.
200. Logunov, D.Y., et al., *Safety and efficacy of an rAd26 and rAd5 vector-based heterologous prime-boost COVID-19 vaccine: an interim analysis of a randomised controlled phase 3 trial in Russia.* Lancet, 2021. **397**(10275): p. 671–81.

201. *CanSinoBIO's COVID-19 vaccine 65.7% effective in global trials, Pakistan official says.* Reuters. February 8, 2021. Retrieved June 29, 2021. https://www.reuters.com/world/china/cansinobios-covid-19-vaccine-657-effective-global-trials-pakistan-official-says-2021-02-08/.
202. *Novavax COVID-19 Vaccine Demonstrates 90% Overall Efficacy and 100% Protection Against Moderate and Severe Disease in PREVENT-19 Phase 3 Trial.* ir.novavax.com. (Press release). June 14, 2021. Retrieved June 29, 2021.
203. Wadhwa, A., et al., *Opportunities and challenges in the delivery of mRNA-based vaccines.* Pharmaceutics, 2020. **12**(2): p. 102.
204. Karikó, K., et al., *Suppression of RNA recognition by Toll-like receptors: the impact of nucleoside modification and the evolutionary origin of RNA.* Immunity, 2005. **23**(2): p. 165–75.
205. Baden, L.R., et al., *Efficacy and safety of the mRNA-1273 SARS-CoV-2 vaccine.* N Engl J Med, 2021. **384**(5): p. 403–16.
206. "Science Brief: Emerging SARS-CoV-2 Variants". www.cdc.gov. January 28, 2021. Retrieved June 25, 2021.
207. Korber, B., et al., *Tracking changes in SARS-CoV-2 spike: evidence that D614G increases infectivity of the COVID-19 virus.* Cell, 2020. **182**(4): p. 812–27.e19.
208. Galloway, S.E., et al., *Emergence of SARS-CoV-2 B.1.1.7 lineage – United States, December 29, 2020–January 12, 2021.* MMWR Morb Mortal Wkly Rep, 2021. **70**(3): p. 95–9.
209. Abu-Raddad, L.J., H. Chemaitelly, and A.A. Butt, *Effectiveness of the BNT162b2 Covid-19 vaccine against the B.1.1.7 and B.1.351 variants.* N Engl J Med, 2021. **385**(2): p. 187–9.
210. Pokhrel, P., C. Hu, and H. Mao, *Detecting the Coronavirus (COVID-19).* ACS Sens, 2020. **5**(8): p. 2283–96.
211. WHO. Summary of available protocols to detect SARS-CoV-2. 2020. https://www.who.int/docs/default-source/coronaviruse/whoinhouseassays.pdf.
212. Diagnostic testing for SARS-CoV-2. WHO Interim guidance. 2020. https://www.who.int/publications/i/item/diagnostic-testing-for-sars-cov-2.
213. Kashir, J. and A. Yaqinuddin, *Loop mediated isothermal amplification (LAMP) assays as a rapid diagnostic for COVID-19.* Med Hypotheses, 2020. **141**: p. 109786.
214. Nicol, T., et al., *Assessment of SARS-CoV-2 serological tests for the diagnosis of COVID-19 through the evaluation of three immunoassays: two automated immunoassays (Euroimmun and Abbott) and one rapid lateral flow immunoassay (NG Biotech).* J Clin Virol, 2020. **129**: p. 104511.
215. Mojsoska, B., et al., *Rapid SARS-CoV-2 detection using electrochemical immunosensor.* Sensors (Basel), 2021. **21**(2): p. 390.
216. Ranjan, P., et al., *Rapid diagnosis of SARS-CoV-2 using potential point-of-care electrochemical immunosensor: Toward the future prospects.* Int Rev Immunol, 2021. **40**(1–2): p. 126–42.
217. Rashed, M.Z., et al., *Rapid detection of SARS-CoV-2 antibodies using electrochemical impedance-based detector.* Biosens Bioelectron, 2021. **171**: p. 112709.
218. Strickler, J.H., B.A. Hanks, and M. Khasraw, *Tumor mutational burden as a predictor of immunotherapy response: is more always better?* Clin Cancer Res, 2021. **27**(5): p. 1236–41.
219. Overview of IVD Regulation. U.S. Food & Drug Administration (FDA). October 18, 2021. https://www.fda.gov/medical-devices/ivd-regulatory-assistance/overview-ivd-regulation.
220. *Commercialization of unapproved in vitro diagnostic devices labeled for research and investigation (draft compliance policy guide).* U.S. Food & Drug Administration (FDA). Center for Devices and Radiological Health. Rockville, MD. August 3, 1992.
221. Gajria, D. and S. Chandarlapaty, *HER2-amplified breast cancer: mechanisms of trastuzumab resistance and novel targeted therapies.* Expert Rev Anticancer Ther, 2011. **11**(2): p. 263–75.
222. García-García, E., et al., *Hybridization for human epidermal growth factor receptor 2 testing in gastric carcinoma: a comparison of fluorescence in-situ hybridization with a novel fully automated dual-colour silver in-situ hybridization method.* Histopathology, 2011. **59**(1): p. 8–17.
223. Hofmann, M., et al., *Assessment of a HER2 scoring system for gastric cancer: results from a validation study.* Histopathology, 2008. **52**(7): p. 797–805.
224. Meric-Bernstam, F., et al., *Advances in HER2-targeted therapy: novel agents and opportunities beyond breast and gastric cancer.* Clin Cancer Res, 2019. **25**(7): p. 2033–41.
225. Fitzgibbons, P.L., et al., *Prognostic factors in breast cancer. College of American Pathologists Consensus Statement 1999.* Arch Pathol Lab Med, 2000. **124**(7): p. 966–78.
226. Wolff, A.C., et al., *American Society of Clinical Oncology/College of American Pathologists guideline recommendations for human epidermal growth factor receptor 2 testing in breast cancer.* J Clin Oncol, 2007. **25**(1): p. 118–45.
227. Wolff, A.C., et al., *Recommendations for human epidermal growth factor receptor 2 testing in breast cancer: American Society of Clinical Oncology/College of American Pathologists clinical practice guideline update.* J Clin Oncol, 2013. **31**(31): p. 3997–4013.

228. Wolff, A.C., et al., *Human epidermal growth factor receptor 2 testing in breast cancer: American Society of Clinical Oncology/College of American Pathologists clinical practice guideline focused update.* J Clin Oncol, 2018. **36**(20): p. 2105–22.
229. Peters, S. and S. Zimmermann, *Targeted therapy in NSCLC driven by HER2 insertions.* Transl Lung Cancer Res, 2014. **3**(2): p. 84–8.
230. Gow, C.H., et al., *Comparable clinical outcomes in patients with HER2-mutant and EGFR-mutant lung adenocarcinomas.* Genes Chromosom Cancer, 2017. **56**(5): p. 373–81.
231. Cousin, S., et al., *Targeting ERBB2 mutations in solid tumors: biological and clinical implications.* J Hematol Oncol, 2018. **11**(1): p. 86.
232. FDA grants accelerated approval to pembrolizumab for first tissue/site agnostic indication. U.S. Food & Drug Administration (FDA). May 23, 2017. Retrieved June 29, 2021. https://www.fda.gov/drugs/resources-information-approved-drugs/fda-grants-accelerated-approval-pembrolizumab-first-tissuesite-agnostic-indication.
233. Ng, S.B., et al., *Targeted capture and massively parallel sequencing of 12 human exomes.* Nature, 2009. **461**(7261): p. 272–6.
234. Griffith, M., et al., *Optimizing cancer genome sequencing and analysis.* Cell Syst, 2015. **1**(3): p. 210–23.
235. Chan, T.A., et al., *Development of tumor mutation burden as an immunotherapy biomarker: utility for the oncology clinic.* Ann Oncol, 2019. **30**(1): p. 44–56.
236. FDA approves pembrolizumab for adults and children with TMB-H solid tumors. U.S. Food & Drug Administration (FDA) (Press release). June 16, 2020. Retrieved May 28, 2021. https://www.fda.gov/drugs/drug-approvals-and-databases/fda-approves-pembrolizumab-adults-and-children-tmb-h-solid-tumors.
237. FDA approves pembrolizumab for first-line treatment of MSI-H/dMMR colorectal cancer. U.S. Food & Drug Administration (FDA) (Press release). June 29, 2020. Retrieved May 28, 2021. https://www.fda.gov/drugs/resources-information-approved-drugs/fda-grants-accelerated-approval-pembrolizumab-first-tissuesite-agnostic-indication.
238. Baretti, M. and D.T. Le, *DNA mismatch repair in cancer.* Pharmacol Ther, 2018. **189**: p. 45–62.
239. Jiricny, J., *Postreplicative mismatch repair.* Cold Spring Harb Perspect Biol, 2013. **5**(4): p. a012633.
240. Luchini, C., et al., *ESMO recommendations on microsatellite instability testing for immunotherapy in cancer, and its relationship with PD-1/PD-L1 expression and tumour mutational burden: a systematic review-based approach.* Ann Oncol, 2019. **30**(8): p. 1232–43.

Application of Modeling and Simulation in the Development of Biologics and Biosimilars

26

Sylvia Nam-Phuong Dinh, Xinyu Pei, and Yihui Shi

Corresponding author: Yihui Shi

Contents

DOI: 10.1201/9780429485626-26

26.1 INTRODUCTION

Biologics are macromolecules synthesized in living systems that copy or improve a naturally occurring product, including a wide range of products, such as vaccines, blood and blood components, allergens, cells, genes, tissues, and recombinant proteins (Ventola 2013; FDA 2018a). Since its introduction, biologics have played a significant role in the treatment of several diseases, including cystic fibrosis, autoimmune disorders, diabetes, cancer, and anemia (Ratih et al. 2021). While the three main classes of biologics are endogenous molecules, monoclonal antibodies (mAbs), and protein mimicking receptors, biologics have expanded to include nanobodies, immunoconjugates, vaccines, and fusion protein (Blandizzi, Meroni, and Lapadula 2017; Kabir, Moreino, and Sharif Siam 2019). Despite its effectiveness, because biologics are expensive and have expiring patents, there has been a rise in investment and development of biosimilars, with an expected 25% rise in compound annual growth rate (CAGR) between 2020 and 2026 (Ratih et al. 2021).

The US Food and Drug Administration (FDA) defines the biosimilars as: "The biological product is highly similar to the reference product notwithstanding minor differences in clinically inactive components, there are no clinically meaningful differences between the biological product and the reference product in terms of the safety, purity, and potency of the product" (FDA 2015). Therefore, according to the FDA, for approval in the United States, general data required includes analytical studies proving similar structures, preclinical studies assessing toxicity, and clinical studies comparing pharmacokinetic/pharmacodynamic (PK/PD) profiles and efficacy (FDA 2017).

The average cost of research and development for biologics is about $391 million, which is $82 million more expensive than the cost for small molecule drugs due to structural complexity and complicated manufacturing process (Lexchin 2020). Biosimilar development has attracted extraordinary attention in the last decade as a result of patent expiration and reduced cost (Reeves et al. 2019; Huang et al. 2020). Despite the cost and time invested into pharmaceutical research and development, only approximately 10% of potential drugs ultimately reach the market (Kim, Shin, and Shin 2018). With 79% of failed potential drugs attributable to issues regarding safety and efficacy, modeling and simulation (M&S) was recognized by the FDA to be a powerful tool to decrease cost and time by improving efficiency and decision making throughout the biologics and biosimilar development process (Wang et al. 2019).

M&S uses mathematical and statistical models that extract and integrate preclinical and clinical data, as well as published industry data to correlate between drug exposure, drug response, and patient outcomes (Gieschke and Steimer 2000). M&S can improve the drug development decision process through all stages of drug research and discovery, from early discovery to post-marketing monitoring studies.

During the preclinical stage, M&S can guide the selection of first-in-human (FIH) dose and dosing regimen, clinical trial design during phase I study, prediction of safety and efficacy in unstudied population in a clinical trial, offering an early view of drug-drug interactions (DDIs). In early clinical stages, M&S can provide guidance in alternate and finalize dosages, optimize drug formulations, predict DDIs and alert other safety concerns, trial design of phase III study, optimization of dose and dosing regimen in special populations, such as pediatric, hepatic- or renal-impaired populations. In the later stage of clinical trials, M&S can guide scaling clinical endpoints from biomarkers, dose adjustment based on patients' age, gender, ethnicity, concomitant diseases, and medications (Milligan et al. 2013; Parekh et al. 2015; Lim 2019; Yellepeddi et al. 2019).

26.2 M&S METHODS USED IN BIOLOGICS AND BIOSIMILARS

The general M&S methods used in small molecules can also be applied in biologics and biosimilars, which include two basic M&S methods: Pharmacokinetic (PK) modeling and PK/PD modeling. As for the PK and PD profiles of biological products, they are substantially different from small molecules.

Therefore, the M&S for biological products have to be performed with some modifications but still use the same concept and output the same information we needed.

26.2.1 Different Performance in PK and PD

Before discussing M&S, it is important to mention the different performances between small molecules and biological products (Table 26.1). Biologics and biosimilars have very unique characteristics in PK profiles and PD properties. The main reason is that the molecular size of small molecules is usually less than 1 kDa, whereas biologics typically range from 1 kDa (e.g., small peptides) to 1000 kDa (e.g., mAbs). Because of their large molecular size, biologics are not well-absorbed by oral administration. Therefore, they are mainly delivered by parenteral administration routes, such as intravenous (IV), subcutaneous (SC), and intramuscular (IM) injections (Zhao, Ren, and Wang 2012). Biologics are absorbed slowly and need to take a longer time to reach peak concentrations (T_{max}). This prolonged T_{max} is associated with slow vascular and lymphatic absorption (Roskos, Ren, and Robbie 2011; Ryman and Meibohm 2017). The bioavailability of biologics after SC administration ranges from 50% to 80%, which is much higher than that of small molecules (Zhao, Ren, and Wang 2012; Anselmo, Gokarn, and Mitragotri 2019). Compared to small molecules, biologics have limited distribution to the central compartment; its volume of distribution (Vd) is similar to the total blood volume (Tang et al. 2004). Biologics are restricted in their initial distribution because of their size, charge, and tight target binding (Tang et al. 2004). Many biologics exhibit target-mediated drug disposition (TMDD) and eliminate through a limited number of target receptors, which causes nonlinear PK (Dua, Hawkins, and van der Graaf 2015; Glassman, Abuqayyas, and Balthasar 2015; Ryman and Meibohm 2017). Other non-target-specific elimination occurs via proteolytic degradation (Tabrizi, Tseng, and Roskos 2006; Kamath 2016). Renal excretion happens when the molecular weight of biologics is smaller than 69 kDa (Zhao, Ren, and Wang 2012). In terms of PD, off-target toxicity effects occur in small molecules because they are not target-mediated binding. Whereas biologics are more specific with their binding, there are limited off-target side effects (Agoram, Martin, and van der Graaf 2007). But the toxicity is related to exaggerated pharmacology effects or immunogenicity (Marshall 2006; Zhao, Ren, and Wang 2012).

TABLE 26.1 Differences between biological products and small molecules

CHARACTERISTICS	*BIOLOGICAL PRODUCTS*	*SMALL MOLECULES*
Molecular weight	1–1000 kDa	Usually less than 1 kDa
Preferred route of administration	Parenteral (e.g., IV, SC, IM)	Oral
Half-life ($t_{1/2}$)	Long (few days to few weeks)	Short (typically few hours to 24 hours)
Absorption	Via lymphatic and vasculature system, high bioavailability	Rapid absorption after oral administration variable bioavailability
Distribution	Limited distribution to central compartment, Vd similar to plasma volume	Normally not restricted to central blood compartment; larger Vd, but widely variable
Metabolism and elimination	Target specific and proteolysis elimination, mostly nonlinear PK	Hepatic cytochrome P450 enzyme metabolism or thru renal filtration
PD	target-mediated binding, specific in action, limited off-target side effects, toxicity caused by exaggerated pharmacology and immunogenicity	non-target-mediated binding, off-target toxicity effects

26.2.2 Pharmacokinetics (PK) Modeling

PK describes changes of drug concentration in body fluids over a period of time which includes the process of absorption (A), distribution (D), metabolism (M), excretion (E), and transport (T). PK modeling is performed by non-compartmental or compartmental methods. The former usually use area under the curve (AUC) to estimate drug exposure (Bulitta and Holford 2014). Whereas the latter uses kinetic models to evaluate the concentration-time curve. In addition to the non-compartmental or compartmental models, PK modeling can be further divided into different sub-modeling types, such as physiologically based PK (PBPK) modeling, quantitative systems pharmacology (QSP) models, and population PK (popPK) modeling, each of which has different focuses and its own advantages.

26.2.3 PBPK Modeling

PBPK modeling consists of multiple compartments, each of which corresponds to different tissues or organs, such as lung, liver, kidney, brain, heart, bone, skin, fat tissues, and muscle tissues. (Figure 26.1). Each compartment is described by specific tissue volume and connected by the circulating blood system. Different from traditional PK models, the PBPK modeling is a mechanism-based approach to predict PK of drugs based on species-specific physiological and anatomical parameters, physical and chemical characteristics of the given drug, as well as the ADME properties that are incorporated in each compartment (Zhuang and Lu 2016). Therefore, PBPK modeling can extrapolate PK across different species, for example, from lab animals to humans, which can also provide a more accurate prediction of PK than empirical approaches (Jones et al. 2006; Lee et al. 2020). PBPK modeling also conducts extrapolation of different physiological conditions, such as elderly, pediatrics, pregnancy, and other diseased populations (Hornik et al. 2017; Marsousi et al. 2017; Dallmann et al. 2018; Mendes and Chetty 2019; Cui et al. 2020). Because biologics are predominantly absorbed via the lymphatic system, lymph flow is used to interconnect each compartment in the PBPK model of biologics, which also differentiates the PBPK models for biologics versus small molecules (Roskos, Ren, and Robbie 2011; Wong and Chow 2017).

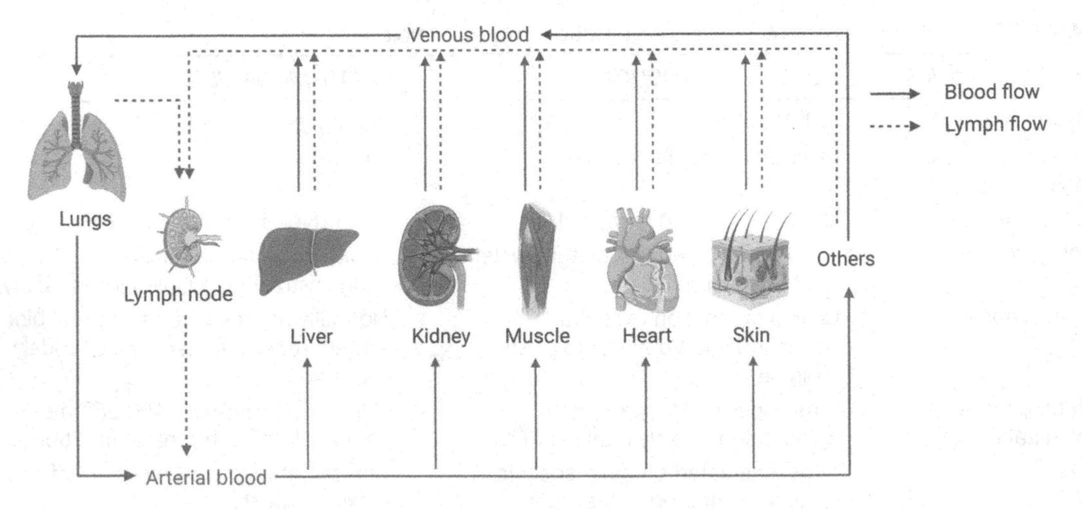

FIGURE 26.1 Schematic of a PBPK model. (Created with permission from BioRender.com.)

26.2.4 QSP Modeling

QSP is an extension of PBPK modeling. PBPK modeling is more focused on predicting PK outcomes, whereas QSP modeling is more focused on predicting PD outcomes and clinical efficacy outcomes (Rostami-Hodjegan 2018). QSP modeling is used to understand the mechanisms of action of novel drugs, identify biomarkers, and predict toxicity (Snelder et al. 2017; Woodhead et al. 2017; Clausznitzer et al. 2018; Schuck et al. 2018). It has been predominantly used in the development of biologics as well. It deepens understanding of the mechanisms of action and how it connects to therapeutic outcomes and regimen optimization in biologics (Hardiansyah and Ng 2019; Milberg et al. 2019).

26.2.5 PopPK Modeling

PopPK uses mathematical and statistical methods to quantify and describe the variability in drug concentrations among individuals (Sheiner, Rosenberg, and Marathe 1977). Following the same dosing regimen, drug concentrations vary significantly among each individual. This variability may be attributed to intrinsic patient factors (e.g., weight, age, gender, ethnicity, genetic phenotype, and the presence of liver or renal impairment) and extrinsic factors (e.g., concomitant medicine and food consumption), all of which can be assessed by popPK analysis to identify factors affecting drug exposure (Ette and Williams 2004; FDA 2019). Additionally, in a popPK model, PK is characterized in target population using sparse samples, which makes it suitable for analyzing sparse data from special populations, such as pediatrics and the elderly (Samara and Granneman 1997). PopPK modeling plays an important role during the regulatory process and supports the FDA in making new drug approval and labeling decisions for both new drug application (NDA) and biological license application (BLA) submissions (Lee et al. 2011a). And in biologics labels, the impact of intrinsic factors is provided by popPK analysis (Ogasawara and Alexander 2019).

26.2.6 PK/PD Modeling

PD describes the relationship between drug concentrations at the site of action and the effects. It emphasizes the dose-response curve. To be simple, "Pharmacokinetics is what the body does to the drug; pharmacodynamics is what the drug does to the body" (Holford and Sheiner 1982). Under PK steady-state conditions, dose-response relationships can be described by five simple models: fixed-effect model, linear model, log-linear model, E_{max} model, and sigmoid E_{max} model (Meibohm and Derendorf 1997).

Under non-steady-state conditions, PK and PD are tightly interconnected with each other. Firstly, changes in drug concentrations in body fluids will demonstrate a PK process over time. Secondly, the time course of drug concentration and effect might dissociate as a result of a time-consuming process that is involved in mediating the observed effect after interaction of the drug with its physiological response structure (Meibohm and Derendorf 1997). This integrates the PK modeling and PD modeling to better describe the time course of the effects resulting from a given dosage (Figure 26.2).

Under the M&S study of biologics and biosimilars, it is even more necessary to integrate the analyses of PK and PD. In most small-molecule PK/PD analyses, it is assumed that the amount of the drug that interacts with a target/receptor is negligible compared to its overall dose. In consequence, it is considered that the PK profile would not be altered by such interactions, and sequentially, the PD profile can be analyzed (Wang et al. 2010). However, this assumption becomes invalid when a TMDD is saturated as the drug concentration excesses the target amount. Under this circumstance, the pharmacological action would lead to a change in the number of targets/receptors in a time-dependent manner (Krzyzanski et al. 2016). In return, it results in a time-dependent PK (Wang et al. 2010; Wang et al. 2019). This PK and PD interplay is known as PD-mediated drug disposition (PDMDD).

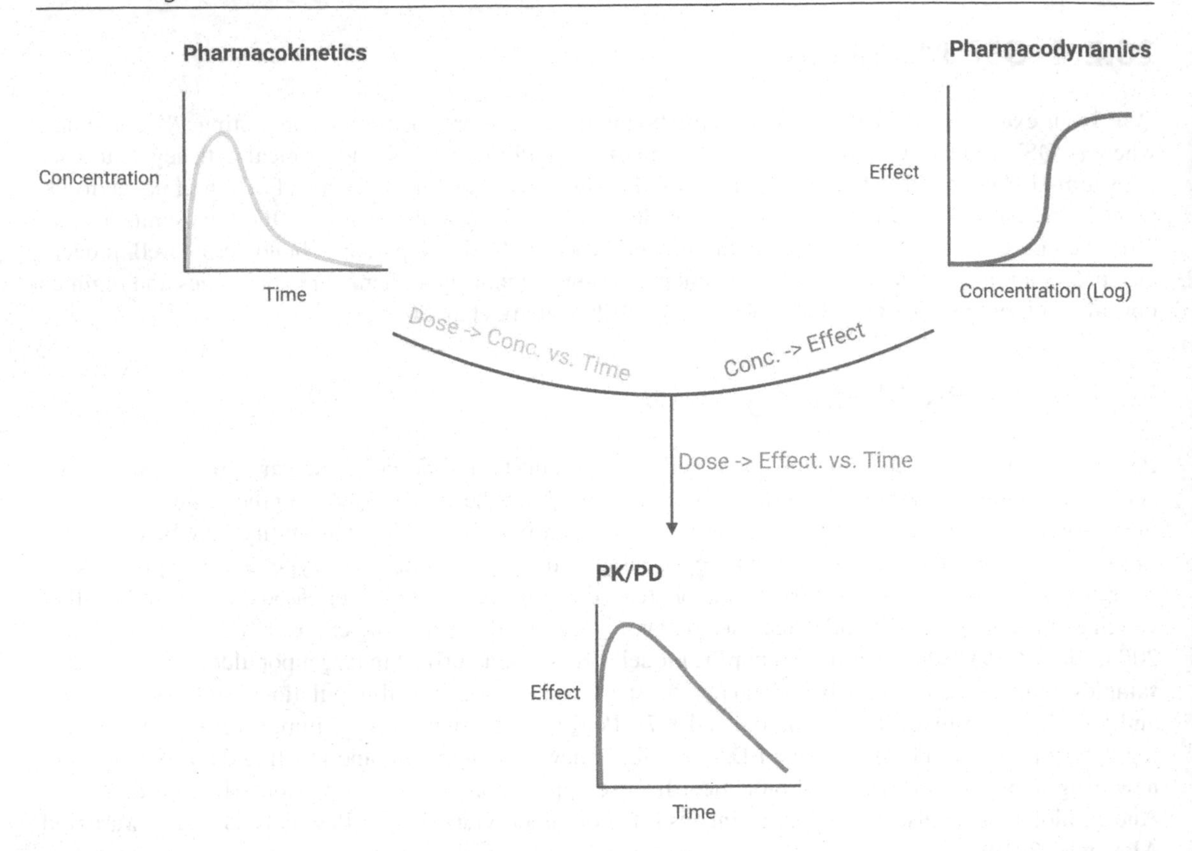

FIGURE 26.2 Pharmacokinetic/pharmacodynamic (PK/PD) modeling builds the bridge between two main branches of pharmacology. Regenerated from (Meibohm and Derendorf 1997). (Created with permission from BioRender.com.)

26.2.7 Use of M&S in Biosimilars

According to the definition of biosimilars described in the introduction, comparable PK, PD as well as the immunogenicity with the reference biologics product in a relevant population need to be demonstrated, followed by a phase III study to confirm similar efficacy and safety in a special population (Berghout 2011; FDA 2016; Cazap et al. 2018). Those similarity studies are where M&S, such as PK modeling, PK/PD modeling, and popPK modeling, are applied. Particularly, popPK modeling is now highly involved in 351(a) BLA submissions (Wang et al. 2019). M&S will help identify dose for PK similarity and PK/PD similarity, optimize study design, and select PD biomarkers (Wang et al. 2019).

26.2.8 Comparison of Different M&S Software

There are many commercially available software and open source tools that have been developed to facilitate PK and PD M&S and have been widely used during the drug development process, such as MATLAB, SAAM II, WinNonLin, NONMEM, Simulation Plus, GastroPlus, Phoenix™ platforms, Simcyp™, SAS, R, and more (Table 26.2). It is worth noticing that due to the difference in PK and PD between small molecules and biologics described previously, biologics and biosimilars should use a special implement tool for their M&S analyses.

TABLE 26.2 Different software developed for PK and PD modeling

SOFTWARE	*DESCRIPTION*	*BIOLOGICS*
ADAPT	A software package developed for PK/PD modeling and individual data analysis (simulation, parameter estimation, and sample schedule design) and population analysis.	Has been used in biologics.
GastroPlus	A physiologically based software package which simulates and models PK and PD with 10 modules that can predict DDI, ADMET (absorption, distribution, metabolism, excretion and toxicity), nonlinear PK for CYP metabolism, different routes of absorption, PK/PD of biologics, and more.	Has a PBPK module designed for biologics.
SimBiology	A software tool that provides programmatic tools for M&S of QSP, PBPK, PK/PD applications using the MATLAB® language.	Has a biologics PBPK model builder that provides a two-pore PBPK to describe the tissue distribution and elimination of biologics.
SAAM II	A modeling, simulation, and analysis software tool for PK studies.	Has not been seen for applications in biologics so far.
Phoenix™ WinNonlin	A software tool can be used in non-clinical to clinical studies, and the non-compartmental analysis and PK/PD modeling can be used to analyze PK, PD, DDI, FIH dosing, toxicokinetic (TK), bioequivalence (BE), and more.	The same as above.
Phoenix™ NLME	A population M&S software that uses non-linear mixed-effect modeling.	The same as above.
NONMEM	A non-linear mixed-effect modeling software for popPK modeling.	The same as above.
Phoenix® IVIVC Toolkit™	A predictive mathematical tool that describes the relationship between the *in vitro* drug dosage form and an *in vivo* PK response for *in vivo* human BE studies.	The same as above.
Simcyp™ PBPK Simulator	A PBPK-based M&S software tool that extrapolates FIH dosing, optimize clinical study design, evaluates new drug formulations, and predicts DDI.	Has a specialized module for biologics – Simcyp Biologics Simulator.
DILIsym®	A QSP-based modeling system designed for drug-induced liver injury, non-alcoholic fatty liver disease, idiopathic pulmonary fibrosis, and acute kidney injury.	Has not been seen for applications in biologics so far.

DDI, drug-drug interaction.

26.3 APPLICATIONS OF M&S

26.3.1 Drug Discovery and Development

M&S at the early stage of drug discovery and development can be an effective tool employed to identify new drug targets with confidence, characterize drug targets quantitatively, and provide a better understanding of drug efficacy (Agoram, Martin, and van der Graaf 2007; Kim, Shin, and Shin 2018). Models can be comprehensive, empirical, or mechanistic; the type of model chosen is dependent on the availability and quality of information from the literature as well as the question of interest. (Visser et al. 2014; Lin, Li, and Lin 2020). While comprehensive systems pharmacology models can provide insight into biological pathways and translatable pharmacological interventions, empirical and mechanistic models can predict ADME properties and guide compound optimization (Visser et al. 2014; Lucas et al. 2019).

QSP integrates modeling of biological and pharmacologic systems to assess the biological and toxicological perturbation caused by a hypothetical compound (Benson et al. 2013; Bradshaw et al. 2019). This predictive approach has been applied in academia and industry at the early stages of drug discovery and development to identify and prioritize druggable targets, define their mechanisms of action, and predict their toxic and therapeutic effects via simulations (Topp, Trujillo, and Sinha 2019). In 2013, Benson et al. constructed a systems pharmacology model of the nerve growth factor (NGF) pathway and conducted a sensitivity analysis, identifying NGF as the most sensitive druggable target and tropomyosin receptor kinase A (TrkA) as another possible druggable target to control diphosphorylated extracellular signal-regulated kinase (dppERK). dppERK is used as a biomarker of pain in this study because its accumulation in the neuronal nucleus, in response to NGF:TrkA complex, activates downstream pathways related to neuronal survival and pain. After the druggable targets were identified, they introduced a hypothetical anti-NGF mAb and TrkA inhibitor in the systems model to simulate its effects, analyze its pharmacologic characteristics, and predict drug dosages required for complete inhibition of dppERK response. The model predicted that for total inhibition of dppERK, a plasma concentration of approximately 1000-fold Kd was required. The model also predicted that >99% of TrkA must be inhibited for equivalent efficacy as anti-NGF mAb. Both of these have since been supported by data from clinical trials on Tanezumab, an NGF-binding mAB and PF-06273340, a pan-TRK inhibitor (Lane et al. 2010; Loudon et al. 2018; Benson 2019).

PBPK M&S has also played an integral role at the early stages of drug discovery and development because of its ability to provide early access to preclinical, *in vitro* ADME data, allowing lead optimization and reducing the need for animal studies (Zhuang and Lu 2016). In a study by Polli et al., PBPK M&S was used to explore the PK/PD properties of 10H6, an anti-carcinoembryonic antigen (CEA) "catch-and-release" (CAR) mAb (Polli, Engler, and Balthasar 2019). Because CAR mAbs have pH-dependent binding affinity and become available for neonatal Fc receptor (FcRn) recycling when released within the acidified endosome, the PBPK model was developed to consider competitive mAb binding with FcRn and its target. Simulations were first used to predict PK data for 10H6, which were found to be comparable to those observed in 10H6-treated colorectal cancer mouse models. Simulations and sensitivity analysis were then used to determine key parameters influencing the disposition of CAR mAb. These factors include the high rate of target dissociation, the high expression of tumor antigen, and the high rate of antigen-mAb complex internalization. Simulations with the model predicted that for minimal FcRn recycling of mAb, the rate of target dissociation at pH 6.0 must be greater than 0.32 min^{-1} and have a dissociation half-life less than the endosomal transit time. Lastly, simulations with the model predicted that CAR mAb had higher receptor occupancy and tumor selectivity than non-CAR mAb.

26.3.2 Preclinical Development

According to the FDA, the findings from the preclinical development stage should provide information regarding the dosing and toxicity of the proposed drug (FDA 2018b). M&S applied at this stage can be used to determine the relationship between dose and response, identify markers for efficacy or toxicity, and form a better understanding of the PK and PD characteristics (Chien et al. 2005; Kim, Shin, and Shin 2018). Allometric scaling, PK/PD, and PBPK modeling are commonly used to study these effects (Kim, Shin, and Shin 2018). Allometric scaling is an empirical model that uses differences in body surface area to describe PK changes that are often used for interspecies conversions (Nair and Jacob 2016; Choi, Lee, and Cho 2019). And PBPK is a mechanistic model that incorporates both physiological and drug-specific parameters to best represent a drug's ADME (Kuepfer et al. 2016).

In a study on MCLA-128 (Zenocutuzumab), an IgG1 bispecific mAb targeting HER2 and HER3 receptor tyrosine kinases, de Vries Schultink et al. utilized allometric scaling and translational PK/PD modeling to predict clinical efficacious exposure and dose from preclinical data on cynomolgus monkey and a mouse xenograft tumor growth model (de Vries Schultink et al. 2018). Using the cynomolgus monkey, PK data allometrically scaled to a 70 kg human, and it was predicted that a flat dose of MCLA-128, ranging from 10 to 480 mg, every three weeks, had a safety margin of greater than 10 folds. And flat doses ≥360 mg every three weeks would be effective, with greater than 99% receptor occupancy. The proposed effective human dose was then evaluated for anti-tumor activity. Because the non-perturbed tumor growth model was evaluated in untreated mice and there was no PK data in mice, the PK data on cynomolgus monkeys had to be allometrically scaled to a 0.02 kg mouse to study the anti-tumor effect. The allometrically scaled PK modeling suggested that MCLA-128 has profound anti-tumor activity in mice, with doses ≥2.5 mg/kg showing decreased tumor growth rate.

PK/PD modeling can also be used to establish a way to measure the efficacy of an investigational drug by assessing and predicting the change in biomarker concentration in response to a drug (Colburn and Lee 2003; Netterberg et al. 2019). For therapeutic mAbs, while it has been suggested that the concentration of free ligand can be a marker of efficacy and help make informed decisions about dose selection in clinical trials, it is often difficult to measure free ligand levels in the presence of high mAb-ligand complexes; PK/PD modeling is an approach that can be used to mitigate this challenge (Lee et al. 2011b; Kielbasa and Helton 2019). In a study by Wang et al., a PK/TE/PD modeling, a type of PK/PD modeling that integrates downstream effects of target engagement (TE), was used to understand the dose-response relationship of Siltuximab, an anti-interleukin-6 (IL-6) mAb, and to predict free IL-6 levels at high dosages (Wang et al. 2014). A low dose PK/TE modeling was first created from measured concentrations of Siltuximab, free IL-6, and Siltuximab/IL-6 complex in cynomolgus monkeys treated with a very low dose of Siltuximab to estimate the *in vivo* equilibrium dissociation constant (K_D). The derived K_D helped establish a high dose PK/TE modeling to predict free IL-6 levels at high, clinically relevant doses of Siltuximab. To establish the *in vivo* relevance of free IL-6 and to verify the ability of PK/TE modeling to accurately predict free IL-6 levels, a TE/PD modeling, describing the relationship of free IL-6 and acute C-reactive protein (CRP), a downstream biomarker of IL-6 activity, predicted CRP levels that were later confirmed in high-Siltuximab-dose-treated cynomolgus monkeys (Wang et al. 2014).

26.3.3 Clinical Development

At the early stages of clinical development, one of the main roles of M&S is to confirm predictions from the preclinical stage, such as those regarding PK/PD properties, and to guide optimal dosing regimen for clinical trials (Chien et al. 2005; Bonifacio et al. 2020). This is commonly done using PK, PK/PD, PBPK, and popPK modelings. Findings from these modelings and simulations can then be used to design efficient and informative subsequent clinical trials and help guide recommendations on dosing and route of administration (Wang et al. 2019). For example, Cai et al. used a TMDD model, a type of PK/PD model describing the nonlinear clearance commonly seen in the use of mAb, to predict the PK, PD, and TE responses to different dosage regimens of BMS-986184, an anti-interferon-γ-induced protein 10 (IP-10) mAb and to guide therapeutic dosing in healthy individuals (Cai et al. 2020). The developed TMDD model was confirmed using data on the total drug, serum unbound IP-10, and serum total IP-10 concentrations, collected from a clinical study on BMS-986184 in 54 healthy participants (NCT02864264). Simulations were then performed on virtual healthy participants to predict the PK and TE responses for six SC, every other week, dose panels for 25 dosing intervals. From M&S, the study made two conclusions to improve dose selection: approximately eight dosing intervals were required to reach a steady-state and a dosing regimen of 150, subcutaneously administered every other week, would be optimal.

At later stages of clinical development, M&S can play an essential role in characterizing DDIs, drug-disease interactions, and tolerance (Ghahramani 2016). In an example by Parrott et al., a PBPK study was used to study the impact of food and gastric pH on the PK of Entrectinib, a tyrosine kinase inhibitor (TKI) (Parrott et al. 2020). A PBPK modeling was built from preclinical and clinical data of patients treated with Entrectinib. Via simulations, the effects of food and the proton pump inhibitor (PPI) on the gastric pH, transit time, and intestinal concentration were predicted. From M&S, it was concluded that when Entrectinib is co-administered with a PPI, there would be an incomplete dissolution of the dose due to gastric pH elevation.

A model-based approach can also be used throughout clinical development to prove that a biosimilar has no clinically meaningful difference from its reference biologic (Ventola 2013; Brekkan et al. 2019). In a study by Candelaria et al., a popPK model was used to compare the pharmacologic properties of RTXM83, a proposed Rituximab biosimilar, to those of its reference product (Candelaria et al. 2018). A popPK modeling was built from PK/PD data from a previous phase III clinical study, where patients with diffuse large B-cell lymphoma (DLBCL) were administered the same dose of IV RTXM83-CHOP or Rituximab-CHOP regimen every three weeks for six cycles. M&S demonstrated PK and PD similarity of RTXM83 to its reference product. Statistical analysis showed comparable PK/PD properties, including B-cell depletion/recovery, AUC, and predicted maximum concentration (C_{max}) for RTXM83 and Rituximab.

26.3.4 Post-Market Monitoring

While M&S is not as commonly used for post-market monitoring as in other parts of the drug development, M&S can still be used to study extended-release formulation and to refine dosing, especially in specific populations where clinical trials are hard to perform (Ghahramani 2016). In a study by Combes et al., PBPK, popPD modeling, and linear mixed regression models were used to predict the short- and long-term efficacy of Everolimus, an mTOR inhibitor, in children aged between six months and two years old, as it had only been approved for the treatment of tuberous sclerosis (TSC)-associated partial-onset seizure (POS) in patients ≥2 years old (Combes et al. 2020). Because recruitment of the targeted population was impossible for a clinical trial, a PBPK modeling incorporating demography, physiology, and CYP3A4-catalyzed metabolism, was used to predict plasma though concentration (C_{min}) after administration of an Everolimus dose of 6 mg/m^2 and a decrease in seizure frequency. Using the generated data, popPD and a linear mixed-effect model were used to determine the efficacy of Everolimus in adults and children. Model-based simulations later concluded that Everolimus is effective in children between six months and two years old, with approximately a 77.8% reduction in seizure frequency seen upon administration of Everolimus at 6 mg/m^2.

26.4 SUMMARY AND FUTURE PERSPECTIVES

In recent years, M&S has become increasingly more important in the drug development process of not only small molecules but also biologics and biosimilars. It has been commonly used among scientists and many biotech companies because of its ability to provide invaluable guidance when making pivotal decisions regarding novel drug development. The usage of M&S in drug development has become more acknowledged. Recently, the FDA recognized the importance of applying M&S during the regulatory process and highlighted the beneficial roles that M&S could play in addressing development gaps. Especially, the popPK analysis is becoming a component of 351(a) BLA submissions.

Due to the larger molecular weight and target-mediated binding, it is necessary to integrate the unique PK, PD characteristics of biologic products with PK/PD, PBPK, and popPK models when simulating them. This has made M&S an indispensable tool in identifying potential receptor/ligand as a feasible antibody-drug target, predicting efficacy and toxicity, determining FIH, evaluating biologics PK profiles in renal or hepatic impairment patients, and making informed decisions when designing clinical trials. The value of M&S in biologics drug development will undoubtedly become even more valuable as the modeling database continues to be updated and refined, making simulations more accurate and precise.

REFERENCES

Agoram, B. M., S. W. Martin, and P. H. van der Graaf. 2007. The role of mechanism-based pharmacokinetic-pharmacodynamic (PK-PD) modelling in translational research of biologics. *Drug Discovery Today* 12 (23–24):1018–1024.

Anselmo, Aaron C., Yatin Gokarn, and Samir Mitragotri. 2019. Non-invasive delivery strategies for biologics. *Nature Reviews Drug Discovery* 18 (1):19–40.

Benson, N. 2019. Quantitative systems pharmacology and empirical models: Friends or foes? *CPT Pharmacometrics and Systems Pharmacology* 8 (3):135–137.

Benson, Neil, Tomomi Matsuura, Sergey Smirnov, et al. 2013. Systems pharmacology of the nerve growth factor pathway: Use of a systems biology model for the identification of key drug targets using sensitivity analysis and the integration of physiology and pharmacology. *Interface Focus* 3 (2):20120071.

Berghout, Alexander. 2011. Clinical programs in the development of similar biotherapeutic products: Rationale and general principles. *Biologicals* 39 (5):293–296.

Blandizzi, C., P. L. Meroni, and G. Lapadula. 2017. Comparing originator biologics and biosimilars: A review of the relevant issues. *Clinical Therapeutics* 39 (5):1026–1039.

Bonifacio, Laura, Michael Dodds, David Prohaska, et al. 2020. Target-mediated drug disposition pharmacokinetic/pharmacodynamic model-informed dose selection for the first-in-human study of AVB-S6-500. *Clinical and Translational Science* 13 (1):204–211.

Bradshaw, Erica L., Mary E. Spilker, Richard Zang, et al. 2019. Applications of quantitative systems pharmacology in model-informed drug discovery: Perspective on impact and opportunities. *CPT: Pharmacometrics and Systems Pharmacology* 8 (11):777–791.

Brekkan, Ari, Luis Lopez-Lazaro, Elodie L. Plan, Joakim Nyberg, Suresh Kankanwadi, and Mats O. Karlsson. 2019. Sensitivity of Pegfilgrastim pharmacokinetic and pharmacodynamic parameters to product differences in similarity studies. *The AAPS Journal* 21 (5):85–85.

Bulitta, Jürgen B., and Nicholas H. G. Holford. 2014. Non-Compartmental Analysis. *Wiley StatsRef: Statistics Reference Online*.

Cai, Weiguo, Tarek A. Leil, Leonid Gibiansky, et al. 2020. Modeling and simulation of the pharmacokinetics and target engagement of an antagonist monoclonal antibody to interferon-γ-induced protein 10, BMS-986184, in healthy participants to guide therapeutic dosing. *Clinical Pharmacology in Drug Development* 9 (6):689–698.

Candelaria, M., D. Gonzalez, F. J. Fernández Gómez, et al. 2018. Comparative assessment of pharmacokinetics, and pharmacodynamics between RTXM83™, a rituximab biosimilar, and rituximab in diffuse large B-cell lymphoma patients: A population PK model approach. *Cancer Chemotherapy and Pharmacology* 81 (3):515–527.

Cazap, Eduardo, Ira Jacobs, Ali McBride, Robert Popovian, and Karol Sikora. 2018. Global acceptance of biosimilars: Importance of regulatory consistency, education, and trust. *The Oncologist* 23 (10):1188–1198.

Chien, J. Y., S. Friedrich, M. A. Heathman, D. P. de Alwis, and V. Sinha. 2005. Pharmacokinetics/pharmacodynamics and the stages of drug development: Role of modeling and simulation. *AAPS Journal* 7 (3):E544–E559.

Choi, G. W., Y. B. Lee, and H. Y. Cho. 2019. Interpretation of non-clinical data for prediction of human pharmacokinetic parameters: In vitro-in vivo extrapolation and allometric scaling. *Pharmaceutics* 11 (4):168.

Clausznitzer, Diana, Cesar Pichardo-Almarza, Ana Lucia Relo, et al. 2018. Quantitative systems pharmacology model for Alzheimer disease indicates targeting sphingolipid dysregulation as potential treatment option. *CPT: Pharmacometrics and Systems Pharmacology* 7 (11):759–770.

Colburn, W. A., and J. W. Lee. 2003. Biomarkers, validation and pharmacokinetic-pharmacodynamic modelling. *Clinical Pharmacokinetics* 42 (12):997–1022.

Combes, F. P., H. J. Einolf, N. Coello, T. Heimbach, H. He, and K. Grosch. 2020. Model-informed drug development for everolimus dosing selection in pediatric infant patients. *CPT: Pharmacometrics and Systems Pharmacology* 9 (4):230–237.

Cui, Cheng, J. E. Valerie Sia, S. Tu, et al. 2020. Development of a physiologically based pharmacokinetic (PBPK) population model for Chinese elderly subjects. *British Journal of Clinical Pharmacology* 87:2711–2722. https://doi.org/10.1111/bcp.14609

Dallmann, André, Juri Solodenko, Ibrahim Ince, and Thomas Eissing. 2018. Applied concepts in PBPK modeling: How to extend an open systems pharmacology model to the special population of pregnant women. *CPT: Pharmacometrics and Systems Pharmacology* 7 (7):419–431.

de Vries Schultink, A. H. M., R. P. Doornbos, A. B. H. Bakker, et al. 2018. Translational PK-PD modeling analysis of MCLA-128, a HER2/HER3 bispecific monoclonal antibody, to predict clinical efficacious exposure and dose. *Invest New Drugs* 36 (6):1006–1015.

Dua, P., E. Hawkins, and P. H. van der Graaf. 2015. A tutorial on target-mediated drug disposition (TMDD) models. *CPT: Pharmacometrics and Systems Pharmacology* 4 (6):324–337.

Ette, E. I., and P. J. Williams. 2004. Population pharmacokinetics I: Background, concepts, and models. *Annals of Pharmacotherapy* 38 (10):1702–1706.

FDA. 2015. Scientific Considerations in Demonstrating Biosimilarity to a Reference Product. Available from https://www.fda.gov/media/82647/download

———. 2016. Clinical Pharmacology Data to Support a Demonstration of Biosimilarity to a Reference Product. Available from https://www.fda.gov/media/88622/download

———. 2017. Biosimilar Development, Rcview, and Approval. Available from https://www.fda.gov/drugs/biosimilars/biosimilar-development-review-and-approval.

———. 2018a. Step 2: Preclinical Research. Available from https://www.fda.gov/patients/drug-development-process/step-2-preclinical-research.

———. 2018b. What Are "Biologics" Questions and Answers. Available from https://www.fda.gov/about-fda/center-biologics-evaluation-and-research-cber/what-are-biologics-questions-and-answers.

———. 2019. Population Pharmacokinetics. Available from https://www.fda.gov/media/71364/download#:~:text=Population%20pharmacokinetics%20is%20the%20study,drug%20of%20interest%20(2)

Ghahramani, Parviz. 2016. Chapter 3 – Modeling and Simulation Applications in Drug Development Process. In *Translational Medicine*, edited by A. Shahzad. Boston: Academic Press.

Gieschke, R., and J. L. Steimer. 2000. Pharmacometrics: Modelling and simulation tools to improve decision making in clinical drug development. *European Journal of Drug Metabolism and Pharmacokinetics* 25 (1):49–58.

Glassman, Patrick M., Lubna Abuqayyas, and Joseph P. Balthasar. 2015. Assessments of antibody biodistribution. *The Journal of Clinical Pharmacology* 55 (S3):S29–S38.

Hardiansyah, D., and C. M. Ng. 2019. Quantitative systems pharmacology model of chimeric antigen receptor T-cell therapy. *Clinical and Translational Science* 12 (4):343–349.

Holford, Nicholas H. G., and Lewis B. Sheiner. 1982. Kinetics of pharmacologic response. *Pharmacology & Therapeutics* 16 (2):143–166.

Hornik, Christoph P., Huali Wu, Andrea N. Edginton, Kevin Watt, Michael Cohen-Wolkowiez, and Daniel Gonzalez. 2017. Development of a pediatric physiologically-based pharmacokinetic model of clindamycin using opportunistic pharmacokinetic data. *Clinical Pharmacokinetics* 56 (11):1343–1353.

Huang, H. Y., D. W. Wu, F. Ma, et al. 2020. Availability of anticancer biosimilars in 40 countries. *The Lancet Oncology* 21 (2):197–201.

Jones, Hannah M., Neil Parrott, Karin Jorga, and Thierry Lavé. 2006. A novel strategy for physiologically based predictions of human pharmacokinetics. *Clinical Pharmacokinetics* 45 (5):511–542.

Kabir, Eva Rahman, Shannon Sherwin Moreino, and Mohammad Kawsar Sharif Siam. 2019. The breakthrough of biosimilars: A twist in the narrative of biological therapy. *Biomolecules* 9 (9):410.

Kamath, Amrita V. 2016. Translational pharmacokinetics and pharmacodynamics of monoclonal antibodies. *Drug Discovery Today: Technologies* 21–22:75–83.

Kielbasa, William, and Danielle L. Helton. 2019. A new era for migraine: Pharmacokinetic and pharmacodynamic insights into monoclonal antibodies with a focus on galcanezumab, an anti-CGRP antibody. *Cephalalgia: An International Journal of Headache* 39 (10):1284–1297.

Kim, Tae Hwan, Soyoung Shin, and Beom Soo Shin. 2018. Model-based drug development: Application of modeling and simulation in drug development. *Journal of Pharmaceutical Investigation* 48 (4):431–441.

Krzyzanski, Wojciech, John M. Harrold, Liviawati S. Wu, and Juan Jose Perez-Ruixo. 2016. A cell-level model of pharmacodynamics-mediated drug disposition. *Journal of Pharmacokinetics and Pharmacodynamics* 43 (5):513–527.

Kuepfer, L., C. Niederalt, T. Wendl, et al. 2016. Applied concepts in PBPK modeling: How to build a PBPK/PD model. *CPT: Pharmacometrics and Systems Pharmacology* 5 (10):516–531.

Lane, Nancy E., Thomas J. Schnitzer, Charles A. Birbara, et al. 2010. Tanezumab for the treatment of pain from osteoarthritis of the knee. *New England Journal of Medicine* 363 (16):1521–1531.

Lee, J. Y., C. E. Garnett, J. V. Gobburu, et al. 2011a. Impact of pharmacometric analyses on new drug approval and labelling decisions: A review of 198 submissions between 2000 and 2008. *Clinical Pharmacokinetics* 50 (10):627–635.

Lee, J. W., M. Kelley, L. E. King, et al. 2011b. Bioanalytical approaches to quantify "total" and "free" therapeutic antibodies and their targets: Technical challenges and PK/PD applications over the course of drug development. *AAPS Journal* 13 (1):99–110.

Lee, Jong Bong, Simon Zhou, Manting Chiang, Xiaowei Zang, Tae Hwan Kim, and Leonid Kagan. 2020. Interspecies prediction of pharmacokinetics and tissue distribution of doxorubicin by physiologically-based pharmacokinetic modeling. *Biopharmaceutics & Drug Disposition* 41 (4–5):192–205.

Lexchin, Joel. 2020. Affordable biologics for all. *JAMA Network Open* 3 (4):e204753.

Lim, H. S. 2019. Evolving role of modeling and simulation in drug development. *Translational and Clinical Pharmacology* 27 (1):19–23.

Lin, Xiaoqian, Xiu Li, and Xubo Lin. 2020. A review on applications of computational methods in drug screening and design. *Molecules (Basel, Switzerland)* 25 (6):1375.

Loudon, P., P. Siebenga, D. Gorman, et al. 2018. Demonstration of an anti-hyperalgesic effect of a novel pan-Trk inhibitor PF-06273340 in a battery of human evoked pain models. *British Journal of Clinical Pharmacology* 84 (2):301–309.

Lucas, A. J., J. L. Sproston, P. Barton, and R. J. Riley. 2019. Estimating human ADME properties, pharmacokinetic parameters and likely clinical dose in drug discovery. *Expert Opinion on Drug Discovery* 14 (12):1313–1327.

Marshall, Eliot. 2006. Lessons from a failed drug trial. *Science* 313 (5789):901.

Marsousi, Niloufar, Jules A. Desmeules, Serge Rudaz, and Youssef Daali. 2017. Usefulness of PBPK modeling in incorporation of clinical conditions in personalized medicine. *Journal of Pharmaceutical Sciences* 106 (9):2380–2391.

Meibohm, B., and H. Derendorf. 1997. Basic concepts of pharmacokinetic/pharmacodynamic (PK/PD) modelling. *International Journal of Clinical Pharmacology and Therapeutics* 35 (10):401–413.

Mendes, M. De Sousa, and Manoranjenni Chetty. 2019. Are standard doses of renally-excreted antiretrovirals in older patients appropriate: A PBPK study comparing exposures in the elderly population with those in renal impairment. *Drugs in R&D* 19 (4):339–350.

Milberg, Oleg, Chang Gong, Mohammad Jafarnejad, et al. 2019. A QSP model for predicting clinical responses to monotherapy, combination and sequential therapy following CTLA-4, PD-1, and PD-L1 checkpoint blockade. *Scientific Reports* 9 (1):11286.

Milligan, P. A., M. J. Brown, B. Marchant, et al. 2013. Model-based drug development: A rational approach to efficiently accelerate drug development. *Clinical Pharmacology and Therapeutics* 93 (6):502–514.

Nair, Anroop B., and Shery Jacob. 2016. A simple practice guide for dose conversion between animals and human. *Journal of Basic and Clinical Pharmacy* 7 (2):27–31.

Netterberg, Ida, Chi-Chung Li, Luciana Molinero, et al. 2019. A PK/PD analysis of circulating biomarkers and their relationship to tumor response in atezolizumab-treated non-small cell lung cancer patients. *Clinical Pharmacology and Therapeutics* 105 (2):486–495.

Ogasawara, K., and G. C. Alexander. 2019. Use of population pharmacokinetic analyses among FDA-approved biologics. *Clinical Pharmacology in Drug Development* 8 (7):914–921.

Parekh, A., S. Buckman-Garner, S. McCune, et al. 2015. Catalyzing the critical path initiative: FDA's progress in drug development activities. *Clinical Pharmacology and Therapeutics* 97 (3):221–233.

Parrott, N., C. Stillhart, M. Lindenberg, et al. 2020. Physiologically based absorption modelling to explore the impact of food and gastric pH changes on the pharmacokinetics of entrectinib. *AAPS Journal* 22 (4):78.

Polli, Joseph Ryan, Frank A. Engler, and Joseph P. Balthasar. 2019. Physiologically based modeling of the pharmacokinetics of "catch-and-release" anti-carcinoembryonic antigen monoclonal antibodies in colorectal cancer xenograft mouse models. *Journal of Pharmaceutical Sciences* 108 (1):674–691.

Ratih, Ratih, Mufarreh Asmari, Ahmed M. Abdel-Megied, Fawzy Elbarbry, and Sami El Deeb. 2021. Biosimilars: Review of regulatory, manufacturing, analytical aspects and beyond. *Microchemical Journal* 165:106143.

Reeves, P., K. Edmunds, A. Searles, and J. Wiggers. 2019. Economic evaluations of public health implementation-interventions: A systematic review and guideline for practice. *Public Health* 169:101–113.

Roskos, Lorin K., Song Ren, and Gabriel Robbie. 2011. Application of Modeling and Simulation in the Development of Protein Drugs. In *Clinical Trial Simulations: Applications and Trends*, edited by H. H. C. Kimko, and C. C. Peck. New York, NY: Springer New York.

Rostami-Hodjegan, Amin. 2018. Reverse translation in PBPK and QSP: Going backwards in order to go forward with confidence. *Clinical Pharmacology and Therapeutics* 103 (2):224–232.

Ryman, Josiah T., and Bernd Meibohm. 2017. Pharmacokinetics of monoclonal antibodies. *CPT: Pharmacometrics & Systems Pharmacology* 6 (9):576–588.

Samara, E., and R. Granneman. 1997. Role of population pharmacokinetics in drug development. A pharmaceutical industry perspective. *Clinical Pharmacokinetics* 32 (4):294–312.

Sheiner, L. B., B. Rosenberg, and V. V. Marathe. 1977. Estimation of population characteristics of pharmacokinetic parameters from routine clinical data. *Journal of Pharmacokinetics and Biopharmaceutics* 5 (5):445–479.

Snelder, N., B. A. Ploeger, O. Luttringer, et al. 2017. Characterization and prediction of cardiovascular effects of fingolimod and siponimod using a systems pharmacology modeling approach. *Journal of Pharmacology and Experimental Therapeutics* 360 (2):356–367.

Tabrizi, Mohammad A., Chih-Ming L. Tseng, and Lorin K. Roskos. 2006. Elimination mechanisms of therapeutic monoclonal antibodies. *Drug Discovery Today* 11 (1):81–88.

Tang, Lisa, Adam M. Persky, Günther Hochhaus, and Bernd Meibohm. 2004. Pharmacokinetic aspects of biotechnology products. *Journal of Pharmaceutical Sciences* 93 (9):2184–2204.

Topp, Brian, Maria E. Trujillo, and Vikram Sinha. 2019. Industrialization of quantitative systems pharmacology. *CPT: Pharmacometrics and Systems Pharmacology* 8 (6):356–358.

Ventola, C. L. 2013. Biosimilars: Part 1: Proposed regulatory criteria for FDA approval. *P&T* 38 (5):270–287.

Visser, S. A., D. P. de Alwis, T. Kerbusch, J. A. Stone, and S. R. Allerheiligen. 2014. Implementation of quantitative and systems pharmacology in large pharma. *CPT Pharmacometrics and Systems Pharmacology* 3 (10):e142.

Wang, Y. M., W. Krzyzanski, S. Doshi, J. J. Xiao, J. J. Pérez-Ruixo, and A. T. Chow. 2010. Pharmacodynamics-mediated drug disposition (PDMDD) and precursor pool lifespan model for single dose of romiplostim in healthy subjects. *AAPS Journal* 12 (4):729–740.

Wang, W., X. Wang, R. Doddareddy, et al. 2014. Mechanistic pharmacokinetic/target engagement/pharmacodynamic (PK/TE/PD) modeling in deciphering interplay between a monoclonal antibody and its soluble target in cynomolgus monkeys. *AAPS Journal* 16 (1):129–139.

Wang, Yow-Ming C., Yaning Wang, Sarah J. Schrieber, et al. 2019. Role of modeling and simulation in the development of novel and biosimilar therapeutic proteins. *Journal of Pharmaceutical Sciences* 108 (1):73–77.

Wong, H., and T. W. Chow. 2017. Physiologically based pharmacokinetic modeling of therapeutic proteins. *Journal of Pharmaceutical Sciences* 106 (9):2270–2275.

Woodhead, J. L., P. B. Watkins, B. A. Howell, S. Q. Siler, and L. K. M. Shoda. 2017. The role of quantitative systems pharmacology modeling in the prediction and explanation of idiosyncratic drug-induced liver injury. *Drug Metabolism and Pharmacokinetics* 32 (1):40–45.

Yellepeddi, V., J. Rower, X. Liu, S. Kumar, J. Rashid, and C. M. T. Sherwin. 2019. State-of-the-art review on physiologically based pharmacokinetic modeling in pediatric drug development. *Clinical Pharmacokinetics* 58 (1):1–13.

Zhao, L., T. H. Ren, and D. D. Wang. 2012. Clinical pharmacology considerations in biologics development. *Acta Pharmacologica Sinica* 33 (11):1339–1347.

Zhuang, Xiaomei, and Chuang Lu. 2016. PBPK modeling and simulation in drug research and development. *Acta Pharmaceutica Sinica B* 6 (5):430–440.

Machine Learning Applications in Biologics and Biosimilars

Disruptive Innovation

27

DeVon Herr, Erika Young, and Catherine F. Yang

Corresponding author: Catherine F. Yang

Contents

DOI: 10.1201/9780429485626-27

27.1 INTRODUCTION

Artificial intelligence (AI) and machine learning (ML)-based technologies, fields of computer science centered around mimicking the biological process of learning, have gained significant momentum in the healthcare industry in recent years.[1,2] With the growing quantity and availability of healthcare and biomedical data, AI and ML can be used to leverage new insights that would otherwise be impossible to discover at such a scale.[3] Real-world data from electronic health records and patient biometrics are common candidates for generating statistical insights in healthcare. Beyond the sheer magnitude, such data are complex, messy, and can be incredibly difficult to use through ordinary computation. On the contrary, ML and AI perform well with large data, particularly with specialized algorithms designed for scalability, in tandem with modeling procedures and computational libraries suited for wrangling data into a more workable form.[2]

Statistics and ML, at their core, center around deriving information from a finite sample of data and thus often overlap. Whereas statistics focuses on inference, creating and/or assessing a mathematical model, ML more closely studies prediction, forecasting future behaviors, and classifying not-yet explicitly classified data. From a computational perspective, statistical algorithms often work better on data where a specific model has already been specified or considered, and data for every possibility is plenty. However, ML makes far weaker assumptions, perhaps none, depending on the algorithm, and is hence a better fit for complex data that may not even fit any sensible model. In practice, this means ML models can be applied to data even without significant domain expertise; one only needs to know the form and features of the data.[4] Predictive models have their use in patient diagnostics and drug discovery,[3] well fit healthcare data's complexity and large volume.[2]

27.2 BIG DATA

The term "big data" is defined as data sets with such scale, complexity, and variance in structure that standard storage, retrieval, analysis, and visualization methods become infeasible.[5] DNA sequencing information,[6] patient electronic health records,[7,8] and patient biomedical data [9] have all been described as instances of big data and have been subject to big data analytics.

27.2.1 Big Data in Healthcare

Real-life data[10] collected regularly from healthcare professionals with respect to treatments, patient information, or clinical outcomes is opportune for observing major changes in prescriptions and medical practices,[11] assessing clinical practice costs and medical decision making,[10] and identifying predictive response parameters among the various heterogeneous responses to treatment[11] compared to clinical trials which are expensive[12] and often narrow in scope. Note that the analysis of real-life data should support and inform decisions that will eventually lead from or to clinical trial rather than in place of a clinical trial. Real-life data, which can be obtained from hospital information systems and medical administrative systems from payers,[7,13] fall under the definition of big data considering magnitude, heterogeneity, and complexities of data access and flow.[8] Yet another source of real-life data for clinical interests are general databases themselves, biomedical or otherwise, alleviating the need to collect data in some circumstances.[11] Such data can be analyzed to determine if the concomitant use of biologics and biosimilars and other medications might cause an increase in unwanted immunogenicity.[14,15]

However, data analysis, especially for ML and AI methods, is not always immediate, often requiring the need of preprocessing the data to address concerns regarding missing or malformed information in the datasets. Individual points of data can differ substantially within the same dataset and this heterogeneity is only amplified when multiple sources of data are considered, such as datasets from multiple healthcare establishments.[16] Conventional statistical algorithms, which often assume some explicit mathematical model and prediction model parameters from data,[7] can struggle to deal with data that is unprocessed, characteristic of real-life data.[10] AI and ML methods, which do not make such modeling assumptions, can often work better for answering questions from the healthcare community with respect to biosimilars and their multiple components.[15]

27.2.2 Machine Learning and Complex Biological Data

To fully understand complex biological phenomena such as human diseases or quantitative traits in animals and plants, massive amounts and multiple types of "big" data are generated from complicated studies.[17] With increases in both computational power and accessibility, parts of the data analysis process become trivial and new problems spring up.

In the past, biologists collecting and aggregating data served as the largest bottleneck; now, with automated, regular, and digitized systems, this obstacle has been drastically reduced.[10] Instead of finding and creating the datasets, the key difficulty has shifted to data mining of the aforementioned datasets, the practice of extracting and discovering the patterns in large, complicated data sets.[18] In the past decade, technological advances in data generation have significantly advanced studies of complex biological phenomena. Next-generation sequencing (NGS) technologies have allowed researchers to screen changes at varying biological scales, such as genome-wide genetic variation, gene expression, and small RNA abundance, epigenetic modifications, protein binding motifs, and chromosome conformation in a high-throughput and cost-efficient manner.[19]

Figure 27.1 depicts machine learning using complex biological data as inputs into a model based on prior knowledge, outputting new knowledge and predictive models. Some high-throughput data generation techniques for different biological aspects are *ATAC-seq* assay for transposase-accessible chromatin using

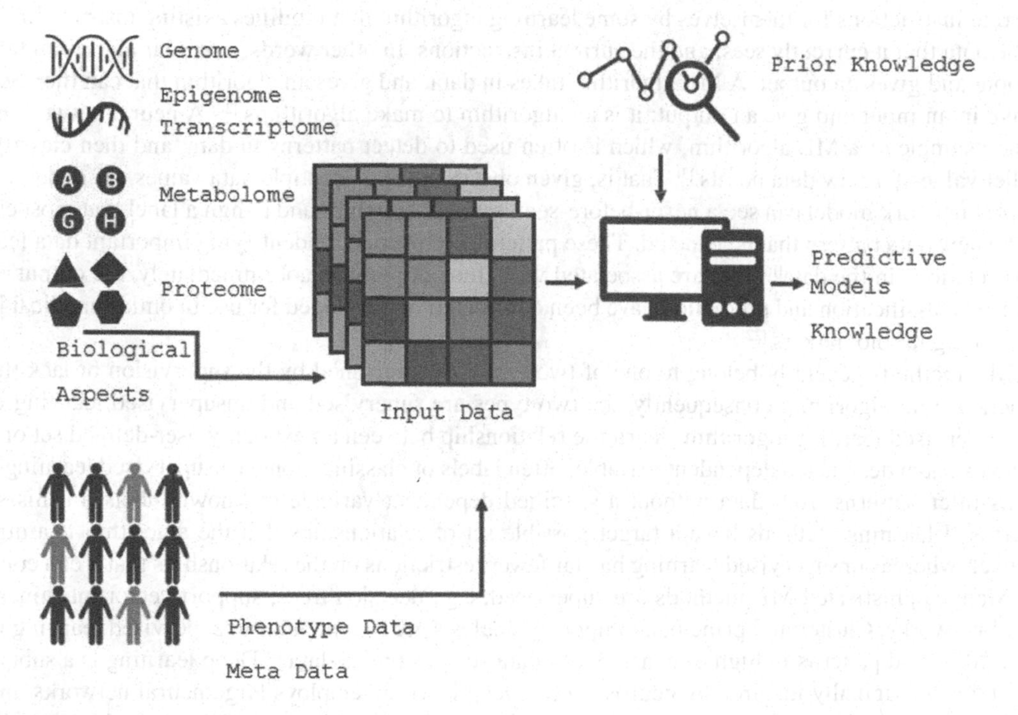

FIGURE 27.1 Machine learning of complex data, a pipeline of biological aspects, phenotype, and metadata.

sequencing; *ChIP-seq* chromatin immunoprecipitation sequencing; *DNase-seq* DNase I hypersensitive sites sequencing; *GC-MS* gas chromatography-mass spectrometry; *LC-MS* liquid chromatography-mass spectrometry; *lncRNA-seq* long non-coding RNA sequencing; *NMR* nuclear magnetic resonance, *RNA-seq* RNA sequencing; *smRNA-seq* small RNA sequencing; *WES* whole-exome sequencing; *WGBS* whole-genome bisulfite sequencing; *WGS* whole-genome sequencing; *Hi-C* chromatin conformation capture combined with deep sequencing; and *iTRAQ* isobaric tags for relative and absolute quantification.[19]

27.2.3 Challenges and Future Outlooks

Importantly, ML and AI methods are not without problems, even in purely analytical and computational contexts. For instance, firstly, interpretation of models derived from some sophisticated ML approaches such as deep learning can be daunting if not impossible. The result of a neural network, a deep learning AI, is a predictive algorithm comprised of an often-unintelligible series of arithmetic computations.[20] In many cases, researchers are more interested in model-building procedures to yield biological insights by introspecting the model itself, rather than the usage of the model. The black-box nature of ML and AI creates a powerful model that cannot be translated outside of its predictions. The information from the model may need further inspection and should be carefully interpreted with corresponding deep biological knowledge.[19]

27.3 MACHINE LEARNING APPLICATION IN HEALTH CARE

From a technical point of view, a regular computer program uses a set of instructions to produce an outcome. Normally, these instructions are explicitly written by a computer programmer. ML programs generate instructions for themselves by some learning algorithm that modifies existing instructions based on the data that it currently sees, and the current instructions. In other words, a regular algorithm takes in an input and gives an output. A ML algorithm takes in data, and gives an algorithm that can then be used to take in an input and give an output; it is an algorithm to make algorithms.[20] A neural network model is one example of a ML algorithm, which is often used to detect patterns in data, and then classify and predict values for new data points.[3] That is, given observations of multiple data values, all under a label, a neural network model can see a never-before-seen set of data values and assign a label that most closely fits the new data pattern that is detected. These patterns correspond to identifying important data features or correlations in the data[16] that are associated with, though perhaps not immediately, the output.[21] For instance, classification and prediction have been considered or developed for use in omics, medical imaging, and digital biomarkers.[22]

ML methods generally belong to one of two types, distinguished by the supervision or lack thereof in the learning algorithm; consequently, the two types are supervised and unsupervised learning methods. Supervised learning algorithms learn the relationship between an explicitly user-defined set of input variables and a designated dependent variable, often labels or classifications. Unsupervised learning algorithms infer patterns from data without a specified dependent variable or known labels. In this sense, supervised learning methods have a target possible set of relationships that the algorithm is aiming to discover, whereas unsupervised learning has far fewer restrictions on the relationships that it can consider.

Many sophisticated ML methods are supervised, e.g., decision trees, support vector machines, and neural networks. Cluster and principal component analysis are two popular unsupervised learning methods used to find patterns in high dimensionality data such as omics data.[1] Deep learning is a subtype of ML and was originally inspired by neuroscience. Deep learning employs large neural networks and has become one of the most popular means for ML. It has been applied in many fields, largely driven by the massive increases in both computational power and big data. Deep learning being both supervised and unsupervised, has revolutionized fields such as image recognition and shows promise for applications in genomics, medicine, and healthcare.[22]

ML has also been applied broadly in biology for predictions and finding insights.[1] With the increasing availability of more and different types of omics data, the application of ML methods, especially deep learning approaches, has become more frequent.[19] One area of opportunity for ML approaches is in the prediction of genomic features, particularly those that are hard to model using conventional approaches, such as regulatory regions. ML has been used to predict the sequence specificities of DNA- and RNA-binding proteins, enhancers, and other regulatory regions on data generated by one or multiple types of omics approach, such as DNase I hypersensitive sites (DNase-seq); formaldehyde-assisted isolation of regulatory elements with sequencing (FAIRE-seq); assay for transposase-accessible chromatin using sequencing (ATAC-seq), and self-transcribing active regulatory region sequencing (STARR-seq).[23,24] ML models can predict regulatory elements and non-coding variant effects *de novo* from a DNA sequence that can then be tested and validated for their contribution to gene regulation and ultimately to observable traits/pathologies.[24]

In addition to the prediction of regulatory regions, supervised learning has recently been used in solving population and evolutionary genetics questions, such as the identification of regions under purifying selection or selective sweeps, as well as more complicated spatiotemporal questions.[25] Further examples of ML approaches are the prediction of transcript abundance,[26,27] imputation of missing SNPs and DNA methylation states,[21,27] variant calling,[28] disease diagnosis and classification, and many different biological questions using datasets from different biological aspects such as genomes, epigenomes, transcriptomes, and metabolomes.[19]

27.4 MODELS OF MACHINE LEARNING

ML models use learning algorithms or processes to detect patterns and predict outcomes. This procedure describes the steps to process the data of the model which it is training on, and which, if any, relationships to consider. This instruction is focused on enhancing the performance of the model to produce the correct output that reflects the available input data; the internal workings of the model will change to optimize its performance on the data that has been seen so far.

The learning can be broken down into three main steps: training, validation, and testing – each associated with a different set of data. In the training step, the model considers the training data set, using its learning procedure to generate instructions to analyze the data and come up with a guessed prediction based on the previous data seen. Using the same learning procedure, it compares the prediction to the actual answer and then modifies its instruction set to account for this. The validation set is the second set of data, in which the model that has been obtained from the training set can be partially tested, and parts of its instructions can be modified for performance. The testing set is then a completely unbiased evaluation set of data to test the data on. It is from the validation and test sets that one can conclude the success or weaknesses of the output.[3]

Although several methods have been developed for interpreting and understanding complicated models, such as perturbation-based methods and gradient-based methods for the interpretation of convolutional neural networks (CNNs), the interpretation of many complicated cases is generally challenging. Joint analysis of multiple biological data types has the potential to further our understanding of complex biological phenomena; however, data integration is challenging due to the heterogeneity of different data types. For example, an expression profile is a list of numbers, while the genetic variants are categorial and of different lengths. Research in this area exists but establishing a best practice has yet to happen.[17,29]

Another challenge is the curse of dimensionality, which refers to the change in volume of a problem when the dimension increases where the volume change is often much larger, often exhausting resources or making some methods infeasible. This difficulty is amplified in a data analysis situation when the actual quantity of data is low compared to the dimension, usually the number of parameters for each observation. Problems such as sparsity, multicollinearity, and overfitting are difficult to avoid in high-resolution studies such as in omics datasets, although the larger sample size and modern ML methods can partially mitigate these problems.[29] To increase the number of samples, it may be necessary to combine data from multiple sources, which may be feasible for qualitative data as single-nucleotide polymorphisms (SNPs). It can be difficult to handle quantitative data such as gene expression data due to confounding factors from laboratory procedure, measurement, or when assuming all data from the different sources describe the same phenomenon, with no clear answer on properly handling the hidden confounding factors even if known.[19]

While improved ML methods and the increasing number of available samples show great promise to better our understanding of complex biological phenomena, building proper machine-learning models can still be challenging due to hidden biological factors such as population structure among samples or evolutionary relationship among genes that are not explicitly enumerated in data. Biological datasets should be carefully curated to remove or control for confounders. Without properly accounting for such factors, the models can be overfitted by learning the noise in the data instead of potentially more explanatory relationships of such factors, leading to false-positive discovery. The model essentially remembers the training data but not any of the actual relationships between the data. Rather than deriving some insight, it performs informed guessing, an undesirable property. To build proper models, the biological and technical factors specific to the modeling scenario must be considered. For example, biological data are often imbalanced, such as the case in some diseases or traits that occur only in a small fraction of a population.[19] It is usually more meaningful to access metrics as F-score, precision, and recall for the non-major class rather than simple accuracy to evaluate model performance for imbalanced classes in the data. That simple accuracy can be conditioned by itself on different classes is also important, as model performance on some classes may be more important than others, depending on the data. It may be more important for a model to handle an infected patient or uninfected patient, depending on the base prevalence in the population. More involved ratios such as efficacy are also a good evaluation metric.[30]

27.5 REGULATORY FRAMEWORK/CLINICAL TRIAL DATA

AI and ML methods also have their place in statistical applications, such as considering clinical trial data. Here, their capabilities for prediction and classification can be utilized to identify and predict the effect of treatment and increase overall productivity.

27.5.1 Machine Learning for Clinical Trials in the Era of COVID-19

Under pandemic time, clinical trials present a unique set of challenges in three key areas: (1) ongoing clinical trials for non-COVID-19 drugs, (2) clinical trials for repurposing drugs to treat COVID-19, and (3) clinical trials for new drugs to treat COVID-19. In each of these three areas, we identify opportunities where we believe that ML can provide important insights and can help address some of the challenges faced in clinical trials.[31]

For the data quality, e.g., in a typical trial condition with a highly controlled environment, extensive data can be collected and monitored for patients throughout the trial. However, the pandemic and its effects disrupt data collection in ongoing trials. Existing ML methods can be used to impute missing data and/or produce estimates less influenced by missing data. ML methods can also be used to flexibly model and uncover biases introduced by changing conditions over the course of the pandemic.[32] Many ongoing (non-COVID-related) clinical trials face temporary suspension. Unplanned interim analyses may present the opportunity to adapt recruitment strategies, in a blinded or unblinded fashion, to increase the likelihood that re-started trials succeed. Further, if a trial is fully suspended, ML methods can be used for the discovery of (heterogenous) treatment effects and for the assessment of uncertainty.[33] Hence some research work can be done before the clinical trials are resumed. The current COVID pandemic provides optimal conditions for existing ML methods for response-adaptive randomization. The time to the clinical endpoint is relatively short, allowing frequent adaptation with a constant stream of new patients, therefore, quick action is key.[34]

27.5.2 Analysis of Data from Ongoing Clinical Trials

During the pandemic, on-site assessment of patients may be less frequent, which will lead to missing data.[31] Moreover, changes in visitation frequency, regularity, and times may lead to inconsistencies in both generating and sampling patient data. ML and AI methods exist to address these exact problems, with some assumptions. It may not be apparent where and when missing data occurs in a data set, for while patient records track when a patient visits and not periods of absences, it is not easy to hand-calculate all differences to detect changes. Assuming that missing data is random and independent, ML methods on temporal data streams, such as using multi-dimensional recurrent neural networks[35] and Generative Adversarial Imputation Nets[36] substantially out-perform previous methods, including multiple imputations by chained equations, matrix completion, and expectation maximization on a variety of datasets.[31]

27.6 ML APPLICATIONS FOR BIOLOGICS AND BIOSIMILARS

Immune reactions which potentially involve large numbers of parameters and large quantities of observations are infeasible to analyze with standard statistical means due to data quantity and complexity. As a simple example, if one has a model of immune reactions with parameters known as *a priori*, ML algorithms can be used to estimate corresponding parameters, and then use the model to predict future immune reactions.[15] Such inputs could be clinical care data or biologics prescribed for predicting adverse events and clinical outcomes as outputs. Such a model can assist with decision-making on substitution and dose regulation.[37] To assess unwanted immunogenicity, one begins with a single clinical question, such as determining if a given substitution increases a certain set of unwanted reactions when biosimilars are used in real-life data, which can then be undertaken using ML and AI methods.

Supervised learning can be readily used to address clinical questions for biosimilars. For a data source, all instances of a patient taking some reference biological drug and then treated with its biosimilar as well any relevant medical record and clinical outcome data suffice as completely labeled data. In particular, all relevant data should be considered in the preliminary model building phase, such as drugs prescribed; indication; biosimilar and reference biologic drugs, batch number; manufacturer and glycosylation rate relevant input factors, as clinical outcomes, adverse events; and severity and non-compliance outputs.[8] These explicitly labeled data can then be used to create a supervised ML model for predicting various relevant data when varying biosimilar treatments. The model attempts to process the patient data to classify them into their predefined categories, based on input and output, respectively. Hence selection of biosimilar related features is particularly important, such as the absence of immune reactions, weak immune reactions, ADA production, and fulminant reactions.[15]

Neural nets have advantages as the model for biosimilars, much from their interaction with new data. In the case where previously irrelevant or unrecorded data is suddenly important, such as the clinical response of hemophilia A depending on the genetic defect affecting factor VIII, which causes an unwanted immune reaction.[38] Neural nets can consider this new data while still maintaining the analysis and training accuracy from the old data. Hence there is a particular time advantage in that a single neural net can be used for multiple different investigations without having to retrain neural nets multiple times, a time-intensive process.

27.6.1 AI and ML Application in Drug Development for Biologics and Biosimilar

Beyond the numerous computational speedups that ML and AI methods provide on big data, AI-based solutions have the capability of generating biological insights in understanding drugs binding to targets to improve the specificity of medicines. If undertaken at the drug discovery stage, AI solutions can increase research and development efforts. ML and AI methods have the potential to speed up the discovery and preclinical stages by a factor of 15,[39] increasing the accuracy of predictions on the efficacy and safety of drugs and diversifying drug pipelines. Currently, AI's role in developments is primarily in chemical, small molecule research. AI has been used to identify the most accurate animal models for different diseases.[40] Research has also been conducted in the binding and structure of antibodies and biologics.[41]

Further applications of AI and ML for drug development include automating document analysis using natural language processing and generation for regulatory filling and safety assessments. By classifying safety reports for a produced drug, one can automatically consider reports of utmost importance first, increasing productivity. Image analysis from automating pathology images in animal studies and clinical trials is another tried method.[40]

27.6.2 Predicting the Long-Term Outcomes of Biologics in a Therapy Using Machine Learning

In the pre-biologic era, skin clearance was difficult to achieve and required a combination of multiple skin-directed therapies and systemic agents, with a significant risk of cumulative toxicities. The unprecedented success and wide implementation of biologics in the therapy of psoriasis has changed the landscape of the medical need in this disease over the last decade. Today, biologics allow us to achieve PASI 75 responses in up to 90% of the patients, and real-world evidence from registries confirms that a comparable proportion of psoriasis patients achieve excellent control of skin disease.[42] Despite the vast amount of data on the efficacy of biologics, therapeutic decision-making (i.e., deciding which treatment to administer for each patient) is still based on a trial-and-error approach. The initial choice of therapy is not always optimal, as reflected by data documenting that over 50% of patients need dose optimization during the therapy[43–46] and 20%–50% of patients experience a relapse of the disease and require a switch to another medication. ML and AI methods have been used for personalized low-dose adjustment and optimization per patient. A feed-forward neural network regression model was trained to predict volumetric computed tomography dose index for a given patients' biomedical data.[47] Such an approach has also been implemented in the Spark framework, making such a model especially conducive to big data.[48] Deep learning has also been used for dose optimization for magnetic resonance imaging (MRI)[49] and an extra-trees regression model predicts lamotrigine dosage particularly well.[50] Hence there exists an opportunity to use similar approaches for potential dose optimization ML and AI models. Note that medical experts must still verify the findings and predictions of models, even with good model performance.

27.7 ECONOMIC INCENTIVES

The high cost of pharmaceuticals, especially biologics, has become an important issue in the battle concerning ever-increasing healthcare costs. Economic incentives to use biosimilars and questions from the clinical community regarding unwanted immunogenicity can be incompatible. Real-life data could therefore be used with support from ML to identify the risk factors that increase immune risks or non-compliance. This assessment could support economic incentives and improve the protection of public health.[15]

With faster algorithms with potentially more expressive power, and using untapped resources, many parts of drug discovery can be made more productive. Successful biologics must satisfy multiple properties including activity and particular physicochemical features. AI and ML methods can aid by using prediction to identify a target if a given drug design is viable without the need of testing, allowing for a smaller initial selection of designs to then test, as well as discern cheaper manufacturing procedures. Hence ML and AI methods can together be used to make numerous processes for biologics and biosimilars less costly and time-intensive.

27.8 ETHICAL AND LEGAL ISSUES

As interacting with biomedical and healthcare expenditure revolves around the personal information of individuals, any ML application needs to comply with the ethical guidelines laid down by national ethics committees for such data. Hence one's data processing methodology must not lead to the profiling of patients from their information. Complying with data privacy regulations in each country could be challenging, but aggregate and anonymous data reduce the risk of violating those regulations. Yet another recent option is blockchain technology to anonymize this data.[15] In particular, some data taken to be input can be legally inaccessible or unusable in other countries,[41] and hence care must be taken in the actual ML architecture to avoid the use of such data.

Neural nets famously lack transparency, which can raise questions about how the ML systems arrived at their conclusions while processing patient data. Neural networks use individually meaningless sequences of arithmetic operations that feed into other operations, and each of these folds involves a leap to a new level of interpretation whose workings may be hidden from clinicians trying to understand the reasoning process. The more complex a model is, the less likely it is that biological insights can be generated from the ML model. Hence there exists some incompatibility in transparency, biological knowledge, and model complexity.[15]

In addition to immunogenicity studies, ML models could automate pharmacovigilance processes, detect safety signals, make risk assessments to facilitate decision-making on dose regimens or switching, and compare switching initiated by a clinician *vs.* substitution at the pharmacy, informing policy on substitution.[15] However, these systems should only be considered after standard operating procedures and processes have been undertaken and should only be used after clinical trial and regulation have verified the efficacy of the biosimilar in question.

27.9 CONCLUSION REMARKS

The application of ML in healthcare creates tremendous opportunities to support biologics and biosimilars development and research in many ways that were not considered possible. Such technology should be seen as a highly informative and efficient tool to learn more about biological drugs after clinical trials and the medical practices during their uses. ML can also track and interpret the information from the immediate reactions to specific biosimilars or experience adverse events only after repeated treatments due to the complex nature of biosimilars. However, ML can be leveraged to its potential, the economic and medical context as well as consequences of using biosimilars that must be carefully understood. There are indeed several limitations that have been described previously that need to be addressed at the earliest stages of any application. Furthermore, other key elements should also be discussed, such as the involvement of multiple stakeholders from the public and private sectors. In addition, the application of ML to other services and products supplied by the pharmaceutical industry and healthcare providers should also be considered.

REFERENCES

1. Shailaja, K., Seetharamulu, B., & Jabbar, M. A. Machine Learning in Healthcare: A Review," 2018 Second International Conference on Electronics, Communication and Aerospace Technology (ICECA), 2018, pp. 910–914, doi: 10.1109/ICECA.2018.8474918.
2. Miotto, R., Wang, F., Wang, S., Jiang, X., & Dudley, J. T. (2018). Deep learning for healthcare: review, opportunities and challenges. Briefings in Bioinformatics, 19(6), 1236–1246. https://doi.org/10.1093/bib/bbx044.
3. Doupe, P., Faghmous, J., & Basu, S. (2019). Machine learning for health services researchers. Value in Health, 22(7), 808–815. https://doi.org/10.1016/j.jval.2019.02.012.
4. Bzdok, D., Altman, N., & Krzywinski, M. (2018). Statistics versus machine learning. Nature Methods, 15(4), 233–234. https://doi.org/10.1038/nmeth.4642.
5. Sagiroglu, S. & Sinanc, D. "Big data: A Review," 2013 International Conference on Collaboration Technologies and Systems (CTS), 2013, pp. 42–47, doi: 10.1109/CTS.2013.6567202.
6. Celesti, F., Celesti, A., Carnevale, L., Galletta, A., Campo, S., Romano, A., Bramanti, P., & Villari, M. (2017). Big data analytics in genomics: The point on Deep Learning solutions. 2017 IEEE Symposium on Computers and Communications (ISCC). https://doi.org/10.1109/iscc.2017.8024547.
7. Rajkomar, A., Oren, E., Chen, K., Dai, A. M., Hajaj, N., Hardt, M., Liu, P. J., Liu, X., Marcus, J., Sun, M., et al. (2018). Scalable and accurate deep learning for electronic health records. NPJ Digital Medicine, 1, 1–26. doi: 10.1038/s41746-018-0029-1.
8. Crown, W. (2019). Real-world evidence, causal inference, and machine learning. Value Health, 22, 587–592. doi: 10.1016/j.jval.2019.03.001.
9. Luo, J., Wu, M., Gopukumar, D., & Zhao, Y. (2016). Big data application in biomedical research and health care: a literature review. Biomedical Informatics Insights, 8(1–10), https://doi.org/10.4137/bii.s31559.
10. Makady, A., de Boer, A., Hillege, H., Klungel, O., Goettsch, W., & (on behalf of GetReal Work Package 1). (2017). What is real-world data? A review of definitions based on literature and stakeholder interviews. Value in Health: The Journal of the International Society for Pharmacoeconomics and Outcomes Research, 20(7), 858–865. https://doi.org/10.1016/j.jval.2017.03.008.
11. Berger, M. L., Sox, H., Willke, R. J., Brixner, D. L., Eichler, H. G., Goettsch, W., Madigan, D., Makady, A., Schneeweiss, S., Tarricone, R., Wang, S. V., Watkins, J., & Mullins, C. D. (2017). Good practices for real-world data studies of treatment and/or comparative effectiveness: recommendations from the joint ISPOR-ISPE special task force on real-world evidence in health care decision making. Value Health, 20(8), 1003–1008. doi: 10.1016/j.jval.2017.08.3019. Epub 2017 Sep 15. PMID: 28964430.
12. Steinwandter, V., Borchert, D., & Herwig, C. (2019). Data science tools and applications on the way to Pharma 4.0. Drug Discovery Today, 24, 1795–1805. doi: 10.1016/j.drudis.2019.06.005.
13. Wang, S. V., Schneeweiss, S., Berger, M. L., Brown, J., de Vries, F., Douglas, I., Gagne, J. J., Gini, R., Klungel, O., Mullins, C. D., Nguyen, M. D., Rassen, J. A., Smeeth, L., Sturkenboom, M., & Joint ISPE-ISPOR Special Task Force on Real World Evidence in Health Care Decision Making (2017). Reporting to improve reproducibility and facilitate validity assessment for healthcare database studies V1.0. Value in Health: The Journal of the International Society for Pharmacoeconomics and Outcomes Research, 20(8), 1009–1022. https://doi.org/10.1016/j.jval.2017.08.3018.
14. De La Forest Divonne, M., Gottenberg, J.E., Salliot, C. (2017). Revue systématique des registres de polyarthrites rhumatoïdes sous biothérapie dans le monde et méta-analyse sur les données de tolérance. Revue duRhumumatisme,84, 199–207. doi: 10.1016/j.rhum.2017.01.003.
15. Perpoil, A., Grimandi, G., Birklé, S., Simonet, J. F., Chiffoleau, A., & Bocquet, F. (2020). Public health impact of using biosimilars, is automated follow up relevant? International Journal of Environmental Research and Public Health,18(1), 186. Published 2020 Dec 29. doi: 10.3390/ijerph18010186.
16. Berdaï, D., Thomas-Delecourt, F., Szwarcensztein, K., table ronde « recherche clinique et méthodologie » des Ateliers de Giens XXXIII, d'Andon, A., Collignon, C., Comet, D., Déal, C., Dervaux, B., Gaudin, A. F., Lamarque-Garnier, V., Lechat, P., Marque, S., Maugendre, P., Méchin, H., Moore, N., Nachbaur, G., Robain, M., Roussel, C., Tanti, A., … Thiessard, F. (2018). Demandes d'études post-inscription (EPI), suivi des patients en vie réelle: évolution de la place des bases de données. Therapie, 73(1), 1–12. https://doi.org/10.1016/j.therap.2017.12.001.
17. Zitnik, M., Nguyen, F., Wang, B., Leskovec, J., Goldenberg, A., & Hoffman, M. M. (2019). Machine learning for integrating data in biology and medicine: principles, practice, and opportunities. Information Fusion, 50, 71–91.

18. Jothi, N., Rashid, N. A., & Husain, W. (2015). Data mining in healthcare – a review. Procedia Computer Science, 72, 306–313. doi: https://doi.org/10.1016/j.procs.2015.12.145.
19. Xu, C., & Jackson, S.A. (2019). Machine learning and complex biological data. Genome Biology, 20, 76. https://doi.org/10.1186/s13059-019-1689-0.
20. Abiodun, O. I. et al. (2019). Comprehensive review of artificial neural network applications to pattern recognition. IEEE Access, 7, 158820–158846, doi: 10.1109/ACCESS.2019.2945545.
21. Baker, R. E., Peña, J. M., Jayamohan, J., & Jérusalem, A. (2018). Mechanistic models versus machine learning, a fight worth fighting for the biological community?. Biology Letters, 14(5), 20170660. https://doi.org/10.1098/rsbl.2017.0660.
22. Hutchinson, L., Steiert, B., Soubret, A., Wagg, J., Phipps, A., Peck, R., Charoin, J. E., & Ribba, B. (2019). Models and machines: how deep learning will take clinical pharmacology to the next level. CPT: Pharmacometrics & Systems Pharmacology, 8(3), 131–134. https://doi.org/10.1002/psp4.12377.
23. Libbrecht, M. W., & Noble, W. S. (2015). Machine learning applications in genetics and genomics. Nature Reviews Genetics, 16, 321–332.
24. Zou, J., Huss, M., Abid, A., Mohammadi, P., Torkamani, A., & Telenti, A. (2019). A primer on deep learning in genomics. Nature Genetics, 51, 12–18.
25. Schrider, D. R., & Kern, A. D. (2018). Supervised machine learning for population genetics: a new paradigm. Trends in Genetics, 34, 301–312.
26. Washburn, J. D., Mejia-Guerra, M. K., Ramstein, G., Kremling, K. A., Valluru, R., Buckler, E. S., et al. (2019). Evolutionarily informed deep learning methods for predicting relative transcript abundance from DNA sequence. Proceedings of the National Academy of Sciences of the U. S. A., 116, 5542–5549.
27. Sun, Y. V., & Kardia, S. L. R. (2008). Imputing missing genotypic data of single-nucleotide polymorphisms using neural networks. European Journal of Human Genetics, 16, 487–495.
28. Poplin, R., Chang, P. C., Alexander, D., Schwartz, S., Colthurst, T., Ku, A., et al. (2018). A universal SNP and small-indel variant caller using deep neural networks. Nature Biotechnology, 36, 983–987.
29. Altman, N., & Krzywinski, M. (2018). The curse(s) of dimensionality. Nature Methods, 15, 399–400. https://doi.org/10.1038/s41592-018-0019-x.
30. Reich, Y., & Barai, S. (1999). Evaluating machine learning models for engineering problems. Artificial Intelligence in Engineering, 13(3), 257–272. doi: 10.1016/s0954-1810(98)00021-1.
31. Zame, W. R., Bica, I., Shen, C., Curth, A., Lee, H., Bailey, S., Van der Schaar, M. (2020). Machine learning for clinical trials in the era of COVID-19. Statistics in Biopharmaceutical Research, 12(4), 506–517. doi: 10.1080/19466315.2020.1797867.
32. Cruz-Roa, A., Gilmore, H., Basavanhally, A. et al. (2017). Accurate and reproducible invasive breast cancer detection in whole-slide images: a deep learning approach for quantifying tumor extent. Scientific Reports, 7, 46450. https://doi.org/10.1038/srep46450.
33. Filipovych, R., Resnick, S. M., & Davatzikos, C. (2011). Semi-supervised cluster analysis of imaging data. NeuroImage, 54(3), 2185–2197. doi: 10.1016/j.neuroimage.2010.09.074.
34. Segar, M. W., Patel, K. V., Ayers, C., Basit, M., Tang, W. W., Willett, D., Berry, J., Grodin, J. L. & Pandey, A. (2020). Phenomapping of patients with heart failure with preserved ejection fraction using machine learning-based unsupervised cluster analysis. European Journal of Heart Failure, 22, 148–158. https://doi.org/10.1002/ejhf.1621.
35. Yoon, J., Zame, W. R., & van der Schaar, M. (2019). Estimating missing data in temporal data streams using multi-directional recurrent neural networks. IEEE Transactions on Biomedical Engineering, 66(5), 1477–1490. doi: 10.1109/TBME.2018.2874712.
36. Yoon, S., & Sull, S. (2020). GAMIN: Generative adversarial MULTIPLE imputation network for highly missing data. 2020 IEEE/CVF Conference on Computer Vision and Pattern Recognition (CVPR). doi: 10.1109/cvpr42600.2020.00848.
37. Tourdot, S., Abdolzade-Bavil, A., Bessa, J., Broët, P., Fogdell-Hahn, A., Giorgi, M., Jawa, V., Kuranda, K., Legrand, N., Pattijn, S., Pedras-Vasconcelos, J. A., Rudy, A., Salmikangas, P., Scott, D. W., Snoeck, V., Smith, N., Spindeldreher, S., & Kramer, D. (2020). 10th European immunogenicity platform open symposium on immunogenicity of biopharmaceuticals, mAbs, 12(1), doi: 10.1080/19420862.2020.1725369.
38. Doevendans, E., & Schellekens, H. (2019). Immunogenicity of innovative and biosimilar monoclonal antibodies. Antibodies, 8(1), 21. https://doi.org/10.3390/antib8010021.
39. Zhavoronkov, A., Ivanenkov, Y. A., Aliper, A. et al. (2019). Deep learning enables rapid identification of potent DDR1 kinase inhibitors. Nature Biotechnology, 37, 1038–1040. https://doi.org/10.1038/s41587-019-0224-x.
40. Poperzi, F., Taylor, K., Steedman, M., Ronte, H., & Haughey, J. (2019). Intelligent drug discovery Powered by AI (Rep.). Deloitte Insights, https://www2.deloitte.com/us/en/insights/industry/life-sciences/artificial-intelligence-biopharma-intelligent-drug-discovery.html

41. Sogaard, M. (2019, April 08). How Artificial Intelligence Will Impact Drug Development [Interview by 1384839161 1011388988 K. Kelleher]. Bio-IT World, https://www.bio-itworld.com/news/2019/04/08/how-artificial-intelligence-will-impact-drug-development
42. Seymour, K., Benyahia, N., Hérent, P., & Malhaire, C. (2019). Exploitation des données pour la recherche et l'intelligence artificielle: Enjeux médicaux, éthiques, juridiques, techniques. Imagerie De La Femme, 29(2), 62–71. doi: 10.1016/j.femme.2019.04.004.
43. Mason, K. J., Barker, J. N. W. N., Smith, C. H., et al. (2018). Comparison of drug discontinuation, effectiveness, and safety between clinical trial eligible and ineligible patients in BADBIR. JAMA Dermatol, 154, 581–588.
44. Gniadecki, R., Kragballe, K., Dam, T. N., & Skov, L. (2011). Comparison of drug survival rates for adalimumab, etanercept and infliximab in patients with psoriasis vulgaris. British Journal of Dermatology, 164, 1091–1096.
45. Egeberg, A., Ottosen, M. B., Gniadecki, R., et al. (2018). Safety, efficacy and drug survival of biologics and biosimilars for moderate-to-severe plaque psoriasis. British Journal of Dermatology, 178, 509–519.
46. Sbidian, E., Mezzarobba, M., Weill, A., et al. (2019). Persistence of treatment with biologics for patients with psoriasis: a real-world analysis of 16 545 biologic-naïve patients from the French National Health Insurance database (SNIIRAM). British Journal of Dermatology, 180, 86–93.
47. Meineke, A., Rubbert, C., Sawicki, L.M. et al. (2019). Potential of a machine-learning model for dose optimization in CT quality assurance. European Radiology, 29, 3705–3713. https://doi.org/10.1007/s00330-019-6013-6.
48. Alla Takam, C., Samba, O., Tchagna Kouanou, A., & Tchiotsop, D. (2020). Spark architecture for deep learning-based dose optimization in medical imaging. Informatics in Medicine Unlocked, 19, 100335. doi: 10.1016/j.imu.2020.100335.
49. Jung, K., Park, H., & Hwang, W. (2017). Deep learning for medical image analysis: applications to computed tomography and magnetic resonance imaging. Hanyang Medical Reviews, 37(2), 61. doi: 10.7599/hmr.2017.37.2.61.
50. Zhu, X., Huang, W., Lu, H. et al. (2021). A machine learning approach to personalized dose adjustment of lamotrigine using noninvasive clinical parameters. Scientific Reports, 11, 5568. https://doi.org/10.1038/s41598-021-85157-x.

Index

Note: Locators in *italics* represent figures and **bold** indicate tables in the text.

B

C

D

I

J

K

L

M

Q

R

S

T

X

Y

Z

Printed in the United States
by Baker & Taylor Publisher Services